Chassis Dynamometer Testing

Addressing the Challenges of New Global Legislation (WLTP and RDE)

Other SAE Books of Interest:

Data Acquisition from HD Vehicles Using J1939 CAN Bus
Richard P. Walter and Eric P. Walter
(Product Code: R-446)

Sensors: Miniaturization and Testing (DVD)
Spotlight on Design
(Product Code: SOD-007)

Fuel Efficiency: Fuel Economy Testing (DVD)
Spotlight on Design
(Product Code: SOD-003/1)

Automotive Fuels Reference Book, Third Edition
Paul Richards
(Product Code: R-297)

Chassis Dynamometer Testing

Addressing the Challenges of New Global Legislation (WLTP and RDE)

Eduardo Galindo, David Blanco, Chris Brace, Ed Chappell, Richard Burke

SAE INTERNATIONAL®

Warrendale, Pennsylvania, USA

400 Commonwealth Drive
Warrendale, PA 15096-0001 USA
E-mail: CustomerService@sae.org
Phone: 877-606-7323 (inside USA and Canada)
724-776-4970 (outside USA)
Fax: 724-776-0790

Library of Congress Catalog Number 2016955855
SAE Order Number R-452
http://dx.doi.org/10.4271/R-452

ISBN-Print 978-0-7680-8278-4
ISBN-PDF 978-0-7680-8412-2
ISBN-epub 978-0-7680-8414-6
ISBN-prc 978-0-7680-8413-9

To purchase bulk quantities, please contact:

SAE Customer Service
E-mail: CustomerService@sae.org
Phone: 877-606-7323 *(inside USA and Canada)*
724-776-4970 *(outside USA)*
Fax: 724-776-0790

Visit the SAE International Bookstore at books.sae.org

Contents

I

Introduction

The use of the chassis dynamometer test cell has been an integral part of the vehicle development and validation process for several decades. Developments in global climate science and environmental health monitoring have driven a wide-ranging debate in the fields of CO_2 reduction and air quality improvement. This discourse has focused on road transport and, especially, passenger cars because they are a very visible and tangible sector requiring improvement. Events have moved the type approval testing of passenger cars from being a somewhat niche topic understood by a small group of engineers and regulators into a widespread topic of conversation. The deficiencies in historical testing regimes and the failure of some companies to adequately regulate their own work have focused the attention of regulators, policy makers, and the public on laboratory procedures and instrumentation methods in an unprecedented manner.

For practicing engineers, it is important to understand our wider role in society, as well as in our own organizations. In chassis dynamometer testing, these responsibilities come together and are focused on the delivery of clean efficient vehicles that meet legislative requirements, as well as customer aspiration. These requirements must be met by vehicles that are robust and durable in the real-world. In addition, the vehicles must be economically viable in order for the industry to continue providing the affordable mobility which has driven so much economic, personal, and social progress over the last century. The pervading environmental lobby presents new challenges to our industry, and the precise experimental validation of vehicle performance on modern capable chassis dynamometers under realistic operating conditions will be fundamental in overcoming this challenge.

The chassis dynamometer test cell design involves specialists in many different fields—such as mechanics, ventilation, and refrigeration—who may not be experts in automotive vehicle engineering. In addition, automotive engineers may not have in-depth knowledge in all aspects of testing facilities.

As the demands on chassis dynamometer testing become more exacting and at the same time more diverse, the challenge of delivering effective installations and operating procedures becomes ever greater. This book sets out to gather knowledge from both groups of specialists to achieve a better understanding on the testing uncertainties associated with the vehicle chassis dynamometer test cell, and to enable informed design and the use of these facilities.

I.1 General Objective

The intention of this book is to analyze the main factors affecting a vehicle in order to reproduce them as close as possible in a testing facility specifically designed for this purpose.

The first objective is to give guidance and recommendations on how vehicle tests should be performed.

The second objective is to obtain a deeper knowledge on the requirements of chassis dynamometer testing facilities so that they can be optimized, reducing unnecessary cost because of overdimensioning or underperformance.

The last objective is to give the reader an overview of the current emissions legislation, its limitations, and its evolution.

I.2 Book Structure

This book has been divided into three parts:

- *Part I: The Vehicle:* The objective of this section is to analyze the main parameters of influence on a vehicle when it is running on the road and look at how to consider these parameters when designing a chassis dynamometer test cell. The key aspects affecting the vehicle system performance and powertrain operating principles must be considered at all stages when designing the facility and test procedures if a realistic validation environment is to be achieved.
- *Part II: The Test Cell:* The objective of this section is to study the different components of a chassis dynamometer test cell, the characteristics, and the design criteria of the test cell, which are relevant to the execution of effective vehicle tests. The chassis dynamometer system is a large and complex installation with many interdependencies. The designer or test engineer must understand enough of the essential behavior of each part of the facility to form an effective overview.
- *Part III: Vehicle Homologation Process and Tests:* This section is largely governed by legislative requirements for today and the foreseeable future.

  The main objective of this section is to give the reader an understanding on the main legislation worldwide. It explains briefly the enforcement programs in the EU and the U.S., analyzing in more detail how effective it has been in Europe until 2017.

  It introduces the reasons for the revision of the EU legislation and the new testing procedures known as Worldwide Harmonized Light-Duty Vehicle Test Procedure and real-driving emissions.

Acknowledgments

This work is the result of the effort to fully understand a complex topic: chassis dynamometer testing. The investment to pursue this project was provided by AVL Spain, a company led by two entrepreneurs, Alberto Zumeta (Managing Director) and Jesus Sinde (Engineering Manager). AVL Spain is a place where people are not only allowed, but encouraged to bring their ideas to work and to develop them beyond the current state of the art. It is the result of implementing a philosophy that questions current beliefs to the ultimate consequences, never giving up until everything is understood and justified. It is the outcome of a long-term strategy that proves that committing new ideas and good technical knowledge can definitely make a company successful.

I would like to thank Alberto and Jesus for creating and sharing these unique values.

All the data and knowledge gathered in this book are the result of team work. It would have not been possible without the help of many colleagues from AVL Spain, especially David Blanco, who believed in the project from the first day and has been a great contributor.

I am also grateful for the opportunity to work for AVL UK. I had the chance to learn from great engineers such as David Moore, Neil Avery, and Colin Hussell and, through them, meet Dr. Chris Brace, Dr. Richard Burke, and Dr. Edward Chappell from the University of Bath. They have been instrumental in complementing our knowledge and working on the material now presented.

We have also had the privilege of having very experienced reviewers such as Ken Barnes, Tim Beatty, Fernando Moreno, and Rodolph Belleux, to whom I want to express special gratitude. The readers will gain immensely from their input.

I also want to thank Carmen Muñoz for the work she has done in integrating the various chapters of the book and Enrique Fradejas for the excellent job with the images we needed.

Finally, I want to thank my parents, Julio Galindo and Carmiña Ferreiro, for teaching me the enormous value of perseverance, and my wife, Sara, for her support and patience during all the hours I have dedicated to writing this book.

Eduardo Galindo

Part I

The Vehicle

Chapter 1
Parameters of Influence on a Moving Vehicle

1.1 Equations of Motion

During real-world on-road driving, the vehicle powertrain must produce enough tractive force at its driven wheels to overcome the losses acting on the vehicle, namely, aerodynamic drag, tire rolling resistance, gravitational pull, and drivetrain losses. It must also, when required, provide force to overcome the vehicle inertia and, therefore, provide the acceleration demanded by the driver. During testing on a chassis dynamometer, the machine is controlled in such a way as to replicate these real-world loads. To understand how this is achieved, it is necessary to consider the forces acting on the vehicle in a free-body diagram, as shown in Figure 1.1. The equations of motion for a vehicle driving along a road can then be constructed.

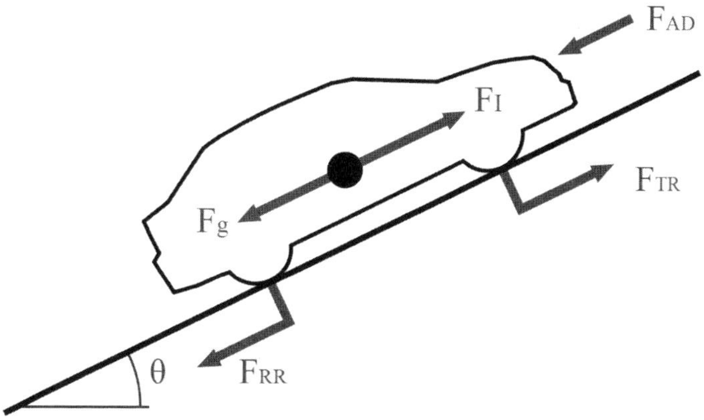

Figure 1.1 Free-body diagram for a vehicle driving up a gradient.

The main forces of concern when considering longitudinal motion of a vehicle are as follows:

- F_{TR} = tractive effort transmitted from the powertrain, through the tires, to the road
- F_I = force acting on the mass of the vehicle to overcome inertia effects (when the rate of acceleration is nonzero)
- F_g = component of force because of gravity acting in line with the road
- F_{AD} = force because of aerodynamic drag
- F_{RR} = force because of tire rolling resistance

These forces can be summed in Eq. 1.1, which will always be balanced. In steady state (zero acceleration), the inertial force F_I is zero and tractive effort will be equal to the sum of the drag terms. If tractive effort exceeds the sum of all the resistive terms, then the vehicle will accelerate at the rate determined by the magnitude of F_I:

$$F_{TR} = F_I + F_g + F_{AD} + F_{RR} \tag{1.1}$$

The mechanisms that govern each of these terms are now introduced.

1.1.1 Vehicle Inertial Loads F_I

In the case where the vehicle is accelerating, the inertias of the vehicle structure and the rotation parts come into play, and the equation of motion is derived from Newton's second law, as shown in:

$$F_I = Ma = m\frac{dV}{dt} \tag{1.2}$$

where M is the equivalent vehicle mass, including an equivalence for rotating assemblies such as wheels, whose rotational velocities are linked to road speed, a is the vehicle acceleration, and V is the vehicle speed.

It is important to note that when a vehicle accelerates, it is usually as a result of torque from the engine rather than a tractive force. Going back one step further, the torque from the engine is actually produced from pressure forces acting on the individual pistons within the engine. The pressure forces create a resultant torque through the arrangement of the connecting rods and crankshaft in the engine. This engine output torque is transmitted through a series of gears and shafts to the wheels where the contact of the tire with the road results in a tractive force in the direction of motion of the vehicle. When the engine accelerates the vehicle, it not only has to accelerate the total mass of the vehicle, but all of the components of the engine (pistons, conrods, crankshaft, etc.) and all of the components of the driveline (gearbox shafts, gears, brakes, wheels, and tires) will also be accelerated and therefore contribute to the overall inertia of the vehicle. Although the masses of these components are small in comparison with the overall vehicle mass, their rotational acceleration can be quite large because of the gear ratios, and should therefore be accounted for.

Figure 1.2 shows the typical layout of a vehicle driveline consisting of an engine, gearbox, prop shaft, differential, axle, and wheel assembly. The different rotational speeds are related to each other through:

$$N_{\text{wheel}} = N_{\text{axle}} = \frac{N_{\text{GB out}}}{GR_{\text{FDrive}}} = \frac{N_{\text{eng}}}{GR_{\text{FDrive}} \times GR_{\text{GBox}}} \tag{1.3}$$

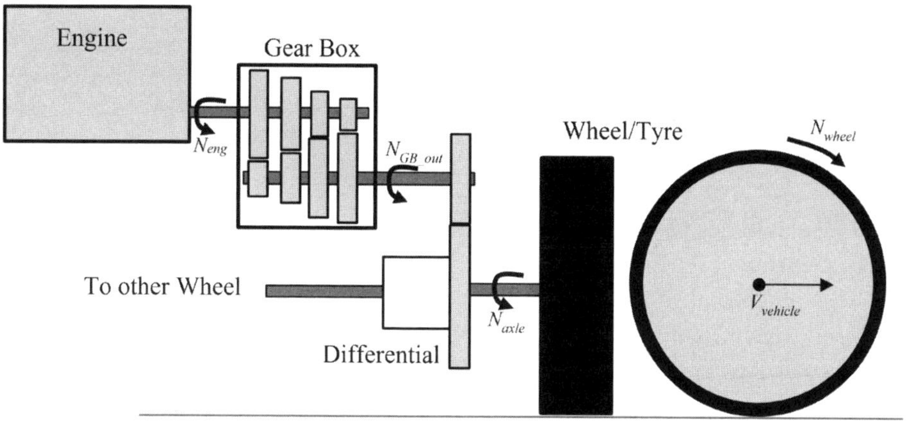

Figure 1.2 Typical vehicle driveline layout showing key inertias and speeds.

It is important to understand the effect of gear ratios on inertia. In a pair of meshing gears such as those shown in Figure 1.3, it is common to refer to the smaller gear as the pinion (the larger gear having no such special name). When the mating gears are operating, the speeds of the two gears are related by the gear ratio as a function of speeds (ω), pitch diameters (d), or the number of teeth (N).

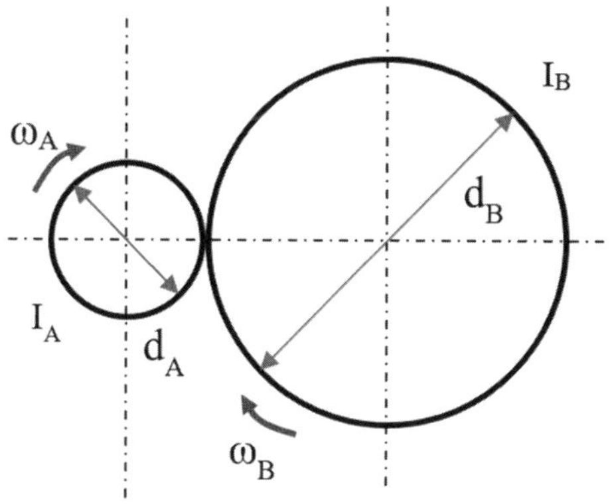

Figure 1.3 Mating external gears.

For meshing gears A and B in Figure 1.3, the inertia perceived by shaft B will be composed of the inertia of shaft B itself and a reflection of the inertia of shaft A onto shaft B, sometimes referred to as an equivalent inertia. If two shafts A and B, with total inertias I_A and I_B, support mashing gears with N_A and N_B teeth, respectively, then the total inertia on shafts A and B will be given by:

$$I_{t(A)} = \frac{I_B}{\left(\frac{N_B}{N_A}\right)^2} + I_A \tag{1.4}$$

$$I_{t(B)} = \frac{I_A}{\left(\frac{N_A}{N_B}\right)^2} + I_B \tag{1.5}$$

The principle also applies to converting a rotational inertia into an equivalent mass. For example, when a wheel is in motion the mass of the wheel must be accelerated in a longitudinal direction, but the inertia of the wheel must also be overcome for it to rotate. The inertia can be converted into an equivalent mass or the mass converted into an equivalent inertia using:

$$I_{eq} = m \text{ or } m_{eq} = \frac{I}{r_w^2} \tag{1.6}$$

The inertias of the driveline can be reflected to give a total equivalent vehicle mass or the vehicle mass, and the inertias can be reflected to give an equivalent total inertia at the engine flywheel. In both cases, these values will be a function of the gear ratios, and therefore change with each gear in the vehicle. For a vehicle with a mass of 800 kg, with an engine, prop shaft, and axle inertias of 0.4, 0.05, and 2.7 kg·m^2, respectively, and a final drive ratio of 3.74, the equivalent masses for five gear ratios are given in Table 1.1 and equivalent inertias at the flywheel in Table 1.2.

Table 1.1 Equivalent Masses for Each Component of the Driveline of a Typical Small Passenger Car

Gear	Gear ratio	$m_{eq\ eng}$ (kg)	$m_{eq,prop}$ (kg)	$m_{eq,\ axle}$ (kg)	$m_{eq,\ veh}$ (kg)	$m_{eq,\ tot}$ (kg)
1	3.7	797	7.3	28.1	800	1632
2	2.01	235	7.3	28.1	800	1071
3	1.2	84	7.3	28.1	800	919
4	0.8	37	7.3	28.1	800	873
5	0.6	21	7.3	28.1	800	856

Table 1.2 Equivalent Inertia at the Flywheel for Each Component of the Driveline of a Typical Small Passenger Car

Gear	Gear ratio	$I_{eq\ eng}$ (kg·m²)	$I_{eq,prop}$ (kg·m²)	$I_{eq,\ axle}$ (kg·m²)	$I_{eq,\ veh}$ (kg·m²)	$I_{eq,\ tot}$ (kg·m²)
1	3.7	0.4	0.004	0.01	0.4	0.82
2	2.01	0.4	0.012	0.05	1.36	1.82
3	1.2	0.4	0.035	0.13	3.82	4.39
4	0.8	0.4	0.078	0.3	8.59	9.37
5	0.6	0.4	0.139	0.54	15.3	16.34

The results for the equivalent mass show that in first gear the inertia of the driveline will more than double the equivalent mass that needs to be accelerated. Conversely, when considering reflected inertia at the flywheel, high gears amplify the mass of the vehicle.

This shows that the inertia is an important consideration when analyzing vehicle performance. On a chassis dynamometer, the vehicle powertrain is part of the vehicle being tested, and therefore does not need to be accounted for in the simulated inertia; however, it is useful to be aware of this aspect of the vehicle system.

1.1.2 Force because of Gravity F_g

As a vehicle ascends a gradient, the force of gravity imposes an additional energy demand equal to the potential energy being invested in the vehicle because of the change in elevation. As the vehicle descends, this energy is returned. The gradient of a hill is typically described in terms of its inclination from the horizontal in degrees or radians, referred to as theta (θ).

The force of gravity is determined by the mass of the vehicle and the gravitational constant. The downward force on an object exerted by gravity is simply its weight:

$$F_g = Mg \tag{1.7}$$

As we are only interested in the component of the weight that is acting against vehicle progress up the gradient, we need to find F_q, the component of F_g that acts in the direction of θ. A free-body diagram will show that $F_q = F_g \sin\theta$. The gravity force equation is then given in Eq. 1.8. It should be noted that the angle θ is directional and a positive value indicates that the vehicle is driving up a hill. If the vehicle traveled down an incline, θ would be negative, and the resultant force would also be negative, meaning its direction would change in Figure 1.1:

$$F_q = Mg \cdot \sin\theta \tag{1.8}$$

As expected, a level gradient ($\theta = 0°$) translates into zero F_q as $\sin 0 = 0$. Gradient has historically been taken as zero when estimating fuel economy. For performance testing and more naturalistic real-world driving, this can no longer be assumed.

On the road, gradient values are usually expressed as a percentage term, whereas for use in the equations of motion above, an angle in radians is required. A gradient in percentage expresses the number of meters gained in altitude per the number of horizontal meters traveled in the forward direction:

$$\theta = \tan^{-1}\left(\frac{\text{Gradient (\%)}}{100}\right) \tag{1.9}$$

1.1.3 Aerodynamic Drag F_{AD}

Aerodynamic drag is created by the displacement of air as the vehicle moves, the friction between the moving air with the vehicle surface, and turbulence in the wake of the vehicle. The force, because of aerodynamic drag, depends mainly on the size and shape of the vehicle, the density of the surrounding air, and the velocity of the vehicle.

Vehicle shape is described by the use of a dimensionless drag coefficient (C_D) and the projected frontal area of the vehicle (A_F). C_D was historically determined experimentally, while A_F is a simple area measurement that was often made photographically. Both parameters can now be determined very accurately using digital techniques well in advance of a physical prototype becoming available.

The following equation is the one describing force because of aerodynamic drag in still air:

$$F_{AD} = \frac{1}{2}\rho \cdot C_D \cdot A_F \cdot V^2 \tag{1.10}$$

where ρ = density of air (kg/m^3)

C_D = drag coefficient of the vehicle (dimensionless)

A_F = projected frontal area (m^2)

V = vehicle velocity (m/s)

1.1.4 Force because of Tire Rolling Resistance F_{RR}

A rolling tire will generate a force that opposes motion called the rolling resistance. This rolling resistance is mainly caused by internal friction within the tire carcass [1-1], [1-2], [1-3], [1-4], [1-5]. The losses within the tire carcass occur because, in the contact area between the tire and road, there is deformation of the tire and the energy dissipated to deform the tire is not completely recovered when it springs back as the tire rolls forward. This energy is lost as heat into the tire, which is why tires become warm while driving—this will be explained in more detail in Chapter 4.

Given a tire design and a standard inflation pressure, a coefficient of rolling resistance (C_{RR}) can be found. C_{RR} is a dimensionless quantity that describes resistance to forward motion per unit of vehicle weight. Normally, the resistance and weight terms are expressed in the same unit, such as pounds or newtons, so that the quotient is dimensionless.

The equation for F_{RR} becomes:

$$F_{RR} = C_{RR} \cdot m \cdot g \cdot \cos(\theta) \tag{1.11}$$

where C_{RR} = tire rolling resistance coefficient (lbf/lbf or N/N, i.e., dimensionless)

g = gravitational constant (9.8 m/s^2)

θ = road gradient (degrees from the horizontal position)

1.1.5 Tractive Power and Energy Equation

Now that the forces acting on the vehicle are defined, we have the information necessary to calculate tractive power and energy demands.

Power is energy per unit time. Because energy is force multiplied by distance and velocity is distance per unit time, it can be shown that power is force multiplied by velocity. Thus, tractive force Eq. 1.1 can be multiplied by velocity V to yield tractive power:

$$P_{TR} = \left(F_{RR} + F_{AD} + F_I + F_G\right) \cdot V \tag{1.12}$$

where

$P_{tractive} = P_{TR}$ = tractive power demand (W).

Substituting the definitions of component forces Eqs. 1.2–1.11 and the mean velocity, the tractive power demand equation becomes:

$$P_{TR} = \left(C_{RR} \cdot M \cdot g^{*} \cos(\theta) + \frac{1}{2} \cdot \rho \cdot C_D \cdot A_F \cdot V^2 + M \cdot \frac{dv}{dt} + m \cdot g \cdot \sin(\theta)\right) \cdot V \tag{1.13}$$

1.1.6 Alternative Arrangement of Vehicle Load Terms

In many cases, it is convenient to divide the forces acting on the vehicle into two categories. The first category is the collection of terms relating to aerodynamic drag, rolling resistance, and gradient. These forces are collectively known as "road load" in that they are the load exerted by the vehicle, which must be matched by the powertrain in order to travel at a steady speed on that road. These terms can be expressed as in:

$$F_{roadload} = F_g + F_{AD} + F_{RR} \tag{1.14}$$

where $F_{roadload}$ is the new concept of road load. In this case, we consider inertial loads as a separate category because it can be argued that the "normal" case is equilibrium, where the vehicle will continue at a constant velocity if the road load is matched by tractive effort. Any surplus or deficit in tractive effort will lead to an acceleration or deceleration. Thus, the road load terms are useful to describe the special case where the vehicle is in equilibrium and distinct from inertial forces, which come into play only when the vehicle is not in equilibrium.

This may be viewed as a somewhat academic division of terms, but as we shall see later, these two categories of load are dealt with by different aspects of the dynamometer control system.

1.2 Vehicle Energy Use and Distribution of Losses

1.2.1 Vehicle Performance and Efficiency: Energy Distribution

Improving vehicle fuel efficiency is a key market driver for any vehicle as many incentives have been placed for both the manufacturer and the customer. In the search for higher efficiency, it is natural to consider how the internal combustion engine (ICE) may be improved.

The real operating efficiency of an ICE when installed in a vehicle is affected by many factors, not least the fact that it seldom operates at its peak efficiency. The typical losses are summarized in Figure 1.4 and quantified for particular driving conditions in Tables 1.3, 1.4, and 1.5. The root of this effect is in the need to install an engine large enough to deliver the power demanded by the driver in all circumstances. Most of the time, however, the power required is well below the maximum installed power. As in most machines, the greater the ratio of maximum power to average power, the less favorable the installation will be from an efficiency perspective. Under typical operating conditions, only about 14% to 26% of the calorific energy of the fuel is converted into mechanical power that is used to move the car along the road. The rest of the fuel energy is lost as waste heat or is used to power auxiliary vehicle systems.

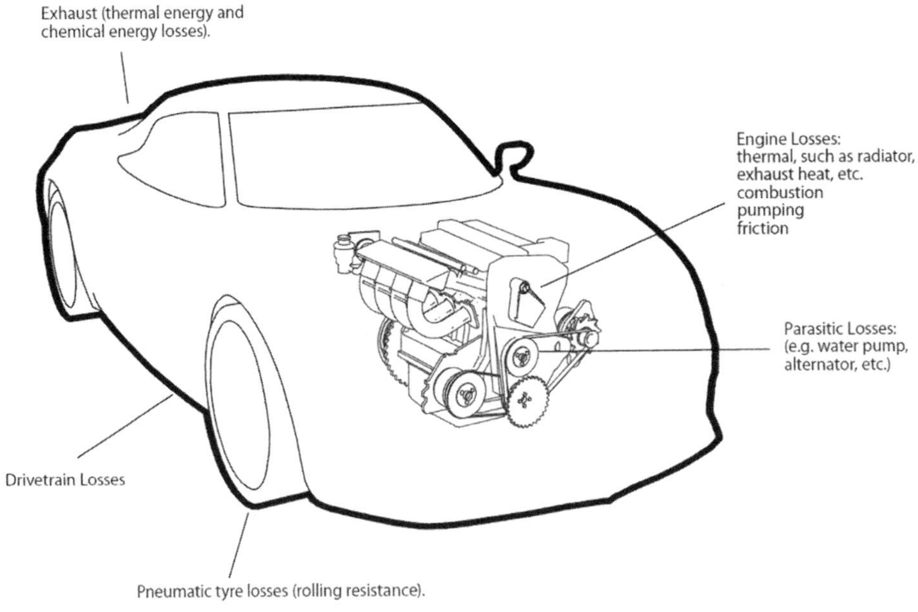

Figure 1.4 Energy requirements.

Table 1.3 Energy Requirements for Highway Driving

Highway driving									
Engine 65–69%				Parasitic 5–7%	Wheels 20–26%			Drive-line 3–4%	Idle 0%
Thermal	Com-bustion	Pump-ing	Friction	-	Wind resis-tance	Rolling resis-tance	Braking	-	-
56–60%	3%	3%	3%	-	13–16%	6–8%	2%	-	-

Table 1.4 Energy Requirements for City (Stop and Go)

City (stop and go)									
Engine 74–75%				Parasitic 6–7%	Wheels 14–16%			Drive-line 4–5%	Idle 6%
Thermal	Com-bustion	Pump-ing	Friction	-	Wind resis-tance	Rolling resis-tance	Braking	-	-
63–64%	3%	5%	3%	-	4%	4–5%	6–7%	-	-

Table 1.5 Energy Requirements for Combined City/Highway Driving

Combined city/highway driving									
Engine 70–72%				Parasitic 5–6%	Wheels 17–21%			Drive-line 3–6%	Idle 3%
Thermal	Com-bustion	Pump-ing	Friction	-	Wind resis-tance	Rolling resis-tance	Braking	-	-
60–62%	3%	4%	3%	-	8–10%	5–6%	4–6%	-	-

It is important to consider such a distribution of energy losses when designing a chassis dynamometer facility and the testing regime. Specifically, the main factors that affect this distribution must be reproduced correctly because they affect both performance and efficiency.

1.2.2 Main Factors Affecting a Vehicle

Figure 1.5 introduces the factors being analyzed in this chapter and upcoming chapters of this book. These chapters deal with the following:

1. wind (interaction between air and vehicle)
2. engine (engine performance and efficiency—engine losses)
3. accessory drives (engine power on accessory drives—parasitic losses)

4. wheels (engine power on wheels—the pneumatic tire rolling resistance)
5. exhaust emissions (exhaust emissions, aftertreatment systems, and parameters of influence)

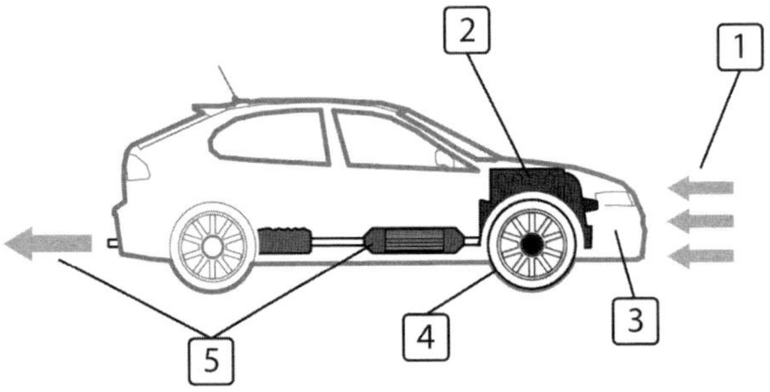

Figure 1.5 Factors to be analyzed.

1.3 Wind: Interaction between Air and Vehicle

When a vehicle is running at a certain speed on the road, the air around it has a great influence on its performance. On one side, the vehicle speed together with its geometry will create a drag load for which the engine will have to compensate, as expressed in Eq. 1.10.

On the other side, the air speed into the vehicle will have a strong influence in the heat exchange with the surrounding air. This heat exchange is crucial for cooling many parts of the vehicle and has a strong influence on the performance of the vehicle under certain conditions.

On the road, vehicles head into an effectively infinite section of air in front of the vehicle, and this air then surrounds the vehicle, as illustrated in Figure 1.6, as it passes through the space.

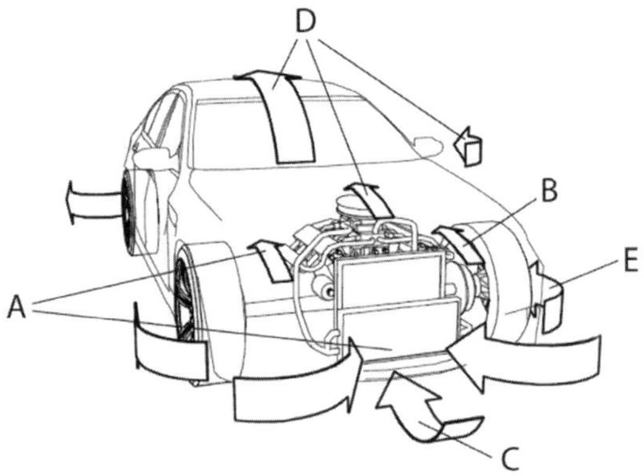

Figure 1.6 Airflow vehicle cooling classification. A—radiator cooling, engine cooling. B—brake cooling. C—underneath the vehicle flow. D—airflow around the vehicle (heat dissipation). E—airflow on tires.

1.3.1 Underhood Air Ducting

The thermal cooling requirements of all the components installed in a vehicle require the manufacturer to design complex air cooling ducts inside the vehicle. A typical location of frontal air ducting is shown in Figure 1.7. Vehicles are designed with air entries in the front, which allow the following:

- air duct for vehicle coolers such as engine radiator, intercooler and oil coolers
- combustion air engine intake
- engine cooling
- exhaust cooling

Figure 1.7 Vehicle front air inlet grid.

Vehicle frontal entry takes air through the vehicle heat exchanger and distributes it. Figure 1.8 shows an idealized path through the heat exchanger and engine compartment to the wheel spaces. The flow of air through these spaces is a complex process, and is typically analyzed using computational fluid dynamics (CFD) in order to ensure adequate cooling performance with a minimum additional drag load. This type of simulation allows for the detailed breakdown of air velocity all around the engine and powertrain to be analyzed, as illustrated in Figure 1.9.

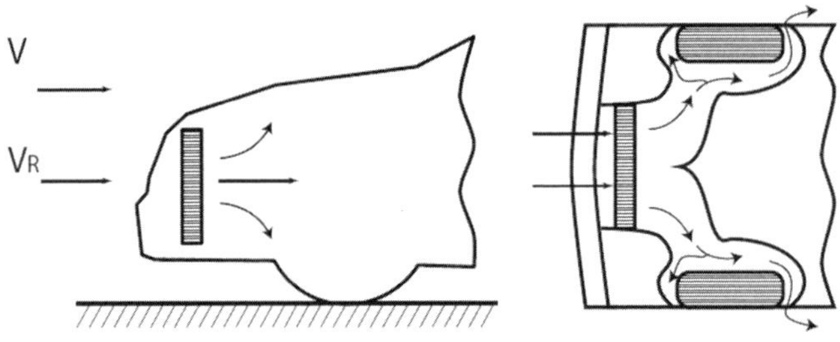

Figure 1.8 Typical vehicle air hood distribution.

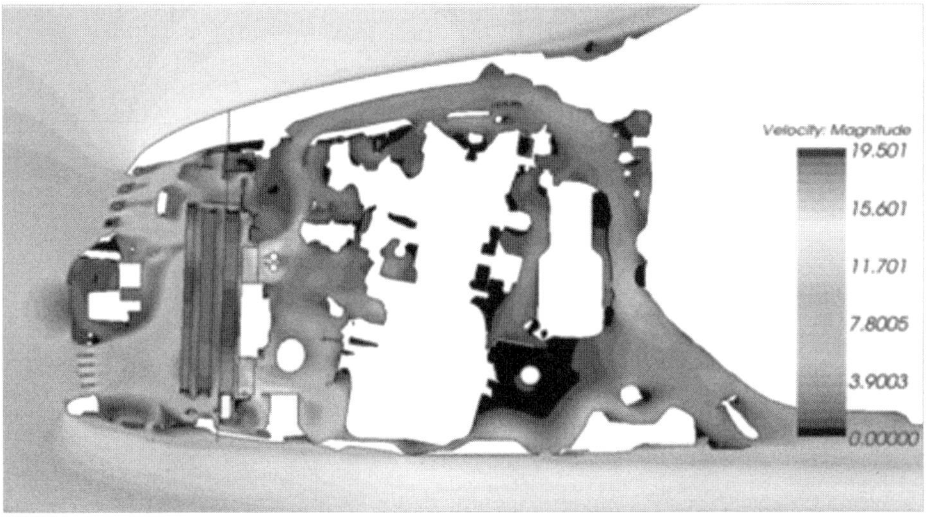

Figure 1.9 Vehicle air hood distribution CFD analysis.

1.3.2 Brake Cooling

Brakes are required to stop the vehicle in a very short time for safety reasons. When a vehicle is traveling at high speeds, all the kinetic energy of the vehicle must be dissipated by the brakes. This is done by friction forces, which convert energy into heat. The power dissipation capacity within the braking system is typically several times that of the propulsion system. This leads to very high rates of heat dissipation into the braking components. Maintaining adequately low temperatures in the brakes is critical for good brake performance. In order to dissipate all this energy, brake cooling ducts are installed in vehicles with high braking-power requirements, as illustrated in Figure 1.10.

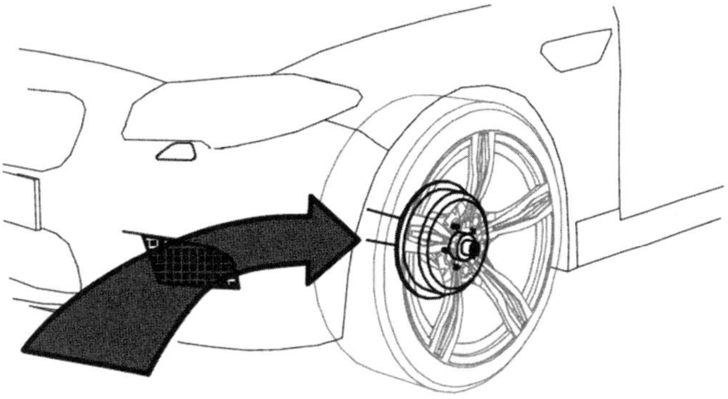

Figure 1.10 Schematic of brakes and air ducts.

1.3.3 Flow underneath the Vehicle

A portion of the air impinging onto the vehicle is redirected to the underside, giving an aerodynamic effect as well as cooling the exhaust system (see Figure 1.11). Cooling of the exhaust system will have an effect on the aftertreatment system, as will be explained in Chapter 5. It is crucial to consider such effects when aiming to replicate on-road performance on a chassis dynamometer.

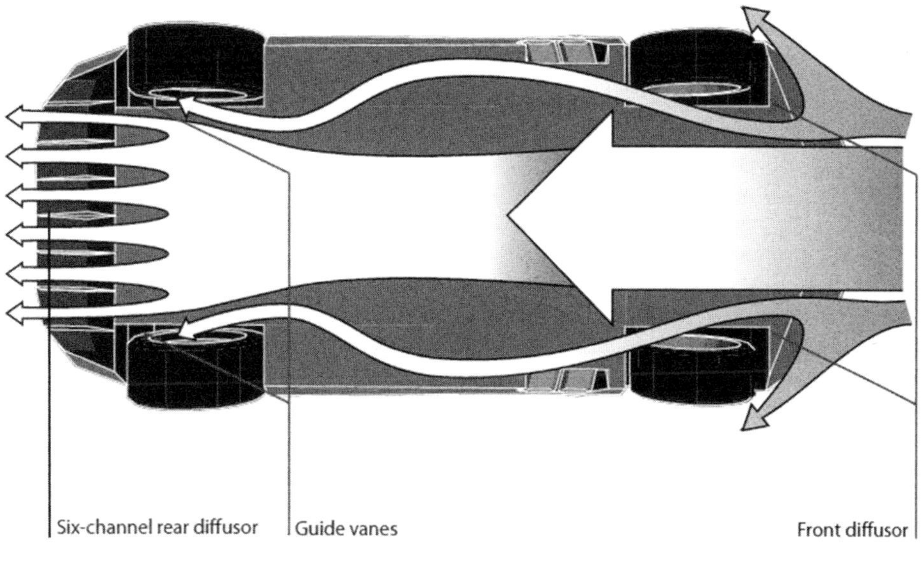

Figure 1.11 Airflow along the underfloor of the Mercedes-Benz SLR McLaren.

1.3.4 Airflow around the Vehicle

The airflow to the vehicle cabin and mainly over the windscreen and windows has a great influence on the cabin temperature and, therefore, on the requirements for the air conditioning. Higher air speeds will increase energy transfer between cabin and surrounding air. This will be explained in detail in Chapter 3.

1.3.5 Airflow around the Tires

As will be explained in Chapter 4, airflow on vehicle tires will have a great influence on the final tire temperature. This will influence both tire temperature and rolling resistance.

References

1-1. Wong, J. Y., 2008. *Theory of Ground Vehicles.* John Wiley & Sons, Inc: New Jersey.

1-2. Clark, S. K., 1971. *Mechanics of Pneumatic Tyres.* U.S. Department of Commerce: Washington, DC.

1-3. Gerresheim, M. "Dependence of vehicle fuel consumption on the tyre-wheel assembly," Proceedings of the Conference Automobile Wheels and Tyres, Sutton Coldfield, UK, IMechE 12:51–63, 1983.

1-4. Genta, G. 2006. *Motor Vehicle Dynamics: Modelling and Simulation.* World Scientific: London.

1-5. Transportation Research Board. "Tires and Passenger Vehicle Fuel Economy: Informing Consumers, Improving Performance," National Research Council. Special Report 286, 2006.

Chapter 2
Engine Architecture Performance and Efficiency

2.1 Introduction

An internal combustion engine (ICE) is a heat engine that converts chemical energy into mechanical work. In this process, a large portion of the energy is wasted and only a small part of it reaches the output shaft. This book is not focused on this topic, and the interested readers are directed to [2-1] and [2-2]. Here, a short overview is presented in order to alert the chassis dynamometer designer or test designer as to what principal effects must be considered. The second law of thermodynamics dictates that in a heat engine, some heat must be rejected to the surroundings. The Carnot cycle defines the maximum efficiency that can be obtained by a theoretical heat engine operating over a closed cycle between any two temperatures. The Carnot cycle describes a closed system with perfect heat exchange between a working fluid and the high and low-temperature reservoirs alternately.

In a real ICE, these temperatures may be considered to be analogous to the combustion temperature and the temperature of the surrounding air. The ICE seeks to approximate the Carnot cycle in a practical, convenient, and affordable machine. There are many compromises that degrade the actual engine efficiency, but a well-designed ICE can still exceed 50% thermal efficiency. This is a significant achievement that can only be achieved under optimum conditions.

The fundamental limitations of practical engine efficiency are well described by the ideal air standard thermodynamic cycles (Otto, diesel, and dual cycles). These closed cycles can be used to estimate the pumping losses present in real engines, which are incurred as we induce the air into the engine and pump the burned gases out. There are many incidental sources of inefficiency that degrade the practical performance of an ICE that can be described by the ideal air standard cycles. For example, the chemical energy (fuel) is not optimally converted into heat during the combustion and, even in an Otto engine, the rate of energy conversion is not instantaneous. Consequently, the heat released during the reactions is lower than the theoretical maximum—this discrepancy can be referred to as the combustion efficiency.

The chemical energy released from the fuel as heat causes a rise in temperature and pressure, which is then converted into mechanical work by the cranktrain. Engine internal components such as the oil pump, coolant pump, and the camshaft consume power from the available shaft power. Friction between the rubbing surfaces generates resistive forces against the engine rotation. Vehicle accessories, such as the air conditioning compressor, alternator, and power steering pump, also use power from the engine accessory drive. The remaining power at the engine flywheel is left to provide tractive power for vehicle motion.

2.2 Engine Energy Distribution

The principal energy flows in an ICE are listed below and illustrated in Figure 2.1:

1. heat input from the combustion of the fuel
2. mechanical work in the crank shaft
3. heat losses to the engine cooling system and the engine oil system that are ultimately lost to ambient, principally through the radiator
4. energy loss in the exhaust pipe in the form of exhaust gas enthalpy and unburnt fuel
5. heat loss from the cooling of intake and/or EGR gases

Of these energy vectors, only a relatively small fraction of the total fuel energy is available to produce mechanical work when installed in a typical nonhybridized passenger car. An indicative distribution of energy flows over a New European Drive Cycle (NEDC) (described in Chapter 12, Section 12.3.2.3) performed from a cold soaked condition is presented in Figure 2.2 for a 2.4-L diesel engine operating in a light commercial vehicle. Around 40 MJ of energy is released from the fuel as heat, which is equivalent to around 1 L of diesel fuel. Less than 20% of this results in mechanical work at the engine output shaft, and a further percentage of this would be lost in the vehicle drivetrain and tires. The other two major routes for energy flows are to the exhaust and the heating of the engine structure and fluids.

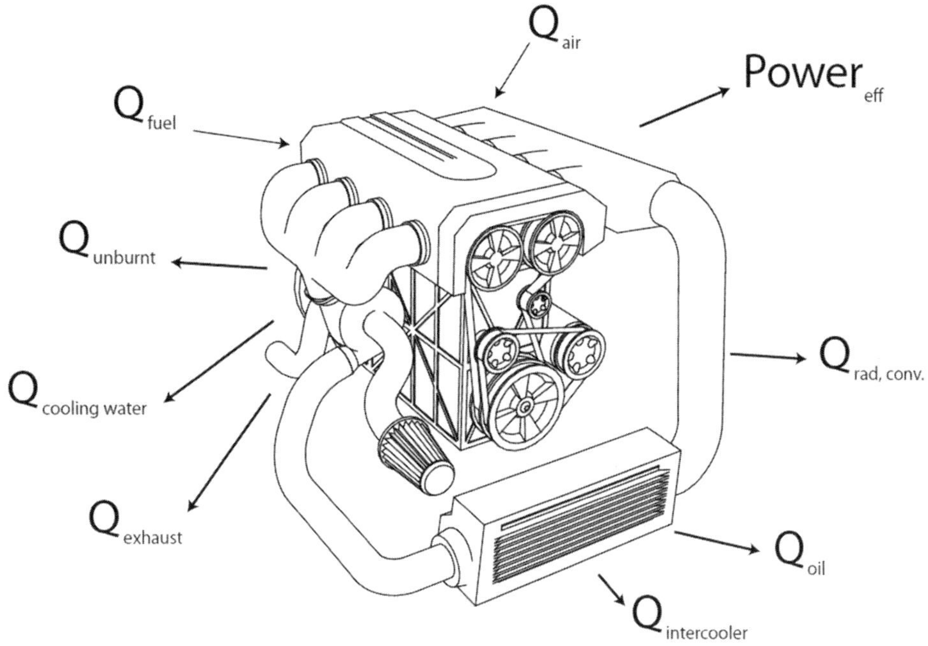

Figure 2.1 Principal engine energy flow vectors.

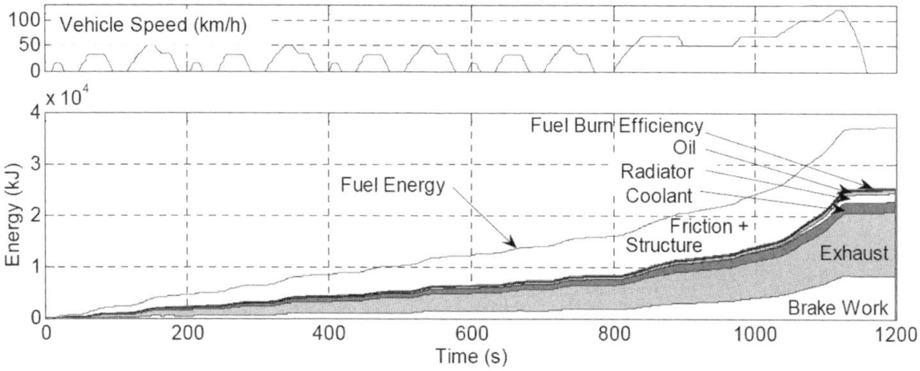

Figure 2.2 Typical energy distribution in ICE powertrain over an NEDC from cold. Note the significant proportion of the fuel energy that goes to heat the structure and overcome friction.

2.2.1 Gasoline Engine Energy Distribution

The thermal efficiency of a typical homogeneous charge, direct fuel-injected gasoline engine varies greatly depending on the operating condition. This is shown in Figure 2.3, which shows how efficiency varies with engine speed and load. Efficiency is represented as specific fuel consumption, which is the ratio of the mass of fuel consumed by the engine to produce a unit of mechanical energy at its crankshaft. The main factors that affect the overall efficiency are the relative magnitude of the friction work and the pumping work as engine speed and load change. Friction increases strongly with engine speed and load. Pumping work increases strongly as the homogeneous mixture preparation requires that the engine is throttled to reduce its power output at part load. In addition, the variations in combustion efficiency, optimum spark phasing, and observed flame speed occur across the operating range. The consequent phasing of the heat release relative to the crankshaft has a major effect on the cycle efficiency. Taken together these factors tend to degrade engine efficiency greatly in the part load condition, where the engine spends a significant part of its operational time in a conventional passenger car powertrain. Engine protection measures, such as mixture enrichment and retardation of spark timing, tend to further reduce efficiency at high power, but these operating points are less significant to overall efficiency as they are rarely visited in normal driving.

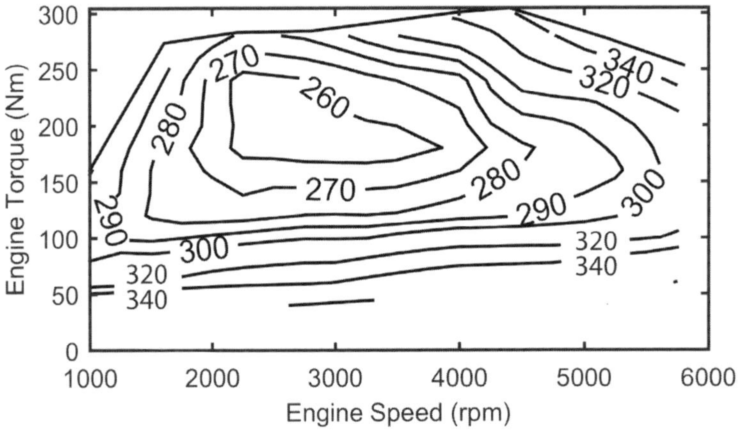

Figure 2.3 Typical fuel consumption contours (BSFC in grams per kilowatt-hour) for a 2.0-L turbocharged direct injection gasoline engine.

2.2.2 Diesel Engine Energy Distribution

The diesel or compression ignition engine adopts a different combustion approach to the spark ignition (SI) engine, allowing much leaner operation and avoiding the need to throttle the air flow at part load. This greatly reduces the pumping losses at part load. In addition, the combustion is tolerant to much greater compression ratios than the typical SI system, and can therefore operate on a more efficient thermodynamic cycle. Modern air motion and fuel injection technologies allow rapid combustion, which can improve combustion phasing and thermodynamic efficiency. The principal compromise in terms of efficiency is the requirement to delay the combustion phasing in order to reduce the in-cylinder production of oxides of nitrogen. This allows a smaller and more efficient NO*x* aftertreatment system but at the expense of engine thermal efficiency. A system-level optimization is required to obtain the best balance between in-engine and after-treatment NO*x* reductions. More recently, there has been a trend to reduce the geometric compression ratio and sacrifice some thermodynamic efficiency for reduced engine friction and mass. Figure 2.4 shows typical brake-specific fuel consumption (BSFC) contours for a modern diesel engine.

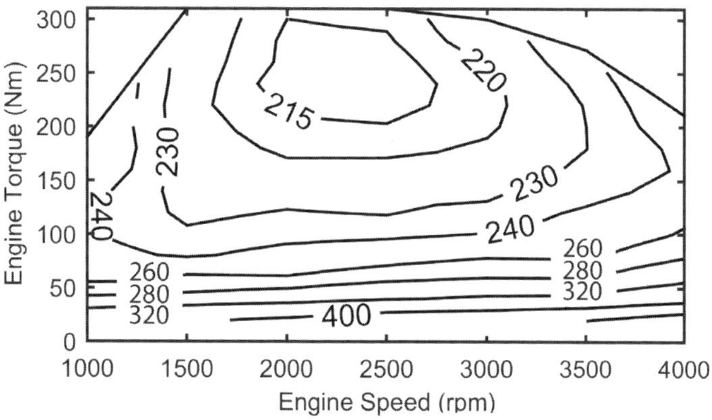

Figure 2.4 Typical fuel consumption contours (BSFC in grams per kilowatt-hour) for a 2.0-L turbocharged direct injection diesel engine.

2.3 Effects of Ambient Conditions on Engine Performance

Considering an engine as a device that converts chemical energy into mechanical work via a combustion process, it becomes obvious that its performance is directly dependent on its capacity to burn fuel. Because of the finite volume of the engine cylinders, the engine performance is strongly dependent on the air density and therefore on the ambient conditions, especially pressure and temperature.

Considering air as an ideal gas, we can calculate the influence of ambient parameters on air density.

2.3.1 Effect on Temperature and Pressure

The ideal gas law is presented in the following equation, which relates pressure (p), volume (V), temperature (T), and the number of moles (n):

$$pV = nRT \tag{2.1}$$

where R is the universal gas constant, which is 8.314 J/K·mol. The equation can also be represented in molar form. Defining the molar mass as the mass of a single mole:

$$M = \frac{m}{n} \tag{2.2}$$

and introducing density as:

$$\rho = \frac{m}{V} \tag{2.3}$$

we can substitute Eq. 2.2 and Eq. 2.3 into Eq. 2.1 to yield the following equations:

$$pV = \frac{m}{M}RT \tag{2.4}$$

$$p = \rho\frac{R}{M}T \tag{2.5}$$

Now, defining the specific gas constant as:

$$R_{\text{specific}} = \frac{R}{M} \tag{2.6}$$

we obtain the same equation in molar form:

$$p = \rho R_{\text{specific}} T \tag{2.7}$$

The advantage of this form of the ideal gas law is that it allows one to calculate the influence of pressure and temperature on density:

$$\rho = \frac{p}{R_{\text{specific}} T} \tag{2.8}$$

where R_{specific} for air is 287,058 J/(kg·K).

Figure 2.5 shows the variation in air density as a function of pressure and temperature. It is evident that density will increase with pressure and decrease with temperature.

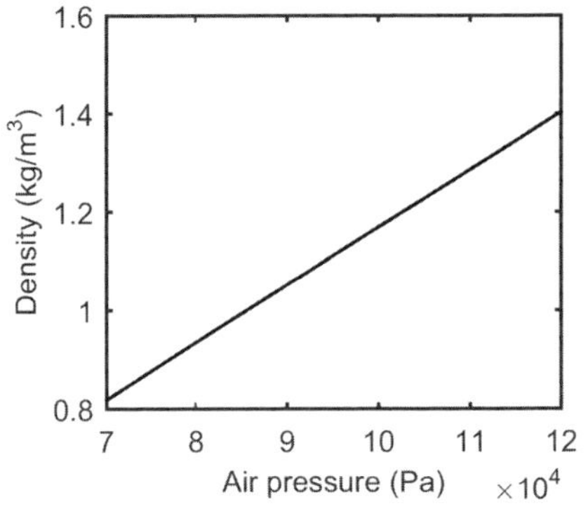

(a)

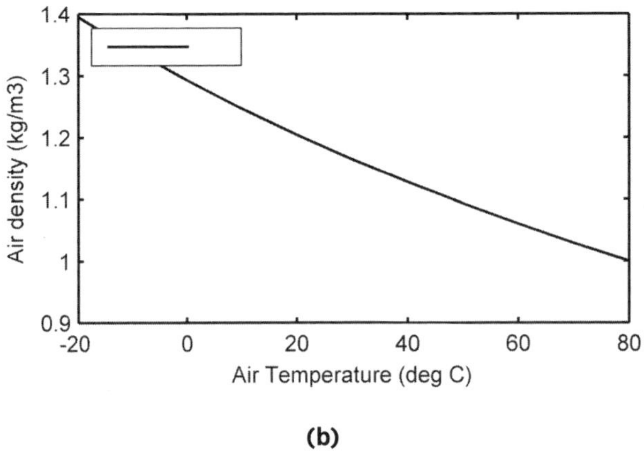

(b)

Figure 2.5 Air density variation at (a) constant temperature (25°C) and (b) constant pressure (101.325 kPa).

Air density is important because engine power is directly linked to the amount of fuel that can be burned in the engine cylinders. The amount of fuel that can be burnt depends on the amount of oxygen present in the cylinders. Since these cylinders are of fixed volume, a higher density of air trapped in the cylinders results in a higher availability of oxygen. Therefore, more fuel can be burned resulting in higher mechanical power output. As engine friction losses are only loosely linked to the pressures in the cylinders, an increased power from a particularly sized cylinder usually results in an increased efficiency because frictional losses remain roughly constant. This is the principle of engine downsizing and turbocharging, which increases the density of air fed into the cylinders.

Assuming engine power is directly proportional to the amount of oxygen that can be introduced in the engine cylinder and therefore to air density, we can plot the influence of pressure and temperature versus engine power as shown in Figure 2.6.

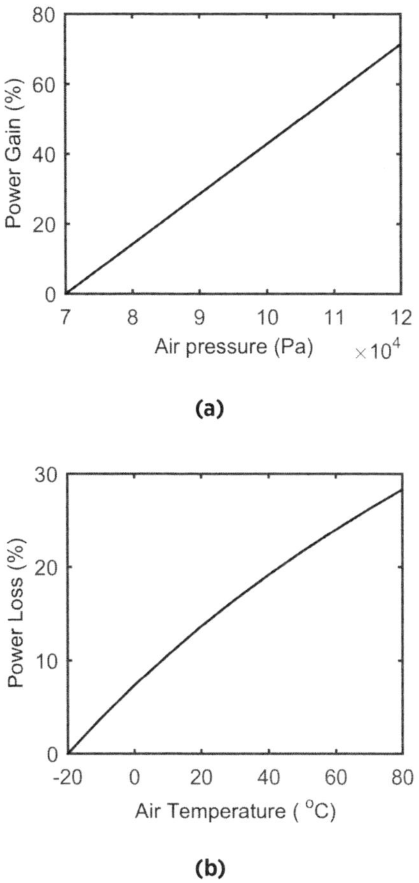

Figure 2.6 (a) Engine power gain (%) versus pressure with constant temperature and (b) engine power loss (%) versus temperature increase with constant pressure.

As shown in Figure 2.6, the influence of both atmospheric pressure and temperature on engine power is very significant. Atmospheric pressure at sea level varies because of atmospheric conditions, for instance, the lowest atmospheric pressure recorded at sea level is around 870 mbar during a tropical cyclone. In the U.S. and UK, the lowest recorded pressures are around 925 mbar. Ordinarily, the effect of changes in altitude is more significant. Although the oxygen concentration is a constant of 20.9% at all altitudes, for a naturally aspirated engine, the reduction in atmospheric pressure will reduce the mass of oxygen ingested by the engine as altitude increases. At a 1000-m altitude, there is already a 10% reduction in the mass of oxygen ingested by an engine.

For supercharged or turbocharged engines, the effect of altitude is different as the amount of oxygen ingested into the combustion chambers is linked to the intake manifold pressure, which is, in any case, above atmospheric pressure. Vehicle manufacturers usually have some scope for the turbocharger system to maintain engine power when at higher altitudes with lower ambient pressure. However, the cost of this is that the turbocharger turbine or mechanical supercharger must work harder as a greater pressure ratio exists between ambient and inlet pressure. This will usually manifest itself through an increase in exhaust backpressure, which will contribute to either increased fuel consumption or eventually, if the ambient pressure is sufficiently low, an inevitable drop in engine power.

In fact, the engine control system monitors mass flow into the engine, and in most cases, attempts to control the mass flow by either opening the throttle (SI engine) or closing the wastegate (turbocharged diesel or SI engine). The fuel injection system will then supply the necessary fuel appropriate for the mass of air admitted. In an SI engine, the air fuel ratio must remain stoichiometric in order to ensure proper catalyst operation. In a diesel engine, there is more flexibility, but a minimum air fuel ratio limit must be observed in order to avoid excessive particulate formation.

2.3.2 Effect of Humidity

Avogadro's law states that two given samples of an ideal gas of the same volume at the same temperature and pressure contain the same number of molecules. As an example, equal volumes of molecular hydrogen and nitrogen contain the same number of molecules when they are at the same temperature and pressure and observe ideal gas behavior. In practice, real gases show small deviations from the ideal behavior and the law holds only approximately, but is still a useful approximation for engineering purposes.

The molecular mass of water (18 g/mol) is less than the molecular mass of dry air (around 29 g/mol). Considering Avogadro's law, when a number of water molecules (in vapor form) are added to a given volume of air, the number of dry molecules must decrease by the same number, to keep the pressure or temperature from increasing. Hence, the mass per unit volume of dry air (its density) decreases, and with it, there is a proportional decrease in the mass of oxygen in the volume.

The increase of water vapor, making the air more humid, reduces the air density. This concept is reflected in Dalton's law, which states that the partial pressure of an ideal gas in a mixture is equal to the pressure it would exert if it occupied the same volume alone at the same temperature. Because of this, the total pressure of a mixture of ideal gases is equal to the sum of the partial pressures of the individual gases:

$$p_t = p_d + p_v \tag{2.9}$$

where p_t = total air pressure, p_d = dry air pressure, and p_v = vapor pressure.

The vapor pressure of water can be calculated from the saturation vapor pressure and relative humidity:

$$p_v = \Phi p_{sat} \tag{2.10}$$

where Φ = relative humidity and p_{sat} = saturation vapor pressure.

The saturation vapor pressure of water at any given temperature is the vapor pressure when the relative humidity is 100%. This can be expressed as:

$$p_{sat} = \frac{e^{77.3450+0.0057T-(7235/T)/T}}{T^{8.2}} \tag{2.11}$$

where p_{sat} = water saturation pressure expressed in Pa and T = temperature expressed in °C (Figure 2.7).

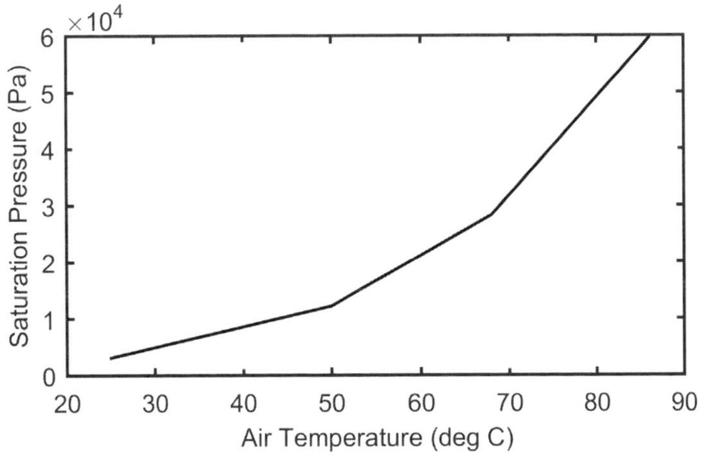

Figure 2.7 Water saturation pressure versus temperature.

As the equation shows, saturation pressure increases exponentially.

The dry air pressure can therefore be expressed as:

$$p_d = p - \Phi\frac{e^{77.3450+0.0057T-(7235/T)/T}}{T^{8.2}} \tag{2.12}$$

Inserting this equation into the ideal gas law in molar form, we can see the influence of air humidity on dry air density, which will influence the oxygen density and engine performance:

$$\rho = \frac{p}{R_{\text{specific}} T} \tag{2.13}$$

$$\rho_d = \frac{p - \Phi \dfrac{e^{77.3450+0.0057T-(7235/T)/T}}{T^{8.2}}}{R_{\text{specific}} T} \tag{2.14}$$

For constant total pressure and temperature, we can see that most of the terms in the previous equation are also constant. By plotting it, we can see the influence of relative humidity on the density of dry air (Figure 2.8).

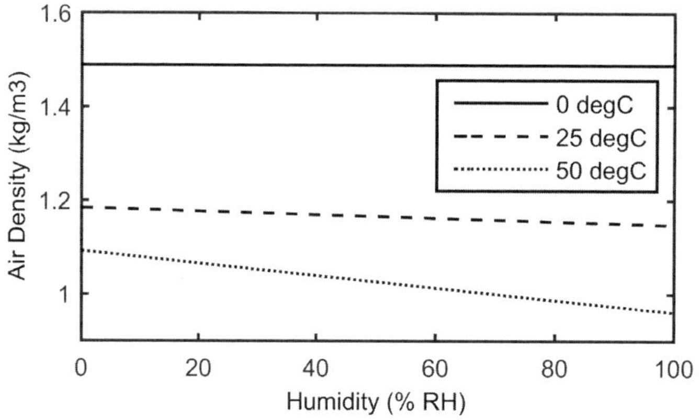

Figure 2.8 Dry air density versus humidity for constant total pressure.

As seen in the graph, the influence of air humidity significantly increases with temperature. This is because the higher the temperature, the larger the mass of water evaporated into the air for a given relative humidity. At 0°C, the influence is insignificant, while at 50°C, there could easily be a 10% reduction in dry air density under prevailing climatic conditions, and so an equivalent reduction in oxygen concentration. In fact, it should be noted that relative humidity is quite a poor indicator of the humidity level for engine and vehicle testing. This is because for the same relative humidity, different ambient temperatures will cause significant differences in specific humidity (the mass of water vapor per mass of dry air).

Considering air density to be directly proportional to engine power (without any compensation by the engine control system), we can estimate engine power loss because of the increase in air humidity for a given air temperature and pressure (Figure 2.9).

As shown previously, the influence for cold air at 0°C is almost zero, while at high temperatures of 50°C, this easily results in over 10% of power loss.

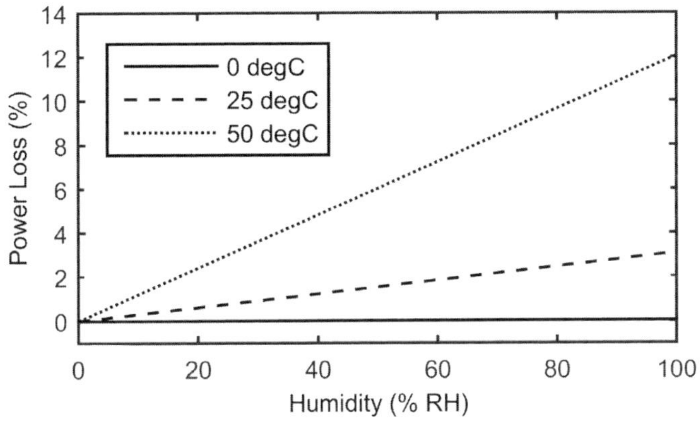

Figure 2.9 Power loss (%) because of relative humidity increase (%) of air.

The effect of humidity does not only manifest itself on engine power, it will also have a significant impact on the emissions formation within the engine. We have seen how humidity displaces oxygen in the cylinder, and this also has a thermal effect because the specific heat capacity of water is greater than that of air. The water vapor therefore has a quenching effect both globally and locally in the cylinder, reducing combustion temperatures. All emissions are sensitive to these temperatures, but NO*x* emissions, which are strongly linked to peak flame temperatures in the cylinder, are affected more significantly than other emissions.

2.3.3 Effect of Altitude

Pressure reduces with altitude uniformly from the Earth's surface to the top of the atmosphere. Although the pressure changes with the weather, NASA data provide a good estimate of the average conditions for all parts of the Earth year-round.

To calculate air density as a function of altitude, the following parameters are required:

- sea-level standard atmospheric pressure $p_0 = 101.325$ kPa
- sea-level standard temperature $T_0 = 288.15$ K
- earth surface gravitational acceleration $g = 9.80665$ m/s^2
- temperature lapse rate $L = 0.0065$ K/m
- ideal (universal) gas constant $R = 8.31447$ J/mol·K
- molar mass of dry air $M = 0.0289644$ kg/mol

Temperature at altitude h meters above sea level is given by:

$$T = T_0 - L \times h \tag{2.15}$$

The following equation can be used to calculate pressure at a given altitude:

$$p = p_0\left(1 - \frac{Lh}{T_0}\right)^{gM/RL} \approx p_0 \exp\left(-\frac{gMh}{RT_0}\right) \tag{2.16}$$

Density can be calculated using the ideal gas law in molar form:

$$\rho = \frac{p}{R_{\text{specific}} T} \tag{2.17}$$

In Figure 2.10, the variation of these parameters is plotted against altitude.

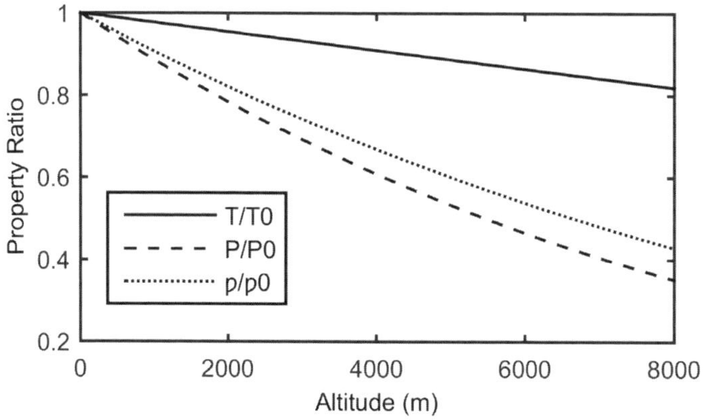

Figure 2.10 Air property ratio versus altitude.

As in the previous cases, we can estimate engine power loss with altitude by assuming this is directly proportional to air density (see Figure 2.11).

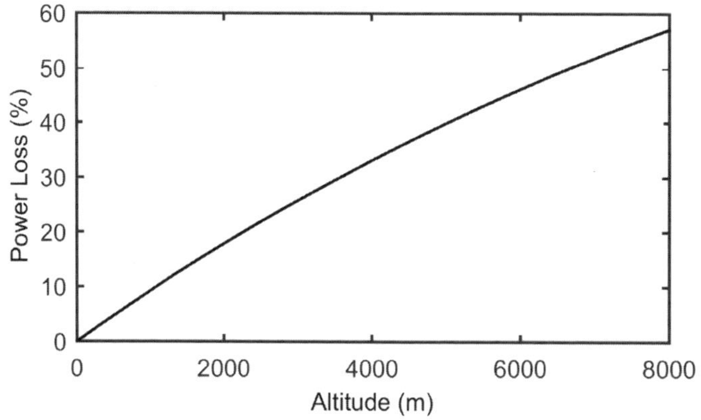

Figure 2.11 Engine power loss (%) versus altitude.

2.3.4 Summary of the Environmental Effect on Engine Power

As a summary of the environmental effect on engine power, we can say the following statements, which are represented graphically in Figure 2.12:

- Air temperature can represent around a 2.5% reduction in engine power for every 10°C of temperature rise.

- Ambient pressure can represent around 1.5% of power decrease for every 1000-Pa fall in atmospheric pressure.
- Humidity can represent around 0.3% of power reduction (for air at 25°C) for every 10% of relative humidity increase.
- Altitude can represent around 3.5% of power reduction for every 500 m increase in altitude.

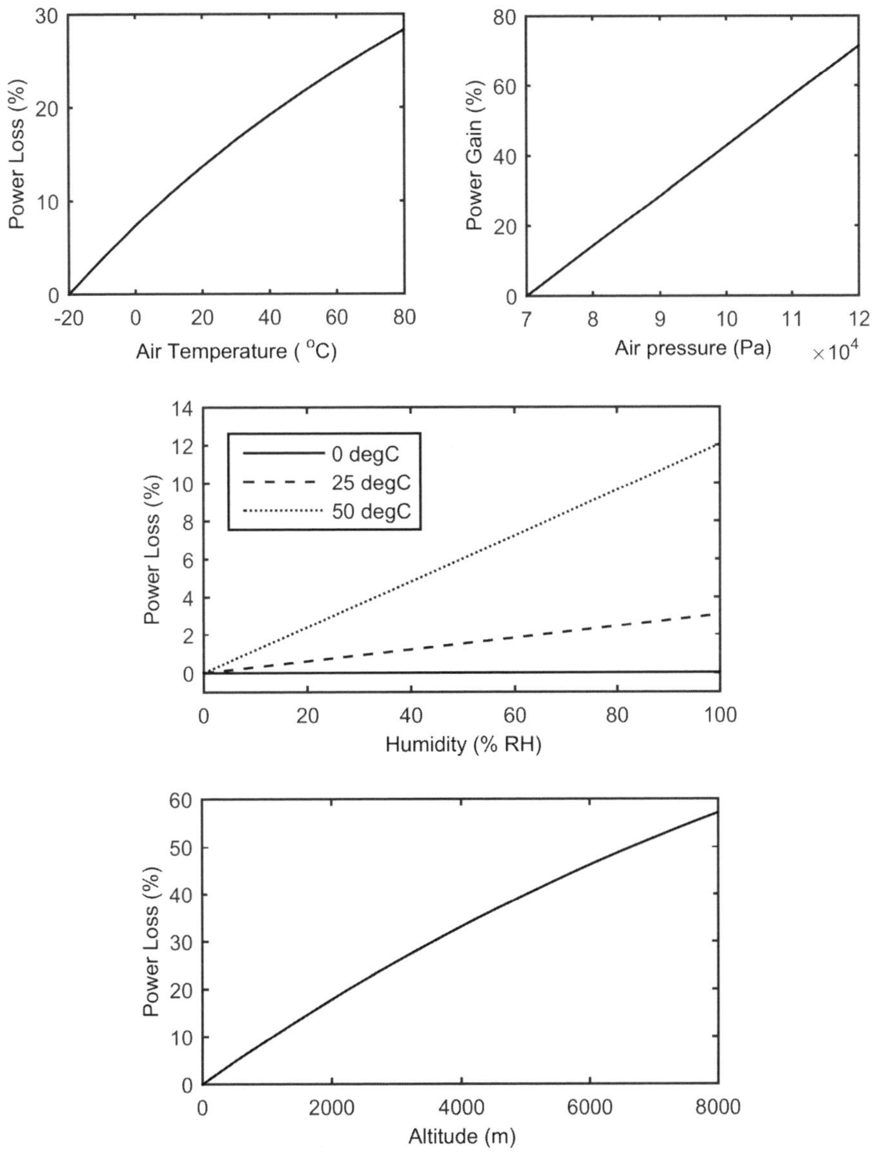

Figure 2.12 Engine power loss (%) versus temperature, engine power gain (%) versus pressure (constant *T*), power loss (%) because of relative humidity increase, and engine power loss (%) versus altitude.

2.3.5 Standardized Correction Factors

Controlling ambient conditions of pressure, temperature, and humidity within a test cell is a fairly recent development in vehicle testing and prohibitively expensive, meaning that full control is not universally adopted. To allow comparisons between tests undertaken in different locations and on different dates, mathematical correction methods for emissions and engine powers have been established and standardized. In most standards, the correction formula for the engine power is of the basic form given by:

$$P_s = C_f P_m \tag{2.18}$$

where P_s = standard power, C_f = power correction factor, and P_m = measured power.

The general form for the correction factor is written as:

$$C_f = \frac{p}{p_0} = \left(\frac{p - p_v}{p_0 - p_{0v}}\right)^{\alpha} \left(\frac{T_0}{T}\right)^{\beta} \tag{2.19}$$

where sub-index 0 denotes reference conditions according to each standard and the main parameters are defined as in Table 2.1.

Table 2.1 Main Parameters of the Reference Conditions for the Power Correction Factor

Organization parameters	SAE J1349	ISO 1585 (1192)
Ambient pressure p_0 (bar)	0.99	1
Vapor pressure p_{0v} (bar)	0.013	0.01
Air supply temperature (K)	302.4	298
α	1	1.2
β	0.5	0.6
Recommended range temperature (K)	288.5 → 316.3	288 → 308
Recommended range pressure (bar)	0.95 → 1.010	0.8 → 1

As the equation shows, the effect of temperature and pressure is taken directly into the formula. The effect on humidity is calculated using measured vapor pressure p_v. Other correction factors are described in standards, such as DIN 90020, that do not take into account vapor pressure and therefore the effect of humidity.

Correction factors outside the recommended range are not accurate and should only be used as an indication.

2.4 Engine Warm-Up and Cooling

In the previous sections, we have seen typical engine efficiencies. The numbers introduced in those sections are measured for engines running in stabilized warm conditions. The energy efficiency of the engine changes significantly during transient cycles such as the engine warm-up period. Therefore, the temperature is a factor that influences engine efficiency significantly.

2.4.1 Cold Start Combustion

In order to start an SI engine, it is necessary to supply a combustible mixture of fuel and air near the spark plug, which can be burned easily—ensuring sufficient fuel evaporation is critical for this task. When an engine is started in cold temperatures, the gasoline fuel has a relatively low rate of evaporation and it enters the combustion chamber in the form of vapor and droplets, which degrade the burning process. To facilitate this process, more volatile fuels would be desirable, but within current legislation, there are strong limits on volatility to reduce evaporative emissions from the fuel tank and inlet manifold when the vehicle is standing idle.

The flammability limit of a typical air and gasoline mixture is about 8 to 1 (8 parts air to 1 part fuel by mass) for rich mixtures and 20 to 1 (20 parts air to 1 part fuel) for lean mixtures. The most practical option to facilitate cold starts is usually to use a far richer air–fuel mixture than normal, to ensure that enough fuel vapor is mixed with the air to guarantee combustion. This causes an increase of unburned gases that are evacuated through the exhaust.

All these factors lead to higher emissions and poor fuel economy directly following a cold start. Any measures that improve fuel/air mixing, such as high-pressure direct fuel injection, will alleviate this effect to an extent. Even so, an additional problem is that the exhaust system catalyst is inactive at low temperatures. The catalyst typically starts to become active when feed gas temperatures are above 150°C. Minimizing the time to achieve catalyst "light-off" is therefore a critical calibration task.

Diesel engines have a much greater ability to achieve adequate combustion at low temperatures because of the significantly higher fuel injection pressures used, leading to more consistent mixture preparation. Nevertheless, glow plugs are typically installed to preheat the cylinder charge for the first combustion event to assist with evaporation and ignition of the fuel.

After starting, the quenching effect of the cold engine structure leads to poor CO and HC emissions on cold start, and to increased heat transfer losses as the engine warms up. The oxidation catalyst and some de-NO*x* systems used for diesel engine systems also require a warm-up period. This is compounded by the much lower exhaust temperature in a diesel engine because it is able to run under very lean conditions, meaning that there is a large mass of unburnt air that can absorb heat from the combustion. This leads to long warm-up times for the exhaust catalysts.

2.4.2 Engine Friction Losses Versus Temperature

Oil viscosity is significantly affected by temperature. When running an engine at low temperatures, the losses that occur in the oil circulation pump and the hydrodynamically lubricated components increase significantly. A 40% increase in friction because of a thicker oil can correspond to an increase in fuel consumption of about 12%. Changing to a lower-viscosity grade lubricant would lead to a decrease in engine friction that could correspond to an improvement of 4% in fuel consumption. Figure 2.13 shows the reduction in engine frictional losses from a change in oil grades from baseline to candidate oils.

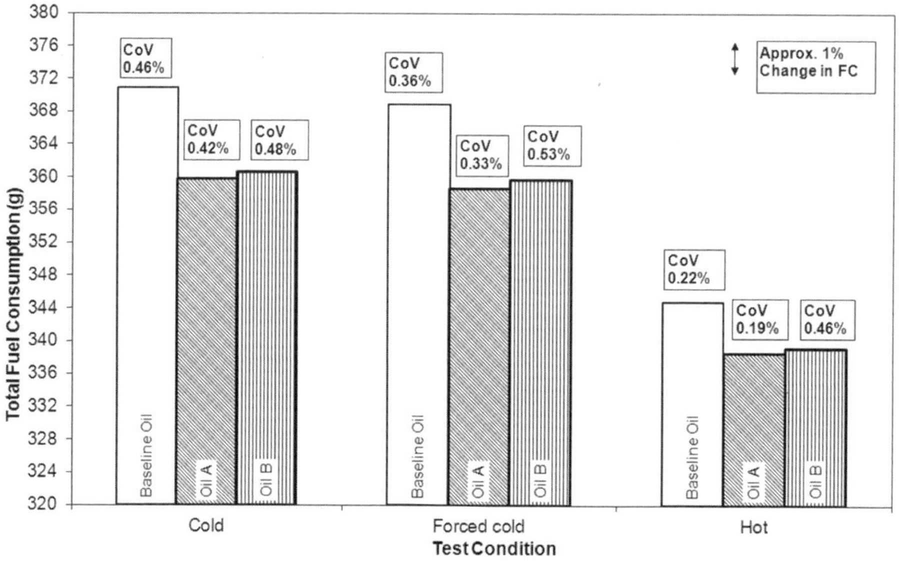

Figure 2.13 Comparison of engine friction power losses for different types of oils.

Figure 2.14 illustrates the reduction in engine friction during a cold start NEDC test cycle. The graph shows friction mean effective pressure (FMEP) for a repeated section of the duty cycle. Frictional losses gradually reduce as the engine warms up. When the engine is fully warm, frictional losses have approximately halved compared to when the engine started at 25°C.

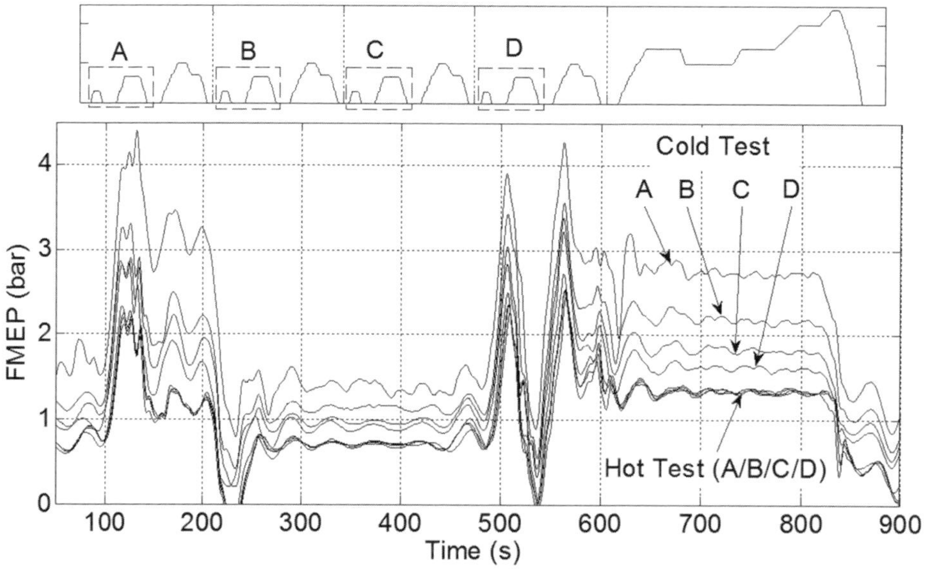

Figure 2.14 Frictional power loss as a function of the lubricating oil sump temperature and engine speed and drop in FMEP over repeated ECE cycle during cold start test.

2.4.3 Engine Warm-Up

The soak temperature of the engine and the ambient conditions all influence the warm-up rate of the engine. Figure 2.15 shows the warm-up profile from a 2.4-L diesel engine over the NEDC cycle (which starts following an overnight soak at 25°C). The coolant warms up faster than the oil for most vehicles and for certain cooling system layouts, the water temperature can rise substantially faster than the oil temperature.

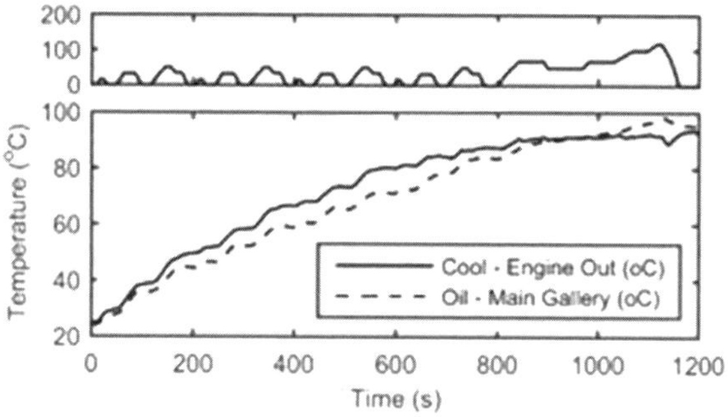

Figure 2.15 Variation of lubricating oil and water temperatures for a typical diesel engine over NEDC.

The cooling system itself will be covered in Chapter 3, but for the purpose of this chapter, it is important to note the following:

1. Significant changes in oil temperature exist that will affect the engine and vehicle performance because of the effect on friction.
2. The warm-up phase of a particular trip can be quite significant; however, the shorter the trip is, the greater the cold start period will affect the overall fuel use. Some studies have correlated the engine fuel consumption with the distance of the trip in order to try to normalize for this warm-up phase:

$$\frac{FC_{COLDSTART}}{FC_{WARMEDUP}} = 1 + L_t^{-0.8} \tag{2.20}$$

where $FC_{COLDSTART}$ = fuel consumption at cold start, $FC_{WARMEDUP}$ = fuel consumption when the engine is already warmed up, and L_T = distanced traveled in miles.

2.5 Engine Lubrication

It is beyond the scope of this text to present a full overview of the science of friction (tribology). Rather, an overview of the main factors affecting the operation and efficiency of the lubrication are given such that the implications for test cell design and operation may be considered.

The basic function of the lubricant is to reduce friction and prevent wear of moving parts. However, in the engine, there are many other functions that the lubricant has to fulfill. One major aspect is cooling of engine components. Although a water cooling system is installed, a significant amount of heat is evacuated by the lubricant from the combustion chamber, the valve train, and the turbocharger. Modern downsized engines now include oil water heat exchangers to cool the oil, as convection from the sump alone is not sufficient. During warm-up, these oil coolers can be used as oil heaters to improve the warm-up rate of the engine oil and reduce friction from cold start. The engine oil is also required to be compatible with all the components of the engine and should not attack any of the surfaces. This can become particularly problematic when the oil becomes contaminated with combustion by-products. The engine oil needs to be stable over its full temperature operating range, which can be from a temperature low as −40°C for extreme cold starting up to a temperature of 150°C for full load operation in warm climates. The oil is required to provide sufficient viscosity over this range, but also all other aspects including flowability at low temperatures.

It is desirable to have a single lubricant for the whole engine system; however, this means that the lubricant must cope with different lubrication regimes (boundary, hydrodynamic, and elastohydrodynamic). The oil must also preserve its characteristics for long periods to ensure long service intervals. An overview of an engine lubricating circuit is shown below. A typical engine oil circuit is shown in Figure 2.16 and illustrated in the sketch in Figure 2.17. Oil is pumped from the sump through a filter to a main gallery. From here, it is distributed to the main engine components (crankshaft, valvetrain, piston cooling jets, and turbocharger). Internal oil passages are drilled into the crankshaft to deliver oil from the main bearings between the shaft and the engine block to the big-end bearings at the base of the connecting rods.

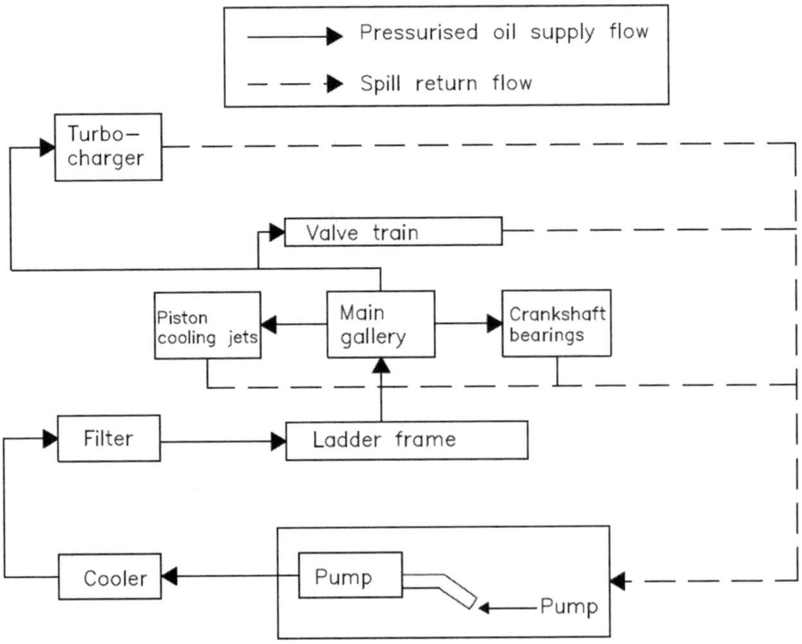

Figure 2.16 Typical gasoline engine lubrication circuit.

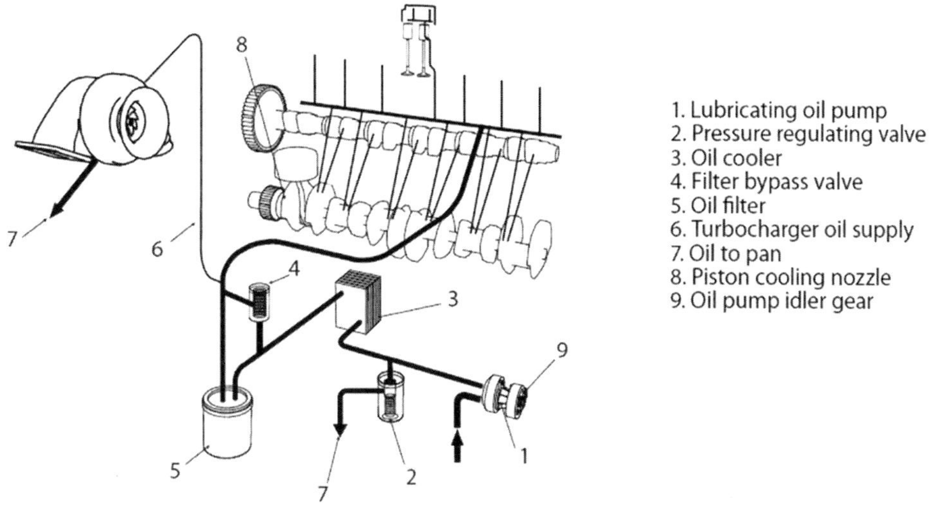

Figure 2.17 Engine oil lubrication circuit.

2.5.1 Oil Pump

The oil pump is powered by the vehicle engine and pumps the oil through the oil filter to the engine and bearings. The oil pump is usually a gear-type or rotor-type, driven directly from the crankshaft. Oil pressure varies widely during the operation because of the relatively crude pressure control systems employed, which must balance sufficient pressure when hot conditions and low speeds are encountered with acceptable pressures in cold conditions at high engine speed. Lower temperature and higher engine speeds increase the pressure to a maximum of around 4.5 bar.

An oil pump is dimensioned for the worst-case operating conditions which is hot idling (low pump speed and low oil viscosity). In all other operation areas, the oil pressure has to be regulated and limited by a pressure control or relief valve, which causes losses and a reduction in efficiency. Typically, a state-of-the-art oil pump has an efficiency of about 60% to 80% on component level. Nevertheless, the efficiency on the system level is lower because of the above-described characteristics. State-of-the-art oil pump systems can vary speed or displacement to match oil delivery with operating demands. This minimizes the excess flow (and power) wasted to sump. The graphs of Figure 2.18 show the oil pump power demand from such a modern system at two different constant pressures, 2 and 3 bar. For comparison, the power absorbed by a fixed displacement system is plotted along with that at maximum displacement of the variable delivery system. Because of the related pump speed to engine speed, the oil flow at high speeds exceeds by far the engine requirements resulting in high losses.

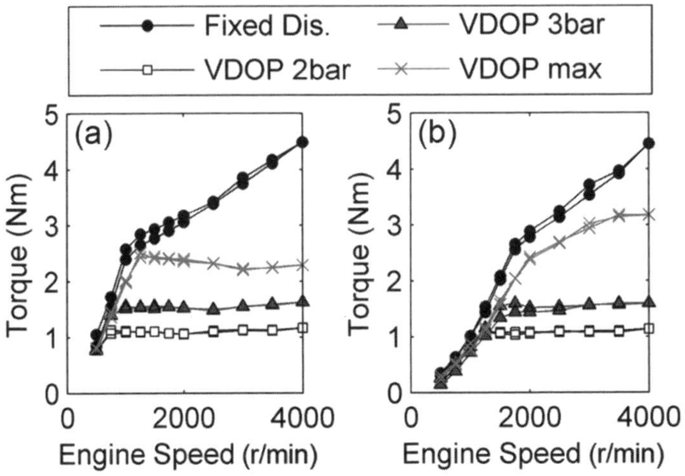

Figure 2.18 Effect of variable displacement oil pumps on parasitic power requirements for pumps with (a) 2-bar and (b) 3-bar delivery pressures compared with a fixed displacement system.

2.5.2 Lubrication Oil Properties

2.5.2.1 Viscosity

The most important property of a lubricant is its viscosity. In a journal bearing, the viscosity must be high enough to avoid being squeezed out from between the surfaces under load; however, a too-high viscosity will result in higher hydrodynamic friction and overall reduced efficiency.

In the first part of the 1900's, monograde oils were available with different viscosity levels measured depending on the performance at 100°C, which was deemed indicative of fully warm engine operation. These were classified using SAE Standard J300 using a numerical system where a higher number indicates a high viscosity. However, using oils with a high viscosity at 100°C resulted in excessively high viscosity at cold start temperature, which affects both fuel economy and performance because of higher frictional losses. Therefore, motorists would typically switch between oil grades between summer and winter months. In 1952, the standard was updated to include winter grade designations using the letter "W" and determined by the oil viscosity at temperatures ranging from 0°C to −30°C. Later, multigrade viscosity oils were introduced with cold and hot grades.

The trend in viscosity grades used in passenger car engines shows a distinct reduction in viscosity since the turn of the 21st century, and is forecast to continue. In 2000, the dominant oil grades were SAE 10W-30 and 10W-40, whereas in 2020, the prediction is that 5W-20 and 5W-30 will dominate. It is also expected that 0W-XX grades will emerge emphasizing the trend for thinner oils as material and manufacturing technologies allow for better precision in the friction components. This trend has been driven by a need to improve fuel economy and cold start fuel economy. When cold, the lubricant with lower viscosity will circulate more easily and reduce friction.

2.5.2.2 Viscosity Measurement Methods

There are two methods used to describe viscosity using two different units.

2.5.2.2.1 Dynamic Viscosity

It is closely related to the "thickness" of oil and does not depend on the density of the sample. It is expressed using the units of centipoise as below, using the symbol μ:

$$\frac{\text{kg}}{\text{m}\cdot\text{s}} = \text{mPa}\cdot\text{s} = \text{cP} \tag{2.21}$$

2.5.2.2.2 Kinematic Viscosity

It is equal to dynamic viscosity divided by the fluid density and is expressed as shown by:

$$\upsilon = \frac{\mu}{\rho} \tag{2.22}$$

Viscosity is characterized at different conditions according to standardized tests designed to represent the different conditions encountered in automotive applications. Viscosity is typically characterized by measuring the flow of oil under its own weight in a capillary tube. When tested in this way at 100°C this determined the SAE viscosity grade (see Table 2.2). Other measurement techniques exist that measure viscosity under high shear conditions, for example, the cold cranking simulator (CCS), which is used to determine the winter or "W" SAE grading or the high-temperature high-shear (HTHS) viscosity, which is more representative of conditions seen on the engine.

Table 2.2 SAE Viscosity Classification

SAE grade	CCS viscosity (cP)	100°C kinematic viscosity (capillary tube) (cSt)	HTHS viscosity (cP)
0W	3250 (−30°C)	3.8	
5W	3500 (−25°C)	3.8	
10W	3500 (−20°C)	4.1	
15W	3500 (−15°C)	5.6	
20W	4500 (−10°C)	5.6	
25W	6000 (−5°C)	9.3	
20		5.6	2.6
30		9.3	2.9
40		12.5	2.9–3.7
50		16.3	3.7
60		21.9	3.7

The low-temperature cranking viscosity test uses a close-fitting rotor and a stator at a controlled temperature. The oil is placed in the gap and the amount of energy required to keep the motor running at a reference speed is recorded. The thicker the oil, the greater the viscous drag will be, and so the energy required will indicate the viscosity.

The low-temperature pumping viscosity test determines the temperature at which the oil reaches a 60.000-cP viscosity, and watches for any signs of the oil beginning to solidify.

The low-shear kinematic viscosity test determines a minimum viscosity at 100°C. Unlike the other tests, it is measured in centistokes reflecting the influence of fluid density.

The combination of these viscosity tests demonstrates the ability of an oil to avoid excessive viscosity when starting at low temperatures such that the oil pump cannot pump it through the oil galleries, and that the oil has sufficient viscosity to guarantee lubrication when reaching high temperatures of 100°C.

Testing of oil performance in engine, rather than a simplified test piece, is very difficult, even with well-developed statistical techniques, because differences tend to be smaller than the uncontrolled variations of engine performance from test to test. Therefore, comparing and quantifying the actual benefit of friction modifiers is challenging.

2.5.2.3 Viscosity Index

Viscosity index (VI) is an indication of the change in oil viscosity with temperature. It is typically measured using two viscosity measurements from the capillary tube method, at 40°C and 100°C. It is an arbitrary scale based on a comparison of the oil viscosity change with that of two reference oils. It is expressed as a percentage (although the % sign is often omitted) with 100 representing the same behavior as the best reference oil and 0 the worst reference oil. This scale was established before the use of additives that are capable of modifying the temperature/viscosity behavior and was initially used as a method for comparing base oil stocks. However, in recent times with the engineering of lubricants through viscosity modifiers, oils with VI greater than 100 are available. Figure 2.19 shows the difference in viscosity/temperature for a single-grade and a multigrade oil.

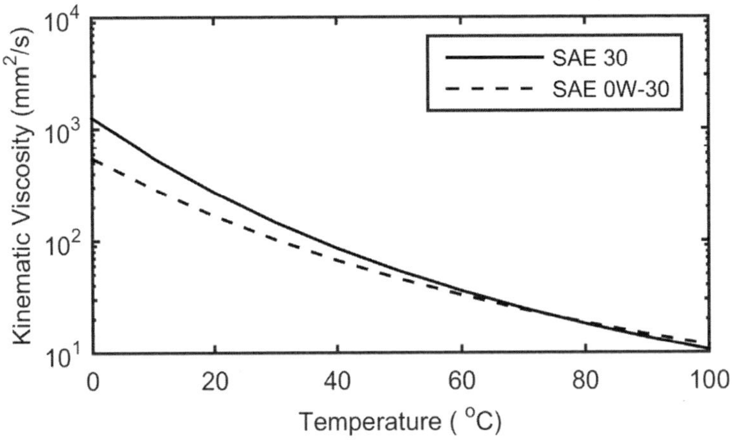

Figure 2.19 Single-grade and multigrade oil viscosities as a function of temperature.

2.5.2.4 Additives

A number of additives can be added to base oil stocks to improve their properties for particular applications and, in particular, the ICE. These can allow for features such as reducing the minimum operating temperature of the lubricant (pour point depressants), modifying the viscosity/temperature relationship (viscosity modifiers or VI improvers), avoiding oxidation of the lubricant, avoiding the formation of deposits on parts, notably bearings and piston liners (detergents), avoiding contamination from soot and water (dispersants), ensuring compatibility with bearing materials (anti-wear), improving boundary friction (friction modifiers), avoiding corrosion of engine parts (rust and corrosion inhibitors), and avoiding the formation of foams (anti-foam).

It should be noted that some additives can fulfill a number of the aims described above, and there is not necessarily one additive for each function. Some of the more pertinent additives for friction are as follows:

- *VI Improvers:* These additives are molecules that increase the viscosity of the base stock at higher temperatures while having little effect on the viscosity at low temperatures. The additive is made of polymer chains that interact heavily with the oil at high temperatures and become completely surrounded by oil molecules.
- *Friction Modifiers:* These additives aim to reduce the friction coefficient between two surfaces when the boundary lubrication regime prevails and conditions are such that a fluid film cannot be sustained. Such conditions occur regularly in the valve train but can also exist between the piston and the liner at top and bottom ring reversals, and in the crank journals under the high loads at firing conditions.

2.5.3 Minimum Starting Temperature

Under cold ambient conditions, lubricants increase in viscosity to the point that they will not flow into the oil pump through the oil galleries. This is especially critical in engines that require good lubrication flow to turbochargers.

As explained above, SAE defines some specifications to assign a winter grade to an engine by performing a cranking simulation test and a low-temperature pumping test. Figure 2.20 shows a study performed by ASTM [2-3] in 1990 to check the correlation between SAE grades and minimum starting temperatures (MSTs) for different types of engines. MST is defined as the temperature at which an engine could not start after six consecutive cranking attempts. Oil viscosity, along with fuel volatility, engine control system, and calibration are all critical factors in achieving acceptable MST.

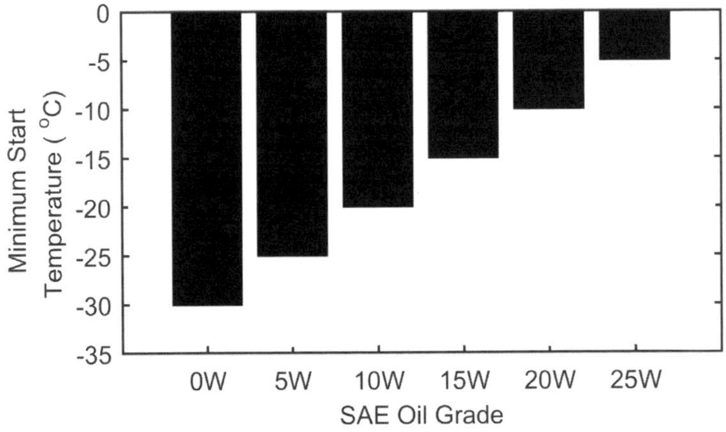

Figure 2.20 Average MSTs for different engines with different oil grades.

References

2-1. Heywood, J. 2011. *Internal Combustion Engine Fundamentals*. McGraw Hill Education: New York.

2-2. Stone, R. 2012. *Introduction to Internal Combustion Engines*. Oxford Press: Oxford.

2-3. Ricardo. "Study of Methods for the Assessment of Mobile Air Conditioning Efficiency." Project Number Q50642. Document Number RD10/215501.2.

Chapter 3
Engine Ancillary Devices

3.1 Introduction

Mechanical power supplied by the engine is delivered at the crankshaft, but not all of this power is available for vehicle traction, as some of it must be supplied to engine accessories such as the alternator, air conditioning compressor, power steering pump, and water pump. In automobiles, these devices are typically driven by belt drives placed in front of the engine, and for this reason, they are called the front engine accessory drive (FEAD), as shown in Figure 3.1.

The power supplied to the accessories is directly subtracted from the engine and is often referred to as parasitic losses. Any vehicle will incorporate some or all of these components, and if a meaningful estimate of vehicle performance is the object of a test campaign, it is important to have a good understanding of how these systems operate and the magnitude of power they are likely to consume. In the following section, we will introduce a number of these accessories. Some are vital to the vehicle health such as the coolant pump, while others are optional luxury items that could be seen as a requirement of the driver (power steering and air conditioning). The engine oil pump does not feature in this chapter as it was covered in Chapter 2 because of its close links with engine performance.

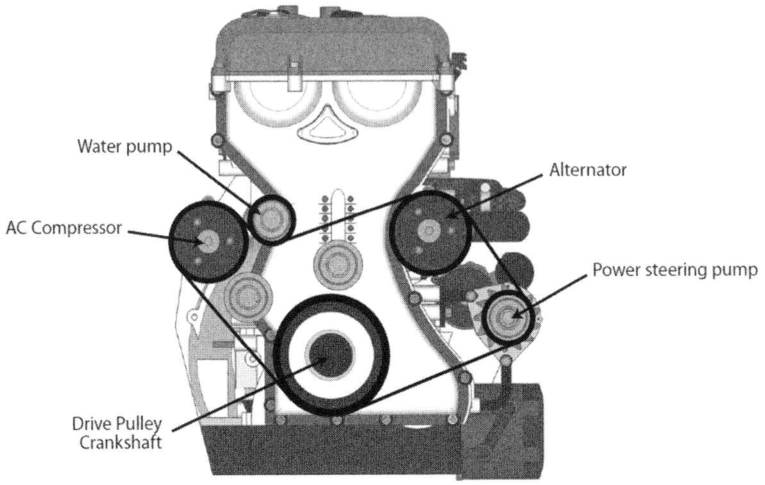

Figure 3.1 Typical FEAD arrangement.

3.2 Power Steering

3.2.1 Introduction to Power Steering Systems

Conventional power steering systems obtain the power from a hydraulic pump, which is connected to the engine accessory drive. These systems are called hydraulic power-assisted systems (HPAS).

In a hydraulic power steering system, a hydraulic pump coupled to the engine shaft supplies the necessary oil pressure to either side of a valve, as indicated in Figure 3.2. When the steering wheels turn clockwise or counterclockwise, oil pressure builds up, more on one side of the valve than the other, thus assisting the turn. In the straight-ahead position, the pressure on both sides of the valve is the same, thus neutralizing any assistance.

The moment at which the greatest steering servo assistance is required, mainly when parking, is when the engine speed is the lowest. The pump capacity is designed for this case. The faster the engine speed, the higher the pump flow that is delivered, even though this is usually not required. In this condition, the excess of oil flow is dumped by a bypass valve. The functional performance of such systems is very good, but its energy efficiency at system level is very low. Over the course of a drive cycle, these losses may account for 0.5% of the total fuel used in an average passenger car.

To reduce this energy consumption, engine manufacturers have developed different systems that adapt steering pump flow supply to specific requirements not depending on engine speed. One increasingly popular method is to drive the power steering pump using an electric motor, mechanically disconnected from the engine. This is known as an electrically powered hydraulic system (EPHS) (Figure 3.3). Such a system will be less efficient at the component level due to the inefficiencies in generating and storing electrical energy on board, and then converting it back to mechanical power. At a system level, however, this approach allows significant power reduction because the system is only switched on and loading the engine when it is needed, and can be completely isolated during straight-line and higher speed driving. Typically, electrically powered hydraulic steering reduces the power consumption by 80% (from around 500 to 100 W) for a typical passenger car application. This advantage is increased further because techniques such as "smart" battery charging on overrun are employed.

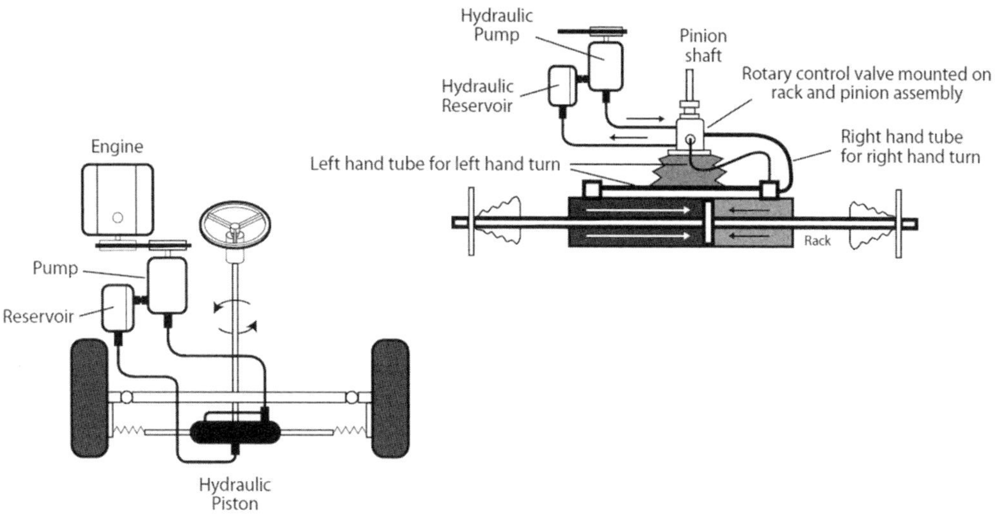

Figure 3.2 Typical schematic of an HPAS.

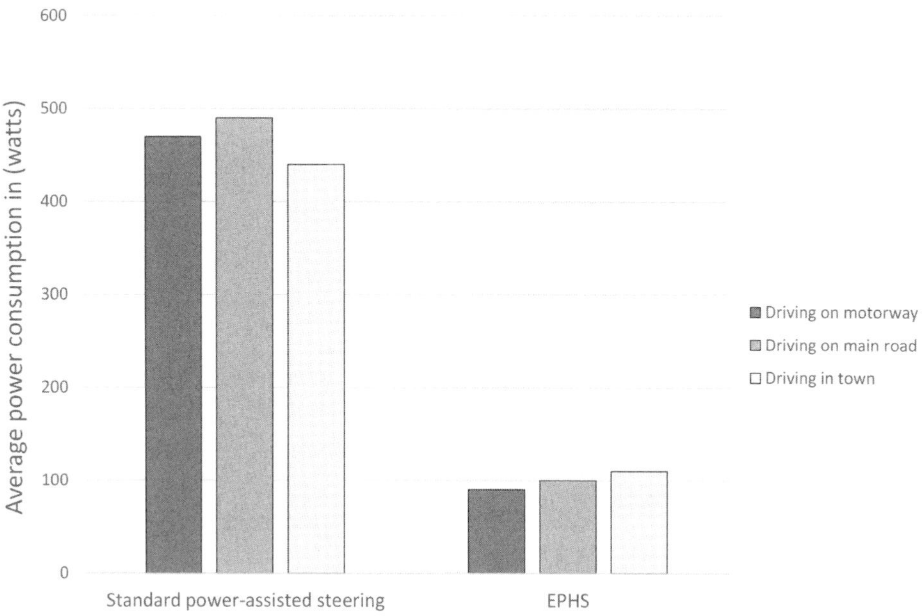

Figure 3.3 Advantages of EPHS.

3.2.2 Factors Affecting Hydraulic Power Steering Efficiency

Apart from the efficiency of the power steering system itself, it is important to know what factors are affecting the efficiency of a given system.

Hydraulic pumps convert the mechanical energy of the prime mover into fluid flow. Pressure is generated by the restriction of this flow in the system. Unfortunately, the hydraulic pump cannot be 100% efficient in this energy conversion. A typical hydraulic pump is 80% to 90% efficient, with energy lost in two main forms:

- *Mechanical Losses:* Energy lost to fluid friction.
- *Volumetric Losses:* Energy lost as the result of internal fluid leakage within the pump.

Mechanical losses are highest when fluid viscosity is high and volumetric losses are highest when fluid viscosity is low. Viewing these two curves on the graph in Figure 3.4 illustrates a viscosity range for optimal efficiency. Because hydraulic fluid viscosity is high at low temperatures and decreases as the fluid temperature rises, staying within this optimal operating range is not simple.

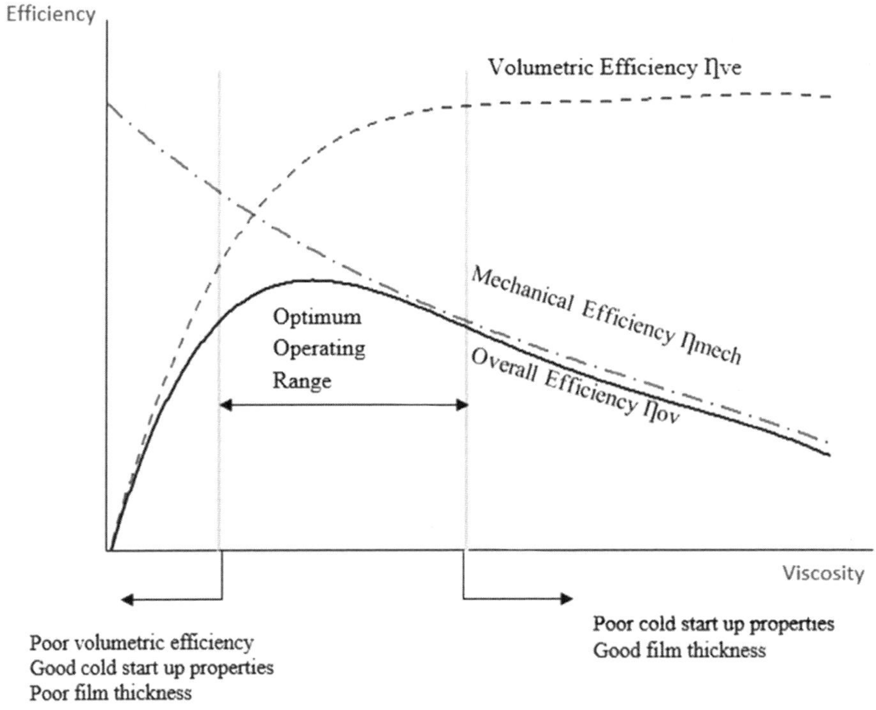

Figure 3.4 Hydraulic pump efficiency versus oil viscosity.

Oil temperature will therefore be an important parameter when reproducing testing conditions for a power steering system and its influence on the overall system efficiency.

3.3 Air Conditioning Compressor

3.3.1 The Refrigeration Cycle

Conventional vehicle air conditioning systems work under the same principle as any standard refrigeration system. They are composed of a gas compressor, a condenser where the gas is cooled down and transformed into liquid, an expansion valve, and an evaporator where air passing to the passenger car compartment is cooled. A typical layout is shown in Figure 3.5.

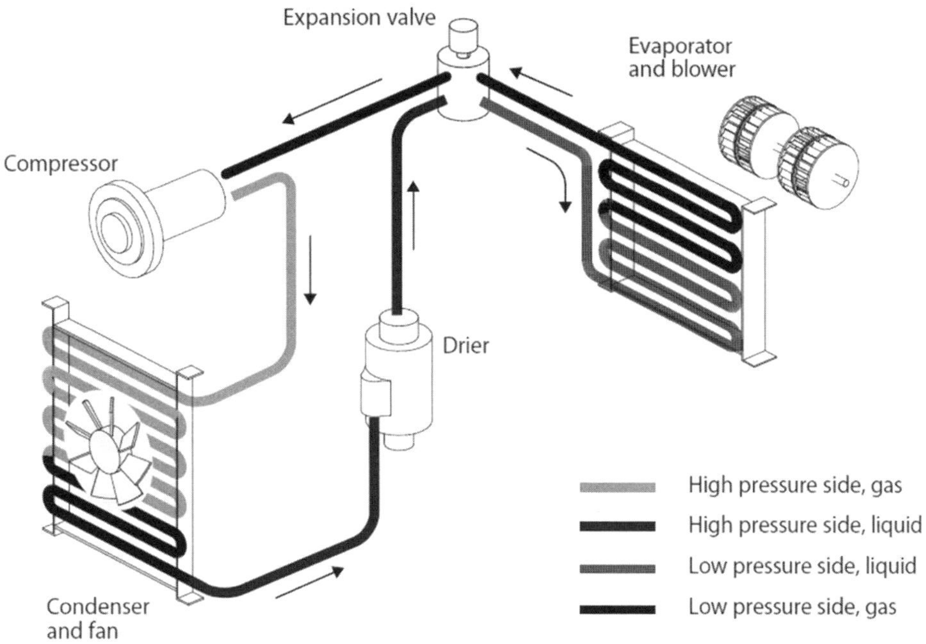

Figure 3.5 AC loop components.

3.3.2 Vehicle Air Conditioning

The automotive application places very special demands on the air conditioning system. A typical vehicle system has a similar cooling capacity to that required for the air conditioning of a small house despite the great size difference.

The first reason for this is that the overall heat transfer coefficient in the vehicle is much greater due to the forced convection between the air and the vehicle created by the relative vehicle speed. Second, the proportion of passenger compartment enclosed by glass is very high and has a great effect on heat input to the vehicle from solar radiation. In addition, it is usually an operational requirement to cool down the vehicle in a very short time after the vehicle has been soaked to high temperatures due to solar radiation during summer conditions. The external heat transfers between the vehicle and its environment are primarily by radiation and convection, as summarized in Figure 3.6.

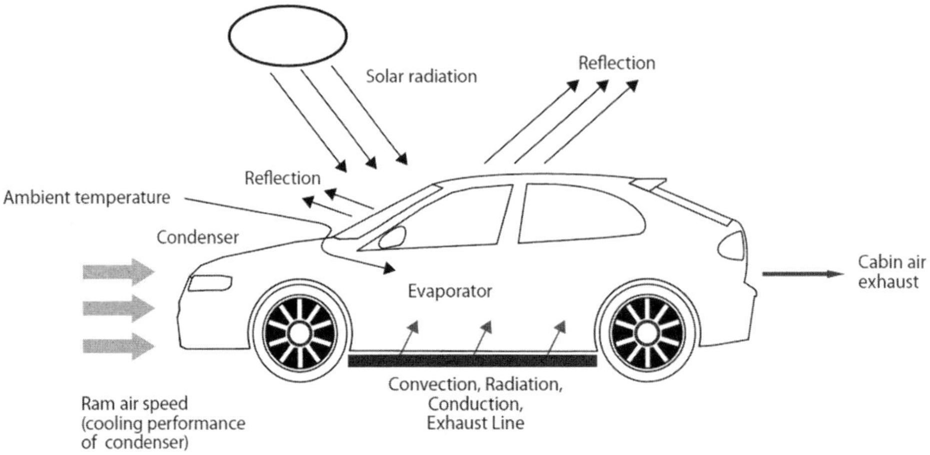

Figure 3.6 Heat and air flow balance into the vehicle affecting air conditioning.

3.3.3 Vehicle Air Conditioning Control

As shown in Figure 3.1, the most common powering method of a vehicle air-conditioning compressor is by connecting the compressor shaft to the FEAD. The main problem resulting from this is how to adjust cooling power requirements in the cabin when the compressor is connected to the variable speed engine, and using a constant displacement compressor, which are the most common. The simplest solution comes from the installation of a clutch in the compressor pulley, which can be connected and disconnected depending on the power requirements. Typically, the air conditioning will only operate if any of the following conditions occur:

- The passenger cabin blower is active.
- The AC is requested by the driver.
- A thermostatic switch based on cabin temperature indicates a high cabin temperature.
- The pressure in the AC loop is neither too high nor too low.
- The ECU is switched on.

In Figure 3.7, the cooling power is controlled by a thermostatic switch that senses cabin temperature, and a low-pressure switch that controls suction pressure of the compressor, which gives a measure of the refrigerant that can be evaporated.

With these two signals and the manual AC switch, the compressor clutch is activated or deactivated, controlling power demand.

Other signals are implemented as safety measures, such as the blower switch to guarantee that there is air blowing on the condenser, and a high-pressure switch for protection of the compressor. This kind of control is simple to integrate but results in the compressor working in a range that is usually not the most efficient. Other systems include variable displacement compressors and even compressors powered by an electrical motor, which allows a wider control range and adjustment in a more efficient area, at a penalty in cost and complexity.

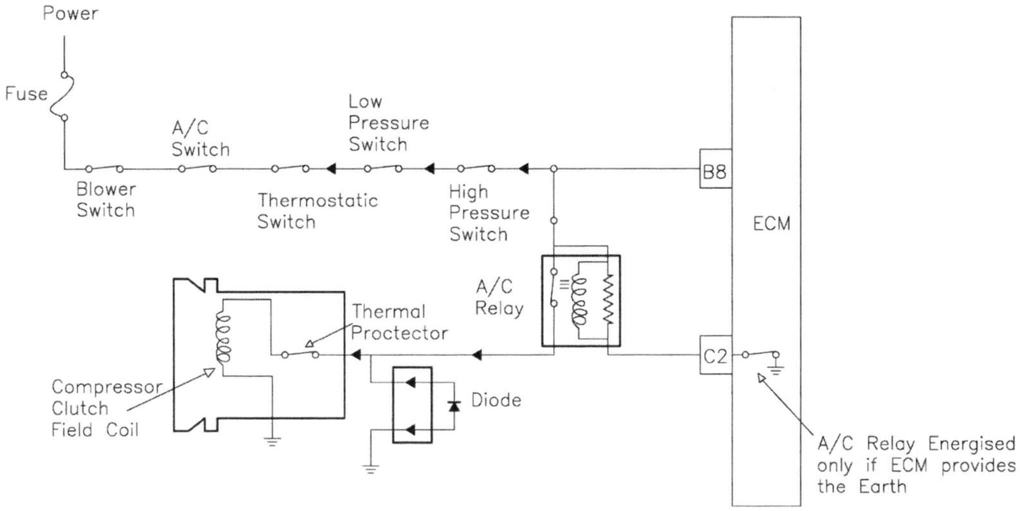

Figure 3.7 Diagram of the cooling power control with a clutch.

3.3.4 Dimensioning a Vehicle Air Conditioning System

The following section will show an example of the design of a vehicle air conditioning system to give an idea of the magnitude of the main factors affecting the component dimensioning. The example is taken from [3-1].

The first task is to define the ambient and vehicle parameters—these are summarized as an example in Table 3.1.

Table 3.1 Example of the Air Conditions for the Ambient and Vehicle Parameters

Ambient temperature	52°C
In-car temperature	24°C
Solar radiation rate	900 W/m^2
Driving speed	55 mi/h (88.5 km/h)
Fuel mileage	28 mpg
Air density at 52°C	1.1 kg/m^3
Air density at 24°C	1.3 kg/m^3
Thermal conductivity of glass	1.4 W/m·K
Thermal mass of windshield	0.25 in (6.35 mm)
Area of windshield (vehicle sample)	1.40 m^2 (100 kg)
Total window area (vehicle sample)	3.41 m^2
Maximum AC pulldown time	120 s

The vehicle heat load is dominated by two large heat sources: solar radiation and body conductance. It is a surprising result that when the car is moving, the largest single heat load on the system is not solar but rather heat conduction through the vehicle windows. Solar irradiation at the surface of the earth is considered as 900 W/m^2. The total window area of the vehicle considered is 3.41 m^2. Considering that at most only half of the vehicle can be exposed to the sun, we have a maximum window area of 1.7 m^2. When the sun is at an angle of 30°, this is reduced to 1.5 m^2. Considering glass transmittance equal to 1, we get a solar heat of 1.35 kW.

Now consider the heat load on the vehicle caused by air convention as a result of flow over the body panels and windows of the vehicle. For moderate vehicle speeds, the body conductance varies roughly linearly with vehicle speed, as illustrated in Figure 3.8. This heat transfer coefficient captures the total heat transfer that occurs from the air inside the passenger cabin to the ambient air flowing over the vehicle. It will include convective terms on the inner and outer face as well as conduction through the body structure. Local values on the vehicle could be very different, such as those on windows or body panels, but the values presented here give an overall approximation for the whole vehicle. When this vehicle is traveling at a steady speed of 88.5 km/h (55 mi/h), this body conductance has a typical value of 80 W/K.

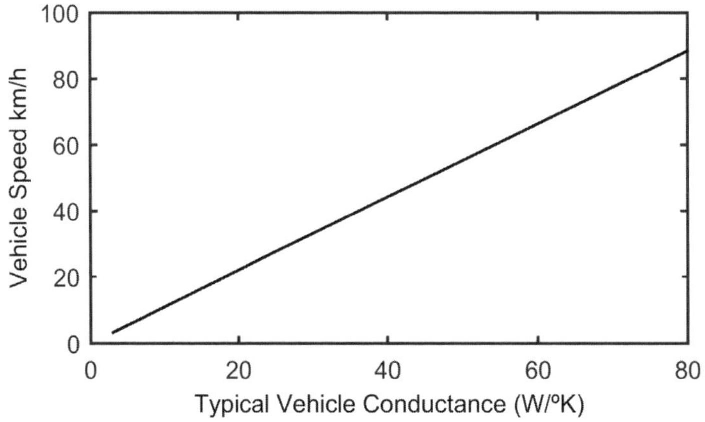

Figure 3.8 Typical vehicle conductance (km/h) versus speed (W/K).

Considering an extreme temperature difference between vehicle and outdoors of 28°C, and the vehicle traveling at 88.5 km/h, yielding a heat transfer coefficient of 80 W/K, we calculate a heat load of 2.24 kW.

There is an additional heat load that must be accounted for in the energy balance due to heat generated by the passengers and the induction of fresh air from outside the vehicle. Out of these two, the most important one is the amount of fresh air taken from outside. This load is very variable, and can go from 100% fresh air to 0% fresh air depending on the adjustment selected by the driver. A conservative amount of air replacement is 0.25 m^3/min, which will be what considered in this example.

Considering outside at 52°C and inside air at 24°C, we have to cool down the fresh air 28°C, which represents 0.12 kW. The passengers can be considered as releasing 120 W per person. Assuming two people in the car, we have 240 W. Finally, we have heat passing from the engine and exhaust into the passenger compartment by conduction, which we can consider as 0.50 kW. Gathering all these factors, we obtain a total of 4.45 kW.

We can summarize the heat sources affecting the sizing requirements of the air conditioning system in Figure 3.9. We can see the importance of solar radiative load and the vehicle conductance because of the speed of the air relative to the vehicle. When simulating air conditioning loads, it is important to bear in mind the source of the main factors.

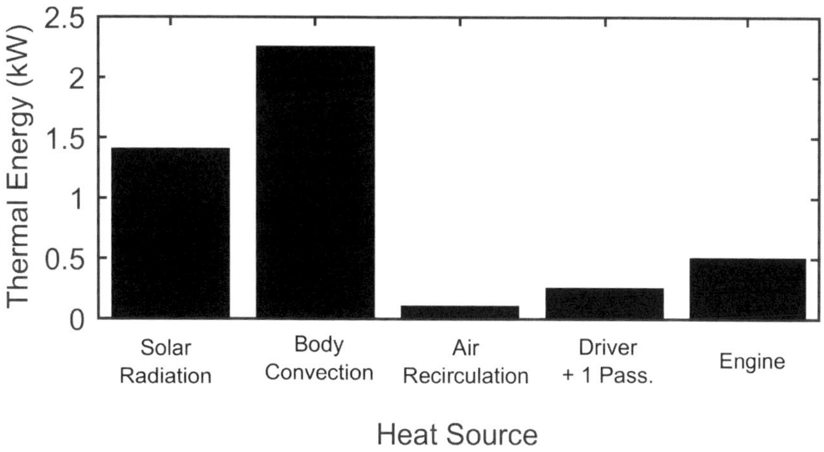

Figure 3.9 Example of AC heat load distribution.

3.3.5 Factors Affecting Vehicle Air Conditioning Performance

3.3.5.1 Influence by Solar Radiation

The design of a car air conditioner is greatly influenced by the maximum cabin (soak) temperature attained when the car is parked on a hot sunny summer day. The soak temperature of the car is strongly influenced by solar gain through its windows and opaque shell. In the past decade, researchers and car manufacturers have used near-infrared reflective windows to reduce solar gain, lowering the soak temperature by several degrees. Using exterior paints with high solar reflectance can further reduce soak temperature.

The sun radiation spectral density showing the amount of energy at particular wavelengths is shown in Figure 3.10. About 37% of the sun's energy lies in the near-infrared band, while another 44% lies in the visible light spectrum.

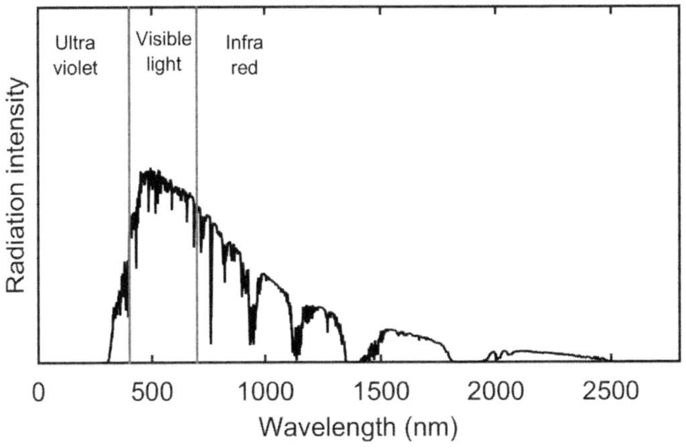

Figure 3.10 Emission of the sun spectrum.

A strategy to reduce heat load in the vehicle consists in filtering the heat load in the near infrared while maintaining visible light spectrum without strong filters. This can be achieved by having a transmittance spectrum as shown in Figure 3.11.

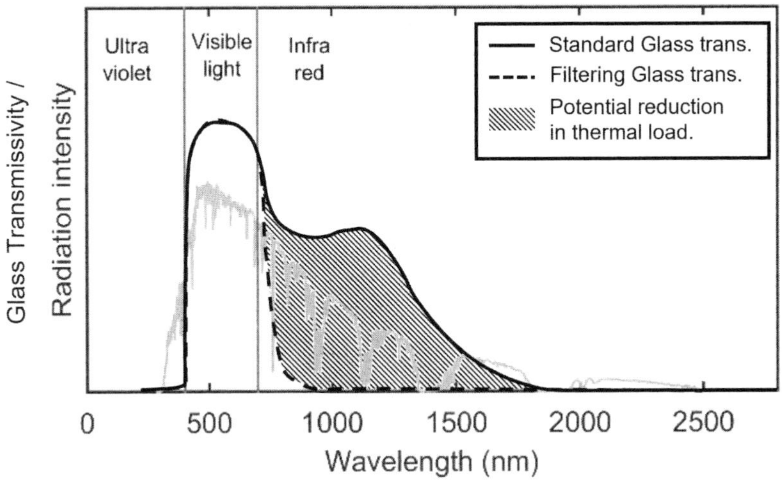

Figure 3.11 Transmittance of solar-reflecting windshield reducing transmissibility outside the visible spectrum.

The color of the body paint also plays a role in the heat load transferred to the vehicle, as each color has a different reflectance over the light spectrum.

3.3.5.2 The Vehicle Cabin Size

The cabin size has a direct impact on the solar radiation and convection load because of a bigger heat exchange surface area. Figure 3.12, taken from a simulation study, shows the cabin temperature evolution of a vehicle in idle with a starting temperature of 60°C (because of solar radiation) with different cabin volumes. A larger cabin volume takes longer to cool because of a larger thermal inertia of the cabin, but it also settles at a higher temperature because the thermal load is larger.

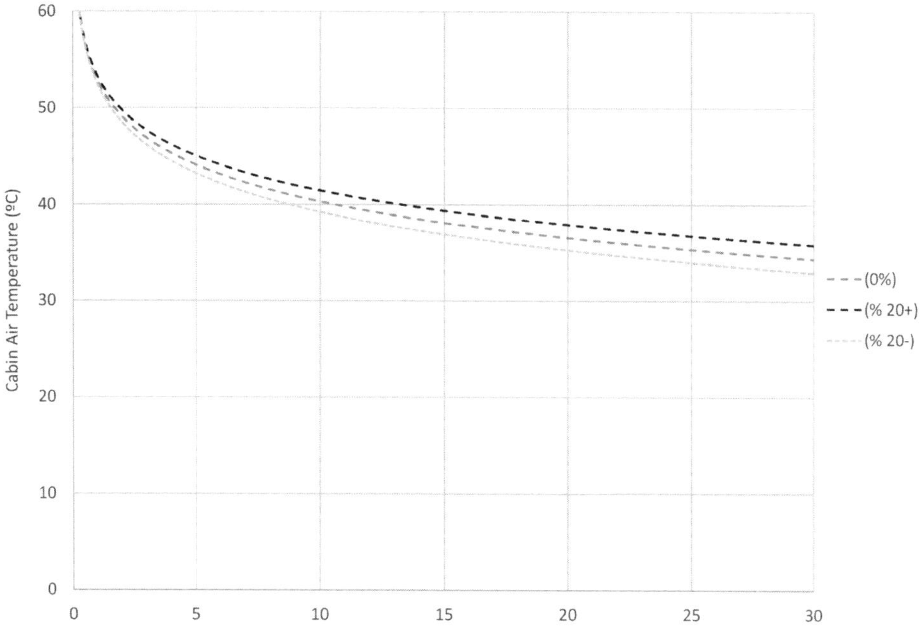

Figure 3.12 Variation of cabin air temperature with time (minutes, displayed in the *X*-axis) for various cabin air volumes (modeling results) for a 4100-W tested evaporator during idling conditions.

3.3.5.3 Fresh Air Ratio (Air Recirculation)

Most air conditioning systems allow the user to select the source of air from fresh air and recirculated air. Some others perform this adjustment automatically depending on the quality of the air inside the vehicle. The amount of fresh air introduced and its temperature and humidity have a direct impact on the energy requirements for the cooling system because ambient air needs to be conditioned to cabin requirements, in terms of temperature and humidity, and cabin air is already conditioned.

Figure 3.13 shows the impact on energy required depending on the amount of recirculated air and ambient conditions including temperature and humidity. Irrespective of whether the outside air is hotter or colder than the cabin air, there is always a higher power requirement for using outside air rather than recirculating. The disadvantage of recirculating air is the buildup of humidity within the cabin.

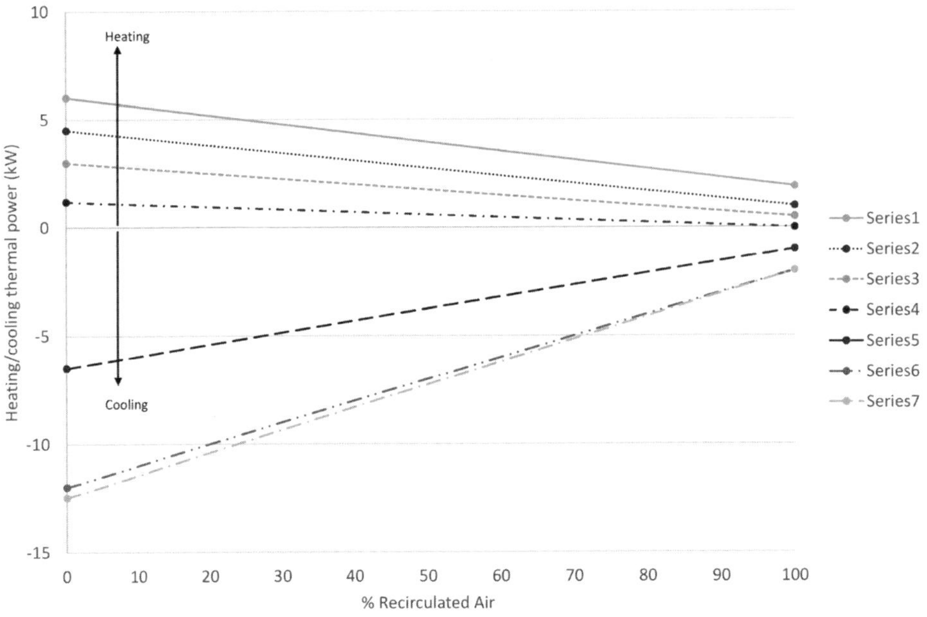

Figure 3.13 Heating/cooling thermal power as a function of percent in recirculated air and ambient conditions.

3.3.5.4 Vehicle Speed

Vehicle speeds have different opposing effects on the cabin temperature. High air speed blowing over the air conditioning condenser will reduce condensation temperature and, therefore, improve air conditioning efficiency.

Conversely, high air speed around the vehicle cabin forces rapid heat transfer between the surrounding air and the vehicle. This effect can be positive when cabin temperatures are above ambient temperatures, as usually happens because of solar radiation, and negative when cabin temperature is below ambient temperature.

Heat transfer through the vehicle windows is driven by the temperature difference between the inner face of the window and the outer face. The temperature gradient through the window varies with air speed because of the increase in glass-to-air heat transfer coefficient as air speed rises.

Figure 3.14 shows an example in an extreme condition where the outside temperature is considered to be 52°C and the inside temperature is 24°C. As shown, when the vehicle speed is zero, the outside window temperature is close to the inside temperature, however, when travelling at maximum speed, the outside window temperature reaches 52°C, causing much more heat dissipation.

The same graph can also be read in reverse for cold conditions, exchanging outside and inside position. In this case, window conductivity will help to cool down the vehicle after being heated by the sun when driving at high speeds in a mild temperature.

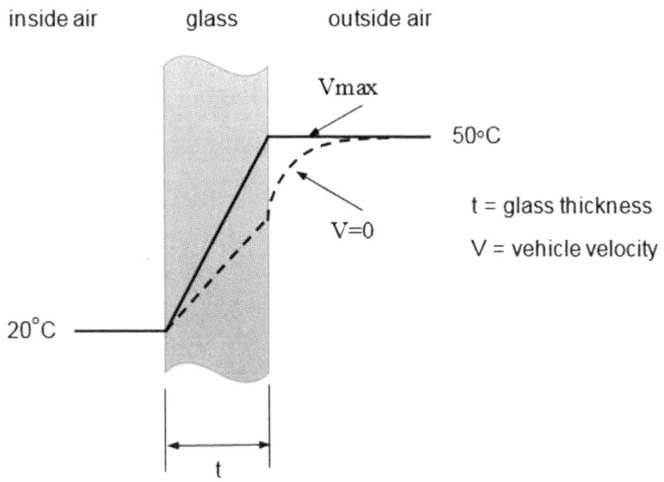

Figure 3.14 Temperature profiles through the vehicle windows, showing the velocity effect.

Vehicle speed is the principal source of air flow over the air conditioning condenser. Higher vehicle speeds will increase heat transfer to the ambient and allow for an overall reduction in the condensation temperature, thus improving the performance of the AC system. This results in a reduced power consumption for a given cooling capacity, or an overall increased cooling capacity should the driver choose to use it.

High vehicle speeds usually demand high engine speeds. As the AC compressor is directly linked to the engine, it will also be driven faster and, therefore, increase the flow of refrigerant in the system, which again improves heat transfer in the condenser. There is therefore a double effect of vehicle speed.

Conversely, the cooling speed at idle is very slow. The air conditioning compressor runs at the engine speed (direct connection) and thus delivers a very small amount of refrigerant to the evaporator. The condenser is only cooled by its fan with a small air flow compared with a moving vehicle. The forced convection of ambient air at 43°C does not help vehicle cooling.

As the vehicle speed increases, the compressor speed increases as does the cooling air flow onto the condenser. The influence of forced convection will be positive in the beginning when cooling the vehicle down to ambient temperature, but negative afterwards.

In a test cell environment, the selection on the vehicle air speed fan has a great influence on the amount of refrigerant that can be evaporated. This will influence the power consumption of the compressor, as well as the vehicle cooling rate.

3.3.5.5 Number of Passengers

The number of passengers in the vehicle has two opposing effects. From one side, we have the thermal load supplied by the passengers in the vehicle, and from the other side, we have a reduction in cabin air volume because of the occupied space. It is assumed that the occupant temperature is governed by the internal regulatory mechanism of the human body, and is not significantly affected by the cabin air temperature. Therefore, the thermal mass of the occupants is neglected.

A detailed analysis on the thermal load supplied by humans can be read in [3-2].

During the cooling down phase, increasing the number of passengers reduces the cooling time because of the reduction in air volume. This effect will be the opposite when the vehicle has already achieved temperature set point, and the factor of the passenger thermal load will have a greater influence than the reduction in cabin volume.

3.4 Alternator

A generator converts mechanical energy into electrical energy, generally using electromagnetic induction. The generator is mechanically driven through a belt at a fixed ratio. Typically, a state-of-the-art automotive generator has an efficiency of about 60%. However, because of the low electrical power demand and the efficiency characteristics of the electrical machine, the average efficiency of a standard generator is only about 40% or 45% (Figure 3.15).

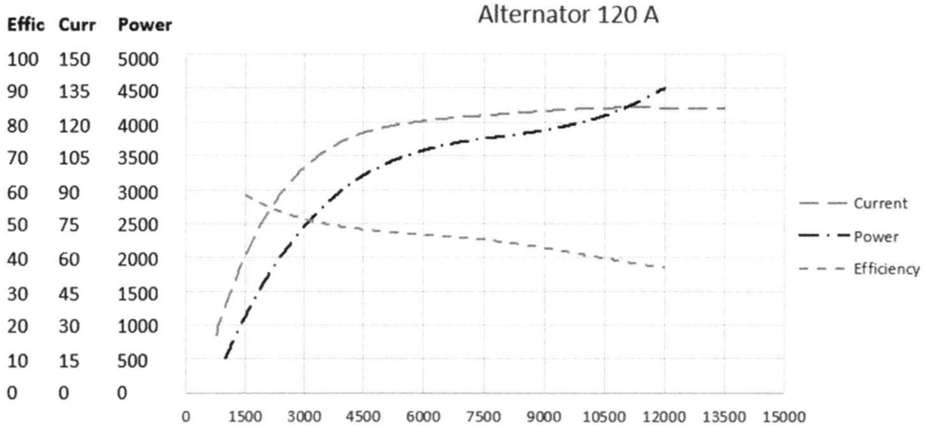

Figure 3.15 Alternator performance curves for a typical passenger vehicle.

The main factors influencing the generator operation are the generator speed, the current level, and the temperature. A simple alternator control strategy regulates output dependent on the inferred state of charge of the battery. This results in a larger load at startup, gradually declining over the operating period. More modern systems preferentially charge the battery during overrun, where they can use vehicle kinetic energy, otherwise dissipated in the brakes, to allow battery charging for almost no fuel cost.

Table 3.2 lists the different electrical loads and typical power consumptions, with their proportion of use.

Table 3.2 Electrical Auxiliaries' Typical Consumption

Auxiliary	Electrical consumption (W)	Use of auxiliary (time proportion)
Dipped headlight	160	During night
Full headlight	170	During night
Turn indicator/stop light	40	1%
Fresh air ventilator	60	50%
Wipers	60	
Radio	15	85%
Rear window defroster	150	50% if outside temperature < 0°C
Seat heating	150	1%

3.5 Water Pump

The water pump is used to supply the engine with sufficient coolant mass flow to maintain engine temperature within a controlled range. As shown in Figure 3.1, it is usually belt driven and its speed is therefore proportional to engine speed.

A typical cooling circuit layout is shown in Figure 3.16. The only feature missing from this circuit is a cabin air heat exchanger, which would apply an additional cooling load to the circuit. As the coolant pump is often mechanically driven from the engine, a thermostat is usually used to control coolant temperature.

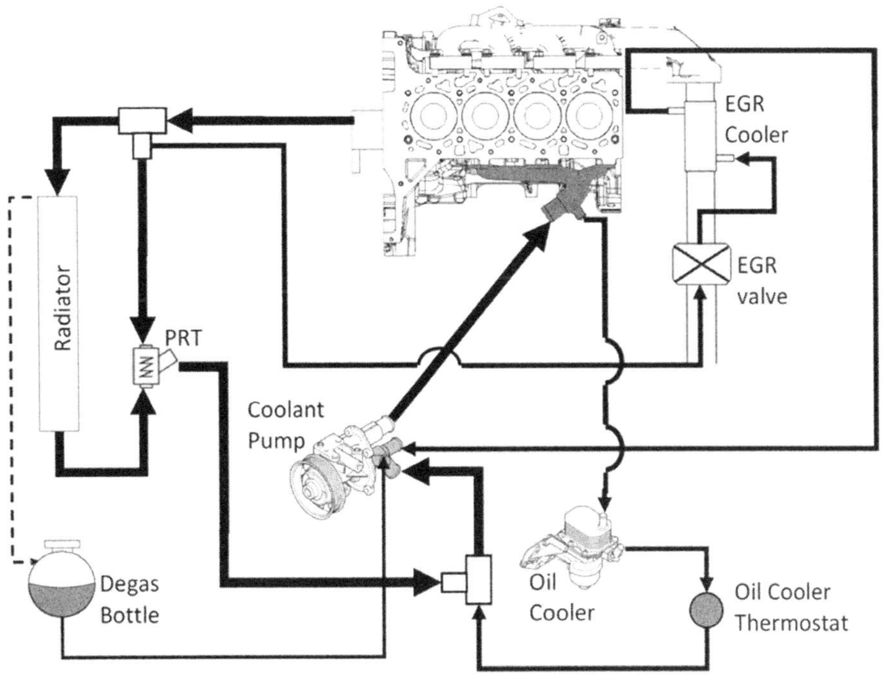

Figure 3.16 An advanced variable flow water cooling system for a four-cylinder diesel engine.

Typically, a state-of-the-art water pump has an efficiency of about 35%. The efficiency on the system level is even lower because of the characteristics of the cooling circuit and the belt efficiency.

A water pump is designed to guarantee its function over a wide range of speed. Hence, in most operation points, the component runs in a less efficient operation mode. To improve the efficiency of water pumps, some modern vehicles have been integrating electrically driven pumps, beginning with BMW since 2004. Again, component efficiencies are lower because of additional energy conversion steps, but the system efficiency can be significantly greater. As the power dissipation to the coolant is roughly proportional to engine power rather than engine speed, the traditional fixed ratio arrangement delivers excess flow for much of the operating range. With an electrically driven pump or other variable flow device, cooling flow can be decoupled from engine speed to allow significant energy savings to be achieved, as well as a more controllable cooling regime.

During the warm-up phase, the role of the cooling system is to promote fast warm-up to reduce frictional losses in the shortest possible timeframe. In some systems, this can involve suppressing cooling flow in the first minutes following startup. However, the cooling system also has to provide cabin heating and demisting capabilities, which presents an additional thermal load on the circuit. Once the engine has reached its operating temperature, this is a useful route to evacuate heat; however, during warm-up, it is an unwanted burden. In some vehicles, an additional fuel or electrically powered cabin heater may be installed to provide a faster supply of hot air to the cabin, but these will have a penalty on vehicle power consumption, which will ultimately result in higher fuel consumption.

References

3-1. Farrington, R., and Rugh, J. *Impact of Vehicle Air-Conditioning on Fuel Economy, Tailpipe Emissions, and Electric Vehicle Range*. National Renewable Energy Laboratory: Golden, Colorado (September 2000).

3-2. Simon Wrigley, Bending, S., Rouaud, C., and Marchi, A. "Study of Methods for the Assessment of Mobile Air Conditioning Efficiency." Ricardo Project Number Q50642. Document Number RD10/215501.2. (June 2010).

Chapter 4

The Pneumatic Tire and Rolling Resistance

4.1 Introduction

The tires are the final interface between the vehicle and the road, and are responsible for the transfer of forces from one to the other. The power transmission chain is shown graphically in Figure 4.1, in which the tire is a critical element as it is partially responsible for the total vehicle performance and efficiency. Rolling resistance, caused by the behavior of the tires, typically uses around 4% of fuel energy [4-1].

This energy consumption depends on several factors, such as the vehicle speed and tire type. Figure 4.2 shows the different energy use paths depending on driving scenario for a VW Golf.

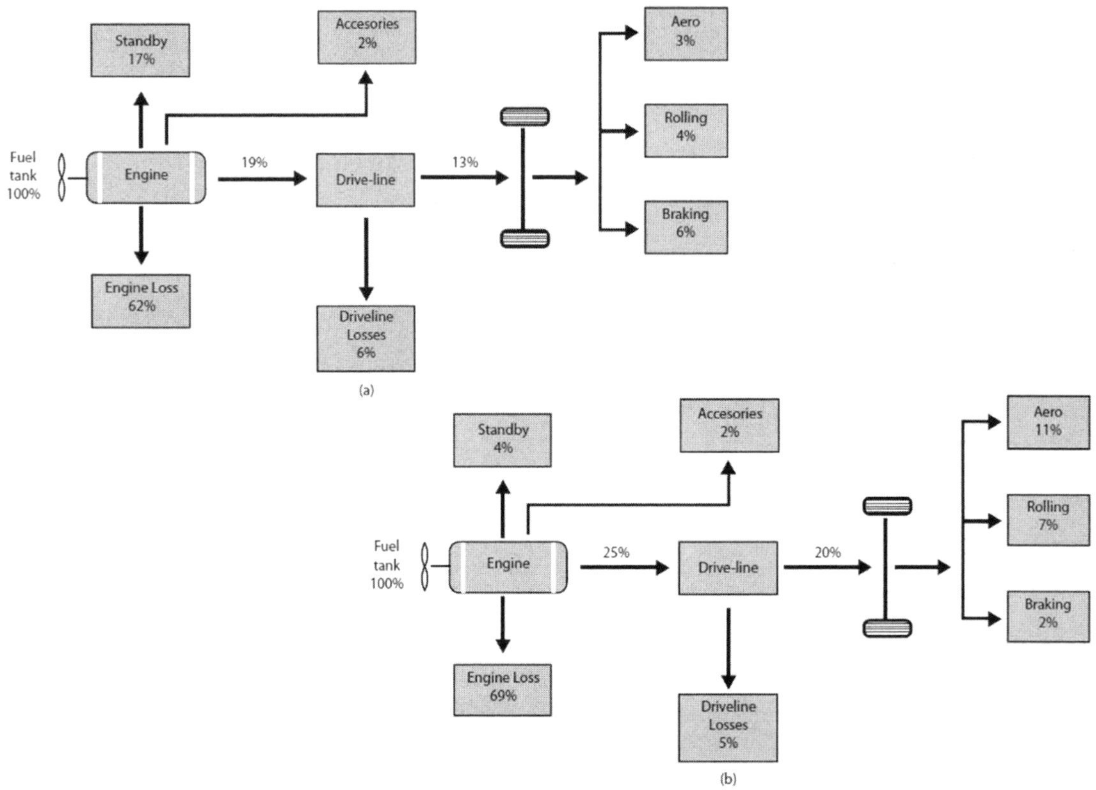

Figure 4.1 Example energy flows for a late model midsize passenger car. (a) Urban driving. (b) Highway driving.

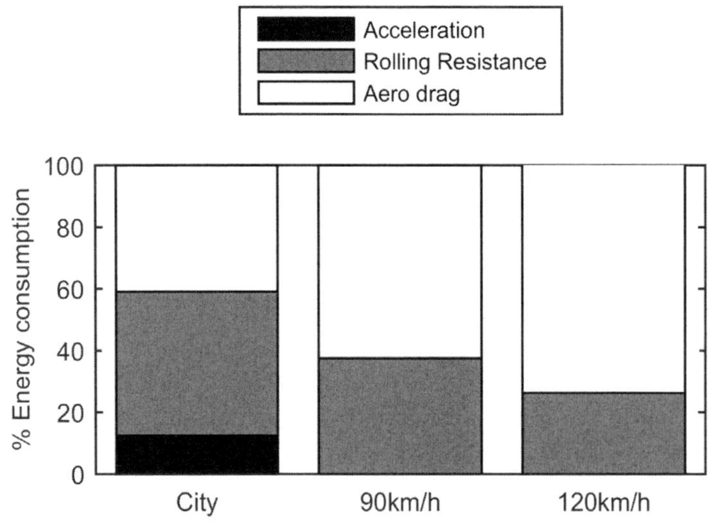

Figure 4.2 Contribution of tire rolling resistance to vehicle fuel economy versus speed.

In town, when a vehicle is traveling at low speeds, up to 40% of the energy to move the vehicle is determined by the rolling resistance, while at high speeds the most critical factor is aerodynamic drag [4-1].

The intention of this chapter is to analyze parameters that affect tire performance, and therefore vehicle performance, to understand their influence when testing a vehicle in a chassis dynamometer. This will allow for a better simulation of real road conditions or at least compensation for these factors when necessary.

4.2 The Concept of Rolling Resistance

When a tire is inflated (with no load), it takes a toroidal shape. When a load is placed on a tire, it flattens the sector in contact with the road. As the tire rolls forward, the part that is deformed moves around the tire such that there is a constant "flow" of the tire rubber through the deformed section. The elasticity of the tire allows for the deformation much like compressing and expanding a spring. In addition, as the tire is being flattened, there are extensive geometry changes around the tire–road contact [4-2]. This is illustrated graphically in Figure 4.3.

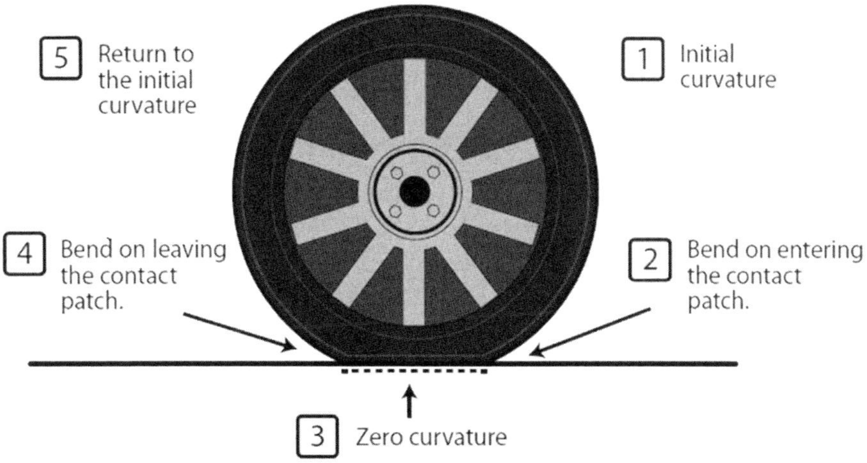

Figure 4.3 Deformation of a pneumatic tire in contact with a road with zero curvature.

As the tire rolls, each part must go through a distorted sector taking a certain amount of energy in the deformation. Only a part of this energy is recovered, while a great part is converted into heat. Figure 4.4 shows a thermal imaging camera shot of a vehicle running on a chassis dynamometer. The tires stand out as the hot spot in this image as a result of heat generation from the deformation of the tire and rolling resistance. In other words, energy required to overcome rolling resistance becomes heat in the tires.

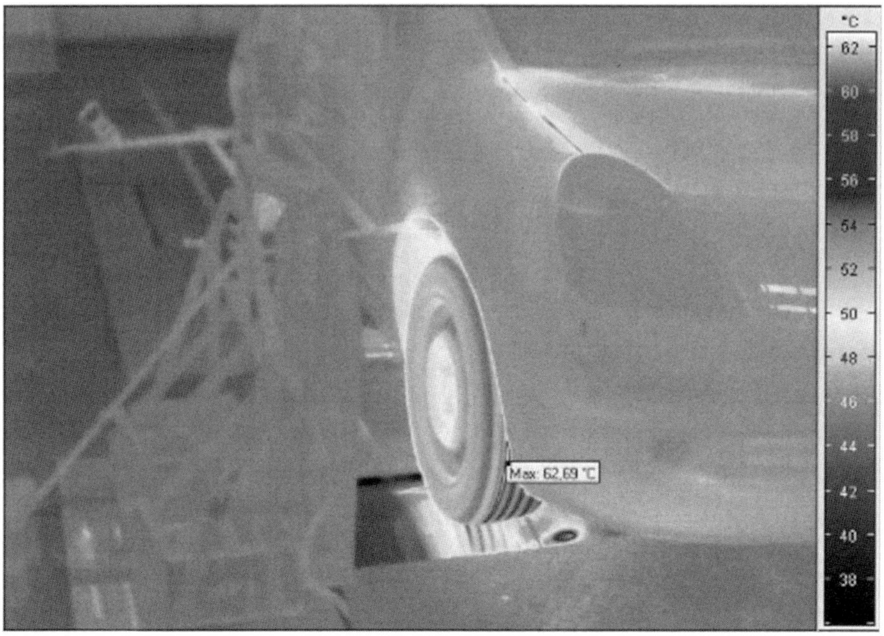

Figure 4.4 Thermography showing energy heat dissipation by a tire operated in a chassis dynamometer.

4.3 Rolling Resistance Coefficient

Rolling resistance is often captured through an empirical constant named the "rolling resistance coefficient," which expresses the amount of force opposing motion to the magnitude of the normal force. For a vehicle, the normal force is the reaction force at the ground because of the vehicle weight. On a perfectly flat road, this would be the vehicle weight. On an incline, it would be the component of the weight acting normally to the ground. The rolling resistance coefficient is expressed in the following equation and the forces are illustrated in Figure 4.5:

$$C_{RR} = \frac{F_{RR}}{N} \tag{4.1}$$

where F_{RR} = rolling resistance force, C_{RR} = dimensionless rolling resistance coefficient or coefficient of rolling friction (C_{RF}), and N = normal force.

Rolling resistance depends primarily on the type and condition of the tire and its inflation pressure. As a guide, the rolling resistance coefficient for a variety of vehicle tire types can be taken from the following ranges [4-3]:

- all-season: 0.005–0.0062
- touring: 0.0058–0.0075
- performance: 0.0065–0.0083
- high performance: 0.009

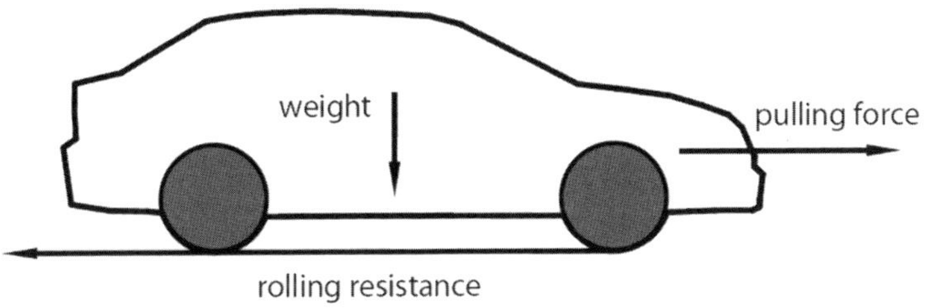

Figure 4.5 Rolling resistance forces.

4.4 Rolling Resistance Dependence

Tire rolling resistance depends on the design and manufacturing process and varies significantly between different manufacturers and models. A tire's rolling resistance is not constant and is affected by operating conditions, which are influenced by several factors. The following sections will evaluate the factors that are important for chassis dynamometer testing.

4.4.1 Influence of Inflation Pressure

Increasing the inflation pressure of a tire is a direct way of reducing rolling resistance. Figure 4.6 shows how rolling resistance decreases with increasing tire pressures.

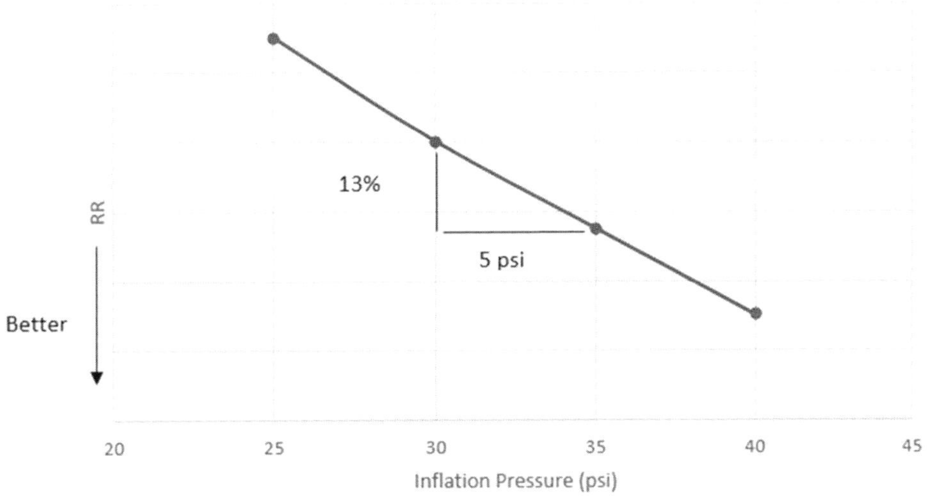

Figure 4.6 Rolling resistance versus inflation pressure.

A rough estimation for rolling resistance is provided by assuming an inversely proportional relationship to the inflation pressure raised to a power of around 0.7 as shown in the following equation:

$$F_{RR} = \frac{C_{tire}}{p^{0.7}} \tag{4.2}$$

where C_{tire} is a constant used to describe the performance of the particular tire without the effect of pressure [4-4].

The 0.7 power relationship is relative to inflation pressure measured while the tire is in operation. Usually, inflation pressure is not measured while in operation, but before the tire is put into operation. This is called "cold inflation" pressure. The relationship for rolling resistance to "cold inflation" pressure is slightly different with a 0.5 exponent as given in:

$$F_{\text{RR}} = \frac{C_{\text{tire}}}{p_c^{0.5}} \approx \frac{C_{\text{tire}}}{p^{0.7}} \tag{4.3}$$

However, as the tire starts to move at the beginning of a trip, the power consumption would be closer to the "cold inflation" state. As the tire is driven, it will make a transition to the in-use relationship. A tire that starts out with a lower air pressure will consume more power than the same tire set to a higher inflation pressure. Consequently, as the tire heats with use, the air pressure increase will be greater.

The increase in air pressure as the tire cavity heats is:

$$\Delta p_a = p_a * \Delta T / T \tag{4.4}$$

where P_a = absolute pressure in the tire and T = temperature and ΔT is the temperature increase expressed in kelvin.

Figure 4.7 shows how the tire and tire pressure affect the coast-down time of a vehicle. This is the time it takes for the vehicle to decelerate between two speeds only under the influence of aerodynamic and rolling resistance forces. The figure also shows how this affects fuel consumption for a particular passenger vehicle.

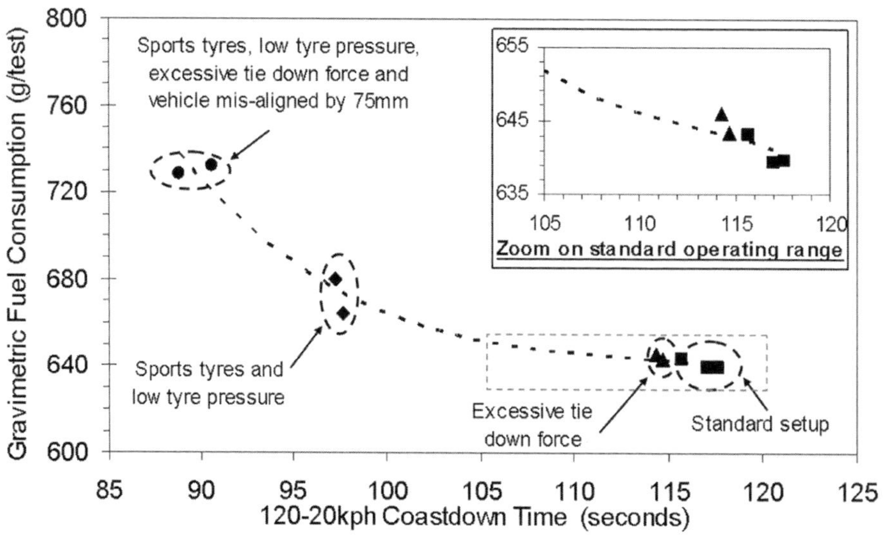

Figure 4.7 Rolling resistance versus tire pressure for a variety of tire types.

4.4.2 Influence of Tire Temperature

Typically, vehicle tires operate at temperatures between 20°C and 90°C depending on the tire type as well as the ambient conditions.

Tire temperature has a major effect on rolling resistance in two direct ways and one indirect way:

1. By changing the amount of heat dissipation from tire to ambient. When the temperature difference increases between the tire and surrounding air, heat dissipation is increased, and therefore, energy losses are also increased.
2. By changing the stiffness and hysteresis of the rubber. The rubber of a cold tire is relatively stiff. This takes more energy to distort it and therefore loses more energy to heat the tire. A tire starting off in cold weather has several times the rolling resistance of a tire running in warm weather. This is mainly because of the change in rubber properties and partly due to the increase in air pressure because of the increase in temperature.
3. By changing the air temperature in the tire and therefore the inflation pressure.

As a consequence of all these factors, as ambient temperature rises, rolling resistance decreases. Rolling resistance dependency on temperature is not linear; however, between 10°C and 40°C, an increase of 1°C corresponds to a reduction in rolling resistance of around 0.6%.

Rolling resistance measurements are frequently adjusted with the following equation:

$$F_{\mathrm{RR}(T\mathrm{ref})} = F_{\mathrm{RR(amb)}} * \left(1 + k\left(T_{\mathrm{amb}} - T_{\mathrm{ref}}\right)\right) \quad (4.5)$$

where $k = 0.006°\mathrm{C}^{-1}$ normal force. Other values of k are sometimes used, but its value is typically in the range of $0.005°\mathrm{C}^{-1}$–$0.008°\mathrm{C}^{-1}$.

Plotting this equation with $k = 0.006°\mathrm{C}^{-1}$, in Figure 4.8, we can see the influence of temperature on the rolling resistance. As shown, the rolling resistance can be up to 50% higher at −30°C and around 10% lower at 50°C, relative to 25°C conditions.

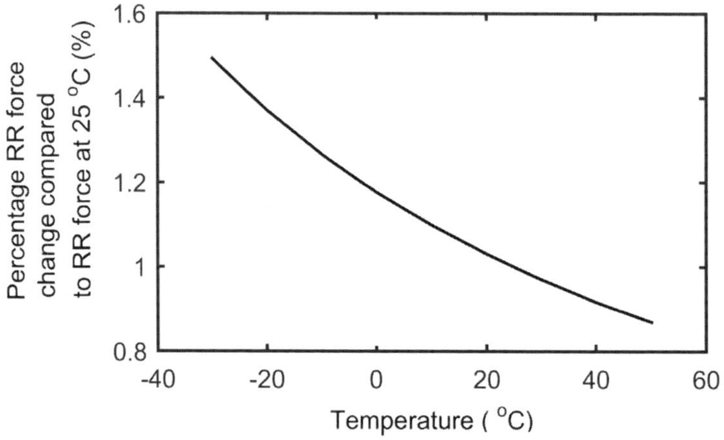

Figure 4.8 Rolling resistance versus ambient temperature compared with the RR at 25°C.

4.4.3 Influence of Vehicle Speed on Tires

In principle, we can think that rolling resistance increases when speed rises because an increase in tire deformation produces an increase in heat dissipation. Nevertheless, for speeds in between 20 and 120 km/h, a tire's rolling resistance is relatively constant.

Increased tire deformations produce higher temperatures and therefore higher pressures inside the tire, which reduce rolling resistance. Opposing this factor is cooling of the tire by increased air speeds over the outer surface of the tire.

Above 120 km/h, the rolling resistance of most vehicles begins to increase significantly. At high speeds, the formation of standing waves in the tires becomes more critical. Centrifugal forces cause the tire's radius to increase in the upper region, opposite to the ground, while the deformations associated with the flattening of the contact patch remain nearly constant. The result is that relative displacements around each revolution become higher and result in higher energy dissipation. To avoid this situation, high speed rated tires are equipped with a reinforced crown to reduce the displacement at speed.

As can be seen in Figure 4.9, at moderate speeds, the resistance coefficient follows a linear function. However, the relationship is closer to a speed squared function when the tire is rolling at higher speeds; in general, we can assume that:

$$f_r = f_0 + 3.24 f_s \left(\frac{v}{62.15} \right)^{2.5} \tag{4.6}$$

where f_0 = basic coefficient of rolling resistance [-], f_s = speed effect coefficient [-], and v = speed (km/h).

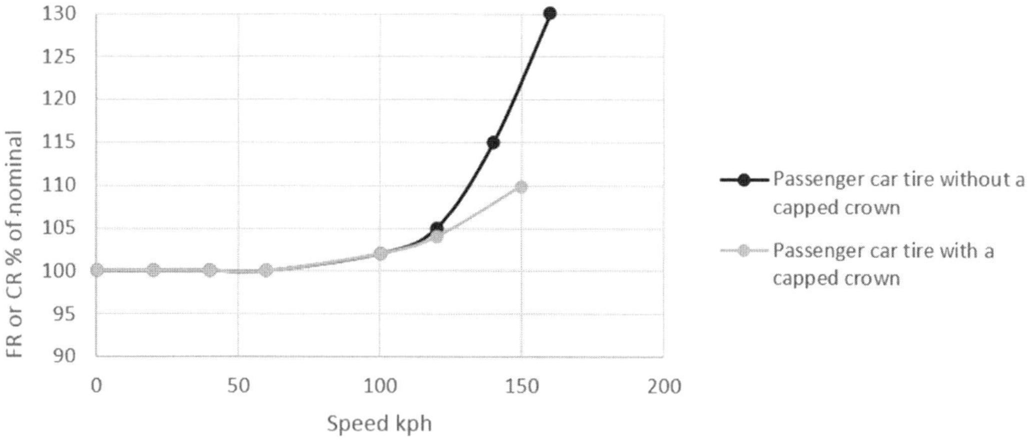

Figure 4.9 Effect of speed on rolling resistance. Rolling resistance plotted as a percentage of that measured at 80 km/h.

When it comes to testing a vehicle in a chassis dynamometer, most of the previous factors are physically occurring in the test cell by turning the tire against the roller, at least for the driven wheels. The factor that is not well replicated is the air cooling effect because of vehicle motion.

Taking a sample vehicle for a real test, we can compare tire performance on the road and on a drum test, this can give us an indication on the influence of air speed across the tire. Figure 4.10 shows the tire sidewall temperature for a highway (on-road) test and on a chassis dynamometer drum. The measured temperatures are considerably higher on the chassis dynamometer because there is much less airflow over the wheels to allow convective cooling. We have seen above how this may affect the rolling resistance of the tire.

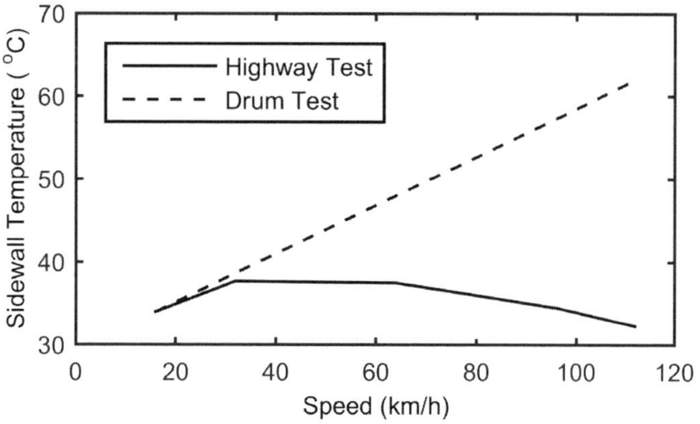

Figure 4.10 Comparison of sidewall temperature from the drum test with that from the highway operation. Conditions: inflation pressure 2.2 bar, road temperature 43.3°C, and room temperature drum test 37.7°C.

If we consider tire sidewall temperature to be indicative of inflation pressure (which is a simplification), applying the ideal gas law (Eq. 2.1), we can calculate air pressure in the tire. This is shown in Figure 4.11.

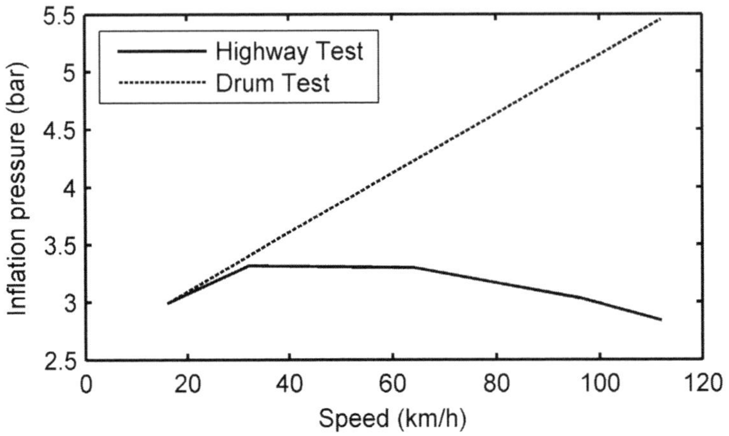

Figure 4.11 Comparison of the calculated inflation pressure from the drum test with that from the highway operation.

Taking the equation that relates rolling resistance to inflation pressure, we can also compare the rolling resistance of a tire on a highway test and on a drum test (see Figure 4.12).

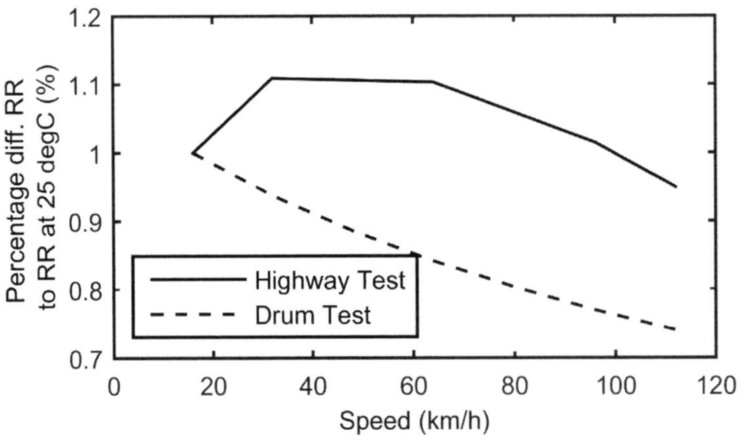

Figure 4.12 Rolling resistances for the drum test and highway test versus the rolling resistance at 20 km/h.

Figure 4.13 plots the error; we see that the difference of testing a tire on a drum can be quite significant, up to 25% in the worst case.

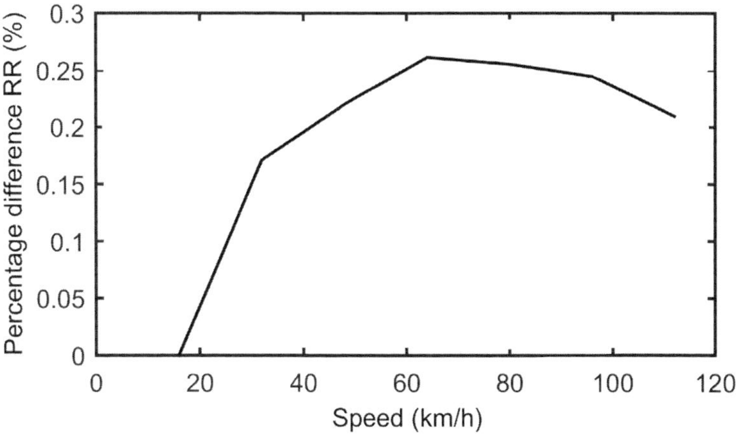

Figure 4.13 Rolling resistance error for the drum test versus that for highway.

A similar analysis can be performed using Eq. 4.5, assuming that ambient temperature equals tire temperature such that there is no heat exchange in between air and the tire. Again, this is a simplification, but illustrative of the factors involved. The effect of low ambient temperatures is significant as illustrated in Figure 4.14.

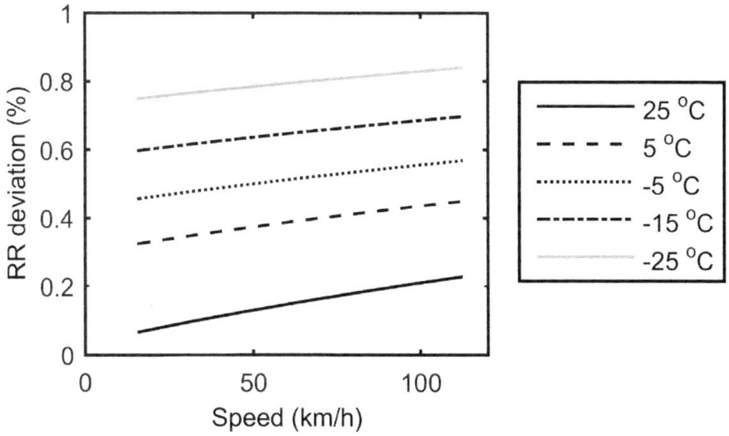

Figure 4.14 Rolling resistance for the drum test versus RR of the highway test at different temperatures.

4.4.4 Influence of Warm-Up Time

When a vehicle is started for the first time, the tires and the air contained within the tires are at a thermal equilibrium with the ambient. As the vehicle is driven, the deformation of the tire generates heat and causes a temperature rise of the tire structure, which convects into the air within the tire. This heats the air, and as it is contained within the approximately fixed inner volume of the tire, the pressure increases. This pressure rise tends to reduce the rolling resistance. The heating process takes around 30 minutes for passenger car tires to reach stable conditions. Emissions tests, for example, do not use warm tires, so this effect must be considered.

4.4.5 Tire Aerodynamic Drag

The tire's aerodynamic drag has an influence on a moving vehicle in two ways (Figure 4.15).

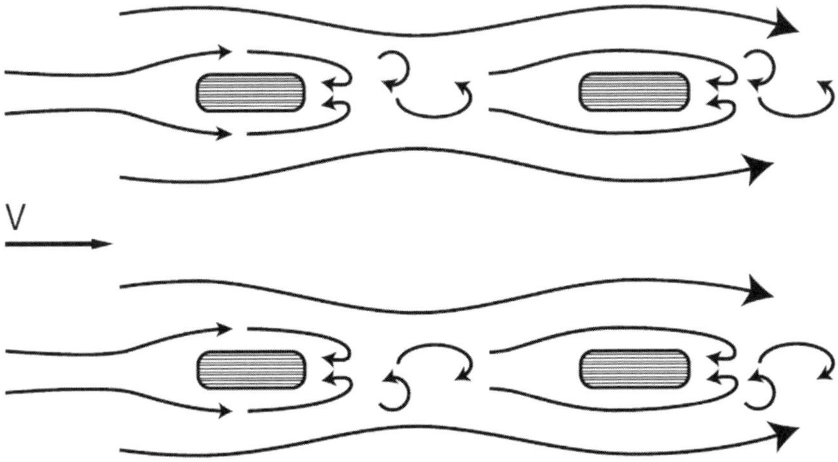

Figure 4.15 Tire's influence on a vehicle aerodynamic drag.

4.4.5.1 Affecting the Vehicle's Properties

The cross-sectional area and the frontal area become part of the vehicle causing an obstacle to the air and therefore causing an aerodynamic drag.

4.4.5.2 Effects of Tire Rotation

The rotation of the tire having air around it causes a force in the opposite direction of the tire turn (Figure 4.16).

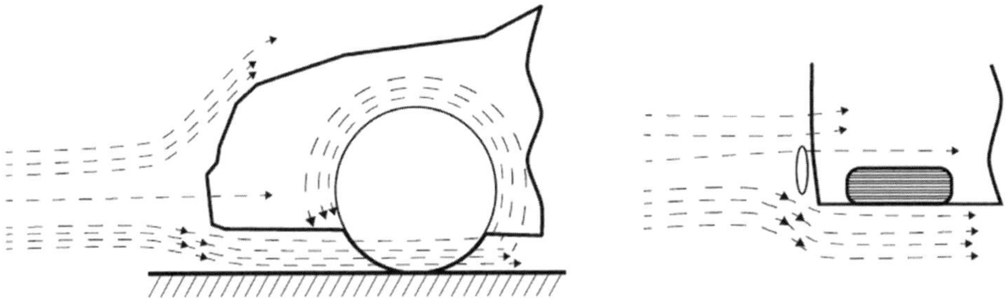

Figure 4.16 The effect of tire's rotational drag.

Other parameters that affect rolling resistance have not been considered in this chapter as they do not represent a factor to be simulated in chassis dynamometer vehicle testing. However, these would include the following:

- effect of vehicle load
- changes in rolling resistance throughout tire life
- effect of applied torque
- surface roughness

4.5 Tire Rolling Resistance Measurements

Rolling resistance measurement for tires is usually performed on a dedicated test rig with no vehicle [4-1]. The equipment consists of a drum coupled to an electrical motor and a cylinder, which applies a given force to the tested wheel, as shown in Figure 4.17.

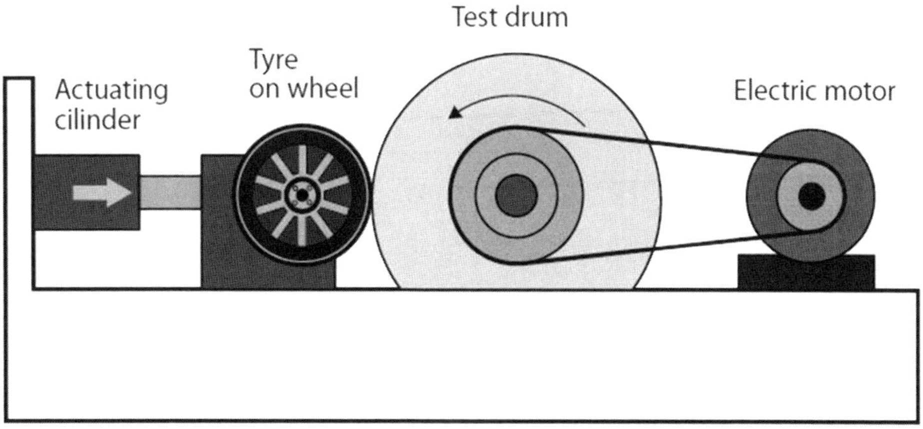

Figure 4.17 Typical test configuration for rolling resistance measurements.

In this type of test, it is possible to measure the different sources of energy loss in the tire that account for the rolling resistance. There are a wide variety of tire test machine configurations, including flat belt designs as well as designs that can accurately recreate the attitudinal changes of the tire relative to the road surface, which strongly influence the losses incurred.

References

4-1. NHTSA. U.S. Department of Transportation. DOT HS 811 154. *Tire Fuel Efficiency Consumer Information Program Development: Phase 2—Effects of Tire Rolling Resistance Levels on Traction, Treadwear, and Vehicle Fuel Economy*. By Larry R. Evans, James D. MacIsaac, Jr., John R. Harris, et al. August 2009. Accessed October 02, 2016. https://www.nhtsa.gov/DOT/NHTSA/.../Vehicle%20Research%20.../Tires/811154.pdf.

4-2. LaClair, T.J. "Rolling Resistance." In *The Pneumatic Tire*, 475–532. DOT HS 810561. February 2006. Accessed October 02, 2016. https://www.nhtsa.gov/staticfiles/safercar/pdf/PneumaticTire_HS-810-561.pdf.

4-3. National Research Council of the National Academies. Transportation Research Board. TRB Special Report 286. *Tires and Passenger Vehicle Fuel Economy: Informing Consumers, Improving Performance*. Compiled by Committee for the National Tire Efficiency Study. 2006. Accessed October 02, 2016. http://onlinepubs.trb.org/onlinepubs/sr/sr286.pdf.

4-4. Padmanabha, Pillai S. "Effect of tyre overload and inflation pressure on rolling loss (resistance) and fuel consumption of automobile and truck/bus tyres." *Indian Journal of Engineering & Material Sciences* 11, no. 5 (October 2004): 406–12.

Chapter 5
Exhaust Emissions and Aftertreatment Systems

5.1 Introduction

The major regulatory bodies around the world, the Environmental Protection Agency (EPA) in the U.S. and the EU have, for three decades, introduced increasingly stricter limits on pollutant emissions from light road vehicles. The most notable recent changes have focused on emissions of particulates and oxides of nitrogen. Also striking is the more recent tendency to use the same test procedure to measure and regulate CO_2 emissions, which, although not toxic, are of concern due to climate change.

For each vehicle type, different standards apply. Compliance is determined by running the engine at a standardized test cycle (Chapter 14). Noncompliant vehicles cannot be sold in the market covered by the regulatory body, but new standards do not apply to vehicles already on the roads. No use of specific technologies is mandated to meet the standards, though available technology is considered when setting the standards. A more detailed examination of the regulatory framework is included in Section 5.3.

5.2 Exhaust Components

A hydrocarbon fuel, if burnt completely, will result in only CO_2 and water vapor as products of combustion, along with the residual fraction of the air unaffected by the combustion. In practice, a wide range of different exhaust gas components are obtained as a result of the combustion process, which is dependent on the engine fuel properties and the combustion process. Fuel types can be classified into two broad groups: spark ignited, including gasoline and compressed natural gas (CNG), and compression ignited, principally diesel and other similar blends.

Exhaust components can be classified as follows.

5.2.1 Nonharmful

- Nitrogen gas (N_2) is an inert gas existing in ambient air in a concentration of approximately 79% and harmless to breathe. This passes through the engine unaffected, except for its role in the production of NO_x.
- Oxygen (O_2) is found in ambient air with a concentration of 20.9% and is harmless, and is indeed required by animal life. It is consumed in the engine as the fuel is oxidized.
- Carbon dioxide (CO_2) is not harmful for human beings and is required by plants, and is of interest as a pollutant only because of its role in global warming.
- Water vapor (H_2O) is harmless for human beings and is also found in ambient air depending on the humidity levels.

5.2.2 Harmful

- Hydrocarbons (HC) or unburnt hydrocarbons are actually a wide range of unwanted hydrocarbons resulting from incomplete combustion. Depending on the molecular structure, the HC pollutants can cause different harmful effects. Benzene, for example, is poisonous, and human exposure to this gas can cause irritation in skin and eyes. Benzene can also cause cancer.
- Oxides of nitrogen (NO and NO_2, referred to together as NO_x) cause respiratory irritation to human beings but are also responsible for the generation of acid rain. The strength of the medical evidence of health concerns has strengthened since 2014, leading to increased motivation to reduce levels emitted from all anthropogenic sources.
- Carbon monoxide is poisonous and can cause human death very rapidly when found in great concentrations.
- Sulfur dioxide (SO_2) is the main industrial pollutant responsible for acid rain.
- Particulate matter (PM) is referred to microscopic solid or liquid matter suspended in the engine exhaust flow. PM has been implicated in a wide range of respiratory and cardiovascular effects, especially in those who already suffer from a respiratory or cardiac condition. In addition to the PM, the particle number is also of concern. The Euro 6 legislation puts a restriction on the number of solid particles a vehicle may emit.

The breakdown of exhaust emissions for a typical gasoline and diesel engine are shown in Figure 5.1. The main point to note is that in both cases the nonharmful emissions make up nearly 99% of the total exhaust flow. A major difference is that the gasoline engine operates at a stoichiometric ratio, meaning that there is just enough oxygen in the cylinder to burn all of the injected fuel. This means there is no, or very little, oxygen in the exhaust flow. Diesel engines always run lean, meaning there is always excess oxygen in the combustion chambers and, therefore, there will always be a proportion of oxygen in the exhaust.

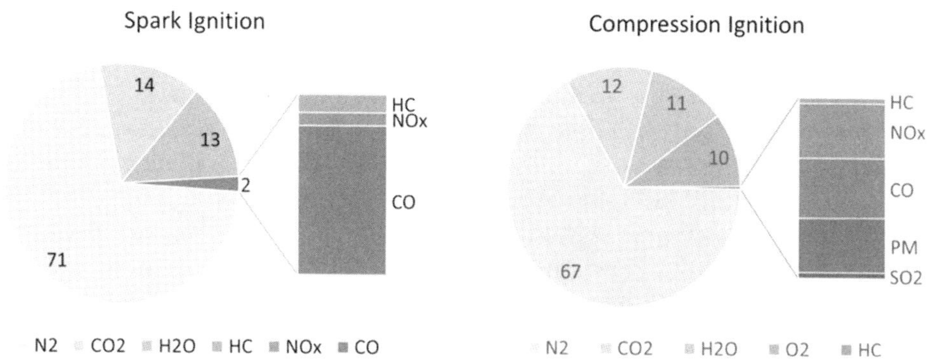

Figure 5.1 Composition of tailpipe exhaust gases for gasoline engines (left) and diesel engines (right).

Another major difference is in the breakdown of harmful emissions. Although they represent only a small amount of total exhaust flow, their effects on human health and the environment are such that even this small amount is important. In a gasoline engine, carbon monoxide emissions tend to be a large proportion because of the scarcity of oxygen for the combustion. In a diesel engine, NO_x emissions tend to be more problematic because the presence of oxygen in the exhaust means that traditional catalytic convertors cannot reduce these components. Also, PM is problematic in diesel engines: although overall the combustion chamber contains a lean mixture of fuel and air, locally there is a very rich combustion which, when combined with the lower temperatures in the diesel engine cylinder, results in this problematic emission.

In a spark ignition engine, pollutant formation is closely linked with air/fuel ratio. This is shown in Figure 5.2.

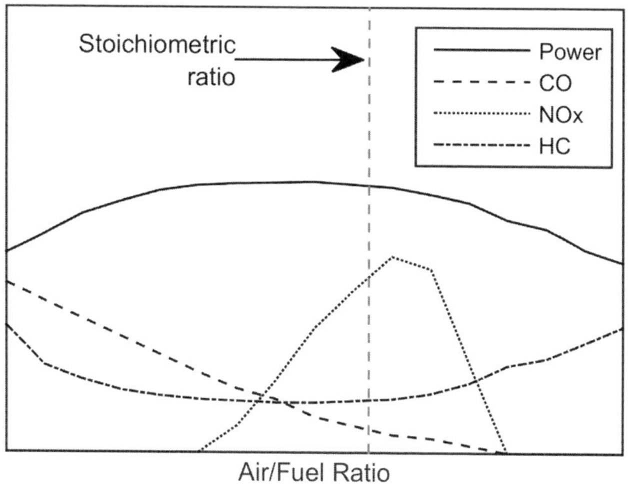

Figure 5.2 Percentage emissions of pollutants as a function of air/fuel ratio.

In diesel engines, especially in small-volume diesel engines for passenger cars, all processes inside the combustion chamber (injection, fuel mixture generation, ignition, combustion, and pollutant formation) are strongly influenced by factors affecting the mixture preparation and combustion. One of the largest factors is the injection strategy, that is, injection timings, quantities, and the number of injections. In addition, the air path and use of exhaust gas recirculation (high and/or low pressure) has a significant effect on the combustion process. The topic continues to be the subject of extensive research and development, but the principal areas of interest are inlet port geometry optimization to introduce turbulence (usually swirl); location, design, and operating pressure of the fuel injector nozzle to promote rapid mixture preparation; design of the piston bowl to promote rapid combustion; and consideration of the influence of fuel composition. Note that this last consideration has to be accomplished across a whole market as a collaborative effort among fuel refiners, vehicle manufacturers, and regulators in order to be practical, such as the desulfurization of diesel in the 1990s, which led to significant reductions in particulate formation.

5.3 Exhaust Emission Component Creation

Exhaust emission components are the result of a combustion process and are strongly influenced by the way the combustion is performed. This section will analyze the influence of the combustion process on the creation of exhaust components.

5.3.1 NO_x

The formation of NO_x is highly influenced by high temperatures in an atmosphere containing nitrogen (as all air breathing engines exhibit) and a moderately high content of oxygen. Oxygen atoms are necessary to initiate the reactions. At high temperatures above 1800–2000 K, oxygen atoms are dissociated, starting consecutive reactions. Indicative reaction rates are shown in Figure 5.3. Note that the NO_x formation rate increases with excess oxygen initially, but falls again in very lean mixtures.

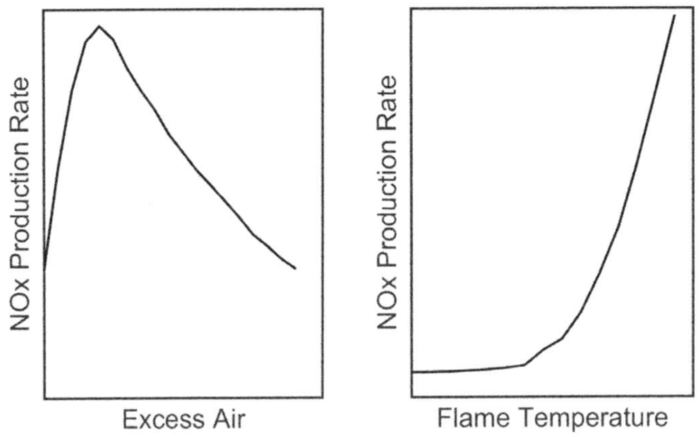

Figure 5.3 Formation rate for NO_x as a function of temperature and oxygen concentration.

Air humidity also influences the specific heat of the mixture. An increase in air intake humidity reduces the formation of NO_x because it lowers the temperature by absorbing a greater amount of heat, particularly in the combustion flame.

5.3.2 Carbon Monoxide

The formation of carbon monoxide is closely linked to air/fuel ratio. It increases in the regions of rich air–fuel mixtures such as seen in spark ignition engines running in component protection regions, or diesel engines burning with low homogeneity. Although CO emissions from the engine cylinders can be high, catalytic convertors are very effective at converting CO into CO_2, making them less problematic for overall vehicle behavior.

5.3.3 Hydrocarbons

Hydrocarbon emissions result when fuel is not completely burned in the engine. Such conditions can be as a result of rich air fuel ratios, or by quenching of the flame by cold surfaces, fuel trapped in crevice volumes or post-injection fuel injector leakage. There are a variety of hydrocarbons emitted into the atmosphere, most notable of which are volatile organic compounds (VOCs) for their impact on health and the environment. These compounds are precursors of ozone and some, such as benzene, formaldehyde, and acetaldehyde, have high toxicity for humans.

5.3.4 Particulate Matter

PM emissions are formed through a complex process that begins in the combustion chamber as the piston expands and the gas temperature drops. Below a certain temperature, the chemical reactions are stopped, effectively freezing the chemical composition at this stage. If the combustion involves a very rich combustion region, such as in diesel engines, a variety of partially burnt products will exist. Many of these are the precursors of soot and they will agglomerate together as they leave the cylinder and travel down the exhaust pipe to form the black smoke that is often seen in diesel exhausts.

5.4 Exhaust Emissions Control Systems

There are a wide range of technical solutions to the problem of reducing exhaust gas emissions. They can be divided into two main groups:

- *Reduction of emissions at source:* These are measures that attempt to reduce the emissions that are formed in the engine's cylinders. Examples of such systems involve the use of exhaust gas recirculation, ignition timing, and fuel injection.
- *Aftertreatment system:* which reduces pollutant's emission levels in the exhaust pipe before they are emitted to the ambient. Examples of such systems include three-way catalysts, particulate filters, and selective catalytic reduction (SCR) catalysts.

This chapter will only focus on the second principle: evaluating the behavior and performance of exhaust aftertreatment systems depending on the working conditions.

The exhaust gas aftertreatment systems of combustion engines can be classified into five main areas:

- *Thermal reactors:* They are based on the combustion of the unburned gases by maintaining high temperatures in the exhaust system. They were used in the U.S. during the 1960's but have now fallen out of use and are not considered further.
- *Catalytic converters:* Using specific metals in the exhaust system, which have a catalytic effect, allows the performance of useful chemical reactions to be accelerated.
- *Chemical reactors:* By adding a new component, allow the execution of a chemical reaction that removes some of the pollutant components.
- *Chemical filters:* Systems that retain or accumulate some of the pollutant components. The saturation of these systems requires a periodic regeneration.
- *Physical filters:* Filters that retain solid or liquid particles contained in the exhaust gases. Saturation requires cleaning or periodic regeneration.

5.4.1 Catalytic Converters

A catalytic converter is a device installed toward the start of the exhaust system to accelerate the oxidation or reduction reactions that have not been completed during the combustion. The presence of a catalytic substance promotes the completeness of chemical reactions at lower temperatures.

Figure 5.4 shows a typical gasoline three-way catalytic (TWC) converter. Exhaust gases are fed into the device and the catalytic substance is usually placed on a substrate around which the exhaust gases circulate. The catalyst typically relies on heat from the engine exhaust gases to reach its activation temperature and is therefore usually located close to the engine. The catalyst being composed of many small passages will create a backpressure on the engine, meaning that there will be a pumping loss associated with the aftertreatment and, ultimately, a fuel consumption penalty associated with this form of emissions control.

There are three main types of catalytic converters classified according to the chemistry that can be promoted within them:

- *Oxidation:* Active against CO and HC. These catalysts are used when there is enough oxygen in the exhaust gases to complete the combustion process. Crucially, they cannot promote reduction chemistry to any significant extent, and so are ineffective for NO_x reduction.
- *Reduction:* Active against NO_x. They are used when the exhaust gas comes from a rich mixture with no oxygen but molecules with a reducing potential such as CO and H_2.
- *Three-Way:* Oxidation and reduction can be promoted within the same exhaust stream, and are therefore active against CO, HC, and NO_x. They can be used when the exhaust gases have little to no oxygen content, such as those produced from the combustion of a stoichiometric mixture.

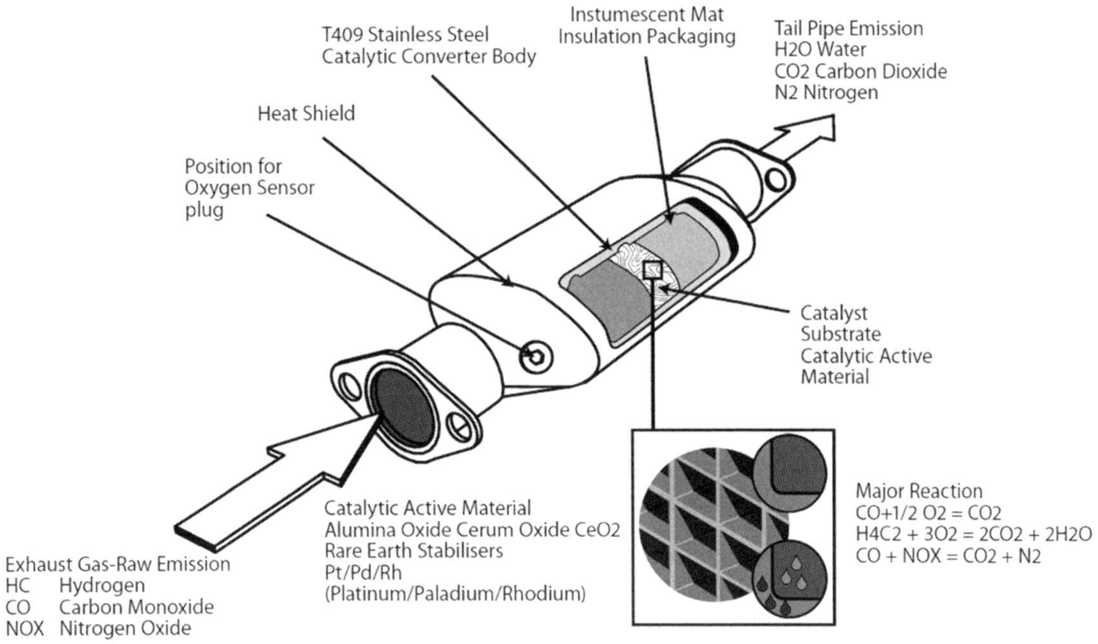

Figure 5.4 TWC converter.

Out of these devices, the most commonly used are the oxidation catalytic converter used for diesel engines and the TWC converter used for spark ignition engines. Only these two will be analyzed in this section.

Catalytic converters are usually located underneath the vehicle (Figure 5.5), near the engine. Air circulation around it will therefore have an influence on the thermal state of the device, so also on its performance.

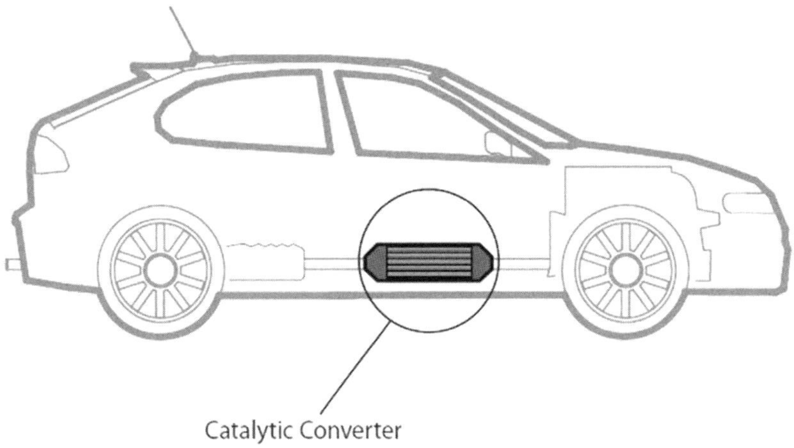

Figure 5.5 Location of the catalytic converter in a vehicle.

5.4.1.1 Diesel Oxidation Catalyst

It is important to remember that diesel engines always run with lean air–fuel mixtures, having an excess of oxygen in the exhaust as shown in the previous sections. The main outcome from this is that the catalytic converter can only oxidize CO and HC, and will not be able to reduce NO_x. This was the principal deficiency in pollution control from diesel engines until the advent of de-NO_x devices described later.

Another important feature of diesel engines is that the exhaust gases are cooler than spark ignition engines falling to a temperature of barely 100°C at idle. Temperature of the exhaust gases is a matter of great importance since the effectiveness of the catalytic converter efficiency is dependent on this variable. Figure 5.6 shows the typical conversion efficiencies as a function of exhaust temperature for a diesel oxidation catalyst (DOC).

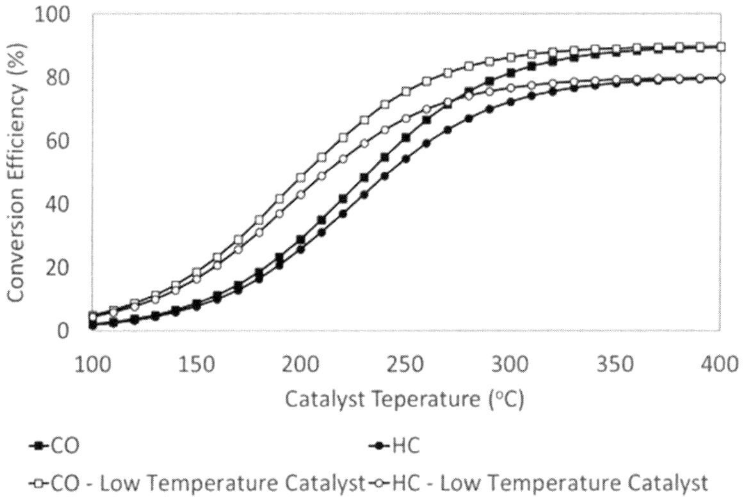

Figure 5.6 Effect of temperature on the operation of a diesel oxidation catalytic converter.

5.4.1.2 Three-Way Catalytic Converter

One of the main differences between spark ignition engines and diesel engines is the air–fuel ratio. While diesel engines always run with a lean mixture and can operate under extremely lean conditions, spark ignition engines may run somewhat lean, a little rich, or stoichiometric. The air–fuel ratio has a major influence on the efficiency of the catalytic converter (see Figure 5.7). Catalytic converters on engines running with a lean mixture are essentially only effective against CO and HC, and are therefore described as "two-way." If the engine runs rich, the catalytic converter will be good at reducing NO_x but will show little effect on CO and HC. When the engine runs with a stoichiometric mixture the catalytic converter will exhibit high conversion efficiency over all three major pollutant groups, and will therefore be known as "three-way." In addition, if the catalyst is exposed to nonstoichiometric conditions for long periods, its operation may be temporarily or permanently degraded.

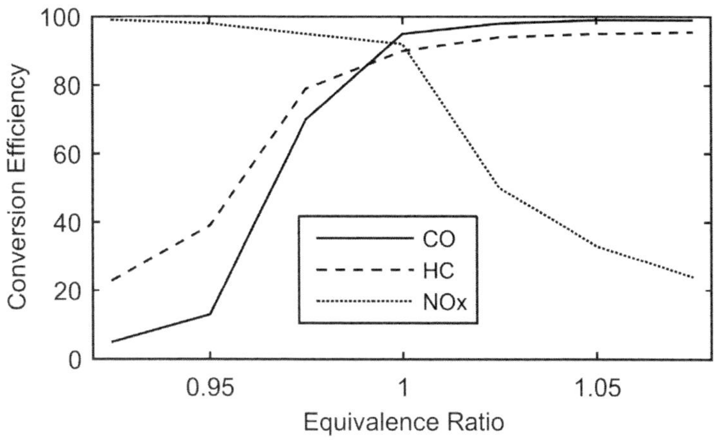

Figure 5.7 Effect of air-fuel ratio on the operation of a catalytic converter.

For the proper performance of the catalytic converter, spark ignition vehicles are equipped with a Lambda sensor, which senses the reducing potential of the exhaust gases, therefore allowing the engine control system to adjust the fuel injection quantity to maintain a stoichiometric mixture.

As explained before, catalysis is greatly influenced by the temperature of the exhaust gases; however, as exhaust gases are typically hotter in a gasoline engine because of the stoichiometric combustion, this is usually less problematic than for diesel engines.

5.4.2 Chemical Reactors

Chemical reactors are based on reducing the concentration of certain exhaust components by causing a chemical reaction. This is usually done by adding a component that is mixed with the exhaust gases. The most commonly used systems are known as selective catalytic reduction (SCR) to reduce NO_x when the conventional catalytic reduction is not possible because of an excess of oxygen. They are used in diesel engines, usually fitted after an oxidation catalytic converter, and use urea as the reduction agent. These systems started to appear on production vehicles in Europe after 2010, and are being used on a much larger scale to meet EURO VI emissions standards (see Section 5.3).

A typical system layout is shown in Figure 5.8. An exhaust fluid dosing system is installed upstream of the catalyst and the role is to maintain a correct level of ammonia storage in the catalyst. The stored ammonia is then able to reduce the NO_x emissions within the catalyst. The challenge with such a system is that a minimum exhaust temperature is required to activate the catalyst (see Figure 5.9), but the ability of the catalyst to absorb ammonia is reduced at higher temperatures. The emission of ammonia is not desirable because of its adverse effects on the environment and, therefore, a careful control system is required. The operation of such a catalyst is more complex than the operation of a DOC or TWC presented above, and the ammonia storage characteristics mean that it has a relatively slow dynamic behavior that must be also accounted for.

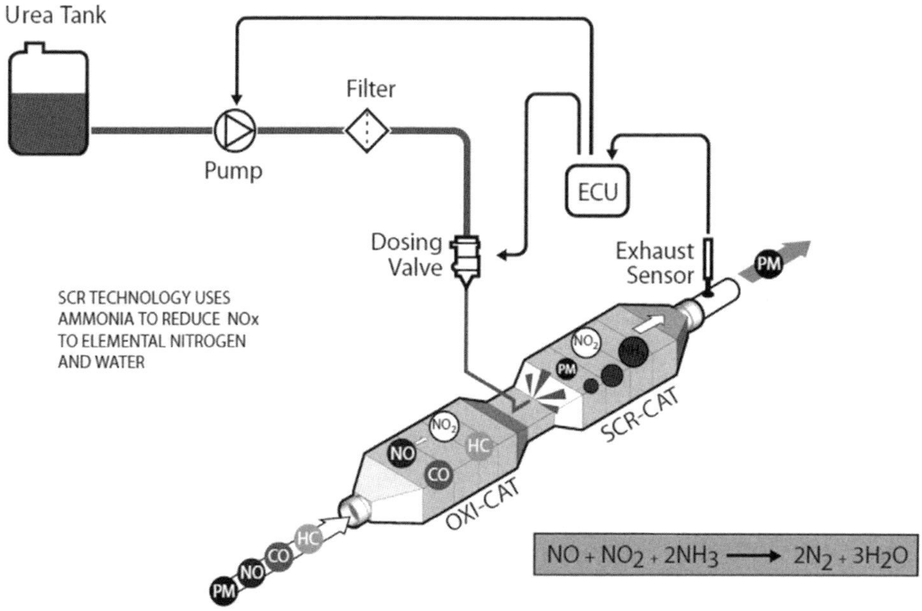

Figure 5.8 Schematic of a combined system using an oxidation catalyst followed by an SCR catalyst using urea.

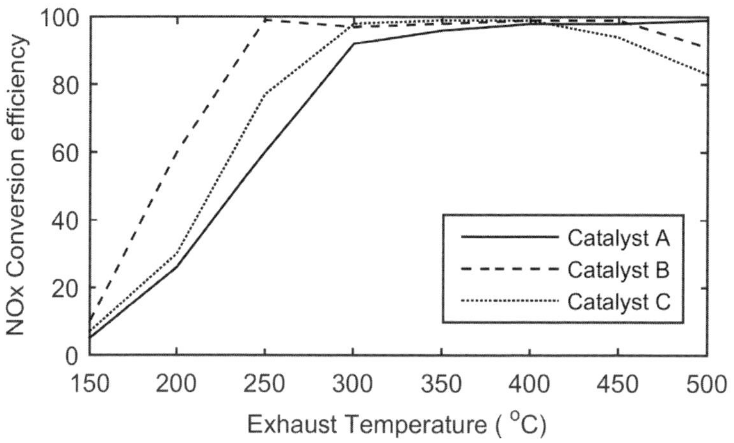

Figure 5.9 Temperature influence on efficiency of SCR catalysts. A and B are zeolite catalysts, while V-SCR is a standard wash coat with vanadium catalyst.

5.4.3 Chemical Filters

An alternative to SCR for engines with lean mixture for the reduction of NO_x is the use of chemical filters. These filters store NO_x during the usual working conditions with an excess of oxygen, which is then released later when there is lack of oxygen.

When the engine is working with a lean mixture, NO is converted into NO_2 in an oxidation catalytic converter, and this is trapped in form of nitrate in an alkaline metallic oxide such as BaO, which is located in the aftertreatment system. When the system is saturated, it is required to regenerate the filter and reduce the stored NO_2 using a rich air–fuel mixture for a short period of time. On a diesel engine, this is achieved at the expense of fuel consumption, through a combination of throttling to reduce the air flow and additional fuel injection into the exhaust or into the cylinder during the exhaust stroke.

Figure 5.10 shows a typical layout of the system. The systems are known as NO_x storage reduction (NSR) or lean NO_x traps (LNTs). A chemical filter's efficiency is also dependent on temperature, but current designs are somewhat more capable at low temperatures than contemporary SCR systems.

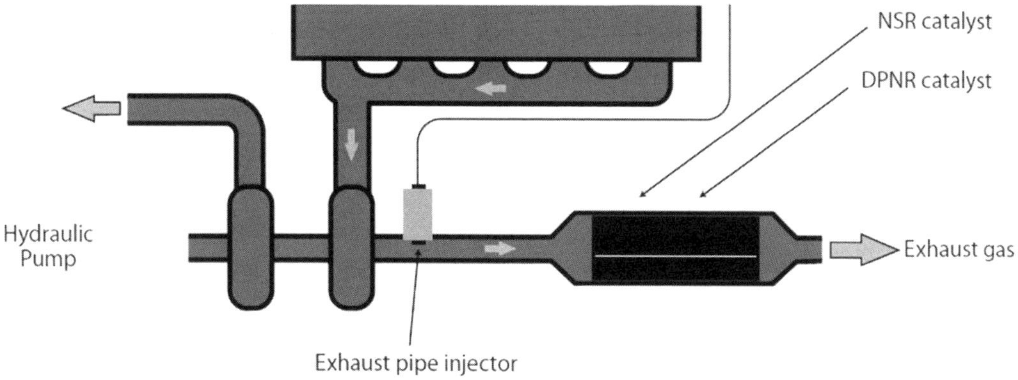

Figure 5.10 Diesel exhaust aftertreatment system using NSR with fuel injection in the exhaust pipe.

5.4.4 Particle Filters

A diesel particulate filter is a device designed to remove diesel PM or soot from the exhaust gases of a diesel engine. They are usually installed after an oxidation catalytic converter (DOC) and SCR. They work based on the physical filtering of the gas through a porous media, which retains the particles. As more and more matter is collected in the filter, the backpressure it creates on the engine increases, increasing the pumping losses. After a certain time, they get saturated and need a regeneration process.

Regeneration is done by burning the particles that have been retained. This can be done by two main technologies (Figure 5.11):

- *Active regeneration:* Requires additional heat supply to reach oxidation temperatures of the PM. This can be done with electrical heaters or fuel injection in the exhaust line. Throttling the air flow into the engine to reduce the air flow also increases the exhaust temperature.
- *Passive regeneration:* Is based on reducing soot oxidation temperature. This is usually done by adding a catalytic substance to the fuel, and by operating the engine with excess air at a flow rate and feed gas temperature sufficient to activate the catalytic reaction.

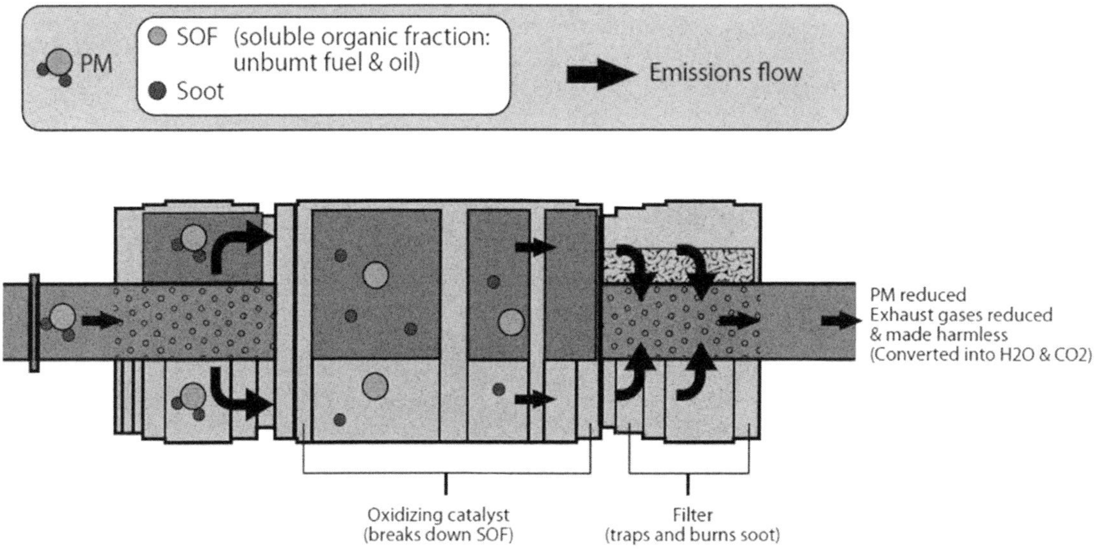

Figure 5.11 Working principle of an exhaust aftertreatment system composed of an oxidizing catalyst and a diesel particle filter.

Temperature is also a crucial variable in the performance of diesel particle filters. Some gasoline engines have started to appear with particulate filters, namely, gasoline particulate. These may become important if or when gasoline combustion systems running with a lean air-to-fuel mixture become widespread.

5.5 Exhaust Pipe Temperatures

As shown in the previous sections, the temperature of the exhaust aftertreatment components is critical for its performance. This temperature is mainly dependent on the engine working conditions, but is also influenced by the operating conditions around the vehicle as indicated in Figure 5.12.

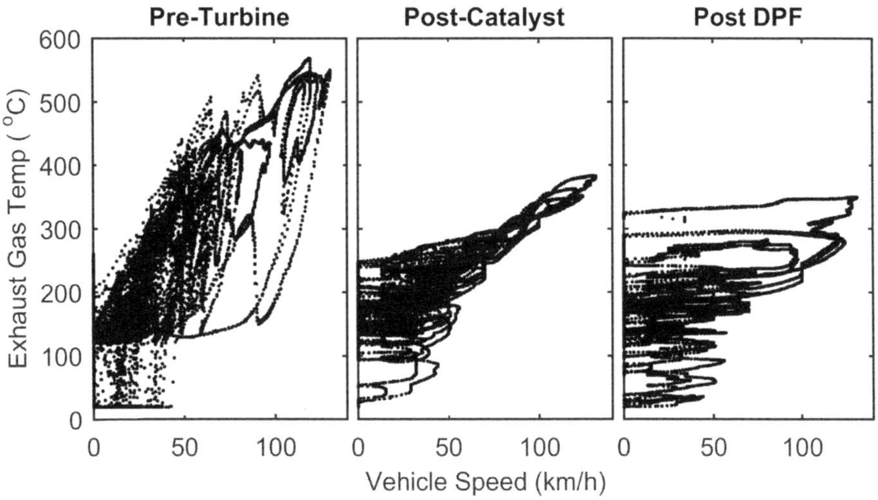

Figure 5.12 Exhaust gas temperature (°C) versus car velocity (km/h).

5.6 Parameters Influencing Emissions

This section shows some parameters influencing emissions in a vehicle. These influences are the summation of all of the individual behaviors introduced above, including the load exerted by the vehicle on the engine, the engine operating state, and the aftertreatment performance. As such, the exact details are highly variable according to the technology under consideration, but must be taken into account when an emissions cycle test is designed and performed.

5.6.1 Influence of Speed

Figure 5.13 shows the typical influence of the vehicle speed on some species of exhaust emissions. It is evident that technological improvements have had a significant impact over time. Fundamentally, though, many emissions increase with the higher powers required at high speed. In addition, emissions tend to be high for low speeds where the systems are operating under adverse conditions and when the vehicle starts.

Although subsequent technology steps have improved the results obtained, the trend is still that dynamic conditions make the task of the aftertreatment system more challenging and lead to a greater fuel use and so CO_2 production [5-1].

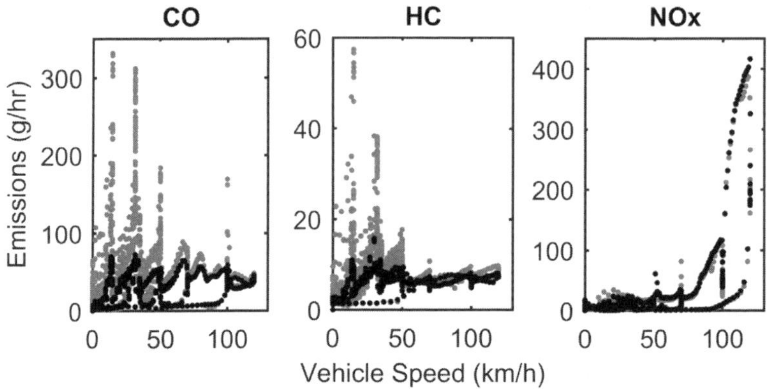

Figure 5.13 Emissions of diesel and petrol light duty vehicles (g/km) versus average speed (km/h).

5.6.2 Influence of Absolute Humidity

Figures 5.14, 5.15, and 5.16 show the absolute humidity influence on pollutants emissions of NO_x, CO, and HC, respectively. The impact of humidity on NO_x is always beneficial, as may be anticipated given that water injection into engines is an effective NO_x control mechanism. For HC and CO, the impact is more complex, depending on the technologies fitted to the vehicle and the fuel type.

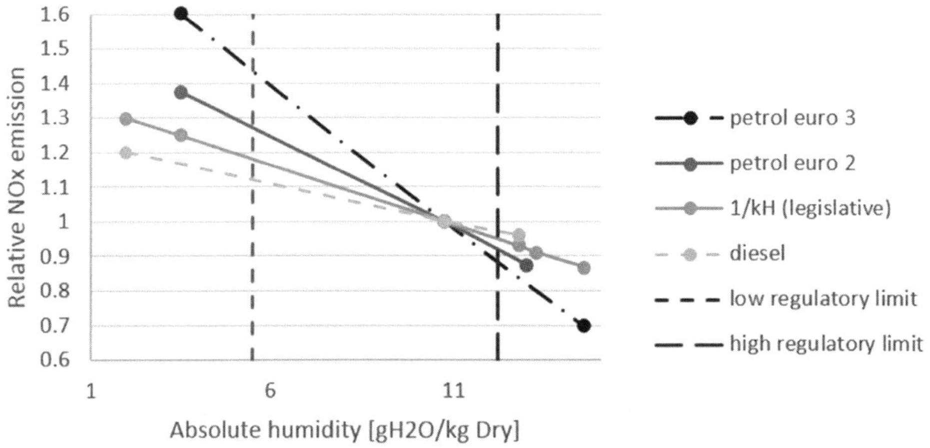

Figure 5.14 Linear models of (uncorrected) NO_x emissions measured in an urban driving cycle, fitted in average values for high, medium, and low humidity.

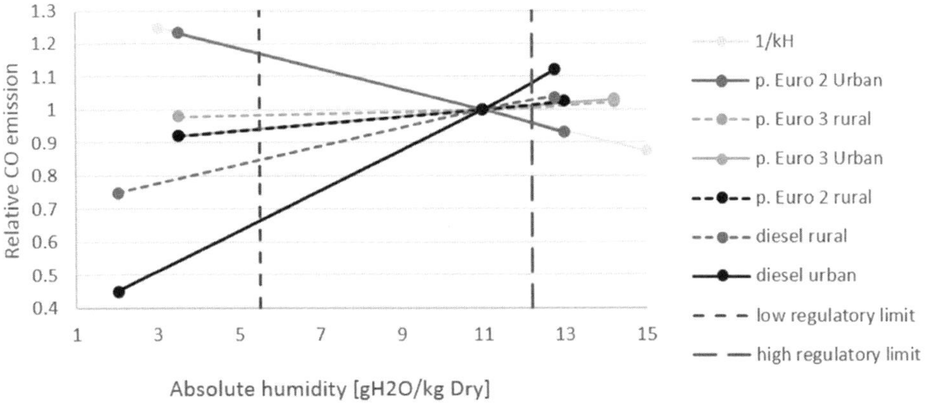

Figure 5.15 Average relative variations of CO emissions according to the absolute humidity.

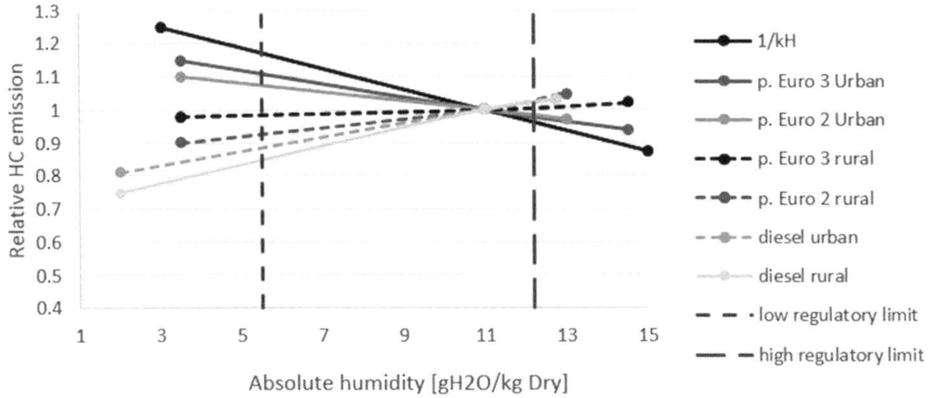

Figure 5.16 Average relative variations of HC emissions according to the absolute humidity.

5.6.3 Influence of Engine Temperature

The influence of temperature on CO_2 is quite straightforward as it is primarily linked to fuel consumption and friction losses. Therefore, CO_2 emissions tend to decrease as the engine temperature rises. The perturbing factor to this rule is the calibration of the engine, which may vary depending on other parameters like combustion timing, exhaust gas recirculation (EGR), and fuel injection strategies to compensate for the effects on emissions. This can affect fuel consumption.

The effect of temperature on all other emissions is more complex. In the combustion chamber, all other parameters being equal, NO_x emissions tend to be lower at colder engine temperatures while CO, HC, and particulates tend to increase. However, the powertrain will be calibrated to use features like EGR, injection, and ignition strategies to compensate for these.

On a chassis dynamometer where the full vehicle is under test, the effect of temperature on tailpipe emissions becomes more complex because the temperature of the after-treatment system also has a strong effect. Colder three-way or dual oxidation catalysts will not function below their light-off temperature, resulting in higher NO_x, CO, and HC emissions. Newer SCR systems have a typical temperature operating window, and the engine controller will take steps to modulate exhaust gas temperature within these windows. Finally, SCR catalysts, particulate filters, and LNTs all have a dynamic behavior of storage of some components (ammonia, particulates, or NO_x) that are temperature dependent and need to be managed. Hence, the effect of engine temperature can have a large impact on the tailpipe emissions both from the formation of emissions in the engine and the treatment of emissions in the exhaust (Figure 5.17).

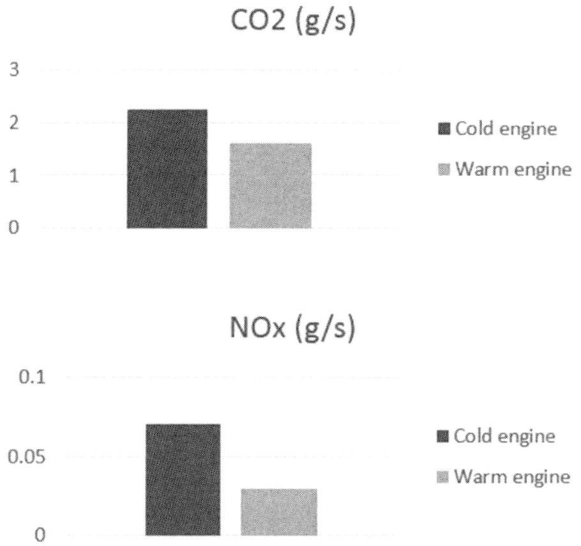

Figure 5.17 CO_2 and NO_x emissions at warm and cold engines.

5.6.4 Influence of Ambient Temperature

At a fundamental combustion level, reducing the ambient temperatures tends to reduce NO_x emissions but increases CO and HC emissions. However, when considering the engine aftertreatment that only operates effectively above a certain temperature threshold, at the tailpipe all three emissions will tend to increase. In practice, when considering a full vehicle system, the situation is much more complex because the powertrain control strategy will have been configured to take steps that mitigate the effect of temperature. Therefore, it is not possible to predict a definite trend of vehicle emissions with ambient temperature.

References

5-1. http://www.kanomax-usa.com/automotive/automotive.html.

Summary of Part I—The Vehicle

In the first part of this section, the forces acting on a moving vehicle on the road are discussed in order that we can reproduce them accurately when running a vehicle in the test cell.

The second part considers the main external parameters affecting the efficiency of a vehicle so that they can be taken into account and reproduced accurately in a test cell.

We have seen the interaction between air and vehicle. The cooling effects on different parts of the vehicle have a strong influence on the efficiency of different components. The main airflows that have an important effect are as follows:

- underhood air flow, which affects vehicle coolers such as radiator, intercooler, and oil coolers, combustion air engine intake, engine cooling and exhaust cooling
- brake cooling, which affects brake temperature and efficiency
- under vehicle flow, which affects exhaust line and exhaust aftertreatment temperature
- airflow around the vehicle, which affects cabin temperature and air conditioning demand
- airflow onto tires, which has a significant influence on rolling resistance

We have also seen the importance of the following factors:

- Engine performance and efficiency, which in the first place is affected by air intake conditions such as temperature, humidity, pressure, and altitude. Temperature also affects other fluids such as lubricating oil, which affects internal frictions, and fuel vaporization, which affects ignition. These parameters are not only impacted by ambient temperature, but also by the engine warm-up process.
- Engine power absorbed by accessory drives. A portion of the mechanical power produced by the engine is not used for the vehicle movement but to power different accessories, such as the power steering pump, the air conditioning compressor, the alternator, and the water pump. Different parameters also affect the power requirements and efficiency of these systems.

- The power steering pump efficiency is influenced by its oil temperature and therefore by ambient temperature.
- The air conditioning compressor load is a complex case. It is affected by the heat load on the vehicle, which is defined by the solar radiation, the size and type of windows, the vehicle color, the cabin size and the number of passengers, the fresh air flow rate, the engine speed, which defines the compressor speed and therefore its efficiency. It is also affected by the vehicle speed, which affects the heat exchange in between the vehicle and surrounding air as well as the flow onto the air conditioning condenser and, therefore, the Carnot cycle efficiency.
- The alternator, which is mainly affected by its speed defined by engine speed, and the load defined by consumers such as lights, fresh air ventilator, wipers, radio, and window defroster.
- The water pump whose efficiency is mainly affected by engine speed.

The pneumatic tire performance directly influences rolling resistance and is highly influenced by the following factors:

- inflation pressure, which affects tire deformation
- tire temperature, which affects material performances and energy dissipation to ambient temperature
- tire and vehicle speed, which affects tire deformation and heat dissipation to the surrounding air
- warm-up time
- aerodynamic drag

Exhaust emissions are highly influenced by the following:

- Combustion conditions are mainly affected by engine type, combustion temperature, combustion air humidity, and air/fuel ratio.
- Exhaust aftertreatment system performance is mainly affected by exhaust gas temperature and composition, as well as the cooling effect of surrounding air. The cooling effect of the surrounding air will be influenced by air speed and temperature around the exhaust aftertreatment component and, therefore, by vehicle speed and ambient temperatures.

Part II

The Test Cell

Chapter 6
The Chassis Dynamometer Test Cell

6.1 Introduction

The objective of this section is to study the characteristics and design criteria of the chassis dynamometer test facilities. Much of the details will be equally applicable to a powertrain test facility that replicates some or all of the vehicle powertrain, but loads the powertrain using directly coupled electrical machines rather than using the tire. The aim of the chassis dynamometer test cell is to subject the engine and/or powertrain to the same conditions as would be experienced when driving on the road. However, as the vehicle is stationary, many large and heavy instrumentation suites can be installed to monitor the behavior of the vehicle, notably for emissions and fuel consumption. The use of a chassis dynamometer also removes the uncertainty associated with on-road driving where the weather conditions are unpredictable, traffic can interfere with the desired duty cycle, and exact test repetition is very challenging.

6.2 The Chassis Dynamometer Test Cell

In order to perform tests in a controlled environment with guaranteed repeatability, a chassis dynamometer test cell must be employed rather than the road. A chassis dynamometer test cell is usually composed of the major items shown in Figure 6.1.

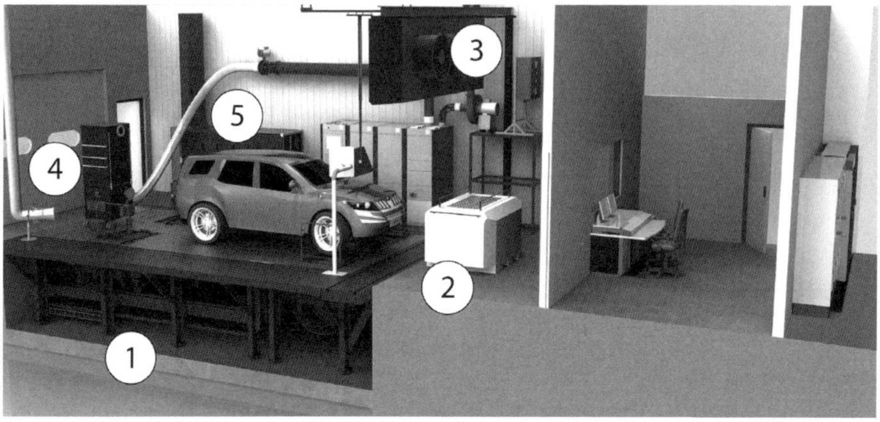

Figure 6.1 Chassis dynamometer test cell. 1—chassis dynamometer. 2—vehicle cooling blower. 3—ventilation or refrigeration system. 4—exhaust extraction system. 5—air intake system.

6.2.1 Chassis Dynamometer

The center of a modern chassis dynamometer consists of a roller drum, coupled to an electric machine that is controlled such that the applied load simulates those encountered by the vehicle during on-road driving. An illustration of such a dynamometer is shown in Figure 6.2. The vehicle is placed on the top of the roller crown, which rotates to emulate the correct vehicle speed. The dynamometer is equipped with instrumentation to measure the tractive effort applied by the vehicle to the roller. This tractive effort forms the input to a vehicle model, which results in a predicted acceleration rate for the vehicle under road load conditions. The speed control loop of the dynamometer is used to achieve this acceleration rate.

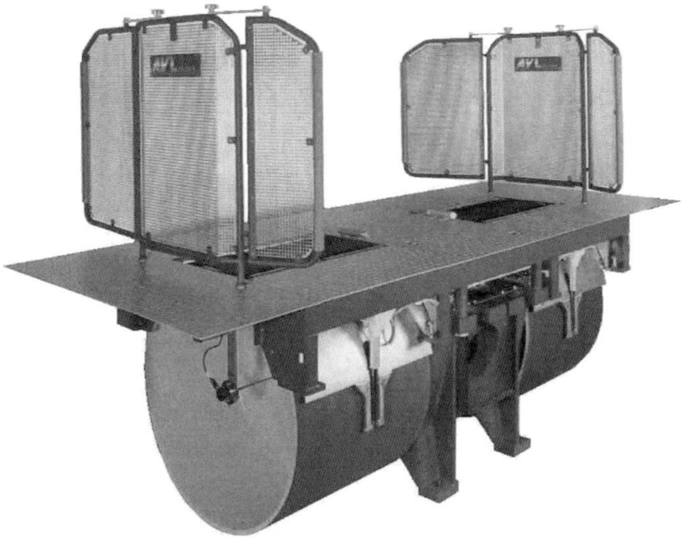

Figure 6.2 Dynamometer. (Reprinted with permission from AVL.)

There is a significant quantity of information regarding the specific design and operation of individual chassis dynamometers supplied by their manufacturers. For more information, the reader is referred to [6-1]. The layout of the test cell and the replication of vehicle loads will be studied in detail in Chapter 7.

6.2.2 Vehicle Cooling Blower

The vehicle cooling blower is placed in front of the vehicle and its rotational speed is controlled to simulate the airflow through the vehicle as if it was on the road. Depending on the test performed, the vehicle areas that must be considered will be different.

Figure 6.3 shows a typical chassis dynamometer blower of medium size. The size of the blower will depend on the requirements of the air flow. In many cases, the blower is only required to provide cooling for the engine and a relatively small blower will be sufficient, focused on the vehicle radiator. In other tests, it may be the thermal behavior of the vehicle cabin that is of interest, and here a much larger blower providing flow representative of real driving over the complete vehicle will be required. In any case, it is important to note the difference between a chassis dynamometer blower, which is intended to provide convective cooling air, and a wind tunnel, which is used to simulate and measure aerodynamic drag.

Figure 6.3 Chassis dynamometer vehicle cooling blower. (Reprinted with permission from AVL.)

Several different designs and control approaches are in common use for the vehicle blowers depending on the type of test to be performed. This issue will be studied in detail in Chapter 8.

6.2.3 Ventilation or Refrigeration System

Because of the heat released from the engine and additional equipment installed inside the test cell, it is necessary to ventilate or refrigerate (depending on test requirements) the test cell to avoid significant changes in the ambient temperature and humidity. Although the vehicle exhaust gases will be extracted through a dedicated system, a substantial amount of heat is lost by convection and radiation from the vehicle and, if allowed to increase the temperature of the air in the test cell, can have significant influences on the engine performance, as detailed in Chapter 2. The ventilation system should be designed to allow control of the environmental conditions inside the test chamber prior to and during a vehicle test.

As it will be explained in Chapter 12, Section 12.3, ambient temperature ranges are often specified in the different legislations both as a required start temperature of the vehicle and a maximum deviation during the test. Controlling the ambient temperature is also vital from a test precision perspective.

The main thermal loads and ventilation/refrigeration design criteria are presented in Chapter 9.

6.2.4 Air Intake System in Chassis Dynamometer Test Cells

The main purpose of this system is to supply combustion air for the vehicle under testing. This air needs to be controlled in terms of temperature and humidity for experimental design purposes as well as to guarantee that there is no condensation or icing inside the chamber. A dedicated system may be specified to condition combustion air such as that shown in Figure 6.4. Alternatively, combustion air can be drawn directly from the test cell, which negates the need for this dedicated system. However, this means the full test cell must be conditioned in terms of pressure, temperature, and humidity to the desired levels we want the engine to be subjected to. Reinforced chambers for operating under pressurized or depressurized conditions can result in especially high costs if pressure control is required. The combustion air intake system will be studied in detail in Chapter 10.

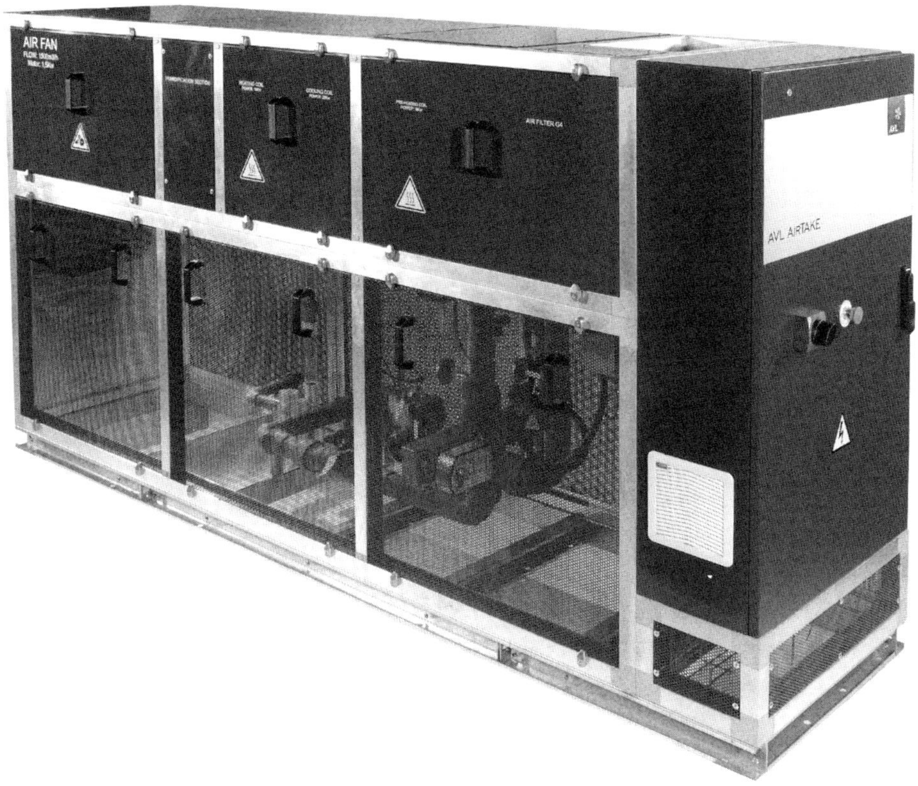

Figure 6.4 Air intake system. (Reprinted with permission from AVL.)

6.2.5 Exhaust Extraction System

Exhaust gases produced during a test must be evacuated from the test cell. Because of the high temperatures at the vehicle tail pipe, these gases must be diluted to reduce their temperature to a range within the limits of the exhaust extraction fan used for this task. The extraction itself will typically be achieved by an electronically driven fan, which mixes fresh dilution air with the exhaust gases and then evacuates these through a chimney as illustrated in Figure 6.5.

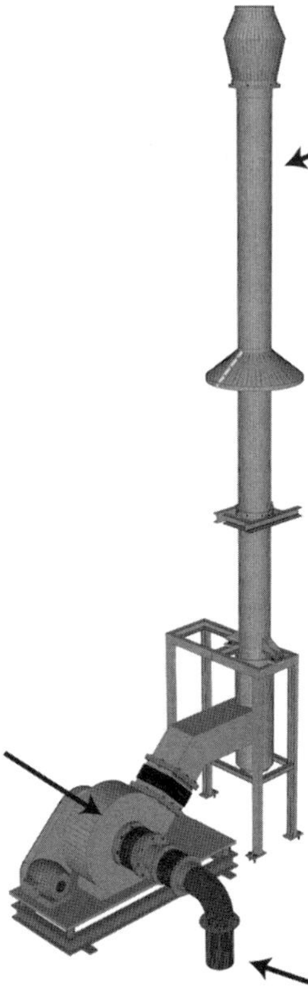

Figure 6.5 Exhaust extraction system.

References

6-1. Verlag Moderne Industrie, "Vehicle Chassis Dynamometers," edited by AVL Zöllner. Verlag Moderne Industrie: Landsberg am Lech, Germany.

Chapter 7
The Chassis Dynamometer

7.1 Introduction

In Chapter 1, the vehicle tractive effort and power demand on a vehicle moving on the road were discussed. In this chapter, this concept is reviewed briefly in the context of replicating these forces on a chassis dynamometer. As the vehicle will not actually be moving, the inertia, rolling resistance, aerodynamic drag, and gradient forces will all need to be reproduced in a way that the engine and powertrain are subjected to the same conditions as if they were driving along the road.

7.2 Vehicle Tractive Energy and Power Demand Review

The equations of motion of a vehicle were developed in full in Chapter 1. However, it is useful to remind ourselves of them here as they are the starting point for calculating the requirements of the chassis dynamometer settings that will replicate these forces while the vehicle is under test. The forces acting on a vehicle as it moves in a straight line on a road are summarized in Figure 7.1.

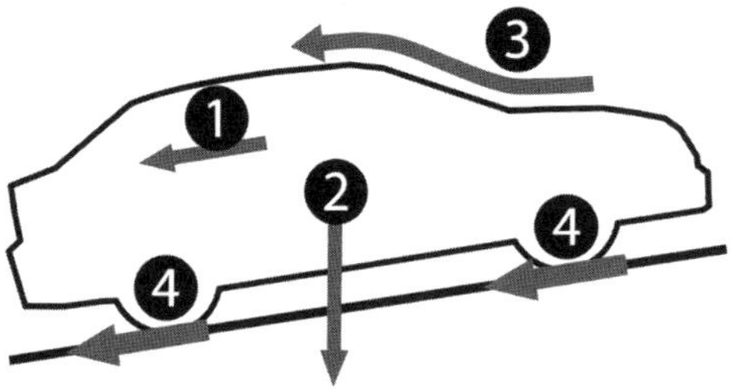

Figure 7.1 Forces acting on a moving vehicle. 1—inertia. 2—gravity. 3—aerodynamic drag. 4—rolling resistance.

The force opposing a vehicle in motion consists of several components:

$$F_{TR} = F_I + F_g + F_{AD} + F_{RR} \tag{7.1}$$

where

F_{TR} = tractive effort transmitted from the powertrain, through the tires to the road

F_I = force necessary to overcome inertia (during acceleration and deceleration)

F_g = force because of gravity when traversing a gradient

F_{AD} = force because of aerodynamic drag

F_{RR} = force because of tire rolling resistance

Analyzing each one of the forces in Eq. 7.1, the general equation that characterizes a vehicle moving on the road is developed in Chapter 1, and reproduced as follows:

$$P_{TR} = \left(M \cdot g \cdot C_{RR} \cos\theta + \frac{1}{2} \cdot \rho \cdot C_D \cdot A_F \cdot V^2 + M \cdot \frac{dv}{dt} + M \cdot g \cdot \sin\theta \right) \cdot V \tag{7.2}$$

where

P_{TR} = tractive power demand (W)

M = vehicle mass equivalent including rotating inertia effects (kg)

g = gravitational constant (9.81 m/s^2)

C_{RR} = tire rolling resistance coefficient

ρ = density of air (kg/m^3)

C_D = drag coefficient

A_F = projected frontal area (m^2)

V = velocity at current time increment (m/s)

θ = road gradient (degrees from the horizontal)

7.3 Simulating Vehicle Tractive Energy and Power Demand on a Chassis Dynamometer

When simulating a vehicle running on the road, most of the forces acting on the vehicle will not be present because of the lack of movement of the vehicle. It is necessary to calculate the equivalent tractive power absorbed by the vehicle motion. This power will then be simulated by the chassis dynamometer. It is necessary to consider the implications of this requirement for each term of the vehicle load.

7.3.1 Inertia Force F_I

Inertia force is the force produced by the change of speed of the vehicle mass. As the vehicle is at zero speed when running in a chassis dynamometer, this force needs to be simulated. To do this, it is only necessary to know the vehicle mass and rate of acceleration.

In a vehicle running on the road, a significant part of the inertia force is associated with the rotating components such as the wheels and transmissions. We saw how to calculate the effective mass of these components in Chapter 1. However, on a vehicle dynamometer, these inertias do not need to be simulated because all these components will be running in similar conditions as on the road. One exception to this is the nondriven wheels in the case of a two-wheel drive chassis dynamometer. In such a test, the nondriven axle is stationary and must be considered in the road load simulation.

7.3.2 Gravity F_g

As the vehicle body is at rest during a chassis dynamometer test, there can be no change in height or potential energy of the vehicle. Therefore, the work done to overcome a gradient will need to be simulated by the chassis dynamometer. As mass is a parameter set by the test procedure and gradient is a property of the test cycle, the calculation of the gravitational term is straightforward. Gradient has traditionally been assumed to be zero when estimating fuel economy. Gradient may be set to other values for specific tests, and will increasingly be considered during vehicle-type approval testing and real-world compliance testing as legislation is updated. Unlike vehicle mass, which will remain approximately constant whilst driving, the gradient is a function of the route driven and will change with time over the course of a test. Such a parameter would need to be programmed into a time-based test sequence, unless the test in question was a constant-gradient hill climb.

7.3.3 Aerodynamic Drag F_{AD}

Even though, all vehicle chassis dynamometers are equipped with a cooling blower, this machine is used for vehicle cooling purposes. It creates no drag force on the vehicle that must be overcome by the powertrain as the vehicle is restrained firmly on the dynamometer. Any load imparted to the vehicle by the air blast will be reacted through the vehicle restraints. Typically, the magnitude of this force is not of interest in a chassis dynamometer as the airflow is not designed to precisely replicate that found in road driving. By comparison, in a wind tunnel, the objective is to replicate real-world flow forces. These are then measured at the restraint system. In such a wind tunnel, the flow stream covers the whole vehicle frontal area. This is not usually the case in a chassis dynamometer; rather, the air flow is designed to ensure realistic cooling for the test piece. Therefore, in a chassis dynamometer, the control system is programmed to apply a representative aerodynamic load to the vehicle via the rollers.

7.3.4 Tire Rolling Resistance F_{RR}

Tire rolling resistance is greatly influenced by the design of the tire, its thermal state, and road surface roughness. The precise type of tire used on a chassis dynamometer test may well be different from that used to conduct the road test, and the road roughness and condition will certainly be different from the chassis dynamometer roller surface.

There will be, however, a significant degree of rolling resistance derived from the tires on a chassis dynamometer test. This resistance force must be evaluated and compensated for by comparing the observed resistance on the dynamometer, with the experimental values observed on road and then computing a revised rolling resistance term, which replicates on-road conditions.

7.4 Measuring Vehicle Characteristic Parameters

When it comes to measuring the values of these forces acting on a real vehicle, it is often useful to take an empirical approach. This is in part because of the experimental challenge in isolating the effect of each physical mechanism as all tend to occur concurrently. While an analytical approach could be taken, the load terms are today seldom calculated individually from theoretical data; instead, the total drag load is measured experimentally.

As was introduced in Chapter 1, the three terms relating to aerodynamic drag, rolling resistance, and gradient are often grouped together and referred to as road load. This is distinct from the remaining component of vehicle load, which comes from inertial forces. A further simplification is usually made at this stage to neglect the effect of gradient when measuring vehicle drag characteristics, because this is trivial to calculate from theory once the vehicle mass and gradient are known.

Conveniently, the sum of the remaining road load terms generally increases with the square of vehicle speed. For convenience, in dynamometer control systems, the road load curve is often expressed in a simplistic manner, with the various physical terms combined into a polynomial expression as in:

$$F_{\text{roadload}} = F0 + F1V + F2V^2 \tag{7.3}$$

Now, the total force acting on the vehicle (equal to tractive effort) is the sum of these components, road load, and inertia load, as shown in:

$$F_{\text{TR}} = F_{\text{roadload}} + F_{\text{inertia}} = F0 + F1V + F2V^2 + M\frac{dV}{dt} \tag{7.4}$$

where F0, F1, and F2 are called the *road load coefficients* and are typically entries to be made in the chassis dynamometer control software. In this expression, the physical meaning of the coefficients becomes indistinct as they are not derived from detailed measurements, and the three coefficients are determined empirically to match on-road behavior. Conceptually, we can say that the following hold:

- The F2 coefficient is dominated by the aerodynamic drag effects.
- The linear coefficient F1 is dominated by bearing friction effects, which are approximately linear with speed [7-1].
- The F0 term is dominated by the tire rolling resistance [7-2].

The task of determining these coefficients is undertaken using the coast-down test.

7.4.1 Coast-Down Test

Coast-down is a fundamental test of a motor vehicle and simply consists of accelerating to some high speed on a test track and then coasting down to rest (or a predetermined lower speed) with the powertrain disengaged. In this way, all the drag terms acting on the vehicle are correctly represented and can be observed using simple measurements. During this coast-down, the speed, deceleration, and distance traveled are recorded until the lower speed point is reached. Theoretically, only vehicle acceleration needs to be measured as time progresses, as this can be integrated to calculate speed and again to give distance. In practice, acceleration and speed would typically be a minimum desirable dataset, measured by an accelerometer and a microwave radar system, respectively. This is usually augmented by the distance derived from high-precision GPS data as shown in Figure 7.2.

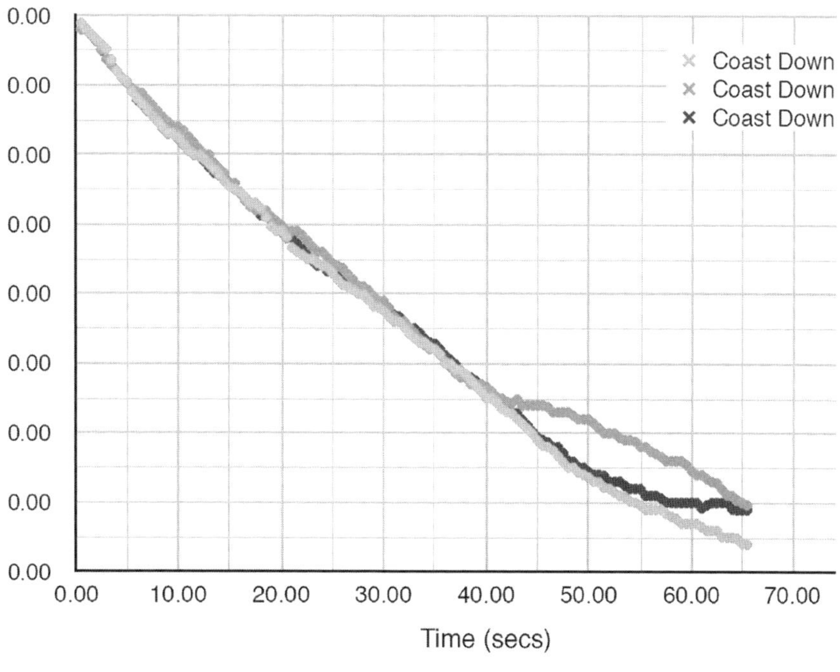

Figure 7.2 Velocity curves for several coasts-down along one particular road strip.

The main aim of this test is to evaluate the values of the resistance forces acting on the vehicle at certain speed and road conditions, so that they may be reproduced on the chassis dynamometer. The test conditions during the coast-down test will have a significant effect on the observed deceleration rates and on subsequent chassis dynamometer tests. The test conditions are carefully defined in legislation for type approval testing. Typically, the data will have some measurement errors because of noise and Eq. 7.4 will be fitted to the experimental data as illustrated in Figure 7.3. Note that by undertaking a mathematical fit, in some cases, the coefficients F0–F2, which were described analogous to real forces, may have some unexpected values. For example, it is not uncommon to have some very small or even negative values in these coefficients, which is a result of the regression approximation rather than any physical effects.

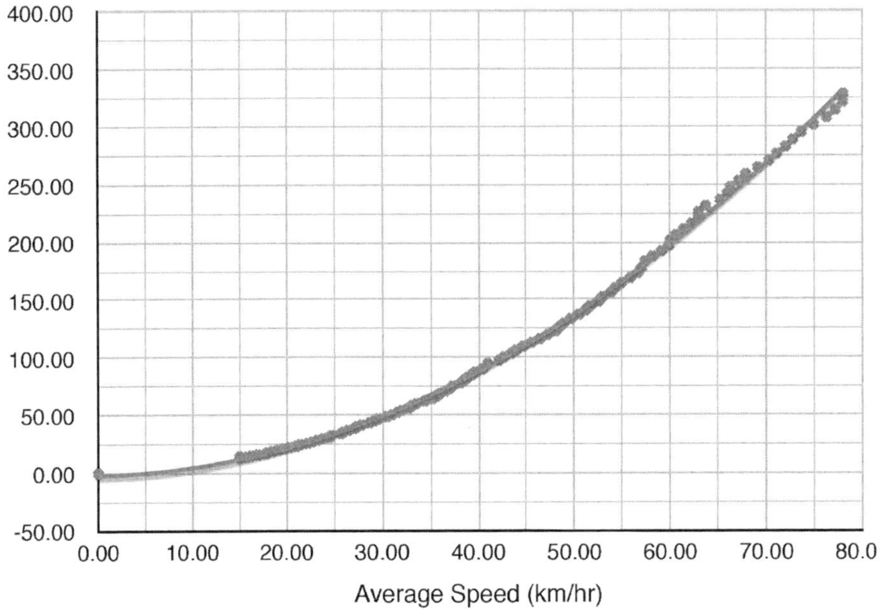

Figure 7.3 Least-squares approximation of experimental data with a second-degree polynomial.

7.4.2 Coast-Down Equation

Once the deceleration of the vehicle has been determined experimentally, we can compare this curve with the theoretical motion equation to identify each term in the equation. Considering again Eq. 7.4, the terms F0–F2 can be determined from measured data as above. The mass in the inertia term must refer to the mass equivalent of both linear and angular moments of inertia of all parts that will not be captured during testing. Typically, this will be composed of the vehicle mass and any nondriven wheels, axles, and brake assemblies. Later, we will see that when considering a chassis dynamometer control algorithm, the angular moment of inertia of the driven wheels does not need to be simulated as this inertia is part of the vehicle under test. When analyzing coast-down data, the apparent mass represented by the rotational inertia does need to be considered as this is part of the energy use of the vehicle. The equivalent mass of the vehicle and nondriven wheels can be determined using the equivalent mass theory presented in Chapter 1, which will give the following expression:

$$\frac{M_{ap} \cdot V^2}{2} = \frac{M \cdot V^2}{2} + \frac{I_w \cdot \omega_w^{\ 2}}{2} \tag{7.5}$$

and $\omega_W = Vr_d$, the rotational speed of the wheels.

The vehicle therefore behaves as having a bigger mass, the apparent mass M_{ap}.

F_{TR} can be expressed as the force delivered by the engine minus frictional losses. $F_{TR} = F_{ENGINE} - F_{FRICTION}$. During a coast-down test, as the vehicle is running freely, with clutch disengaged and transmission in neutral, the force delivered by the engine comes to zero, and the force F_{TR} becomes negative because of the friction in axle mechanisms, drive line, and output assembly of the transmission:

$$F_{TR} = -R_f = \frac{M_f}{r_d} \tag{7.6}$$

where M_f = equivalent friction moment applied to the driving wheels.

Substituting Eq. 7.6 into Eq. 7.4 yields:

$$-R_f = F_{roadload} + F_{inertia} = F0 + F1V + F2V^2 + M_{ap}\frac{dV}{dt} \tag{7.7}$$

which gives the deceleration equation during coast-down:

$$-M_{ap}\frac{dV}{dt} = R_f + F0 + F1V + F2V^2 \tag{7.8}$$

In practice, the driveline friction is generally considered a property of the vehicle under test, which will also be present on the chassis dynamometer. Therefore, the R_f term, which tends to be linear with speed, is usually lumped into the overall road load and Eq. 7.8 can be simplified to:

$$-M_{ap}\frac{dV}{dt} = F0 + F1V + F2V^2 \tag{7.9}$$

A numerical procedure is then used to calculate the three coefficients of quadratic equation (Eq. 7.9) from the experimental data. This quadratic term is used to program the vehicle simulation within the dynamometer control system directly. One important factor to consider is the units of velocity used within the control system and vehicle road load database, m/s and km/h, which are both commonly used in road load libraries. The two units are a factor of 3.6 different, leading to catastrophic errors if the wrong units are assumed.

7.5 Chassis Dynamometer Load Setting Procedure

An on-road coast-down test allows you to calculate the main parameters that define vehicle characteristics. This type of test involves running the vehicle at a test track, driving up to a certain speed, and disengaging the clutch while recording the deceleration of the vehicle. A similar type of coast-down can be performed on a chassis dynamometer, which has already been configured with some base coefficients derived from the test-track coast-down test and calculated according to the previous section. In this case, the vehicle will be driven up to a simulated speed on the chassis dynamometer and then allowed to decelerate against the speed-dependent resistance of the chassis dynamometer.

This resistance will comprise both the programmed road load, which is applied by the electric motor according to Eq. 7.9, and some internal friction inherent in the dynamometer bearings and resulting from the particular contact between the drum and the tires, which may be different from the resistance between the tires and the road. Both these internal frictions must also be taken into account.

For this reason, the most common way to calibrate the dynamometer setting is by running a coast-down test of the vehicle installed on the dynamometer. The most common procedure for achieving this calibration is to perform an iterative test to obtain matching coast-down times on the dynamometer to those observed on the road. In this way, friction losses from the chassis dynamometer and rolling resistance of the vehicle tires on the roller are also taken into account.

7.6 Simulating Road Load on the Dynamometer

To simulate the total force applied to the vehicle tire in Eq. 7.9, the dynamometer must combine its own inherent physical properties with some level of simulated properties. The dynamometer itself is composed of rotating elements as shown in Figure 7.4. This general scheme is representative of all chassis dynamometers; however, older installations may consist of larger flywheels representing significant inertias and a small electrical machine, while newer installations may have no flywheel (only the inertia of the drum and electrical machines) and, instead, rely on a larger electrical machine to simulate a proportion of the inertia effects.

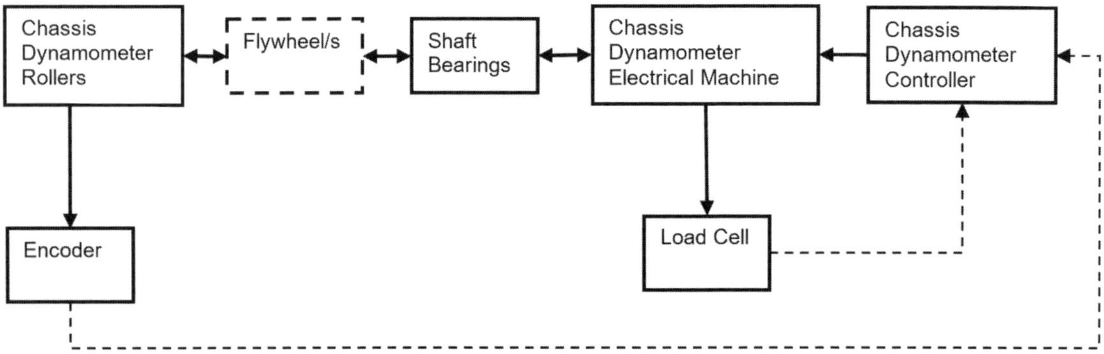

Figure 7.4 Chassis dynamometer functional block diagram. The thick solid lines represent mechanical connections where rotational power is transferred between blocks. The dotted lines represent signal connections. An optional item is shown in the dotted outlined box.

The most common inertia system nowadays is the fully electric system. Fully electric inertia simulation systems will usually consist of one AC motor with a power rating between 150 and 300 kW [7-3] mounted between two rollers (see Figure 7.5). This configuration is known as "motor in the middle" and has substantial advantages as there is no requirement for external shafts or bearings. An alternative layout of the same system is to use two smaller AC motors, one driving the left-hand roller and the other the right-hand roller. The advantage of this setup is that you can simulate curves by making one roller independent from the other. A final alternative within the fully electric inertia simulation systems is to use DC motors in place of AC. These tended to be used in the early days of electrical inertia simulation (most prevalent in the early to mid-1990s), and have effectively been superseded by the AC motor because of its improved response time.

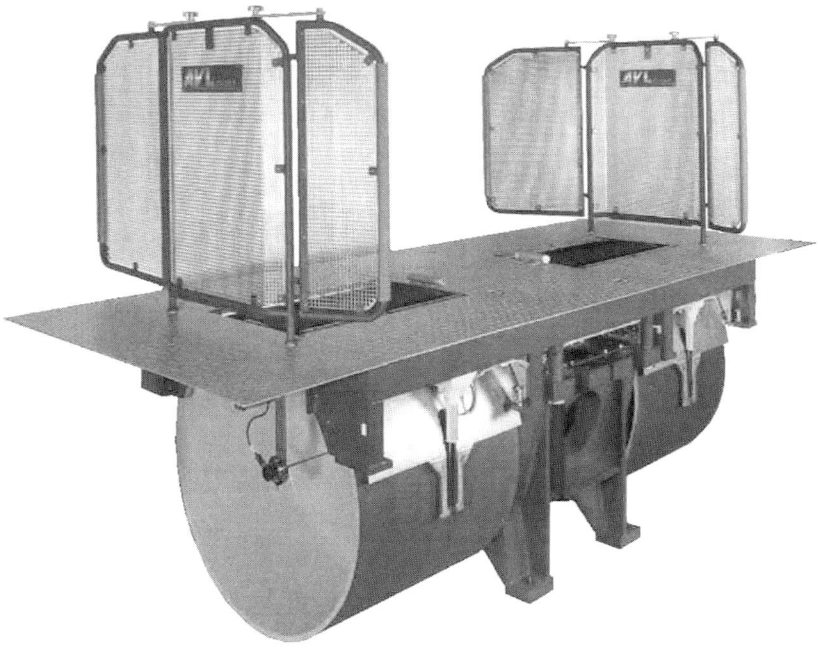

Figure 7.5 A 48" single roller 2WD AVL motor in the middle chassis dynamometer. (Reprinted with permission from AVL.)

In operation, the dynamometer control system is given a definition of the dynamometer road load it must apply to the vehicle during testing. This road load is defined by the equivalent vehicle mass plus the three coefficients, F0–F2 (as defined by Eq. 7.9). The dynamometer controller takes a measure of the roller speed from one or often two encoders positioned on the roller shafts. The force applied by the electrical machine is measured by a load cell positioned between the chassis dynamometer bed frame and a lever arm attached to the electrical machine housing (other techniques can use different instrumentation for the torque measurement). Often, the dynamometer electrical machine will be configured so that the inertia and road load simulation are not active when the rollers are stationary. Consider a vehicle starting from rest on a chassis dynamometer; the vehicle must initially accelerate only the base inertia of the rollers, the vehicle wheels, driveline, and the mechanical couplings in the dynamometer. Once the roller speed is above a certain threshold, the motors become active and the dynamometer controller can adjust the motor current to achieve the desired road load and inertia simulation force for the given speed and acceleration rate. As the load cell gives only a measure of the motor force, the controller must have an estimation of the loss force between the motor output and the roller surface to calculate the load applied to the vehicle. These losses arise primarily from bearing friction, flywheel and roller windage and motor coasting torque losses [7-1], [7-4]. Bearing losses are reported to be approximately linear with increasing speed and vary with viscosity of the lubricating medium and bearing preload, both of which are dependent on bearing temperature variations [7-1]. The estimation of the total losses is typically obtained by a separate test where the rollers are accelerated and decelerated at a fixed rate and the error in the motor force is measured. From this test, the dynamometer losses can be defined by a quadratic equation, as shown by:

$$F_{\text{Dyno\_Loss}} = A0 + A1V + A2V^2 \tag{7.10}$$

where $F_{\text{Dyno\_Loss}}$ is the dynamometer loss force, $A0$–$A2$ are the loss force coefficients, and V is the dynamometer roller surface speed.

References

7-1. Thompson, J. K., et al. 2002. "Inertia Simulation in Brake Dynamometer Testing." SAE Technical Paper 2002-01-2601. SAE International: Pennsylvania.

7-2. Chapin, C. E. 1981. "Road Load Measurement and Dynamometer Simulation Using Coastdown Techniques." SAE Technical Paper 810828. SAE International: Pennsylvania.

7-3. Horiba Ltd. 2013. «VULCAN EMSCD48.» Available from http://www.horiba.com/automotive-test-systems/products/mechatronic-systems/vehicle-test-systems/details/vulcan-emscd48-626/. Accessed 26 November 2013.

7-4. D'Angelo, S., et al. 1996. "Evaluating the Performance of Chassis Dynamometers with Electric Inertia Simulation." SAE Technical Paper 960716. SAE International: Pennsylvania.

Chapter 8
The Vehicle Cooling Blower

8.1 Introduction

The vehicle cooling blower is responsible for cooling the main parts of the vehicle and its characteristics need to be carefully selected according to test parameters.

8.2 Wind Blower Classification Depending on the Kind of Test

Taking the classification of the air flows around the vehicle described in Chapter 1, we can identify the importance of the air flows depending on the kind of test (see Chapter 12). The different air flows around a vehicle during on-road driving are summarized in Figure 8.1. The air flows are used to cool various components of the vehicle that generate heat while driving. The main sources of heat are the engine radiator, the engine, the exhaust pipes, the brakes, and the tires.

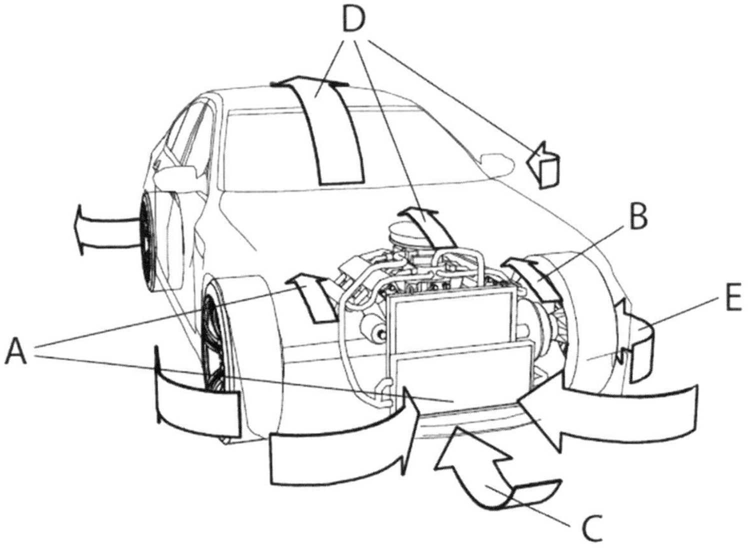

Figure 8.1 Key air flows around a vehicle when driving on the road. A—radiator cooling, engine cooling. B—brake cooling. C—underneath the vehicle flow. D—air flow around the vehicle (heat dissipation). E—air flow on tires.

Table 8.1 shows which of the vehicle cooling flows are important for different types of vehicle tests. The type of test is important because it will influence the size and type of cooling fan that will be employed. In the planning stage, it will also determine how much electrical power is required to operate such a fan and, ultimately, costs. More details and descriptions of the different types of vehicle tests can be found in Chapter 12.

Table 8.1 Air Flow Vehicle Cooling Depending on the Kind of Test

	Air Flow				
Vehicle Test	A	B	C	D	E
Aeroclimatic	✓	✗	✗	✓	✗
Emissions	✓	✗	✗	✗	✗
RDE	✓	✗	✗	✗	✗
RDE-Advanced	✓	✓	✓	✓	✓
Friction	✓	✓	✓	✓	✓
High Speed/Durability	✓	✗	✓	✗	✗

Taking a typical mid-size vehicle, we can identify its main frontal dimensions in Figure 8.2 and see the influence of the different blower nozzle dimensions used for testing. Although the overall frontal area of the vehicle is quite large, the area required for powertrain cooling, roughly the area of the frontal radiator, is usually substantially smaller. On most production cars, the radiators and intercoolers are located centrally at the front of the vehicle; however, care should be taken to identify the locations of all the radiators to ensure that sufficient air flow is provided to them. Usually, the standards specify the size of cooling fans required for emission testing.

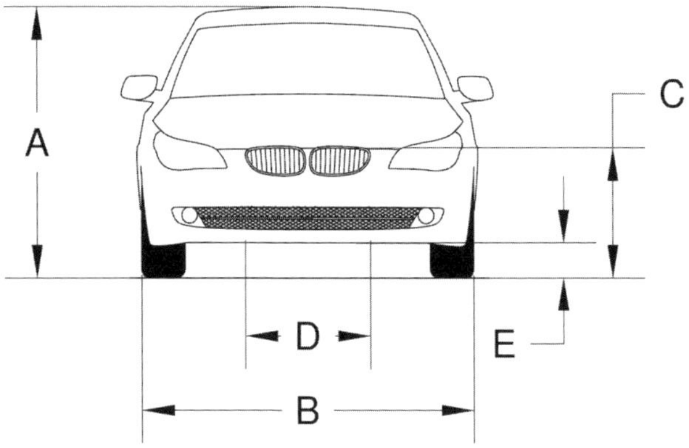

Figure 8.2 Main dimensions of a mid-size vehicle (mm). Approximate lengths for passenger car (mm): A: 1471, B: 1777, C: 710, D: 870, and E: 220.

8.2.1 Aeroclimatic Tests

Aeroclimatic tests are usually performed at highway speeds (within legal limits) to test the interactions between the vehicle cabin and the surrounding air. The most critical areas are the vehicle inlet frontal area, covering the main entries of the car, and the vehicle cabin with special emphasis on the windscreen due to its high heat exchange potential.

The U.S. SC03 test defines the outlet section of the blower as 4.25 ft × 4.25 ft (1290 mm × 1290 mm) with a maximum air speed of 54.8 m/h (88.2 km/h). This area is represented schematically in Figure 8.3.

Figure 8.3 Nozzle coverage for an aero climatic test (U.S. SC03 test).

According to the legislation, the distance from the outlet section to the floor should be between 0 and 6 in (0–15.2 cm); any value within this range would be accepted. The distance from the nozzle to the vehicle is quite large because of the large section of the nozzle (in between 60 and 90 cm in the U.S. SC03).

8.2.2 Emission Tests

Historically, emission tests have been performed within legal speed limits, with a maximum speed of around 130 km/h. The power requirements of the vehicle are not very demanding, and therefore, the exhaust line is not too hot and does not require special cooling. The heat exchange between the vehicle and surrounding air is not subject to study. The only section that requires cooling is the vehicle radiator, which has a relatively small air inlet section as shown in Figure 8.4.

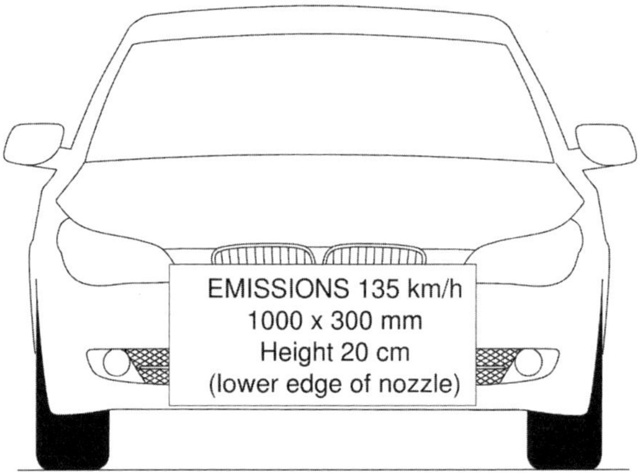

Figure 8.4 Nozzle coverage for an emission test.

The European legislation requires an outlet section of the nozzle of 0.3 m^2 (at least) placed at 30 cm from the vehicle and 20 cm from the ground floor.

8.2.3 Research and Development

Many manufacturers in research and development (R&D) are considering the different types of tests. We will examine two kinds of tests in this group: real-driving emission (RDE) simulation tests and friction tests.

8.2.3.1 Real-Driving Emission Simulation in a Chassis Dynamometer

A new RDE testing procedure is being introduced and is described in detail in Chapter 15. The new test will use a portable emissions monitoring system to measure pollutant emissions directly on roads.

With the move toward RDE legislation, new cooling fans will be required to support verification tests on a chassis dynamometer that better replicate real-world conditions; however, it is important to clarify that RDE testing in chassis dynamometers is an R&D task and cannot be considered as part of the legislation.

Because RDE will be measured on the road, the legislation does not specify the size of the nozzle for chassis dynamometer testing. For RDE testing, in order to achieve reproducibility of on-road conditions, all flows shown in Figure 8.1 should be represented; however, the vehicle underhood air flow is the most critical air flow for this test profile. Thus, the nozzle and blower recommended for emission testing would be suitable for RDE as well, assuming a maximum speed of 160 km/h (which is the limit for an RDE test with extended range (see Part III)). (See Figure 8.5.)

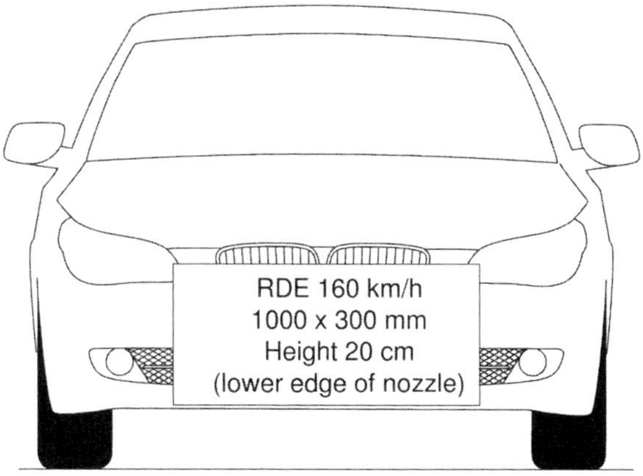

Figure 8.5 Nozzle recommended for an RDE simulation in a chassis dynamometer.

Bigger blowers and nozzles can be adopted for better reproducibility of real-world driving where tire cooling, exhaust cooling, and a higher air flow around the vehicle (to reproduce air conditioning system performance) would be considered. Computational fluid dynamics (CFD) studies have shown that a nozzle of 1.3 m width and a height of between 1 and 1.3 m (the height should be enough to exceed the vehicle hood for every case) can be considered as the optimal area, installed at 1 m from the front of the car.

These considerations lead to a new concept known as RDE-Advanced with nozzle dimensions and speed, and is illustrated in Figure 8.6.

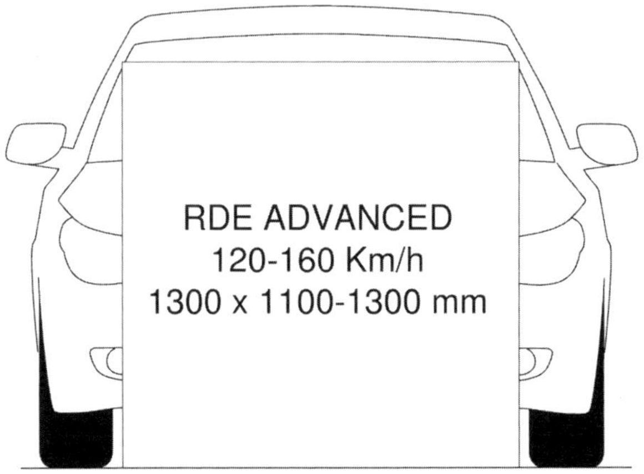

Figure 8.6 Nozzle recommended for an RDE-Advanced simulation in a chassis dynamometer.

8.2.3.2 Friction Tests

Friction tests are performed at high speeds, from 140 to 200 km/h, with special attention paid to exact reproducibility of vehicle mechanical efficiency. Exhaust cooling is required because of the high temperatures obtained in the exhaust line. This requires cooling under the vehicle. Tire cooling is also required to obtain consistent tire performance and, therefore, rolling coefficient. Finally, a representation of the heat exchanged between air and vehicle is also required to test the influence of the air conditioning on the overall vehicle performance. Different configurations can be adopted as shown in Figure 8.7 and Figure 8.8, depending on the maximum speed that has to be reached and the height of the vehicle hood. A nozzle of around 1.3–1.4 m width and 0.7–1.3 m height should cover most of the cars in the market (Figure 8.9). In order to limit maximum blower air flow, many manufacturers opt for an adjustable nozzle that can achieve higher speeds with a reduced section and lower speeds with an extended section.

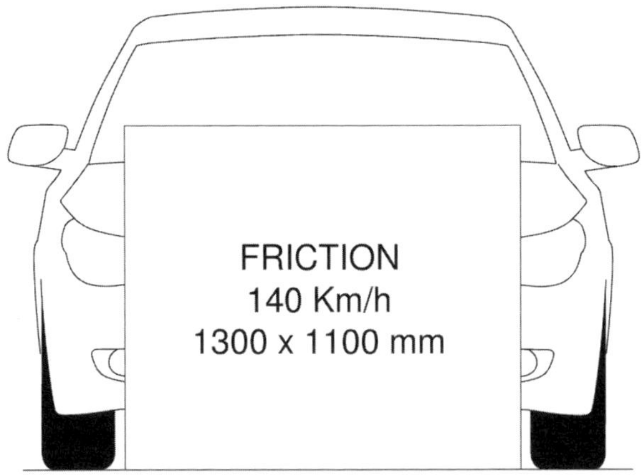

Figure 8.7 Nozzle coverage for a friction test (140 km/h).

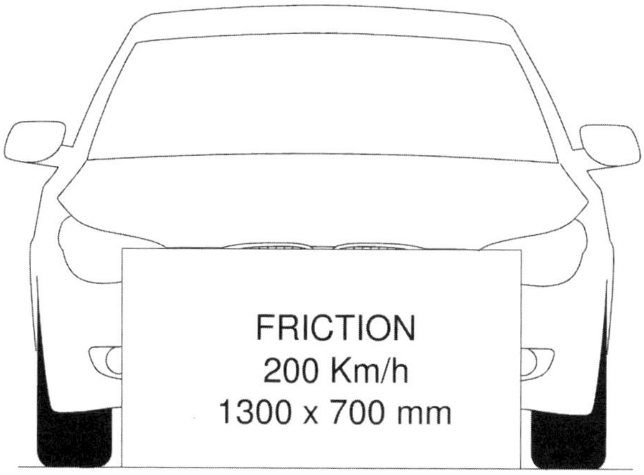

Figure 8.8 Nozzle coverage for a friction test (200 km/h).

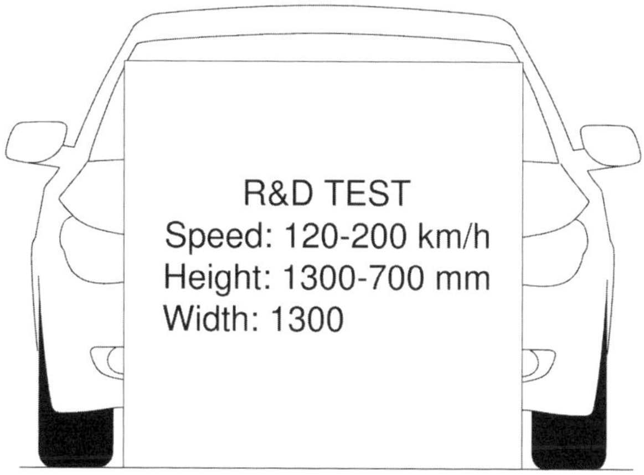

Figure 8.9 Nozzle coverage for every R&D test.

8.2.4 High Speed/Durability Tests

Durability tests are often performed at high speeds, but their purpose is to test vehicle fatigue rather than vehicle performance. Exhaust cooling is required because of the very high speeds achieved, but tire cooling is not normally necessary because of the lower accuracy demands. Air conditioning simulation is not required for such tests. A typical fan area for durability testing is illustrated in Figure 8.10.

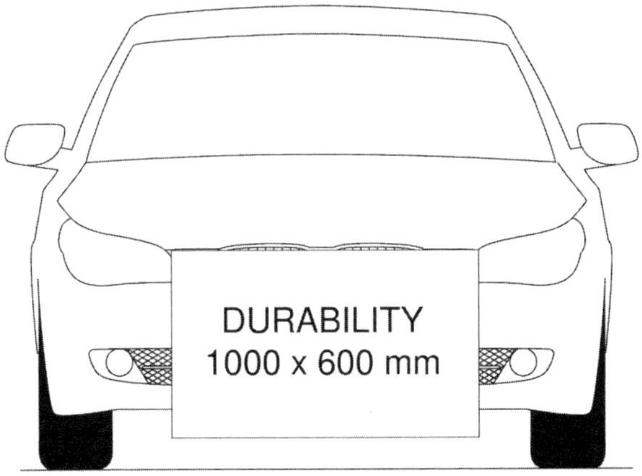

Figure 8.10 Nozzle coverage for a durability test.

Chapter 9
Chassis Dynamometer HVACR

9.1 Introduction

To ensure realistic and repeatable results, environmental conditions need to be controlled when a vehicle is tested on a chassis dynamometer. Furthermore, current legislation sets the temperature range within which the type approval vehicle must be tested. For these reasons, it is important to analyze the thermal loads on the test cell during a driving cycle.

9.2 Main Thermal Loads in the Test Cell

The main thermal loads to bear in mind regarding the test cell are illustrated in Figure 9.1 and listed below:

- thermal dissipation from the engine and vehicle to be tested
- exhaust pipe line radiation
- vehicle wind blower
- test chamber lights
- heat transfer through the walls
- thermal state of new air ingested into the cell

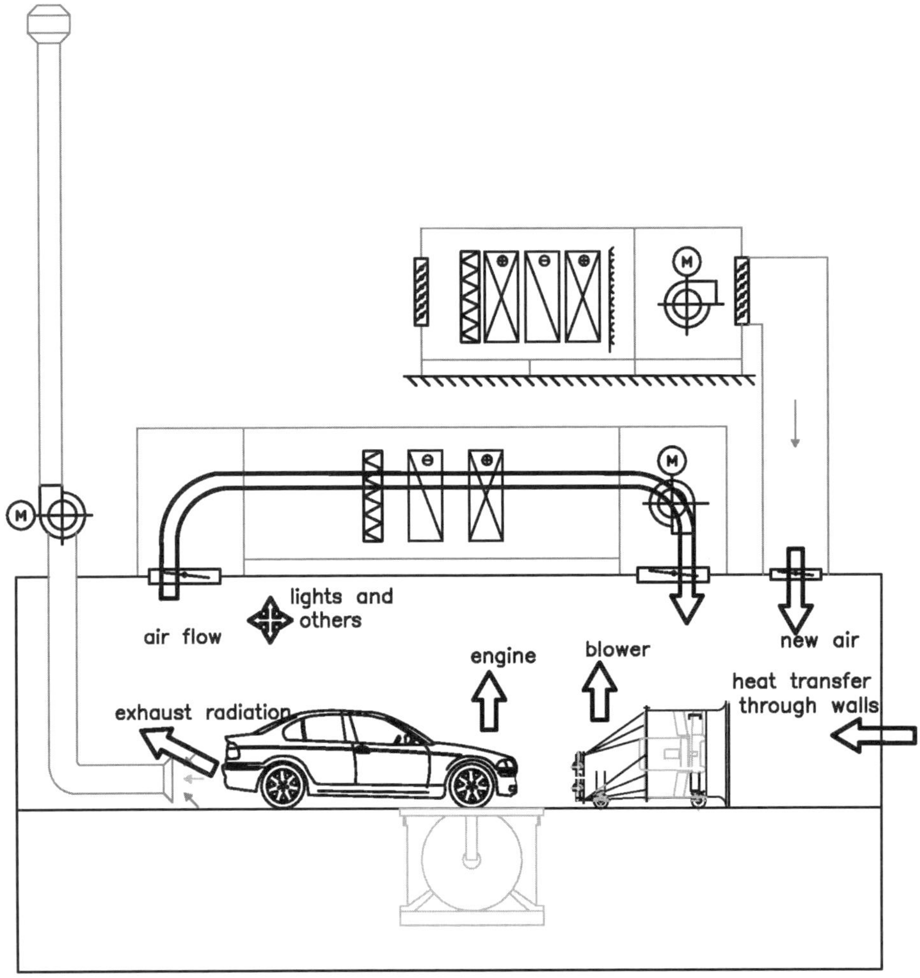

Figure 9.1 Thermal loads in a chassis dynamometer test cell.

Thermal loads in the test cell will mainly depend on the vehicle mass and driving cycle performed and not so much on the temperature set point (although the engine heat rejection and its efficiency will change at low temperatures).

Figure 9.2 shows an example of how the temperature increases when a New European Driving Cycle (NEDC) (further information in Chapter 12, Section 12.3.2.3) test is carried out in a chassis dynamometer without controlled environmental conditions.

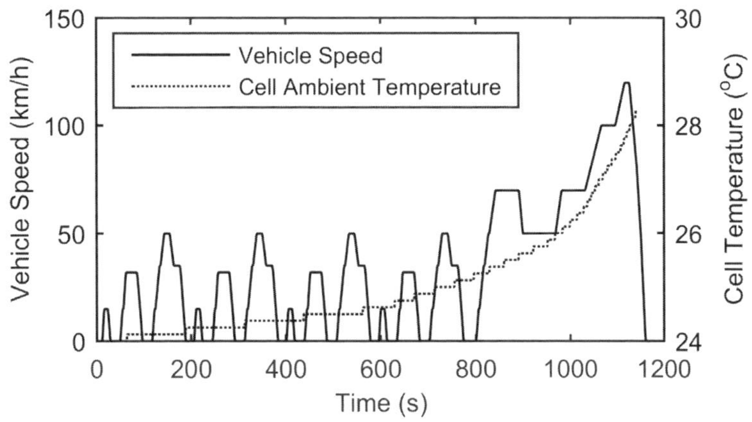

Figure 9.2 Temperature increase in a chassis dynamometer during a test (NEDC).

9.3 HVACR Design Criteria

Based on the different temperature requirements in the test cell, we can classify the chassis dynamometer test cells according to two different temperature ranges:

- ambient temperature
- low temperature

9.4 Ambient Temperature Chassis Dynamometer Test Cell

Ambient temperature controlled test cells usually work in a temperature range from 14°C to 50°C. Temperature control is usually performed by an air handling unit (AHU), which can be located within or above the cell in a technical area. Cooling capacity is usually taken from a central chilled water facility. The selection of the type of AHU depends mainly on two conditions:

- requirement to have humidity control in the test cell
- local climate of the laboratory

9.4.1 Non-Humidity Controlled Test Cell

When the humidity control is not required in a test cell design, the AHU will usually be as shown in Figure 9.3. It consists of a cooling coil, a heater (optionally), fans, and some dampers to control the degree of fresh air and recirculated air in the system. The air to cool the vehicle is discharged via a directed nozzle through the cooling fan. The air passes around the vehicle increasing the temperature. The extraction fan of the AHU takes the hot air at the rear part of the vehicle and then is drawn through the coils to be cooled. The air is afterward impelled by means of the impulsion fan to the front of the cell. Finally, the supplied air is drawn into the blower suction side and discharged at the vehicle again.

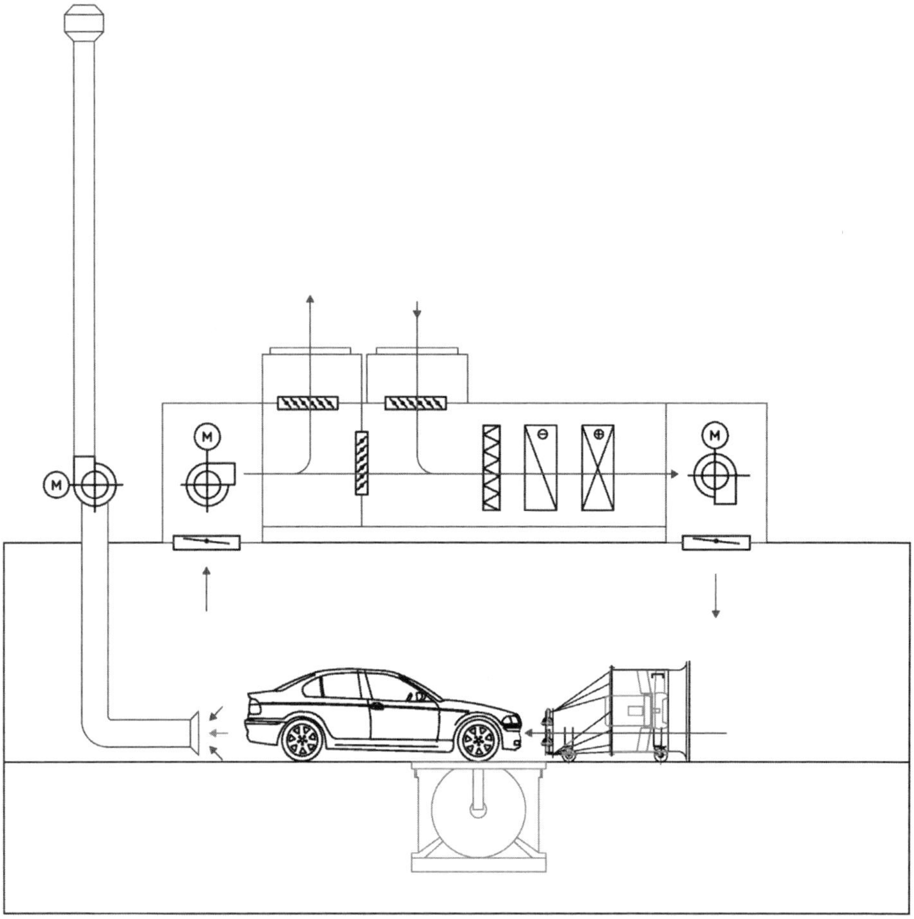

Figure 9.3 Scheme of an AHU without humidity control.

When the ambient temperature is lower than the set point required for the test, fresh air flow can be used to cool down the cell, saving energy consumption from the chiller plant. If the ambient temperature is higher than the set point required but lower than the extraction temperature, regulation flaps control the percentage of the fresh air to minimize the cooling requirements. A minimum percentage of fresh air has to be reached to supply combustion air to the vehicle and to guarantee a minimum number of air changes in the test cell.

9.4.2 Humidity Controlled Test Cell

Depending on weather conditions, two different configurations can be selected depending on the energy consumption and installation costs.

9.4.2.1 Single-Stage AHU

A typical AHU is equipped with preheating, cooling, and post-heating coils together with a humidifier and a mixing section, as shown in Figure 9.4.

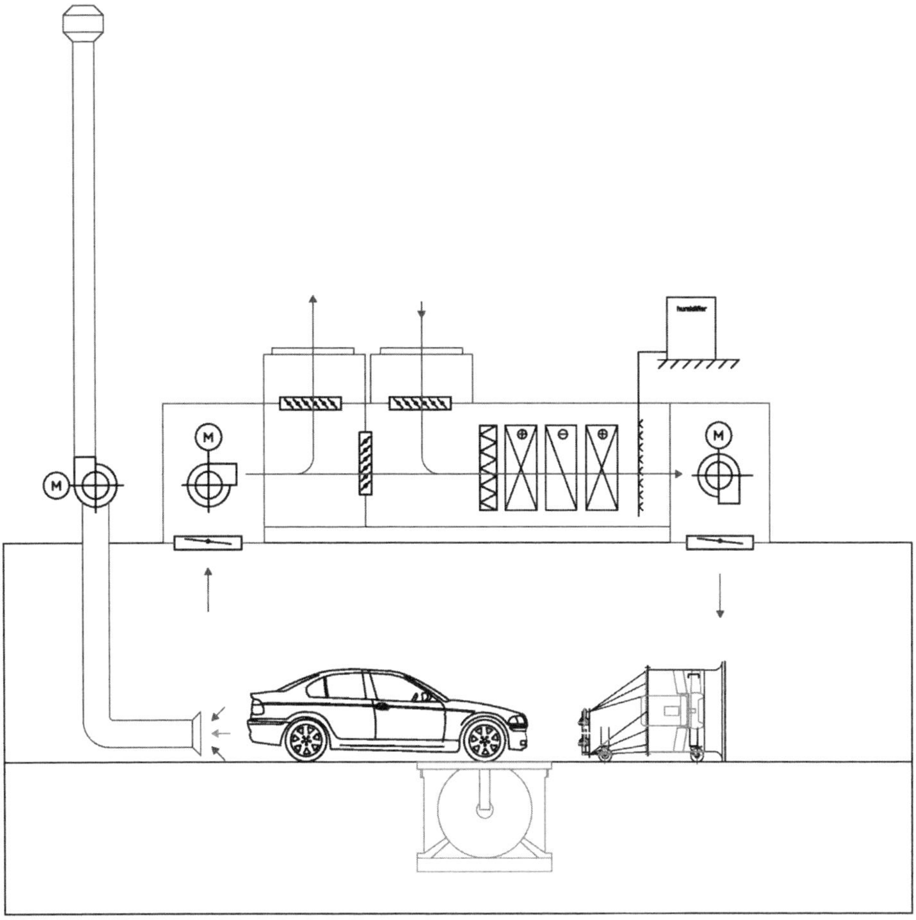

Figure 9.4 A single-stage AHU with humidity control.

The pre-heating coil is designed to protect the cooling coil from freezing if required. The cooling and post-heating coil adjust temperature and humidity when there is a need to dry the fresh air. Finally, the humidifier adds humidity to the air flow when required.

9.4.2.2 Two-Stage AHU

In humid conditions, when there is a strong requirement for air drying, a single-stage AHU needs to cool all the air flow to the requested dew point and heat it again to obtain the desired temperature and humidity. If this process is performed within the AHU flow, it is very expensive in terms of energy consumption.

To reduce the energy consumption, the AHU is divided into two units: 1) one unit that treats all the air flow in the test cell, and 2) the other unit that preconditions only the fresh air (see Figure 9.5).

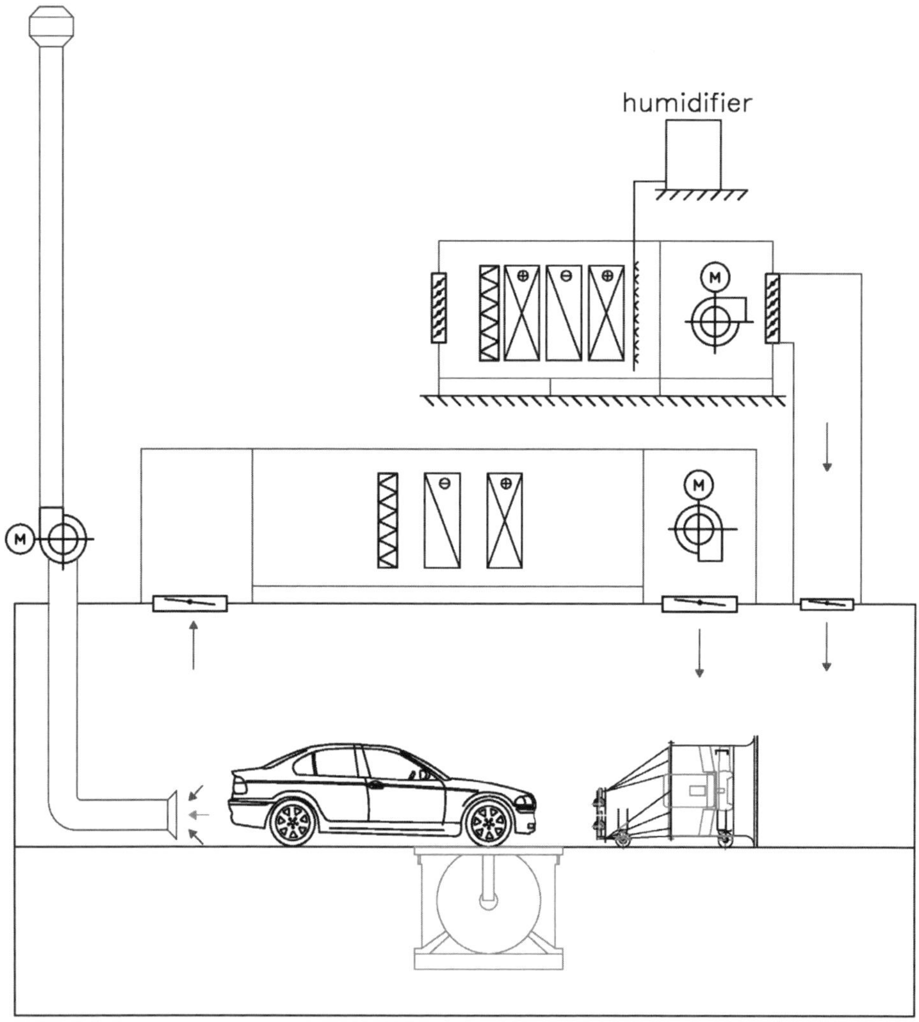

Figure 9.5 A two-stage AHU with humidity control.

The requirement to dry the fresh air can be fulfilled by the fresh AHU with a much smaller amount of air flow. The main AHU needs then to perform very little humidity control because the air recirculated in the test cell is already at the target humidity level, if we can guarantee that there is no condensation in the cooling process.

The main AHU can be installed inside the test cell (at the ceiling) through an auxiliary frame where the filters, coils, and fans are installed. This configuration is especially advantageous when the available space above the cell is limited, or even if there is not technical floor considered in the building. Figure 9.6 shows the solution described.

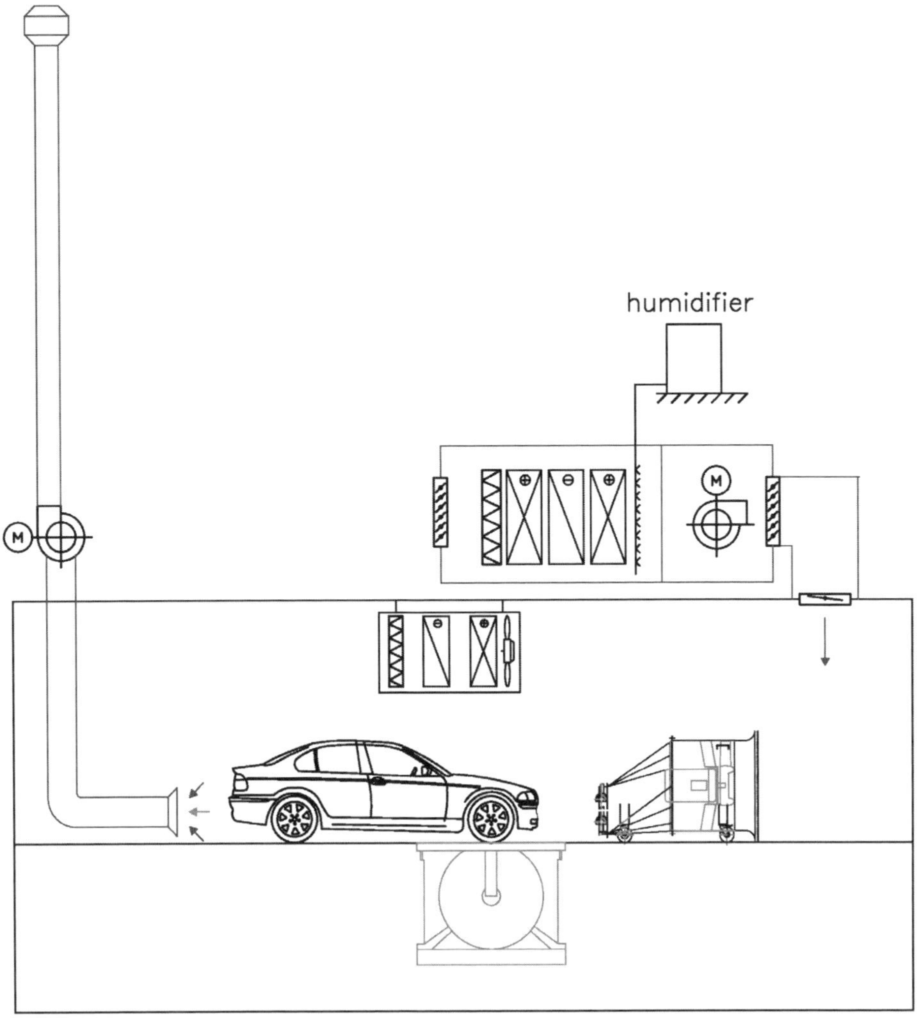

Figure 9.6 In the cell—AHU solution.

9.4.3 How to Control the Condensation in the Cooling Coil

Water condensation on a cooling coil depends on the coil surface temperature. When this temperature is below the air dew point, moisture held in the air will condense. If this is above the dew point, the air will not condensate.

There are cases in which air needs to be cooled but condensation is not desired because humidity is already at the required level, as it is the case of the main AHU in a two-stage system. In some other cases, condensation is desired as in the fresh AHU in a two-stage system.

Two different strategies are used depending on the requirement:

- constant temperature with variable flow
- constant flow with variable temperature

9.4.3.1 Constant Temperature with Variable Flow

A chiller plant provides a flow of water at constant temperature, as shown in Figure 9.7. A control valve controls the amount of this fluid that flows through the cooling coils, and the rest of the flow is bypassed, thus controlling the temperature of the cooled air leaving the coils. If this temperature is below the dew point for the intake air, condensation will occur, thus reducing the specific humidity of the air.

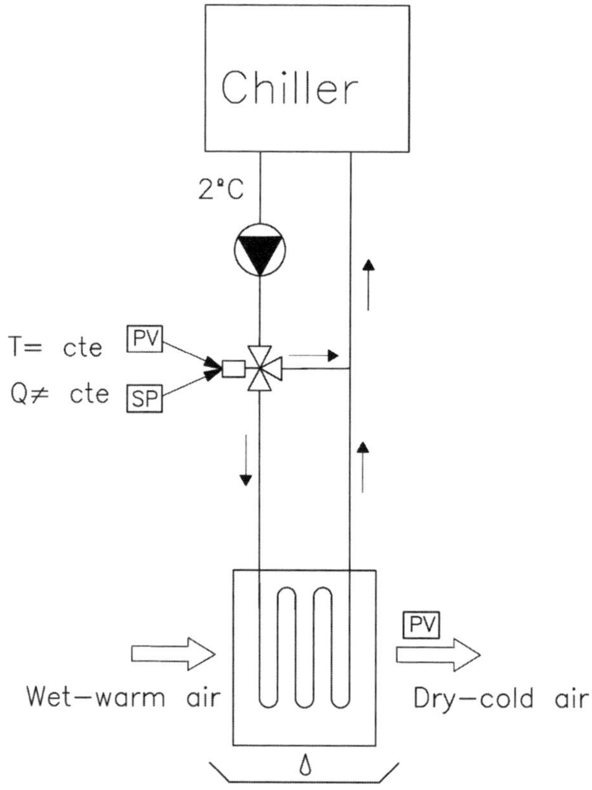

Figure 9.7 Constant temperature and variable flow cooling coil regulation.

9.4.3.2 Constant Flow with Variable Temperature

Figure 9.8 shows a constant flow, variable temperature system where a pump is installed to provide a constant flow through the chiller coils. The system still relies on a chiller unit to provide a source of cooling fluid at constant temperature, but in this case, a control valve is used to provide a mix of cooled (from the chiller) and non-cooled (recirculation in the coils) fluid. This system allows water temperature control, giving the possibility of adjusting coil surface temperature easily.

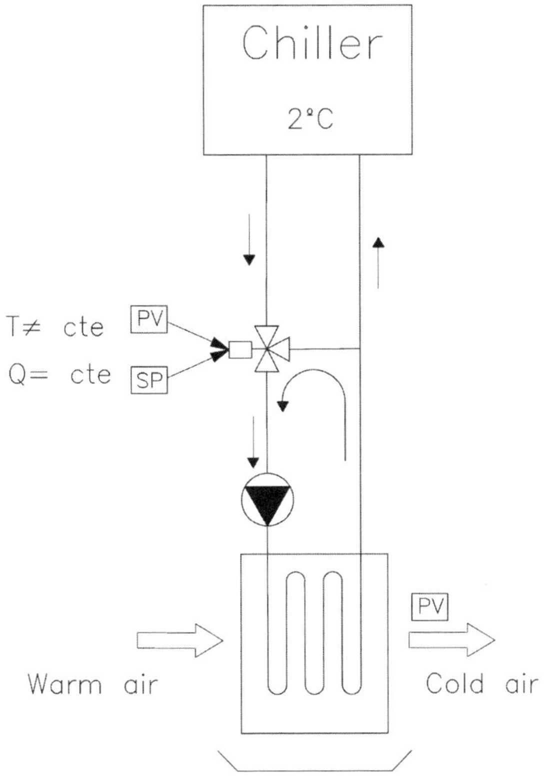

Figure 9.8 Constant flow and variable temperature cooling coil regulation.

9.5 Low Temperature Chassis Dynamometer Test Cell

Refrigeration systems are used for low-temperature tests. They work with a refrigerant gas following the reversed Carnot thermodynamic cycle.

Low-temperature-controlled test cells usually work in a temperature range from −10°C (or −30°C) to 40°C. To reduce heat losses to the environment, refrigeration is performed by a coil or evaporator placed on the ceiling of the test cell. Cooling energy usually comes from a specific chiller or refrigerant plant used only for this chamber; however, sometimes a testing facility with several chassis dynamometer test cells is considered. In that case, all the cooling coils from different cells can be connected to the general low-temperature chilled water generation plant (usually −30°C).

9.5.1 Introduction to the Refrigeration Machine

The basic configuration of this type of refrigeration system consists of a compressor, a condenser, an expansion device, and an evaporator (see Figure 9.9).

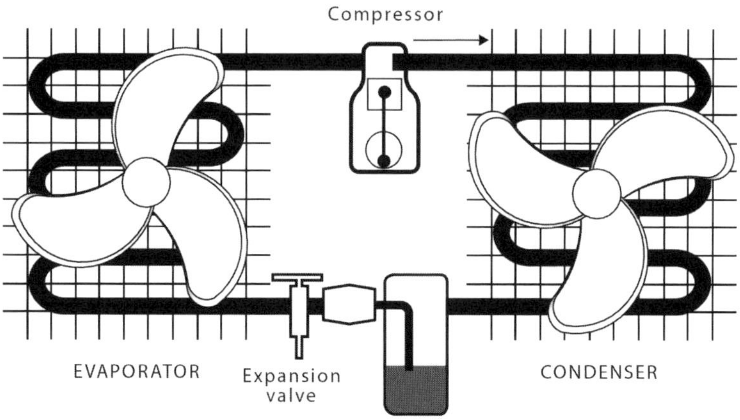

Figure 9.9 Vapor compression refrigeration system.

The circulating refrigerant enters the compressor in the thermodynamic state known as a saturated vapor, and is compressed to a higher pressure, resulting in a higher temperature as well. The hot compressed vapor is then in the thermodynamic state known as superheated vapor, and it is at a temperature and a pressure at which it can be condensed with either cooling water or cooling air flowing across the coil or tubes. This is where the circulating refrigerant rejects heat from the system, and the rejected heat is carried away by either the water or air.

The condensed liquid refrigerant, in the thermodynamic state known as a saturated liquid, is next routed through the expansion valve where its pressure is drastically reduced. That pressure reduction results in adiabatic flash evaporation of a part of the liquid refrigerant. This process lowers the temperature of the liquid and vapor refrigerant mixture to where it is colder than the temperature of the enclosed space to be refrigerated.

The cold mixture is then routed through the coil or tubes in the evaporator. A fan circulates the warm air in the enclosed space across the coil or tubes carrying the cold refrigerant liquid and vapor mixture, in which warm air evaporates the liquid part of the cold refrigeration mixture. At the same time, the circulating air is cooled and thus lowers the temperature of the enclosed space to the desired temperature. The evaporator is where the circulating refrigerant absorbs and removes heat, which is subsequently rejected in the condenser and transferred elsewhere by the water or air used in the condenser. To complete the refrigeration cycle, the refrigerant vapor from the evaporator is again a saturated vapor and is routed back into the compressor.

9.5.2 Performances and Efficiencies

The refrigeration cycle performance is usually described by the coefficient of performance (COP). It is the ratio between useful energy acquired (heat extraction by the evaporator) and energy applied (compressor work) and can be expressed as:

$$\text{COP} = \frac{Q_{\text{evap}}}{W_{\text{comp}}} \tag{9.1}$$

The COP represents, therefore, the amount of thermal power obtained from a refrigeration machine compared with the electrical power transferred to the compressor.

The COP of real systems will mainly depend on the temperatures of the cold source and the hot sink. When designing a climatic chamber, it is critical to bear in mind that dissipating a given amount of power from the test cell, at a lower temperature, will require more energy. Figure 9.10 shows the characteristic drop in COP as the ratio of ambient temperature T_c (condensation temperature) to environmental chamber temperature T_e (evaporation temperature) rises.

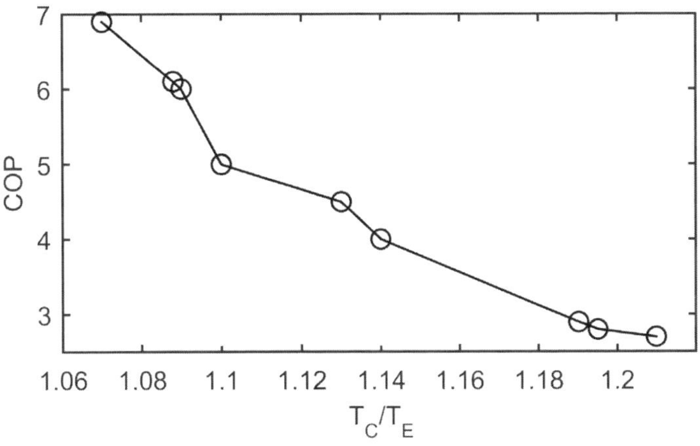

Figure 9.10 Compressor COP against different T_c(K)/T_e(K) values.

9.5.3 Refrigeration Systems in Chassis Dynamometers

Climatic chambers for chassis dynamometers represent a very specific situation due to the nature of the heat load to be dissipated.

Unlike most refrigeration chambers, designed to cool constant heat loads, vehicle climatic chambers have a highly variable heat load coming from the variable conditions of the vehicle during a test. Additionally, it is commonly necessary to control air humidity in the test chamber. Temperature and humidity need to be controlled very accurately and dynamic compensation of vehicle heat dissipation must be achieved almost instantaneously.

The main classification of the chassis dynamometer (CD) refrigeration systems in the market can be made according to the cooling coils used:

- direct expansion
- indirect expansion systems using thermal fluids

9.5.3.1 Direct Expansion System

In these systems, cooling is done by evaporating refrigerant gas directly in the climatic chamber. The evaporator is installed inside the chamber where temperature and humidity control is required, and the chiller plant is installed outside and connected to the evaporator through the piping. Refrigerant gas condensation can be made by surrounding air or by cooling or chilled water (thermal fluid depends on working temperature of the chamber). A schematic is shown in Figure 9.11.

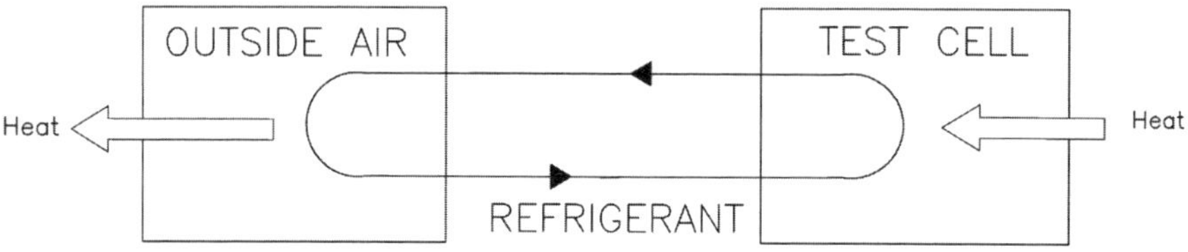

Figure 9.11 Scheme of a direct expansion system.

9.5.3.2 Indirect Expansion Systems Using Thermal Fluids

In this system, the main refrigerant exchanges heat with an intermediate thermal fluid. Then the cooled thermal fluid absorbs heat from the test cell (see Figure 9.12).

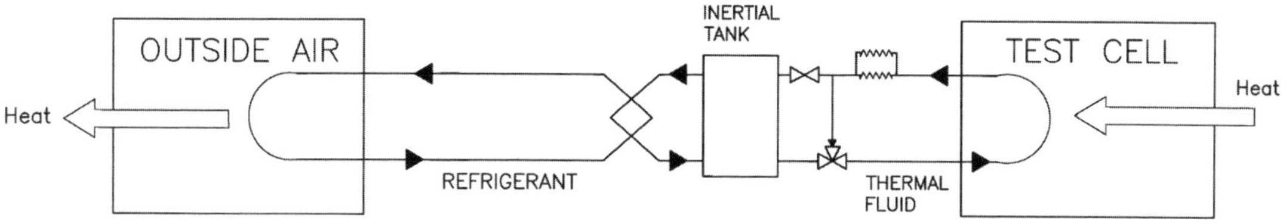

Figure 9.12 Scheme of an indirect expansion system using thermal fluid.

Once cooling energy is transferred from the refrigerant to a thermal fluid, temperature control in the climatic chamber is done in a similar way to ambient temperature chambers (further information is provided in Section 9.4).

9.5.4 Working Principle of a Refrigeration Plant

The basic practical configuration of a refrigeration plant is shown in Figure 9.13.

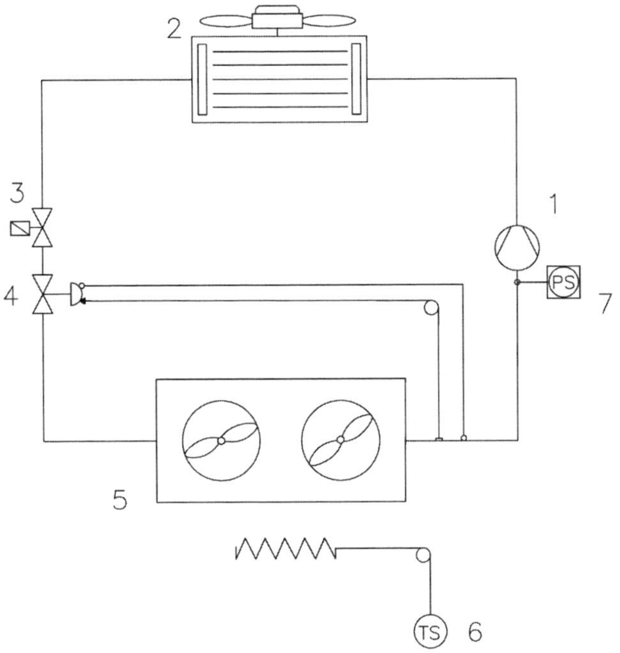

Figure 9.13 Refrigeration plant components. 1—compressor. 2—condenser. 3—solenoid valve. 4—expansion valve. 5—evaporator. 6—thermostat. 7—low pressure switch.

The refrigerant is compressed in the compressor (1) and condensed into liquid in the condensers (2). It then passes through the solenoid valve (3), which opens the refrigerant flow. If the solenoid valve is open, a refrigerant is injected in the evaporator (5) by the expansion valve (4), which adjusts the amount of liquid injected by sensing pressure and temperature at the outlet of the evaporator (5). Expanded gas is then drawn into the suction port of the compressor (1) to start the cycle again.

When the temperature in the chamber reaches the set point as sensed by the chamber thermostat (6), the solenoid valve (3) is closed, closing the suction side to the compressor (1). The compressor (1) draws in refrigerant until it reaches a low pressure, which triggers low pressure switch (7) and removes electrical power to the compressor.

When the temperature in the chamber rises above a preset value, the thermostat (6) opens the solenoid valve (3) and starts the compressor (1) to start the cycle again.

A temperature control system like this continually cycles the compressor on and off at a rate dependent on the thermal loading of the chamber (see Figure 9.14).

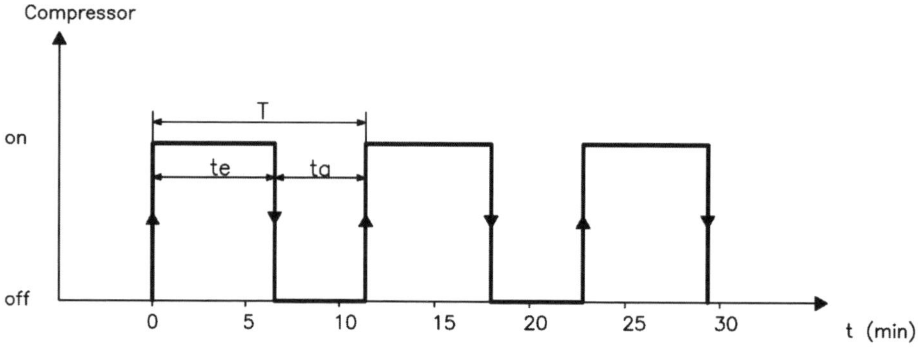

Figure 9.14 The compressor switching diagram, where: T = Compressor switching time, T_e = time the compressor is on, and T_a = Time the compressor is off.

Therefore, temperature control in the chamber is subject to temperature oscillations associated with the compressor start and stop cycles (see Figure 9.15).

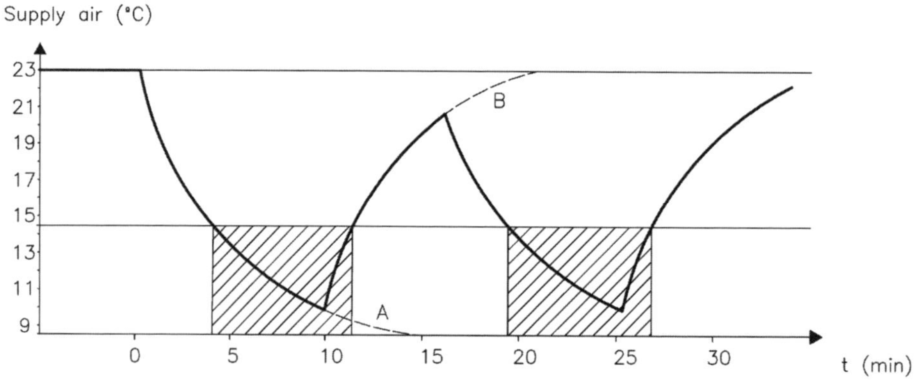

Figure 9.15 Supply air temperature progression, where A = cooling curve and B = reheating curve.

This principle and control strategy is widely used in stationary applications, but it has some limitations when used in dynamic systems.

When the re-heating curve (B) slope is too high and the control accuracy requested has a tight tolerance, the compressor switching cycle (T) will be very low, forcing the compressor to start and stop with a very high frequency. This causes overheating of the electrical motor driving the compressor, which leads to motor failure.

This control strategy can be used when the following conditions hold:

- control accuracy requested has a very wide tolerance (high temperature fluctuations)
- re-heating curve (B) slope is very low for systems with very high inertia or low heat dissipation

It is very common to find this control strategy in soak rooms where the vehicles are not started or in refrigeration systems used to cool thermal fluid tanks with a high inertia.

Another critical issue when using this principle is the lack of control on the evaporator temperature, which is directly related to the compressor suction pressure. A very low temperature on the evaporator may lead to condensation or icing, which might be critical in some applications.

To solve these issues in climatic chambers with a high dynamic process or humidity control, some other strategies need to be applied.

9.5.5 Evaporator Control Capacity

A mechanical thermostatic expansion valve works by sensing the refrigerant pressure and temperature at the outlet of the evaporator, trying to inject as much liquid as possible while guaranteeing that all of it can be evaporated. It works on the principle of adjusting a constant superheat of the refrigerant at the outlet of the evaporator. This means that the evaporator will always be used at 100% of its capacity.

There are other applications thought, where the capacity needs to be adjusted to the heat load to have a more accurate control.

In these applications, superheat controllers allow a variable superheat set point depending on the desired capacity of the evaporator at each instant (Figure 9.16). They measure pressure and temperature at the outlet of the evaporator and comparing the conditions with the refrigerant table give a set point to an electric injection valve, which adjusts the level of superheat at the outlet of the evaporator.

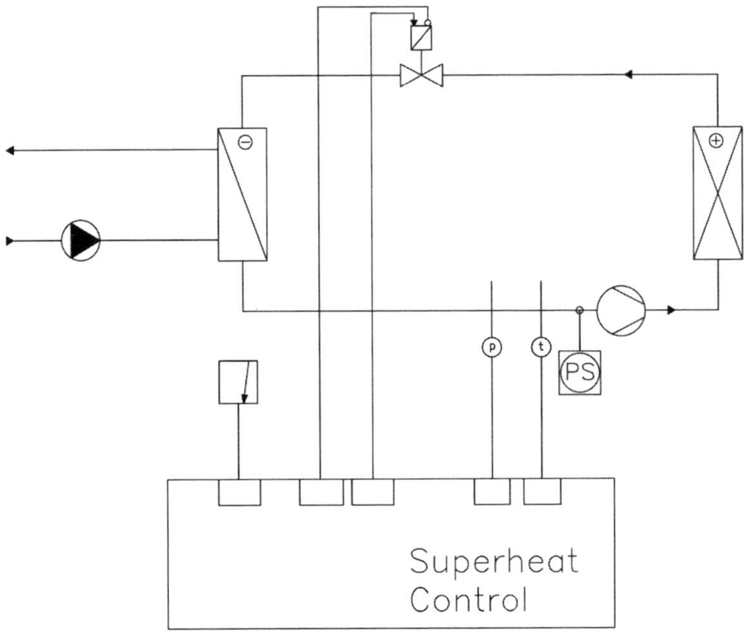

Figure 9.16 Superheat controller.

This allows capacity control within a certain range, but as the injection valve is closed, the suction pressure at the compressor is reduced reaching a point where the low-pressure switch is triggered, and the compressor is stopped.

Once the compressor is stopped, it needs a certain time to start again. During this time, the climatic chamber has no cooling capacity and temperature can drift out of the tolerance requested if there is significant heat dissipation during this period. That means we need additional control strategies to get an accurate control in applications with high dynamic loads.

9.5.6 Compressor Capacity Control

To adjust suction pressure at compressor inlet, most compressors are equipped with a capacity control system. In screw compressors, suction pressure is adjusted by a sliding valve, which bypasses a part of the compressed gas back to suction, as shown in Figure 9.17.

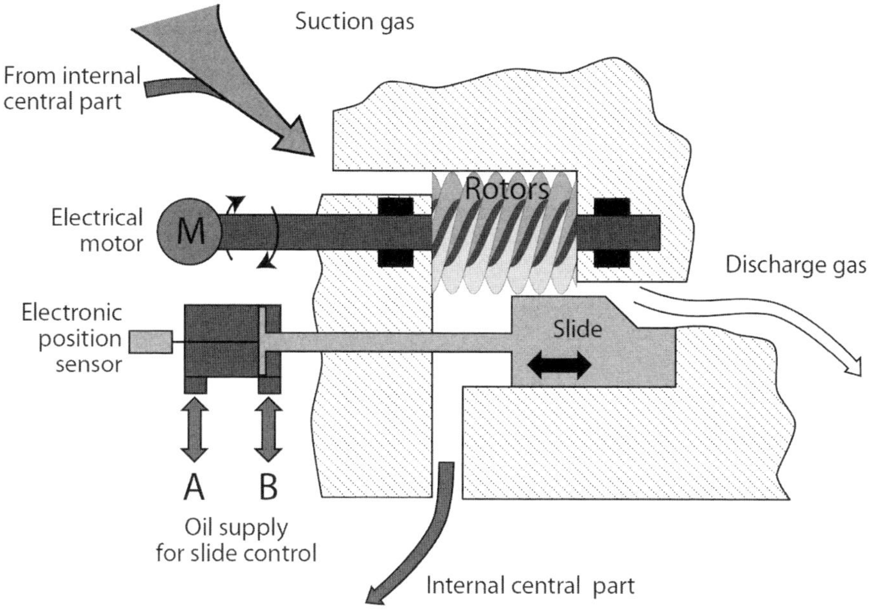

Figure 9.17 Screw compressor capacity control by sliding valve.

This capacity control allows compressor regulation typically in the range of 25% to 100%. However, reducing the capacity control further down is not possible with this technology.

To do this other control strategies are required. Some systems are equipped with multiple stages of compressor that can be sequentially brought into operation as demand increases. This is effective but costly. The most common method used in chassis dynamometer climatic chambers is the hot-gas bypass.

9.5.7 Hot Gas Bypass Capacity Control

The most common solution to avoid compressor suction pressure dropping below the operation limit, without evaporating refrigerant in the condenser, consists of bypassing the hot gas at the high-pressure side of the compressor to the suction side.

This can be achieved by installing a hot gas bypass valve as shown in Figure 9.18. A post-injection valve is usually considered as well to avoid overheating at the compressor suction side.

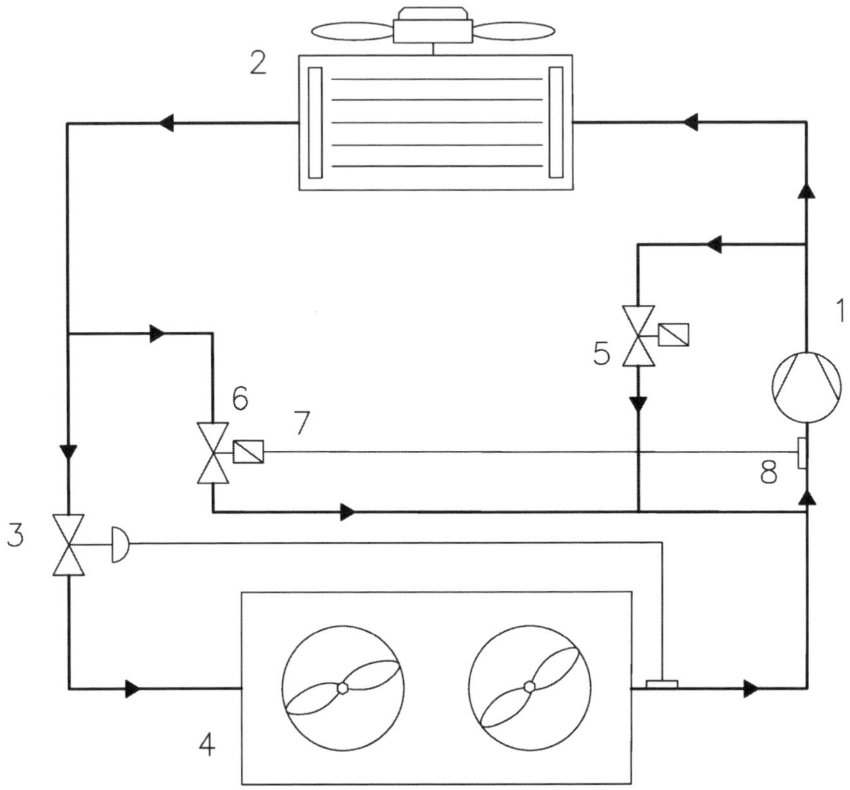

Figure 9.18 Direct hot gas bypass control. 1—compressor. 2—condenser. 3—expansion valve. 4—evaporator. 5—bypass control valve. 6—post-injection valve. 7 and 8—sensor.

If full cooling capacity is required, the hot-gas bypass control valve (5) and the automatic thermostatic post-injection valve (6) are closed. The compressor transports the total gas volume around the original cycle through the condenser, expansion valve, and evaporator. If the demand falls, the control valve (5) opens continuously, drawing off a portion of the compressed gas and mixing it with the suction gas. To prevent excessive heating of the suction gas in the partial load range, a cool refrigerant is mixed in via the post-injection valve (6) to the suction gas temperature.

9.5.8 Evaporator Temperature Control

When the evaporator outlet is directly connected to the compressor inlet, both have the same pressure. This pressure is directly related to the evaporation temperature.

In a process where air humidity needs to be controlled, it is important to guarantee that the evaporator surface temperature is above the air dew point to avoid water condensation or icing. The simplest way to do this is by controlling compressor inlet pressure, but this is not always possible due to the working range of the compressors.

Compressor evaporating temperatures are usually constrained to limits determined by their manufacturers. Figure 9.19 shows a typical compressor working range. As can be seen in the figure, evaporation above 8°C is not allowed.

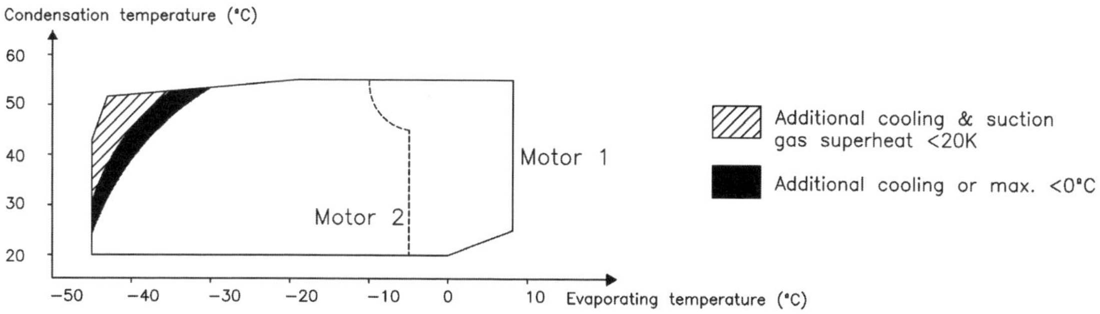

Figure 9.19 Example of a compressor working range.

A common approach to achieve this is to install a suction pressure control valve as shown in Figure 9.20.

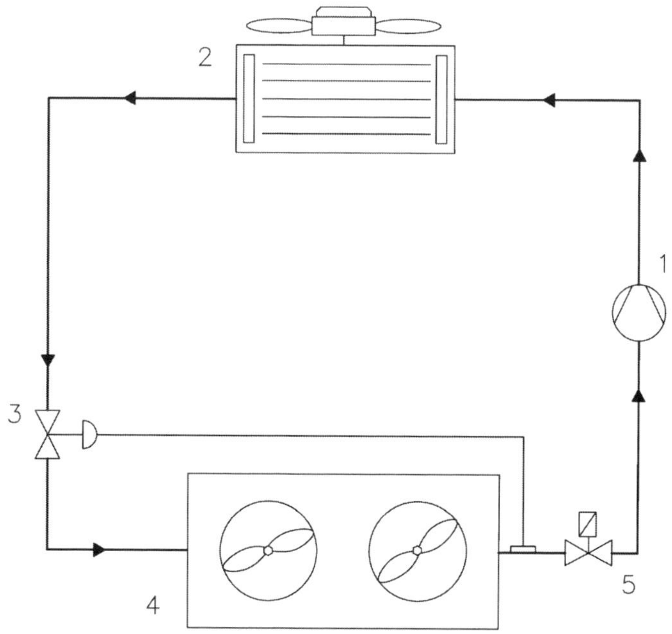

Figure 9.20 Simple suction gas pressure control. 1—compressor. 2—condenser. 3—expansion valve. 4—evaporator. 5—suction throttle valve.

The pressure and the vaporization temperature increase in the evaporator to avoid condensation and freezing. The control valve and the compressor capacity control are used together to allow the compressor to stay within its operating range.

9.5.9 Chassis Dynamometer Refrigeration Plant with Direct Expansion

Considering all the control strategies introduced above, a typical chassis dynamometer refrigeration plant is equipped with the components shown in Figure 9.21.

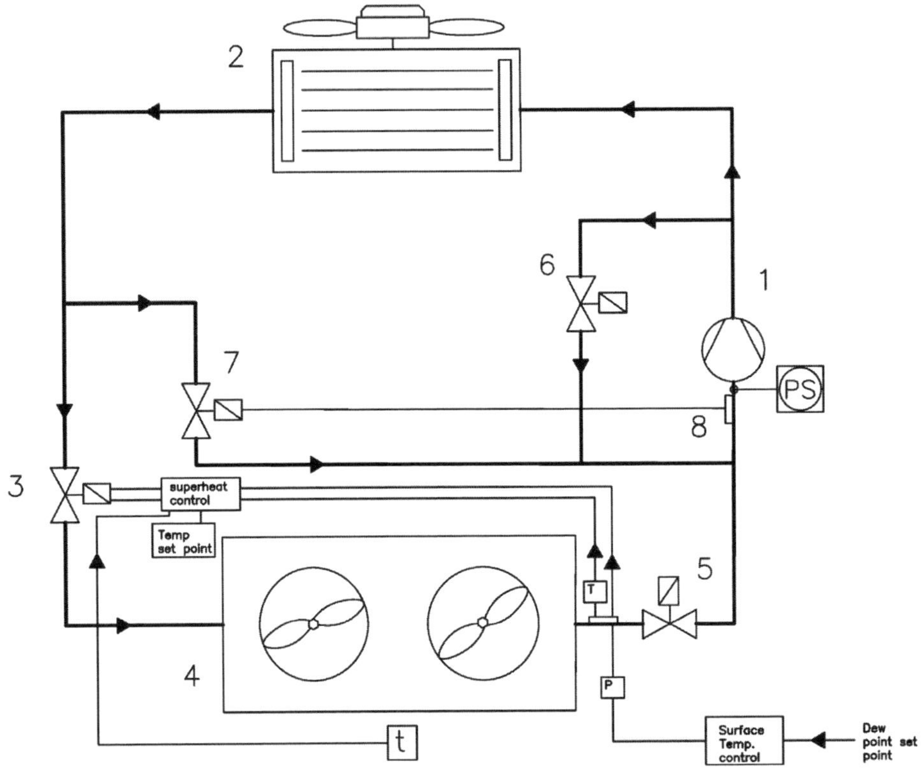

Figure 9.21 Typical chassis dynamometer refrigeration plant components. 1—compressor. 2—condenser. 3—expansion valve with superheat control. 4—evaporator. 5—suction throttle valve. 6—bypass control valve. 7—post-injection valve. 8—high- and low-pressure switch.

The refrigerant is compressed in the compressor (1) and condensed into liquid in the condenser (2). It then goes through the expansion valve, which adjusts the amount of refrigerant injected in the evaporator (4) by a superheat controller measuring test cell temperature and refrigerant gas temperature and pressure after the evaporator. The suction throttle valve (5) adjusts the pressure at the outlet of the evaporator to avoid condensation and icing.

To avoid stopping the compressor due to a very low pressure at the suction side, a sliding gate in the compressor (internal) is adjusted following a low-pressure set point. Additionally, when the sliding gate reaches its minimum position, a bypass control valve (6) lets hot gas from the high-pressure to the suction pressure side to maintain minimum pressure at compressor inlet. Finally, a post-injection valve (7) expands liquid refrigerant at compressor inlet to maintain compressor inlet temperature at desired values.

9.5.10 Indirect Expansion Chassis Dynamometer Refrigeration Plant with Thermal Fluids

Some chassis dynamometer refrigeration plants use a thermal fluid cooling coil instead of direct expansion. The thermal fluid is cooled by a chiller, and then pumped into a cooling coil installed inside the climatic chamber (Figure 9.22).

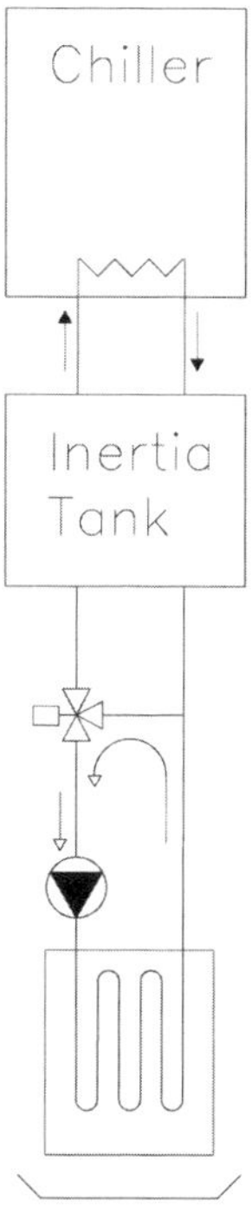

Figure 9.22 Scheme of a chassis dynamometer refrigerant plant using and intermediate fluid.

The advantage of these systems is that the fast time response is supplied by a valve installed in the thermal fluid flow. The chiller is working against a tank with high inertia, which allows a much simpler control of the refrigerant chiller.

Additionally, by the installation of a constant flow loop on the cooling coil, the system allows control of the surface temperature of the chamber cooling coil, independently of the chiller evaporation temperature and thus allows the avoidance of condensation and icing inside the coil. The main disadvantage is that the chiller is forced to work at a lower temperature and therefore has a lower COP.

9.5.11 Combined Systems

It is possible to combine both cooling strategies using direct expansion for low temperatures to avoid the compressors from evaporating to even lower temperatures, and from using thermal fluids at high temperatures for easier temperature control of the cooling coil surface without using a suction throttle valve. This is shown in Figure 9.23.

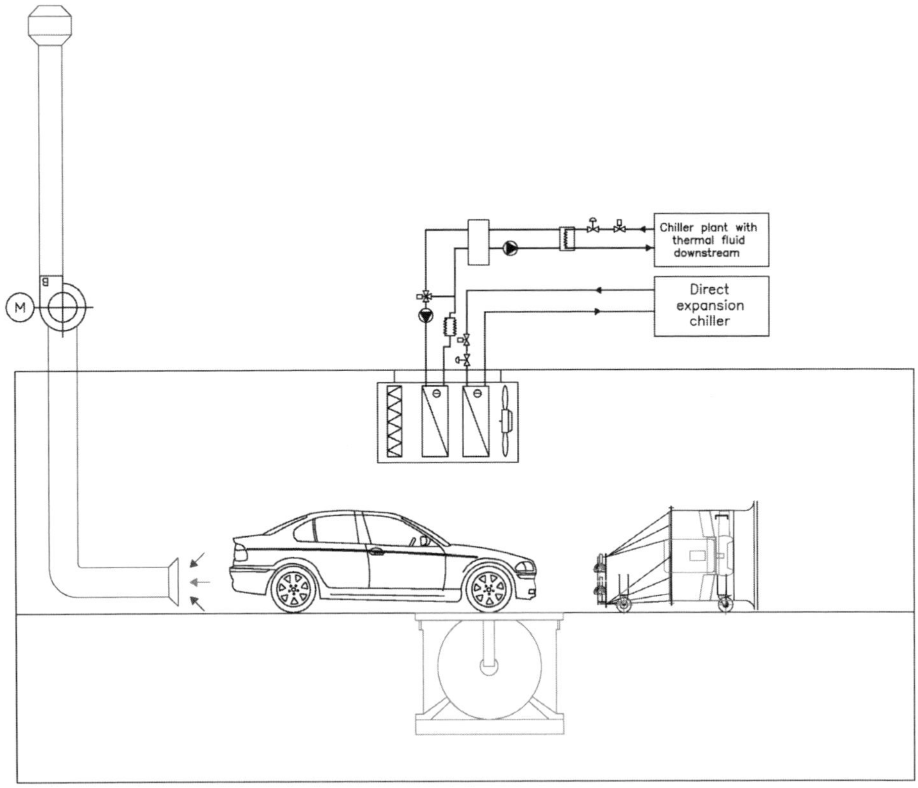

Figure 9.23 Combination of cooling strategies.

Chapter 10

Air Intake in Chassis Dynamometer Test Cells

10.1 Introduction

When testing a vehicle in a confined environment, it is necessary to supply air for the combustion process and extract the exhaust gases produced.

As it was explained in Chapter 9, in ambient temperature test cells where humidity control is not required, this combustion air is supplied by the test cell air handling unit (AHU). If humidity control is required, or in the case of low-temperature test cells, there is a specific AHU to supply the fresh air at suitable conditions in terms of temperature and humidity. The risk of icing is especially important for low-temperature applications if ambient air is drawn in and cooled at these low points. An overview schematic of an external air handling unit is shown in Figure 10.1.

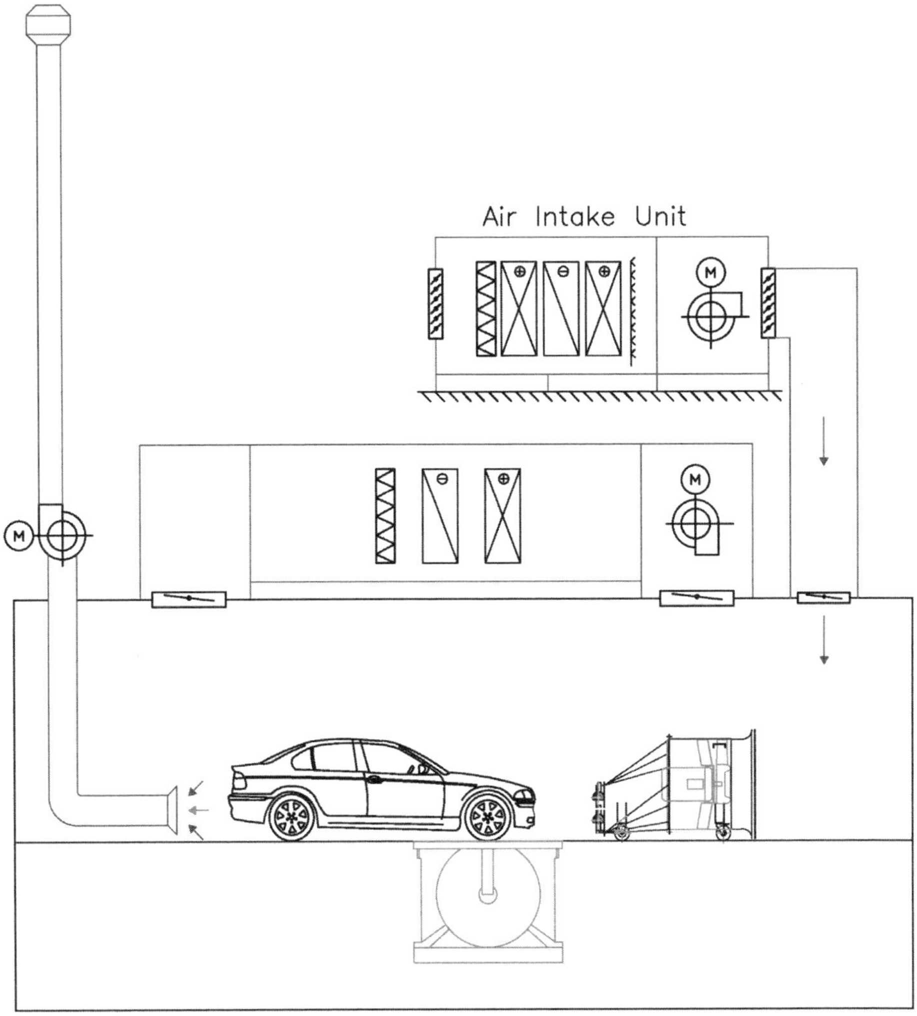

Figure 10.1 Air intake chassis dynamometer test cell.

10.2 Basic Functions of an Air Intake Handling Unit

The air intake handling unit is usually the only air supply into low-temperature and ambient temperature test cells with humidity control.

It has two main functions:

- humidity control
- avoid condensation or icing inside the chamber (working together with coil surface temperature control)

It is responsible for the humidity control in the test cell. When a test requires positive humidification, it will need to add humidity into the fresh air, but it will also be required to supply dry air into the room when tests are performed at low temperatures to avoid condensation and freezing of moisture from humid air entering the chamber.

Different technologies, which are explained in this chapter, are used for drying and humidification technologies.

10.3 Humidity Control

10.3.1 Air Dryers

There are two technologies used to dry air:

- dehumidification by condensation
- desiccant dehumidifiers

10.3.1.1 Dehumidification by Condensation

The basic concept consists of cooling down the air with a cooling coil at low temperature until the air reaches saturation and condenses, as shown in Figure 10.2.

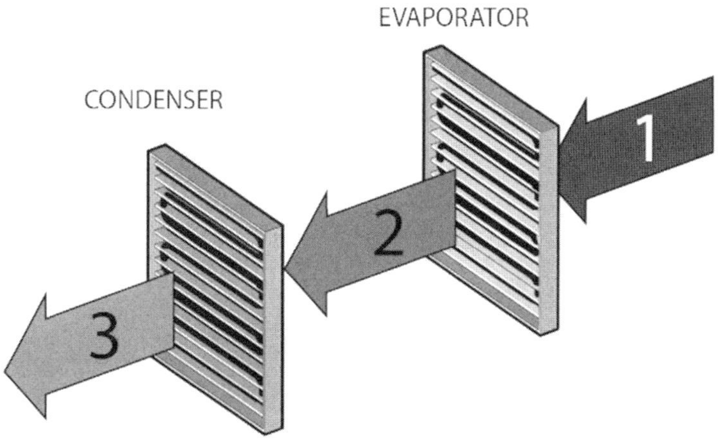

Figure 10.2 Dehumidification by condensation. 1—warm humid air. 2—cool saturated air. 3—warm dry air.

Once the desired dew point is achieved, the air can be heated up to obtain the desired temperature. The entire process is illustrated in Figure 10.3.

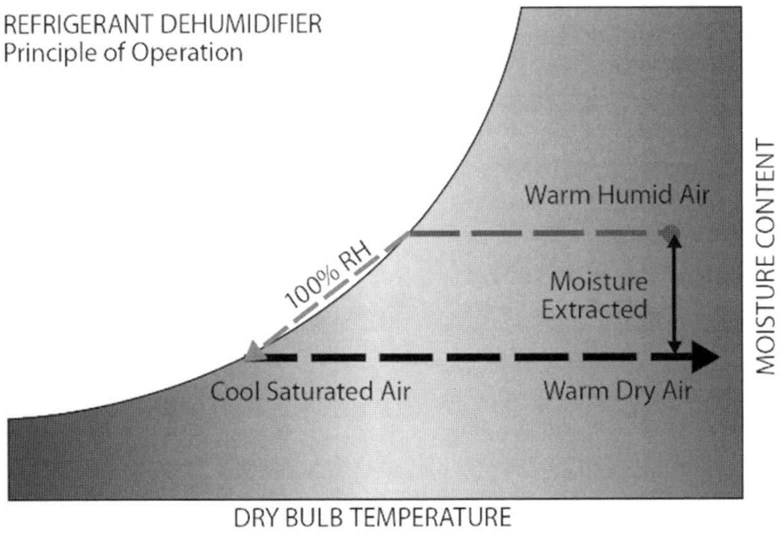

Figure 10.3 Air conditioning process by condensation.

If the dew point of the air is already low, the coil temperature necessary to create condensation may be subzero. At these conditions of temperature and humidity, ice can form on a refrigeration dehumidifier's coils, and this phenomenon can prevent a proper air flow through the coil (and eventually form a solid block encasing the coils). This issue prevents the dehumidifier from operating effectively, and in extreme cases, the ice can deform mechanical elements, causing permanent damage.

Because of the limitations of refrigerant dehumidifier, if very low humidity is desired, it is recommended that a desiccant dehumidifier be selected.

10.3.1.2 Desiccant Dehumidifier

The working principle is that some active components can absorb water molecules from the air when they are in contact with it.

The system consists of a rotor or a desiccant wheel with two different areas through which the air is blown by means of one or more fans. The rotor is constructed of alternate layers of flat and corrugated sheets, which are impregnated with an active component (desiccant). These sheets are arranged so that they allow for air flow perpendicular to the rotor. This rotor or desiccant wheel functions to remove moisture from incoming air.

The humid process air passes through 75% of the rotor face area, and the process air is dry after crossing the rotor because moisture has been extracted by the absorption process to the active component. Simultaneously, reactivation airflow is heated and crosses through the regeneration zone of the rotor. The active component transfers the moisture into the reactivation airflow by means of evaporation of water into the hot air flow that has increased capacity to be saturated with water vapor. The warm, humid airflow is then exhausted to the outside of the desiccant dehumidifier. As these two opposing airflows pass through the rotor simultaneously, a continuous and automatic dehumidification process is achieved. Drying capacity is normally controlled by regulation of the reactivation heater power. The process is shown in Figure 10.4.

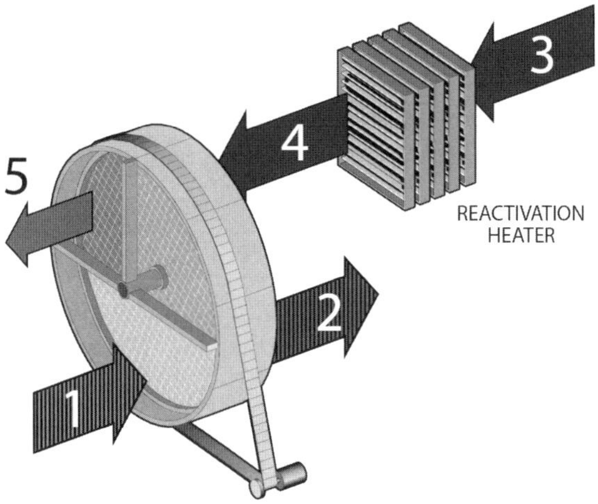

Figure 10.4 Desiccant dehumidifier. 1—cool humid air. 2—warm dry air. 3—fresh air in. 4—hot reactivation air. 5—wet air out.

One of the main benefits of desiccant humidifiers is that they perform well when a low dew point is required. As there is no actual water produced during process, these units can work effectively at subzero temperatures.

10.3.2 Humidification Technologies

There are usually two different technologies for humidification in chassis dynamometer climatic chambers; both are classified as isothermal because air temperature is not modified during the humidification process.

They are classified as follows:

- atmospheric pressure humidifiers
- pressurized humidifiers

10.3.2.1 Atmospheric Pressure Humidifiers

These systems usually consist of a water vessel with a heating element (usually an electrical heater) that boils water at atmospheric pressure. The electrical heater power is directly controlled by the air humidity measured in the air flow (Figure 10.5).

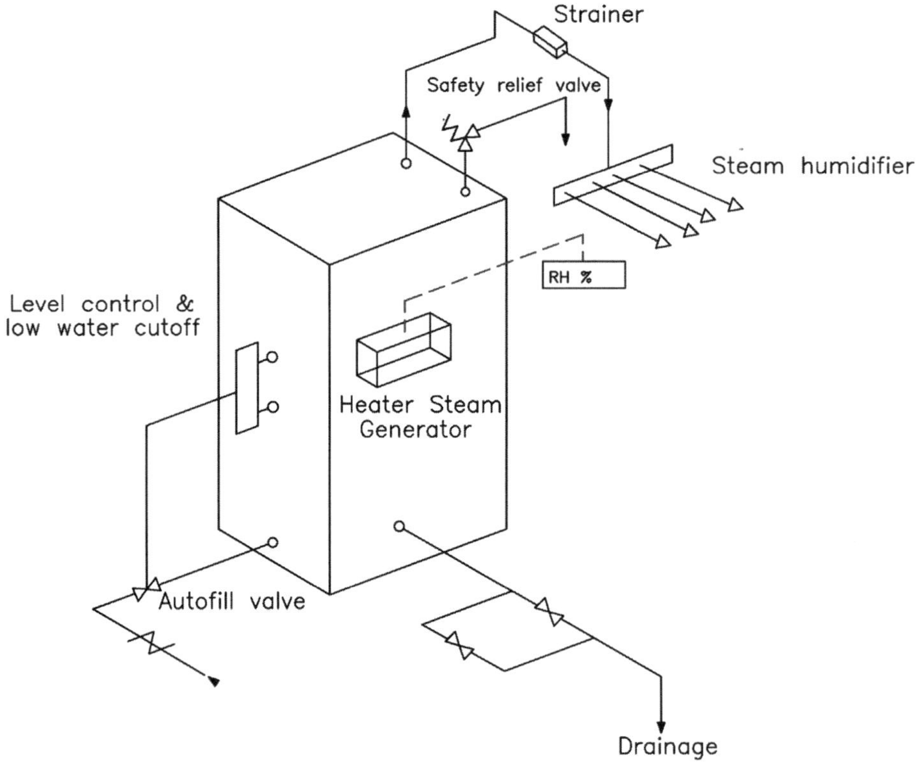

Figure 10.5 Direct control of an isothermal humidification.

The main disadvantage of this system is that the response is quite slow because of the high inertia of the water. Some units also have issues during the water refilling process because incoming water cools down the boiling water.

Slow response is usually not a problem in chassis dynamometer test cells because of the large volume of the test cells and the high inertia of the test cell humidity.

10.3.2.2 Pressurized Humidifiers

These systems usually consist of a water vessel with a heating element (usually an electrical heater) and a valve that regulates steam flow into the process air. Because of the valve restriction to the flow of steam, boiling occurs at positive pressures.

The electrical heater is controlled to maintain a constant pressure in the water vessel and the valve is controlled to achieve the desired humidity in the process air (Figure 10.6).

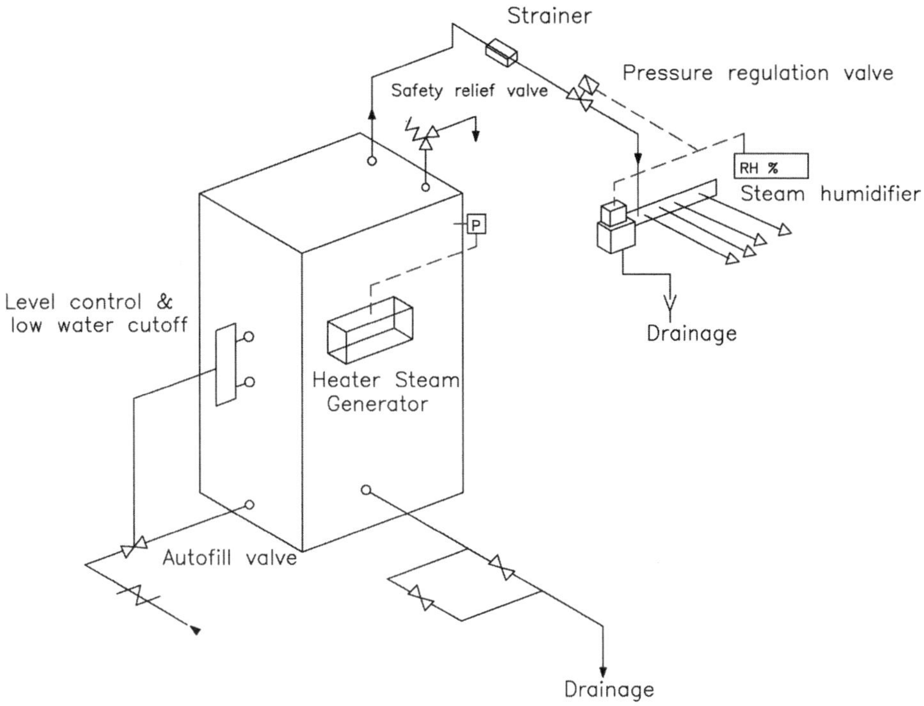

Figure 10.6 Pressure control of an isothermal humidification.

Having pressurized steam in the vessel guarantees an available steam reservoir and allows a much faster response from the regulation valve.

10.4 Operating Principle of an Air Intake Handling Unit

A typical air intake handling unit is represented in Figure 10.7.

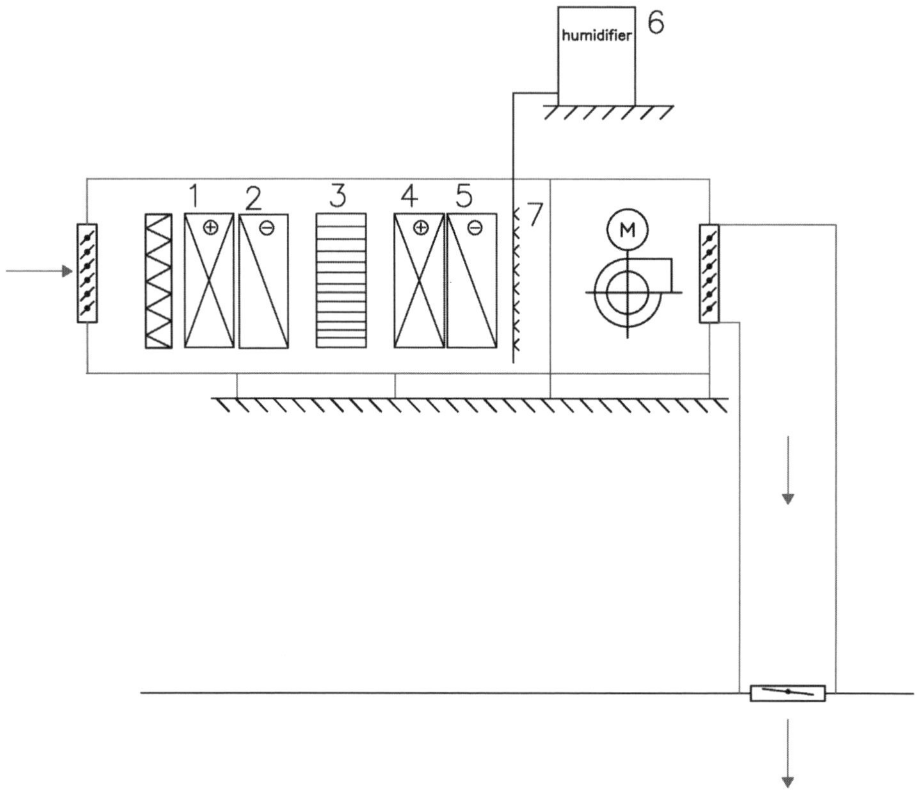

Figure 10.7 Typical arrangement of an intake air handling unit. 1—preheating. 2—cooling coil. 3—desiccant air dryer. 4—heater. 5—post cooler. 6—humidifier. 7—steam injection.

Ambient air is taken and heated in the preheating coil (1) if required to avoid possible freezing of the cooling coil (2). The cooling coil (2) cools the air to the desired temperature of dew point without icing. The desiccant air dryer (3) reduces the humidity content further below the dew point in the testing chamber. The heater (4) and post-cooler (5) adjust the temperature to the test cell temperature, and finally, the humidifier (6) produces steam, which is injected (7) into the process air.

Chapter 11
Instrumentation and Control

11.1 Introduction

The final purpose of a chassis dyno test cell is to measure vehicle performance. The main parameters to be measured in a chassis dyno test cell are as follows:

- vehicle power, torque, and speed
- exhaust emissions
- fuel consumption
- other engine and vehicle parameters

Once we have all the instrumentation in place, we need to perform the tests. Section 11.6 will explain how this is done.

11.2 Vehicle Power, Torque, and Speed

Torque and speed are the fundamental parameters to be measured in a chassis dynamometer test cell. Power can be easily calculated afterward.

Chassis dynamometer motors are usually mounted in a cradle configuration with a load cell measuring the tractive force. The load cell works in both directions of rotation and load (forward/backward and compression/tension), and uses compensated strain gauges. The configuration is illustrated in Figure 11.1.

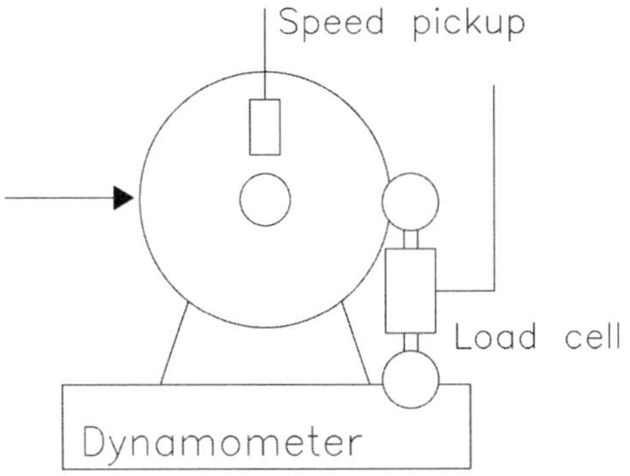

Figure 11.1 Chassis dyno torque and speed measurement.

Dyno speed is measured using an encoder that detects each turn. From this signal speed, distance can be calculated knowing the roller diameter.

11.3 Exhaust Emissions

Exhaust emissions have become one of the most critical factors to be measured in a chassis dynamometer because of regulatory purposes. They have to be measured in the development of the vehicle, in the homologation phase and during the vehicle life, as it will be explained in Chapter 12.

To have comparable results, regulations define the measuring principles of the equipment and test procedures to be used in certification. This has created a well-accepted standard for the measurement of regulated pollutants, which include CO, HC, NO_x, and PM (all measured in units of mass). Some regulations, such as Euro 5, also include a solid particle number (PN) limit. CO_2 is also measured to calculate fuel consumption and regulation limits.

Some chassis dyno test cells also include the measurement of not regulated pollutant. When this is the case, different testing methodologies and tools are used creating a bigger disparity from different testing facilities.

11.3.1 Emissions Analyzers

The first step to measure emissions is to have analyzers capable of detecting the chemical components in the sample.

The standardized measuring principle specified in the legislation is summarized in Table 11.1.

Table 11.1 Measurement Principles of Exhaust Gas Analyzers

Gas Component	Measurement Principle
CO, CO_2	NDIR
HC	FID
NO_x	Chemiluminescence detector

Other nonstandardized techniques frequently use Fourier transform infrared (FTIR), which will also be explained below.

11.3.1.1 Nondispersive Infrared Analyzers

The nondispersive infrared (NDIR) analyzers work under Beer–Lambert's law by measuring the degree of absorption of infrared radiations when passing through a column of gas (Figure 11.2). Infrared wavelengths are selected depending on the kind of gas to be analyzed; for example, CO has a very strong absorbance in the range of 4–5 μm.

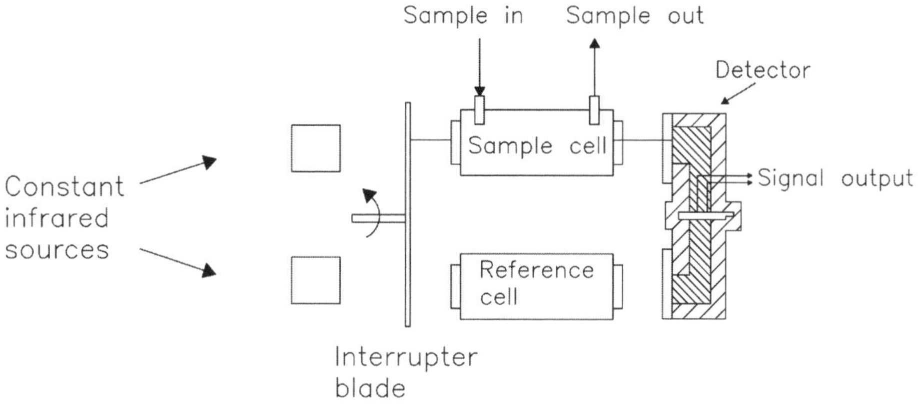

Figure 11.2 Schematic of an NDIR analyzer for the measurement of CO and CO_2 concentrations.

A detector capable of measuring differential pressure is installed in between two chambers, one filled with the gas to be measured and another one with a reference gas and a very low absorbance in that infrared range, usually N_2. An infrared beam from a single source is then applied to both chambers; the chamber with the measured gas will absorb this energy, while the reference gas will not. This difference in energy absorption will generate a differential pressure, which will be measured by the sensor.

11.3.1.2 Flame Ionization Detector Analyzers

They work under the principle that pure H_2–air flames are practically ion free, but on the introduction of HC, the flames cause ionization and become electrically conductive (Figure 11.3). The ionization current is proportional to the number of C atoms present in HC molecules. Flame ionization detector (FID) analyzers are therefore carbon atoms counters.

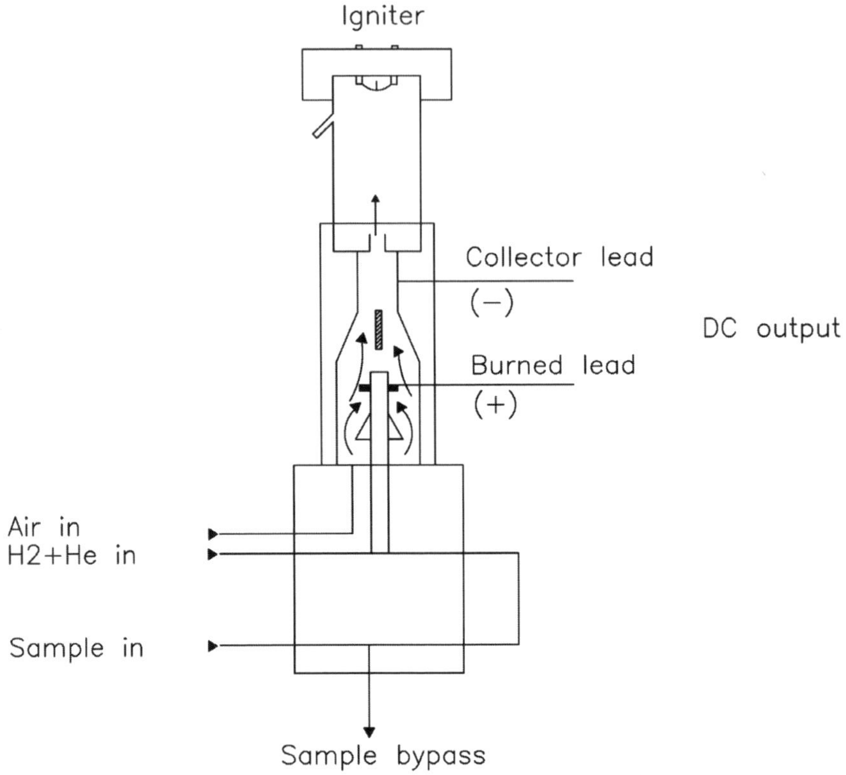

Figure 11.3 Schematic of an FID for HC measurement.

FID analyzers are basically an H_2 burner where the sample gas is introduced. They supply a voltage to the burning flame and measure current through the flame. H_2 is usually mixed with helium to decrease flame temperature and increase stability.

For measuring HC in the diesel exhaust, the sample line is heated to temperatures around 190°C to minimize condensation of heavy hydrocarbons.

FID analyzers are usually calibrated with propane or methane mixtures in H_2.

11.3.1.3 Chemiluminescence Detector Analyzers

They work by measuring the light emitted by NO when reacting with ozone (O_3) (Figure 11.4). To measure all NO_x, NO_2 is first converted into NO. This is then mixed with ozone (O_3) coming from a generator and the intensity of the light emitted, which is proportional to the NO particles, is measured.

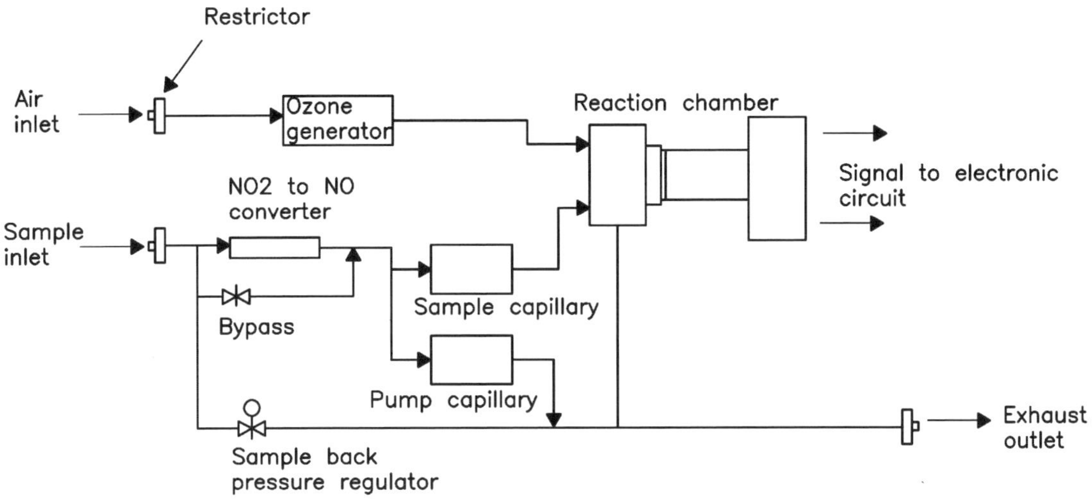

Figure 11.4 Schematic and flow diagram of a Chemiluminescence NO_x detector.

11.3.1.4 Fourier Transform Infrared

FTIR works by emitting infrared light that passes through the gas to be evaluated. Some of the radiation is absorbed by the sample, while some is transmitted. The resulting spectrum shows the composition of the gas sample. FTIR technology allows the measurement of NO, NO_2, N_2O, NH_3, CO, CO_2, CH_4, as well as alcohols.

11.3.2 Diesel Particle Emission Measurement (PM)

PM is not a clear substance because it is a mixture of unstable solid and liquid components changing composition during the sampling and measurement process. It is defined by the measuring principle, as any material deposited on the filter from the dilute exhaust gases is sampled at a temperature of approximately 52°C.

The filter papers are weighed to measure the number of particles in the filters and compared with the filters that have been used with clean reference air. Filters need to be conditioned before and after the filtration to guarantee no contamination with other particles. It is also important to control the temperature and humidity of the room where weight is measured so that ambient temperature and humidity do not change the content of the filters by vaporization.

11.3.3 Diesel Particle Counter (PN)

Particles within a size range of 23 nm to 2.5 μm have been proved to be dangerous for humans. Because of their very small size, they are usually not detected by conventional optical techniques. For this reason, they need to be enlarged first. This is done by a condensation particle counter where particles are enlarged using them as nucleation centers because of the condensation of butanol, and counted using the light scattering method.

During the sampling process, it is important to guarantee that particles are maintained with the same characteristics avoiding agglomeration and deposits. This is done by dilution sampling, which will be explained in the following chapters.

11.3.4 Smoke Opacity Meters

Smoke opacity meters measure the visual black smoke. Two different principles are used.

The filter-paper method consists in passing a fixed amount of gas sample through a clean filter to capture a certain concentration of soot. The filter soot causes blackening of the paper, which is then measured by photoelectric sensor.

Another common technique that allows continuous measurement consists in measuring the light beam extinction when going through a sample.

11.3.5 Portable Equipment Measurement

A portable emissions measurement system (PEMS) is a complete set of emission measurement instruments that can be carried on board the vehicle to record instantaneous emission rates of selected pollutants with good levels of accuracy (see Figure 11.5). A PEMS unit usually comprises a set of gas analyzers with sample lines directly connected to the tailpipe. An exhaust gas flow meter (PEMS) gives the emission as a concentration. An exhaust gas flow meter is needed to convert the emission measurement into a value expressed in grams per kilometer. An engine diagnostics scanner is designed to connect with the on-board diagnostic link of the vehicle.

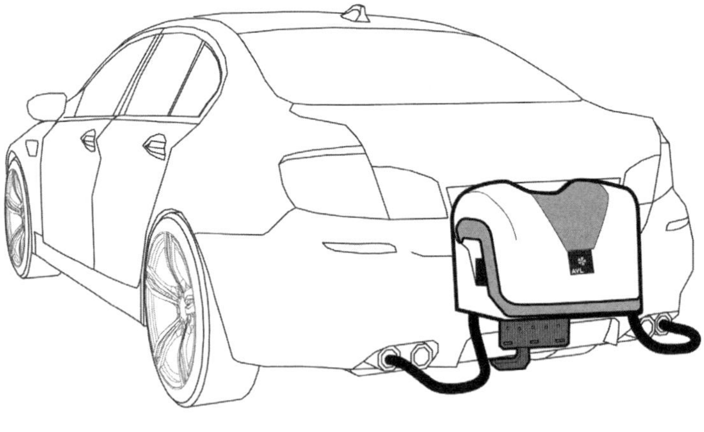

Figure 11.5 Passenger car instrumented with the PEMS. (Reprinted with permission from AVL.)

PEMS technology is relatively new, but has undergone major development over the past two decades, with improved gas measurement principles and significant reduction in size, weight, and overall complexity. PEMSs are simple and relatively inexpensive to purchase and maintain, in comparison with a full chassis dynamometer test cell. In recent years, PEMSs have also been utilized for regulatory purposes.

PEMSs typically measure raw exhaust emissions of CO_2, CO, NO_x, and THC. Still, the range of pollutants that can be measured with PEMS is limited in comparison to laboratory measurements.

Some alternative gas analyzing techniques are used to allow the system to be mobile and reduce the need for references gases measurement carried out using an ultraviolet analyzer (which can measure NO and NO_2).

11.3.6 Gas Sampling

Once we have the required analyzers, we need to take the sample from the exhaust and precondition it to the requirements of the measurement instruments, without altering the composition of the sample.

One limiting factor is usually temperature because many analyzers are designed to work at ambient temperatures while exhaust samples are at high temperatures.

Another critical factor is the fact that exhaust emissions contain volatile materials such as water and sulfur compounds, and cooling the sample would lead to saturation of many of this species, resulting in condensation and nucleation. To avoid this, three different sampling principles can be distinguished: raw (undiluted), partial dilution, and full flow dilution.

11.3.6.1 Undiluted Analytics (Raw Sampling)

Raw emissions sampling presents a difficulty in that instruments must be designed for the high temperatures at the outlet of the engine exhaust. It is also important to guarantee that exhaust content does not condensate during the sample. They have an advantage, though, in that they can measure very low content of a certain unit because concentration is very high. Considering that the regulations are lowering the limits of pollutants, this is a considerable advantage.

Undiluted techniques must work with an exhaust flow measuring device to be able to calculate overall emissions in the exhaust from the measured sample.

11.3.6.2 Partial Flow Dilution

Partial dilution only takes a certain sample of the exhaust and dilutes it to adapt it to the requirements of the measuring instruments. By diluting the exhaust with air, partial pressure of unburned HC and water is reduced. This reduces the risk of condensation of HC and risk of water condensing and trapping particles. Dilution proportion is dependent on the humidity content of dilution air.

When doing partial dilution, it is critical to guarantee that the sample has the same concentration as the exhaust flow. To prevent aerodynamic effects, which may lead to a sample with higher concentration of certain components such as particles, exhaust samples must be taken with the same velocity as the main flow stream. This is referred to as isokinetic sampling.

11.3.6.3 Fully Diluted Sampling (CVS)

In full dilution, all the exhaust flow is diluted in a constant resulting flow. This technique is referred to as constant volume sampling (CVS).

11.3.7 Constant Volume Sampling

CVS is the accepted method by legislation for sampling and diluting the exhaust for homologation vehicles at the same time, as it maintains a constant flow rate of exhaust plus dilution air.

As an alternative to the flow measurement of the exhaust gas in raw conditions, which requires high-temperature flow meters, the CVS technique consists in diluting the exhaust gas in air resulting in a constant volume. In this way, pollutants are proportional to the exhaust gas flow rate. At low engine loads, exhaust gas flow will be low and dilution air will be high, which will lead to low polluted samples. At high loads, the exhaust gas flow will be high and the dilution air low, which will lead to much polluted samples.

This is what is achieved in a CVS instrument, as shown in Figure 11.6. A CVS usually consists of a dilution tunnel, which leads to a venturi and an extraction fan at the outlet. Air is suctioned by the extraction fan from the venturi, guaranteeing that the flow is always constant.

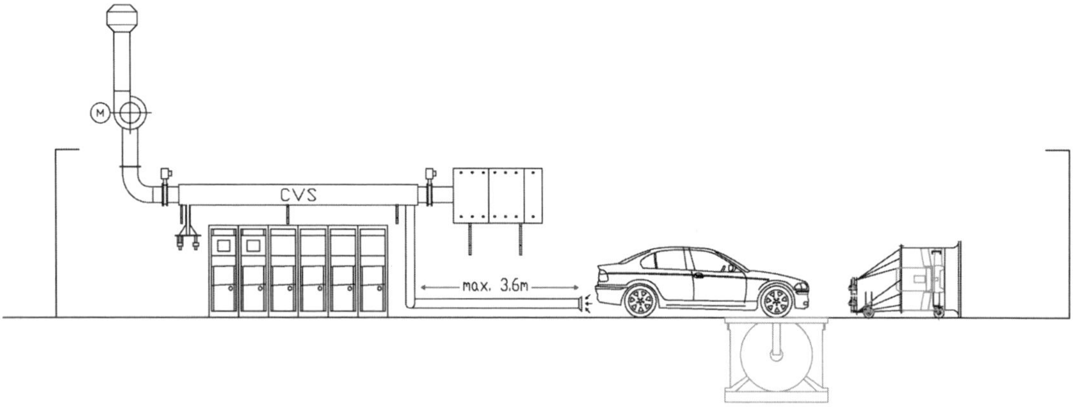

Figure 11.6 Emission CVS setup.

The proportion of dilution air needs to be selected to guarantee that the mixture will not create condensation, and this proportion can be adjusted by reducing the humidity content of the dilution air. If dilution air is dried, the dilution ratio can be lowered leading to a higher content of concentration of exhaust in the mixture.

A sample is continually taken from the dilution tunnel and collected in Teflon bags. Different bags are used per section of the driving cycle covering the whole cycle. This gas from the bags is then fed to the analyzer to evaluate pollutant gas concentrations.

11.3.8 Complete Emission Setup

All the previous measuring techniques are integrated in the test cell depending on the type of tests that will need to be performed. As much of the instrumentation is highly expensive, specialized test cells will be constructed for dedicated purposes and typical instrumentation suites are shown in Figure 11.7.

Light Duty Vehicles– Emission Certification

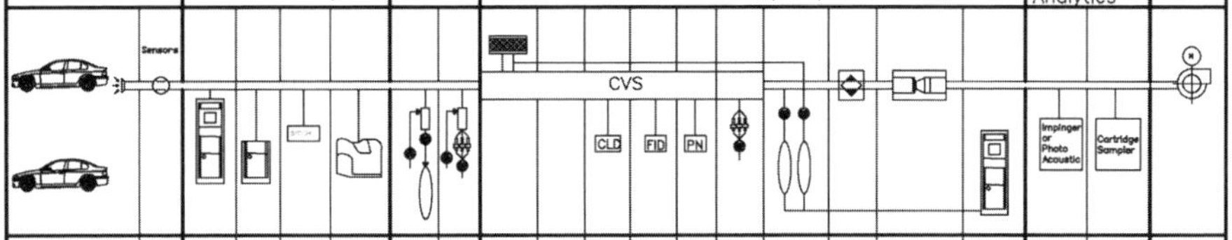

	Testbed		Undiluted Analytics				Partial Flow Dilution		Full Flow Dilution (CVS)										Special Analytics		Blower
	Load Unit	Exh. Flow	Raw Bench	Raw FTIR	Smoke	PEMS	BMD	PFD PM	Mixing Device	Dil. Tunnel	Tunnel CLD	Tunnel H-FID	PN	PM	Bags	Heat Exch.	Venturi	Diluted Bench	Alcohol Sampler	Aldehyde Sampler	Blower
EU	Vehicle Chassis Dyno				✓				✓	✓		✓	✓	✓	✓		✓	✓			✓
EU – RDE	Vehicle on road	✓				✓															
US	Vehicle Chassis Dyno								✓	✓		✓		✓	✓		✓	✓	✓	✓	✓
Alternative		✓					✓	✓										✓	✓	✓	✓
Alternative		✓	✓					✓											✓	✓	✓
Japan	Vehicle Chassis Dyno								✓	✓		✓		✓	✓	✓	✓	✓			✓

Note: The alternatives for the US Standards relate to CFR–1066 and CFR–1065. Smoke measurement accordingly to UN–ECE R–24 on engine test bed or chassis dyno.

Figure 11.7 Light duty emissions certification instrumentation [11-1].

11.3.8.1 Undiluted Analytics

- *Gaseous emissions:* An emission bench is used to measure components such as CO_2, CO, THC, CH_4, NO_x, O_2, and sometimes also components such as NH_3, NO_2, and N_2O. Additionally, a multicomponent analyzer, such as an FTIR, can be used for research and development (R&D) and for NH_3 and N_2O for the legislation purpose.
- *Smoke emissions:* Particulates, smoke, and opacity are measured by various devices, such as smoke meters, opacity meters, and micro-root sensors.
- *Real-driving emissions:* The PEMS also apply raw emission measurement.

11.3.8.2 Partial Flow Dilution

There are two partial flow dilution systems:

- partial flow dilution and particulate sample systems for particulate measurement (PM), such as the AVL smart sampler
- bag mini diluter (BMD) systems for gaseous emissions

11.3.8.3 Full Flow Dilution (CVS)

An emission bench for diluted concentrations analyzes the samples of diluted exhaust and dilution air in the CVS bags. Optionally, the diluted exhaust can be analyzed continuously direct from the CVS (modal diluted).

For compression ignition engines (such as diesel), a heated FID is used to measure the hydrocarbons directly and continuously from the dilution tunnel.

If NO_2 is measured, it has to be measured directly from the diluted exhaust, or if it is determined by NO_x minus NO, then NO has to be measured directly and continuously from diluted exhaust, while NO_x can be measured from the CVS bags.

11.4 Fuel Consumption Measurement

Fuel consumption is one of the fundamental measurements for any vehicle test on a chassis dynamometer. Different principles can be used:

1. carbon balance of exhaust emissions
2. interrogation of engine controller (engine control unit (ECU)) signals
3. external fuel combustion measuring devices

11.4.1 Carbon Balance

Carbon balance is the preferred method when having exhaust emissions measuring systems in a chassis dyno test cell. It has the advantage that gives good precision with no intrusion on the vehicle fuel system.

An indirect estimate of fuel consumption can be obtained by equating the mass of carbon in the exhaust to the mass of carbon in the burnt fuel. The mass of carbon in the exhaust is estimated from exhaust concentrations of CO_2, carbon monoxide (CO), and unburned hydrocarbons (THC). The carbon content of the fuel is obtained from the fuel supplier or measured using chemical analysis. The emissions can be sampled at different points in the exhaust system on a continuous basis or collected and analyzed post-test. While the latter will not give continuous fuel consumption over a testing sequence, it gives the overall fuel consumption, which is the final value desired. This is the standard method adopted for homologation.

11.4.2 Engine Control Unit

The ECU is responsible for the fuel injection into the engine. Having access to this information allows a simple estimation of fuel consumption. This signal is not a measure but a demand signal, and requires calibration for each engine to yield accurate results.

The ECU method relies on interpreting the numbers programmed into the engine ECU to infer fuel consumption. These numbers are not the result of any measurement, but rather tabulated values used within the control algorithms of the ECU. Their level of accuracy is therefore not always known. However, they do have the advantage of providing a cycle-by-cycle breakdown of the amount of fuel that the engine was intending to inject, and, if calibrated against a real fuel consumption measurement, can provide useful information.

11.4.3 External Fuel Combustion Measuring Devices

External fuel combustion measuring devices such as fuel balances or Coriolis principle meters used in engine test cells allow a very accurate measurement of fuel consumption, but have the main problem that they need to supply fuel to the engine directly bypassing the vehicle fuel tank.

This is not acceptable for homologation because it modifies the original vehicle. It can be used for R&D in installations not equipped with emissions measurement technologies.

11.5 Other Engine and Vehicle Parameters

R&D chassis dyno test cells can be designed to study any parameter in the vehicle including the engine or any other parameter in the vehicle. It is therefore common to find additional sensors installed in the vehicle, such as temperature and pressure sensors or flow meters to monitor vehicle conditions.

11.6 Chassis Dyno Control

Chassis dyno control is usually not fully automated because it requires the interaction with a human driver who is performing the tests.

Once the vehicle is put in place and all the instrumentation is adjusted, the driving cycle, which is going to be simulated, is loaded into the automation system. This will be the demand for the roller motors control, which will adjust speed and torque according to the vehicle equation of motion.

Coordination with the driver is critical so that the driver follows the same driving cycle as the roller is trying to simulate. To do this, the automation system needs to give the driver the speed set point, which he needs to follow. This is done by a computer screen called "driver's aid," which shows the profile that the driver must follow with time in the y-axis and speed in the x-axis. As the screen moves the driving cycle in the y-axis (time), the driver needs to adjust the speed, which, as a result, moves the cursor right and left to stay within the acceptable tolerances.

As an alternative, robot drivers can also be installed in the car to control the gas pedal and the brake and follow the speed profile. This technique is not allowed for homologation, but can be used for other kind of tests such as durability, where it is the preferred because of the time of the tests.

References

11-1. Engeljehringer, K. 2016. AVL Emission Testing Handbook. AVL: Graz, Austria.

Summary of Part II—The Test Cell

This section has studied the different components in a chassis dynamometer test cell.

We have covered the following:

- A chassis dynamometer, which consists of a roller drum coupled to a dynamometer, which simulates the vehicle tractive forces and power demands.
- A vehicle blower, which is placed in front of the vehicle and simulates the airflow which the vehicle faces on the road due to relative speed. We have seen the importance of the wind blower outlet section selection depending on the test to be performed. We have also seen the different wind speed measurement philosophies and the homogeneities requirements.
- Chassis dynamometer HVACR, which is responsible for regulating the testing chamber in temperature and humidity while heat is dissipated by the vehicle under tests. Different technologies are used depending on the tests to be performed and parameters to be regulated. Two main classifications have been established for low-temperature and high-temperature controlled environments.
- Air intake, which supplies air to the chamber in controlled conditions to be used as engine combustion air. Different temperature and humidity control technologies have been analyzed.
- Exhaust extraction, which evacuates the exhaust gases produced by engine combustion.
- Exhaust emission measurements check the performance of the engine and the exhaust aftertreatment systems in terms of pollutants. Different techniques have been analyzed during this part of the book.

This section has analyzed different technologies for simulating vehicle on-road conditions and for checking vehicle performance, taking into account all the parameters studied in the previous sections.

Part III

Vehicle Homologation Process and Tests

Chapter 12
Chassis Dynamometer Test Standards

12.1 Introduction

Vehicle tests on chassis dynamometers can have many different purposes, and we will not be able to make an exhaustive list of all possible test scenarios within this book. First, we will summarize some of the principal types of tests that are performed on a chassis dynamometer. Following this, we will review the testing standards that exist for emissions testing up to 2017 in the legislation in Europe and the U.S. before discussing the limitation of these tests and the future evolution.

Chassis dynamometers can be used to perform any test that automotive engineers can design. Even though it is not possible to cover every alternative, we will try to highlight the tests most commonly performed among vehicle manufacturers (most of which were mentioned in Chapter 8). The size of the wind blowers depends on the type of test being performed:

- *Friction Tests:* Friction tests study the mechanical efficiency of the engine and drivetrain. Tests are performed with vehicle speeds up to 200 km/h and with temperatures usually varying from 10°C to 40°C. It is crucial to reproduce engine and vehicle conditions, including transmission and tire performance. They do not measure exhaust emissions from the vehicle.
- *Durability Tests:* Durability tests, or mileage accumulation tests, are intended to reproduce vehicle age. In this kind of test, strict reproducibility of vehicle engine conditions is usually not required. For this reason, they are frequently performed in a chassis dynamometer installed outdoors with no temperature regulation. Air flow over the vehicle needs to be sufficient to guarantee that all vehicle components work properly at the tested speeds. They do not measure exhaust emissions from the vehicle.

- *Aeroclimatic Tests:* Aeroclimatic tests typically study vehicle performance in a temperature range of −30°C to +50°C. The most typical test consists of irradiating a vehicle with solar simulation for a time period to approximate the vehicle being parked in the sun. The vehicle is then started and the capacity of the air conditioning to cool down the vehicle at relatively high speeds is tested. Due to the interaction of the vehicle and the surrounding air in the vehicle cabin, wind simulation around the vehicle is very important. One example of an aeroclimatic test that is defined in the U.S. emissions legislation to test the air conditioning is U.S. SC03.
- *Cold Starts Tests:* Intended to test only the cold start, chambers are usually conditioned to temperatures down to −30°C and a cold start is performed. Temperature in the test cell does not need to be conditioned after the cold start. The refrigeration system is designed to cool down the engine in a certain amount of time (usually around 12 h), but not to compensate heat dissipation from the engine. A chassis dynamometer and blower are not required.
- *Emissions Tests:* These tests are the only ones that are defined by different legislation and are used to certify vehicles in different countries. Speeds are a relatively modest representative of normal usage. Temperature control is important. Humidity is controlled directly. A vehicle blower is used for engine cooling, but does not have a large enough outlet to cover the entire frontal area of the vehicle.

12.2 The History of Emission Testing Procedures

Since the 1950's, governments have been well aware of the importance of vehicle emissions and have been establishing increasingly strict regulations with the development of new technologies.

The first step was established in California in 1961 with the requirements to install the positive crankcase ventilation system. This draws fumes-heavy in unburned hydrocarbons—from the crankcase into the engine's intake track so they are burned, rather than released unburned from the crankcase.

The first legislation for exhaust emissions was established in California in 1966, followed by an extension in 1968 applying the laws to the rest of the U.S. [12-1].

Considering all the different working conditions of a vehicle and its engine, which depend on multiple factors such as speed, road, route, and traffic, one of the major problems legislation has faced is how to set a reference measurement.

In order to resolve this issue, different governments created reference driving cycles (e.g., New European Driving Cycle (NEDC) in Europe and FTP75 in the U.S.).

The first cycles were developed in the 1950s by the Los Angeles County Air Pollution Control District for emissions measurement of typical Los Angeles driving. At that time, researchers characterized a real-world driving database according to the proportion of time spent in specific engine speed-intake manifold pressure bins. These bins were used to define "modes" of driving. Based on a survey of 1956 driving, the seven-mode drive cycle, the first certification drive cycle, was developed.

The seven-mode drive cycle was used for 13 years; however, this test did not accurately represent day-to-day driving behaviors.

In 1969, work began on a more realistic drive cycle, designed to represent "typical" morning (home-to-work) driving in Los Angeles. A specific 12-mile route, called the LA 4-road route, beginning and ending at the California Emissions Laboratory was selected as the basis of this new drive cycle. In 1969, six different Environmental Protection Agency (EPA) personnel drove the route. From this data set, a speed trace was selected and shortened to represent the average commute length in Los Angeles at that time. Due to the mechanical limitations of dynamometers then, all accelerations and decelerations were cut back to 3.3 mi/h/s (1.5 m/s^2).

The result was a new drive cycle forming the Urban Dynamometer Driving Schedule, also referred to as the LA4-S3, the LA4, or the FTP 72 [12-2].

The cycle simulates an urban route of 7.5 mi (12.07 km) with frequent stops. The maximum speed is 56.7 mi/h (91.25 km/h) and the average speed is 19.6 mi/h (31.5 km/h).

The same driving cycle is known in Australia as the Australian Design Rules 27 (ADR 27) cycle, and in Sweden as the A10 or constant volume sampling (CVS) cycle.

In Europe, vehicle emission regulations were first introduced in the 1960s and 1970s. At that time, the first version of a European drive cycle for vehicle testing under laboratory conditions was developed. This test cycle included only urban driving and only a maximum speed of 50 km/h. CO_2 emissions were not tested, only air pollutant emissions. This test is what we now refer to as the Urban Driving Cycle ECE-15. In the 1990s, the EU amended the test cycle to also include an extra urban segment (Extra Urban Driving Cycle (EUDC)), which reached a maximum speed of 120 km/h. In 2000, the idle period considered before the tests was removed and the vehicle was started from cold conditions. The test was then called the New European Driving Cycle (NEDC).

Since the initial creation of these driving cycles, various countries, governments, and private organizations have created different driving cycles to try to better represent driving conditions in different cities or countries, as detailed below. Emission limits certified over these driving cycles have been reduced year after year, but reference cycles have not changed significantly.

12.3 Worldwide Emission Testing Procedures and Standards

Emission testing standards are defined in four different parts:

1. *Emissions limits:* Define maximum tailpipe emissions allowed, typically per unit distance traveled.
2. *Test cycles:* Define how an engine or vehicle is operated during an emissions test.
3. *Test procedures:* Describe how to perform the tests, test specifications, and measurement systems.
4. *Equations:* Define how to calculate final results.

Emissions limits (1) define two different limits for emissions pollutants and CO_2. Usually, exhaust pollutant limits are set per vehicle, while CO_2 emission limits are set by vehicle fleet where the manufacturer is responsible for the overall production. They can compensate for the case of heavier vehicles over lighter ones. These emission limits are usually reviewed and made stricter while maintaining the test cycles and procedures.

Emission test cycles (2) define the driving characteristics, but it is the test procedure (3) where the specifications for the test, such as test temperature, humidity, and preconditioning requirements are defined.

There are three main emission legislations defined by Europe, the U.S., and Japan. Most countries worldwide follow these legislations or a combination of them, as we can see in Figure 12.1.

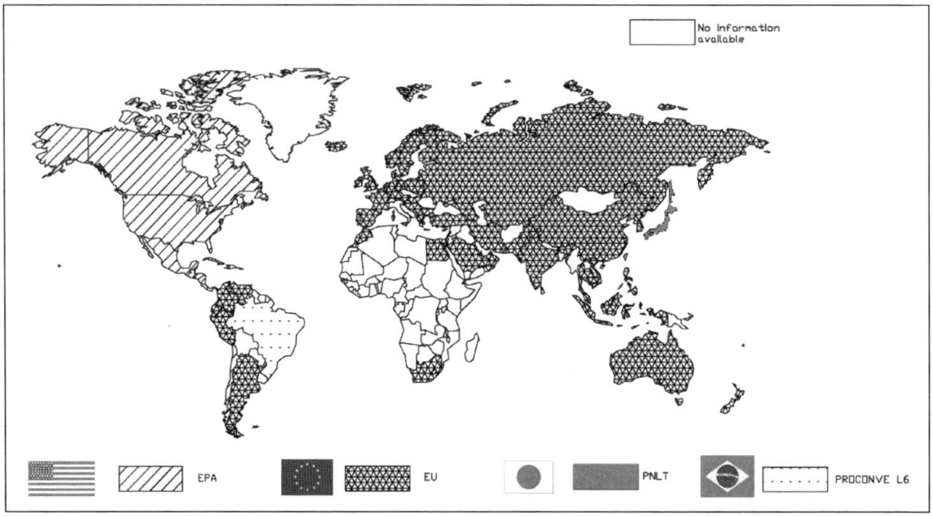

Figure 12.1 The world map defining regions following different legislations.

Clearly, for a global business such as the automotive sector, it is inefficient and inconvenient to be subject effectively the same product to different standards around the world. The cost of these inefficiencies is ultimately paid by car buyers. In an attempt to harmonize the different tests, experts (mainly from the European Union, Japan, and India) under the guidelines of the United Nations Economic Commission for Europe (UNECE) World Forum for Harmonization of Vehicle Regulations collaborated to design the Worldwide Harmonized Light Vehicles Test Procedure (WLTP), the text of which was released in 2013 in order that individual legislative regions could adopt it to the legislation according to their own timelines.

12.3.1 United States

The EPA is the agency responsible for protecting human health and the environment by writing and enforcing regulations based on laws passed by Congress. They have the authority to ensure that all vehicles meet standards, not only when manufactured, but also during the vehicle life.

For historical reasons, California has its own regulatory agency called California Air Resources Board (CARB).

EPA pollutants limits are set by tier standards, while those of CARB are set by Low Emission Vehicle (LEV) standards. Exhaust pollutant legislation is defined by CFR-86 in transition to CFR-1066, while for fuel consumption, the U.S. has two parallel standards, Corporate Average Fuel Economy Standards adopted by NHTSA and Green House Gas standards adopted by EPA.

12.3.1.1 U.S. Vehicle Testing and Enforcement Program

All engines and vehicles sold in the U.S. must have a valid certificate of conformity issued by the EPA before entering the market. Manufacturers must certify the most polluting vehicle out of a group of engines or vehicles with similar characteristics, such as engine displacement and cylinder number. These vehicles must be tested to determine emissions levels according to the Federal Test Procedure (FTP 75), High Speed Driving US06, and air conditioning testing (SC03).

For fuel consumption, the focus is different, as CO_2 production limits are set on the vehicle fleet instead of the single vehicle. The selection criteria of the test sample aim to cover a representative value of the average consumption of all the vehicles produced. For this, manufacturers take a sample of at least 90% of each model year. Fuel economy is calculated based on the five-cycle method result considering a Federal Test Procedure (FTP) cycle at regular temperature, an FTP at cold temperature, a highway fuel economy test (HWFET), high-speed tests (US06), and air conditioning tests (SC03).

Apart from the manufacturer's tests, the EPA can also perform confirmatory tests to validate emissions and fuel economy results reported by the manufacturer.

As explained in Chapter 7, one of the most critical factors for chassis dynamometer testing is the definition of the road load parameters. In the U.S., the manufacturer tests vehicles and declares the parameters that become publicly available via the EPA website. The EPA regularly performs its own test to validate these values.

Once the prototype vehicle is tested and certified, some deviations may occur during production. In order to guarantee that vehicles still perform within the same emissions levels, the EPA can perform different tests. The Selective Enforcement Audit allows the EPA to take any vehicle from the production line without prior notice to test its accordance to the standards.

Vehicle life and aging may also influence emissions levels, but manufacturers must still guarantee a certain level of performance. The In-Use Verification Program requires manufacturers to perform chassis dynamometer tests of in-use vehicles with low mileage (10.000 to 16.000 km) and high mileage (50.000 to 80.000 km). Even though the results allow 30% more than the standard limit, they are controlled.

12.3.1.2 U.S. Test Cycles

Table 12.1 summarizes the standard U.S. test schedules.

Table 12.1 Characteristics of U.S. Test Cycles

Driving schedule attributes	Test Schedule				
	City	Highway	High speed	AC	Cold
	EPA FTP	EPA Highway fuel economy	US06	SC03	EPA FTP
Trip type	Low speeds in stop-and-go urban traffic	Free-flow traffic at highway speeds	Higher speeds; harder acceleration and braking	AC use under hot ambient conditions	City test with colder outside temperature
Top speed	56 mi/h	60 mi/h	80 mi/h	54.8 mi/h	56 mi/h
Average speed	21.2 mi/h	48.3 mi/h	48.4 mi/h	21.2 mi/h	21.2 mi/h
Maximum acceleration	3.3 mi/h/s	3.2 mi/h/s	8.46 mi/h/s	5.1 mi/h/s	3.3 mi/h/s
Simulated distance	11 mi	10.3 mi	8 mi	3.6 mi	11 mi
Time	31.2 min	12.75 min	9.9 min	9.9 min	31.2 min
Stops	23	None	4	5	23
Idling time	18% of time	None	7% of time	19% of time	18% of time
Engine Startup*	Cold	Warm	Warm	Warm	Cold
Laboratory conditioned ambient temperature	68–86°F			95°F	20°F
Vehicle air conditioning	Off	Off	Off	On	Off

12.3.1.2.1 The U.S. FTP-75 Test Cycle

The Federal Test Procedure-75 (FTP-75) has been used for emission certification and fuel economy of light duty vehicles in the U.S. model since 2000. Due to the limitations of these tests, vehicles certified with this procedure on pollutant emissions also need to be evaluated in two additional tests. The US06 to simulate high-speed driving and aggressive acceleration, while the SC03 test simulates the air conditioning system. Fuel economy is calculated based on the five-cycle method considering an FTP cycle at regular temperature, an FTP at cold temperature, an HWFET, high-speed tests (US06), and air conditioning tests (SC03).

The FTP-75 consists of three segments [12-2]:

- cold start phase
- transient phase
- hot start phase

After the second phase is completed, the engine is stopped for a 600 s soak and restarted. The following are the basic parameters of the cycle:

- distance travelled: 11.04 mi (17.77 km)
- duration: 1874 s
- average speed: 21.2 mi/h (34.1 km/h)
- top speed: 91.3 km/h (56.7 mi/h)
- number of stops: 23

The vehicle speed trace against time of the FTP-75 cycle is shown in Figure 12.2.

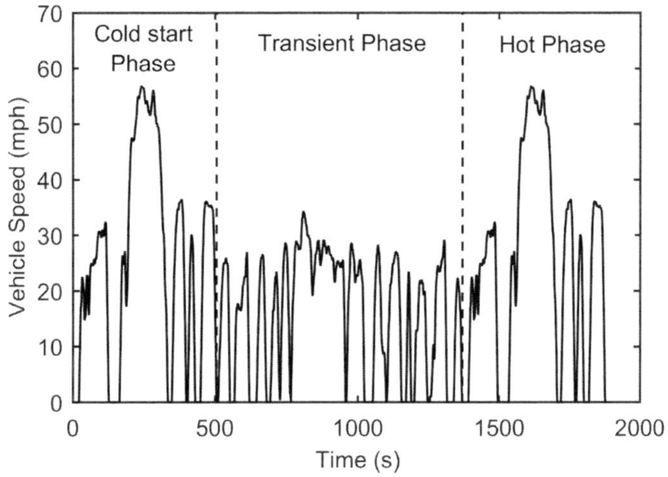

Figure 12.2 The FTP-75 cycle (Ref. [12-2]).

The emissions from each phase are collected in separate bags, analyzed, and expressed in grams per mile (grams per kilometer) (Table 12.2). The weighting factors are 0.43 for the cold start, 1.0 for the transient phase, and 0.57 for the hot start phase.

The FTP-75 cycle is known in Australia as the ADR 37 (Australian Design Rules 37) cycle.

The main parameters to control in the testing chamber are as follows:

- ambient temperature: 20 to 30°C (68 to 86°F) and humidity: not specified
- cold temperature: −6.7°C ± 1.7°C (20°F ± 3°F) and humidity: not condensing
- vehicle cooling air flow constant, at least 5300 ft^3/min (9005 m^3/h) squarely positioned within 12 in (30.5 cm) of the vehicle and with the vehicle hood open

12.3.1.2.2 EPA Cycle: The Highway Fuel Economy Test Driving Schedule

The HWFET cycle is a chassis dynamometer driving schedule, developed by the U.S. EPA for the determination of fuel economy of light duty vehicles [12-2].

The following are some characteristic parameters of the cycle:

- duration: 765 s
- total distance: 10.26 mi (16.45 km)
- maximum speed: 96.5 km/h (59.9 mi/h)
- minimum speed: 45.7 km/h (28.4 mi/h)
- average speed: 48.3 mi/h (77.7 km/h)

The vehicle speed trace against time of the cycle is shown in Figure 12.3.

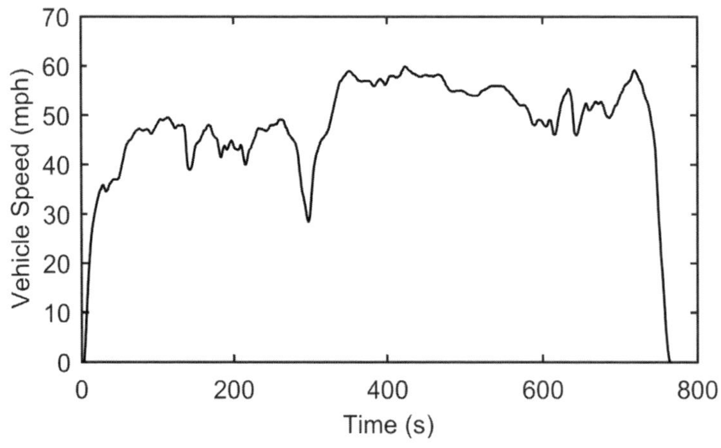

Figure 12.3 The EPA HWFET cycle.

The main parameter to control in the testing chamber is as follows:

- ambient temperature: 20 to 30°C (68 to 86°F) and humidity: not specified

12.3.1.2.3 The US06 Supplemental Federal Test Procedure

The US06 Supplemental Federal Test Procedure (SFTP) was developed to address the shortcomings with the FTP-75 test cycle in the representation of aggressive, high speed, and/or high acceleration driving behavior, rapid speed fluctuations, and driving behavior following startup.

The following are the basic parameters of the cycle [12-2]:

- length: 596 s
- distance: 12.8 km (8.0 mi)
- top speed: 129.2 km/h (80.3 mi/h)
- average speed: 77.44 km/h (48.4 mi/h)
- number of stops: 5

The vehicle speed trace against time of the US06 cycle is shown in Figure 12.4.

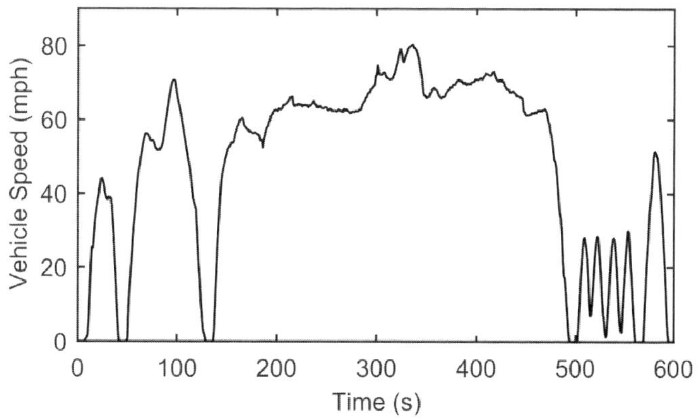

Figure 12.4 The US06 cycle.

The main parameters to control in the testing chamber are as follows:

- ambient temperature: 20 to 30°C (68 to 86°F) and humidity: not specified
- vehicle cooling air flow: constant, at least 15.000 ft^3/min (25.485 m^3/h) positioned within 24 in (61 cm) and vehicle hood open

12.3.1.2.4 The U.S. SC03 Speed Correction Driving Schedule

The U.S. SC03 Speed Correction Driving Schedule is used in addition to the above-mentioned FTP-75. It simulates urban driving and engine load with the air conditioning unit turned on for the entire duration of the test [12-2]. The vehicle speed trace against time of the cycle is shown in Figure 12.5.

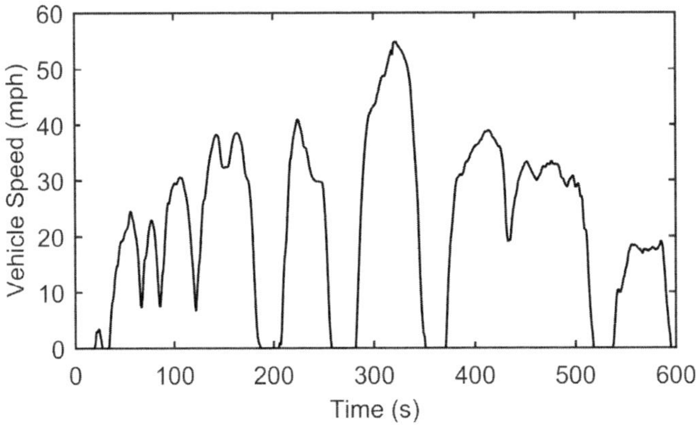

Figure 12.5 SC03 speed correction driving schedule.

The air conditioning system controls must be set as follows:

- air conditioning mode setting at maximum
- airflow setting at recirculate, if so equipped
- fan setting at highest setting
- air conditioning temperature setting at full cool (for automatic systems set at 72°F)
- air conditioning controls should be placed in the "on" position prior to vehicle starting so that the air conditioning system is active whenever the engine is running.

The following are the basic parameters of the cycle:

- length: 596 s
- distance: 5.8 km (3.6 mi)
- top speed: 88.2 km/h (54.8 mi/h)
- average speed: 34.8 km/h (21.6 mi/h)
- number of stops: 6

The main parameters to control in the testing chamber are as follows:

- ambient temperature: 35°C (95°F ± 2°F)
- 100 ± 5 grains of water/pound of dry air (approximately 40% relative humidity)
- solar heat load intensity of 850 W/m^2
- vehicle cooling air flow proportional to vehicle speed, which is equal or exceeds vehicle frontal area. Optimum 18 ft^2 (4.25 × 4.25) positioned in between 2 and 3 ft (61–91 cm) and 0 to 6 in from ground.

12.3.2 Europe

In Europe, pollutant emission limits up to September 2017 are defined by Euro Standards designated as Euro X (where X is a number in chronological order or a number plus a letter) and the regulation UN-ECE-R-83. CO_2 and fuel consumption are defined by UN-ECE-R101.

In September 2017, legislative changes come into effect, known as Euro-6d, which incorporates the new Global Technical Regulation GTR-15 (also known as WLTP) to define the test procedure. Euro-6d also incorporates Test 1 (average emissions after a cold start) previously defined in UN-ECE-R-83 and CO_2 emissions clauses previously defined in UN-ECE-R101. The older UN-ECE-R-83 will still cover some other vehicle tests such as type VI low ambient temperature (–7°C) exhaust emissions. It is predicted that GTR-15 will later be expanded to cover the rest of these tests.

Additionally, a new test, Real-Driving Emissions (RDE), will be implemented in September 2017 for exhaust pollutant emissions, as is explained in Chapter 15.

Table 12.2 EU Pollutant Emissions Legislations

Date	1995	Jan 98	Jan 01	Jan 05	Sep 09	Sep 11	Sep 14	Sep 17	Jan 21
Regulation	ECE R-83 w/o cold start		ECE-R-83 with cold start			ECE-R-83 +PMP		GTR 15	
Drive cycle	ECE 15.04	ECE 15.05	NEDC					WLTC+RDE	
Limits	Euro 1	Euro 2	Euro 3	Euro 4	Euro 5a	Euro 5b	Euro 6b	Euro 6d TEMP	Euro 6d

12.3.2.1 *European Union Vehicle Testing and Enforcement Program*

Before a vehicle goes into production, vehicle manufacturers need to kick off the type approval process, also known as homologation of the vehicle, to guarantee that the vehicle meets all the requirements to be sold in the EU. The type approval process consists of several steps including homologation of individual components such as mirrors, tires, and brakes, as well as CO_2 and pollutant emissions production, which is the central focus of this book.

Each EU member has its own type approval authority, which is also recognized by all members' states. Appointed by the type approval authority, there are technical services authorized to carry out type approval tests.

In order to homologate a vehicle, manufacturers can select any type approval authority within all the member states and an authorized technical service. They can then perform the homologation tests in the technical service site or in their own facilities witnessed by the technical service.

For exhaust pollutant emissions and CO_2 output, manufacturers need to guarantee that all vehicles are below the required limits. Not all variants will be tested because this would lead to a prohibitive regulatory burden where may be hundreds of similar variants. Instead, the example representative of each major type of vehicle configuration will be certified.

It is interesting to note that emissions and CO_2 could, theoretically, be tested under different type approval authorities and different technical services.

For pollutant emissions certification, the main tests to perform on a chassis dynamometer are described as the Type 1 Average exhaust emissions after a cold start, which is performed in between 20°C and 30°C and Type VI low ambient (–7°C) temperature emissions. For CO_2 and fuel emissions, a test is performed at temperatures between 20°C and 30°C.

As previously stated, the definition of the road load coefficients is crucial for the chassis dynamometer tests. In the EU, in contrast to the U.S., road load coefficients are the subject of confidentiality and are not declared publicly. A manufacturer has the option of performing a coast-down test with the supervision of the technical service or may take the default values defined in the procedure. The fact that these values are confidential and cannot be compared or contrasted causes a lot of uncertainty of their accuracy.

Once the manufacturer has performed their internal tests on the chassis dynamometers, they can declare an emissions value, which must then be achieved in any subsequent conformity test within a specified tolerance. Tests are then repeated on later samples of production vehicles, in the presence of the technical services witness and if passed, the manufacturer has complied with the conformity of production requirements.

In-service compliance currently only needs to be demonstrated for exhaust pollutants and does not cover CO_2 emissions. Every two years, the manufacturer needs to demonstrate compliance of vehicles within its fleet that are aged between 0 and 5 years and have travelled between 15.000 and 100.000 km. The results from the vehicle tests need to be reported to the independent authorities.

12.3.2.2 European Union Test Cycles

European emissions limits are defined with respect to tests performed over defined driving cycles. The main cycles to be performed in chassis dynamometers are listed in Table 12.3.

Table 12.3 European Test Cycles to be Performed in Chassis Dynamometers

EU test cycles	Positive ignition	Compressed ignition	Test cycle	Temperature	Humidity
Type I	YES	YES	4x ECE +1XEUDC	20–30°C	5.5 g ≤ H ≤ 12.2 g H_2O/kg dry air
Type VI	YES	NO	4x ECE	–7°C ± 2°C	Not condensing

For both types of tests, the cooling fan requirements are specified as follows:

- speed proportional from 10 km/h to at least 50 km/h
- area: at least 0.2 m^2
- height of the lower edge above ground: approximately 20 cm
- distance from the front of the vehicle: approximately 30 cm
- as an alternative, the blower speed shall be at least 6 m/s (21.6 km/h).

12.3.2.3 The New European Driving Cycle (ECE 15 + EUDC)

The NEDC is constructed from two shorter cycles: the European Commission for Europe (ECE) and the EUDC. It is sometimes referred to as NEDC, and therefore as the ECE + EUDC cycle. It is also known as the MVEG-B. The cycle is performed on a chassis dynamometer used for certification of light-duty vehicles in Europe [12-3]. The cycle was originally developed in the 1970's for comparing emissions on a standard-duty cycle, suitable for the testing technologies of the day. However, it has been adopted for emissions and fuel consumption certification while receiving relatively few updates to the actual procedure:

- *Test type I, average exhaust emissions after a cold start:* The entire cycle includes four Elementary Urban Driving Cycle ECE segments (see Figure 12.6) repeated without interruption, followed by one EUDC segment (see Figure 12.7). Before the test, the vehicle is allowed to soak for at least 6 h at a test temperature of 20 to 30°C. It is then started and emissions are measured directly from cold start.
- *Test type VI:* Consists of four Elementary Urban Driving Cycle ECE segments (see Figure 12.6), performed at –7°C ± 2°C.

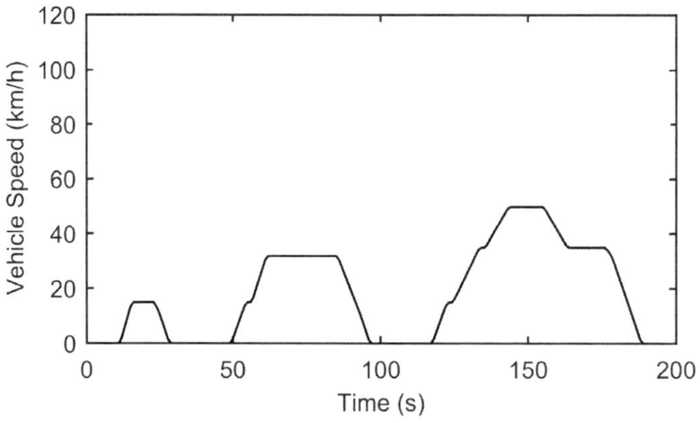

Figure 12.6 The ECE cycle.

Emissions are sampled during the cycle according the CVS technique (which was explained in Chapter 11), analyzed, and expressed in grams per kilometer for each of the pollutants.

The EUDC segment has been added after the fourth ECE cycle to account for higher speed driving modes representative of highway driving (see Figure 12.7). The maximum speed of the EUDC cycle is 120 km/h. An alternative EUDC cycle for low-powered vehicles has been also defined with a maximum speed limited to 90 km/h (Figure 12.8).

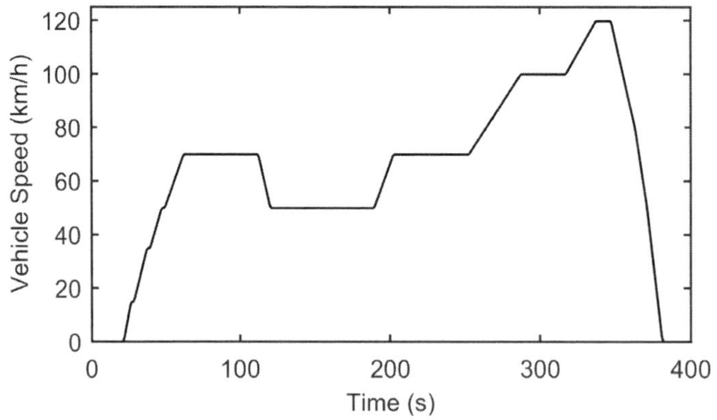

Figure 12.7 The EUDC cycle.

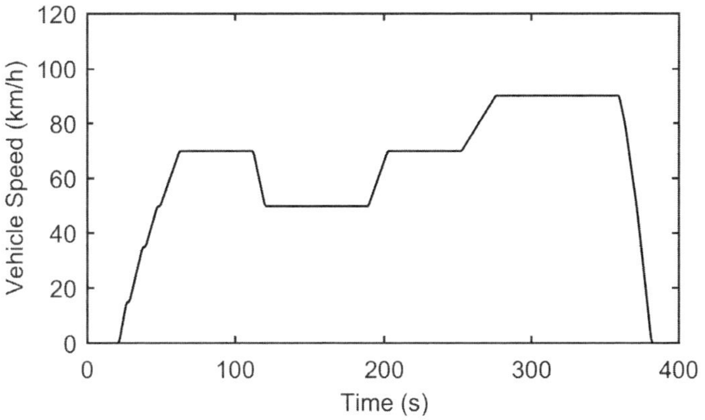

Figure 12.8 The EUDC cycle for low-power vehicles.

Table 12.4 includes a summary of selected parameters for the ECE, EUDC, and NEDC cycles.

Table 12.4 Selected Parameters for the ECE, EUDC, and NEDC Cycles

Characteristics	Unit	ECE	EUDC	NEDC
Distance	km	4 × 1.013 = 4.052	6.955	11.007
Duration	s	4 × 195 = 780	400	1180
Average Speed	km/h	18.7 (with idling)	62.6	33.6

12.4 Environmental Requirements for the Emissions of the Chassis Dynamometer Test Cell

Based on the previous tests, we can summarize the requirements for the test cell. We can make a classification according to standard tests and complementary tests, as summarized in Table 12.5 and Table 12.6.

Except for the air conditioning testing SC 03, it is possible to create a chassis dynamometer test cell that can simultaneously fulfill the requirements of European, U.S., and WLTP legislation, if it is capable of the following specifications:

- ambient temperature test controlling temperature in between 20 to 30°C with humidity in between $5.5 < H < 12.2$ g of water/kg of dry air
- low temperature tests at −7°C ± 2°C
- high temperature tests at 35°C ± 2.8°C (40% RH)
- air blower up to 39.000 m^3/h

Table 12.5 Characteristic of Standard Tests

	Temperature	Humidity	Cooling Fan
EU	293–303 K	Absolute 5.5 g ≤ H ≤ 12.2 g H_2O/kg dry air	Speed proportional from 10 km/h to at least 50 km/h Area: at least 0.2 m^2 Height of the lower edge above ground: approximately 20 cm Distance from the front of the vehicle: approximately 30 cm As an alternative, the blower speed shall be at least 6 m/s (21.6 km/h)
U.S.	298 ± 5K	NA	Fixed speed fan (not exceed 5300 ft^3/min)
WLTP (see Chapter 15)*	296 ± 5K	Absolute 5.5 g ≤ H ≤ 12.2 g H_2O/kg dry air	Speed proportional above speeds of 5 km/h (measurements made with no vehicle) Area: at least 0.3 m^2 Width/diameter of at least 0.8 m Height of the lower edge above ground: approximately 20 cm Distance from the front of the vehicle: approximately 30 cm The air temperature and humidity shall be measured at the vehicle cooling fan outlet at a rate of 1 Hz

*At the time of publication, WLTP requirements are not fully confirmed

Table 12.6 Characteristic of Complementary Tests

	Temperature	Humidity	Cooling fan
EU Type VI (Low temperature)	−7°C ± 2°C	Not condensing	Speed proportional from 10 km/h to at least 50 km/h Area: at least 0.2 m Height of the lower edge above ground: approximately 20 cm Distance from the front of the vehicle: approximately 30 cm As an alternative, the blower speed shall be at least 6 m/s (21.6 km/h)
U.S. Low Temp	20°F ± 3°F −6.7°C ± 1.7°C	Not Condensing	Fixed speed fan (not exceed 5300 ft^3/min) Distance to vehicle 12 in (30.5 cm)
U.S. US06	68–86°F (20–30°C)	Not Specified	Fixed speed 15.000 ft^3/min (25.485 m^3/h) positioned within 24 in (61 cm)
U.S. SC03	35°C (95°F ± 2°F) + Solar heat of 850 W/m^2	100 ± 5 grains of water/ pound of dry air (approximately 40% relative humidity)	Vehicle cooling air flow proportional to vehicle speed, which is equal to or exceeds vehicle frontal area. Optimum 18 ft^2 (4.25 × 4.25) positioned in between 2 and 3 ft (61–91 cm) and in between 0 and 6 in

References

12-1. Irene Berry, B.S. 2010. *The Effects of Driving Style and Vehicle Performance on the Real-World Fuel Consumption of U.S. Light-Duty Vehicles.* MIT Press: New York.

12-2. DieselNet. 2014. "Engine & Emission Technology." DieselNet: Ontario, Canada.

12-3. EEC Directive 90/C81/01. 1999. DieselNet: Ontario, Canada.

Chapter 13
The Need of a New Emission Regulation

13.1 Is the EU Legislation in 2016 Fulfilling Its Final Goal?

Now, we have seen the legislation of 2016 (UN-ECE-R-83 for pollutants and UN-ECE-R101 CO_2) and we will analyze as to how well it fulfilled its final purpose, which is reducing exhaust pollutant emissions and CO_2 into the environment.

In 2015, the International Council on Clean Transportation (ICCT) did an analysis on the divergence between real-world and manufacturer approval cycles in Europe measuring CO_2 and NO_x emissions [13-1]. Data on almost 600.000 vehicles from 11 data sources and 6 countries were analyzed. This study reveals a clear trend over time: the divergence (or "gap") between real-world and official CO_2 emissions increased from approximately 8% in 2001 to 50% in 2014 (see Figure 13.1). Each data source used for the study includes a unique set of vehicles and drivers, so estimates of the divergence of real-world from official values vary among them. However, the increase in the gap cannot be explained based on driving behavior or differences between the data sources, but is instead a result of increasingly unrealistic type approval values.

Figure 13.2 shows the measured emission values of CO_2 and NO_x for European vehicles compared with the limits for CO_2 and NO_x for European legislations EU-5 and EU-6.

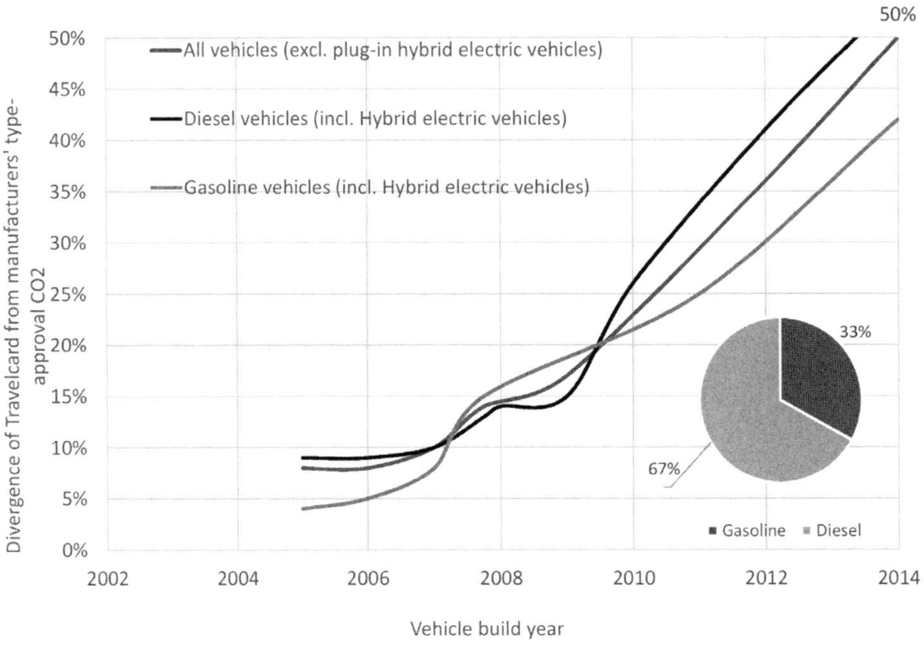

Figure 13.1 Divergence of real-world emissions versus manufacturers' type-approval CO_2 emissions (Ref. [13-1]).

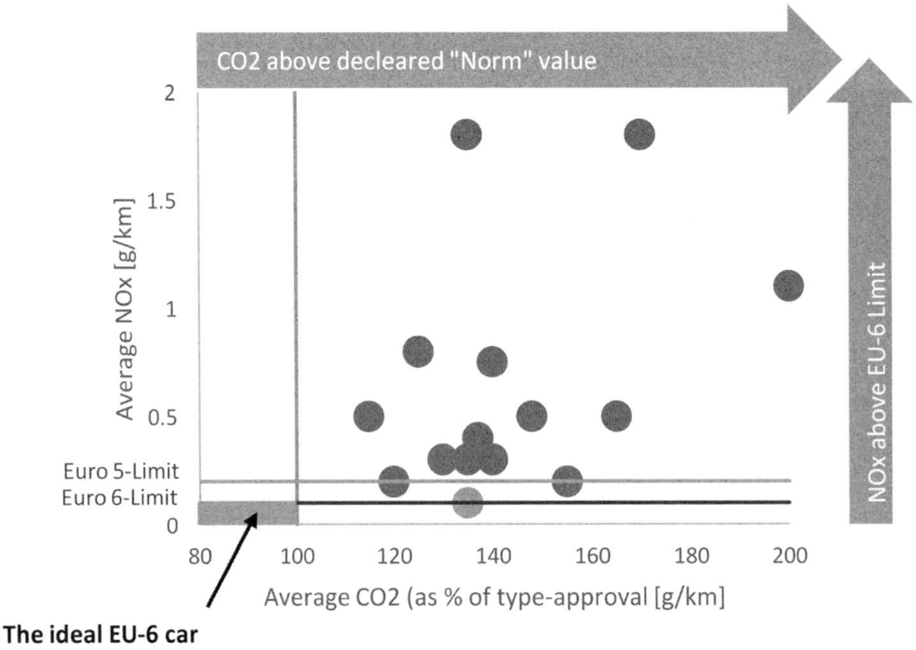

Figure 13.2 Divergence of real-world emissions versus manufacturers' type-approval CO_2 and NO_x emissions.

As it can be seen, all the vehicles tested are greater than the authorized values when tested in "real-world" conditions, even though they are all fulfilling the existing legislation when tested according to the legislative standard.

This evidence indicates that the existing type approval tests are not a good indication of the real-world driving conditions. This does not mean that the legislation has been without merit because it has been an important step and motivation for manufacturers to develop new technologies to reduce emission pollutants.

The increasing deviation starting in 2002 as shown in Figure 13.1 has led to a recognition within the industry and regulatory bodies that a review is needed to make the legislation more representative of real-driving conditions. This is expected to incentivize technologies and designs that perform better in real-world conditions, rather than the historically dated conditions represented by the NEDC.

Before defining any new legislation, it is important to understand the limitations of the existing procedures and where they deviate from real-world conditions.

13.2 Inaccuracies of Existing Legislations

13.2.1 What Is the Reason for these Deviations?

In 2015, the ICCT investigated the reasons why certifications testing results for fuel consumption results are not representative of real-world usage in Europe [13-1]. Four key factors were identified:

1. *Technology deployment:* Some technologies, such as start–stop systems and hybrid powertrains, have a different effect on CO_2 emissions in the approval procedure than they do during real-world driving. The specific driving cycle that is used for approval testing differs from everyday vehicle operation. Some of these technologies will perform better in the real-world than in the laboratory and vice versa. An example would be any technology that promotes a rapid engine warm-up. Real-world driving is often shorter in duration and so would see more benefit from such measures than is captured by the type approval process.
2. *Road load determination:* To develop a vehicle certification test in the laboratory, the coefficients for road load must be determined to characterize the forces (mainly aerodynamic drag and rolling resistance) that a car needs to overcome as it is driven on the road. These coefficients are determined through a series of coast-down tests on an outside track. There are a number of aspects of this load determination where the manufacturers have some tolerances and flexibilities. These include tire selection and preparation, selection of the test track, ambient test conditions, and preconditioning of the vehicle. It is estimated that about one-quarter of the overall gap observed in 2014 is explained by exploitation of tolerances and flexibilities in the methods required by the EU regulation for determining road load.

3. *Chassis dynamometer testing:* Under the EU regulation, there are areas in which flexibility could affect the results seen during chassis dynamometer testing. These include break-in periods of the test vehicle, tolerance regarding laboratory instruments, the state of charge (SoC) of the battery, optimized test driving techniques, and the use of pre-production parts that are not representative of mass-produced vehicles. There have been accusations that manufacturers could use this flexibility to their advantage, and while this is possible, there is considerable risk of noncompliance and poor customer satisfaction if this approach is taken. On balance, the biggest problems stem squarely from the unrepresentative nature of the cycle. It is this fact that has driven the industry and regulators to update the test procedures, along with the introduction real-driving emissions (RDE) measurements as a validation step.
4. *Other parameters:* Operating equipment such as air conditioning systems and entertainment systems increase fuel consumption during real-world driving. Nonetheless, these devices are either switched off or are not fully taken into account during the type-approval emissions test, leading to unrealistically low CO_2 emission values.

All these factors explain the increasing divergence between type-approval and real-world exhaust emission levels. The weights of these factors are estimated in Figure 13.3.

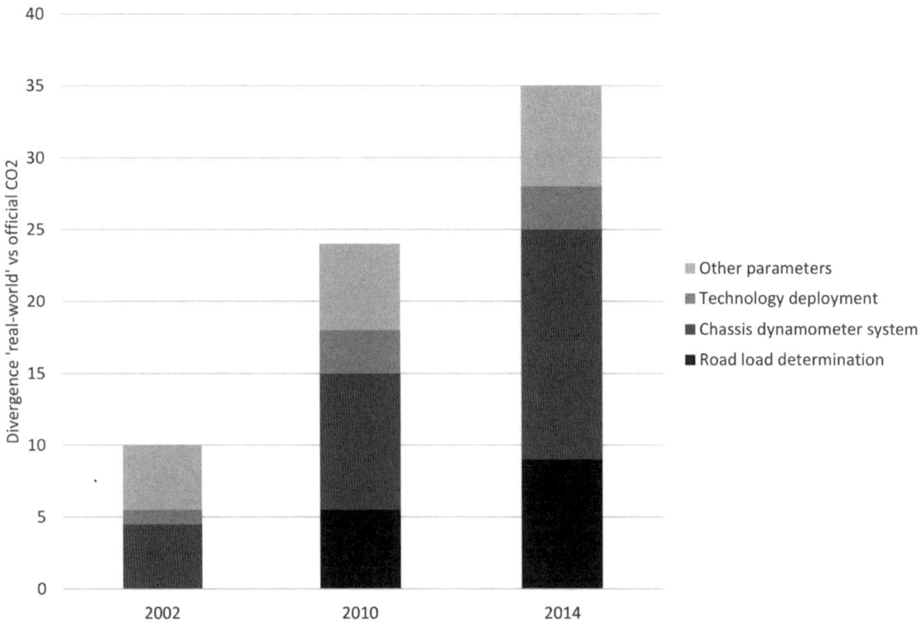

Figure 13.3 Estimate of the reasons for the divergence between type approval and real-world CO_2 emission levels for new passenger cars.

We will now try to analyze all these factors in detail.

13.2.2 Technology Deployment

Technology deployment analyzes how well the tests performed on the chassis dyno are representative for real on-road conditions.

13.2.2.1 The EU Driving Cycle

As we have seen in the previous sections, the NEDC cycle consists of four elementary urban driving cycles (ECE) (Figure 12.6) followed by an extra urban driving cycle (EUDC) (Figure 12.7).

Comparing the existing driving patterns with the driving cycle, it has been seen that the driving cycle has a relatively low mean velocity as well as low acceleration. Peak velocities are also set to 120 km/h, which are not representative of the actual driving patterns, where maximum speeds in some countries are 130 km/h such as in France or even higher, such as in Germany where there are zones with no speed limit.

The cycle represents urban driving and highway driving, but does not represent rural driving.

The cycle also has 14 stop phases that are very relevant for congested city driving, but probably not representative of real conditions in more free-flowing cities. This has led to the development of stop–start technologies, which result in reduced fuel usage in the tests and in real driving in congested cities, but maybe not in real on-road conditions depending on the usage of an individual driver.

The fact that the ECE cycle is repeated four times also reduces the variability of on-cycle working conditions. Figure 13.4 shows a comparison between engine speed and load operating points visited in the real-world and those visited during an NEDC test.

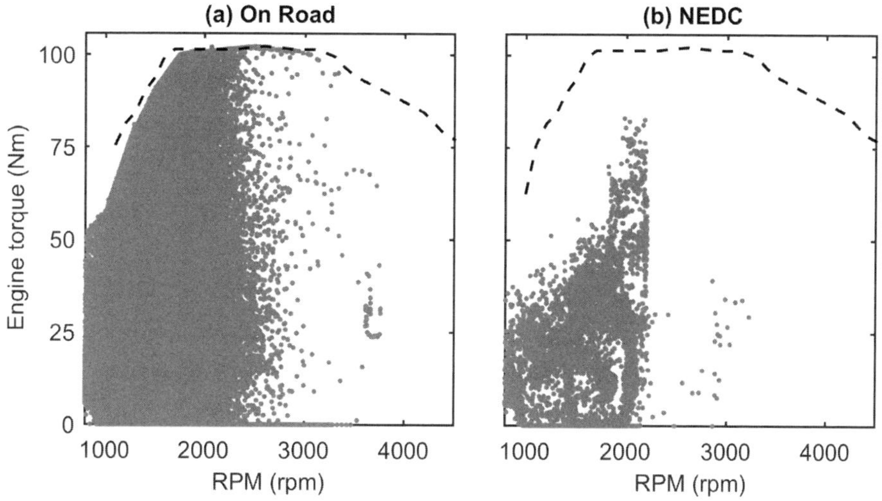

Figure 13.4 Engine working conditions. Left: real-world. Right: NEDC cycle.

As it can be seen, the NEDC represents a much smaller operating range than the real-world data.

13.2.2.2 The Problem of a Defined Cycle

An emissions test cycle is just one set of carefully controlled operating conditions that are tested to make results comparable across the vehicle type and manufacturer. Ideally, the results would be broadly representative of any working condition, implying a random or unregulated cycle comprising real-driving maneuvers in uncontrolled environmental conditions. In reality, it is very challenging to use any data from such maneuvers as the basis of a limit-based legislative instrument, because the conditions encountered vary very significantly in different regions, climates, and geographies.

Nevertheless, a further motivation for the adoption of a truly random test cycle would be to render it impossible to implement a "defeat device," which can detect the driving cycles based on the speed profile adjusting the vehicle to low emission mode. There have been very few high-profile cases of such manipulation, and a random real-world validation test would remove this vulnerability.

13.2.3 Road Load Determination Factor

In Chapter 1, we have seen the equation of motion of a vehicle on the road, and in Chapter 7, we have seen how this equation is replicated in the chassis dynamometer (Eq. 7.3). It is now possible to analyze how accurate is the replication of this equation in the chassis dynamometer compared with the road:

$$F_{\text{roadload}} = F0 + F1V + F2V^2 \tag{13.1}$$

The accuracies of the factors F0, F1, and F2 will directly influence the representative of the tests compared with the road.

In the EU legislation in 2016 (UN-ECE-R-83), there are two accepted methods for defining the road-load coefficient factors. The first one was explained in Chapter 7, it is known as coast-down testing and is defined across two SAE standards, J1263 [13-3] and J2263 [13-4]. The alternative method is to use the loads defined by ECE regulation 83-06 known as "cookbook."

When providing the "cookbook" data, the regulators must represent a middle band road load, which means that they typically take an "average" in between heavier and lighter vehicles for the type approval vehicle. This means that lighter vehicles loads are often overestimated, while heavier vehicles loads are typically underestimated.

As we said at the beginning of this section, in the EU the road load coefficients are confidential due to competition issues, while in the U.S., they are public and frequently checked by the EPA. In the EU, while the loads are confidential, they are checked by the type approval authority at the time of the test, via an end of coast-down test.

If we compare published values of road load coefficients in the U.S. with the values provided in the "cookbook" supplied by the EU, we can see that there are significant differences. This is a clear indication of the variability allowed in the selection of these parameters in the EU, which have a great influence in the declared vehicle performances and fuel consumption.

When the coast-down method is used in the EU, manufacturers have a certain degree of freedom when preparing the vehicle and selecting the road for the tests:

- The tires can be selected to give lowest rolling resistances using special tires or optimizing tire pressure.
- The test track can be selected to have an especially low rolling resistance. Additionally, the legislation allows 1.5% gradient. Regulations insisting that runs are performed in both directions, and they do not preclude the design of a test track that slopes downward in both directions [13-2], [13-3].
- *The vehicle:* When it comes to the optimization of the vehicle, mass manufacturers are free to optimize the vehicle mass so that the lightest possible type approval vehicle is tested at the track. The type approval regulations state that the test vehicle must have no optional extra equipment fitted [13-3].

The derivation of road load coefficients will have a direct influence on fuel consumption. Even this will change from one vehicle to another, and some values can be given as an estimation. A change of 1% in the road load force can represent a change of 0.63% in CO_2. A 1% change on vehicle inertia can represent 0.16% in an NEDC test (Figure 13.5).

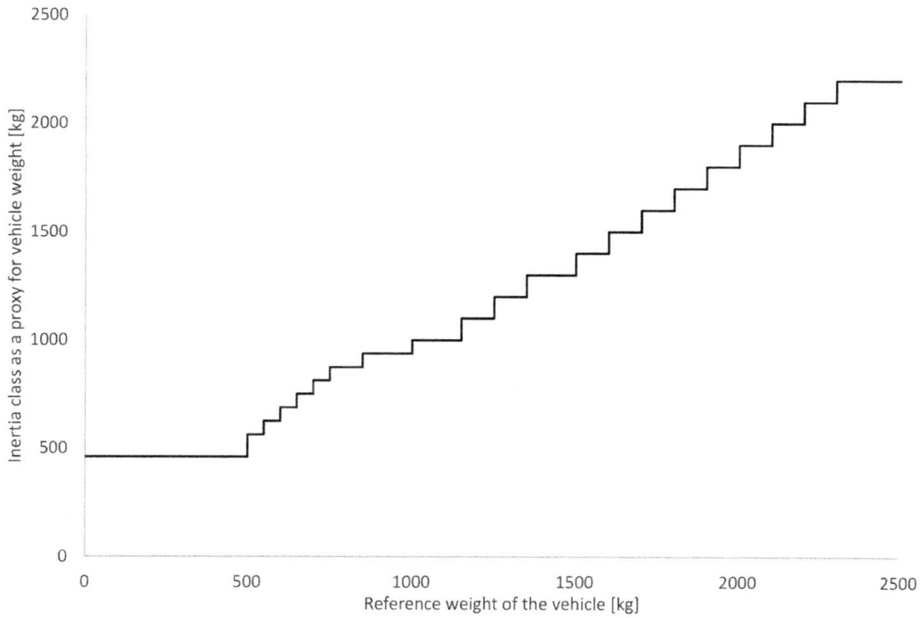

Figure 13.5 Inertia class steps as used in the NEDC to determine the emissions level of a light-duty vehicle.

Coast-down tests are performed at a given temperature and so the results obtained are representative of such a temperature. In the equations presented in the previous section, we see that the aerodynamic drag coefficient is dependent on air density. This means that, if required, this factor may be recalculated in the chassis dynamometer control system, if performing the test at a different temperature or altitude. Because the vast majority of testing has historically been focused on type approval testing at fixed temperatures at sea level, this is not a widespread procedure.

13.3 Chassis Dynamometer Testing

Imprecision coming from testing in a chassis dynamometer represents one of the most significant factors to consider. Even though some of these factors can be controlled or improved, some others are inherent to operating principle in a chassis dynamometer, such as the errors from the instruments used or the errors induced by some of the equipment such as the curvature of the dyno.

Figure 13.6 shows the effect on fuel consumption measurement that 12 vehicle configuration variables could have. These are put into context against a change in oil viscosity and highlight how the perturbing factors could mask real effects that need to be demonstrated, if they are not controlled properly.

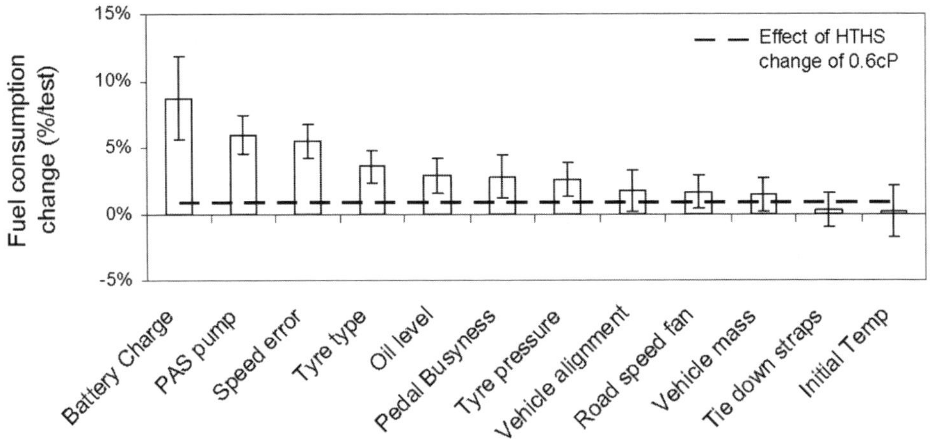

Figure 13.6 Absolute effect on fuel consumption measured, ordered, and compared with the typical effect of change of oil's high-temperature high-shear value from 2.9 to 3.5 cP (Ref. [13-4]).

To analyze these factors in more detail we will distinguish in between three different groups:

- tolerance of instruments
- tolerance of vehicle preparation
- driving techniques

13.3.1 Tolerance of Instruments

13.3.1.1 The Roller

13.3.1.1.1 The Influence of the Roller Diameter

As explained in Chapter 4, the influence on the deformation of a tire in contact with a roller instead of a flat surface creates a greater rolling resistance. Chassis dynamometers force tires to work on a non-flat surface adapting the contact area to the roller. This causes higher deformations in the tire, which result in higher energy losses.

A simplified expression developed by W.K. Klamp is commonly accepted for this correction in the rolling resistance forces (Figure 13.7):

$$F_{Rd} = F_{Rf}\left(1+\frac{r}{R}\right)^{1/2} \tag{13.2}$$

When a coast-down is performed in a chassis dynamometer, this correction is automatically achieved, but when high precision is required, such as a friction test, larger diameter rollers are preferred. Historically, twin roller dynamometers were common, with the tire resting between two small diameter rollers. Such machines were convenient to manufacture and install, but the tire losses were significantly larger due to the excessive deformation of the tires. Today, almost all new dynamometers used for emissions testing follow the EPA standard of a single roller of diameter 48".

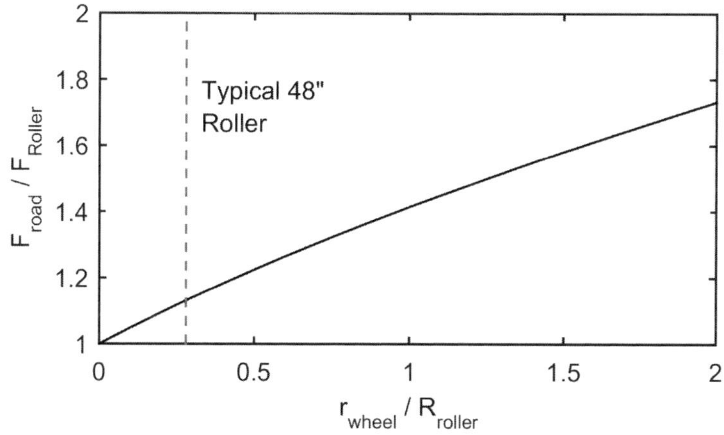

Figure 13.7 Influence of the roller diameter on the rolling resistance force.

13.3.1.1.2 Vehicle-Roller Alignment

When a vehicle is installed onto a chassis dynamometer, it needs to be aligned perpendicularly to the rollers. There are several procedures to perform this task.

If this alignment is not done properly, the resulting misalignment will cause a certain error. This is shown in exaggerated form in Figure 13.8. Studies on vehicle setup have shown than a misalignment of 75 mm can cause a 1.7% increase in vehicle fuel consumption over the NEDC cycles [13-4].

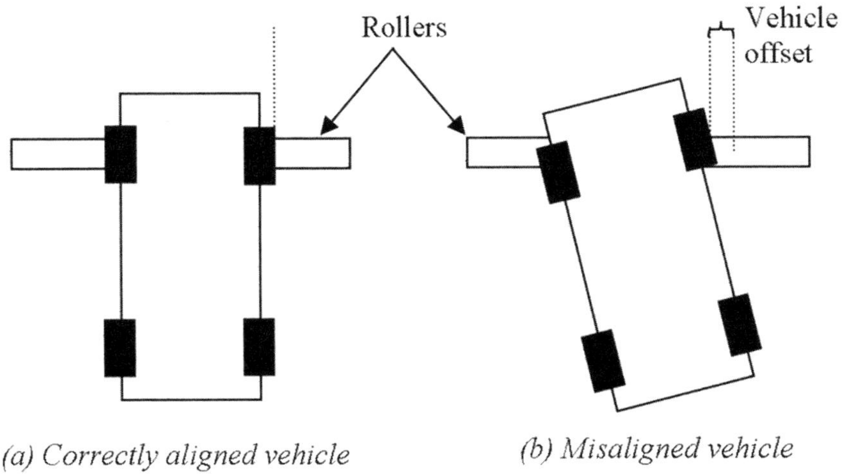

Figure 13.8 The diagram of vehicle misalignment on a chassis dynamometer (Ref. [13-4]).

13.3.1.1.3 The Influence of the Roller Temperature

Section 7.6 also explained the equation for internal dyno losses:

$$F_{\text{Dyno\_Loss}} = A0 + A1V + A2V^2 \tag{13.3}$$

Dyno friction losses are usually temperature dependent. Performing a test at a different temperature from the dynamometer load settings may lead to inaccurate results.

The results show that lower temperatures lead to higher internal friction from the chassis dynamometer and, therefore, higher CO_2 emissions during a specific test. Some chassis dynamometers have friction mitigation measures, such as motorized outer bearing races or friction stabilization methods such as heated bearings, in an attempt to minimize errors from such sources.

13.3.1.1.4 Vehicle-Roller Fixation

When the vehicle is installed in the dynamometer, it needs to be fixed properly to avoid movement and misalignment. When this is done, some additional vertical load can be set on the tires apart from the vehicle weight.

An increase of the vertical load imparted on the vehicle through the tie-down straps will result in an increase of the rolling resistance of the tires. It is therefore important to minimize this force or avoid it as some of the new restraint systems allow. It is somewhat difficult to quantify the magnitude of vehicle restraints unless the magnitude of the strapping force is directly measured. What can be said is that, by reducing the opportunity for the restraints to deform the load on the vehicle, is more constant with an equivalent improvement in the vehicle CO_2 emission precision.

13.3.1.2 The Vehicle Blower

13.3.1.2.1 The Blower Dimensions

Chapter 8 explained the different airflows that a vehicle experiences when driving on the road.

Regulation UN-ECE-R-83, for European type approval testing, defines a vehicle blower of at least 0.2 m^2 outlet section, positioned 30 cm from the front of the vehicle and aimed at the main vehicle radiator. In this way, cooling is focusing mainly on the engine compartment and virtually no cooling flow reaches the brakes, tires, or vehicle body.

Not cooling the tires will cause them to heat up, which will increase the tire pressure, and therefore affect the rolling resistance, as we explained in Chapter 4.

In addition to the lack of airflow over the vehicle powertrain systems, the lack of air flow over the vehicle body will in turn reduce the cooling (or heating) effect on the vehicle cabin, meaning that it is not possible to simulate the working conditions of the air conditioning system correctly, as explained in Chapter 6.

Ideally, vehicle cooling blowers should cover the whole body of the vehicle to replicate on road conditions. However, such hardware is very expensive, and the resulting requirements to upgrade test cells would likely have a significant financial impact on the industry. Furthermore, the increased electrical power required to drive a large cooling fan could have an environmental and financial impact.

13.3.1.2.2 Wind Blower Speed Measurements

Measuring the air speed at the outlet of a wind blower is typically carried out using an anemometer. However, it is not a simple task to get accurate and repeatable measurements because the readings are affected by many other parameters such as the orientation of the anemometer and the selection of the measuring points at the fan outlet and disturbances created by the vehicle in front of the blower. UN-ECE-R-83 requires that measurements are carried out with no vehicle in front of the blower to eliminate the disturbance created by that vehicle.

A calibration of air speed is performed to guarantee the linear velocity of the air at the blower outlet shall be within ±5 km/h of the corresponding roller speed within the range of 10 to 50 km/h. At the range over 50 km/h, the linear velocity of the air shall be within ±10 km/h of the corresponding roller speed. At roller speeds of less than 10 km/h, the air velocity may be zero.

This calibration is usually fulfilled by a linear correlation between the chassis dynamometer speed and blower speed. Once it is performed with no vehicle, it is used for any vehicle model. Because the vehicle is not placed in front of the blower when performing this calibration, a certain error is introduced when installing the vehicle.

Placing a vehicle in front of the blower can cause a drop in air speed; however, this can vary, vehicle to vehicle. To avoid this error, some manufacturers prefer to control air speed, continuously measuring speed at the outlet of the blower.

13.3.1.2.3 Blower Speed Control

Different philosophies are also applied for the speed control of the blower.

UN-ECE-R-83 allows three different conditions:

- proportional speed between 10 and 50 km/h
- proportional speed between 10 km/h and the maximum speed
- fixed speed constant at 6 m/s

Even though the legislation considers all possibilities for the wind speed control, some studies have been done comparing the different strategies showing that they have different results on exhaust temperatures and emissions.

Despite the flexibility offered by the legislation, most OEMs are controlling blower demand to achieve air speed proportional to roller speed up to the maximum speed of the cycle tested because this is the closest to real-life on-road conditions.

13.3.1.3 The HVACR System

13.3.1.3.1 Soak Temperature

In Chapter 4, we examined the influence of environmental conditions on engine performance, and in Chapter 5, the influence on exhaust emissions aftertreatment systems was explained.

Typical temperatures for engine coolants and lubricants during the regular operation of a vehicle range between 90°C and 110°C. After engine operation, the engine temperature will slowly decrease toward the ambient temperature as the vehicle soaks with its surroundings, and after restarting it will take some time for the engine to heat up again. When a vehicle is started after several hours of soaking, such that the powertrain and ambient temperatures are approximately the same, it is known as a cold start. During cold starts, the friction losses are higher than during normal operation and therefore fuel consumption and CO_2 emissions also are higher. A cold start is required on the NEDC that forms part of the current type approval regulations. To achieve repeatable cold start condition, the test vehicle is preconditioned in a soak area at constant ambient temperature before the actual test takes place.

The effect of this cold start depends on the starting temperature of the engine and the overall driven distance in the test cycle.

For the NEDC, studies are available quantifying the effect of start temperature on emission levels (EPA and ICCT Web sites). Figure 13.9 [13-5], [13-6] shows the influence of the engine start temperature on the CO_2 emissions. A starting temperature of 23°C is set as the baseline, and from this baseline, the CO_2 emissions for cold start increase by

19% (i.e., higher CO_2 emissions) as the starting temperature drops to −7°C and the CO_2 emissions decrease by up to 12% (i.e., lower CO_2 emissions) for starting the vehicle with a hot engine.

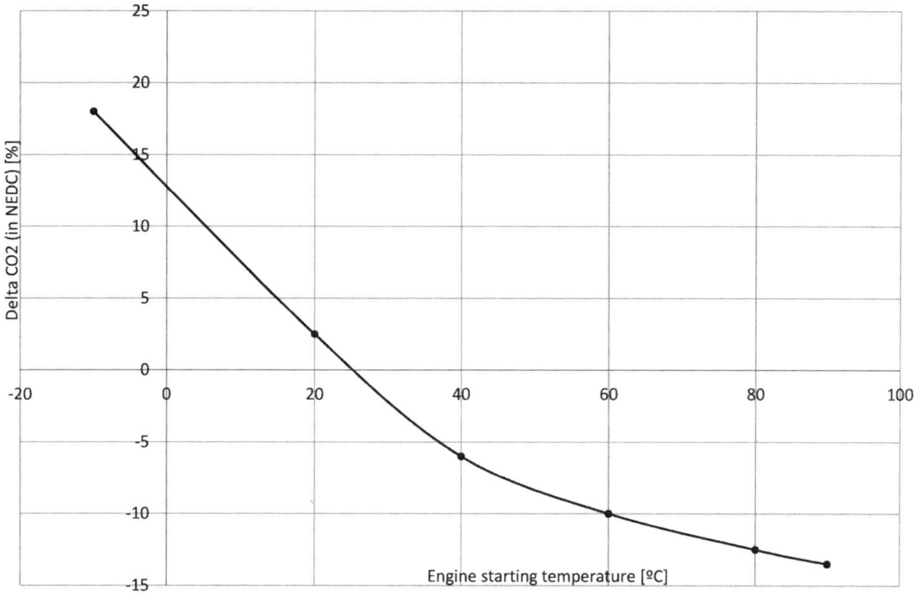

Figure 13.9 Impact on engine starting temperature on CO_2 emissions based on NEDC results.

As seen in Chapter 5, exhaust aftertreatment systems efficiency is also greatly influenced by temperature. Soak temperature will therefore affect pollutant emission results significantly.

UN-ECE-R-83 allows a soak temperature range of 25°C ± 5°C (Figure 13.10). Any values within this range are acceptable, and any variability within this range is acceptable. This also allows soaking at 28°C ± 1°C as many manufacturers do to gain advantage of the benefit of higher temperatures.

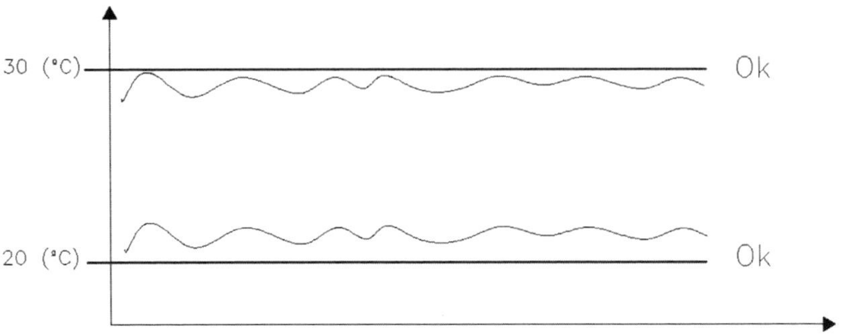

Figure 13.10 Soak temperature range for NEDC tests.

13.3.2 Tolerance of Vehicle Preparation

13.3.2.1 Status of Electrical Ancillaries

For type approval legislative emission testing, electrical loads should be turned off. The effects of switching on ancillary loads can be very large and this is therefore a key factor in these differences between real-world and legislative emissions. Studies attempting to quantify the differences have recorded changes to the CO_2 emissions in the range of +5% to +50% depending on the number of accessories and magnitude of the electrical load.

13.3.2.2 Battery State of Charge

The effects of battery SoC on fuel consumption or CO_2 emissions are well known. A lower SoC during vehicle operation will result in a demand from the alternator to recharge the battery which, depending on the driving mode, will put more load on the engine and cause a higher fuel consumption. For type approval emissions, UN-ECE-R-83 only specifies that the vehicle is run through a conditioning cycle prior to be soaked for between 6 and 30 h without any battery charging [13-3]. It is therefore possible to enter the test with a range of battery SoC depending on the vehicle behavior during the conditioning cycle. Traditionally, the easiest way to control battery SoC is to ensure it is always fully charged at the start of the test by ensuring the battery is on charge whenever not in use. However, with the advent of modern vehicle electronic control systems, this is no longer universally true, because it is now possible to control the alternator output so that no or limited load is placed on the engine during acceleration events, and instead increase the alternator during deceleration fuel shut off modes to reduce fuel consumption.

To quantify the possible effects of battery SoC, some authors have carried out vehicle tests with differing levels of initial SoC. One study showed that discharging the battery by leaving the vehicle headlamps on for 90 min prior to an NEDC emission test caused an 8.7% increase in fuel consumption during the subsequent test. Another study showed that larger discharges toward the limit of capacity (to the point where the battery can only just ensure starting of the car) of the batteries have shown up to 30% increase in fuel consumption [13-7].

The effect on CO_2 varies significantly from vehicle to vehicle. For example, Table 13.1 shows the increase in CO_2 emission for four different gasoline/diesel vehicles over the NEDC with near depletion of battery SoC. For all vehicles, the impact in percentage terms on fuel consumption is larger in the urban phase rather than the extra urban phase of the cycle. This is a result of the urban phase occurring first (when the battery charge is lowest), and that the engine powers are lower, meaning that the electrical load is a higher percentage of total engine load. It should also be noted that the overall effect on fuel consumption varies between 8% and 30%.

Table 13.1 Percentage Deviation from Basic NEDC CO as a Result of Initial Battery Discharge for a Range of Vehicles (Ref. [13-7])

Vehicle	Urban phase of NEDC	Extra urban phase of NEDC	Total NEDC
1	+35%	+9%	+21%
2	+15%	+4%	+9%
3	+46%	+14%	+30%
4	+31%	+5%	+27%

13.3.2.3 Tire Pressure

Chapter 8 has explained the influence of the tire conditions on its rolling resistance.

UN-ECE-R-83 states that the tire pressure must be set before testing when the vehicle is cold; however, no specific temperature is stipulated. It is possible therefore that by setting the tire pressure when the vehicle is very cold, it is possible to ensure the highest pressure and therefore the lowest rolling resistance during the test, when the vehicle is hot. From a precision perspective, it is therefore vital that tire pressures are set and checked at a repeatable soak temperature. One study has shown that a 7 psi decrease in the tire pressure of the conventional tire caused a 2.6% change in the measured fuel consumption over the NEDC cycle [13-4]. This gives a sensitivity of 0.4% increase in fuel consumption per 1 psi tire deflation.

13.3.3 Driving Techniques

13.3.3.1 The Type Approval Driver

During an emission test, a driver is shown on real time a display of the vehicle speed against a target speed trace on a device known as a driver's aid, as explained in Chapter 11. The center point of the display is always a moving real-time graphical representation of the real-time dynamometer speed against the target. An example of the type of graphical display being described is shown for the first 40 s of an NEDC test in Figure 13.11.

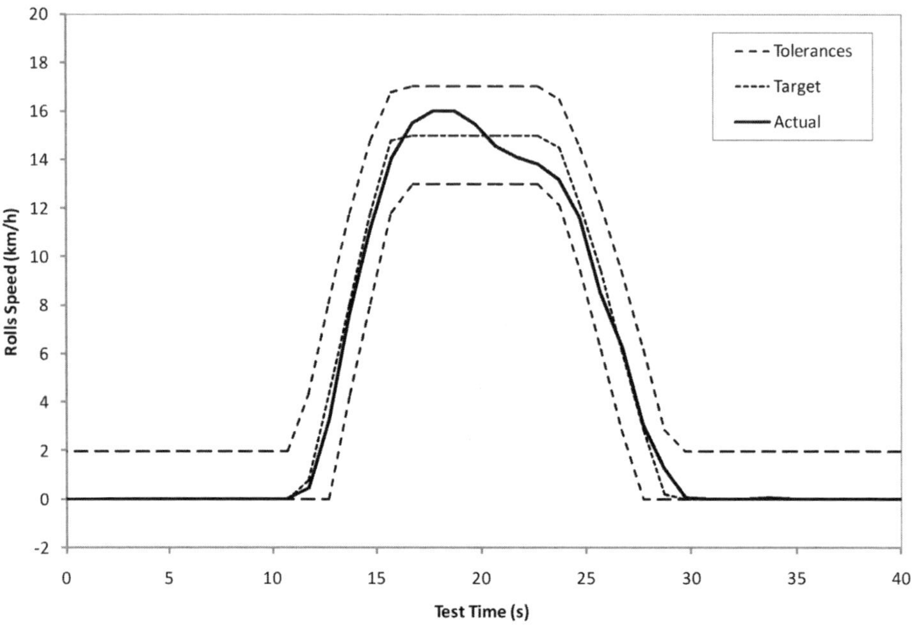

Figure 13.11 Example driver's aid speed display with tolerance lines shown.

During the test, the driver must control the vehicle speed to ensure that it stays within the prescribed tolerance, and ensure that they change gear when prompted to do so by the display on the driver's aid, unless the vehicle is fitted with the gear shift indicator (GSI). The driver must therefore control the accelerator and brakes, and change gear and start/stop the engine at the beginning/end of the test. For a vehicle with an automatic transmission, the gearbox must be operated in fully automatic mode. For a TA emissions test, the target speed trace is subject to a speed tolerance of ±2 km/h and ±1 s.

The ways in which the driver can affect the CO_2 emissions and fuel consumption through the way the vehicle is driven, include the following [13-8]:

- vehicle speed error compared to target duty cycle
- throttle operation
- engine starting procedure
- clutch control
- gear changing

13.3.3.2 Speed Error

The speed tolerance represents quite a wide window in which a variety of driving styles can be accommodated. In fact, one study showed that driving variation within the tolerances can cause changes to CO_2 emissions of up to 4% [13-9]. It is therefore possible to have two or more particular driving styles, within the tolerances of type approval, that yield very different values in total fuel consumption. Table 13.2 shows the change in fuel consumption for six gasoline and diesel engines [13-7]. The benefits are more pronounced in the high-speed extra-urban phase of the cycle.

Table 13.2 Test Results Showing the Effect on Fuel Consumption of Optimized Driving within the Type I Emissions Test Legal Tolerances across Six Test Vehicles. Vehicles 1–3 are spark ignition and vehicles 4–6 are compression ignition (Ref. [13-7])

Vehicle	Urban phase of NEDC	Extra urban phase of NEDC	Total NEDC
1	−0.7%	0%	−0.1%
2	+0.6%	−0.5%	−0.5%
3	+0.9%	−1.2%	−0.8%
4	−0.1%	−1.8%	−1.5%
5	−0.1%	−0.3%	−0.6%
6	+1.2%	−3%	−1.5%

Another study showed that a constant speed error of +3 km/h during the cruise modes of the NEDC can result in a 5.5% change in the vehicle fuel consumption [13-4].

13.3.3.3 Accelerator Pedal Busyness

Some authors have developed a new metric, known as pedal busyness for quantifying driver behavior based on their throttle pedal inputs [13-4]. This is defined in the following as the integral of the absolute rate of change of pedal position:

$$\text{Pedal Busyness} = \int abs\left(\frac{d}{dt}\text{Pedal}\right) \tag{13.4}$$

This is a useful metric for quantifying driver behavior because it is a measure of how oscillatory the driver's accelerator pedal movements are when integrated for the length of the test cycle. The use of this metric has been demonstrated using a robot driver where the pedal busyness was changed by tuning the robot's PID controller to force oscillatory behavior. When doubling the pedal busyness, the NEDC fuel consumption was seen to increase by 2.8%. This offers a way of quantifying pedal activity after a test has been completed, and so is useful for test validation and explaining any observed imprecision.

13.3.3.4 Gear Shifting

It has also been found that the timing of the gear shift points in the cycle can cause significant changes in the fuel consumption and emissions [13-9], [13-10]. The gear shift points for a vehicle without a GSI are fixed, with respect to vehicle speed, by UN-ECE–R-83. However, many new vehicles have dashboard mounted GSIs that tell the driver when to change up and down gear. In this case, for legislative TA emissions tests, the driver must change gear when the vehicle indicates. This allows manufacturers to tune their vehicle shift indicators to provide the best possible gear change strategy for the NEDC. When tuning a GSI, calibrators will have to make assumptions about what loads the driver places on the engine during a TA test. Clearly, it is possible to calculate the engine load using the target NEDC speed trace. However, we have seen that there is a wide range of driving styles that can be adopted within the TA speed tolerance. Naturally, this results in a certain amount of interaction among the vehicle, the GSI, and the driver. For example, a driver might need to minimize their speed and acceleration within the TA tolerance to maximize the benefit of the GSI. If they do not, the GSI's effect will be limited.

13.3.4 Other Factors

One of the most critical factors studied in Chapter 3, Section 3.3 was the air conditioning system. NEDC test is performed with the air conditioning off. Due to the limited area of the wind blower nozzle, the air cannot properly cool down all the vehicle and having the air conditioning on would not be representative. However, as we have seen in Section 3.3, the power requirements of the air conditioning system can be quite significant.

13.4 The Need for New Emission Legislation

As it has been seen, UN-ECE-R-83 has contributed a lot to the improvement of different technologies to optimize vehicles and reduce pollutant emissions. It has also been proved that it does not represent the real-world with the desired accuracy.

The need of a revised legislation becomes obvious. The first alternative is to review the existing standards trying to improve the inaccuracies, but as we can see, some of these inaccuracies are inherent in the philosophy of testing a vehicle on chassis dyno test cells, such as the differences in airflow over the vehicle and the fact that the non-driven wheels might be stationary (for a 2WD Dyno).

The second alternative is to change the testing philosophy and test vehicles on the road. New portable emissions measurement system (PEMS) equipment would allow this option. Comparability of results would, however, be very difficult due to changes in traffic and climatic conditions.

The best solution seems to take a compromise in between both options with some improved tests to be done in a chassis dynamometer plus some tests on the roads. Tests on the road would replicate any possible driving conditions, and test on the chassis dynamometer would guarantee the accuracy of the measurement comparing results from the road and the chassis dynamometer, as well as a reference cycle.

13.4.1 A Global Legislation for Global Vehicles

In additional to the improvement of the testing methodologies, it is desirable to have a worldwide standard, which can help manufacturers to focus the efforts in one direction without spending time to understand and fulfill the legislation of different countries.

Harmonizing testing legislation across the world should reduce vehicle development costs to benefit both manufacturers and consumers.

In 2007, the World Forum for Harmonization of Vehicles Regulations of the United Nations Economic Commission for Europe established an informal group to prepare a road map for the development of a world-harmonized light-duty vehicle test procedure (WLTP). This road map was proposed in 2009 and called for the development of a global technical regulation (GTR) for light-duty vehicles.

13.4.2 Legislation Beyond September 2017

To fulfill the requirements for revised legislation, a new testing procedure is implemented in Europe, in September 2017, known as the WLTP (GTR-15) and is used for the certification of CO_2 and exhaust pollutants. Additionally, a new road testing legislation called Real-Driving Emissions (RDE) is adopted in September 2017, where vehicles are tested on exhaust pollutants on the roads using the PEMS.

Correlation factors are defined for chassis dyno results under WLTP and on-road tests on RDE.

References

13-1. ICCT (2015), "From laboratory to road: A 2015 update of official and 'real-world' fuel consumption and CO2 values for passenger cars in Europe."

13-2. Kadijk, G. and N. Ligterink (2012). TNO 2012 R10237: Road load determination of passenger cars, TNO.

13-3. United Nations (2013). ECE 83-06. Uniform Provisions Concerning the Approval of: Vehicles with Regard to the Emission of Pollutants According to Engine Fuel Requirments, InterRegs Ltd.

13-4. Brace, C. J., et al. (2009). «Increasing accuracy and repeatability of fuel consumption measurement in chassis dynamometer testing.» Proceedings of the Institution of Mechanical Engineers, Part D: Journal of Automobile Engineering 223(Compendex): 1163–1177.

13-5. The Handbook Emission Factors for Road Transport (HBEFA). (July 2014). Database issued by INFRAS, Bern, Switzerland. Version 3.2.

13-6. Ligterink, N., van Mensch, P., Cuelenaere, R., Hausberger, S., Leitner, D. & Silberholz, G. (5 August 2014). "Correction algorithms for WLTP chassis dynamometer and coast-down testing."

13-7. Schmidt, H. (2010). Factors influencing NEDC CO2 emissions during type approval of passenger cars, TuV Nord.

13-8. Bielaczyc, P. and A. Szczotka (2007). "Analysis of Uncertainty of the Emission Measurement of Gaseous Pollutants on Chassis Dynamometer." SAE Technical Paper 2007-01-1324, SAE International.

13-9. Kadijk, G., et al. (2012). Supporting Analysis Regarding Test Procedure Flexibilities and Technology Deployment for Review of the Light Duty Vehicle CO2 Regulations TNO.

13-10. Vagg, C., et al. (2012). "Development of a new method to assess fuel saving using gear shift indicators." Proceedings of the Institution of Mechanical Engineers, Part D: Journal of Automobile Engineering 226(12): 1630–1639.

14.2.1.1.3 Class 1 Cycle

Class 1 with the lowest power-to-mass ratio represents vehicles driven in India (Figure 14.4 and Table 14.4).

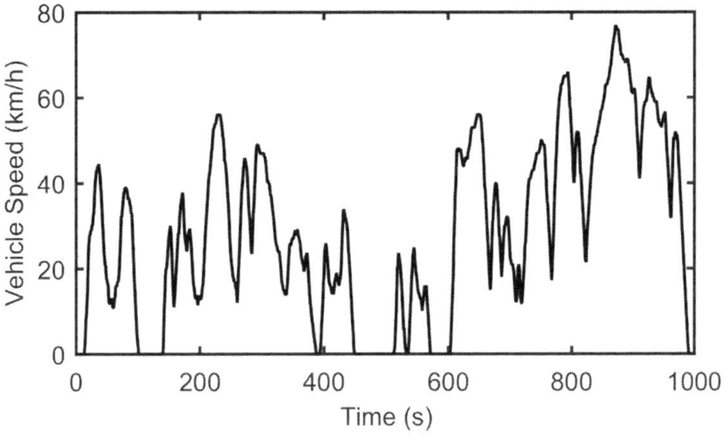

Figure 14.4 WLTP cycle for class 1 vehicles.

Table 14.4 WLTP Class 1 Cycle: Selected Parameters

Phase	Duration s	Stop duration s	Distance m	p_stop	v_max km/h	v_ave without stops km/h	v_ave with stops km/h	a_min m/s^2	a_max m/s^2
Low	589	155	3324	26.3%	49.1	27.6	20.3	−1.00	0.76
Middle	433	48	4767	11.1%	64.4	44.6	39.6	−0.57	0.63
Total	**1022**	**203**	**8091**						

Table 14.5 compares the NEDC test with the new WLTP to show main deviations.

Table 14.5 Comparison between the NEDC and the WLTP (Main Parameters)

	Units	NEDC	WLTP
Start conditions		Cold	Cold
Duration	s	1180	1800
Distance	km	11.03	23.27
Mean velocity	km/h	33.6	46.5
Maximum velocity	km/h	120	131.3
Stop phases		14	9
Durations:			
Stop phases	s	280	226
Constant driving	s	475	66
Acceleration	s	247	789
Deceleration	s	178	719
Shares:			
Stop phases		23.7	12.6
Constant driving		40.3	3.7
Acceleration		20.9	43.8
Deceleration		15.1	39.9
Mean positive acceleration	m/s^2	0.59	0.41
Maximum positive acceleration	m/s^2	1.04	1.67
Mean positive velocity*acceleration (acceleration phases)	m^2/s^3	4.97	4.54
Mean positive velocity*acceleration (whole cycle)	m^2/s^3	1.04	1.99
Maximum positive velocity*acceleration	m^2/s^3	9.22	21.01
Mean deceleration	m/s^2	−0.82	−0.45
Min deceleration	m/s^2	−1.39	−1.5

Analyzing the table, we can observe the following:

- *Increase of test duration:* An increase in test duration will reduce the influence of the pollutants emitted during the first part of the test, when the vehicle is cold and the vehicle aftertreatment system is not working to the full effect. Even though the pollutant emissions will be similar, the proportional influence on the overall tests will be reduced.
- *Stop time:* In the WLTC (12.6% stop time), there are less stationary phases than in the NEDC (23.7% stop time). Stop–start systems stop the engine during phases where the vehicle is stationary to eliminate vehicle emissions in these phases. As the WLTC has a greatly reduced stop time compared with the NEDC, this technology will have a lesser effect on reducing the overall CO_2 emissions result.
- *Higher speeds and accelerations:* This will have a direct impact on the working conditions of the engine. Comparing the NEDC with the WTLC and real-driving

conditions, we can see how the engine working map in WLTC is getting closer to real-world (Figure 14.5).

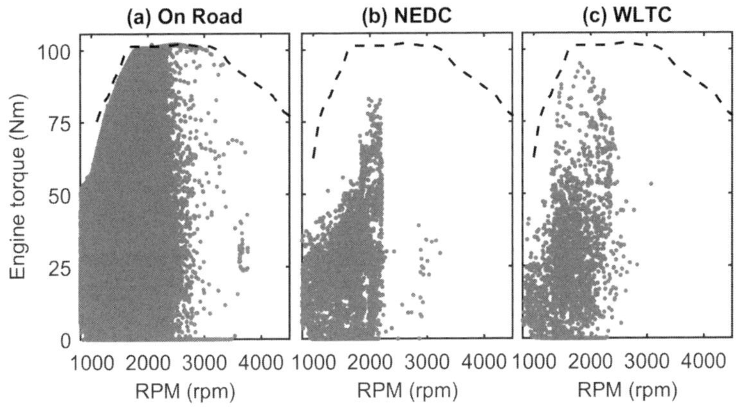

Figure 14.5 Engine working conditions in real-world, NEDC, and WLTP.

14.2.2 Road Load Determination Factor

The WLTP will introduce improvements around the road load requirements and tolerances have been tightened in this area via the implementation of a few key changes. The basic process for road load determination, using vehicles and a test track, will remain the same.

Some of the improvements are as follows:

1. Instead of declaring one CO_2 value for an entire family of vehicles (as required by NEDC legislation), each individual vehicle within a vehicle family will receive a CO_2 value based on its individual mass, rolling resistance, and aerodynamic drag, as determined by its standard and optional equipment.
2. Instead of the "table of running resistances" (the "cookbook" of road load values that can be used if the road load for a vehicle has not been determined by track tests), a formula for calculating road load is provided, based on a relevant vehicle characteristic.
3. Use of on-board anemometers will be permitted, and the correction method applied for wind during the coast-down method (for both stationary wind measurement and on-board anemometry) is improved.
4. Air drag will be compensated for the deviation of air density taking into account pressure, density, and humidity.
5. The test track characteristic (e.g., road inclination) will be more stringent to reduce positive influences on the road load determination.
6. Tire specifications must be similar to those of the vehicle that will be manufactured.

a. Test tire preconditioning (tread depth, tire pressure, run-in, shape, no heat treatment allowed, etc.) will be more stringent, to more closely match the tire conditions to those on production vehicles.

b. Tire pressure must be set according to vehicle specifications.

c. Special brake preparation to avoid parasitic losses from brake pads touching brake discs will be prevented.

7. Inertia simulation for rear wheels will be considered for one-axle chassis dynamometers (+1.5%).

14.3 Chassis Dynamometer Testing

14.3.1 Tolerance of Instruments

14.3.1.1 The Vehicle Blower

14.3.1.1.1 The Blower Dimensions

The vehicle blower dimensions have been increased to 0–3 m^2 compared with 0–2 m^2 for the NEDC. This will cover a larger section of the vehicle frontal area and will, therefore, replicate vehicle cooling closer to on-road conditions.

In addition, the regulation also defines the width/diameter dimension of at least 0.8 m.

However, this relatively small increase will not be sufficient to provide fully realistic airflow to powertrain components, such as the tires and vehicle body.

14.3.1.1.2 Blower Speed Control

The three different types of fan control allowed in the previous NEDC regulations are no longer permitted. Blower speed now needs to be proportional to roller speed for all speeds above 5 km/h within ±5 km/h or ±10%, whichever is greater.

This eliminates the uncertainties of blowers working at a fixed speed or with a maximum speed of 50 km/h.

14.3.1.2 The HVACR System

14.3.1.2.1 Soak Temperature

The test temperature in the laboratory is lowered from a range of 20°C to 30°C to a set point of 23°C.

In the NEDC regulations, the test temperature is set to range between 20°C and 30°C during soak and testing. For the WLTP, there is a more stringent definition, setting the soak temperature at 23°C ± 3°C and test temperature at ±5°C. Emphasis is placed on the average to be at 23°C stating that a clear offset can be rejected (Figure 14.6).

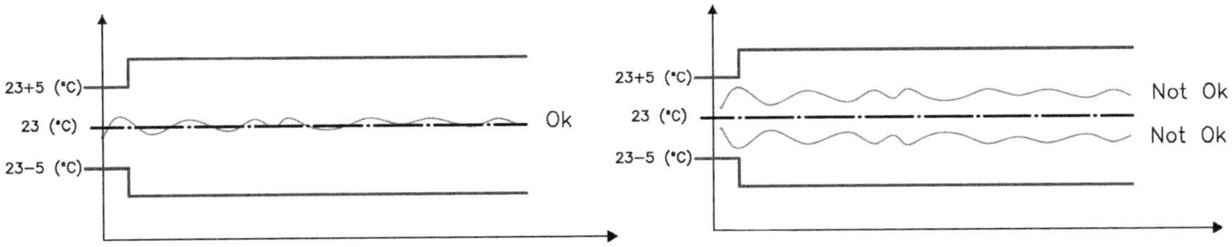

Figure 14.6 Soak temperature limits for WLTP. Left: acceptable. Right: not acceptable.

With this, they avoid the possibility to soak vehicles at higher temperatures like 28°C ± 1°C.

Additional considerations are being made to adapt an ATCT were the vehicle is soaked and tested at 14°C, which is much more representative of the average temperature in Europe.

14.3.2 Tolerance of Vehicle Preparation

14.3.2.1 Battery State of Charge

The difference in battery stage of charge over the cycle is monitored and the consumption corrected as needed, based on changes in battery state of charge over the cycle. Battery state of charge at the start of the test is changed from fully charged (NEDC) to a representative start value (the fully charged battery will be partially depleted by first driving a WLTC as the preconditioning cycle).

14.3.2.2 Tire Pressure

Stricter control on tire pressure on the dynamometer test is requested.

14.3.2.3 Gear Shifting

A gear shift strategy for manual transmission has been created. This strategy simulates the representative gear shift operation as a function of engine operating point. This was one of the most elaborate descriptions of gear shift strategy, and takes into account an analysis of the vehicle tractive force requirements compared with the torque available in a particular gear. This means the gear shifting strategy is related to the power available for a specific vehicle, not just a specific engine speed range.

14.4 Expected Results From WLTP

With the implementation of the new WLTP, the legislator expects to reduce divergence from real on-road conditions in the short term. However, as manufacturers' emission targets are pushed down, with the possible findings of new loopholes, divergence may increase again as shown in Figure 14.7.

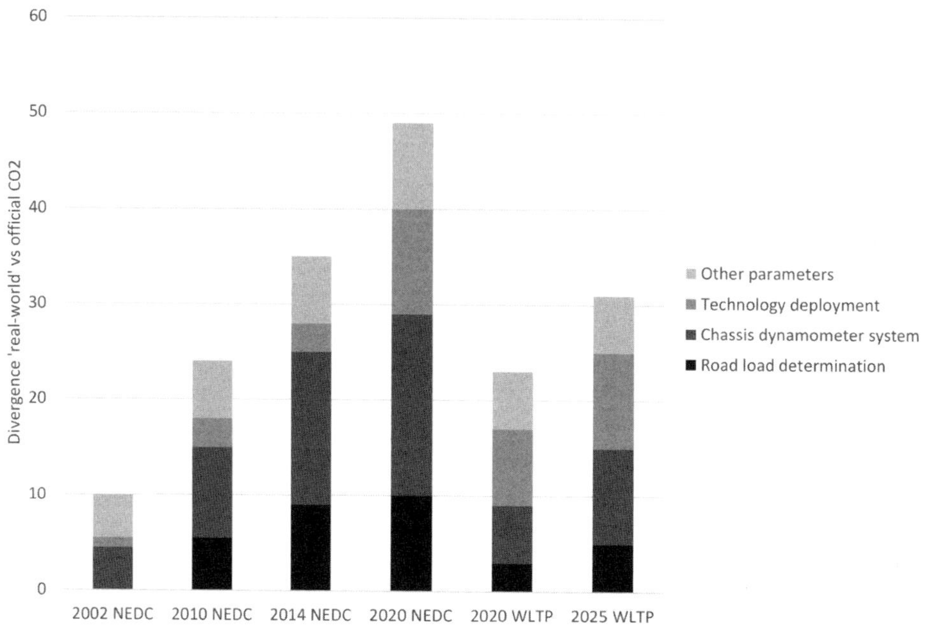

Figure 14.7 Existing and future expected divergence in between real-world and official CO_2 emissions (%).

As we have seen in the previous chapters, some of the divergences in between testing procedures and real road can be reduced, but some others are inherent to testing in a chassis dynamometer. The final step to guarantee a good representation of real-road conditions is to test on the road, as it will be explained in Chapter 15. This will only be applied so far for exhaust pollutants and not for CO_2.

References

14-1. Economic and Social Council. 2015. Proposal for Amendments to Global Technical Regulation No. 15 pm Worldwide Harmonized Light Vehicles Test Procedure (WLTP). UNECE: Geneva, Switzerland.

Chapter 15
Real-Driving Emissions

15.1 Introduction

As we have explained in Chapter 11, Section 11.3.5, the development of new technologies has allowed the introduction of emissions measurements on the road using portable equipment measurement systems.

Testing on a chassis dyno test cell allows one to have the vehicle under stationary conditions. This also allows one to increase the repeatability and the comparability of results. However, it is also an artificial way of measuring emissions, and its results may differ from the actual on-road emissions because it eliminates several factors that influence emissions (e.g., road gradient, hard accelerations, use of air conditioning, and traffic or weather conditions).

The combination of improved tests performed on a chassis dynamometer with on-road Real-Driving Emissions (RDE) tests seems to be the solution that the legislators implement from September 2017 onward. This offers a better position in terms of accuracy between real-world and laboratory testing.

15.2 Requirements for Real-Driving Emissions

Theoretically, RDE could result in vehicle testing under any on-road conditions. However, we must set some boundaries to make results comparable among tests and be able to set limits to pollutant emissions.

The regulators therefore define certain conditions for an RDE test such as vehicle payload considering the number of passengers, the vehicle working conditions such as the air conditioning or auxiliary devices, which must replicate normal conditions, and the trip distance and the environmental conditions.

In order for an RDE test to be representative of different driving conditions, but consistent, it must be composed of urban driving (speeds up to 60 km/h and approximately

34% distance) followed by rural driving (speeds between 60 and 90 km/h and approximately 33% distance) and motorway driving (speeds above 90 km/h and approximately 33%). Some other conditions are set on the trip such as the maximum speed of 145 km/h (160 km/h in extended conditions), average speed, and duration of stops [15-1].

As temperature also has a significant effect on emissions, two temperature conditions are defined: 1) a moderate temperature condition, which is defined as temperatures in between 0°C and 30°C, and 2) an extended condition, which is defined from −7°C to 0°C and 30°C to 35°C [15-1].

The altitude is also defined as moderate for altitudes below 700 m above sea level, and extended for altitudes up to 1300 m [15-1].

Many other boundary conditions will be defined by the legislator in order to make the test as representative as possible to real-driving conditions, yet practically possible to achieve and suitable for comparison test to test.

15.3 Preparing Vehicles for RDE

Even though there are some boundary conditions to an RDE test, the variability in a test is still very significant as there are many aspects on the road, which cannot be controlled or reproduced, such as the traffic, the number of stops, wind blowing into the vehicle, and the ambient temperature.

This poses a difficult question: "How can manufacturers study and improve their vehicles in such variable conditions?" The solution that most seems to be applying is making them go back to chassis dyno test cells where they can control all these conditions in a repeatable fashion and study the performance of the vehicle.

The test cells needed in the future are, therefore, likely to be more capable than traditional certification test cells, as they need to reproduce nearly all on-road conditions as close as possible.

One component that has a significant importance on this issue is the wind blower as it is responsible for many of the deviations from a test cell and the road. A wind blower for RDE replication on a chassis dyno test cell should allow close simulation of the on-road airflow over the tires and, similarly, allow for accurate operation for the air conditioning system.

A proposal for vehicle wind blower to simulate RDE in a chassis dyno test cell was done in Chapter 8.

Another issue that has been detected by some researchers is that it is very difficult to obtain comparable results using a human driver in a chassis dyno test cell trying to replicate an RDE on-road tests due to the speed profile error. Best results have been obtained by replicating exactly the same engine conditions giving this demand directly to the engine [15-2].

References

15-1. Official Journal of the European Union. 2016. Amending Regulation (EC) No 692/2008 As Regards Emissions from Light Passenger and Commercial Vehicles (Euro 6). Official Journal of the European Union. http://eur-lex.europa.eu/legal-content/EN/TXT/?uri=CELEX%3A32016R0427.

15-2. Dr. Günter Fraidl, Dr. Paul Kapus, and K. Vidmar. "From RDE towards RDD, Does RDE Require Real-Driving Diagnostics?" Internationales Symposium für Verbrennungsdiagnostik, 2016.

Summary of Part III—Vehicle Homologation Process and Tests

In Chapter 12, Section 12.1, we have seen the different tests that can be done in a chassis dyno test cell and how different legislations started to standardize testing procedures to control pollutants and CO_2 emissions.

We have seen how the main organizations worldwide (from the United States and Europe) control pollutants and CO_2 limits by defining legislations that control the requirements for the vehicle manufacturers. An important part of these legislations is the vehicle testing protocols and test cycles to measure vehicle emissions.

We have then analyzed how legislation is fulfilling its final goal observing that there is a significant deviation up to 2017, from expected emission values coming from laboratory tests to on-road results. This had led to the analysis of this deviation in four main points:

- Technology deployment, which indicates how good the tests of some technologies on a chassis dyno test cell correlate to real road conditions.
- Road load determination, which analyzes how good the road load equation is replicated in a chassis dyno test cell compared with the real road.
- Chassis dynamometer testing, which analyzes how good a chassis dynamometer tests cell operated under certain procedures replicates real conditions.
- Other parameters such as air conditioning or entertainment devices, which are usually not simulated.

As a conclusion, we have seen that there are deviations that are inherent in the philosophy of testing a vehicle in a chassis dynamometer, and not much can be done without changing this testing concept. Some others are limitations on the existing testing procedures that can be improved.

We have then seen how some limitations are being addressed by a new WLTP.

The limitations inherent in testing a vehicle in a chassis dynamometer are being addressed by a new testing methodology testing vehicles on the road with portable emissions measurement equipment.

Notation(s)

Reference	Description
WLTP	Worldwide Harmonized Light-duty Vehicle Test Procedure
F_I	Force necessary to overcome inertia
F_g	Force due to gravity
F_{AD}	Force due to aerodynamic drag
F_{RR}	Force due to tire rolling resistance
MST	Minimum starting temperature
HVACR	Heating, ventilation, air conditioning, and refrigeration
PM	Particle measurement
PN	Particle number
PEMS	Portable equipment measurement system
CVS	Constant volume sampling
ECU	Engine control unit
RDE	Real-driving emissions
EU	European Union
U.S.	United States
F_{tot}	Total (resultant) force
F_{TR}	Tractive force
a	Vehicle acceleration
N_{wheel}	Wheel rotational speed
N_{axle}	Axle rotational speed
$N_{\mathrm{GB\ out}}$	Gear box output rotational speed
$\mathrm{GR}_{\mathrm{FDrive}}$	Flywheel drive gear ratio
$\mathrm{N}_{\mathrm{eng}}$	Engine rotational speed
$\mathrm{GR}_{\mathrm{GBox}}$	Gearbox gear ratio
ω	Speed
d	Pitch diameter
N	Number of teeth
m_{eq}	Equivalent mass
r_W	Wheel radius
M	Vehicle mass

Reference	Description
dv/dt	Acceleration (or deceleration) rate
Θ	Level grade
C_D	Drag coefficient
A_F	Projected frontal area
ρ	Density of air
v	Vehicle velocity
q	Road gradient
$\mathbf{g}$	Gravitational constant
C_{RR}	Tire rolling resistance coefficient
F_{roadload}	Road-load force
I_{CE}	Internal Combustion Engine
C_{FD}	Computational Fluid Dynamics
EGR	Exhaust Gases Recirculation
P_{Tractive}	Average tractive power demand during time increment
$\dot{Q}_{\mathrm{fuel}}$	Fuel power
$\dot{Q}_{\mathrm{air}}$	Air power
$\dot{Q}_{\mathrm{intercooler}}$	Intercooler power
$\dot{Q}_{\mathrm{exhaust}}$	Exhaust power
$\dot{Q}_{\mathrm{oil}}$	Oil power
$\dot{Q}_{\mathrm{coolingwater}}$	Cooling water power
$\dot{Q}_{\mathrm{unburnt}}$	Unburnt power
$Q_{\mathrm{rad,conv}}$	Radiation and convection power
$\mathrm{Power}_{\mathrm{eff}}$	Effective power for mechanical work production
NEDC	New European Driving Cycle
CI	Compression Ignition
SI	Spark Ignition
BSFC	Brake-specific fuel consumption
p	Pressure of the gas
V	Volume of the gas
T	Temperature of the gas
R	Universal gas constant
n	Amount of substance of gas
M	Molar mass
R_{specific}	Specific gas constant
p_t	Total air pressure
p_d	Dry air pressure
p_v	Vapor pressure

Reference	**Description**
Φ	Relative humidity
p_{sat}	Saturation vapor pressure
p_0	Sea-level standard atmospheric pressure
T_0	Sea-level standard temperature
L	Temperature lapse rate
h	Altitude
P_s	Standard power
C_f	Power correction factor
P_m	Measured power
P_0	Ambient pressure
P_{0v}	Vapor pressure
CoV	Coefficient of variation
FMEP	Friction mean effective pressure
ECE	Elementary urban driving cycles
$FC_{COLDSTART}$	Fuel consumption at cold start
$FC_{WARMEDUP}$	Fuel consumption when the engine is already warmed up
L_t	Distanced travelled
μ	Dynamic viscosity
v	Kinematic viscosity
SAE International	Society of Automotive Engineers
CCS	Cold cranking simulator
HTHS	High-temperature high-shear
VI	Viscosity index
ASTM	American Society for Testing and Materials
FEAD	Front engine accessory drive
HPAS	Hydraulic power assisted systems
EPHS	Electrically powered hydraulic system
ECM	Engine control module
PRT	Pressure reduction terminal
C_{tire}	Constant to describe the performance of the tire without pressure effect
Pa	Absolute pressure in the tire
T_{amb}	Ambient temperature
T_{ref}	Reference temperature
RR	Rolling resistance
f_r	Resistance coefficient
f_0	Basic coefficient of rolling resistance
f_s	Speed effect coefficient

Reference	**Description**
CAT	Contained air temperature
FR	Rolling force
EPA	Environmental Protection Agency
CNG	Compressed natural gas
SCR	Selective catalytic reduction
DOC	Diesel oxidation catalyst
TWC	Three-Way Catalytic converter
NSR	NO_x Storage Reduction
LNT	Lean NO_x traps
DPNR	Diesel Particulate-NO_x Reduction
SOF	Soluble organic fraction
GPF	Gasoline particulate filters
F0	Road load coefficient
F1	Linear road load coefficient
F2	Quadratic road load coefficient
F_{inertial}	Inertial force
M_{ap}	Mass in the inertia, apparent mass
ω_w	Rotation speed of the wheels
I_w	Wheel inertia
Mf	Equivalent friction moment applied to the driving wheels
R_f	Resistance generated by the drivetrain frictions
AC motor	Alternating current motor
DC motor	Direct current motor
$F_{\text{Dyno\_loss}}$	Dynamometer loss force
A0	Loss force coefficient
A1	Linear loss force coefficient
A2	Quadratic loss force coefficient
AHU	Air handling unit
COP	Coefficient of performance
Q_{evap}	Heat extraction by the evaporator
W_{comp}	Compressor work
T_{c}	Condensation temperature
T_{e}	Evaporation temperature
CD	Chassis dyno
NDIR	Nondispersive infrared
FID	Flame ionization detector
CLD	Chemiluminescence detector

Reference	Description
FTIR	Fourier transform infrared
OBD	On-board diagnostics
PFD	Partial flow dilution
BMD	Bag mini diluter
PCV	Positive crankcase ventilation
UDDS	Urban dynamometer driven schedule
EUDC	Extra urban driving cycle
UNECE	United Nations Economic Commission for Europe
CARB	California Air Resources Board
LEV	Low emission vehicle
CAFÉ	Corporate average standards
NHTSA	National Highway Traffic Safety Administration
GHG	Greenhouse gas
FTP	Federal test procedure
HWFTE	Highway fuel economy test
SEA	Selective enforcement audit
IUVP	In-use verification program
TBC	To be confirmed
ICCT	International Council on Clean Transportation
RW	Reference mass of vehicle
HTHS	High-temperature high-shear
PAS	Power assisted steering
F_{RRd}	Rolling resistance on the roller
F_{RRf}	Rolling resistance on the road
r	Tire radius
R	Roller radius
OEM	Original equipment manufacturer
SoC	State of charge
TA	Type approval
GSI	Gear shift indicator
GTR	Global Technical Regulation
UN	United Nations
PMR	Power-to-mass ratio
WLTC	Worldwide harmonized Light-duty vehicle Test Cycle

About the Authors

Eduardo Galindo, a native of Valladolid, studied in Spain, USA, and UK. He graduated from the University of Valladolid with a Master's degree in Industrial Engineering and, concurrently, obtained another Master's degree in Vehicle Testing from UCLM (University of Castilla La Mancha).

He joined AVL in 2001, focusing his career on testing facilities, and developing an engineering team specialized in fluids and HVAC for engine and vehicle testing.

He has worked on multiple international projects in various countries, including France, Germany, Sweden, Poland, Hungary, Romania, Italy, Russia, China, India, Thailand, Iran, and Brazil.

Still at AVL, he is now working with a European group specialized in chassis dynamometer testing, whose goal is to rapidly improve vehicle performance and emissions.

David Blanco studied in Spain. He graduated from the University of Valladolid with a Master's degree in Industrial Engineering. He worked in the ICE department focused on combustion research and predictive computational algorithms. He started his professional degree in Madrid as project leader in a French company specializing in IT facilities.

He joined AVL in 2012, as an engineer specialized on HVAC and fluids systems for turbocharger and engine testing. He worked as a designer for AVL Turbocharger test beds specialized in turbocharger mapping and thermal shock tests. He subsequently worked full time in the BMW Steyr Turbocharger Technical and Production Centre for a year as AVL residence engineer. His projects were developed in many countries, including Germany, Hungary, Netherland, Spain, Thailand, USA, and Austria where he lived for two years.

Currently, he continues his professional degree at AVL as a member of new group focused on chassis dynamometer testing.

Chris Brace is Professor of Automotive Propulsion and Deputy Director of the Powertrain Vehicle Research Centre at the University of Bath. He is a past Chair of the Automobile Division of the Institution of Mechanical Engineers (IMechE) (2012 to 2014), and is member of the Technical Committee of FISITA, and Members Council of LowCVP. He has published over 110 conference and journal papers with over 50 industrial co-authors. He attracted over £10 million for collaborative research funded by EPSRC, DTi, TSB/Innovate UK, APC, and directly from industry.

After receiving his Bachelor's degree in Engineering from the University of Bath (UK) in 1990, Chris worked at Massey Ferguson for a short period designing gearboxes. He returned to the University of Bath and completed his PhD in 1996. He then joined the Department of Mechanical Engineering at the same university, progressing through reader to professor in 2014. During his time at the University of Bath, Chris has taken charge of the building of the university's research chassis dynamometer facility and its subsequent upgrade in 2014.

Chris now leads a wide portfolio of powertrain systems-based research projects including diesel and SI engine systems, advanced transmission systems and HEV, BEV, and REEV concepts. The research has a common theme around the intensive measurement of the thermodynamic and mechanical processes, leading to greater insight into system behavior under dynamic operating conditions. This research is done in collaboration with industry, most notably Ford Motor Company and Jaguar Land Rover.

Ed Chappell is a Research Associate at the University of Bath. He received his Master's in Automotive Engineering from the University of Bath (UK) in 2009 and, subsequently, completed a PhD at the same university in 2015. His PhD research focused on identifying and investigating the causes of imprecision in CO_2 emissions during legislative emissions testing on chassis dynamometers. During his doctoral program, he worked on a number of collaborative projects with lubricant manufacturers and the Ford Motor Company. He subsequently worked full time in the Ford Dunton Emissions Laboratory at the Ford Motor Company's UK site in Essex.

Ed returned to the University of Bath in 2014 and was responsible for delivering a £2.2 million upgrade to the research chassis dynamometer facility, including the upgrade to 4-wheel drive system, improving the instrumentation suite, exhaust analysers, and implementing a robot vehicle driver. His role currently involves overseeing all experimental test programs on the University of Bath Chassis Dynamometer facility, from vehicle setup to data analysis.

Richard Burke studied at the University of Bath (UK) and obtained his Master's in Automotive Engineering in 2008 and his PhD in 2011. His doctoral research focused on understanding the interactions between diesel combustion, friction, and engine controllers with a particular objective of optimizing emissions and fuel consumption during warm-up. This research also included the topics of high-precision and high accuracy dynamic testing, and design of experiments methodologies for transient cycles.

He joined the Department of Mechanical Engineering at the University of Bath as a research fellow in 2011 and became a lecturer in 2014. Over this period, he worked on a number of projects related to turbocharging and engine downsizing. In particular, he developed new experimental techniques to characterize the turbocharger under transient conditions.

As a lecturer since 2014, he has managed a range of experimental research projects on engine and chassis dynamometers, always with a focus on transient measurement precision, transient testing methods, and reproducibility of real-driving conditions within the test cell environment.

He has published 15 journal papers and over 30 conference papers, is a chartered engineer with the Institution of Mechanical Engineers, and a Distinguished Associate of the Internal Combustion Engine Division of the American Society of Mechanical Engineers (ASME).

Index

Note: Page numbers followed by *'f'* and *'t'* represent figures and tables respectively.

grundsätzlich davon ausgehen, dass die Leistungen der ausgewählten Person erinnerungswürdig sind.[37] Zudem attestieren die an der Gestaltung des Mythos beteiligten Einzelpersonen oder Gruppen der Person eine heldische „Aura“ und damit eine Verkörperung und Konkretisierung von in der Regel abstrakten Wert- und Begriffsvorstellungen.[38] Der mythische Held wird zum historischen Exempel, zu einer Ikone.[39]

Ob von einem Personen-Mythos gesprochen werden kann, hängt somit maßgeblich davon ab, ob sich ein „Kanon“ ausmachen lässt, also ausreichend Darstellungen vorhanden sind, auf die die genannten Kriterien zutreffen.[40] Erst die Kombination sogenannter „Mytheme“, die kleinsten semantischen Einheiten eines Mythos, konstituiert den Mythos.[41] Dabei rekurrieren Mytheme in der Regel auf bereits vorhandene Mythen, im Falle der Fliegerei z. B. auf den Ikarus-Mythos, oder sie stellen die Person in eine „erfundene“ Traditionslinie mit anderen Helden.[42]

Politische Mythen manifestieren sich vor allem durch Gedenktage, z. B. Geburts- oder Todestage, Gedenkschriften wie Zeitungsartikel oder eine Familienbiografie sowie durch (rituelle) Zeremonien oder Kulthandlungen in der Gesellschaft. Ferner erfolgt die Verbreitung des Mythos auch über ikonografische Erinnerungsformen wie Denkmäler oder Bilder, so dass er zu „ein[em] komplexe[n] System von Zeichen oder vielmehr von auslegbaren und deutbaren Symbolen“ wird,[43] die von seinen Anhängern gezielt initiiert werden.[44]

Als Sinnstifter und Erzeuger kollektiver Identität sind Mythen Teil der Erinnerungskultur und der Versuch, Vergangenes zu vergegenwärtigen, zu verstehen und für die Zukunft nutzbar zu machen. Dabei verstehen sich die meisten Mythen lediglich als Sinnangebot, das man annehmen kann, aber nicht muss, und deswegen als „grundsätzlich unanstößig und legitim“.[45] Zugleich sind

37 Vgl. Wulf Wülfing, Die heilige Luise von Preußen, in: Jürgen Link/ders. (Hg.), Bewegung und Stillstand in Metaphern und Mythen, Stuttgart 1984, S.233–275, hier S.235f., und Wulf Wülfing/Karin Bruns/Rolf Parr, Historische Mythologie der Deutschen 1798–1918, München 1991, S.3.

38 Bernhard Giesen, Die Aura des Helden, in: Anne Horner/Ronald Kurt/Jo Reicherts (Hg.), Diesseitsreligion. Zur Deutung der Bedeutung moderner Kultur, Konstanz 1999, S.437–444.

39 Siehe zum Potenzial einer Analyse eines historischen Exempels: Mischa Meier, Rezension zu: Albertz, Anuschka, Exemplarisches Heldentum. Die Rezeptionsgeschichte der Schlacht an den Thermopylen von der Antike bis zur Gegenwart, München 2006, in: H-Soz-Kult, 5.5.2008, http://www.hsozkult. de/publicationreview/id/rezbuecher–9825 (16.6.2014).

40 Vgl. Wülfing/Bruns/Parr, Historische Mythologie, S.4.

41 Claude Lévi Strauss, The Structural Study of Myth, in: The Journal of American Folklore 68 (1955), S.428–444.

42 Vgl. Wülfing/Bruns/Parr, Historische Mythologie, S.5f. Zu den „erfundenen Traditionen“ siehe Eric Hobsbawm, Introduction: Inventing traditions, in: ders./Terence Ranger (Hg.), The Invention of Tradition, 7. Aufl., Cambridge 2009, S.1–14.

43 Speth, Nation, S.32.

44 Siehe hierzu: Münkler, Die Deutschen, S.477–491; Dörner, Politischer Mythos, S.256ff.

45 Tepe, Entwurf, S.57.

Mythen elementarer Bestandteil jeder Gesellschaft, befriedigen sie doch das menschliche Grundbedürfnis, sich zu orientieren und Vergangenheit handhabbar zu machen.[46] Auch der Mythos um Josef Suwelack versuchte einem Grundbedürfnis nach Heldentum und Orientierung nachzukommen. Der Kern dieses Mythos besteht in seiner Lebensgeschichte. Doch erst in der Verbindung von Dekonstrukion (des Mythos) und Rekonstruktion (seines realen Lebens) gewinnt seine Biografie historische Kontur.

Biografie

Wolfram Pyta konstatierte 2009, dass die „deutsche Geschichtswissenschaft ... auf breiter Front die Biographie wiederentdeckt [habe und man] geradezu von einem biographischen Boom" sprechen könne.[47] Doch obschon sich die Biografik in den letzten Jahren wieder großer Beliebtheit erfreut, zählt die historische Biografie weiterhin „zu den schwierigsten Arten der Geschichtsschreibung".[48] Das ist unter anderem dem Umstand geschuldet, dass trotz aktueller Theoriebestrebungen – die lange Zeit ein Desiderat der Biografik darstellten – Definitionen fehlen, die wissenschaftliche und semi- bzw. nicht-wissenschaftliche Biografik eindeutig voneinander abgrenzen.[49]

Die Theorieresistenz hat dieser Gattung von mehreren Forschungsdisziplinen wie den Sozial- und Politikwissenschaften und auch vonseiten der Geschichtswissenschaft Kritik eingebracht.[50] Vor über einer Dekade unter Wissenschaftlern noch als „akademischer Selbstmord"[51] verschrien, gewinnt sie jedoch zunehmend wieder an Akzeptanz, da sich die neuere wissenschaftliche Biografik interdisziplinären Ansätzen öffnet und versucht, die an der Mehrzahl der Biografien geübte

46 Jan Assmann zufolge hat jede Gesellschaft „ihre Mythomotorik, d. h. einen Komplex narrativer Symbole, fundierender und mobilisierender Geschichten, die gegenwartsbedeutend und zukunftsweisend wirken". Jan Assmann, Frühe Formen politischer Mythomotorik. Fundierende, kontrapräsentische und revolutionäre Mythen, in: Dietrich Harth/ders. (Hg.), Revolution und Mythos, Frankfurt a. M. 1992, S. 39–61, hier S. 40.

47 Wolfram Pyta, Biographisches Arbeiten als Methode. Geschichtswissenschaft, in: Christian Klein (Hg.), Handbuch Biographie. Methoden, Traditionen, Theorien, Stuttgart 2009, S. 331–338, hier S. 331.

48 Jacques Le Goff, Ludwig der Heilige, Stuttgart 2000, S. 4.

49 Vgl. die Ausführungen bei Anita Runge, Wissenschaftliche Biographik, in: Klein, Handbuch Biographie, S. 113–121.

50 Texte zur Biografie-Kritik finden sich bei Bernhard Fetz/Wilhelm Hemecker (Hg.), Theorie der Biographie. Grundlagentexte und Kommentar, Berlin/New York 2011. Zu nennen sind in diesem Zusammenhang auch Siegfried Kracauer, Die Biographie als neubürgerliche Kunstform, in: ders., Das Ornament der Masse, Frankfurt a. M. 1970, S. 75–80; Hayden Whites These von der „Fiktion des Faktischen": Hayden White, Der historische Text als literarisches Kunstwerk, in: ders., Auch Klio dichtet, S. 101–122; Michel Foucault zum „Tod des Autors": Michel Foucault, Was ist ein Autor?, in: ders. (Hg.), Schriften zur Literatur, Frankfurt a. M. 1988 [1969], S. 7–31 sowie Bourdieu, Illusion, S. 75–81.

51 Deirde Bair, Die Biographie ist akademischer Selbstmord, in: Literaturen (2001), H. 7/8, S. 38–39.

Kritik konstruktiv zu nutzen. Dementsprechend ist es kein Zufall, dass dieser „Bastard der Geisteswissenschaften" kein eindeutig nur einer Fachwissenschaft zugeordnetes, klar umrissenes Genre darstellt,[52] sondern sich „an der Schnittstelle von Historiographie, Literaturwissenschaft und Literatur" befindet.[53] Spätestens seit den *cultural turns* stehen der Geschichtswissenschaft zudem neue Kategorien wie Geschlecht oder Identität und weitere Analysekonzepte zur Verfügung,[54] die als Ausgangspunkte für eine reflektierte Biografie dienen können. Dabei wird nicht nur der Biografierte, sondern auch der Biograf in den Blick genommen. Von Letzterem wird gefordert, dass er sich selbst Rechenschaft über sein methodisches Vorgehen und seine eigene Zeitgebundenheit ablegt.[55] Nur so kann er sich in eine kritische Distanz zum biografischen Gegenstand bringen, „sich den Helden also vom Leib halten".[56]

Eine explizit theoretische Selbstvergewisserung des eigenen Standpunktes und des methodischen Vorgehens schärft das Bewusstsein dafür, dass die Struktur der Biografie von individuellen Sinnbildungsprozessen, aber auch von den Deutungsabsichten des Verfassers geprägt ist. Diese Transparenz und Offenlegung des eigenen Erkenntnisinteresses zwingt den Biografen zu einer vorüberlegten und fragengeleiteten Wissensentnahme aus den vorliegenden Quellen und Darstellungen über den Untersuchungsgegenstand.[57] Entsprechend verweigern sich die meisten derart theoriegeleiteten modernen Biografien einem chronologischen Ansatz, der vermeintlich geradlinige Entwicklungen in scheinbar zielgerichteten Lebensläufen rekonstruiert.[58]

Einen Gegenentwurf zur klassischen Biografie bildet die sogenannte Anti-Biografie, die über die multiperspektivische Darstellung einer Persönlichkeit „a set of irreconciliable propositions about [a] ... possible character" entwickelt.[59] Thomas Etzemüller zufolge werde dadurch das „biographische Paradox" hervorgehoben:[60]

52 Christian Klein, Einleitung. Biographik zwischen Theorie und Praxis. Versuch einer Bestandsaufnahme, in: ders. (Hg.), Grundlagen der Biographik. Theorie und Praxis des biographischen Schreibens, Stuttgart/Weimar 2002, S.1–22, hier S.1.

53 Ebd., S.12.

54 Doris Bachmann-Medick, Cultural Turns. Neuorientierungen in den Kulturwissenschaften, 5. Aufl., Reinbek bei Hamburg 2014.

55 Vgl. dazu Simone Lässig, Die historische Biographie auf neuen Wegen?, in: Geschichte in Wissenschaft und Unterricht (GWU) 60 (2009), S.540–553.

56 Christian Meier, Die Faszination des Biographischen, in: Frank Niess (Hg.), Interesse an der Geschichte, Frankfurt a.M. 1989, S.100–111, hier S.101.

57 Vgl. Bernhard Fetz, Grundfragen biographischen Schreibens. Zur Bedeutung der Quellen, in: Christian Klein (Hg.), Handbuch Biographie. Methoden, Traditionen, Theorien, Stuttgart 2009, S.433–438.

58 Das Potenzial dieses Verfahrens führt Karl Heinrich Pohl am Beispiel von Gustav Stresemann vor. Pohl versteht die neueren biografischen Ansätze als Chance, den bisherigen Forschungsstand über Stresemann zu erweitern. Karl Heinrich Pohl, Zur Konstruktion einer neuen Biographie, in: GWU 60 (2009), S.554–567.

59 Ein Beispiel für eine solche Anti-Biografie ist: David E. Nye, The Invented Self. An anti-biography,

„Eine Biographie gilt als Konstruktion, die gleichwohl eine nicht fassbare Realität abbildet; die ‚anti-biographische' Dezentrierung bedarf eines organisierenden Eigennamens; – der multiple Charakter einer Biographie kann nur mithilfe eines Kohärenz stiftenden Genres dargestellt werden".

Gleichwohl ist Kohärenz nicht gleichbedeutend mit der Konstruktion einer geschlossenen „Ganzheitlichkeit", wenngleich dies häufig genug eine Vorstellung ist, von der sich Biografen leiten lassen.[61]

Die Sozialwissenschaften haben mit ihren Untersuchungen zu Generationszugehörigkeit und Milieubildung wesentlich zum Verständnis von Handlungsmöglichkeiten, von Chancen und Grenzen beigetragen. Gerade bei der Konstruktion einer modernen Biografie sind die Bedingungen, unter denen die Individuen agieren (können), nicht zu unterschätzen. So resümiert Frank Becker in Bezug auf den beruflichen und gesellschaftlichen Status einer Person: „Karriereverläufe hängen ganz wesentlich von kollegialen, freundschaftlichen oder verwandtschaftlichen Kontakten ab, jedenfalls von Personen, die bereit sind, zu helfen, zu fördern, ein Wort einzulegen."[62]

Aus diesem Grund scheint es geboten, ein Analyseinstrument zu verwenden, das der Ambivalenz von struktureller Determiniertheit und kontingenten Handlungsspielräumen Rechnung trägt.. Wenngleich der bzw. die BiografIn versucht, theoretisch gerahmt Handlungen und Ereignisse zu erklären und ein Leben zu erzählen, kann er bzw. sie sich letztlich allerdings auch nicht dem Vorwurf entziehen, eine „andere Art der Kontingenzbekämpfung" – wie sie wohl jeder Form der Geschichtsschreibung letztendlich eigen ist – zu praktizieren.[63]

Doch das Potenzial einer reflektierten und selbstreflexiven Biografie kann gerade darin liegen, dem Leser Einblicke in die „Konstruktions- und Dekonstruktionsprozesse und die mühevolle Annäherung an Wahrnehmung und Erinnerung, Vergessen, Umdeutung und Überschreibung" zu gewähren.[64] Gelingt es,

from documents of Thomas A. Edison, Odense 1983, Zitat S.23–24. Siehe auch den Rückblick Nyes: David E. Nye, Nach Thomas Edison. Rückblick auf eine Anti Biographie [2003], in: Bernhard Fetz/Wilhelm Hemecker (Hg.), Theorie der Biographie. Grundlagentexte und Kommentar, Berlin/New York 2011, S.347–360; zudem den Betrag von Bernhard Fetz, Struktur statt Psychologie. Die „Anti-Biographie" als biographisches Modell, in: ebd., S.361–366.

60 Siehe auch nachfolgendes Zitat: Thomas Etzemüller, Biographien: lesen – erforschen – erzählen, Frankfurt a. M. 2012, S.164.

61 Vgl. Caitríona Ní Dhúill, Widerstand gegen die Biographie: Sigrid Weigels Ingeborg-Bachmann-Studie, in: Wilhelm Hemecker (Hg.), Die Biographie – Beiträge zu ihrer Geschichte, Berlin/New York 2009, S.43–68, hier S.43.

62 Frank Becker, Perspektiven einer Carl-Diem-Biographie, in: BIOS 18 (2005), H. 2, S.157–168, hier S.161.

63 Christian Klein, „Biographismus ist eine andere Art der Kontingenzbekämpfung". Der Literaturkritiker Hubert Winkels über die „gute Biographie", ihr Verhältnis zu Roman und Realität sowie den Erfolg des Genres, in: Non Fiktion. Arsenal der anderen Gattungen 8 (2013) H. 1, S.105–111.

64 Lässig, Die historische Biographie auf neuen Wegen?, S.549. Ähnlich äußert sich auch der Literaturkritiker Hubert Winkels über die „gute Biographie": „Biographien sind gut, wenn der Prozess der Selbst- und Fremderfindung, der Selbst- und Fremdmythologisierung als Zugangsbedingung

das Individuum in seinen Strukturen sichtbar zu machen, ein Leben als Summe von Selbst- und Fremdzuschreibungen zu analysieren und sich gleichzeitig als Teil dieser Konstruktion zu begreifen, kann eine solche biografische Darstellung sowohl veranschaulichen, wie sich beispielsweise Josef Suwelack vor dem zeitgenössischen gesellschaftlichen und politischen Hintergrund „erfand", als auch, wie er in der zeitgenössischen Wahrnehmung und kollektiven Erinnerung damals und heute memoriert wurde und wird.

Das Habitus-Konzept Pierre Bourdieus

Das Konzept des Habitus gehört zu den wesentlichen Kategorien der Sozialwissenschaften – ebenso wie Vorstellungen vom sozialen Raum und Feld, vom ökonomischen und kulturellen Kapital sowie von der symbolischen Gewalt –, die Pierre Bourdieu als Analyseinstrumente für soziales Handeln entwickelt hat. Insbesondere durch die Kategorie des Habitus erfolgte ein Paradigmenwechsel in den Sozialwissenschaften, eine „Abkehr von einer Vorstellung vom sozialen Handeln, die dieses als Resultat bewusster Entscheidungen bzw. als das Befolgen von Regeln begreift."[65]

Als erste Annäherung an die Bedeutung des Begriffs lässt sich feststellen, dass der Habitus Gewohnheiten umfasst, die in Interaktion mit der sozialen Umwelt erlernt wurden.[66] Durch sein physisches In-der-Welt-Sein agiert das Individuum von Geburt an in der Gesellschaft, ist somit in gesellschaftliche Zusammenhänge integriert. „Gesellschaft und Individuum erzeugen sich gegenseitig",[67] wodurch sich soziale Realität „in den Sachen und in den Köpfen, in den Feldern und in den Habitus, innerhalb und außerhalb der Akteure" manifestiere. Auf diese Weise wird das handelnde Subjekt sowohl von der Gesellschaft geprägt und geformt, umgekehrt interagiert es aber auch mit der Welt, indem es sie in seinem Inneren rezipiert, die Welt für sich selbst subjektiv verständlich macht.[68]

Um diese Interaktionen analysieren zu können, führte Bourdieu unter anderem die Begriff „sozialer Raum", „Feld", „Klasse" und „Geschlecht" ein. Dabei platziert Bourdieu das Individuum in einen sozialen Raum, der zum einen geprägt ist von Unterschieden, die sich in ökonomischem, hier vor allem im Besitzstand, und kulturellem Kapital, z. B. im Bildungsgrad, niederschlagen. Zum

und Erkenntnisregel durchschaut und diskutierbar gemacht wird" und sich der Biograf eine „narrative Skepsis" bewahre. Klein, Kontingenzbekämpfung, S.107 und 108.

65 Beate Krais/Gunter Gebauer, Habitus, 3. Aufl., Bielefeld 2010, S.5.

66 Vgl. Gerhard Fröhlich/Boike Rehbein, Bourdieu Handbuch. Leben – Werk – Wirkung, Stuttgart/Weimar 2009.

67 Ebd., S.78.

68 Vgl. Krais/Gebauer, Habitus, S.74–76.

anderen ist der soziale Raum ein „Raum von Beziehungen",[69] der die Laufbahn einer Person für den gesamten Zeitraum ihres Lebens umfasst. Es geht dabei um Aufstieg oder Abstieg, um den Lebensstil, der die Klassen voneinander trennt.

Klassen sind für Bourdieu, im Unterschied zu Marx, nicht einfach vorhanden, sondern werden erst durch das ritualisierte Alltagshandeln von Individuen konstruiert.[70] Insofern können Klassen durch Unterschiede in der Lebensführung variieren, die Bourdieu zufolge wiederum auf den Habitus zurückzuführen und ohne ihn nicht erklärbar seien.[71]

> „Wer in der Wohlhabenheit, in ökonomischem und kulturellen Reichtum, in der damit gegebenen Sicherheit und Freiheit aufgewachsen ist, entwickelt nicht nur einen anderen Geschmack, sondern auch ein anderes Verhältnis zur Welt als jemand, der von frühester Kindheit an mit Not und Notwendigkeit ... ebenso wie mit der Notwendigkeit des Sich-Wegduckens und Stillhaltens gegenüber den Einflussreichen und Mächtigen konfrontiert war."[72]

Der soziale Ort formt den Habitus und bringt innere Haltungen, sogenannte Dispositionen, und Handlungsweisen sowie bestimmte Wünsche und Umgangsformen mit der Welt hervor, die wiederum zur Reproduktion der eigenen Klasse führen.[73] Ferner impliziert das Eingebundensein in einen sozialen Raum spezifische Möglichkeitsräume.[74]

Das Habitus-Konzept ist ein Konstrukt, das erlaubt, die Interaktion von Individuum und Gesellschaft zu beschreiben. Mit seiner Hilfe können Handlungen von Individuen als Resultat von gemachten Erfahrungen und zukünftigen Erwartungen betrachtet werden, die erst im Wechselspiel mit der sie umgebenden Gesellschaft entstehen und an sie rückgekoppelt sind. Der Habitus ist gewissermaßen ein „Sammelbecken" von Erfahrungen, die abgespeichert und auf neue Situationen in dann anderen sozialen Feldern erfolgreich oder erfolglos angewendet werden. Dabei versteht Bourdieu diese sozialen Felder als „Kräftefelder",[75] die jeweils eine eigene Logik besitzen und in denen es um Macht- und Einflussverteilung geht. Ähnlich wie in einem Spiel gibt es Spielregeln und bedarf es gewisser Fähigkeiten, es existieren Ziele und Einsätze. Man tritt in ein soziales Feld jedoch nicht sofort mit einem kompatiblen Habitus ein, sondern muss erst lernen, nach den Regeln des Feldes zu spielen. Das bedeutet, „dass man unzählige Akte der Anerkennung erbringt, dass man in das Spiel investiert, dass man sich anstrengt alles richtig zu machen, dass man versucht dahinter zu kommen, ... welche Positionen wichtig sind, welche weniger, welche expliziten und impliziten Regeln gelten".[76]

69 Vgl. ebd., S.36.
70 Vgl. ebd., S.35–36.
71 Vgl. ebd., S.41.
72 Ebd., S.43.
73 Zum Begriff der Disposition siehe Fröhlich/Rehbein, Bourdieu Handbuch, S.73–75.
74 Vgl. Krais/Gebauer, Habitus, S.46.
75 Vgl. ebd., S.56; Fröhlich/Rehbein, Bourdieu Handbuch, S.100.
76 Krais/Gebauer, Habitus, S.62.

Dieses Spiel entsteht über die Praxis der Akteure und kann durch hinzukommende Akteure oder Veränderungen in der Machtstruktur im „Spielverlauf" beeinflusst werden: „Die Gewichte im rationalen Gefüge verschieben sich, neue Konventionen bilden sich heraus, das *Spiel* wird anders gespielt."[77] Trotz ihrer Verschiedenheit verbindet die Akteure der praktische Glaube, die *illusio*, an das Spiel. Sie akzeptieren die Logik des Feldes und lernen über ihre Teilnahme den Einsatz und Kriterien für „elegantes oder auch nur erfolgreiches Spiel kennen".[78] Über diese Praxis werden die Regeln des soziale Feld Teil des Habitus und nach und nach als selbstverständlich erfahren und somit Teil der Identität der Person.[79]

Der Habitus rüstet die Akteure allerdings nur für bestimmte Kontexte und Handlungsmöglichkeiten, auf deren Grundlage neue Situationen interpretiert werden. Passt sich der Habitus dem Feld nicht an, ist Fehlverhalten – Hysteresis – die Folge. Die Karten – um beim Spiel zu bleiben – werden von Zeit zu Zeit neu gemischt; eine Mischung aus Reproduktion und Variation ist die Folge.[80] Da Bourdieus Theorie „als relativ offenes Konzept angelegt" ist,[81] lassen sich über die verschiedenen Kategorien kulturelle Dimensionen integrieren, die „nicht nur als geschlossene Bedeutungssysteme, sondern als soziale Praxis" betrachtet werden können.[82]

Das Habitus-Modell ist aus mehreren Gründen aufschluss- und hilfreich bei der Arbeit an einer Biografie. Mit seiner Hilfe ist es möglich, die Krux der Chronologie, zumindest stellenweise, zugunsten einer an sozialen Feldern orientieren Analyse aufzulösen und damit auch die rückblickend scheinbar zwangsläufigen (teleologischen) Kausalitäten aufzubrechen. Im Fall von Josef Suwelack sind beispielsweise vier soziale Felder denkbar: „Familie", „Fluglehrer-, Konstrukteur- und Flugzeugfirmen-Zeit", „Flugwettbewerbe" und „Fliegerei im Ersten Weltkrieg/Feldfliegerabteilung". Anhand dieser Felder kann jeweils ein Teil der Persönlichkeit – des Habitus – Suwelacks untersucht und mit den zeitgenössisch wirksamen Strukturen verknüpft werden. Ausgehend von der Eingebundenheit eines Individuums in seine Zeit, kann die Lebensgeschichte des westfälischen Piloten Suwelack überdies als Fenster in eine andere Zeit dienen.[83]

77 Ebd., S.59, Kursivsetzung im Original.
78 Ebd., S.62.
79 Ebd., S.59.
80 Vgl. ebd., S.80f.
81 Schwingel, Bourdieu, S.59.
82 Welskopp, Väter, S.194.
83 Simon Karstens, Die Summe aller Wahrheiten und Lügen. Ein Erfahrungsbericht zur geschichtswissenschaftlichen Biographie, in: BIOS 24 (2011), H. 1, S.78–97, hier S.81. Karstens führt den Wert einer Biografie, die das Individuum in Interaktion mit seiner Umwelt analysiert, am Beispiel des Publizisten, Juristen und Staatsmannes Joseph von Sonnenfels vor: Simon Kastens, Lehrer – Schriftsteller – Staatsreformer. Die Karriere des Joseph von Sonnenfels (1733–1817), Köln/Weimar/Wien 2011.

2. Fragestellungen und Forschungsstand

Ziele

Das Innovative des hier beschriebenen Ansatzes liegt in der Herstellung einer reflektierten Distanz zwischen BiografIn und Biografiertem sowie in einer Auseinandersetzung mit dem biografischen Genre. Dennoch ist es sicherlich nicht von der Hand zu weisen, dass sich auch die vorliegende Biografie an zeitbezogenen, d.h. heutigen Wertmaßstäben orientiert. Diese Arbeit verfolgt aber insbesondere das Ziel, die Mythen über die ersten Flieger zu dekonstruieren und die dahinterstehenden Erzählabsichten zu analysieren. So ist beispielsweise zu klären, inwiefern das Narrativ vom „ritterlichen" und „ehrenvollen" Luftkampf an ältere Erzähltraditionen aus dem deutsch-französischen Krieg 1870/71 anknüpft, während durch das anonyme Massensterben des Grabenkrieges der im hohen Maße technisierten Weltkriege der Einzelne namen- und bedeutungslos wurde. Die Flieger verstanden sich weiterhin als Einzelkämpfer, die sich mit messbaren Erfolgen, etwa durch Bombenabwürfe getroffene Ziele oder Abschüsse von Flugzeugen, „einen Namen machen" konnten.[84]

Durch den Vergleich verschiedener Narrative in den Texten über Josef Suwelack können zudem typische erzählerische Muster und Mythen aufgezeigt und beschrieben werden, die neue und andere Perspektiven auf nicht erzählbare oder in den Augen der jeweiligen VerfasserInnen nicht erzählenswerte Aspekte der Lebensgeschichte Suwelacks ermöglichen.

Ebenso ermöglicht der an Handlungsfeldern ausgerichtete Ansatz, Josef Suwelack nicht nur auf seinen Kriegseinsatz zu reduzieren, sondern gleichermaßen sein Vorkriegsleben und sein außerhalb des Krieges verlaufenes Leben (Ingenieurtätigkeit, Wettbewerbsteilnahme, Heimaturlaub, Familie etc.) zu betrachten. Zudem erlaubt diese Vorgehensweise, Handlungen und Denkmuster sowohl in den jeweiligen Feldern zu beschreiben, als auch Verbindungen zwischen ihnen aufzuzeigen. Jeder dieser sozialen Räume bildet einen eigenständigen Bereich ab, der nach bestimmten Gesetzmäßigkeiten organisiert ist.[85] So verläuft die Tätigkeit des Flugpioniers zwischen einem weitgehend isolierten Handeln als Konstrukteur und Pilot und einer dazu parallel bestehenden bedeutenden Flugschau-Wettbe-

84 Siehe zum Verlust des traditionellen Heldenmodells und zu den neuen Fliegerhelden die Ausführungen bei Manuel Köppen, Das Entsetzen des Beobachters. Krieg und Medien im 19. und 20. Jahrhundert, Heidelberg 2005, S.228f.; vgl. auch Jörg Bernig, Anachronistisches Kriegsbild, Selbstinszenierung und posthume Heroisierung. Manfred von Richthofen: Der rote Kampfflieger (1917), in: Thomas F. Schneider/Hans Wagener (Hg.), Von Richthofen bis Remarque: deutschsprachige Prosa zum 1. Weltkrieg, Amsterdam/New York 2003, S.97–111, hier S.101ff.

85 Siehe zu diesem Gedanken Reinhart Koselleck, „Erfahrungsraum" und „Erwartungshorizont" – zwei historische Kategorien, in: ders., Vergangene Zukunft. Zur Semantik geschichtlicher Zeiten, Frankfurt a.M. 1995, S.349–375.

werbskultur. Das Feld „Fliegerabteilung“ ist wiederum über private Feldpostbriefe und Fotos von der Front mit Suwelacks Familie verknüpft, so dass unterschiedliche Schwerpunkte innerhalb, aber auch zwischen den Feldern gesetzt werden können.

Ein weiterer Schwerpunkt soll auf die Analyse der verschiedenen Kontexte gelegt werden, in denen die Fliegerei den Piloten sowohl Frei- wie auch Spielräume zur Verfügung stellte, in denen sich die Piloten beruflich und privat verwirklichen konnten.

Neben den privaten Aufzeichnungen und sonstigen Überlieferungen zur Flugleidenschaft von Suwelack soll dem Fliegerumfeld und den dazugehörigen Publikationsorganen des Flugwesens, wie flugtechnischen Zeitschriften oder dem speziellen „Flugschau-Marketing“, größere Aufmerksamkeit geschenkt werden. Denn erst durch die ausführliche Berichterstattung und Werbung in den entsprechenden Zeitungen und Zeitschriften sowie durch das feingliedrige Beziehungsgeflecht zwischen den Piloten, den Flugzeugfirmen, den Sponsoren und den Flugzeugabnehmern, so die These, wurde die Fliegerei zu einem lukrativen Geschäft nicht nur für die Piloten, sondern auch für Firmen, Investoren und Sponsoren.

Eine Analyse dieser Strukturen und Beziehungsgeflechte, des „Flieger-Feldes“, hilft zu verstehen, nach welchen Regeln ein sozialer oder wirtschaftlicher Auf- oder Abstieg erfolgte, welche Formen der Konkurrenz und Kooperation unter den Piloten, Konstrukteuren und Flugzeugfirmen bestanden und welche Faktoren ausschlaggebend waren, um in der Fliegerbranche zu reüssieren. In diesem Zusammenhang stellt sich auch die Frage, inwiefern die Fliegerei und deren Akteure im hohen Maße durch das Prinzip der hegemonialen Männlichkeit definiert wurden.

Da sich die Lebensdaten Josef Suwelacks (1888–1915) vornehmlich auf die Zeit vor dem Ersten Weltkrieg beziehen, wird sich ein Großteil dieser Arbeit mit seiner Zeit als Flugpionier, -lehrer und Konstrukteur befassen. Für die Zeit seines militärischen Einsatzes im Ersten Weltkrieg bietet der vorgestellte biografische Ansatz Gelegenheit, Einblicke in Arbeitsabläufe und -routinen einer Feldfliegerabteilung zu erhalten. Anhand der Feldpostkorrespondenz Josef Suwelacks und einer umfangreichen Fotosammlung können Mentalitätswandel und Brüche, aber auch Kontinuitäten in seiner Einstellung nachvollzogen und diese auf Bezüge zur politischen Kultur des Deutschen Kaiserreiches hin untersucht werden. Von der „Front“ geben die Quellen hingegen nur bedingt Auskunft, ebenso über die Logistik der Fliegertruppen sowie über Beziehungsgefüge und Aktivitäten jenseits des eigentlichen Kampfgeschehens. Zahlreiche Fotos Suwelacks von seiner Militärzeit vermitteln hingegen einen guten Einblick in die Freizeitgestaltung der Flieger an der Front, Fotos, die zum Teil touristische Motive zeigen, aber auch aus dem „Wunsch nach Dokumentation und Illustration der Worte“ entstanden.[86]

86 Zur Vorstellung vom Krieg als touristisches Motiv siehe Bernd Hüppauf, Kriegsfotografie und die

Der biografischen Studie sind durch die Konzentration auf *ein* Leben erkenntnistheoretische Grenzen gesetzt: Zum einen kann Josef Suwelack nur stellvertretend für einen bestimmten Flieger-Typus des Kaiserreiches „gelesen" werden; zum anderen gehörte Suwelack einer spezifischen Generation an, einer „Gründergeneration", die in einer Zeit technischer Neuerungen und einer Kultur der Machbarkeit aufwuchs.[87] Dennoch können im Vergleich mit anderen Fliegerkarrieren sowohl Besonderheiten wie Merkmale aufgezeigt werden, die die Fliegerszene im Allgemeinen charakterisierten. Suwelacks Leben lässt sich somit als eine Art Sonde lesen, an deren Beispiel sich zeitgenössische Vorstellungen und Maßstäbe veranschaulichen lassen.

Ziel dieser Studie ist es, Josef Suwelack als Akteur in seinen Strukturen sichtbar zu machen und die Wechselwirkungen zwischen Individuum und Gesellschaft zu rekonstruieren, um den Piloten Suwelack in seiner Zeit zu positionieren und generelle Charakteristika eines „Flieger-Feldes" der Jahrhundertwende auszumachen.

Zum Stand der Forschung

Anlässlich des einhundertsten Jahrestages des Ersten Weltkrieges wurden in den unterschiedlichen (Fach-)Medien neue oder bereits bekannte Aspekte über die „Urkatastrophe des 20. Jahrhunderts" veröffentlicht.[88] Mit dem Projekt „europeana 1914–1918" wurden darüber hinaus über 700.000 digitalisierte Objekte aus Museen, Bibliotheken und Archiven online kostenlos bereitgestellt.[89] Seit Ende 2008 in Betrieb, hat sich diese virtuelle Bibliothek zur Aufgabe gemacht, das kulturelle Gedächtnis Europas zu bewahren. Das Angebot umfasst über 660 Stunden Filmmaterial zum Ersten Weltkrieg sowie weitere von Privatpersonen zur Verfügung gestellte Zeitdokumente wie Tagebücher, Fotoalben, Briefe etc. und archivarische Quellen. Unter ihnen befindet sich auch ein Fotoalbum eines deutschen Fliegers, der bei einer Fliegerabteilung an der Westfront stationiert war.[90]

Erfahrung des Ersten Weltkriegs, in: Barbara Naumann (Hg.), Vom Doppelleben der Bilder. Bildmedien und ihre Texte, München 1993, S. 29–50, Zitat S. 31.

87 Vgl. Kurt Möser, Fahren und Fliegen in Frieden und Krieg. Kulturen individueller Mobilitätsmaschinen 1880–1930, Heidelberg/Ubstadt-Weiher/Neustadt a. d. W./Basel 2009, S. 79f.

88 George F. Kennan, The Decline of Bismarck's European Order, Franco-Russian relations, 1875–1890, Princeton 1979, S. 3.

89 Vgl. www.europeana.eu, online seit November 2008 (Stand Juli 2017).

90 Joachim Leu, http://www.europeana1914-1918.eu/en/contributions/1191#prettyPhoto, online seit 2012. Das Album enthält unter anderem Luftaufnahmen, Bilder von Flugzeugen und Mannschaften (12.8.2014). Vgl. auch die Ausstellung der Stiftung Zollverein „1914 – Mitten in Europa" (30.4.–26.10.2014): LVR (Hg.), http://www.rheinland1914.lvr.de/de/ausstellungen/essen.html, online seit 2013 (12.8.2014). Siehe auch Heinrich Theodor Grütter/Walter Hauser (Hg.), 1914 – Mitten in Europa. Die Rhein-Ruhr-Region und der Erste Weltkrieg. Katalogbuch zur Ausstellung

Im Vergleich zu den umfangreichen Forschungen zum Heer und zur Marine findet sich allerdings bis heute eine nur marginale wissenschaftliche Aufarbeitung des Flugwesens und der Luftkriegsgeschichte dieser Jahre.[91] Nach wie vor dominieren auf diesem Gebiet vor allem populärwissenschaftliche Arbeiten, die enthusiastisch und detailliert, jedoch zumeist unreflektiert lediglich die Kriegstechnik rekonstruieren. Allein in den letzten Jahren entstanden zahlreiche Studien zu einzelnen Piloten, fliegenden Verbänden, zur Luftkriegsführung und zu Flugzeugen aus nationalen und transnationalen Perspektiven.[92] Typisch ist eine Vielzahl unkommentierter Illustrationen in diesen Publikationen sowie bei biografischen Arbeiten ein Hang zur Legendenbildung. Christian Kehrt, der sich in seiner erfahrungsgeschichtlichen Studie erstmals ausführlich mit Militärpiloten auseinandergesetzt hat, spricht in diesem Zusammenhang von einer „Verengung des Blickes auf die Helden des Luftkrieges",[93] wobei eine wissenschaftliche Aufarbeitung bislang ausgeblieben sei.

Reflektierte historische Forschungen nutzen hingegen Ansätze der neuen Kulturwissenschaft und gelangen zu weit differenzierteren und aussagekräftigeren Urteilen.[94] Als ertragreich haben sich dabei unter anderem erinnerungskulturelle

des LVR-Industriemuseums und des Ruhr Museums auf der Kokerei Zollverein 30. April bis 26. Oktober 2014, Essen 2014. Belgien erinnert z. B. von 2014 bis 2018 an den Ersten Weltkrieg und organisiert Veranstaltungen zu Kriegseintritten, Schlachtverläufen etc. Siehe z. B. be14–18.be, online seit 2014 (12.8.2014). Auch in Deutschland folgten ab 2015 weitere Ausstellungen: Volksbund deutscher Kriegsgräberfürsorfe e.V. (Hg.), http://100-jahre-erster-weltkrieg.eu/nc/veranstaltungen.html, online seit 2014 (12.8.2014).

91 James R. Hansen, Aviation in the wider view, in: Technology and Culture 30 (1989), S.643; Edward L. Homze, German Military Aviation. A guide to the literature, New York 1984, S.111; Harald Potempa, Die Königlich-Bayerische Fliegertruppe: 1914–1918, Frankfurt a. M. 1997; ders., Die Königlich-Bayerische Fliegertruppe als Teil der Deutschen Luftstreitkräfte bis 1918, in: Blätter zur Geschichte der Deutschen Luft- und Raumfahrt XIX (2013), S.11–22; ders., Krieg in der Luft, in: Markus Po?hlmann/Harald Potempa/Thomas Vogel (Hg.), Der Erste Weltkrieg 1914–1918. Der deutsche Aufmarsch in ein kriegerisches Jahrhundert, München 2013, S.89–109; Lothar Saupe, Palästina von oben. Luftbilder der bayerischen Fliegerstaffel 304 im Ersten Weltkrieg aus Beständen des Bayerischen Hauptstaatsarchivs. Begleitausstellung des Bayerischen Hauptstaatsarchivs zur Ausstellung „Mission Palästina", 18. Juni bis 15. September 2006, München 2006; Dieter Storz, Kriegsbild und Rüstung vor 1914, Herford/Berlin/Bonn 1992; Helmuth Trischler, Luft- und Raumfahrtforschung in Deutschland 1900–1970, Frankfurt a. M./New York 1992, S.31; Militärgeschichtliches Forschungsamt (Hg.), Die Militärluftfahrt bis zum Beginn des Ersten Weltkrieges, 2 Bde. Frankfurt a. M. 1965/66. Oliver Frei, Bilder für den Krieg. Die Tagebücher des Lichtbildoffiziers Erich Ewald 1914–1918/19. Ein Beitrag zur Geschichte der Luftaufklärung und des militärischen Luftbildwesens, Berlin 2011.

92 Folgende Titel geben einen Eindruck: Bryce D. Gibby, Valiant Young Men: heroes of flight, Wilmington 2006; Patrick de Gmeline, Baron rouge et Cigogne blanche: Manfred von Richthofen et René Fonck, Paris 2011; Walter M. Pieters, Above Flanders' Fields: a complete record of the the Belgian fighter pilots and their units during the Great War, 1914–1918, London 1998; Georges Pagé, L'aviation francaise. 1914–1918, Paris 2011; David W. Lewis, Eddie Rickenbacker, an American Hero in the Twentieth Century, Baltimore 2008; Bernard Palmirie, René Fonck. L'as des as, l'homme, Paris 2007.

93 Kehrt Moderne Krieger, S.14.

94 Siehe z. B. Christian Kehrt, „Schneid, Takt und gute Nerven". Der Habitus deutscher Militärpilo-

Fragestellungen erwiesen, die durch eine Dekonstruktion vorhandener Narrative auf die Bildung von Mythen und die nachhaltige Wirkmächtigkeit der Kriegspropaganda seit dem Ersten Weltkrieg hinweisen konnten.[95]

So spielt beispielsweise der ritterliche Zweikampf in der Luft ein zentrales Motiv in den Narrationen über den Ersten Weltkrieg, obgleich er nicht unbedingt den Kriegserfahrungen der Mehrheit der Militär- und Zivilpiloten entsprach, war doch deren Hauptaufgabe „im Ersten Weltkrieg nicht der Luftkampf, sondern die Durchführung der Luftaufklärung."[96] Nicht nur an diesem Beispiel lässt sich zeigen, dass zwischen der fach- und der populärwissenschaftlichen Forschung ein Spannungsfeld besteht, das alle Nuancen zwischen reflektierter Urteilsbildung und unreflektierter Legendenbildung aufweist.[97]

Auch die neuere Kulturgeschichte interessiert sich für den Heldenkult als Forschungsfeld. In diesem Kontext liegen auch bereits einige Studien zur Symbolik und medialen Darstellung von Piloten vor.[98] Peter Fritzsche hat auf einer breiten

ten und Beobachter im Kontext technisch strukturierter Handlungszusammenhänge, 1914–1918, in: Technikgeschichte 72 (2005), S.177–201; ders., „A horribly beautiful picture". Aesthetic dimensions of aerial warfare, in: Reinhard Heil/Andreas Kaminiski/Markus Stippak et al. (Hg.), Tensions and Convergences. Technological and aestetic transformations of society, Bielefeld 2007, S.337–349; Kurt Möser, Schlachtflieger 1918 – ein technisches Waffensystem im Kontext, in: Technikgeschichte 77 (2010), S.185–230; John H. Morrow, Knights of the Sky. The Rise of Military Aviation, in: Frans Coetzee/Marilyn Shevin-Coetzee (Hg.), Authority, Identity and the Social History of the Great War, Oxford/Providence 1995, S.305–324; Stefanie Schüler-Springorum, Vom Fliegen und Töten. Militärische Männlichkeit in der deutschen Fliegerliteratur, 1914–1939, in: Karen Hagemann (Hg.), Heimat – Front. Militär und Geschlechterverhältnisse im Zeitalter der Weltkriege, Frankfurt a. M. 2002, S.208–233; Bernig, Anachronistisches Kriegsbild, S.97–111; Jörg Lehmann, Imaginäre Schlachtfelder. Kriegsliteratur in der Weimarer Republik, o. O. 2014.

95 Vgl. z. B. Garry Campion, The Good Fight. Battle of Britain in wartime propaganda and the few, Basington 2008; Möser, Fahren und Fliegen. Einen guten Überblick über aktuelle Forschungskonzepte und Methoden der Zeitgeschichte, die auch für andere Epochen nutzbar gemacht werden können, bieten: Frank Bösch/Jürgen Danyel (Hg.), Zeitgeschichte – Konzepte und Methoden, Göttingen 2012.

96 Kehrt, Schneid, S.177.

97 Exemplarisch die Diskussion um den Tod des amerikanischen Piloten Frank Luke im Ersten Weltkrieg: Tobias Arand, Der Nachlass des Leutnants Bernhard Mangels aus Münster. Exemplarische Überlegungen zur inhaltlichen und formalen Erschließung von Kriegsfotoalben deutscher Veteranen des Ersten Weltkrieges, in: Westfälische Forschungen 58 (2008), S.421–439; Stephen Skinner, The Stand. The final flight of Lt. Frank Luke jr., o. O. 2008.

98 Burkhard Fuhs, Fliegende Helden. Die Kultur der Gewalt am Beispiel von Kampfpiloten und ihren Maschinen, in: Rolf Brednich/Walter Hartinger (Hg.), Gewalt in der Kultur. Vorträge des 29. Deutschen Volkskongresses, Bd. 2, Passau 1992, S.705–720; René Schilling, Der Körper des „Helden". Deutschland 1813–1945, in: Bielefelder Graduiertenkolleg Sozialgeschichte (Hg.), Körper Macht Geschichte, Geschichte Macht Körper. Körpergeschichte als Sozialgeschichte, Bielefeld 1999, S.119–140; Florian Schnürer, „Nun hat der große Flieger dich geholt ..." – Die Totenfeiern für die „Ritter der Lüfte" als transnationales Medienereignis, in: Kulturwissenschaften exemplarisch. Gießener Forschungsbeiträge zu acht Kernkompetenzen, Trier 2011, S.167–192; Schüler-Springorum, Vom Fliegen und Töten, S.208–233; Martin Stern, Heldenfiguren im Wagnissport. Zur medialen Inszenierung wagnissportlicher Erlebnisräume, in: Thomas Alkemeyer/Bernhard Boschert/Robert Schmidt et al. (Hg.), Aufs Spiel gesetzte Körper. Aufführungen des Sozialen in Sport und populärer Kultur, Konstanz 2003, S.37–54.

Quellenbasis von Zeitungen, Wochenmagazinen und spezieller Fliegerliteratur die kulturelle Konstruktion der Fliegerei untersucht, der er eine besondere Wirkmächtigkeit bei der Ausbildung des Nationalismus zuspricht. Nach Fritzsche sind Flugzeuge und Luftschiffe in besonderer Weise Symbole des neuen deutschen Nationalismus, mit denen sich die – nach Benedict Anderson so definierten – *imagined communities* identifiziert und Luftschiffer wie Ferdinand Graf von Zeppelin zu nationalen Identifikationsfiguren erhoben hätten.[99] Sabine Höhler konnte zudem in ihrer Studie über Luftfahrtforschung nachweisen, dass die Luftfahrt am Beginn des 20. Jahrhunderts dazu genutzt wurde, Männlichkeitsbilder zu generieren, die komplementär zur „schwachen Weiblichkeit" entworfen wurden und (hegemoniale) „Männlichkeit" als Symbol der Beherrschung, im Sinne von Kontrolle und Überlegenheit sowie Macht, setzten.[100] Kurt Möser verbindet in seiner interdisziplinären Studie „Fahren und Fliegen in Frieden und Krieg" kulturelle und technische Prozesse miteinander.[101] Dabei nimmt er die individuelle Mobilität mit Rädern wie Fahrrädern oder mit Flugzeugen in den Blick und untersucht als einer von wenigen die Wettbewerbskultur der Fliegerei. So thematisiert Möser den Zusammenhang zwischen Mobilität und Mode, die Fliegerausbildung sowie die Faszination von Katastrophen und Unfällen auf das Publikum. Zudem fragt er nach den Ursachen für die schon im Jugendalter einsetzende Technikbegeisterung.

Untersuchungen zur Kriegserfahrung von Militärpiloten befassen sich selten mit den technischen Voraussetzungen und den Wechselwirkungen zwischen Pilot und Flugzeug. Insofern stellen Arbeiten zur Technikerfahrung wie allgemein zur Erfahrungsgeschichte von Zivil- und Militärpiloten bislang noch ein Desiderat dar. Als „Genre" lag lange Zeit der Schwerpunkt auf den Jagdpiloten und ihren Geschwadern, wohingegen die Aufklärungs- und „Arbeits-Fliegerei" wenig Beachtung fand.[102] Zudem sind systematische Studien zur Infrastruktur und Logistik des Flugwesens bis heute rar und nur vereinzelt ist von Personalstrukturen, Materialtransporten oder der konkreten Tätigkeit der Beteiligten die Rede.[103] Eine Sozialgeschichte der Luftstreitkräfte, die das Milieu, aus dem sich die Piloten rekrutierten, oder ihre Denk- und Verhaltensweisen in den Fokus stellt, gibt es

99 Vgl. Peter Fritzsche, A Nation of Fliers. German aviation and the popular imagination, Cambridge/Massachusetts/London 1992, S. 1–101. Das Schlagwort der *imagined communities* wird thematisiert bei Benedict Anderson, Immagined Communities. Reflections on the origin and spread of nationalism, Londen/New York 1991, S. 6.

100 Vgl. Sabine Höhler, Luftfahrtforschung und Luftfahrtmythos. Wissenschaftliche Ballonfahrt in Deutschland 1880–1910, Frankfurt a. M./New York 2001, S. 195–206.

101 Möser, Fahren und Fliegen.

102 Diese einseitige Betrachtung kritisiert auch Möser, Fahren und Fliegen, S. 555.

103 Potempa, Fliegertruppe; bislang gibt es nur zwei Studien zu Militärpiloten im Ersten Weltkrieg: Kehrt, Moderne Krieger; Sebastian Rosenboom, Im Einsatz über der „vergessenen Front". Der Luftkrieg an der Ostfront im Ersten Weltkrieg, Potsdam 2013, bes. S. 81–98.

ebenso wenig wie eine wissenschaftliche Aufarbeitung des Luftkrieges aus sozial- oder strukturgeschichtlicher Perspektive.[104] Zwar ist Anfang der 1990er Jahre der Versuch unternommen worden, auch den „Krieg des kleinen Mannes" zu erforschen,[105] doch finden sich immer noch mehr Studien zur Generalität als zu „gemeinen" Flugzeugführern und Mannschaften.

In die jeweilige Regional- bzw. Lokalgeschichtlich eingebettet existieren zur Fliegerei im Ruhrgebiet und in Westfalen wie auch zur Berliner Luftfahrtgeschichte – z. B. zum Flugplatz Berlin-Johannisthal, einer Wirkungsstätte Josef Suwelacks – einige Publikationen, ebenso eine Reihe von Arbeiten zur Pionierzeit der motorisierten Luftfahrt „schwerer als Luft".[106] Gegenstand zahlreicher Publikationen ist die zeitgenössische Rezeption der Fliegerei in den damaligen Medien wie Zeitungen und Magazinen sowie in der Kunst und Literatur, ebenso wie auch die Verarbeitung der neuen Raumwahrnehmung im beginnenden 20. Jahrhundert.[107]

104 Zur Strukturgeschichte, d. h. der Organisation des Flugwesens und den damit verbundenen Einsatzfeldern (Luftbilder, Kartografie, Befehlsstrukturen etc.) liegen wenige Studien vor: Potempa, Die Königlich-Bayerische Fliegertruppe; Rudolf Hafeneder, Überblick über das militärische Geowesen Deutschlands im 19. und 20. Jahrhundert, Euskirchen 2004; Helmut Jäger, Erkundung mit der Kamera. Die Entwicklung der Photographie zur Waffe und ihr Einsatz im 1. Weltkrieg, 2. Aufl., München 2007. Ebenso die Bände des militärgeschichtlichen Forschungsamtes: Die Militärluftfahrt bis zum Beginn des Ersten Weltkrieges 1914, 2 Bde., Frankfurt a. M. 1965/66; und: Die Entwicklung der Heeres- und Marineflugzeuge (technischer Band), 2. Aufl., Frankfurt a. M. 1966. Das Fehlen einer Sozialgeschichte der Luftstreitkräfte beklagt z. B. Dietmar Süß, Tod aus der Luft. Kriegsgesellschaft und Luftkrieg in Deutschland und England, München 2011, S. 23.

105 Wolfram Wette (Hg.), Der Krieg des kleinen Mannes. Eine Militärgeschichte von unten, München/Zürich 1992; vgl. auch Jörg Echternkamp, Militärgeschichte, in: Frank Bösch/Jürgen Danyel (Hg.), Zeitgeschichte – Konzepte und Methoden, Göttingen 2012, S. 293–312.

106 Werner Freitag (Hg.), Geschichte der Stadt Billerbeck, Bielefeld 2012; Westfälischer Heimatbund (Hg.), Jahrbuch Westfalen, Jg. 54, Münster 2000, S. 26–39; Ernst Kühl, Das schöne Münster. 50 Jahre Sportfliegen in Münster, Münster 1960; Hans Langenfeld/Klaus Prange, Münster. Die Stadt und ihr Sport, Münster 2002; Frank Radzicki, Der Traum vom Fliegen im Ruhrgebiet. Bilder der Luftfahrt, Erfurt 2006. Guido Rißmann-Ottow, Glück ab! Frühe Luftfahrt im Revier, Essen 2002; Günter Schmitt, Als die Oldtimer flogen. Die Geschichte des Flugplatzes Berlin-Johannisthal, 3. Aufl., Oberhaching 1995; ders., Fliegende Kisten von Kitty Hawk bis Kiew, 2. Aufl., Berlin (Ost) 1990; ders./Werner Schwipps, 20 Kapitel frühe Luftfahrt, Planegg 1990; Werner Schwipps, Schwerer als Luft. Die Frühzeit der Flugtechnik in Deutschland, Koblenz 1984; ders., Riesenzigarren und fliegende Kisten. Bilder aus der Frühzeit der Luftfahrt in Berlin, Berlin 1984; Hans Vogt, Seidene Kugel und Fliegende Kiste, Krefeld 1993.
Als Fliegerei „schwerer als Luft" wird die motorisierte Luftfahrt mittels dynamischen Auftriebs bezeichnet, nicht zuletzt aufgrund des Eigengewichts des jeweiligen Luftfahrzeuges. Gleit- und Ballonflieger hingegen werden aufgrund ihres statischen Auftriebs als „leichter als Luft" bezeichnet.

107 Siehe z. B. Christoph Asendorf, Super Constellation. Das Flugzeug als kulturelle Erfahrung, in: Wolfgang Ruppert (Hg.), Fahrrad, Auto, Fernsehschrank. Zur Kulturgeschichte der Alltagsdinge, Frankfurt a. M. 1993, S. 188–216; Wolfgang Behringer/Constance Ott-Koptschalijski, Der Traum vom Fliegen. Zwischen Mythos und Technik, Frankfurt a. M. 1991; Hartmut Berghoff, „Dem Ziele der Menschheit entgegen". Die Verheißungen der Technik an der Wende zum 20. Jahrhundert, in: Ute Frevert (Hg.), Das Neue Jahrhundert. Europäische Zeitdiagnosen und Zukunftsentwürfe um 1900, Göttingen 2000, S. 47–77; Fernando Esposito, Mythische Moderne. Aviatik, Faschismus

Zum Stellenwert der Flugzeugtechnologie erschienen in den letzten Jahren einige kulturgeschichtliche Monografien, die meist an einem Fallbeispiel national spezifische Rezeptionsweisen darstellen. Zu nennen sind die Untersuchungen über die Luftfahrtbegeisterung als „geflügeltes Evangelium“ von Joseph J. Corn für die USA,[108] die Analyse der englischen Flugzeugdiskurse durch David Edgerton, die Darstellung der *„early birds“*[109] durch Henry Serrano Villard oder die Studie Rüdiger Haudes über die politische Symbolik der Luftfahrt am Beispiel von Aachen.[110]

Forschungen zur Technikbegeisterung und zur Sensationslust liegen vor allem für das Zeppelin-System vor,[111] wobei hier die Technikfaszination selten explizit thematisiert wird.[112] Zum Thema Geschlecht und Aviatik gibt es jenseits der Studien zur Männerdomäne der Militär- und Kriegsfliegerei biografische Arbeiten zu den wenigen Flugpionierinnen der Vorkriegszeit. Darüber hinaus existieren einige meist populärwissenschaftliche Beiträge zu den sogenannten „Alten Adlern“, denjenigen Fliegerinnen und Fliegern, die vor Ausbruch des Ersten Weltkrieges ihren Pilotenschein erhalten hatten.[113]

Bislang wurde und wird der visuellen Vermittlung der neuen Technik, den handelnden Akteuren und den damit verbundenen Marketingstrategien, beispielsweise Flugwettbewerbsplakaten oder Anzeigen für Flugzeugteile (Motor, Propeller etc.), wenig Beachtung geschenkt. Werbeanzeigen, Fotos oder Postkarten von Piloten in oder vor ihren Maschinen dienen größtenteils lediglich zu Illustrationszwecken, um dem Leser eine Person oder die Technik optisch zu präsentieren. Die

und die Sehnsucht nach Ordnung in Deutschland und Italien, München 2011; Kurt Möser, Fliegerblick 1914, in: Journal of New Frontiers in Spatial Concepts (2009), Vol. 1, S. 99–106, http://ejournal.uvka.de/spatialconcepts/archives/858 (13.9.2013).

108 Joseph J. Corn, The Winged Gospel. America’s romance with aviation, Baltimore/London 2002.

109 Henry Serrano Villard, Contact! The story of the early birds, Washington D.C. 2002.

110 David Edgerton, England and the Aeroplane. An essay on a militant and technological nation, Houndsmills/London 1992; Rüdiger Haude, Grenzflüge. Politische Symbolik der Luftfahrt vor dem Ersten Weltkrieg. Das Beispiel Aachen, Köln/Weimar/Wien 2007.

111 Forschung zur Technikfaszination Flugzeug wurde betrieben von Martina Glanemann, Technikbegeisterung und kleinstädtische Strukturen – Flugtage im Münsterland 1908–1914, unveröffentlichte Masterarbeit, Münster 2012; Helmut Reinicke, Deutsche Auffahrten. Mentalitätsformen der Technikbegeisterung in der Entwicklung der Aviation, in: Wechselwirkung 54 (1992), H. 4, S. 22–28; siehe auch Möser, Fahren und Fliegen, bes. S. 261ff.

112 Exemplarisch einige jüngere Studien zum Zeppelin-Kult: Saskia Frank, Zeppelin-Ereignisse. Technikkatastrophen im medialen Prozess, Marburg 2008; Jeannine Meighörner, „Die Essenz dieses Mannes ist Mut“. Graf Zeppelin und die Presse: ein Heldenepos, in: Wolfgang Meighörner (Hg.), Der Graf. 1838–1917 (Ausstellungskatalog), Friedrichshafen 2000, S. 204–227; Guillaume de Syon, Zeppelin! Germany and the airship, 1900–1939, Baltimore 2002.

113 Siehe z. B. Bettina Gundler, „Nie wird die Fliegerei als Beruf überfüllt sein von Frauen“. Geschlechterverhältnisse und berufliche Perspektiven von Pilotinnen in den Anfängen der Luftfahrt, in: Metis 4 (1995), H. 1, S. 29–40 (dort auch weitere Literatur zu den Fliegerinnen); zu den „Alten Adlern“ siehe Alexander Kauther/Paul Wirtz, Robert Thelen – ein Alter Adler in Berlin-Johannisthal, Berlin/Jülich 2011 (Dokumentationsreihe über den Flugplatz Johannisthal 1909–1914, H. 27) und weitere Titel in dieser Reihe.

Kommentare unter den Bildern beschränken sich häufig auf Informationen zum Erscheinungsjahr oder Herausgeber und schöpfen das Potenzial dieses Mediums als Kommunikationsmittel zwischen Bildgegenstand und RezipientInnen nicht aus.[114] Die wissenschaftliche Erschließung und systematische Analyse von Fliegeralben, Bildpostkarten und der medial inszenierten aviatischen Kultur, die rund um das Flugzeug entstand, sowie die visuelle Aneignung und Verarbeitung der Technik durch die Öffentlichkeit steht für die Zeit vor dem Ersten Weltkrieg erst am Anfang.

Quellen

Die Quellenlage zur Person und zum Umfeld Josef Suwelacks kann als äußert günstig gewertet werden. Der derzeit 18 Archivkartons umfassende, privat verwahrte Nachlass hält umfangreiche Bestände unterschiedlichster Quellengattungen für den Zeitraum von 1902 bis 1915 bereit. Darüber hinaus sind Quellen zu Personen erhalten, die in einem verwandtschaftlichen Verhältnis zu dem Piloten standen, sowie Archivalien, die Auskünfte über die Rezeptionsgeschichte seiner Person und Tätigkeit geben können.

Die Schwerpunkte der Überlieferung liegen bei der zivilen Flugtätigkeit Josef Suwelacks, angefangen bei seinen ersten heimatlichen Flugversuchen in Billerbeck und Münster bis hin zu seiner Tätigkeit bei den Rumpler- und den Kondor-Werken. Der Schriftverkehr Suwelacks mit Motorenfabriken, Flugzeugfirmen, der Militärverwaltung, potenziellen Kunden und Flugschülern, Anfragen, die seine Ausbildung betreffen oder die Bereitstellung von Fotomaterial etc., sind durch Schreiben von Firmen und Kunden dokumentiert; teilweise liegen auch die Antwortschreiben Suwelacks als Eigenbeleg in einem von ihm selbst angelegten Register vor.

Aus seiner Zeit in Johannisthal bei den Rumpler-Werken (1911–1912) und als Pilot und Flugzeugkonstrukteur bei den Kondor-Werken (1912–1914),[115] mit denen er bis zu seinem Tod 1915 in Verbindung stand, ist eine Vielzahl unterschiedlicher Quellengattungen vorhanden: Neben Einladungen zu den Gesellschafterversammlungen der Kondor-Werke und den Protokollen, notariellen Schreiben und Jahresgeschäftsberichten existieren auch privat angelegte Fotoalben aus diesem Lebensabschnitt, ebenso Werbeanzeigen für Flugzeuge, Plakate zu Flugveranstaltungen und Listen mit Erfolgen und Preisen, die auf den Fabri-

114 So sind in den Publikationen von Werner Schwipps oder Kurt Möser zahlreiche Bildquellen in teils sehr guter Qualität abgedruckt, die aber nicht in die Narration mit einfließen. Sie stehen neben dem Text und werden nicht durch Verweise integriert (z. B. „vgl. Abb. 1" o.ä.).

115 Die Jahresangaben beziehen sich auf Suwelacks Verweildauer bei den Fabriken etc. und nicht auf den Zeitraum des Bestehens der Flugzeugwerke etc.

katen erzielt wurden. Darüber hinaus sind zeitgenössische Berichterstattungen über die Flugzeugwerke und selbst- oder fremdverfasste (Zeitungs-)Berichte über die Konstruktionen und Unternehmungen mit den Flugzeugen sowie über Josef Suwelack als Flieger dokumentiert.

Aus seiner Zeit als Zivilflieger im Ersten Weltkrieg sind 95 Feldpostbriefe im Nachlass erhalten geblieben.[116] Sein Kriegserlebnis hat Suwelack außerdem in weit über tausend Fotos festgehalten, die seine Zeit bei der Feldfliegerabteilung dokumentieren. Der fotografische Quellenbestand umfasst Luft-, Landschafts- und Mannschaftsaufnahmen sowie Bilder von Kriegsgerät, Schlachtfeldern, Gefangenen und Toten, von Gräbern, Feierlichkeiten, Pausenzeiten und Freizeitgestaltung.

Daneben sind auch Schriftstücke vorhanden, in denen Suwelack Glückwünsche übermittelt wurden, ferner Belege über die Überweisung von Preisgeldern und Gehältern, z. B. eine Aufstellung der Vergütung für Flüge von September 1914 bis März 1915. Von den abgeschlossenen Verträgen ist sowohl ein Arbeitsvertrag mit den Kondor-Werken im Original überliefert als auch eine Bestätigung des Vertrages Josef Suwelacks mit dem Deutschen Reich vom 2. Februar 1915. Als weiteres offizielles Dokument findet sich die Verleihungsurkunde des Albrechtsordens, datiert auf den 9. Dezember 1914.

Neben den Schrift- und Bildquellen sind einige Sachquellen vorhanden, darunter das Flugzeugführerzeugnis Josef Suwelacks, militärische Orden, seine blutbefleckte Fliegerbinde der Feldfliegerabteilung 24, ein von ihm während des Krieges geführtes Flugbuch mit Aufzeichnungen zur Anzahl der Flüge und den Namen der mitfliegenden Beobachter sowie private Notizen. Ebenso befinden sich im Nachlass Mitgliedsausweise und Aufnahmegesuche, beispielsweise in den Kaiserlichen Aero-Club, oder eine abgeschlossene Unfallversicherung mit der Victoria. Was fehlt, ist der Blick von außen – Aussagen von engen Freunden, Fliegern und Geschäftspartnern über Josef Suwelack. Dies hätte stellenweise eine differenziertere Sicht ermöglicht.

Über den privat verwahrten Nachlass hinausgehend lässt sich die Quellenbasis noch um weitere Bestände aus kommunalen und staatlichen Archiven ergänzen: So verfügt das Archiv des Deutschen Technikmuseums in Berlin im Nachlass Willy Stiasnys über eine umfangreiche Zeitungsausschnittsammlung zur frühen Luftfahrt, über Dokumente zu Rumpler und weitere kleinere Fliegernachlässe etwa von Heinz Erblich. Im Bundesarchiv in Freiburg lagern im Bestand PH 19 Unterlagen über die Heeresflieger der preußischen Armee, im Bestand PH 9 Akten zu Flugvorschriften und zur Luftbildfotografie. Dokumente zur Flieger-

116 Die tatsächliche Anzahl der Briefe liegt weit höher, denn viele wurden auch an Freunde und Verwandte geschrieben, einige der Schreiben erreichten – vor allem in den ersten Kriegswochen – darüber hinaus ihren Bestimmungsort erst gar nicht.

truppe sind hingegen nur unvollständig überliefert, da ein Großteil 1945 im Heeresarchiv in Potsdam verbrannt ist. Der Restbestand ging zunächst als Kriegsbeute nach England. Die erhaltenen Akten wurden nach ihrer Restauration in die Archive der ehemaligen Bundesstaaten zurückgegeben. Teile der Dokumente der ehemaligen sächsischen Armee, zu der auch die Feldfliegerabteilung 24 von Josef Suwelack gehörte, liegen im sächsischen Hauptstaatsarchiv in Dresden.[117] Dort gibt es auch eine schmale Überlieferung zum Flugplatz in Großenhain, auf dem die Fliegerabteilung bis zu ihrer Verlegung an die Westfront stationiert war.[118] Ansonsten erweisen sich die Bestände des Dresdener Hauptstaatsarchivs zur Fliegertruppe als wenig ergiebig, da die Überlieferung äußert fragmentarisch ist und die Dokumente zum Teil schwer zu kontextualisieren sind.

In Westfalen, der Heimatregion Josef Suwelacks, befindet sich im Stadtarchiv Münster ein schmaler Bestand zum Verwaltungsvorgang um den „Josef-Suwelack-Weg", eine Straße in Münster, die 1936 nach dem westfälischen Piloten benannt wurde.[119]

Akten zu den Kondor-Flugzeugwerken, deren technischer Leiter Josef Suwelack war, liegen in Kopie im privaten Familienarchiv von Walter Suwelack vor, genauso wie entsprechende Unterlagen aus dem Billerbecker Stadtarchiv. Weitere Bestände zum Flugplatz Gelsenkirchen-Essen-Rotthausen (1912–1920), dem Stammsitz der Kondor-Werke, lagern im Essener Stadtarchiv, im Historischen Archiv Krupp sowie im Essener Luftfahrtarchiv.

Die öffentliche Seite der Fliegerei, hier vor allem die Wettbewerbskultur, kann über die Lokalberichterstattung in Zeitungen und aviatischen Zeitschriften wie z. B. dem „Flugsport" erschlossen und analysiert werden. Insbesondere das Dortmunder Zeitungsarchiv hält eine Vielzahl zeitgenössischer deutscher Zeitungen bereit, die ausführlich über Flugtage berichteten. Die für die fliegerischen Aktivitäten Josef Suwelacks (1908–1915) relevanten Jahrgänge des „Billerbecker Anzeigers", der lokalen Heimatzeitung Suwelacks, lagern im Privatarchiv von Rotraud Illisch in Billerbeck. Damit ergibt sich ein Quellenbestand, der es ermöglicht, verschiedene Aspekte der Lebensgeschichte Josef Suwelacks detailliert zu betrachten.

Ähnlich umfangreich wie das Quellenmaterial gestaltet sich auch die Literatur

117 Siehe http://www.argus.bstu.bundesarchiv.de (Beständeübersicht Bundesarchiv, aktualisiert im November 2014) (15.1.2015). Edmund Rumpler (1872–1940) war ein österreichischer Flugzeug- und Automobilkonstrukteur. Seit 1913 besaß er die deutsche Staatsangehörigkeit. Er war Namensgeber eines der bekanntesten Flugzeuge der Vorkriegszeit, der Rumpler-Taube (siehe Feld 2, Abschnitt zu den Rumpler-Werken). Heinz Erblich (1891–1932) war Fluglehrer und Ingenieur und hinterließ neben Publikationen zur Fliegerei kolorierte Skizzen- und Notizbücher.

118 Aufarbeitung der Geschichte des Flugplatzes durch Polo Palmen (Hg.), Flugplatz Großenhain. Historischer Abriss, o. O. 2007.

119 Siehe auch http://www.lwl.org/westfaelische-geschichte/nstopo/strnam/Kommune_150.html „Entscheidung des OB (Dez. Konferenz v. 17.8.1936)" (20.1.2015). Ich danke Marcus Weidner für die Übersendung der entsprechenden Akten.

zu dem Piloten Josef Suwelack. Berichte über seinen Tod liegen in 27 Nachrufen vor, die teils im Original, teils in Kopie, darunter auch ein Nachruf auf Spanisch, dokumentiert sind. Des Weiteren sind 25 Dokumente und Berichte zu seinem Tod überliefert, darunter eine französische und eine englische Darstellung der Todesumstände Josef Suwelacks und seines Beobachters Leutnant Oskar Teichmann. Von familiärer Seite gibt es Todesanzeigen und neun Seiten handschriftliche Erinnerungen von Richard, dem jüngeren Bruder von Josef Suwelack.

In Zeitungsberichten der Lokalpresse des Münsterlandes wurde und wird heute noch regelmäßig anlässlich von Jahrestagen an Suwelacks Weltrekord, seine Leistungen für die frühe Fliegerei oder an seinen Todestag erinnert. 1988 veröffentlichten Nachfahren Josef Suwelacks anlässlich seines 100. Geburtstages eine Biografie, die 2001 in überarbeiteter Form unter einem neuen Titel erneut erschien.[120] Weitere Hinweise auf den westfälischen Piloten finden sich sowohl in wissenschaftlichen und populären Darstellungen zur frühen Luftfahrt, in Studien zur Flugzeugkonstruktion und -produktion als auch in lokal- beziehungsweise regionalgeschichtlichen Beiträgen zum ausgehenden Kaiserreich.[121] Darüber hinaus taucht sein Name in einer Übersicht zu Wohnanschriften der ersten Flugpioniere auf der Internetseite „johflug.de“ und auf Seiten zur deutschen Fliegertruppe auf,[122] ferner im Zusammenhang mit der weiter oben bereits angesprochenen Straßenumbenennung.[123] Auch drei Kurzfilme, ausgestrahlt vom WDR, erinnern an den Flugpionier. Die Familie Suwelack hält die Erinnerung an ihn unter anderem durch die Nachlasspflege, private Fahrten zu seinem Kriegsgrab und die finanzielle Förderung von (Forschungs-)Projekten, etwa zur Billerbecker Geschichte, lebendig.[124]

Eine performative Darstellung der Lebensgeschichte des Piloten wurde 1914

120 Winfried und Walter Suwelack, Josef Suwelack und der Traum vom Fliegen. Leben und Tod des Billerbecker Flugpioniers und Weltrekordlers Josef Suwelack, 30. April 1888 bis 13. September 1915, anlässlich der 100. Wiederkehr seines Geburtstages, Dülmen 1988; Rainer A. Krewerth (Hg.), Ein westfälisches Fliegerleben. Josef Suwelack 1888–1915, Münster 2001.

121 Auch im Internet ist der westfälische Pilot unter anderem mit einem Wikipedia-Artikel, verfasst von Walter Suwelack, sowie mit einer eigenen Homepage vertreten. Vgl. http://de.wikipedia.org/wiki/Josef_Suwelack_(Flieger). Homepage, gestaltet von Wolfgang und Walter Suwelack (seit 2011): www.josef-suwelack.de. Siehe auch zur Familie Suwelack heute die Darstellung der Unternehmerfamilie Suwelack und ihrer Beziehung zur Stadt Billerbeck: Wolfgang Suwelack (Hg.), www.suwelack-stiftung.de, online seit 2004 (alle Seiten 28.6.2014).

122 Alexander Kauther/Peter Wirtz (Hg.), www.johflug.de, online seit 2009. Die Seite dokumentiert die Geschichte des Flugplatzes Berlin Johannisthal. Zu den Wohnanschriften der Flieger siehe http://www.johflug.de/pdf/Wohnanschriften_bis_%201914_Dez_2012.pdf. Vgl. auch Thorsten Pietsch (Hg.), www.frontflieger.de (zur Feldfliegerabteilung 24 u. a.), online seit Februar 2001 (alle Seiten 25.6.2014).

123 Marcus Weidner, Die Straßenbenennungspraxis in Westfalen und Lippe während des Nationalsozialismus. Datenbank der Straßenumbenennung 1933–1945, online seit Mai 2013 unter: http://www.strassennamen-in-westfalen-lippe.lwl.org (10.6.2014).

124 http://www.suwelack-stiftung.de (10.6.2014).

mit der Uraufführung eines plattdeutschen Theaterstückes realisiert, das den Flieger Suwelack karikiert und mit einer humoristischen Liebesgeschichte verbindet.[125] Zudem tritt ein nachgebautes Exemplar des von Josef Suwelack konstruierten Kondor-Endeckers seit 2010 regelmäßig bei einem Seifenkistenrennen mit nachgebauten Flugzeugen aus der Frühzeit der Fliegerei auf, bei denen Berliner Grundschulen auf dem Flugplatz Johannisthal als Reenactment seit 2009 jährlich gegeneinander antreten.[126]

Insgesamt steht damit eine große Bandbreite unterschiedlichster Quellengattungen und Darstellungen zur Verfügung, die eine ausreichende Basis sowohl für eine sozial- und kulturgeschichtliche Rekonstruktion wie auch für eine Dekonstruktion des Mythos um den Flieger Josef Suwelack bilden.

Aufbau der Arbeit

Durch eine kritisch reflektierende Analyse bisheriger Geschichtsdarstellung zu Josef Suwelack sollen gängige narrative Konstrukte im Kontext von Flieger-Mythen hinterfragt und auf ihrer Deutungsabsichten untersucht werden (Abb. 1).

Mit Rückgriff auf das Habitus-Konzept des Soziologen Pierre Bourdieu, das eine Verknüpfung individueller und struktureller Momente erlaubt, werden Handlungs-Felder die Struktur der Arbeit vorgeben, unter denen spezifische Aspekte des Lebens und Wirkens Josef Suwelacks thematisiert werden. Die Luftfahrt als neue Technologie des 20. Jahrhunderts setzt den Rahmen für das individuelle Handeln des Piloten. Als Flugpionier, Konstrukteur, Pilot und späterer Kriegsteilnehmer war er unmittelbar an der Entwicklung und Nutzung der neuen Technik beteiligt. Die Raumwahrnehmung aus der Luft führte zu neuen Sinneseindrücken, wurde von den Fliegern und ihren Passagieren in Erlebnis- und Augenzeugenberichten verarbeitet und medial über die Presse, über Bildpostkarten, Briefmarken, Fotos etc. verbreitet. Nutznießer der neuen Technologie wurden zunehmend auch scheinbar unbeteiligte Personengruppen, zunächst die Zuschauer auf Flugwettbewerben, dann im Krieg die Infanteristen, die sich durch den Blick nach oben kurzzeitig vom erstarrten Kriegsgeschehen am Boden ablenken konnten.

Den Rahmen der Arbeit bildet die Re- bzw. Dekonstruktion der historischen Rezeption des Lebens und Wirkens Josef Suwelacks seit den 1920er Jahren. Hie-

125 Abendgesellschaft des Zoologischen Gartens Münster (Westf.), Familie Uhlenhorst off: Buren un Kabeleeren. Großes Volksstück mit Gesang und Tanz in 4 Akten von Franz Falger, Münster 1929ff.

126 Siehe www.johflug.de, Startseite (Stand November 2015).

ran schließt sich die an vier Handlungsfeldern ausgerichtete biografische Rekonstruktion an, die den Hauptteil der Arbeit ausmacht.

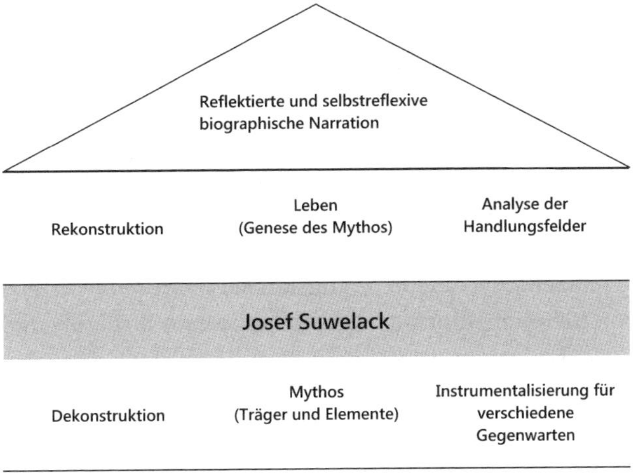

Abb. 1: Methodischer Zugriff auf die Biografie Josef Suwelacks

Im ersten Feld „Familie“ stehen Fragen nach der Sozialisation, dem Milieu sowie nach dem kulturellen und ökonomischen Kapital Josef Suwelacks im Vordergrund, die richtungsweisend für seinen weiteren Lebensweg waren und das Fundament für seine spätere Karriere als Konstrukteur und Pilot legten. In den Blick genommen werden ferner die Beziehungen Josef Suwelacks zu seiner Familie, die ihn finanziell unterstützte, ihm beratend zur Seite stand, ja seine Karriere überhaupt erst möglich machte. Eine Analyse der Strukturen der katholischen Kleinstadt Billerbeck und der Auswirkungen von Modernisierungsprozessen auf den Heimatort von Josef Suwelack verdeutlicht das – von Hoffnungen wie Ängsten begleitete – Eindringen von Technik in das Leben der Bewohner dieser noch ländlichen Kleinstadt. Der individuellen Motivation Josef Suwelacks, den Fliegerberuf zu ergreifen, wird anhand der lokalen Berichterstattung im „Billerbecker Anzeiger“ und der entstehenden aviatischen Kultur – der Praxishandbücher, Zeitschriften, Zukunftsliteratur oder Postkarten – nachgespürt. Durch den Einbezug der Strukturen wird ein multiperspektivischer Erklärungsansatz verfolgt, der die Entscheidung, Flieger zu werden, nicht auf ein „Schlüsselerlebnis“ zurückführt, sondern auf verschiedene Einflussfaktoren verweist.

Im zweiten Feld „Flugpionier und -lehrer, Konstrukteur und Unternehmer“ geht es um die firmeninterne Flieger- und Konstrukteurtätigkeit Josef Suwelacks, zunächst als angestellter Fluglehrer und späterer Chefpilot bei den Rumpler-Werken in Berlin-Johannisthal, dann als leitender technischer Angestellter bei den Kondor-Werken in Essen, an deren Gründung er 1912 beteiligt war. Seine dortigen Tätigkeiten geben sowohl Einblicke in das Entstehen und die Strukturen der Flugzeugfirmen als auch Auskunft über Arbeitsstrukturen sowie die Bedeutung von persönlichen Kontakten zu Industriellen und Militärs. Welche Hierarchien bestanden in den Betrieben, wie viel Kapital musste eine Flugzeugfirma aufwenden, um sich erfolgreich in der Flugzeugindustrie zu behaupten? Welche Personenkreise hatten Interesse am Auf- und Ausbau einer derartigen Firma, wie präsentierten sich die Firmen, z. B. durch Werbeanzeigen oder Schauflüge, und wie positionierte sich Josef Suwelack in diesem Feld? Mit welchen Vorschriften und Auflagen sahen sich die Akteure konfrontiert und welche Personen oder Organisationen regten Innovationen an? Fragen dieser Art tragen grundlegend zum Verständnis des Feldes der Flugzeugproduktion und -vermarktung bei und unterstreichen, dass die handelnden Akteure – angefangen beim Monteur über den Piloten bis hin zur Unternehmensführung – in ein komplexes Struktur- und Beziehungsgeflecht eingebunden waren. Die Regeln dieses Feldes zu skizzieren, dient nicht zuletzt dazu, am Beispiel Josef Suwelacks Gestaltungsmöglichkeiten zu bestimmen und die Fremd- und Selbstwahrnehmung der Akteure zu beleuchten.

Im dritten Feld „Flugwettbewerbe" sollen die Organisation und die Funktionen von Flugwettbewerben auf nationaler, lokaler und militärischer Ebene an ausgewählten Beispielen beschrieben und Veränderungen in der Wettbewerbsausrichtung aufgezeigt werden. Während die ersten Flugveranstaltungen noch in einer „Experimentier- und Spielphase" der Technik stattfanden, wurden die späteren Wettbewerbe zunehmend auf militärische Bedürfnisse zugeschnitten und der sportliche Aspekt allmählich verdrängt. Dies wiederum wirkte sich massiv auf die Flugzeugindustrie aus, war doch das Militär fast alleiniger Abnehmer von Flugzeugen und damit ein wichtiger Impulsgeber der Flugzeugtechnik.

Zur öffentlichen Seite der Fliegerei zählen außerdem die Berichterstattung über die Wettbewerbe und die Reaktionen der Bevölkerung auf das „moderne Wunder" des Flugzeuges und seine Beherrscher.[127] In Zeitungen und Flugzeitschriften wurde detailliert und intensiv über die Fliegerei und die Piloten berichtet. Die Sensationslust der Menschenmassen, von denen die Presse regelmäßig berichtete, erstreckte sich dabei nicht nur auf erfolgreiche Flugversuche; auch Unfälle und Todesstürze weckten gleichermaßen das Interesse und die Neugier der Öffentlichkeit. Diese Faszination für Erfolg und Katastrophe machten sich die Flieger und Militärs zunutze, inszenierten sie sich doch als männliche Technikbeherrscher, die zudem als nervenstarke Dandys gefeiert und noch im Tod als Nationalhelden verehrt wurden.[128] Die Selbst- und Fremdvermarktung der Flieger sowie die Konstruktion von Mythen über Flieger-Helden, die im Ersten Weltkrieg fortgesetzt und noch gesteigert wurden, gilt es zu thematisieren.

127 Rieger, Technology, S. 22.
128 Flugwettbewerbe wurden zu regelrechten „Dandyplattformen". Vgl. Möser, Fahren und Fliegen, S. 138.

Das letzte Feld „Die Fliegerei im Ersten Weltkrieg" bietet schließlich eine andere Perspektive auf den Ersten Weltkrieg – „mit der Kamera bewaffnet"[129] und aus der Luft. Umfangreiches Bildmaterial und die Feldpostbriefe Josef Suwelacks geben Einblicke in die Arbeit bei einer Feldfliegerabteilung, Auskunft über Truppen- und Materialtransporte, über Befehlsstrukturen und Verpflegung etc. Als Beobachtungsflieger an der Westfront, dessen Aufgabe unter anderem das Überfliegen und Fotografieren der feindlichen Stellungen war, hatte Josef Suwelack eine andere Sicht auf das Kriegsgeschehen und die Möglichkeit, seine Eindrücke aus der Luft und am Boden sowohl bildlich als auch schriftlich festzuhalten. Im Gegensatz zu den Soldaten in den Schützengräben hatte er meistens ein Haus- oder Zeltdach über dem Kopf, Ruhepausen und „Freizeit" hinter der Front, die er zum Jagen, Malen oder Briefeschreiben nutzte.

In den Briefen an seinen Bruder Richard, der als Feldmagazininspekteur der Reserve in Laon stationiert war, an seine Eltern, Verwandte und Freunde berichtete Josef Suwelack sowohl von den Annehmlichkeiten bei der Feldfliegerabteilung, von Treffen mit hochrangigen Militärs und den Geschäftsführern der Kondor-Werke, er verschwieg aber auch die Schrecken des Krieges nicht. Wie viele andere Kriegsteilnehmer war auch Suwelack zu Beginn des Weltkrieges davon überzeugt, dass er rasch wieder zuhause in Billerbeck sein würde. Mit zunehmender Kriegsdauer und der Erstarrung der Fronten schlug seine anfängliche Begeisterung jedoch in Ernüchterung um. Überzeugt, einen „gerechten Verteidigungskrieg" gegen den „Erbfeind" Frankreich zu führen, und stolz, sich im Feindesland „häuslich" eingerichtet zu haben, erlebte er „seinen" Krieg bis zu seinem Tod im September 1915 als jemand, der sich freiwillig zur Fliegertruppe gemeldet hatte, bis zuletzt bei der Fliegerei blieb und als „Held", ausgezeichnet mit dem Eisernen Kreuz II. Klasse und dem Albrechtsorden mit Schwertern II. Klasse, im Alter von 27 Jahren für sein Vaterland fiel.

Das Leben von Josef Suwelack endete nach zeitgenössischen englischen und französischen Berichten am 13. September 1915, die Geschichte seines Wirkens aber ging weiter – bis heute. Zahlreiche Nachrufe würdigten ihn als einen erfolgreichen Unternehmer und „leidenschaftlichen" Flieger und sähten die Saat für den Flieger-Mythos um seine Person.

Zum Schluss der Analyse wird der Versuch unternommen, wiederkehrende Inhalte von Fliegermythen aus der Biografie und Rezeptionsgeschichte Josef Suwelacks abzuleiten und für deren Mechanismen zu sensibilisieren. Der letzte Abschnitt resümiert das Biografie-Projekt und die damit verbundenen Chancen

129 So der Titel eines von Anton Holzer herausgegebenen Sammelbandes: Anton Holzer (Hg.), Mit der Kamera bewaffnet. Krieg und Fotografie, Marburg 2003.

und Herausforderungen. Vor allem die eigene Position der Autorin als nicht-fliegende Frau des 21. Jahrhunderts mit dem Wissen um die verheerenden Auswirkungen von Luftkriegen, der einhundertjährige Altersunterschied und die persönliche, sich im Schreibprozess wandelnde Sichtweise auf Josef Suwelack sollen hier zur Sprache kommen.

B. Rezeptionsgeschichte

1. Die Arbeit am Mythos seit den 1920er Jahren

1.1. Quantitative Häufigkeit der Mytheme

Zwischen 1924 und 2012 wurde Josef Suwelack anlässlich von Jahrestagen, etwa seinem Todestag oder Weltrekord, oder im Kontext bestimmter Ereignisse wie den Billerbecker Flugtagen von 1912 alleine oder zusammen mit anderen westfälischen Flugpionieren, porträtiert. So verschieden die Anlässe der Erinnerung waren, so zeigt sich in der Analyse doch, dass bestimmte Elemente seines Lebens im kollektiven beziehungsweise regionalen Gedächtnis erhalten geblieben sind. Die Darstellungen – in Zeitungsberichten, Filmen, der Literatur oder auf der Theaterbühne – spiegeln den medialen Umgang mit dem Billerbecker Flieger im Verlauf fast eines Jahrhunderts. Dabei wird deutlich, dass die unterschiedlichen Rezeptionen doch stets einen Josef Suwelack ins Bild setzten, der Züge der jeweiligen Gegenwart besaß, einen Menschen, der unter wechselnden politischen und gesellschaftlichen Vorzeichen mal zum bewunderungswürdigen Kriegshelden, mal zu einem moralischen Vorbild und Pazifisten stilisiert wurde. Insofern erfüllen die AutorInnen die Lesererwartungen ihrer Zeit und nicht selten auch die der Familie Suwelack, die als Nachfahren des Piloten ein hohes Interesse an seiner privaten und öffentlichen Erinnerung hat und diese auch selbst initiiert. Als „Nachlassverwalter und -hüter" nahmen die Suwelacks häufig direkt oder indirekt Einfluss auf die Art der Berichterstattung, schließlich sind sie es, die den Zugang zum Nachlass des Piloten regeln. Zwar schreiben sie nicht vor, was und wie über ihren Vorfahren berichtet wird, doch ist es naheliegend, dass aus Takt- und Mitgefühl gewisse Elemente seiner Biografie weggelassen wurden und werden, um eine angemessene Form der Erinnerung zu finden. Zugleich nutzen die Autoren die Lebensgeschichte und Persönlichkeit des Piloten als Plattform und Exempel, um ihrer Vorstellung von einer anzustrebenden, gesellschaftlich erwünschten Zukunft Ausdruck zu verleihen.[1]

Da einige Berichte zum Teil identische Passagen enthalten, ist anzunehmen, dass sie sich auf eine gemeinsam Quelle beziehen oder, was wahrscheinlicher ist, schlicht voneinander abschreiben. Überwiegend stammen die Berichte aus

1 Vgl. hierzu die Überlegungen zu einer Biografie über Werner Best bei Ulrich Herbert, Best. Biographische Studien über Radikalismus, Weltanschauung und Vernunft 1903–1989, 3. Aufl., Bonn 1996, S.11–25.

Deutschland, allerdings liegen auch zwei französische Artikel vor, die Josef Suwelack anlässlich seines Weltrekordes würdigen sowie im Kontext des englischen Soldatenfriedhofs von Erquingem-sur-Lys.

Die Narrationen in diesen Texten orientieren sich in der Hauptsache an den Konzepten „Aktualität", „Zwangsläufigkeit", „Erfolg" und „Schicksal". Betont werden Josefs Beharrlichkeit und Zielstrebigkeit als Flugpionier und späterer Unternehmer, seine großen und kleinen Erfolge im Wettstreit mit anderen Fliegern und schließlich die von außen auf ihn einwirkenden Strukturen, denen er sich stellte oder, so scheint es, nicht ausweichen konnte.

Bei den RezipientInnen lassen sich drei Gruppen unterscheiden: Erstens die mediale Öffentlichkeit – d.h. (Zeitungs-)LeserInnen, Luftfahrtinteressierte und LokalpatriotInnen –, zweitens die Gruppe der (Populär-)Wissenschaftler, die Josef Suwelack im Kontext der Fliegerei rezipieren, und drittens die Familie Suwelack, die das Scharnier zwischen öffentlicher und privater Historisierung des Piloten bildet.

1.2. Inhalte und Funktionen des Suwelack-Mythos

Der „Traum vom Fliegen", die „Eroberung der Luft" und das „Schlüsselerlebnis"

Der „Traum vom Fliegen" war und ist Bestandteil zahlreicher Publikationstitel zur Fliegerei und ein typisches Narrationsmuster in Piloten-Darstellungen.[2] Als Leitmotiv strukturiert es die Berichte und betont die tief in der Menschheitsgeschichte verwurzelte „Sehnsucht", Grenzen zu überwinden, frei, losgelöst und ungezwungen die Welt von oben zu entdecken und zu erobern.

„Der alte Menschheitstraum vom Fliegen machte auch vor den Westfalen nicht halt",[3] führt Carsten Krystofiak den Leser in die Lebensgeschichte Josef Suwelacks ein, um dann im letzten Abschnitt seines Porträts den Topos vom – zunächst friedlichen – Flugtraum erneut aufzugreifen, der schließlich durch den Ausbruch des Ersten Weltkrieg ein jähes Ende findet. Zwar habe sich im Krieg der Wunsch

2 Die folgenden Titel stehen beispielhaft für den Traum-Topos: Behringer/Ott-Koptschalijski, Der Traum vom Fliegen; Jürgen Bleibler/Lioba Meyer (Hg.), Der Traum vom Fliegen. Johann Schütte, ein Pionier der Luftschifffahrt (Eine Kooperation des Stadtmuseums Oldenburg mit dem Zeppelin-Museum Friedrichshafen und der Universität Oldenburg aus Anlass einer Ausstellung im Stadtmuseum Oldenburg vom 2. Juli bis 22. Oktober 2000 und im Zeppelin-Museum Friedrichshafen vom 14. Juni bis 9. September 2001), Oldenburg 2000; Radzicki, Der Traum vom Fliegen im Ruhrgebiet; Knut H. J. Röder, Der Traum vom Fliegen. Karl Friedrich Röder: Ein Pionier der Luftfahrt, o.O 2011; Winfried Suwelack/Walter Suwelack, Josef Suwelack und der Traum vom Fliegen.

3 Carsten Krystofiak, Möönsch, he flögg!! Tüftler, Dandys, Seifenkisten: Todesmutige Luftfahrtpioniere machten zu Kaisers Zeiten Münsters Luftraum unsicher, in: Ultimo 22 (2010), S.8–9, hier S.8.

zu Fliegen noch erfüllt, doch nach dem Kriegsende sei der Traum und mit ihm die „Zeit der Pioniere ... endgültig vorbei" gewesen.[4]

Auch in der Familienbiografie dient der Flug-Traum Rainer A. Krewerth als Auftakt für seine Schilderung: „Fliegen – der uralte Menschheitstraum: frei wie ein Vogel die Lüfte erobern und sich ganz von aller Erdenschwere lösen; die Welt von oben besehen und über alltäglich enge Horizonte äugen".[5] In einem Artikel des „Aerokurier" aus dem Jahr 1988 bildet der Jugendtraum vom Fliegen ebenfalls das Intro,[6] wie auch im Internetartikel eines Bloggers die Rede vom „Lebenstraum vom Fliegen" ist,[7] der Josef Suwelack nicht mehr losgelassen habe. Adolf Johanterwage titelte 2001: „Fliegen war sein Lebensinhalt"[8] und 1988 wurde die kurze Lebensspanne des Fliegers unter das Motto gestellt: „Josef Suwelack hat nur wenig mehr als zweieinhalb Jahrzehnte geträumt."[9]

Anscheinend ermöglicht der Traum-Topos, Suwelacks Handeln einen Sinn und ein Ziel zu geben, wenn die AutorInnen Suwelacks Fliegerei als seine alles beherrschende „Leidenschaft" herausstellen, der der „blutjunge wagemutige Visionär" unbeirrt nachgegangen sei.[10] Zugleich erlaubt die Rede vom Flug-Traum, den Flieger als Teil einer größeren Aufgabe zu sehen und seine Leistung bei der Realisierung dieses Traumes zu würdigen. Sowohl der Rückblick in eine oft mythische Vergangenheit wie auch die Traumvorstellung vom Fliegen als uralte Sehnsucht verknüpft, so Hartmut Rosa, die Alltags- mit der Weltgeschichte und legitimiert „die kulturellen und institutionellen Formen des Wandels und der Beharrung ..., wobei sich die Balance zwischen dynamischen und stabilisierenden Kräften, zwischen Bewegung und Beharrung historisch natürlich wandelt."[11] Mit der motorisierten Fliegerei schwerer als Luft war der Traum Realität geworden und in diesem Sinne hat sich Josef Suwelacks Flug-Traum materialisiert.

Keine Bedeutung spielt das Narrativ des Flug-Traumes in Berichten, in denen es darum geht zu erklären, weshalb sich Josef Suwelack für den Fliegerberuf entschied. Stattdessen stützt sich hier die Familienbiografie auf einen Vortrag, den Suwelack 1912 im Münsterischen Verein für Luftschiffahrt vor Publikum hielt

4 Ebd., S.9.

5 Krewerth, Ein westfälisches Fliegerleben, S.6.

6 H. Müller, „Ja, wenn ick un Motörken hädde...". Josef Suwelack vor 100 Jahren geboren, in: Aerokurier 5 (1988), S.625–626, hier S.625: „Seit seiner frühesten Kindheit träumte Suwelack davon, selbst fliegen zu können."

7 Blogger „Moritz", Tollkühne Kerle aus Westfalen erobern den Himmel, online seit dem 3.6.2012 unter http://vivatwestfalia.blogspot.de/2011/09/tollkuhne-kerle-aus-westfalen-erobern.html (15.7.2013).

8 Adolf Johanterwage, Fliegen war sein Lebensinhalt, in: Die Glocke vom 27.9.2001.

9 „Traum vom Fliegen", in: Billerbecker Anzeiger vom 28.10.1988.

10 Krewerth, Ein westfälisches Fliegerleben, S.7.

11 Hartmut Rosa, Beschleunigung. Die Veränderung der Zeitstruktur in der Moderne, ND Frankfurt a.M. 2010, S.35.

und dem er den Titel „Aus meinem Fliegerleben“[12] gab. Darin antwortete er auf die Frage, was ihn bewogen habe, Flieger zu werden, dass bei Gesprächen auf dem örtlichen Tennisplatz in Billerbeck der Name der Brüder Wright gefallen sei, die 1909 mit ihren öffentlichen Motorflügen für Diskussion und Aufsehen gesorgt hatten. Daraufhin habe Josef behauptet, dass er auch fliegen könne, sei aber von seinen Freunden verlacht worden, was ihn weiter angespornt habe, sein Talent unter Beweis zu stellen. Zudem habe er sich schon früh für Technik interessiert.

Das zum Schlüsselerlebnis erklärte Gespräch auf dem Tennisplatz findet sich in ähnlicher Form auch in anderen Fliegerbiografien wieder und wurde bereits von den Zeitgenossen als Erklärungsmodell genutzt, Fliegern wie Suwelack oder nationalen Fliegergrößen wie Max Immelmann und Oswald Boelcke rückblickend eine jugendliche Technikbegeisterung zu attestieren.[13] Stefanie Schüler-Springorum konnte darüber hinaus zeigen, dass häufig auch auf das sportliche Talent oder eben die Faszination für Technik späterer Flieger verwiesen wurde: „Diesem Muster gemäß interessierten sich fast alle späteren ‚großen Flieger‘ schon in jüngsten Jahren für technische Dinge ... mit einer Vorliebe für alles, was sie ‚hoch hinaus‘ führte.“[14] Hellmuth Hirth, mit dem Josef Suwelack 1911 bei den Rumpler-Werken in Berlin-Johannisthal zusammenarbeitete, berichtete in seiner Autobiografie „20.000 Kilometer im Luftmeer“ rückblickend von seinem großen Interesse an allem Technischen: „Zu meinem sechsten Geburtstag bekam ich mein erstes Fahrrad, und damit ging mir ein heftiger Wunsch in Erfüllung, denn jede Maschine interessierte mich von klein auf“. Ebenso betonte Hirth seine Leistungsorientierung: „In jedem sportlichen Gebiet, im Spielen und Turnen, war ich immer erster ...; kein Baum war mir zu hoch, und dreistöckige Häuser wurden mit Fingern und Fußspitzen erklettert.“[15] Karl-Heinz Knubel hingegen stellte in einem Interview über seinen Vorfahren Anton Knubel aus Münster, der sich zeitgleich mit Josef Suwelack im Fliegen versuchte, lediglich fest, dass die Fliegerei „*die* Technik der Zeit“ gewesen sei und sich Anton Knubel deshalb wahrscheinlich auch gefragt habe, ob er nicht auch fliegen könne.[16]

12 Nachlass (NL) J. Suwelack, Vortrag „Aus meinem Fliegerleben“ gehalten am 24.2.1912 vor Mitgliedern des „Luftschiffahrt-Verein Münster für Münster und das Münsterland E.V.“.

13 Vgl. Schüler-Springorum, Vom Fliegen und Töten, S.212. So auch 1936 in der Buerschen Zeitung vom 22.6.1936: „Segelflugzeug ‚Josef Suwelack‘. Buersche Segelflieger ehren einen gefallenen Kriegsflieger. Segelflugtaufe in den Borkenbergen.“, „Von Jugend auf begeisterte er sich für die Fliegerei.“

14 Ebd. Schüler-Springorum macht u.a. auch darauf aufmerksam, dass der Name Immelmann, also „Bienenmann“, für die Zeitgenossen ein Hinweis darauf gewesen sei, dass sich derjenige der Fliegerei zuwenden werde.

15 Hellmuth Hirth, Meine Flugerlebnisse. 20.000 Kilometer im Luftmeer, 2. durchgesehene und erw. Auflage, Berlin 1915, S.9 und 14. Die Erstauflage erschien im Verlag Braunbeck, Berlin 1913 unter dem Titel „20.000 Kilometer im Luftmeer“. In der zweiten Auflage wurden sowohl Kürzungen vom Verlag als auch zeitliche Ergänzungen von Hirth vorgenommen.

Der eigentliche Entschluss Suwelacks, professioneller Pilot zu werden, steht jedoch in der Rezeption seines Wirkens am Ende einer Phase der Unentschlossenheit und des Ausprobierens: „Lange Zeit konnte er sich nicht entscheiden, was er werden sollte",[17] heißt es in einem Kurzfilm des Journalisten Markus Schröder, um schließlich mit dem „Erweckungserlebnis" auf dem Tennisplatz zu enden. Statt Fragen nach den Voraussetzungen und Möglichkeiten beherrschen monokausale Erklärungen die Grundlage der Narrationen, wie z. B.: „Er hatte immer von den Gebrüdern Wright aus Amerika gesprochen und wollte ihnen nachahmen".[18] Eine alternative Darstellung, die auf die Möglichkeit hinweist, dass Josef Suwelack auch die Gelegenheit gehabt hätte, ins elterliche Molkereigeschäft einzusteigen, ist bislang ausgeblieben. Auch thematisiert kein Bericht die Kostspieligkeit des „Hobbys" der Fliegerei und seine (finanziellen) Bedingungen.

Nur selten – die Familienbiografie über Josef Suwelack ausgenommen – kommen die Darstellungen überhaupt auf den familiären Hintergrund zu sprechen. Carsten Krystofiak verweist zwar auf die Wohlhabenheit von Josef Suwelacks Vater als Molkereidirektor, wodurch sich sein Sohn dem „gepflegten Müßiggang" habe hingeben können; als eigentlicher Impuls wird jedoch erneut der Hinweis auf „das Gespräch im elitären Baumberger Tennisclub" genannt. Josef Suwelack sei „sofort begeistert" gewesen und habe sich doch noch entschlossen, „einen Beruf zu ergreifen".[19] Elke Seul umriss 2012 lediglich in zwei Sätzen die Sozialisation Josef Suwelacks: „Als Ältester von zehn Geschwistern wuchs er in Billerbeck auf. Sein Vater Josef leitete die Billerbecker Molkerei."[20] Ansonsten erwähnte sie den Vater nur noch einmal, als sein Sohn ihm sein erstes Flugzeug-Modell vorstellte: „Ein Modell aus Bambusstäben überzeugte seinen Vater. Mit dessen Erlaubnis baute er in der Billerbecker Schreiner-Werkstatt Kleideiter ein Gleitflugzeug."[21] In dem französischen Artikel aus der Zeitung „La Voix du Nord" wird Josef zumindest als drittes von zehn Kindern vorgestellt und auf seine Schul- und Studienzeit hingewiesen:[22]

16 Markus Schröder, Lokalzeitgeschichte – Münsterländische Flugpioniere, in: Lokalzeit des WDR vom 23.9.2009. Hervorhebung im Original.

17 Dazu Markus Schröders Bericht über Josef Suwelack und Anton Knubel in: Schröder, Lokalzeitgeschichte – Münsterländische Flugpioniere.

18 Markus Schröder, Lokalzeitgeschichte – Flugpionier Josef Suwelack, in: Lokalzeit des WDR vom 24.10.2001. Zum Erlebnis auf dem Tennisplatz vgl. u. a. auch Peter Elgaß, der bereits 1989 im „Fliegermagazin" schrieb, dass die „Entscheidung des Josef Suwelack ... 1909 im Tennisclub in Billerbeck" fiel. Peter Elgaß, Josef Suwelack – ein kurzes Fliegerleben, in: Fliegermagazin (1989), Nr. 4, S. 86–88, Zitat S. 86.

19 Krystofiak, Möönsch, S. 8.

20 Seul, Tollkühner Billerbecker.

21 Ebd.

22 Tatsächlich war Josef Suwelack der zweitgeborene Sohn der Suwelacks. Der Erstgeborene Heinrich starb 1887 bereits vier Monate nach seiner Geburt (25.4.–25.8.1887) und Bernhard Suwelack verstarb ebenfalls circa ein halbes Jahr nach seiner Geburt (24.9.1889–10.4.1890).

> „Il était le troisième enfant d'une fratrie de dix, de Joseph Suwelack et de Katharina Brockmann. Josef passa sa scolarité au collège de Burgsteinfurt ..., puis, en 1906, il rejoignit l'école technique supérieure de Darmstadt ..., qu'il quitta peu après pour l'académie des Beau-Arts de Düsseldorf. Mais l'idée de voler lui troittait déjà dans la tête."[23]

Alain Fernagut, Verfasser des dreiteiligen Artikels, verwendet in diesem Artikel das Wort „Traum" (rêve) zwar nicht, doch habe sich, so Fernagut, die Idee zu Fliegen bereits in Suwelacks „Kopf" (tête) befunden.

Der überwiegende Teil der Darstellungen verschweigt hingegen sowohl die Herkunft Suwelacks aus einer wohlhabenden Großfamilie als auch, dass er sich beruflich durchaus in verschiedene Richtungen orientierte. So erscheint der Direktoren-Sohn – aus einer Großfamilie – in den meisten Texten eher als „Einzelkind", das einem vorgefassten Plan folgte.[24]

Verständlicherweise widmen die Suwelacks in ihrer Familienbiografie diesem Thema weit mehr Aufmerksamkeit und stellen Josefs Eltern als warmherzige Menschen und Joseph Suwelack senior als erfolgreichen Unternehmer vor, der die spätere Karriere seines Sohnes überhaupt erst möglich gemacht habe.[25] Darüber hinaus skizzieren sie den schulischen sowie den beruflichen Werdegang des späteren Piloten und widmen sich den Interessen des jungen Erwachsenen:

> „Wir bekommen hier einen ersten Eindruck von seiner beruflichen Lebensplanung, die seinen vielseitigen Begabungen gerecht werden sollte. ... Voller Hingabe betrachtete er sein Leben lang die Malerei, ließ selten eine sich bietende Gelegenheit aus, um die wenigen Stunden der Muße damit zu verbringen. ... Warum er so bald die Kunstakademie [in Düsseldorf] wieder verließ, lässt sich nicht erforschen. Ging ihm schon die Fliegerei durch den Kopf? Vielleicht."[26]

Die prinzipielle Offenheit der Berufslaufbahn Josef Suwelacks verschwindet im folgenden Kapitel allerdings hinter dem „Versprechen auf dem Tennisplatz", also jenem Schlüsselmoment, der aus dem unentschlossenen jungen Mann einen zielstrebigen Flugpionier werden ließ: „Im Jahre 1909 beginnt Josef Suwelack dann, einen mühsamen, steilen Weg zu beschreiten, der ihn buchstäblich in große Höhen führte."[27]

Diese Ursprungserzählung von der heimatlichen Verwurzelung Josef Suwelacks und seinem anschließenden Aufbruch in die Welt der Fliegerei lässt ihn im weiteren Verlauf als mutigen Abenteurer erscheinen, als Sohn eines erfolgreichen Unternehmens, der seinen Weg geht und zum gefeierten Flieger und Konstrukteur wird. Das „Tennisplatz-Erlebnis" bildet hierbei den Ausgangspunkt für seine individuelle Fliegerkarriere.

23 Alain Fernagut, Josef Suwelak, un aviateur allemand inhumé au cimetière militaire britannique d'Erquinghem, in: La Voix du Nord vom 13.3.2007 (Teil 1 von 3).

24 Vgl. Krystofiak, Möönsch, S.8.

25 Vgl. Krewerth, Ein westfälisches Fliegerleben, S.18ff.

26 Ebd., S.24.

27 Ebd., S.25.

Die „westfälische Beharrlichkeit" und der „Schwur auf dem Trümmerhaufen"

Wo liegen die Ursachen und Gründe für den Erfolg Suwelacks, oder anders formuliert: Wie wurde aus dem ländlichen Unternehmersohn ein überregional bekannter Flieger? Für viele AutorInnen erschließt sich die Antwort auf diese Fragen aus seiner westfälischen Herkunft. Insbesondere Texte aus Westfalen über Josef Suwelack unterliegen dabei häufig einem Lokalpatriotismus, der sich in generalisierenden Aussagen zeigt. So wurde z. B. im „Westfalenspiegel", dem Presseorgan der westfälischen Heimatbundes, aus dem Jahres 1960 explizit auf das Durchhaltevermögen und Festhalten an Prinzipien von aus der Region stammenden Flugpionieren verwiesen. Trotz der öffentlichen Präferenz des Zeppelin-Systems, so der Artikel, „gehen westfälische Konstrukteure [unbeirrbar] ihren Weg. Und dieser Weg ist, wie sich schon bald herausstellt, der richtige Weg."[28] Ähnlich argumentiert auch der Verfasser eines Artikels in der Zeitschrift „Aerokurier", der von Josefs ersten vergeblichen Flugversuchen in Billerbeck und Münster berichtete: „Daß das Gleitflugzeug nicht die gewünschten Leistungen erbrachte, entmutigte Suwelack keineswegs. Mit westfälischer Beharrlichkeit konstruierte er ... einen motorbetriebenen Eindecker."[29] Der Kommentar des Bloggers Moritz, Westfalen seien „durch und durch Optimisten",[30] reiht sich ebenfalls in das Narrativ der westfälischen Beharrlichkeit ein, die Flugpioniere wie August Euler, Anton Knubel und Josef Suwelack als Männer mit einem feinen Gespür und sicheren Instinkt für technische Innovationen scheinbar verbindet: „Es ist kennzeichnend für die Aufgeschlossenheit, den Wagemut und die Instinktsicherheit westfälischer Menschen, wenn im gleichen Jahr [der ersten erfolgreichen Motorflugversuche], 1909, also noch in der Pionierzeit ..., der Luftfahrtverein Münster gegründet wird."[31] Neben Zuschreibungen wie Wagemut und Beharrlichkeit werden auf diese Weise über die Flugpioniere – und somit auch über die Person Josef Suwelack – Fähigkeiten wie Ideenreichtum und Pioniergeist zu westfälischen Tugenden erklärt.[32]

Nicht überraschend ist, dass im Mittelpunkt der Familienbiografie – wie das

28 Gerd Schröder, 50 Jahre Flugsport in Westfalen, in: Westfalenspiegel (1960), S.21–24, hier S.21.

29 Müller, Motörken, S.625.

30 Moritz, Tollkühne Kerle.

31 Ebd., S.21.

32 Siehe zur Genese und Tradierung eines Westfalen-Stereotyps: Karl Ditt, Was ist „westfälisch"? Zur Geschichte eines Stereotyps, in: Westfälische Forschungen 52 (2002), S.45–94. Ditt verfolgt die Herausbildung und Verfestigung eines Westfalen-Bildes ausgehend vom 8. Jahrhundert bis ins 21. Jahrhundert. Ihm zufolge sind die allgemein formulierten Stereotypen der „Urtümlichkeit, Stärke und Bäuerlichkeit" etc. kompatibel mit dem Selbstbild der Westfalen: „Die universale und wiederholbare Verwendbarkeit dessen, was als ‚westfälisch' angesehen wurde, machte diese Interpretation zu einem Stereotyp. Es bietet eine bequeme Orientierung für die Kategorisierung, Erklärung und Bewertung – sei es als Selbstlob oder Diskreditierung – der Bevölkerung und ihres Verhaltens." Ebd., S.92–93.

Inhaltsverzeichnis bereits ausweist – Josef Suwelacks Weg „nach oben“ steht. Die Überschriften „Ein kurzes Himmelstürmerleben“, „Aufwärts mit der Rumpler-Taube“ oder „Weltpresse meldet Weltrekord“ betonen hierbei seine Leistungen und seine unaufhaltsamen Karriereschritte.[33] In diesem Sinne heißt es auch am Ende des Vorworts: „Dieses Buch erzählt die atemberaubende Geschichte eines westfälischen Fliegerlebens“.[34]

Eng verknüpft mit dem Topos von der „westfälischen Beharrlichkeit“ liest sich der Mythos vom „westfälischen Blut“ und damit von der „natürlichen“ (vererbten) Veranlagung für die Fliegerei. So bescheinigte Adolf Johanterwage Suwelack intuitives Handeln in der Bildunterschrift zu einem Foto, das den Piloten in Fliegermontur mit den typischen Accessoires zeigt wie Handschuhe, Mantel, Fliegermütze, Brille und Schal : „Josef Suwelack aus Billerbeck war ein Naturtalent, das theoretisch-technisches Wissen und praktisches Können als Pilot in nahezu idealer Weise erfolgreich vereinte.“[35]

Mit diesen Worten greift der Verfasser auf einen Fliegertopos zurück, der schon zu Suwelacks Zeiten in Militärkreisen kontrovers diskutiert wurde: Den Topos vom „fliegerischen Gefühl“, das gleichzusetzen sei mit der Intuition als *airman*, aber auch mit der Eigenstabilität der Maschine, die dem Piloten dann allerdings nur die passive Rolle als „Chauffeur“ zubilligt.[36] Dabei wird dem *tacit knowledge*, dem impliziten Wissen, in der Technikforschung eine große Bedeutung bei die Entwicklung neuer Technologien zugesprochen und auch an Suwelacks Beispiel lässt sich zeigen, dass er seine Erfahrungen beim Fliegen zur Verbesserung und zum Umbau seiner Flugapparate praktisch anwandte.[37] Zugleich erscheint er als Beherrscher der Technik, der seine Nerven[38] – ein weiterer Topos – stets im Griff hatte und seine Fähigkeiten als Pilot stets angemessen einschätzen konnte.[39]

2001 teilte Wolfgang Suwelack in einem Interview mit: „Mich haben in erster Linie fasziniert ... der Wagemut und der Ideenreichtum, den der Josef von seinem Vater Joseph geerbt haben muss.“ Der Kommentator schloss sich diesem Urteil

33 Vgl. Krewerth, Ein westfälisches Fliegerleben, S.4.

34 Ebd., S.7.

35 Johanterwage, Fliegen war sein Lebensinhalt.

36 Vgl. zum „fliegerischen Gefühl“: Kurt Möser, Fahren und Fliegen, S.200ff. und zu den Begriffen *airman* und „Chauffeur“ Christian Kehrt, Vom Luftakrobaten zum Cyborg? Technikerfahrungen deutscher Militärpiloten im 20. Jahrhundert, in: Eberhard Birk/Heiner Möllers/Wolfgang Schmidt (Hg.), Die Luftwaffe zwischen Politik und Technik, Berlin 2012, S.15–37, hier S.20ff. Siehe auch ders., Moderne Krieger, S.107ff.

37 Zum Begriff des *tacit knowledge* in der Technikforschung vgl. Matthias Heymann/Ulrich Wengenroth, Die Bedeutung von „tacit knowledge“ bei der Gestaltung von Technik, in: Ulrich Beck/Wolfgang Bonß (Hg.), Die Modernisierung der Moderne, Frankfurt a. M. 2001, S.106–121. Zu Suwelacks Tätigkeit siehe die Ausführungen zum Feld 2.

38 Joachim Radkau, Das Zeitalter der Nervosität. Deutschland zwischen Bismarck und Hitler, Wien 1998.

39 Siehe die Ausführungen zum Feld 1.

an: „Eine gute Portion Mut und Optimismus müssen dem Flieger Josef Suwelack im Blut gelegen haben. Pioniergeist wächst eben auch in ländlich verträumten Regionen."[40] Erfolg wird hier nicht auf Leistung oder Zufall zurückgeführt, sondern auf familiäres und heimatliches „Blut", und somit auf genealogische Verwandtschaft und überlieferte heimatliche Tradition.[41]

Neben diesen „intrinsischen Fundamenten" sind für die Flieger-Mythen vor allem exzeptionelle Fähigkeiten sowie eine Entdeckerfreude, die die Flugpioniere aus ihrer heimatlichen Enge herausführt, charakteristisch. Auch Biografen von Josef Suwelack betonen dies, wenn sie z. B. schreiben, dass er 1910 „der westfälisch-bürgerlichen Enge entflohen" sei.[42]

Flugpioniere wurden dabei nicht selten als *self-made-men* gezeichnet, die ihren Weg wenn nötig alleine gingen und sich durch nichts aufhalten ließen. Als Josef Suwelack im April 1911 in einem Flugzeug der Deutschen Flugmaschinenbau Gesellschaft (DFG) in Berlin-Rummelsburg seinen Flugschein machen wollte, stürzte die Maschine durch Motorversagen ab und wurde schwer beschädigt, Suwelack hingegen blieb unverletzt. Daraufhin, so berichtete der Flieger in seinem Vortrag von 1912, habe er sich „auf dem Trümmerhaufen" des zerstörten Apparates geschworen, nie wieder mit Menschen ohne die nötige Begeisterung und Leidenschaft für die Fliegerei zusammenzuarbeiten. Josef Suwelack inszenierte sich hier buchstäblich als wiedererstandener Phönix aus der Asche, der gestärkt aus dem Absturz hervorging. Diese retrospektiv entstandene autobiografische Selbstdarstellung des Fliegers nutzten später seine Nachfahren und formten daraus die Losung: „Aus dem Unglück bei der Deutschen Flugmaschinenbau-Gesellschaft, aus den Trümmern seiner Maschine erwächst die steile Karriere des Josef Suwelack."[43] Anlehnend an Suwelacks Selbstbeschreibung schrieb Peter Elgaß 1989 dazu: „Für einen Mann mit fliegerischem Feuereifer waren solche Menschen keine Basis für eine fliegerische Zusammenarbeit",[44] und ein Jahr zuvor konnte man im „Aerokurier" lesen: „Auf den Trümmern des Apparates beschloß er, jetzt bei Hirth in die Lehre zu gehen."[45]

Auch in anderen Berichten werden Suwelacks Fehlschläge relativiert oder sogar gänzlich weggelassen. Zwar finden sich auch hier Formulierungen wie „unvermeidliche... Rückschläge",[46] doch letztlich zahlte sich nach dem *Trial-and-*

40 Interview mit Wolfgang Suwelack, in: Schröder, Lokalzeitgeschichte – Flugpionier Josef Suwelack, Min. 4:57ff. sowie ebd., Min. 5:10ff.

41 Vgl. Ditt, Was ist „westfälisch"?, S.77ff.

42 Buchvorstellung „Josef Suwelack und der Traum vom Fliegen", in: Billerbecker Anzeiger vom 28.10.1988.

43 Krewerth, Ein westfälisches Fliegerleben, S.36.

44 Elgaß, Josef Suwelack, S.87.

45 Müller, Motörken, S.625.

46 Johanterwage, Fliegen war sein Lebensinhalt.

Error-Prinzip seine Beharrlichkeit schließlich aus: „Von Bruchlandungen lässt er sich keineswegs abschrecken". Zentral war hier der Fortschrittsglaube der Autoren, der auch schon den Zeitgenossen Suwelacks geholfen hatte, ihre Erfahrungen zu verarbeiten, um die „Ambivalenzen dieser Moderne" in eine Balance zu bringen, indem sie die technische Entwicklung nicht als Bruch mit der Vergangenheit, sondern vielmehr als deren Weiterentwicklung deuteten.[47]

So wurde auch ein gescheiterter eigenmächtiger Langstreckenflug, den Josef Suwelack im Februar 1913 in seinem Kondor-Eindecker unternahm, von einigen Autoren noch in einen Erfolg verwandelt. Suwelack war bei seinem geplanten Flug von Essen nach London über die Nordsee in eine Nebelbank geraten, hatte die Orientierung verloren und musste schließlich bei Deventer notlanden. Sein „Flug über den Kanal" war in den folgenden Tagen Gegenstand zahlreicher Presseartikel und sorgte für Aufsehen und Aufregung bei den Gesellschaftern der Kondor-Werke, die wenig von Suwelacks riskantem „Aus-Flug" hielten (siehe Feld 2). Suwelack selbst hielt seine Eindrücke und Erfahrungen in seinem Bericht „Mein Flug zur Nordsee"[48] fest, der 1913 in der Zeitschrift „Flugsport" abgedruckt wurde.

Aus einer Kosten-Nutzen-Sicht war der Flug gescheitert, denn der Pilot hatte sein Ziel nicht erreicht und Betriebsmittel verbraucht. Argumentiert man hingegen vor dem Hintergrund der Vermarktung des Piloten und des neuen Flugzeuges, ergibt sich eine differenziertere Bewertung. Josef Suwelack hatte Mut bewiesen und die Zuverlässigkeit des Apparates demonstrieren wollen. Bereits 1909 hatte Louis Blériot den Ärmelkanal überquert und wurde in den folgenden Jahren von anderen Aviatikern nachgeahmt, die ebenfalls nach Ruhm und Aufmerksamkeit strebten. Robert Wohl bezeichnet es als „social drama, which would be played out time and again throughout the twentieth century by a long series of aviators".[49] Insofern erhält der Flug Suwelacks im Kontext der medialen Kontextualisierung, die mit spannungsgeladenen Berichten die Sensationslust der Menschen verstärkte, seine Bedeutung. Darüber hinaus wird der nachträgliche „Erfolg" dieses Unglücks erst vor dem Hintergrund einer Flugzeugindustrie verständlich, die von einem harten Konkurrenzkampf geprägt war und umso mehr das Risiko des „Pioniergeists" als Marketingelement benötigte. In der Rezeption dieses Ereignisses interessierten diese Zusammenhänge allerdings weniger, vielmehr stand die individuelle Leistung des wagemutigen Fliegers im Vordergrund, wenngleich

47 Aleida Assmann, Ist die Zeit aus den Fugen? Aufstieg und Fall des Zeitregimes der Moderne, München 2013, S.58ff. Bedrich Loewenstein, Der Fortschrittsglaube. Geschichte einer europäischen Idee, Göttingen 2009.

48 Josef Suwelack, Mein Flug zur Nordsee, in: Flugsport (1913), Nr. 5, S.173–174.

49 Robert Wohl, A Passion for Wings. Aviation and the western imagination. 1908–1918, Yale 1994, S.66.

auch hier Josef Suwelack als erfolgreiche westfälische Marke inszeniert wurde:[50] „1913 startet ein Flugzeug made in Westfalen by Suwelack mit Kompass und Karte an Bord zum ersten Kontinental-Linienflug: Berlin – Brüssel – Paris – London. Die Internationale Route ergänzen die Münsterländer selbstbewusst durch die Etappe ‚Rotthausen/Westfalen'!"[51]

In der Familienbiografie liegt der Schwerpunkt auf dem Erlebnisbericht des Fliegers, wie er von Suwelack später auf der Schreibmaschine notiert und in leicht geänderter Form in der Zeitschrift „Flugsport" abgedruckt wurde. Drei Seiten sind dem Bericht in der Biografie gewidmet und am Ende des langen Berichts urteilen die Verfasser:

> „Diese Niederschrift gehört ohne jeden Zweifel zu einer der schönsten Darstellungen aus der Pionierzeit der Luftfahrt. ... Sie legt Zeugnis ab von der Einsamkeit eines Menschen, der sich nicht satt sehen kann an den Schönheiten der Natur – zwischen den Wolken am Horizont und den Wellen der Nordsee mit ihren weißen Schaumkronen."[52]

Fliegerberichte wie der von Josef Suwelack waren typisch für die Zeit der frühen Fliegerei und halfen den Akteuren, ihre Erfahrungen und die neue Wahrnehmung aus der Luft, angelehnt an die romantische Erzähltradition eines Joseph von Eichendorff, zu verarbeiten und gleichzeitig die Leser an ihrem Abenteuer teilhaben zu lassen.[53]

Indem Josef Suwelack zu einem „Himmelstürmer" stilisiert wird, der den Luftraum „erobert" – ein Fortschrittsnarrativ, das in seine Zeit zurückreicht –,[54] entsteht das Bild eines Menschen, dem keine Grenzen gesetzt sind und der – wenngleich vornehmlich außerhalb Westfalens – auf seinem Weg als Flugpionier seinen Wurzeln und Tugenden treu bleibt.

50 Vgl. Günter Schweiger, Image und Imagetransfer, in: Bruno Tietz/Richar Köhler (Hg.), Handwörterbuch des Marketings, Stuttgart 1995, Sp. 915–928; zur Werbung mit Westfalen in der Nahrungsmittelindustrie: Hannes Siegrist, Regionalisierung im Medium des Konsums, in: Comparativ 11 (2001), H. 1, S. 7–26, und grundsätzlich: Ditt, Was ist „westfälisch"?, S. 88ff.

51 Krystofiak, Möönsch, S. 9.

52 Krewerth, Ein westfälisches Fliegerleben, S. 71.

53 Siehe die Ausführungen zum Feld 2.

54 Der Topos der „Eroberung der Luft" strukturierte die zeitgenössischen Erzählungen über die Fliegerei und war Ausdruck des nationalen Selbstbewusstseins und der deutschen nationalen „Größe". Vgl. exemplarisch folgende Titel: Rudolf Martin/Gustav Schalk, Die Eroberung der Luft. Kritische Betrachtung über die Motorluftschifffahrt. Mit 5 Abbildungen, Berlin 1907; Ferdinand von Zeppelin, Die Eroberung der Luft. Ein Vortrag gehalten im Saale der Sing-Akademie zu Berlin am 25. Januar 1908, Stuttgart/Leipzig 1908; E. H. Bethke (Hg.), Die Eroberung der Luft von Luftschiffen und Flugmaschinen. Ausgeführte Vortragsabende mit Lichtbildern, Vorträgen, Deklamationen, Liedern und Bühnenspiel, Leipzig 1913; Jules Verne, Robur der Flieger. Ergänzt durch die Eroberung der Luft bis zur Jetztzeit, Berlin 1914; Willi Hackenberger, Deutschlands Eroberung der Luft. Die Entwicklung deutschen Flugwesens an Hand von 315 Wirklichkeitsaufnahmen dargestellt, Berlin 1915.

„Die tollkühnen Männer in ihren fliegenden Kisten" oder: auf den „Spuren großer Männer" – zur Konstruktion stereotyper Charakterzüge

„Die tollkühnen Männer in ihren fliegenden Kisten" ist nicht nur der weithin bekannte Titel einer britischen Filmkomödie aus dem Jahre 1965 über einen internationalen Flugwettbewerb im Jahr 1910, sondern auch ein geflügelter Satz, wenn es um die Charakterisierung von Flugpionieren geht. So titelte 2012 der Blogger Moritz: „Tollkühne Kerle aus Westfalen erobern den Himmel", und Elke Seul überschrieb ihren Beitrag mit: „Tollkühner Billerbecker in fliegender Kiste". Von „tollkühnen Münsteranern in klapprigen Kisten" berichten Krystofiak und Schröder und im Vorwort der Familienbiografie von 2001 schreibt Rainer A. Krewerth: „Tollkühne Kerle aus ihrer Mitte [der Westfalen] wagten bis dahin fast Unvorstellbares. Sie knatterten todesverachtend himmelwärts!"[55] „Tollkühnheit", so scheint es, ist der ideale Begriff, um alle außergewöhnlichen Fliegereigenschaften zu bündeln. Und regelmäßig hoben die AutorInnen in diesem Zusammenhang seine Todesverachtung und die Verwegenheit des Fliegers hervor, wobei sie ihm gleichzeitig auch großes Talent attestierten. Dabei wurde er mal als „forscher Pilot"[56] beschrieben, mal als „einer der Wagemutigsten".[57] In diesen Kontext passt auch eine Charakterisierung von Johannes Loy, der ihn 2001 in einer Rezension zur neu aufgelegten Familienbiografie als einen Mann beschrieb, der anscheinend überall reüssierte und sich „heimisch" fühlen konnte: „Er war nicht nur Ingenieur, Konstrukteur, Unternehmer und Rekordflieger, sondern auch Lebenskünstler, charmanter Plauderer, Student technischer und künstlerischer Institute und Maler, und nicht zuletzt ein überzeugter Katholik."[58]

Charakterisierungen dieser Art evozieren das Bild des Dandys, der seine Fähigkeiten geschickt für seine Zwecke zu nutzen weiß und sich seiner Wirkung auf Andere bewusst ist. So schreibt Elgaß: „Das [Finden von Geldgebern] fiel ihm mit seinem Ruf nicht übermäßig schwer."[59] Er sei in den Berliner Fliegerkreisen ein „gern gesehener Mann"[60] gewesen. Dazu sei er ehrgeizig und zielstrebig gewesen, ein junger Mann, der von Erfolg zu Erfolg geeilt sei. „Kurz entschlossen" sei er zunächst nach Münster,[61] dann nach Berlin gegangen, um seinem Ziel Flieger zu werden, näher zu kommen: „Schon einige Wochen später war Suwelack im

55 Moritz, Tollkühne Kerle; Seul, Tollkühner Billerbecker; Krystofiak, Möönsch, S.8; Schröder, Lokalzeitgeschichte – Münsterländische Flugpioniere; Krewerth, Ein westfälisches Fliegerleben, S.71.
56 Krystofiak, Möönsch, S.9.
57 NL J. Suwelack, „Erstmalig um die Türme von Billerbeck. Vier Stunden bedeuteten damals noch Weltrekord", in: [unbekannter Zeitungstitel] 1961.
58 Johannes Loy, Flugpionier aus dem Münsterland, in: Westfälische Nachrichten vom 27.9.2001.
59 Elgaß, Josef Suwelack, S.88.
60 Ebd., S.87.
61 Ebd., S.86.

Wettstreit mit den besten Piloten seiner Zeit um 10.000 Goldmark bei einem Wettbewerb zur Nationalen Flugwoche.“[62]

Gleich welche Zeit, benutzten die jeweiligen Zeitgenossen Suwelack schließlich auch als Projektionsfläche eigener Heldenvorstellungen. So stilisierte man ihn in der Zeit des Nationalsozialismus zum vorbildhaften Kameraden, dem die Opferbereitschaft das höchste Gut schien:

> „Zum Inhalt der fliegerischen Kameradschaft [...] gehört vor allem auch, daß jeder ohne Neid und ohne Rücksicht auf sich selbst auch die Leistungen des anderen anerkennt. [...] Die Jugend will auch in Bewunderung vor den Heldentaten anderer Waffengattungen stehen, die ihr Leben in gleicher Treue und Opferbereitschaft im Weltkrieg für das Vaterland [eingesetzt haben]“.[63]

Knapp 50 Jahre später dann wurde Suwelack zu einem Held des Widerstandes:

> „Wäre er ein Fliegeras des Zweiten Weltkriegs geworden, ein Konstrukteur genialer Flugmaschinen? Er hätte, mutmaßlich, als phantasiebegabter Techniker und Unternehmer seinen Weg weit fort aus dem Münsterland gemacht, hätte sich dem sogenannten Dritten Reich verweigert (dafür sprechen die ergreifenden Dokumente seiner Glaubenstreue) und stünde heute mit seinem Namen – statt auf einem französischen Grabstein – in den Registern jenes Buches, das die Namen bekannter und erfolgreicher Westfalen festhält.“[64]

So habe sich Suwelack als Katholik „vorm Töten geekelt“, was „glaubhaft“ in der Biografie nachzulesen sei.[65] In beiden Darstellungen wird Suwelack Teil des jeweiligen Geschichtsbildes und zum Vorbild für die jeweilige Gegenwart konstruiert mit dem Appell, es ihm gleichzutun.

Gemeinsam ist den Helden-Narrationen, dass sie in Josef Suwelack einen Mann sehen, der „Geschichte geschrieben“ und sich stets politisch und moralisch korrekt verhalten habe.[66] Nur so kann Josef Suwelack in eine Reihe großer Namen der Fluggeschichte gestellt – an die Seite von Otto Lilienthal, den Wrights oder Manfred von Richthofen –[67] und zu einem ebenbürtigen Mitbegründer der Luftfahrt erhoben werden, der Teil einer Fliegerelite gewesen sei: „Il incarnait une élite parmi les élites. La guerre aérienne fit de lui un guerrier du ciel.“[68]

62 Ebd., S. 87.
63 Rede anlässlich der Taufe eines Segelflugzeuges auf den Namen „Josef Suwelack“: „Segelflugzeug ‚Josef Suwelack'“, in: Buersche Zeitung [1936].
64 „Traum vom Fliegen“, in: Billerbecker Anzeiger vom 28.10.1988.
65 Ebd.
66 So Markus Schröder in der „Lokalzeitgeschichte – Flugpionier Josef Suwelack“.
67 Loy, Flugpionier.
68 Fernagut, Un aviateur, Teil 3.

Heldenbilder und -gedenken im Wandel der Zeit: Josef Suwelack als Sport-, Lokal- und Kriegsheld

Das zeitlose Bedürfnis nach Helden als moralische, kämpferische, gesellschaftliche oder private Vorbilder spiegelt sich auch in der Rezeptionsgeschichte Josef Suwelacks wider.[69] Die jeweiligen Heldenbilder vermitteln dabei viel über die jeweiligen Wertmaßstäbe und die politische Kultur einer Gesellschaft.[70]

Drei Heldentypen kristallisieren sich in den Darstellungen heraus: Erstens der Kriegsheld Suwelack, der sein Leben für das Vaterland opferte, zweitens der Held im Wagnissport,[71] der seine Gesundheit für den Wettbewerb riskierte, und schließlich der lokale Held als Ikone und Werbefigur für seine Region. Der letztgenannte Heldentypus speist sich wiederum vornehmlich aus den beiden anderen Typen, denn sowohl die sportlichen als auch die im Krieg erzielten Erfolge, die bis zur Heimatfront vordrangen, wurden von den Medien verfolgt.

Insbesondere die Fliegerei ist ein Beispiel dafür, wie Erfolge in Wettbewerben auf den Kriegseinsatz übertragen wurden und der sportliche Wettkampf – verstärkt in den letzten beiden Vorkriegsjahren – zu einer nationalen Angelegenheit und Prestigefrage umgedeutet wurde.[72] Paul Dietschy hat dieses Phänomen am Beispiel verschiedener französischer Sportzeitschriften nachgewiesen und gezeigt, dass in Frankreich der Krieg als Sport-Spektakel und *grand match* begriffen wurde.[73] In der Fachpresse sei unter anderem darüber diskutiert worden, ob der Sport die jungen Männer auf den Krieg vorbereite, und Dietschy zufolge sei es vor allem die Presse gewesen, die maßgeblich zur Entstehung des Fliegerhelden, zum Topos des ritterlichen Luftkampfes und des Fairplay beigetragen habe.[74]

Beispielhaft für die Idealisierung Suwelacks zum Sporthelden lässt sich eine Passage aus dem „Fliegermagazin" aus dem Jahr 1989 anführen, in der er zum

69 Siehe dazu: Albrecht von Lucke, Neue Helden braucht das Land, in: Frankfurter Hefte 4 (2008), S.64–65.

70 Siehe zum ambivalenten Heldenbegriff z. B. Michael Naumann, Strukturwandel des Heroismus. Vom sakralen zum revolutionären Heldentum, Königstein/Ts. 1984; Tina Hildenbrandt/Elisabeth Niejahr, „Was ist ein Held?". Verteidigungsminister Thomas de Mazière und der Schriftsteller Uwe Tellkamp im Gespräch über Drohnen, gezieltes Töten, Moral und Tapferkeit, in: Die Zeit vom 14.7.2013; Ausschreibung zum Geschichtswettbewerb des Bundespräsidenten 2008/09: Helden. Verehrt – verkannt – vergessen, in: Spuren suchen 22 (2008), S.3–42, bes. S.24–29 und S.30–31; LWL-Industriemuseum (Hg.), Helden. Von der Sehnsucht nach dem Besonderen. Katalog zur Ausstellung des LWL-Industriemuseum Heinrichshütte Hattingen 12.3.–31.10.2010, Essen 2010.

71 Stern, Heldenfiguren im Wagnissport.

72 Siehe zu diesem Punkt die Ausführungen zum Feld 3.

73 Paul Dietschy, Kriegshelden – Sporthelden? Das Dilemma der französischen Sportpresse im Ersten Weltkrieg, in: Arnd Krüger/Swantje Scharenberg (Hg.), Zeiten für Helden – Zeiten für Berühmtheiten im Sport. Reflektionen der 9. Hoyaer Tagung Sportstars, Helden und Heldinnen. Veränderungen in der Darstellung berühmter Sportler und Sportlerinnen in den Massenmedien, Berlin 2014, S.49–64, Zitat S.54.

74 Vgl. ebd., bes. S.57ff.

Bezwinger der Natur im Kampf Mensch und Maschine gegen die Naturgewalten stilisiert wird: „Trotz des etwas böigen Windes startete er mit seiner Rumpler-Taube mit Aeolus-Motor und Garuda-Propeller zum Dauerflug-Weltrekord.“[75] Ein Jahr zuvor berichtete der „Billerbecker Anzeiger“ von Josef Suwelack als „der Held aus Billerbeck“;[76] 1961 erhielt er posthum im Kontext der Flugtage auf der Münsterschen Loddenheide die Würde des sportlichen „ Held[en] des Tages“.[77]

Obwohl die Mehrzahl der Artikel nach dem Ende des Zweiten Weltkrieges verfasst wurde, konnten sich die Autoren nicht vom Heroischen und von der Darstellung des Fliegers als Kämpfer lösen. Die gängige Narration zu seiner Person verlagerte sich seit den 1960er Jahren inhaltlich zwar mehr zu seiner Konstrukteurtätigkeit, jedoch erscheint er auch in diesem Bereich eher als Einzel-Kämpfer denn als Ingenieur im Team.

Am deutlichsten tritt die legendenhafte Überformung des Billerbecker Fliegers und Unternehmers bei der Skizzierung seiner Todesumstände hervor, die bis heute nicht endgültig geklärt werden konnten – ein Umstand, den bislang jedoch nur die Familienbiografie und drei weitere Autoren hervorgehoben haben:[78] „Die Umstände seines Todes haben seine Nachfahren bis heute beschäftigt. Wahrscheinlich wurde er 1915 nach einer geglückten Landung von britischen Soldaten am Boden erschossen“,[79] schrieb Loy 2001.[80] Dabei stützen sich die Autoren vor allem auf zwei Berichte über die Todesumstände Suwelacks, die kurz nach seiner Vermisstenmeldung erschienen. Zum einen war dies ein englischer Militärbericht, in dem es über den 13. September 1915 heißt, dass es zu einem Luftkampf zwischen einem englischen und einem deutschen Flugzeug gekommen sei, infolgedessen Josef Suwelack und sein Beobachter Leutnant Oskar Teichmann angeschossen wurden und die Maschine wegen des Beschusses hinter den feindlichen Linien musste:

> „Points of interest: – 2nd Lt/ Shield and Corporal Bennett (No 16 Sq) on a B.E. 2c. attacked a German Albatross in the vicinity of ARMENTIERES.Lt. Shield dived from 10000 to 7000 ft. and whilst circling round the hostile machine emptied two drums at it. The Albatross dived and eventually landed without any harm to the machine near STEENWERCK.“[81]

75 Elgaß, Josef Suwelack, S.88.
76 „Traum vom Fliegen“, in: Billerbecker Anzeiger vom 28.10.1988.
77 StdA Münster Bestand Dok-Ztg, Serie Stadt-Dok. Nr. 806, 1960–61.
78 Seul, Tollkühner Billerbecker; Fernagut, Un Aviateur; Loy, Flugpionier.
79 Loy, Flugpionier.
80 Die Frage am Ende der Familienbiografie „Wie starb der Pilot wirklich?“ zeigt den Wunsch nach Gewissheit, der trotz intensiver Nachforschungen letztlich nicht befriedigt werden konnte. Vgl. Krewerth, Ein westfälisches Fliegerleben, S. 106–111, Zitat S.106.
81 Familienarchiv Suwelack, Bericht des Royal Flying Corps vom 13.9.1915 (Imperial War Museum, London, AIR1/2166/209/11/11), http://www.josef-suwelack.de/der-ersteweltkrieg/sein-todestag-am-13-september-1915 (16.6.2013).

Weiter hieß es im Kriegstagebuch des Royal Flying Corps:

> „The pilot Lt SUWELACK, 24th Flieger Abteilung attached to XIX (Saxon) Corps, and the observer, Lt OSKAR TEICHMANN, 2nd Photographic Section, were both killed by bulllets. The machine which was undamaged, except by bullets, was a new Albatross of the small type with 160 H.P. MERCEDES engine, its number was C 60/15, makers No. 853. There was a machine gun in it on a good type mounting, admitting of all-round fire. Three cameras were found in the aeroplane.“[82]

Diese Meldung, so stand es in einem späteren Schreiben der Feldfliegerabteilung 24 an Josefs Vater, sei von deutschen Stellen ebenfalls bestätigt worden.[83] Die Auflistung der Ausstattung des Albatros-Doppeldeckers lässt ebenfalls vermuten, dass die Maschine relativ unversehrt hatte landen können, denn an Bord wurde die komplette Ausstattung des Flugzeuges gefunden, die von den Engländern eingehend inspiziert wurde.[84]

Im französischen Bericht der Zeitung „Le Matin“ vom 16. September 1915, der sich auf einen englischen Bericht in der „Daily Mail“ bezog, war hingegen zu lesen, dass der deutsche Flugzeugführer und sein Beobachter, nachdem sich die Flugzeuge einige Minuten umkreist hätten, in der Luft getötet worden seien – eine Kugel habe den Benzintank getroffen und ihn zur Explosion gebracht, während das englische Flugzeug unbeschadet entkommen sei:

> „Quartier Géneral Anglais, 13. Septembre. Après un emouvant combat aérienn aéroplane allemand d' un des tout derniers modèıes a été abattu ce matin par un avion anglais; les deux aviateurs qui l'ocoupaint ont été tués. ... De grandes croix noires peintes sur la surface inferieure des ailes indiquaient sa nationalilté. L'arrivéé de l'avion ennimi suscita aussi-tôt un grand remue-ménage dans le camp ou rertentit bientôt le brûit des canons anti-aéronals qui de tous côtée tiraient sur lui. En méme temps un avion anglais prenait de la hauteur à grande allure et se dirgeait. sur l'aéronef ennemi. ... Tout d'abord les deux avions décrivaient des cercles l'un autour de l'autre, mois tout à coup on vit l'aéroplane ennemi piquer brusquement de l'avent; il réussit pourtant à se rétablir, mais pour un instant seulement et bientôt aprés il venait s'ecraser sur le sol ou 1'on constata que ses réservoires d'eau et pétrola avaient été percés par les balles. Ses deux oacupants avaient été tués. L'aeroplane anglais par contre était complétement indemme.[85]

Die widersprüchlichen Darstellungen – die vollständige Zerstörung des Albatros-Doppeldeckers und die Tötung Suwelacks in der Luft auf der einen, eine „undamaged machine“ auf der anderen Seite – bereiteten den Nährboden für Spekulationen. Es erschien durchaus möglich, dass Suwelack und Leutnant Teichmann den Absturz überlebt hatten, wofür sowohl der englische Bericht über die Inspektion des deutschen Flugzeuges als auch der Umstand sprachen, dass die Habseligkeiten der beiden Insassen, die von den Engländern als militärisch nicht relevant eingestuft wurden – so auch Josef Suwelacks Fliegerarmbinde –, an die Feldflie-

82 Ebd.

83 Vgl. NL J. Suwelack, Brief Hauptmann Barths an Josef Suwelack senior, 14.9.1915, 2 Seiten.

84 NL J. Suwelack, „Inventory of Albatross“.

85 Un Duel Aérien sur les Lignes Anglaises, in: Le Matin vom 16.9.1915. Siehe auch die Dokumentation des Todestages Josef Suwelacks auf der von der Familie Suwelack ins Leben gerufenen Homepage zu seinem Leben: http://www.josef-suwelack.de/der-erste-weltkrieg/sein-todestag-am–13-september–1915 (16.6.2013).

gerabteilung 24 übersandt wurden. Ferner erreichte das deutsche Fliegerkorps eine Nachricht des britischen Majors der 16. Fliegerstaffel Hugh C. T. Dowding mit folgendem Inhalt:

> „The German Flying Corps. Herewith some of the personal effects of the pilot and observer, who were killed near Steenwerk a few days ago. They are buried on the Eastern edge of the village ERQUINGHEM-SUR-LYS.If any of your pilots have an engine failure or are brought down near here, they will be kindly treated. I can't tell you where it is, but I daresay you know.[86]

Dowdings Biograf Robert Wright schrieb ihm darüber hinaus eine Aussage zu, der zufolge die britische Staffel im September 1915 einen deutschen Flieger „heruntergeholt" habe, der jedoch sicher hinter den britischen Linien gelandet sei. Dann aber wären Pilot und Beobachter von „vorbeikommenden Soldaten kaltblütig erschossen" worden, ein Vorfall, so Wright, der Dowding „noch heute zornig" mache.[87]

Eine endgültige Antwort zu den genauen Todesumständen Josef Suwelacks existiert (noch) nicht und soll in dieser Arbeit auch nicht erfolgen. Stattdessen lässt sich gerade an diesem Thema und den zur Verfügung stehenden Quellen und Darstellungen die Konstruktion des Mythos um den Tod Suwelacks besonders deutlich veranschaulichen.[88] Die (Aus-)Wahl zwischen zwei Narrationen, dem Tod in der Luft und dem „kaltblütigen" Tod am Boden, hat sowohl Konsequenzen für die Art der kollektiven und individuellen Sinnstiftung als auch für die Form der Erinnerung. Während die erste Version jegliche Hoffnung auf Rettung in letzter Sekunde jäh zerstört und Suwelacks Körpers vollständig auslöscht, versetzt die zweite Version den Kriegsflieger vermeintlich in die Lage des noch aktiv Handelnden, der noch im Angesicht des Feindes und im Moment des Absturzes Ruhe, seine Nerven und Haltung bewahren konnte,[89] zumal diese Charakterzüge, so wird noch zu zeigen sein, wesentliche Bestandteile der zeitgenössischen Pilotentugenden darstellten.

Nicht selten bot gerade der letzte Flug eines Aufklärers oder der letzte Kampfeinsatz eines Jagdpiloten damals wie heute Raum für Spekulationen und Gerüchte und bahnte damit den Weg für Legendenbildungen. Das Mysterium des Todes bildete somit immer auch einen wichtigen Bestandteil des posthumen Ruhmes.[90] So lautete auch nicht selten das Narrativ für diese Flieger: Als „Märtyrer" starben

86 NL J. Suwelack, Nachricht des englischen Fliegerkorps von Major Dowding an das deutsche Fliegerkorps vom 22.9.1915. Diese Nachricht wurde später von Alfred von Back von den Kondor-Werken an Richard Suwelack übersandt. Vgl. NL J. Suwelack, Brief Alfred von Backs an Richard Suwelack vom 1.10.1915.

87 Robert Wright, Der vergessene Sieger. Die wahren Hintergründe der Schlacht um England, Bergisch-Gladbach 1970, S.29.

88 Siehe dazu Feld 4.

89 Vgl. Fritzsche, A Nation of Fliers, S.97f.

90 Vgl. Schnürer, Die Totenfeiern.

sie den „Heldentod fürs Vaterland“ auf dem „Feld der Ehre“[91] im vermeintlich ritterlich-fairen Zweikampf[92] – Topoi, wie sie auch auf Totenzetteln von Gefallenen häufig zu finden waren.[93]

Im Unterschied zu den täglich zu Tausenden Getöteten besaßen die Piloten nach wie vor eine Identität, eine Identifikation mit ihrer Maschine und nicht selten einen ebenso klar identifizierbaren Gegner. In der Luft fand – so zumindest in der Vorstellung vieler Zeitgenossen – noch ein Duell auf Augenhöhe statt. Der „Fliegertod“ als „letzter Akt der Ritterlichkeit“[94] vermochte in einer Zeit, in der angesichts des Massensterbens von Ritterlichkeit eigentlich nicht mehr die Rede sein konnte,[95] den Glauben an ältere Kampftraditionen aufrecht zu erhalten. Diese Vorstellung von den „Rittern der Lüfte“ fand bereits während des Ersten Weltkrieges, aber auch noch danach über populäre Schriftenreihen und Erinnerungsliteratur mit fiktiven und autobiografischen Schilderungen von Piloten öffentlich Verbreitung.[96] Die messbaren Erfolge lenkten kurzfristig vom Schrecken und der Gewalterfahrung des Krieges ab, bestärkten die Siegesgewissheit[97] und erlaubten nicht selten eine Trivialisierung des Krieges.[98]

Insbesondere die Erzählung vom Absturz Suwelacks und seiner Tötung am Boden schuf zudem Feindbilder und vermischte sich mit der Dolchstoßlegende, nach der Deutschland an Land, zu Wasser und „in der Luft unbesiegt“ gewesen und nur durch hinterlistiges Verhalten in die Knie gezwungen worden sei.[99] Zusätzlich gewann Suwelacks Tod an mythischer Qualität durch den Todeszeitpunkt. Nur einen Tag zuvor, am 12. September 1915, hatte er vorsorglich einen

91 George L. Mosse, Gefallen für das Vaterland. Nationales Heldentum und namenloses Sterben, Stuttgart 1993, S.13.

92 Vgl. Fritzsche, A Nation of Fliers, S.87f.; Schnürer, Die Totenfeiern, S.173: Der Realität der ersten anderthalb Kriegsjahre entsprach vielmehr das gegenseitige Umkreisen und der Versuch, den Gegner von oben mit Bomben zu bewerfen oder dem Gegner zu entkommen.

93 Dazu Stephan Fuchs, „Vom Segen des Krieges“. Katholische Gebildete im Ersten Weltkrieg. Eine Studie zur Kriegsdeutung im akademischen Katholizismus, Wiesbaden 2004, S.100–106; Monika Schulte, Die Totenzettel aus dem Ersten und Zweiten Weltkrieg. Soziale Bedeutung für das Image der Männlichkeit, o.O. 2011 sowie Beate Sturm, „Heldentod fürs Vaterland“. Totenzettel als Quelle für Einzelschicksale aus der Zeit des Ersten Weltkrieges, in: Jörg Engelbrecht (†)/Simone Frank/Christian Krumm/Holger Schmenk (Hg.), Rhein-Maas Geschichte, Sprache und Kultur. Der Erste Weltkrieg, Bd.5, Oberhausen 2014, S.62–75.

94 Vgl. Dietmar Süß, Tod aus der Luft. Kriegsgesellschaft und Luftkrieg in Deutschland und England, München 2011, S.33f.

95 Vgl. Mosse, Gefallen für das Vaterland, S.172; Schilling, Kriegshelden, S.256; Schnürer, Die Totenfeiern, S.186.

96 Siehe z.B. „Fliegertod“, Helden der Luft. Episoden aus Fliegerfahrten und Kämpfen in Feindesland, Nr.2, Berlin 1915, S.11–13, online verfügbar auf der Seite der Staatsbibliothek Berlin, Permalink: http://resolver.staatsbibliothek-berlin.de/SBB00006C3C0000 0000 (16.6.2013).

97 Vgl. Esposito, Mythische Moderne, S.242f.

98 Vgl. Schnürer, Die Totenfeiern, S.186.

99 Georg Paul Neumann, In der Luft unbesiegt. Erlebnisse im Weltkrieg erzählt von Luftkämpfern, München 1923; Gerd Krumeich, Die Dolchstoß-Legende, in: Etienne François/Hagen Schulze (Hg.), Deutsche Erinnerungsorte 1, München 2001, S.585–599.

Abschiedsbrief an seine Familie verfasst, mit der Bitte an seine Abteilung, den Brief seinen Eltern zu überbringen, sollte er fallen.[100] Zudem hatte er am 5. September 1915 seine Eltern in einem Feldpostbrief gebeten, ihm ein „kleines dünnes Gebetbuch" zuzusenden, „wo Messe, Beichte, Kommunion, Kreuzweg usw. drin ist".[101] Diese Bitte in Verbindung mit dem Abschiedsbrief ließ den Eindruck einer Vorsehung oder Vorahnung entstehen. Hatte es zuvor Luftkämpfe im eigentlichen Sinne aufgrund unzureichender Bewaffnung nicht gegeben, da die Trefferwahrscheinlichkeit äußerst gering war,[102] wurden ab Frühjahr 1915 verstärkt mit dem Propeller synchronisierte Maschinengewehre eingesetzt, so dass das Fliegen um ein Vielfaches gefährlicher wurde. All diese Einzelheiten spielten jedoch für die typisierende Rezeption seines Todes eine eher marginale Rolle. Entscheidend war hingegen der Topos der Tapferkeit, oder wie es Hauptmann Barth in einem Brief an Josef Suwelacks Vater formulierte: Er habe „wacker ein- oder zweimal mit feindlichen Flugzeugen herumgeschossen" und sei „nach übereinstimmenden Meldungen von zahlreichen Beobachtungsstellen ... im *Gleitflug* (also nicht abgestürzt) ca. 12 km hinter der feindlichen Front niedergegangen".[103]

Die Diktion zur Erinnerung an Suwelacks Tod im Jahre 1936 liest sich bereits wie ein Vorgriff auf die Nachrufe während des Zweiten Weltkrieges, wenn von „treuer Pflichterfüllung" die Rede war: „Am 13. September 1915 starb er zusammen mit seinem Begleiter, Leutnant Teichmann, bei Armentières den Fliegertod. Im Verlaufe des Luftkampfes mit einem Engländer wurde er von einer Kugel tödlich getroffen, worauf das Flugzeug abstürzte."[104] Ein zweiter Artikel aus dem gleichen Jahr, der von der Taufe eines Segelflugzeuges auf den Namen „Josef Suwelack" berichtete, sprach sogar von einem „Kampf mit mehreren Engländern",[105] was den Opfertod Suwelacks zusätzlich aufwertete, da er hier als Einzelkämpfer gegen zahlenmäßig überlegene Gegner erschien. Das Heldennarrativ ermöglichte es, auch über Zeiten hinweg den Durchhaltewillen an der Front zu forcieren und die Soldaten daran zu erinnern, welch hohen Preis viele ihre Mitstreiter zu bezahlen bereit gewesen waren, deren Erbe sie nun antraten.

Auch für die Familie Suwelack spielt der Helden-Topos eine zentrale Rolle im Gedenken an ihren Vorfahren. In den Familienbiografien lassen sich auf diese Weise gleichermaßen seine Verdienste für die Zivil- wie für die Militärfliegerei

100 NL J. Suwelack, Abschiedsbrief Josef Suwelacks an seine Familie, datiert auf den 12.9.1915.
101 NL J. Suwelack, Feldpostbrief Josef Suwelacks an seine Eltern, Lille, 5.9.1915.
102 Vgl. Georg Paul Neumann, Die gesamten deutschen Luftstreitkräfte im 1. Weltkrieg, Berlin 1920 [ND Bremen 2011], S.232 (hier Ausgabe von 2011); Ernst von Hoeppner, Deutschlands Krieg in der Luft, Leipzig 1921 [ND Barsinghausen 2013], S.17 und S.23 (hier Ausgabe von 2013).
103 NL J. Suwelack, Brief Hauptmann Barths an Joseph Suwelack senior, 14.9.1915, 2 Seiten. Hervorhebung und Klammern im Original.
104 „Münster ehrt Pioniere", in: Münsterischer Anzeiger vom 18.10.1936.
105 „Segelflugzeug ‚Josef Suwelack'", in: [unbekannter Zeitungstitel] vom 22.6.1936.

würdigen. Illustriert werden die Erzählungen mit zahlreichen Fotos aus seinem Nachlass und mit ihm zuteil gewordenen Auszeichnungen: „Der Held dieses Buches war vielseitig begabt“[106] lautet die Unterschrift zu einem Genrebild, das Josef als jungen Mann an der Staffelei zeigt. Über die Präsentation von Fotos gewonnener Preise und Auszeichnungen – Ehrenbecher der Stadt Essen 1913, Pokal und Ehrenteller als Preise beim Prinz-Heinrich-Flug 1913 oder auch die Kriegsauszeichnungen Josef Suwelacks, der Albrechtsorden mit Schwertern und das Eiserne Kreuz II. Klasse – wurde das leistungsorientierte Narrativ der Erfolgsgeschichte, das das Bild Josef Suwelacks nachhaltig geprägt hat, konsequent verfolgt und reproduziert.[107]

Das Narrativ der Erfolgsgeschichte und der Umgang mit Misserfolg

Dass das heroische Bild des westfälischen Piloten bis heute ungebrochen besteht, hat seinen Grund unter anderem in der Konstruktion einer scheinbar kohärenten Struktur seines Lebens, die in den jeweiligen Text(-sorten) immer wieder abgerufen wird.

Das Synonym für Suwelacks Erfolg ist vor allem sein Weltrekord, der in allen Darstellungen Erwähnung findet.[108] So wurde er z. B. im Jahr 1936 in einer Zeit des erstarkenden Nationalismus und Chauvinismus als der erste Weltrekord für Deutschland gewürdigt.[109] 1957 wurde Suwelacks Weltrekord dann mit dem damals zeitgenössisch herausragenden Ereignis der ersten Raumfahrt verglichen und als „Sputnik-Sensation im Jahre 1911“[110] bezeichnet. Die Idee vom erfolgreichen Lebensweg, den – einmal einschlagen – Josef Suwelack zielstrebig gegangen sei, bildete stets das Fundament der Narration. Dabei dient seine Lebensgeschichte auch der westfälischen Identitätsstiftung und als Beleg für die Willenskraft der Westfalen, indem die individuelle Identität Suwelacks zu einer kollekti-

106 Krewerth, Ein westfälisches Fliegerleben, S.24.

107 Siehe ebd., S.66, 74 und 76 und www.josef-suwelack.de. Siehe für ähnliche Narrationen z. B. den Film über Josef Suwelack, in: Schröder, Lokalzeitgeschichte – Münsterländische Flugpioniere; Moritz, Tollkühne Kerle; Fernagut, Un Aviateur, 3. Teil.

108 Mit Ausnahme von Berichten, die sich einem bestimmten Ereignis wie den Billerbecker Flugtagen 1912 zuwenden, vgl. z. B. NL J. Suwelack, „Erstmalig um die Türme von Billerbeck. Vier Stunden bedeuteten damals noch Weltrekord“, in: [unbekannter Zeitungstitel] (1961); Müller, Motörken, S.626; „Vom Weltrekord im Passagierflug“, in: Münsterische Zeitung vom 30.3.1988.

109 Vgl. NL J. Suwelack, „Vor 25 Jahren. Ein Streifzug durch den November des Jahres 1911“, in: [unbekannter Zeitungstitel] (1936).

110 „Sputnik-Sensation im Jahre 1911“, in: Münsterische Zeitung vom 1.5.1958. Siehe zu den folgenden Ausführungen zu Kontingenz: Michael Makropoulos, Modernität und Kontingenz, München 1997. Die Kontingenz betont die prinzipielle Offenheit der Geschichte beziehungsweise das Denken in Alternativen: „Kontingent ist, was auch anders möglich ist. Und es ist auch anders möglich, weil es keinen notwendigen Existenzgrund hat“. Ebd., S.13.

ven Identität erhoben wird und auf diese Weise der „Glanz“ des Piloten auf einen allen Westfalen innewohnenden Charakter zurückgeführt wird.

Kontingenz, d.h. Umwege, Rückschläge und Momente des Zweifels, verschwindet vor diesem Hintergrund in den Darstellungen zugunsten von Erklärungen, die immer dann auf gesellschaftliche Strukturen zu sprechen kommen, wenn Josef Suwelack vom vorgesehenen Weg abwich. In der Eigenlogik des Mythos ist die Einsicht, dass ein Leben nicht planbar ist, sondern von persönlichen Kontakten, dem zur Verfügung stehenden ökonomischen Kapital und vom Zufall beeinflusst wird, ein Problem, dem durch Strategien der (unterbewussten) Kontingenzbekämpfung und Komplexitätsreduktion entgegengewirkt wird.

Misserfolge hingegen werden lediglich auf externe Faktoren zurückgeführt. So verkehrte Peter Elgaß den aufgrund schlechten Wetters missglückten Flug Suwelacks über den Kanal durch den Einschub eines „trotzdem“ in einen Erfolg: „Trotzdem – die Bewährungsprobe war geglückt.“[111] Externe Faktoren – das sind in den Artikeln in der Hauptsache ungünstiges Wetter, die Technik, das Militär, andere, mit den Kondor-Werken in Konkurrenz stehende Flugzeugfirmen oder der Ausbruch des Ersten Weltkrieges – bilden eine Projektionsfläche für Misserfolge, ihnen wird implizit die Schuld an ausbleibenden Aufträgen (Firmen/Militär) oder Bruch- beziehungsweise außerplanmäßigen Landungen (Technik/Wetter) angelastet.

Diese externen Faktoren werden ebenso als Ursache für Krisenzeiten oder -momente in Suwelacks Fliegerbiografie benannt, denen er „ausgeliefert“ gewesen sei oder nur wenig entgegenzusetzen gehabt habe. Als sich die Kondor-Werke beispielsweise im Laufe des Jahres 1913 im Zuge wachsender Konkurrenz, höherer Anforderungen an Flugzeuge und eines sich verschärfenden nationalen Konkurrenzwettbewerbs unter den Flugzeugwerken mit Absatzschwierigkeiten und vielen Schulden konfrontiert sahen, ergriffen die Verfasser der Familienbiografie – verständlicherweise – für Josef Suwelack Partei:

> „Trotz seiner hervorragenden Konstruktionsleistung und der durch den Neunstundenflug bewiesenen Zuverlässigkeit [des Kondor-Eindeckers] hatte er bei der Heeresverwaltung keinen Erfolg. Wochenlang bemühte er sich vergeblich um Aufträge, hatte Änderungen vorgenommen, die ganz oben gewünscht wurden. Immer neue Hoffnungen hatte man ihm zwar gemacht, zu einem Vertrag kam es dennoch nicht. Die Heeresverwaltung bestellte eben bei den großen Fabriken.“[112]

Die zitierte Passage ist keineswegs als unzutreffend anzusehen, stützt sie sich doch auf Quellen wie die Gesellschafter-Protokolle der Kondor-Werke, auf ein Schreiben des Kuratoriums der Nationalflugspende von 1914 und auf private Schriftwechsel Josef Suwelacks mit den Gesellschaftern der Kondor-Werke.[113]

111 Elgaß, Josef Suwelack, S.88.
112 Krewerth, Ein westfälisches Fliegerleben, S.78.
113 Siehe die Ausführungen zum Feld 2.

Jedoch bleibt die Perspektive des Militärs unberücksichtigt: Die Heeresverwaltung musste die Flut an eingereichten Innovationsvorschlägen und Patentanmeldungen der Firmen ohne ausreichend zur Verfügung stehendes Personal bewältigen. Zudem existierten in den ersten Jahren der Fliegerei keine Prüfstellen oder zuverlässigen Prüfmethoden, da der militärische Einsatz von Flugzeugen noch in einer Experimentierphase begriffen war und der schelle Kauf von wenig erprobten Maschinen in finanzieller und sicherheitstechnischer Hinsicht hohe Risiken barg.[114] Das im Zitat zum Vorschein kommende Bild von „die da oben – wir da unten" vermittelt den Eindruck, die Kondor-Werke und mit ihnen Josef Suwelack wären der Macht des Militärs ausgeliefert gewesen und gleichsam zwischen den Strukturen zerrieben worden, eine Argumentation, die auch von späteren Autoren erneut aufgegriffen wurde. So ist bei Moritz und Krystofiak z. B. die Rede davon, dass der Starrsinn des Militärs, das unerfüllbare Forderungen aufgestellt und falsche Hoffnungen geweckt habe, die weiteren Pläne der Kondor-Werke durch die Unverbindlichkeit der Verhandlungen zum Scheitern gebracht habe.[115]

Bis auf wenige Darstellungen,[116] die körperliche Schwäche (z. B. Krampf im Arm aufgrund der Kälte) als Ursache für eine vorzeitige Landung angeben, wird Suwelack selbst selten für Flugabbrüche verantwortlich gemacht. Stattdessen werden auch hier externe Faktoren als Gründe genannt: entweder setzte der Motor aus oder der zur Verfügung gestellte Apparat war ungeeignet oder Sturm und Regen machten einen „sicheren" Flug unmöglich.[117] Interessanterweise werden selbst spontan und unüberlegt anmutende Entscheidungen des Piloten mithilfe zeitgenössischer, als typisch für Flieger erachteter Charakterzüge in das kausale Erklärungsmodell eingepasst, wie sich an den folgenden Ereignissen exemplarisch verdeutlichen lässt.

Josef Suwelack hatte Anfang des Jahres 1911 einen Vertrag mit den Rumpler-Werken in Berlin-Johannisthal geschlossen und war in der Folgezeit als Fluglehrer und Pilot beschäftigt gewesen. Für Rumpler hatte er vertragsgemäß an den hoch dotierten Flugwettbewerben teilzunehmen, bei denen er sich im Laufe des Jahres erfolgreich platzieren konnte.[118] Während der Weihnachtsfeiertage 1911, die Josef bei seiner Familie in Billerbeck verbrachte, war die Idee zu einem eige-

114 Siehe zum Erwartungshorizont des Militärs in der Vorkriegszeit: Storz, Kriegsbild und Rüstung, S.303, und Möser, Fahren und Fliegen, S.496ff.

115 Vgl. Elgaß, Josef Suwelack, S.88 sowie Krystofiak, Möönsch, S.9.

116 Siehe Schröder, 50 Jahre Flugsport, S.22, und NL J. Suwelack, „Erstmalig um die Türme von Billerbeck", in: [unbekannter Zeitungstitel] (1961).

117 Siehe z. B. „Aus ‚Billerbecker Großflugtagen' wurden 1912 begrabene Hoffnungen", in: Billerbecker Anzeiger, Ostern 1982 (Aussetzen des Motors, schlechtes Wetter); Elgaß, Josef Suwelack, S.87 (mangelhafter Apparat); z. B. auch Moritz, Tollkühne Kerle.

118 Siehe den Abschnitt zu den Rumpler-Werken in Feld 2 und die Abschnitte zu den genannten Flugwettbewerben in Feld 3.

nen Flugtag in Billerbeck entstanden, der dann Ostern 1912 auch realisiert wurde.[119] Am 15. Januar 1912 hatte Josef seinen Arbeitgeber davon unterrichtet, dass er beabsichtige, den heimatlichen Flugtag mit der Rumpler-Taube zu bestreiten. Parallel dazu hatte er sich heimlich und vertragswidrig jedoch zwei Renneindecker der Firma Aviatik gekauft, die ihren Sitz in Mühlhausen im Elsass hatte. Ebenfalls Anfang 1912 schloss er einen privatrechtlichen Dienstvertrag mit dem Deutschen Reich, in dem er die Verpflichtung einging, sich im Kriegsfall als Flieger dem Militär zur Verfügung zu stellen.[120] Hatte Josef Suwelack am 14. Februar 1912 noch mündlich gegenüber Edmund Rumpler erklärt, er wolle sich zwei Eindecker von ihm kaufen,[121] so nahm er wenig später wieder Abstand davon, kündigte noch am gleichen Tag bei Rumpler und leitete rechtliche Schritte gegen ihn ein. Er warf Edmund Rumpler „unsittliches Verhalten“ vor und forderte ausstehende Zahlung aus seiner Teilnahme an Wettbewerben. Sein Arbeitgeber wiederum wies die Anschuldigungen als „schroff und unmotiviert“ zurück und erkundigte sich am 23. Februar schriftlich,[122] weshalb Josef seinen „vertragsmässigen verpflichtungen [sic!] seit einigen Tagen nicht mehr“ nachkomme.[123] Die Kündigung Suwelacks erregte Aufsehen in Johannisthal und zog zahlreiche Anfragen anderer Flugzeugfirmen nach sich, die ihm eine gute Stellung und ein lukratives Gehalt anboten:[124] „Wo ich am I. III. von Rumpler fort gehe[,] fallen die Firmen wie die Raben über mich her. Ich habe Angebote von 2000 Mk monatlich e.t. c.“[125] Er lehnte jedoch alle Anfragen ab, denn er hatte sich bereits entschieden, künftig für die Aviatik AG tätig zu sein, mit der er schließlich am 23. Februar einen Anstellungsvertrag schloss.[126] Ein persönlicher Versuch Rumplers, Josef Suwelack zum Bleiben zu bewegen, scheiterte.[127]

Soweit die Kurzfassung des Fortgangs Josef Suwelacks bei den Rumpler-Wer-

119 Siehe Feld 3: Billerbecker Flugtag.

120 Diesen Vertrag ließ er sich 1915 erneut bestätigen. Vgl. NL J. Suwelack, Bestätigung des Vertrags zwischen dem Deutschen Reich und Josef Suwelack vom 2.2.1915. Der dreiseitige Vertrag enthält mehrere verschiedene Datumsangaben, was vermutlich darauf zurückzuführen ist, dass er von mehreren Behörden bzw. zuständigen Personen gegengezeichnet wurde.

121 NL J. Suwelack, Schreiben der E. Rumpler Luftfahrzeugbau G.m.b.H. an Josef Suwelack vom 15.2.1912.

122 Siehe NL J. Suwelack, Schreiben Suwelacks an Rumpler vom 5.3. und Antwort Rumplers vom 7.3.1912. Worauf sich der Vorwurf des „unsittlichen Verhaltens“ genau bezog, geht aus dem Schreiben nicht hervor, vermutlich ging es dabei um Arbeitszeiten und eine schlechte Bezahlung.

123 Eigentlich hatte Suwelack eine sechswöchige Kündigungsfrist und hätte vertragsmäßig bis zum 31. März bleiben müssen. Da er allerdings am 18. Februar ein neues Abkommen geschlossen hatte, konnte er bereits zum 15. März aus dem Unternehmen ausscheiden. Vgl. NL J. Suwelack, Schreiben Rumplers an Rechtsanwalt Dr. Kelk vom 23.2.1912.

124 Siehe den Abschnitt zu den Rumpler-Werken im Feld 2.

125 NL J. Suwelack, Brief Suwelacks an Hubert Ahlers vom 13.2.1912. Siehe für weitere Angaben auch den Abschnitt zu den Rumpler-Werken im Feld 2.

126 NL J. Suwelack, Vertrag mit der Aviatik AG vom 23.2.1912.

127 Vgl. NL J. Suwelack, Schreiben Rumplers an Suwelack vom 9.3.1912.

ken,[128] die aufgrund einer lückenhaften Quellenlage einige Fragen nicht eindeutig klären kann. Für die Familienbiografen Josef Suwelacks erschien sein Fortgang als „völlig überraschend“ und nicht nachvollziehbar:[129] „Es ist schwer, diesen plötzlichen Wechsel zu verstehen. Die Rumpler-Taube war das erfolgreichste Flugzeug der Jahre vor dem Ersten Weltkrieg ... Das Konstruktionsprinzip der Taube wurde überall in der Luftfahrt nachgeahmt.“[130]

Diesem Bruch in seiner Biografie begegnen die Nachfahren Suwelacks, indem sie über den Rückgriff auf einen stereotypen Charakterzug, die Exzentrik der frühen Flieger, auf die Sprunghaftigkeit Suwelacks verweisen: „Auch Rumpler machte jene Erfahrung ..., derzufolge Flieger eben durch die Bank exzentrische Menschen sind.“[131] Ferner machten sie auf den Vortrag Josef Suwelacks vom Februar 1912 aufmerksam, in dem er vom Kauf der Aviatik-Renneindecker berichtet hatte. Für sie stand fest: „Man spürt es, Josef Suwelack will Rennen fliegen“ und sich an den Geschwindigkeitswettbewerben beteiligen.[132] Mithilfe dieser Argumentation wurde versucht, eine Gradlinigkeit in Suwelacks Handeln zu konstruieren.

2. In memoriam I: Öffentliche und private Formen der Erinnerung an Josef Suwelack

Die bisher vorgestellten Erinnerungen an den Flieger beschränkten sich auf schriftliche Darstellungen. Darüber hinaus existieren aber noch weitere Formen des Gedenkens an Josef Suwelack, die im privaten beziehungsweise öffentlichen Raum Verbreitung fanden. Dazu zählen die Straßenbenennung nach Josef Suwelack in Münster, ein plattdeutsches Theaterstück über den Piloten sowie ein Berliner (Flug-)Seifenkisten-Rennen, ferner die Würdigung Suwelacks in der Literatur zur Fluggeschichte, digitale Darstellungen im Internet und schließlich private und öffentliche Formen der personalen Denkmalpflege durch die Familie Suwelack.

128 Siehe den Versuch einer differenzierten Betrachtung der Arbeitssituation Josef Suwelacks in Johannisthal im Feld 2.

129 Krewerth, Ein westfälisches Fliegerleben, S.60.

130 Ebd. Diese Aussage ist durchaus zutreffend, denn tatsächlich galt die Rumpler-Taube als zuverlässiges Flugzeug der Vorkriegszeit. Schätzungen zufolge haben mehr als fünfzig Firmen der Taube nachempfundene Modelle gebaut. Vgl. John A. de Vries, Taube, Dove of War. Historical aviation album publication, o.O. 1974, S.IV, vgl. auch Jörg Armin Kranzhoff, Edmund Rumpler, S.362f., und „Die erfolgreichsten Flugzeuge des Jahres 1911“, in: Flugsport (1911), Nr.23, S.836–837.

131 Krewerth, Ein westfälisches Fliegerleben, S.60.

132 Ebd.

2.1. Entstehung eines Münsteraner Fliegerviertels – zur Geschichte des „Josef-Suwelack-Wegs“

Neben einer topografischen Orientierungsfunktion haben Straßennamen in der Regel eine traditionsbildende und identitätsstiftende Absicht, die sich in den Namen für bedeutend erachtete Persönlichkeiten oder Orte manifestiert.[133] Das Straßennamensbild einer Stadt oder einer Region lässt Rückschlüsse auf den politischen Diskurs zu, denn die jeweiligen Entscheidungsträger praktizieren durch Namensänderungen eine „symbolische ... Besetzung des öffentlichen Raums“,[134] das heißt, sie bringen ihre Vorstellungen und Wertmaßstäbe in Namen und Bezeichnungen zum Ausdruck. Rainer Pöppinghege beschreibt die Straßenneubeziehungsweise Umbenennungspraxen daher als „eine höchst selektive Angelegenheit“, da die Auswahl dessen, wer oder was auf einem Straßenschild erinnert werden sollte, meist „den Vorlieben bestimmter gesellschaftlicher Gruppen geschuldet“ war bzw. ist.[135] Insofern spiegeln sich in Straßennamen bestimmte Geschichtsbilder wider, die wiederum Veränderungen unterliegen können.[136]

In der Zeit des Nationalsozialismus wurden in fast allen deutschen Städten und Gemeinden die Namen von Straßen, Plätzen und Wegen geändert, obwohl eine Notwendigkeit durch bauliche oder polizeiliche Maßnahmen mehrheitlich nicht bestand. Ausnahmen bildeten Umbenennungen an einem bestimmten Stichtag, die meist ein ganzes Stadtviertel betrafen.[137] Das Ziel der Umbenennungen bestand für die nationalsozialistischen Machthaber vor allem darin, durch Straßenbezeichnungen ihre politischen Führer und „Märtyrer“ „sichtbar und die Bevölkerung mit Erinnerungsorten und Vorbildern ... vertraut zu machen“.[138] Personennamen, die der NS-Ideologie widersprachen, wurden zugunsten von Orts- und Personennamen beseitigt, welche die künftige politische Marschrichtung unterstreichen sollten: So traten an die Stelle von Weimarer DemokratInnen, linken ParteienvertreterInnen sowie christlichen oder jüdischen Personennamen vor allem Personennamen aus dem Umfeld der NSDAP, militärische Personennamen, sofern sie für eine völkisch-nationalistische Ideologie instrumentalisiert wer-

133 Vgl. Rainer Pöppinghege, Geschichtspolitik per Stadtplan. Kontroversen zu historisch-politischen Straßennamen, in: Matthias Frese (Hg.), Fragwürdige Ehrungen!? Straßennamen als Instrument von Geschichtspolitik und Erinnerungskultur, Münster 2012, S.21–40, hier S.23–24.

134 Peter Stachel, Stadtpläne als politische Zeichensysteme. Symbolische Einschreibungen in den öffentlichen Raum, in: Rudolf Jaworski (Hg.), Die Besetzung des öffentlichen Raumes. Politische Pläne, Denkmäler und Straßennamen im europäischen Vergleich, Berlin 2007, S.13–60, hier S.23.

135 Pöppinghege, Geschichtspolitik, S.24 (dort beide Zitate) und S.33f.

136 Vgl. ebd., S.33ff., bes. S.35.

137 Vgl. Marcus Weidner, „Wir beantragen...unverzüglich umzubenennen“. Die Straßenbenennungspraxis in Westfalen und Lippe im Nationalsozialismus, in: Frese, Fragwürdige Ehrungen!?, S.41–98, hier S.72.

138 http://www.strassennamen-in-westfalen-lippe.lwl.org (17.6.2014).

den konnten, oder territoriale Namen wie z. B. Schlachtorte des Ersten Weltkrieges.[139] Straßennamen, die mit nationalsozialistischen Vorstellungen konform schienen oder zumindest eine gewisse Schnittmenge aufwiesen, blieben bestehen. Zudem wurde „bis zu einem gewissen Grad Rücksicht auf historisch gewachsene Namen mit einer besonderen [lokalen] Bedeutung" genommen. Die Praxis der Umbenennung erfolgte dabei zumeist nicht, wie die Ausrichtung am „Führer-Prinzip" vermuten lassen könnte, von oben nach unten, sondern hier wurden örtliche Instanzen aktiv, die sich allerdings – oft im vorauseilenden Gehorsam – an Parteivorstellungen orientierten. Ferner gab es Lobbyisten, die Rundschreiben an die zuständigen Behörden richteten oder sich für die Ehrung einer Person einsetzten.[140]

Nach dem Ende des Nationalsozialismus und des Zweiten Weltkrieges setzte 1945 erneut eine Straßenum- beziehungsweise -rückbenennungswelle ein, diesmal im Zuge der Demilitarisierung und Entnazifizierung Deutschlands. Die Umbenennungen beschränkten sich auf Namen von der NS-Ideologie eindeutig zuzuordnenden Personengruppen der NS-Zeit. Dabei handelte es sich vorrangig um NS-Eliten wie Himmler, Göring oder Hitler, daneben um NS-Ideologen und -Funktionäre sowie um die Bezeichnungen und Orte, die mit der NS-Ideologie eng verknüpft waren. Die Entscheidung, ob eine Umbenennung erfolgte, oblag zu diesem Zeitpunkt den alliierten Besatzern.[141] Maßgeblich für die Entscheidungsfindung der Nachkriegszeit wurde die Kontrollratsdirektive Nr. 30 vom Mai/Juli 1946,[142] die die Denazifizierung und Demilitarisierung Deutschlands auf die symbolische Ebene des Gedenkens ausweitete. Um eine gründliche „Säuberung" von altem Gedankengut zu gewährleisten, wurde als Stichtag der 1. August 1914 festgesetzt, so dass militärische Kontinuitäten aus dem Ersten Weltkrieg ebenfalls beseitigt werden konnten.[143] Diese Auswechslung erfolgte in einigen Städten sehr konsequent, wohingegen andernorts reaktionäre Kräfte an alten Bezeichnungen festhielten oder die Namen als unbedenklich einstuften.[144] Eine Wiederherstellung des Status Quo vor 1933 erfolgte somit nicht. Stattdessen gestaltete sich die Neu- beziehungsweise Rück- oder Umbenennung heterogen: Einige Straßen erhielten gänzlich neue Namen, andere bekamen ihren alten Namen zurück und

139 Vgl. ebd., weitere Zitate finden sich ebenfalls dort.

140 Vgl. Weidner, „Wir beantragen ...", S. 69.

141 Vgl. Marcus Weidner, „Mördernamen sind keine Straßennamen". Revision und Beharrung in der Straßenumbenennungspraxis der Nachkriegszeit – Westfalen und Lippe 1945–1949, in: Frese, Fragwürdige Ehrungen!?, S. 99–120, S. 102f.

142 StdA Münster, Kreis C, 467, Direktive Nr. 30, Beseitigung deutscher Denkmäler und Museen militärischen und nationalsozialistischen Charakters, in: Amtsblatt des Kontrollrats in Deutschland, Nr. 7, 31.5.1946, S. 48f., online abrufbar unter http://www.lwl.org/westfaelische-geschichte/nstopo/normen/1946-10-16.pdf (17.6.2014).

143 Vgl. ebd., S. 104f.

144 Vgl. ebd, S. 111ff. und 118f.

wieder andere Straßen erfuhren gar keine Veränderung, obwohl die auf diese Weise geehrte Person durchaus belastet war. Auch heute noch sind circa ein Drittel der NS-Benennungen gültig, wie etwa in Detmold-Hiddesen der „Hindenburgplatz", an dem sich 2012 eine Debatte entzündete, die 2014 zu dem Beschluss führte, ein Hinweisschild anzubringen, auf dem die Bürger über den Diskurs informiert wurden.[145]

Gerade in den letzten Jahren ist in Westfalen eine erneute öffentliche Auseinandersetzung mit geschichtsträchtigen Namen und Bezeichnungen auf Straßenschildern zu beobachten, die nun verstärkt auch Personen kritisch beleuchtet, die keine hohen Repräsentanten des Nationalsozialismus, aber dennoch im sogenannten „Dritten Reich" aktiv waren. Karl Ditt hat in diesem Zusammenhang darauf hingewiesen, dass es sich bei derartigen Diskussionen um „Diskussionen über Maßstäbe" handelt.[146] Ditt stellt die seit den 1980er Jahren wachsende Sensibilisierung der Gesellschaft für die begangenen Verbrechen im Nationalsozialismus heraus und sieht im Kampf gegen Fremdenfeindlichkeit einen Auslöser für die erneute kritische Beschäftigung mit der NS-Vergangenheit.

In Münster wurden im Jahr 1936 im Ortsteil Gremmendorf, nahe dem ehemaligen Flugplatz Loddenheide, vier neue Straßen nach lokalen Fliegern benannt. Mit Beschluss der Dezernentenkonferenz unter Leitung des Oberbürgermeisters erhielten diese Wege am 17. September 1936 die Namen von Paul Engelhard, Anton Knubel, Josef Suwelack und Gustav Tweer.

In anderen Ortsteilen Münsters wurden am gleichen Tag Straßen bzw. Wege nach Oswald Boelcke (Boelckeweg), Manfred von Richthofen (Manfred-von-Richthofen-Straße) und Max Immelmann (Immelmannweg) benannt. Alle Straßennamen, bis auf den Immelmannweg in Münster, der seit 1974 Lindberghweg[147] heißt, bestehen in den Ortsteilen Gremmendorf und Mauritz bis heute.[148]

Der zweiseitige Beschluss vom 17. August 1936 über die Vergabe von neuen Straßennamen in besagtem „Fliegerviertel" gibt als Grund für damalige Namensänderung an: „Entsprechend der fortschreitenden baulichen Entwicklung der

145 Siehe „Hindenburg-Schild soll nun kommen", in: Lippische Landes-Zeitung vom 14.2.2014, siehe auch Marcus Weidner, Die Straßenbenennungspraxis in Westfalen und Lippe während des Nationalsozialismus. Datenbank der Straßenbenennungen 1933–1945, Münster 2013ff., http://www.lwl.org/westfaelische-geschichte/nstopo/strnam/Begriff_2.html (Stand: 12.12.2013) (17.6.2014). Zum „Fall Hindenburg" im Allgemeinen siehe Hans-Ulrich Thamer, Straßennamen in der öffentlichen Diskussion. Der Fall Hindenburg, in: Frese, Fragwürdige Ehrungen!?, S.251–264.

146 Karl Ditt, Karl Wagenfeld (1869–1939). Dichter, Heimatfunktionär, Nationalsozialist?, in: Frese, Fragwürdige Ehrungen!?, S.179–232, Zitat S.181.

147 Charles Lindbergh (1902–1974) überquerte 1927 als Erster im Alleinflug den Atlantik von New York nach Paris, was ihn zu einer der bekanntesten Personen der Luftfahrt machte. Zuvor war es jedoch anderen Fliegern als Mannschaft gelungen, die Strecke ebenfalls im Flugzeug ohne Zwischenstopp zurückzulegen, siehe z.B. http://www.welt.de/kultur/history/article13361494/Der-erste-Atlantikueberflieger-hiess-nicht-Lindbergh.html, online seit dem 16.5.2011 (20.11.2014).

148 Siehe http://www.lwl.org/westfaelische-geschichte/nstopo/strnam/Jahr_1936.html (12.12.2013).

Stadt wird die Festlegung mehrerer neuer Straßennamen erforderlich."[149] Im Anschluss an die Bestimmung der Straßennamen folgen knappe Erläuterungen zu Anton Knubel, Gustav Tweer, Josef Suwelack und Paul Engelhard:

> „Anton Knubel, Vorkämpfer und Konstrukteur im Flugsport, verunglückt bei einem Fluge auf der Loddenheide. Gustav Tweer, einer der ersten Flieger in Osnabrück und Münster. Josef Suwelack aus Billerbeck, einer der ersten Flieger des Münsterlandes. Korvettenkapitän Paul Engelhard, geboren zu Münster, erster Militärflieger, Pilotenzeugnis Nr. 3, verunglückt bei einem Fluge im Jahre 1911."[150]

Einzig Paul Engelhard wurde direkt mit dem Militär in Verbindung gebracht, die übrigen drei Flieger wurden aufgrund ihrer fliegerischen Pionierleistungen für das Münsterland als „Männer der ersten Stunde" geehrt. In einem Zeitungsbericht über die neuen Straßennamen dann erfolgte auch eine Würdigung Suwelacks wegen seines Kriegseinsatzes.[151]

In der bundesrepublikanischen Nachkriegszeit sollte die Straßenbenennung zugunsten von Suwelack nochmals auf der Agenda stehen: Hierzu tagte am 18. Januar 1950 ein Umbenennungsausschuss im Fall „Josef Suwelack" und beriet, ob sein Name durch den des Sportfliegers Ernst-August Schröder ersetzt werden sollte. Eine Entscheidung zugunsten einer Neubenennung erfolgte allerdings nicht, so dass Suwelacks Name weiterhin auf dem Straßenschild verbleiben konnte.[152] Leider existiert kein Protokoll dieses Ausschusses,[153] so dass keine Aussagen zu den Antragsstellern oder zu den Pro- bzw. Contra-Argumenten mehr möglich sind. Festzuhalten bleibt allerdings für die Zeit vor 1945, dass lokale „Helden" wie Josef Suwelack aus dem Münsterland für den NS-Staat vereinnahmt wurden und gleichsam als Aushängeschilder für das Regime dienen sollten.

2.2. „Familie Uhlenhorst off: Buren un Kabeleeren, oder: he wedd flügge" – Josef Suwelack im Theater

Angelehnt an die Lebensgeschichte Josef Suwelacks wurde 1914 das plattdeutsche Theaterstück „Familie Uhlenhorst off: Buren un Kabeleeren" verfasst und uraufgeführt.[154] Weitere Aufführungen folgten 1919, 1929, 1952, 1964 und 1987/88. Als „Großes Volksstück mit Gesang und Tanz in 4 Akten" beworben, brachte es

149 StdA Münster, Akte 1936–08–17a.
150 StdA Münster, Akte 1936–08–17b.
151 Siehe auch: http://www.lwl.org/westfaelische-geschichte/nstopo/strnam/Index_Kategorien.html (17.6.2014).
152 StdA Münster, Amt 23, 23, Dokumentation Straßennamen, S. 422f.
153 Mehrere Nachfragen im Stadtarchiv Münster sowie bei Mitarbeitern am geschichtlichen Straßenprojekt in Westfalen haben dies bestätigt.
154 Abendgesellschaft des Zoologischen Gartens Münster (Westf.), Familie Uhlenhorst off: Buren un Kabeleeren. Großes Volksstück mit Gesang und Tanz in 4 Akten von Franz Falger, Münster 1929, S. 62.

die Abendgesellschaft des Zoologischen Gartens Münster (Westf.) nach der Drehbuchvorlage von Franz Falger aus Münster auf die Bühne.

Zusammengefasst beinhaltet das Stück folgende Geschichte: Heinrich Schulte-Uhlenhorst, ein alteingesessener Gutsbesitzer im Münsterland, und seine Frau Milla haben einen Sohn namens Theodor. „Teddo“, wie er von seinen Eltern genannt wird, tritt als Hoferbe nur zögerlich in die Fußstapfen seines Vaters und widmet sich lieber seinem Hobby: der Fliegerei. Mit seinem Doppeldecker übt er sich im Kunstflug. Eines Tages erwarten die Uhlenhorsts ihren Nachbarn, Baron von Treuberg, einen Rittergutsbesitzer, mit seinen beiden Töchtern als Gäste auf ihrem Hof zu einem Erntefest. Theodors Mutter hofft, dass eine der Baronessen ihren Sohn heiraten werde, auf dass dieser der Fliegerei den Rücken kehren und seinen gesellschaftlich rechtmäßigen Platz einnehmen werde. Herr Uhlenhorst und Baron von Treuberg haben zu diesem Zeitpunkt bereits eine informelle Absprache getroffen, die die Verbindung der Familien als beschlossene Sache sieht. Wenig später reist Theodor nach Berlin, um an einer Meisterschaft im Kunstflug teilzunehmen. Seine technik-skeptischen Eltern wohnen der Veranstaltung bei und ebenso die beiden Baronessen in Begleitung ihres Vaters, der wiederum von der neuen Technik fasziniert ist und Herrn Schulte-Uhlenhorst drängt, Theodor finanziell zu unterstützen. Der Flieger selbst will durch spektakuläre Kunst- und Rekordflüge auf sich aufmerksam machen und sieht sich einer Reihe Verehrerinnen gegenüber. Nach Theodors erfolgreichem Flug trifft man sich in erlesener Gesellschaft im Kabarett Zum blauen Affen. Während Theodors Familie und die von Treubergs an der Tafel zusammensitzen, verbringt der Flieger die Zeit mit seiner Geliebten, der Tänzerin Adele, zurückgezogen in der Loge. Zurück aus Berlin auf dem elterlichen Hof erwarten die Uhlenhorsts und Herr Treuberg mit seinen Töchtern Theodors Entschluss, für welche der Baronessen er sich entschieden hat. Dieser verkündet jedoch, er werde Adele aus Liebe heiraten.

Der Schauplatz des Stückes „Familie Uhlenhorst“ spielt in Billerbeck sowie in anderen Städten des Münsterlandes, so dass anzunehmen ist, dass die Zuschauer wussten, auf wessen Leben und Familienhintergrund angespielt wurde. Die Handlungsorte „Haus Uhlenhorst im Münsterlande“, „Flugplatz Tempelhof bei Berlin“ sowie das Kabarett Zum blauen Affen beispielsweise weisen deutliche Parallelen zu den Handlungsfeldern Josef Suwelacks auf:[155] Der elterliche Wohnsitz mit dem angrenzenden Molkereigebäude entspricht im Theaterstück dem Hof Uhlenhorst mit Molkerei und das auf dem Flugplatz angesiedelte Café Senftleben besagtem Kabarett. Das Stück erfuhr mehrere Aufführungen, so dass von einer großen Resonanz auszugehen ist.

Im Stück wird Suwelack in der Person Theodor Schulte-Uhlenhorsts als

155 Ebd., S. 4.

Emporkömmling und Frauenschwarm karikiert, seine Eltern erscheinen als neureiche, wenngleich „einfache Leute", denen nur aufgrund ihres Vermögens Zugang zur „feinen Gesellschaft", verkörpert durch die von Treubergs, gewährt wird. Theodors Eltern fügen sich, anders als ihr Sohn, nicht in die gehobene Gesellschaft ein und es ist der Flieger, der zwischen den beiden aufeinandertreffenden Welten vermittelt. Im Gegensatz zu den von Treubergs, die weltoffen und technikbegeistert sind, wirken die Eltern des Fliegers konservativ und reserviert gegenüber der Flugzeugbegeisterung ihres Sohnes sowie generell skeptisch gegenüber der Flugzeugtechnologie. Abfällig äußert sich Theodors Vater über die „niemodsken Erfindungen" und meint: „Ick seh dat am besten an mienem Jungen. De verdammte Fleigerie, de se nun erfunnen häfft, is em auk in de Planete stiergen. Nu will he sogar in vetteihn Dagn nao Berlin un sien Weltmeisterschaftsexamen maken." Statt am Steuer eines Flugzeuges sehen sie ihren Sohn lieber als mit einer der Baronessen verheirateten Hoferben, der weiß, wo sein Platz ist und sich seiner Heimat verpflichtet fühlt. Eine ganz andere Ansicht vertritt hingegen Baron von Treuberg, der der Gruppe der Technikbegeisterten zugerechnet werden kann: „Wir wollen uns doch freuen, in dieser Zeit der großen Erfindungen und Rekordleistungen zu leben, und wenn die Sache Ihrem Filius auch mal ein paar braune Lappen kostet, was macht das für Sie aus? Sie haben es doch."[156]

Theodor Uhlenhorst wirkt ehrgeizig und geübt im Umgang mit der gehobenen Klasse. Er rebelliert gegen seinen Vater, der seine Pläne, als Flieger erfolgreich zu sein und Bekanntheit zu erlangen, nicht unterstützt. So äußert er sich gegenüber den von Treubergs: „Über diese Sache [die Fliegerei] ist mit meinem Vater nun einmal nicht zu reden, hoffentlich gelingt es mir, noch seine Ansicht zu ändern".[157] Von allen Seiten wird ihm Bewunderung entgegengebracht. So lobt Baron von Treuberg Theodor nach dessen erfolgreichem Flug mit den Worten: „A la bonheur, sind ja inzwischen schon Berühmtheit geworden. ... Werden noch der Stolz der ganzen Nation"[158]

Das Narrativ des Stolzes, das auch in den Lokalzeitungsberichten über Josef Suwelack immer wieder zum Thema gemacht wird, bildet ebenso einen Leitgedanken im Theaterstück. In der Gaststätte ist der „berühmte Sturzflieger Uhlenhorst" das Gesprächsthema der Tänzerinnen, die mit einem anderen Flieger in exklusiver Gesellschaft teuren Champagner trinken.[159] An späterer Stelle unterhält sich Adele mit ihrem zukünftigen Schwiegervater Herrn Uhlenhorst und spricht ebenfalls auf die Bekanntheit Theodors an: „Aber Herr Schulte, den kennt

156 Ebd., S.17 (dort beide Zitate).
157 Ebd.
158 Ebd., S.24.
159 Ebd., S.37.

doch wohl jedes Kind in Berlin und Sie sind der Vater dieses berühmten Fliegers? Nein, Sie haben aber einen Sohn, auf den können Sie sehr stolz sein".[160]

Theodor ist bei der Mehrheit der Damen aufgrund seines vermögenden Hintergrundes und seines Bekanntheitsgrades beliebt, weniger jedoch wegen seines Charakters. Die Unterhaltungen mit den Baronessen lassen erkennen, dass diese den Flieger einzig wegen seines Wagemuts und seines Geldes achten: „Herr Uhlenhorst, nur eine Frau mit großem Mut kann meiner Ansicht nach die rechte Gattin eines Fliegers sein. Ich würde z. B. den nötigen Mut in mir führen." Vor allem für Rosalinde, die zweite Baronesse, spielt sein ökonomisches Kapital eine Rolle.[161]

Neben den zum Teil deutlichen Verweisen auf reale Personen lebt das Stück von der satirischen Kritik an der gehobenen Gesellschaft, den „Kabeleeren" – Adligen –, und kontrastiert ihre affektierte und luxuriöse Lebensweise mit dem bodenständigen Provinzleben der „Buren", der Bauernfamilie Uhlenhorst. Der Name Uhlenhorst, „Eulennest", in Verbindung mit dem Zusatz „oder he wedd flügge" spielt auf den Flieger an, der das elterliche „Nest" verlässt und „flügge", also selbstständig wird, indem er seinen Weg aus der Provinz hinaus in die Welt (Berlin) macht.

Theodor Schulte-Uhlenhorst hebt sich durch sein aufsehenerregendes Hobby von seinen konservativen, technik-skeptischen Eltern ab und sucht den Kontakt zu reichen, exklusiv-exzentrischen Kreisen. In der Liebe allerdings entscheidet sich Theo statt für die vertraglich verabredete adelige Partie für die Bauerstochter und Tänzerin Adele.

„Familie Uhlenhorst" spielt im Jahr 1910, zu einem Zeitpunkt, da die Fliegerei eine noch junge Disziplin war und nur einer kleinen, vermögenden Schicht offenstand, die die nötige Zeit und das entsprechende Geld besaß, sich dieses kostspielige Hobby zu leisten.[162] Neben der humoristischen Verarbeitung des Aufeinandertreffens zweier sozialer Schichten bedient sich das Stück der Figur Josef Suwelacks und seiner Position als Flugpionier der ersten Stunde, während zudem das ambivalente Verhältnis zur Technik – Begeisterung einerseits, Furcht und Skepsis andererseits – durch die Charakterisierung der beiden gegensätzlichen Familien veranschaulicht wird.

160 Ebd., S.45.
161 Ebd., S.18 (dort beide Zitate).
162 Vgl. Kehrt, Moderne Krieger, S.72.

2.3. Josef Suwelack in analogen und digitalen Fachpublikationen zur Fliegerei

Jenseits der populärwissenschaftlichen Literatur über die „Alten Adler“, wie die Flugpioniere, die vor 1914 ihren Pilotenschein erworben hatten, genannt wurden, und der „Jubiläumsschriften“ zu Josef Suwelack fand sein Name auch Eingang in Publikationen zur Fluggeschichte. Rezipiert wurde Suwelack vor allem im Kontext der Rumpler- oder Kondor-Werke, von Flugtagen sowie im Zusammenhang mit seinem Weltrekord. Im Unterschied zu Zeitungs- und Zeitschriftenartikeln, die sich ausschließlich mit ihm befassen, kann man seinen Namen in einigen Publikationen jedoch leicht überlesen. Bis auf wenige Ausnahmen finden sich meist nur zwei bis drei Sätze oder punktuelle Erwähnungen auf einzelnen Seiten.[163] Nur wenige Publikationen beschäftigen sich eingehender mit Suwelack: Guido Rißmann-Ottow, der sich in seiner Arbeit mit der frühen Luftfahrt in Essen beschäftigt und in diesem Kontext auch die Kondor-Werke in den Blick genommen hat, setzte sich kritisch mit der Rolle Suwelacks im dortigen Unternehmen auseinander.[164] Des Weiteren stellt Jörg Armin Kranzhoff den Piloten in seiner Darstellung über Edmund Rumpler vor. Er zeigt auf, dass Josef Suwelack stark in den Flugbetrieb und die Instandhaltung der Maschinen eingespannt wurde und Edmund Rumpler seine Angestellten trotz steigender Gewinne und wachsender Arbeitsbelastung nicht entsprechend entlohnte, sie sogar finanziell benachteiligte.[165] Suwelack tritt bei Kranzhoff vor allem in seiner Funktion als Pilot und Konstrukteur in Erscheinung, der der Fliegerei, wie die meisten zeitweise bei Rumpler Beschäftigten, ein Leben lang verbunden geblieben sei – bis er „am 13. September 1915 über englischen Stellungen abgeschossen“ wurde.[166] In einer anderen Monografie Kranzhoffs über die Arado-Flugzeuge wird Josef Suwelack als Fluglehrer Heinrich Lübbes und „Chefkonstrukteur der Essener Kondor-Flugzeugwerke“ genannt, Lübbe habe Josef Suwelack 1914 dann zum Aufbau einer Fliegerschule in Düsseldorf geraten.[167]

Siebzig Jahr zuvor berichtete Peter Supf ähnlich wie spätere Zeitungsartikel von dem westfälischen Flieger, indem er vornehmlich sein Erfolge auf Flugwettbewerben und seinen Weltrekord hervorhob. Erwähnt wird zudem seine Zeit bei Rumpler und Aviatik, bevor er schließlich „Gründer und Direktor der Kondor-Werke“ geworden sei. Die Ausführungen enden mit den Worten: „Er fällt im Luft-

163 So zum Beispiel bei Bruno Lange, Das Buch der deutschen Luftfahrttechnik. Textteil, Mainz 1970. Zu Suwelack siehe S.45 mit Fokus auf heimatlichen Flugversuchen und Kondor.

164 Rißmann-Ottow, Glück ab!.

165 Jörg Armin Kranzhoff, Edmund Rumpler – Wegbereiter der industriellen Flugzeugfertigung, Bonn 2004, S.120 und S.132ff.

166 Ebd., S.366.

167 Jörg Armin Kranzhoff, Arado-Flugzeuge. Vom Doppeldecker zum Strahlflugzeug, Bonn 2000, S.21.

kampf 1915 an der Westfront".[168] Generell vermerkt Supf zur „Werdezeit", der Pionierzeit der Luftfahrt:

> „Was eben noch Heldentum war, nun zählt es schon zu Alltäglichkeiten. Die Taten, die von vielen vollbracht werden, die nichts Ungewöhnliches, Außergewöhnliches mehr darstellen, hören auf, Heldentaten zu sein. Das aber waren sie in der Werdezeit der Fluggeschichte, und darum sollte die Geschichte jener Jahre viel tiefer im Bewußtsein des Volkes wurzeln!"[169]

Während die meisten Autoren Josef Suwelack lediglich im Kontext von Flugwettbewerben in Teilnehmer- und Siegerlisten sowie in Rekordlisten[170] erwähnen oder auf den vermeintlichen Luftkampf zwischen einem englischen Flugzeug und der Maschine Suwelacks eingehen,[171] wird Suwelack in der Biografie über Melli Beese im Zusammenhang mit den 1911 in Dresden, Chemnitz und Detmold stattfindenden Schauflügen genannt. Anders als die Suwelacksche Familienbiografie schreiben die Verfasser dieser Biografie, dass die drei Teilnehmer Witte, Beese und Suwelack aufgrund der auf den Flugplatz strömenden Menschen aus Sicherheitsgründen nicht gestartet seien und nicht, wie es in der Familienbiografie dargestellt wird, Flüge von zwanzig Minuten Länge ausgeführt und die Detmolder begeistert hätten.[172]

In regionalgeschichtlichen Darstellungen fand Josef Suwelack sowohl Eingang in die Fachliteratur zur Sportgeschichte als auch in Lexikoneinträge und Aufsätze zu den Kondor-Werken. In dem Sammelband „Münster. Die Stadt und ihr Sport", der über die Anfangsjahre der Fliegerei im Münsterland berichtet, findet auch Josef Suwelack Erwähnung, der, so der Autor, „auf das Fliegen versessen" gewesen sei.[173] Vorgestellt wird Suwelack als Weltrekordler – neben anderen Münsteraner Fliegern wie Karl Rösner und Paul Engelhard.[174] Im Zusammenhang mit den

168 Peter Supf, Das Buch der deutschen Fluggeschichte, Bd. 1: Vorzeit – Wendezeit – Werdezeit, Berlin 1935, alle Zitate S. 419.

169 Ebd., S. 444.

170 Siehe Eileen F. Lebow, Before Amelia: Women Pilots in the Early Days of Aviation, Washington, D.C. 2002, S. 69ff. Lebow bemerkt, die Resultate der Johannisthaler Herbstflugwoche seien von der französischen und amerikanischen Presse als nichts Besonderes abgetan worden, vgl. S. 75; Günter Schmitt, Als die Oldtimer flogen. Die Geschichte des Flugplatzes Berlin-Johannisthal, 3. Aufl., Berlin 1995, S. 120 und 131; Werner Schwipps, Riesenzigarren und fliegende Kisten. Bilder aus der Frühzeit der Luftfahrt in Berlin, Berlin 1984, S. 154; ders., Schwerer als Luft. Die Frühzeit der Flugtechnik in Deutschland, Koblenz 1984, S. 72 (2. Platz Herbstflugwoche Berlin-Johannisthal 1911); Wolfgang Wagner, Der deutsche Luftverkehr – Die Pionierjahre 1919–1925, Koblenz 1987, S. 251.

171 Siehe Richard Dietrich, Im Flug über ein halbes Jahrhundert, Gütersloh 1942, S. 97–98.

172 Vgl. Livia Käthe Wittmann/Barbara Zibler, Melli Beese und die „Flügel am Horizont". Die Geschichte der ersten deutschen Pilotin, Berlin 2009, S. 58 und 62. Die Autorinnen nehmen zudem Stellung zur Darstellung der Suwelacks, ebd., S. 62.

173 Langenfeld/Prange, Münster. Die Stadt und ihr Sport, S. 173.

174 Mit dem gleichen Ereignis und seinen Lebensdaten ist sein Name in einem Lexikonartikel unter dem Stichwort „Flugpioniere" aufgeführt, der ihn als Konstrukteur des Kondors würdigt. Vgl. Detlef Fischer, Münster von A bis Z. Wissenswertes in 1500 Stichworten über Geschichte, Kunst und Kultur, 2. Aufl., Münster 2004, S. 100.

Kondor-Werken erscheint er meist als deren (technischer) Direktor, der für Konstruktionsentwürfe, deren bauliche Realisierung und das Einfliegen und Vorstellen der Maschinen verantwortlich gewesen sei.[175]

Abgerundet wird die lokale Rezeption des Flugpioniers derzeit mit einer Masterarbeit über Technikbegeisterung im Münsterland am Beispiel von Flugtagen sowie einem Werk über die Geschichte der Stadt Billerbeck aus dem Jahr 2012.[176] Beide Schriften rezipieren Josef Suwelack im Kontext der Billerbecker Flugtage 1912 sowie seiner vorherigen Flugversuche in seiner Heimat und in Berlin. Martina Glanemann zeichnet anhand von Zeitungsberichten und Briefwechseln die Vorbereitung und Durchführung von Flugtagen im Münsterland nach. Sie zeigt, dass Josef Suwelack der Stolz Billerbecks war und mit seinen Flugvorführungen, über die die Lokalpresse ausführlich berichtete, zum gefeierten Helden und „Sturmflieger" wurde.[177]

Insgesamt betrachtet kann man zwischen einer „Eckdaten-Narration" – Lebensdaten und Weltrekord – in vornehmlich allgemeinen Darstellungen zur Fliegerei und einer lokalspezifischen Rezeption Suwelacks in Publikationen über das Münsterland differenzieren. Letztere sind durch Charakterisierungen des Piloten und die Zuwendung zu einer bestimmten Phase seines Lebens, z. B. Flugversuche in Billerbeck oder Teilnahme an Wettbewerben, gekennzeichnet. Einzig Glanemann, Kranzhoff und Rißmann-Ottow rücken von der heldenhaften Darstellung ab und zeigen Josef Suwelack sowohl als technischen Angestellten in Flugzeugfirmen sowie im Kontext einer technikbegeisterten Gesellschaft.

Die regionale Geschichtsschreibung erinnert an den Piloten und Konstrukteur, wohingegen Suwelack, zugespitzt formuliert, in der allgemeinen Fluggeschichte zwar „vorkommt", aber nicht besonders herausgestellt wird. In der Literatur zur Fluggeschichte agiert Suwelack eher als Nebenakteur und erfährt nicht die Aufmerksamkeit, wie sie etwa Hellmuth Hirth oder Melli Beese als erste deutsche Pilotin erhielten beziehungsweise erhalten. Insbesondere sein erzielter Weltrekord sicherte ihm jedoch einen Platz unter den „heldenhaften Flugpionieren".

Nichtsdestotrotz wäre es daher unzutreffend davon zu sprechen – auch und gerade in Anbetracht einer bis in die Gegenwart reichenden Rezeption seiner „Geschichte" –, Suwelack sei „fast vergessen".[178] Doch hat er nicht die (inter-)

175 Siehe z. B. Hartmut Hernig/Michael Kraus, Und das ist unsere Geschichte. Gelsenkirchener Lesebuch, Oberhausen 1984, S.146: „Technischer Leiter der Kondor-Werke wurde der Flieger Josef Suwelack, der dann auch einen nach ihm benannten Eindecker konstruierte und baute". Hugo Rieth, Die Kondor-Flugzeugwerke (1912–1918). Ein Kapitel aus der Pionierzeit deutscher Luftfahrtgeschichte, geschrieben in Essen und Rotthausen, in: Die Heimatstadt Essen (1975), S.121–130, hier S.123. Frank Radzicki, Der Traum vom Fliegen im Ruhrgebiet, Erfurt 2005.

176 Martina Glanemann, Technikbegeisterung und kleinstädtische Strukturen. Flugtage im Münsterland 1908–1914, unveröffentlichte Masterarbeit, Münster 2012; Werner Freitag (Hg.), Geschichte der Stadt Billerbeck, Bielefeld 2012, zu Suwelack S.241ff.

177 Siehe Glanemann, Technikbegeisterung, S.30ff., Zitat S.32.

nationale Berühmtheit eines Manfred von Richthofen oder Frank Luke erlangt, denn der Billerbecker war in erster Linie Zivilflieger und Konstrukteur und zählt eher zu den lokalen Fliegergrößen. Zu bedenken ist außerdem, dass er bereits im Alter von 27 Jahren starb und sein Leben im Unterschied zu Jagdfliegern wie Manfred von Richthofen (1892–1917), Max Immelmann (1890–1916) oder Frank Luke (1897–1918), abgesehen von öffentlichen Auftritten auf Flugwettbewerben oder Vorträgen, fernab der Öffentlichkeit verlief. So schrieb 1913 schon die Zeitschrift „Flugsport": „Der bekannte Flieger Suwelack, von dem man in letzter Zeit wenig hörte, hat in aller Stille die ‚Kondor'-Flugzeug-Werke G. m. b. H in Essen gegründet und vor einigen Wochen ... den neuen Eindecker-Typ herausgebracht."[179] Insbesondere die Tätigkeit als Konstrukteur und sein Kriegseinsatz sichern ihm jedoch heute auch im Internet einen Platz.[180] So verzeichnet ihn z. B. die Online-Dokumentation über den Flugplatz Johannisthal in einer Liste zu Wohnanschriften der Berliner Flieger und im Kontext eines Flugkistenrennens, das Berliner Grundschulen im Oktober 2010 mit Nachbauten von Flugzeugen aus der Zeit von 1909 bis 1914 veranstalteten.[181]

Im Jahr 2009 wurde in der Sendung „Lokalzeit Münsterland" des WDR in der Rubrik „Lokalzeitgeschichte" auf das Jahr 1959 im Münsterland zurückgeblickt,[182] in dem ein Festumzug durch Billerbeck stattfand, bei dem auch das von Josef Suwelack 1909 erste selbstgebaute Flugzeug als Nachbildung präsentiert wurde: „Vor fünfzig Jahren, nämlich 1909, hatte der Billerbecker Flugpionier Josef Suwelack sein erstes Flugzeug gebaut, eine Meisterleistung, die es im Festumzug zu würdigen galt."

Darüber hinaus erhält man bei der Internetrecherche mehrere Einträge zu den Schlagworten Suwelack/Kondor, die sowohl Informationen zu verschiedenen Flugzeugmodellen als auch zu Personalzahlen und dem Flugbetrieb bieten.[183] Auch in Foren zur Fliegerei, etwa im *„Great War Forum"* oder auf „theaerodrome.

178 So der Moderator in der „Lokalzeitgeschichte – Münsterländische Flugpioniere".

179 Flugsport (1912), Nr. 24, S. 903.

180 Die folgenden Ausführungen sollen einen Eindruck davon vermitteln, in welchen Kontexten Josef Suwelack in Erscheinung tritt, erheben aber keinesfalls Anspruch auf Vollständigkeit! Ferner variiert die Schreibweise seines Namens, was wiederum bei der Recherche zu verschiedenen Treffern führt, z. B. „Suvelak" oder „Suvelack". Einträge zur Literatur über den Flieger wurden nicht berücksichtigt (z. B. Treffer bei Amazon etc. für die Familien-Biografie der Suwelacks).

181 Vgl. http://johflug.de/pdf/Wohnanschriften_bis_%201914_Dez_2012.pdf, S. 1.; http://johflug.de/pdf/Flugkisten2010.pdf, S. 3. Für ein Video zum Rennen siehe https://www. youtube.com/watch?v=i_OvhqsM_Dk, online seit dem 22.1.2011 (25.5.2013).

182 Schröder, Lokalzeitgeschichte – Münsterländische Flugpioniere, dort auch das folgende Zitat, Min. 12:52–13:08.

183 Siehe zum Beispiel: http://www.gelsenkirchener-geschichten.de/viewtopic.php?t=5557 (Diskussion seit dem 9.10.2008), „Die interaktive Spurensammlung Gelsenkirchener Geschichte – Soziokulturelles von Gestern und Heute"; http://flyingmachines.ru/Site2/Crafts/Craft28883.htm, online seit Dezember 2011 (2.2.2014).

com“, finden sich Hinweise auf den westfälischen Piloten.[184] Spuren seiner Tätigkeit als Zivilflieger bei der Feldfliegerabteilung 24 im Ersten Weltkrieg finden sich auf Seiten zur Fliegertruppe, etwa auf „frontflieger.de“,[185] oder in Foren, die sich mit Weltkriegsopfern und Gefallenen des Krieges befassen.[186] Von familiärer Seite existiert sowohl ein von Walter Suwelack verfasster Web-Eintrag zu Josef Suwelack auf „wikipedia.de“ als auch eine von der Familie gestaltete Homepage, die das Leben des Fliegers mit zahlreichen Dokumenten aus dem Familienarchiv nachzeichnet.[187]

Die öffentliche Erinnerung an den Sohn, Flieger, Unternehmer und Kriegsteilnehmer Suwelack ist geprägt von seinen Leistungen für die motorisierte Fliegerei und wird dominiert von seinen Erfolgen als Wettbewerbsteilnehmer und Konstrukteur und schließlich von seinem „Opfertod“ als 27-Jähriger im Ersten Weltkrieg. Die Inanspruchnahme seiner Person für die jeweilige Gegenwart verstellt hierbei oftmals den Blick auf den Pionier der Fliegerei und seine Zeit. Häufig befriedigen die biografischen Darstellungen das jeweilige Leserinteresse ihrer Zeit, schreiben Josef Suwelack aktuell passende und mit dem jeweiligen Zeitgeschmack konforme Charakterzüge und Eigenschaften zu. Als moralisches Vorbild besitzt die Geschichte des Fliegers nicht selten Appellfunktion, will erziehen und Orientierung für Gegenwart und Zukunft bieten.

2.4. Josef Suwelack und seine Nachfahren: Die Familie Suwelack als Pflegerin der Erinnerung

Für die Tradierung des Flieger-Mythos um Josef Suwelack ist neben der Frage, wie erinnert wird, vor allem die Frage von Bedeutung, durch wen erinnert wird. Für die Erzählform macht es einen bedeutenden Unterschied, ob Journalisten, Historiker oder Familienangehörige sich dem Menschen Josef Suwelack zuwenden. Als Nachfahren des Piloten sind die Suwelacks emotional involviert, fühlen sich

184 Jan den Das, Josef Suwelack/Kondor-Flugzeugwerke, veröffentlicht am 30.10.2006, http://1914–1918.invisionzone.com/forums/index.php?showtopic=62224; Hannes Täger, Teichmann/Suwelack and Dowding, http://www.theaerodrome.com/forum/1999/7262-teichmann-suwelack-dowding.html, Diskussion seit dem 1.11.1999 (2.2.2014).

185 Thorsten Pietsch (Hg.), http://www.frontflieger.de/2-ffa024.html, Angehörige der FfA 24 sowie Gefallenenlisten, online seit Ende Januar 2002 (2.2.2014).

186 Ludger Bäumer (Hg.), http://www.weltkriegsopfer.de/Kriegsopfer-Josef-Suwelack_Soldaten_0_11498 6.html?PHPSESSID=1b58feca48720558713e2feb77517161, online seit 2003 (2.2.2014). Auch auf Seiten zu Kriegsgräbern finden sich Informationen zu Suwelack. Vgl. z. B. das „War Graves Photographic Project“, http://twgpp.org/information.php?id=2016101. Dieses Projekt verzeichnet derzeit (Stand 2013) über 1,7 Millionen fotografierte Kriegsgräber. Ziel ist es, „to record and make available, photographs of as many war graves and memorials as possible so that there is a pictorial record of yesterday's sacrifices for tomorrow's decendants“, online seit Februar 2008 (2.2.2014).

187 Siehe www.josef-suwelack.de.

seinem Andenken verpflichtet und wollen ihn in guter Erinnerung behalten. Gleichzeitig verstehen sie sich als pflichtbewusste und geschichtsinteressierte Nachlasshüter und engagieren sich gesellschaftlich, unter anderem durch die finanzielle Unterstützung und Mitwirkung bei Arbeiten zur westfälischen Geschichte. Mit der Erinnerung an Josef Suwelack treten sie das Erbe des Piloten an und unterstreichen durch ihr Andenken den andauernd hohen Stellenwert ihres familiären Zusammenhalts. Josef Suwelacks Leben hat nicht nur sichtbare Spuren in Gestalt seines Nachlasses hinterlassen, sondern durch die mündlichen Überlieferungen blieb er weit über seinen Tod hinaus Teil der Familie. So äußerte sich Julia Koch-Suwelack,[188] die Tochter von Josefs jüngerem Bruder Richard (1891–1968), im Interview 2001:

> „Onkel Josef Suwelack hat uns die ganze Kindheit über begleitet. Wir haben von seinen Geschichten gehört, wir haben mit seinen Bildern gelebt, seinen Abenteuern. Unser Vater, der ihn unmittelbar begleitet hat nach Johannisthal und [ihn] in anderen Dingen auch beraten hat, als er zum Beispiel dort [in Essen] sein Flugzeugwerk gegründet hat, der erzählte fast tagtäglich von diesen Abenteuern und Geschichten".[189]

Aufgewachsen mit den Geschichten über Josef Suwelack überrascht es nicht, dass sich die Erzählung von der familiären „Leidenschaft für die Fliegerei" bis in die Gegenwart fortsetzt.

Bereits 1917 hatte sich der damals achtzehn Jahre alte Bruder Josefs, Wilhelm Suwelack, als Kriegsfreiwilliger gemeldet und sich in Schwerin zum Flieger ausbilden lassen.[190] Die Passion für die Fliegerei habe auch ihn wie zuvor schon Josef nicht mehr losgelassen und er sei wie sein älterer Bruder in Flugvereinen und auf Wettbewerben aktiv gewesen. Während des Zweiten Weltkrieges kämpfte er zunächst an der West-, ab April 1941 dann an der Ostfront, wo er am 14. Mai 1941 in der Ukraine bei Melitopol fiel und in Nikolajew bestattet wurde. In einem zeitgenössischen Nachruf heißt es:

> „Wie sein Bruder, der im Weltkriege gefallen ist, war auch Wilhelm Suwelack mit Leib und Seele ein Flieger. Im nationalsozialistischen Deutschland stellte er sich voll und ganz der Erziehung des Fliegernachwuchses zur Verfügung Als Flieger erlebte Wilhelm Suwelack die Rückkehr der Sudetendeutschen im Reich. Seit dem ersten Tage dieses Krieges um Deutschlands Zukunft als Flieger tätig, erwarb er sich bei seinen Kameraden und bei seinen Vorgesetzten vollste Liebe und Achtung.

188 Julia Koch-Suwelack (*1921 in Billerbeck) lernte in Münster und München Grafik und Illustration. Nach ihrer Ausbildung war sie zunächst als freie Grafikerin in Havixbeck tätig, unterbrochen durch den Zweiten Weltkrieg. Nach dem Krieg setzte sie ihre Arbeit als Illustratorin vor allem von Schulbüchern und Autorin fort und studierte in den fünfziger Jahren Kunst in Hamburg, um Kunstlehrerin zu werden. Ende der sechziger Jahre schloss sie noch ein Studium an der Düsseldorfer Kunstakademie an, um auch in der Kunstlehrerbildung tätig werden zu können. Ihr publizistisches und künstlerisches Schaffen fand Eingang in Museen und schlug sich in zahlreichen Publikationen nieder. Informationen entnommen aus Julia Koch-Suwelack, Grafiken, Pastelle, Ölbilder. Arbeiten aus 60 Jahren, in: 20 Jahre Baumberger Sandstein-Museum, hg. vom Nord-West Verlag in Kooperation mit dem Baumberger Sandstein-Museum. Melle/Havixbeck 2014, S.32–34.

189 Schröder, Lokalzeitgeschichte – Flugpionier Josef Suwelack, Min. 1:14ff.

190 Die folgenden Ausführungen basieren auf: Krewerth, Ein westfälisches Fliegerleben, S.121.

> Seine Zuverlässigkeit und sein Pflichtbewusstsein wurde sehr geschätzt und trug ihm manchen ehrenvollen und wichtigen Sonderauftrag ein.[191]

Nach Josef und Wilhelm wurde mit ihrem Neffen Winfried Suwelack diese „Tradition" fortgesetzt, der sich im Zweiten Weltkrieg zum Flieger ausbilden und sich seine Qualifikation mit einer erneuten Flugprüfung 1952 bestätigen ließ.[192]

Gegenwärtig widmen sich vor allem Wolfgang und Walter Suwelack, der den privaten Familiennachlass verwaltet und „hütet", sowie auch weiterhin Winfried Suwelack mit großem Engagement der Familiengeschichte.[193] 1988 verfassten sie anlässlich des 100. Geburtstages Josef Suwelacks die genannte Biografie über ihren Onkel,[194] die 2001 in überarbeiteter Form von Rainer A. Krewerth herausgeben wurde.[195] Die Biografie ist explizit als Gedenkschrift konzipiert, die „das Andenken" an Josef Suwelack „an nachfolgende Generationen weitergeben" soll. Die Jubiläumsschriften und weitere von der Familie initiierte Artikel, Filme und geförderte Projekte, dazu zählt auch die hier verfasste Darstellung, tragen zu einer beständigen Auseinandersetzung mit Josef Suwelack und nicht zuletzt auch zur Verfestigung seines Mythos bei.

Im Geburtsort Josef Suwelacks ist sein Name symbolisch auf einem Grabstein der Familiengruft aufgeführt, mit dem Hinweis, dass er in Frankreich gefallen und bestattet worden sei. Sein Name ist ebenfalls auf einer Gedenktafel am Billerbecker Mahnmal für die Opfer von Krieg und Gewalt verzeichnet. Private Fahrten zu seinem Kriegsgrab auf dem englischen Soldatenfriedhof in Erquinghem-sur-Lys in Frankreich, über die auch eine private schriftliche Dokumentation existiert, zeigen auf, dass die Suwelacks ihren Vorfahr als Helden (ver)ehren und bestrebt sind, ihm ein Denkmal zu setzen.

Die folgende Annäherung an sein Leben versteht sich als ein Versuch, Josef Suwelack in seiner Zeit zu verorten, die Strukturen, in denen er sich bewegte, und seine Beziehungen und Prägungen in den Blick zu nehmen. Im Zentrum stehen Fragen danach, durch welche Erfahrungen, von welchen Erwartungen sein Leben bestimmt war und welche Personen Anteil an seinen Erfolgen und Misserfolgen hatten. Dabei möchte diese neue biografische Rekonstruktion keineswegs die „Heldengeschichte(n)" ausblenden, sind sie doch als narratives Konstrukt ein zentraler Bestandteil des Flieger-Mythos um Josef Suwelack.

191 Heimatbrief Billerbeck, 18.10.1941. Die Seite zu den Familienanzeigen zu Billerbeck existiert seit 2007 und wird bearbeitet von Rotraud Ilisch und Sabine Lleschi. http://familienanzeigen.genealogy.net/anzeigen.php?bild=anzeigen/BillerbAnz/T/T–1941–10–12-suwelack-wilhelm.jpg&PID=882 (2.2.1914).

192 Vgl. Krewerth, Ein westfälisches Fliegerleben, S.124f.

193 Schröder, Lokalzeitgeschichte – Münsterländische Flugpioniere, vgl. Min. 0:57ff.

194 Winfried und Walter Suwelack, Josef Suwelack und der Traum vom Fliegen.

195 Krewerth, Ein westfälisches Fliegerleben. Vgl. hierzu Rezensionen: „Josef Suwelack und der Traum vom Fliegen. Ein neues 135 seitiges Buch über den berühmten Billerbecker Flugpionier", in: Billerbecker Anzeiger vom 28.10.1988.

C. Annäherungen an Josef Suwelack

Feld 1: Familie und Prägungen

1.1. Startkapital ins Leben: Josefs Eltern und die Billerbecker Molkerei

Josefs Vater, Joseph Suwelack senior (4.1.1850–9.3.1929), wuchs als drittes von vier Kindern auf dem elterlichen Hof in Greven-Pentrup auf. Bereits mit fünf Jahren wurde er Vollwaise – seine Eltern Anna Elisabeth, geborene Lohaus (1817–1853), und Landwirt Johan Wilhelmus Süverlake (1817–1855), waren früh verstorben. Über seine schulische Bildung existieren keine Quellen, doch wahrscheinlich wird er die Volksschule besucht haben. Nach einer landwirtschaftlichen Ausbildung wurde er Verwalter verschiedener Güter, zuletzt auf dem Gut von Twickel in Havixbeck.

In der lokalen Geschichtsschreibung erscheint Joseph Suwelack sen. im Zusammenhang mit der Gründung der Billerbecker Molkerei als Visionär. Seinem Sohn Richard zufolge, der 1909 anlässlich des 25-jährigen Jubiläums der Molkerei eine Festschrift verfasste, habe Joseph Suwelack 1879 beschlossen, „auch in hiesiger Gegend den Landwirten durch eine gemeinsame Verarbeitung der Milch die Vorteile einer guten Milchviehzucht zu bieten."[1]

Auf der Suche nach Unterstützern für sein Vorhaben trat er an einflussreiche Landwirte der Gegend heran, die ihn im Februar 1884 zu einer landwirtschaftlichen Versammlung nach Billerbeck einluden. Nachdem „umfangreiche Erkundigungen bei bestehenden Molkereien" eingeholt und Skeptiker überzeugt worden waren und sogar bei der Stadt Billerbeck für das Projekt geworben werden konnte, wurde am 29. März des gleichen Jahres in Billerbeck eine „Gelegenheitsgesellschaft mit der Aufgabe der Begründung einer Aktien-Gesellschaft Sammelmolkerei Billerbeck" gegründet.[2] Ferner erfolgte die Wahl einer Molkereikommission,[3]

1 Richard Suwelack, Festschrift anläßlich des 25jährigen Bestehens der Dampfmolkerei Billerbeck Gesellschaft mit beschränkter Haftung, o. O. 1909, S.8. Da vom eigenen Sohn geschrieben, ist die Festschrift und die Darstellung der Entwicklung der Molkerei entsprechend kritisch zu sehen. Der „Billerbecker Anzeiger", die Lokalzeitung Billerbecks, berichtete bis zur Gründung einer eigenen Molkereizeitung 1895 in der Rubrik „Lokales und Provinzielles" über die Entwicklungen im Molkereiwesen. Von der Molkereizeitung selbst ist nur eine einzige Ausgabe von 1911 erhalten, so dass hieraus keine nennenswerten Informationen gewonnen werden konnten. Vgl. zur Quellenlage: Katja Rösler, Die Sammel-Molkerei in Billerbeck im Kontext kleinstädtischer Entwicklung und Agrarmodernisierung um 1900, unveröffentlichte Magisterarbeit, Münster 2012, S.9ff. An dieser Stelle sei Frau Rösler für die Übersendung ihrer Arbeit an mich herzlich gedankt.

2 Ebd., S.8–9. Bereits auf dieser Versammlung wurden fast alle Aktien ausgegeben: „Auf der Versammlung erklärten schon sofort 68 Landwirte mit 79 Anteilen à 300 Mark sich zu beteiligen." Suwelack, Festschrift, S.9.

deren Aufgabe es sein sollte, sich weiter über das Molkereiwesen kundig zu machen.

Um die nötigen Fachkenntnisse zu erwerben, besuchte Joseph Suwelack die Milchwirtschaftliche Versuchsstation und höhere Lehranstalt für Molkereiwesen zu Raden in Mecklenburg-Schwerin – die erste ihrer Art.[4] Hier zeigten sich bereits die Aufgeschlossenheit und der Ehrgeiz Joseph Suwelacks, neue Wege zu beschreiten, sich weiterzubilden und mit der Zeit zu gehen. Zudem besaß der zukünftige Molkereidirektor Überzeugungskraft und Mut, hatte er sich doch aus eigener Initiative an die örtlichen Landwirte gewandt und ihnen seine Idee unterbreitet, Eigenschaften, die auch seinen Sohn Josef ausmachen sollten.

Nachdem ein weiterer Monat verstrichen war, fiel die Entscheidung über den zukünftigen Standort der Molkerei auf Billerbeck, genauer auf die Bauernschaft Holthausen Nr. 26 im Kirchspiel Billerbeck, da an dieser Stelle alle Anforderungen und Auflagen an den Betrieb erfüllt werden konnten.[5] Schließlich erfolgte die Gründung der Sammel-Molkerei Akt.-Ges. Billerbeck am 29. September 1884 in Anwesenheit einer Gerichtstagskommission sowie von 54 Landwirten.[6] Daran schloss sich die Wahl des Aufsichtsrates und des Vorstandes an. Zu letzterem zählten Landwirt Anton Ueding aus Lutum bei Billerbeck, Kaufmann Josef Ahlers und Joseph Suwelack. Dieser wurde mit der Verwaltung der Molkerei beauftragt und somit zu ihrem Direktor ernannt.[7]

Bereits vier Monate später, im Januar 1885, war die Anlage „ziemlich fertiggestellt" und nahm am 1. Februar 1885 ihren Betrieb mit 157 Landwirten als Milchlieferanten und einem verbliebenen Betriebskapital von 8.268 Mark auf.[8] Der Betrieb war der erste seiner Art im Kreis Coesfeld und erst die zweite Sammel-Molkerei in Westfalen.[9] Selbst als andere Orte es Billerbeck nachtaten, blieb die dortige Molkerei diejenige mit den größten Auswirkungen auf die Region.[10]

3 Die Mitglieder der Molkereikommission waren: Ueding, Dr. Potthast, Eißing und Oeing für Billerbeck, Suwelack, Raestrup für Havixbeck, Darup und Wellberg für Darup, Darup-Deiters und von Twickel für Nottuln, Tenhagen und Waßmer für Osterwick und Höping für Darfeld. Vgl. Carl Knüppel, 50 Jahre Molkerei Billerbeck. Ein Stück Heimatgeschichte. Geschrieben anläßlich des 50 jährigen Bestehens der Molkerei Billerbeck G.m.b.H. Billerbeck Westf., Billerbeck/Westf. 1935, S. 9.

4 Suwelack, Festschrift, S. 10.

5 Zum Beispiel eine stetige Kühlung der Molkerei durch Grundwasserversorgung, eine gute Bausubstanz etc., siehe dazu Rösler, Sammel-Molkerei, S. 23.

6 Suwelack, Festschrift, S. 10.

7 Vgl. Rösler, Sammel-Molkerei, S. 24 sowie Suwelack, Festschrift, S. 10.

8 Suwelack, Festschrift, Zitat S. 12, S. 40. Insgesamt 58 Aktionäre hatten 81 Aktien gezeichnet, das Anlagekapital betrug 24.300 Mark, wovon 16.032 Mark für den Anlagenbau verbraucht worden waren.

9 Vgl. ebd., S. 12.

10 Vgl. Constanze Sieger, Die Ludgerusstadt im 19. Jahrhundert, in: Freitag, Geschichte der Stadt Billerbeck, S. 111–260, hier S. 219.

Sah sich die Molkerei in den ersten Jahren mit einigen Problemen wie mangelhafter technischer Ausstattung, ungeschultem Personal oder ausbleibenden Milchlieferungen konfrontiert,[11] rentierten sich Investitionen langfristig und trugen dazu bei,[12] dass das Unternehmen zu einem wichtigen Wirtschaftsfaktor und zum größten Steuerzahler der Region wurde.[13] Die Geschäfte liefen so gut, dass nach einigen Jahren der Molkereistandort zu klein und 1897 ein Grundstück an der Münsterstraße erworben wurde. Hierfür verwandten die Investoren die beachtliche Summe von 100.000 Mark aus dem Reingewinn des Betriebes.[14]

Die Gründung von Aktiengesellschaften (AG) erlebte durch den Wegfall der Konzessionspflicht nach der Reichsgründung von 1870/71 und die erlassene „Aktienrechtsnovelle“,[15] die die Unternehmen aus der bisherigen staatlichen Reglementierung entließ, neuen Aufschwung.[16] Die Unternehmensform der AG veränderte zugleich die ökonomischen und sozialen Beziehungen, denn durch die zunehmende „Trennung zwischen Besitz und Kontrolle“,[17] ein Merkmal moderner Gesellschaften, kam es zu einer Ausdifferenzierung und Spezialisierung einzelner Organisationsfunktionen.

Über die Jahre wuchsen der Kundenstamm und die Lieferantenzahl der Molkerei stetig an. Das Produktangebot wurde an die Bedürfnisse der Kunden, aber auch an die Kapazitäten der Molkerei selbst angepasst, so dass auch in Krisenzeiten gute Gewinne erzielt werden konnten. U.a. weitete die Molkerei ihre Produktpalette immer weiter aus. Neben der Hauptproduktion von Butter und Käse betätigte sich das Unternehmen nun auch auf dem Gebiet der Schweinemast und der Obstverwertung und unterhielt seit 1901 einen gemeinschaftlichen Eierverkauf.[18] Darüber hinaus ging die Molkerei dazu über, Kraftfutter und Düngemittel zu verkaufen So war es möglich, dass trotz einer Schweinepest 1908, die den gesamten Bestand vernichtete, durch die übrigen Geschäftsfelder, vor allem aber durch den angewachsenen Reservefonds der Verlust aufgefangen werden konnte. Wenn-

11 Siehe dazu ausführlich Rösler, Sammelmolkerei, bes. S.25f. und 62ff.

12 Hier sind exemplarisch zu nennen: Anbauten an der Molkerei, Anlegung eines Bohrbrunnens, Anschaffung neuer bzw. weiterer Molkereigeräte, Kauf eines Automobils und Einrichtung einer Kühlanlage 1907 etc.

13 Vgl. Rösler, Sammel-Molkerei, S.76 und 79.

14 Vgl. ebd., S.48.

15 Die Aktienrechtsnovelle von 1870 sollte eine stärkere Liberalisierung des Marktes bewirken, denn der Staat war aus Sicht der Aktionäre seiner Aufsichtsfunktion nur unzureichend nachgekommen und hatte eher geschäftsschädigend denn fördernd gewirkt. Siehe dazu: Werner Schubert, Die Abschaffung des Konzessionssystems durch die Aktienrechtsnovelle von 1870, in: Zeitschrift für Unternehmens- und Gesellschaftsrecht 10 (1981), H. 2, S.285–317.

16 Siehe zur Entwicklung des Aktienrechts: Walter Bayer/Mathias Habersack (Hg.), Aktienrecht im Wandel. Bd. II: Grundsatzfragen des Aktienrechts, Tübingen 2007, bes. S.60ff.

17 Werner Sombart, Der moderne Kapitalismus, S.14, zit. nach Morten Reitmayer, Bankiers im Kaiserreich. Sozialprofil und Habitus der deutschen Hochfinanz, Göttingen 2011, S.83.

18 Vgl. Rösler, Sammel-Molkerei, S.35ff.

gleich sich die Obstdörranlage als unrentabel erwies, macht das gewachsene Produktspektrum dennoch deutlich, dass man bereit war, in die Firma zu investieren und Neues zu riskieren.[19] Dass dabei auch kontinuierlich an einer Verbesserung der Wirtschaftlichkeit gearbeitet wurde, lässt sich an Joseph Suwelacks erfolgreich verlaufenden Versuchen zur Herstellung von Trockenmilch ablesen. Deren Produktion führte er seit 1910 ein, nicht zuletzt da sich die Molkerei in den Sommermonaten mit dem Problem der Konservierung von Milch konfrontiert sah.[20] Katja Rösler schreibt in ihrer Arbeit zur Sammel-Molkerei Billerbeck über die Milchtrocknung: „Die Produkte fanden solchen Absatz, dass bis kurz vor Ausbruch des Ersten Weltkriegs 1914 die Maschinen der Trockenmilchanlage Tag und Nacht arbeiteten, um die angeforderten Milchpulvermengen herstellen zu können."[21] Die Milchpulvergewinnung schuf neue Absatzmärkte und stieß offensichtlich auf großen Zuspruch.[22] Im gleichen Jahr wurde die Nährmittelfabrik GmbH gegründet, woran die Molkerei mit 50 Prozent und der Molkereidirektor und dessen Sohn Richard zu je 25 Prozent beteiligt waren.[23]

Darüber hinaus setzte das Unternehmen auf Weiterbildung des Personals und achtete penibel auf Gesundheit und Hygiene, um hochwertige Produkte garantieren zu können. Dazu wurden Qualitätskontrollen eingeführt und die Lieferanten ermahnt, einwandfreie Ware zu liefern. So bezahlte die Molkerei etwa seit 1906 nicht mehr nach der Menge, sondern nach dem Fettgehalt der gelieferten Milch.[24]

In den ersten fünfundzwanzig Jahren ihres Bestehens entwickelte sich die Molkerei zu einem florierenden Unternehmen, das hohe Gewinne erwirtschaftete. Die Anzahl der Milchlieferanten wuchs von 157 im Jahr 1885 auf 1.151 im Jahr 1909 an.[25] Da die Molkerei das verbliebene Geld in einem Reservefonds sparte und nur ab und an Renditen an die Aktionäre ausschüttete, hatte sich „schon im achten Geschäftsjahre die ganze Anlage bis auf einen Rest von 1.275,14 Mark amortisiert und außerdem noch ein Reservefonds von 14.000 Mark angesammelt".[26] Im Gründungsjahr belief sich der Geldumschlag auf 242.000 Mark und lag 1909 bereits bei 3,7 Millionen Mark.[27] Selbst nach Abzug der Lohn- und Betriebskosten ist anzunehmen, dass immer noch eine hohe Summe übrig blieb. Rösler zufolge schwankte die Mitarbeiterzahl der Molkerei zwischen 16 im Jahr

19 Dazu kritisch Rösler, Sammel-Molkerei, S.43.
20 Vgl. für mehr Informationen die Ausführungen bei Rösler, Sammel-Molkerei, S.44.
21 Ebd.
22 Im Jahr 1912 exportierte die Molkerei das Milchpulver nach Amerika, wenngleich Rösler anmerkt, dass das Ruhrgebiet wichtigster Absatzmarkt für das Milchgeschäft geblieben sei; vgl. ebd., S.35.
23 Vgl. http://www.josef-suwelack.de/ueber-die-familie/joseph-suwelack-sen (3.10.2014).
24 Vgl. Rösler, Sammel-Molkerei, S.29 und S.56ff.
25 Vgl. Suwelack Festschrift, S.40.
26 Ebd., S.24.
27 Vgl. ebd., S.34–51.

1893 und 38 im Jahr 1899. Angesichts der Betriebsgröße hält sie eine durchschnittliche Mitarbeiterzahl von 16 für wahrscheinlich.[28]

Die wirtschaftliche Krise während des Ersten Weltkrieges ging auch am Molkereigeschäft nicht spurlos vorüber: „Waren die angelieferten Milchmengen bis zum Jahre 1915 schon immer weniger geworden, so erfolgte doch besonders mit der Aufrichtung der Zwangswirtschaft ... ein starker Rückgang in der Milchanlieferung."[29] Diesem Umstand begegnete die Molkerei mit der Produktion von Suppen für die Heeresverwaltung. Absatzschwierigkeiten nach dem Ersten Weltkrieg und die Inflation von 1923 machten diesen neuen Geschäftszweig der Nährmittelfabrik allerdings insolvent – lediglich die Molkerei überstand die Krise dank hoher finanzieller Aufwendungen. Allerdings entschied sich die Aktionärsversammlung 1924, die Molkerei zu verpachten, worauf hin Joseph Suwelack und sein Sohn Richard aus der Geschäftsführung ausschieden. Die Lokalzeitung, der „Billerbecker Anzeiger", berichtete am 3. Oktober darüber:

> „In der letzten Versammlung der Aktionäre der Dampfmolkerei Billerbeck am 30. September ist der Beschluss gefasst worden, den Molkereibetrieb vorläufig auf fünf Jahre zu verpachten. Die jetzt herrschenden wirtschaftlichen Verhältnisse, die ein Weiterführen des Molkereibetriebes auf bisheriger Grundlage unrentabel machten, gaben zu dieser Verpachtung Veranlassung. Der Geschäftsführer der Gelsenkirchener Milchhändlergenossenschaft, Herr Karl Henn, ist der Pächter. Er habe, wie uns mitgeteilt wird, lange Jahre mit der Dampfmolkerei in Geschäftsverbindung gestanden und sei ein tüchtiger Fachmann. Mit dem 15. Oktober dieses Jahres übernimmt Herr Karl Henn den Molkereibetrieb. Die Zahlung der Milchgelder erfolgt 14 täglich am 20. und 5. jeden Monats. Herr Schräder, der bislang den Verkehr mit den Milchlieferanten getätigt hat, bleibt nach wie vor in gleicher Weise tätig."[30]

Vollständig aufgelöst wurde die Molkerei erst knapp 40 Jahre später, am 19. März 1963.[31]

Die am wirtschaftlichen Aufschwung des Unternehmens beteiligten Personen hatten oftmals einflussreiche Positionen in der Gemeinde und waren wichtige Entscheidungsträger. Als erfolgreicher Unternehmer heiratete Joseph Suwelack am 24. Mai 1886 die Billerbeckerin Katharina Brockmann. Die Tochter der Eheleute Heinrich Brockmann (1829–1899) und Theresia Brockmann, geb. Thier, wurde am 28. Oktober 1863 als drittes von fünf Kindern geboren. In der Brockmannschen Familienchronik schreibt ihr Vater über sie:

> „Nach ihrer Entlassung aus der Elementarschule war sie etwa 3/4 Jahr im Pensionate der Schwestern Unserer Lieben Frau zu Coesfeld und darauf, nach Aufhebung dieses Klosters, im Pensionate der Franziskanessen in Oldenzaal. Danach war sie ein Jahr in Pension beim Gastwirt Willenbrink zu Herzfeld zur Erlernung des Haushalts. Am 24. Mai verehelichte sie sich mit dem Molkereidirektor Josef Suwelack zu Billerbeck."[32]

28 Vgl. Rösler, Sammel-Molkerei, S.65.
29 Dampfmolkerei Billerbeck. Lutum-Gausing & Co., Gesamtübersichts- und Schlußbericht erstattet auf der Generalversammlung am 23.3.1964, Billerbeck 1964, S.8.
30 Billerbecker Anzeiger vom 3.10.1924.
31 Vgl. http://www.josef-suwelack.de/ueber-die-familie/joseph-suwelack-sen (3.10.2014).
32 Zit. nach Krewerth, Ein westfälisches Fliegerleben, S.18.

Der Chronist selbst gründete 1866 die Sparkasse Billerbeck und verhalf seiner Familie zu Wohlstand und Ansehen.[33] Brockmanns Sohn Clemens (1868–1949) übernahm 1897 die Sparkasse seines Vaters und machte Karriere als Sparkassen-Rendant. Bei der Gemeindeverordneten-Wahl wurde er für die Periode 1900–1903 beziehungsweise 1905 zusammen mit Joseph Suwelack und dem Apotheker Adolf Rave gewählt.

Wie auch Amtmann August von Twickel, Aktionär der Billerbecker Molkerei, engagierte sich Joseph Suwelack gesellschaftlich. So wirkte er etwa bei der Gründung des „Verschönerungsvereins" am 22. November 1899 mit, der das äußerliche Stadtbild z. B. durch Blumenschmuck attraktiver machen wollte, und war Teil des Vorstandes.

In der zweiten Hälfte des 19. Jahrhunderts hatte sich wie im Deutschen Reich insgesamt auch in Billerbeck ein reges Vereinsleben entwickelt,[34] das Traditionen der katholischen Schützenbruderschaften fortführte, aber auch neue Vereinsformen, insbesondere Geselligkeitsvereine schuf. Die örtlichen Wirtschaften dienten dabei als Orte des Austausches und des Beisammenseins.[35] Es ging nicht zuletzt aber auch um Mitsprache und Gemeinschaftsfürsorge, um eine funktionierende Selbstverwaltung Billerbecks durch das örtliche Bürgertum. Die Verflechtungen dieser städtischen Eliten stellt Ria Hänisch als bedeutsam heraus und meint, es sei „in Fragen der Kreditvergabe und Finanzierung für die Molkerei nicht ganz unerheblich, wenn der örtliche Sparkassen-Rendant der Schwiegervater des Molkereidirektors ist".[36]

Mit dem Erfolg der Molkerei zogen in Billerbeck auch nach und nach Errungenschaften der Moderne ein: So plädierten im Zuge des Standortwechsels der Molkerei die Unternehmer, wie schon zehn Jahre zuvor, erneut für einen Bahnanschluss Billerbecks an die Umgebung sowie für die Errichtung eines Elektrizitätswerkes. Beides wurde 1908 realisiert und verhalf sowohl der Molkerei wie aber auch der Billerbecker Bevölkerung zu mehr Mobilität und damit zu größerer Unabhängigkeit.[37] Denn vor dem Anschluss der Stadt an das Schienennetz mussten die Waren zunächst noch mit Kutschen, seit 1907 mit einem Automobil zum nächstgelegenen Bahnhof transportiert werden. Im Neubau an der Münsterstaße hatte die Molkerei zunächst noch einen Dampfdynamo zur Stromerzeugung

33 Die folgenden Ausführungen basieren auf Rösler, Sammel-Molkerei, S.70ff.

34 Thomas Nipperdey, Verein als soziale Struktur in Deutschland im späten 18. und frühen 19. Jahrhundert. Eine Fallstudie zur Modernisierung I, in: ders., Gesellschaft, Kultur, Theorie. Gesammelte Aufsätze zur neueren Geschichte, Göttigen 1976, S.174–205, und ders., Deutsche Geschichte 1800–1866. Bürgerwelt und starker Staat, München 1994, S.266ff.

35 Ria Ha?nisch, Vereine im Wirtshaus. Kleinstadtkultur in Billerbeck um 1900, in: Freitag, Geschichte der Stadt Billerbeck, S.590–619.

36 Ebd., S.71.

37 Gleichwohl entstand der Bahnanschluss an einer für die Molkerei logistisch ungünstigen Stelle. Siehe Rösler, Sammel-Molkerei, S.51.

installiert, bevor sich das Amt Billerbeck 1908 schließlich doch entschloss, der Forderung nach einem Elektrizitätswerk nachzukommen.[38]

Das Engagement der Billerbecker Sammel-Molkerei steht somit beispielhaft für Impulsgeber kommunaler Entwicklungen in dieser Zeit, die sowohl wirtschaftliches Wachstum wie auch gesellschaftlich-technischen Fortschritt generierten. Darüber hinaus trug das Wirken des Unternehmens dazu bei, dem Image der „verschlafenen Kleinstädte", die den „Anschluss an die Industrialisierung" verpasst hätten,[39] Paroli zu bieten.

Die Entstehung und Entwicklung der Billerbecker Sammel-Molkerei, die sich in hohem Maße der Initiative Joseph Suwelacks sen. verdankte, bildet schließlich auch den Rahmen für die spätere Positionierung Josef Suwelacks. Die insgesamt zehn Kinder der Familie wuchsen in unmittelbarer Nähe zur Molkerei auf, denn das Wohnhaus, zunächst an der Beerlager Straße, befand sich seit 1899 an der Münsterstraße im Neubau über der Verwaltung der Molkerei.[40]

Der erstgeborene Sohn, Heinrich Suwelack, verstarb 1887 bereits vier Monate nach seiner Geburt. Ein Jahr darauf, am 30. April 1888, wurde Josef Suwelack geboren.[41] Auf ihn folgten zwischen 1889 und 1904 weitere acht Geschwister, wovon Bernhard, wie schon sein Bruder Heinrich, mit nur sieben Monaten im Jahr 1890 verstarb. Im Januar 1911 verstarb auch Josefs drittjüngerer Bruder Aloysius.[42]

Für Josef als ältester seiner Geschwister hätte es nahegelegen, in die Fußstapfen seines Vaters zu treten und in das lukrative elterliche Molkereigeschäft einzusteigen; dies hat er nicht getan. Gleichwohl hat er sich vom Unternehmergeist seines Vaters inspirieren lassen, da er sich seit 1912 selbst in Essen als Flugzeugunternehmer betätigte, Neues wagte und in diesem Sinne auch in seine Zukunft investierte. Seine Unternehmungen, insbesondere seine spätere Fliegerlaufbahn, wurden allerdings vor allem durch die finanzielle Großzügigkeit seines Vaters überhaupt erst möglich. Sie erlaubte es ihm, sich auszuprobieren und beruflich zu orientieren. Josefs Vater, so scheint es, hat vor allem seinen Söhnen viele Freiräume gelassen, sich zu verwirklichen.[43]

38 Vgl. ebd., S.54f.
39 Detlef Fischer, Chronik des Münsterlandes, Münster 2003, S.388.
40 Vgl. Rösler, Sammelmolkerei, S.46ff.
41 Siehe die Lebensdatenübersicht der Familie Suwelack im Anhang.
42 Vgl. Krewerth, Ein westfälisches Fliegerleben, S.123.
43 Nach Auskunft Walter Suwelacks war Joseph Suwelack senior sehr daran gelegen, seinen Kindern im Rahmen seiner finanziellen Situation stets alles zu ermöglichen. Vor diesem Hintergrund wird verständlich, weshalb er Josef bei seinen heimatlichen autodidaktischen Versuchen des Flugzeugbaus und des Fliegens unterstützte.

Abb. 2: Die Familie Suwelack, Billerbeck 1901: v.l. Aloysius, Vater Joseph, Richard, Josef, Mutter Katharina mit Wilhelm, sitzend: Mathilde und Helene Suwelack (Foto: Familienarchiv Walter Suwelack, NL Josef Suwelack)

1.2. Sozialisation und Lebenswelt: Religion, (Aus-)Bildung und ökonomisches Kapital

Josef Suwelack entstammte einer katholischen Familie. Seine Eltern waren treue Kirchgänger, die ihre Kinder dazu erzogen, ihren religiösen Pflichten nachzukommen. Ein an Josef gerichteter Brief seiner Mutter vom 18. April 1915 – der einzige erhaltene Brief der Mutter an ihren Sohn – verdeutlicht den hohen Stellenwert der Religion für die Familie Suwelack:

> „Lieber Joseph! Zu Deinem Namensfeste sende ich Dir die besten Glück- und Segenswünsche. Möge der liebe Gott geben, daß Du wieder gesund zu uns kommst. Ich will an diesem Tage fleißig für Dich beten; ich habe auch eine neuntägige Andacht für Dich angefangen, auch ist die hl. Kommunion für Dich bestimmt. Auch will ich Dir mitteilen, daß das Fest des hl. Joseph auf den 3. Sonntag nach Ostern verlegt ist. Gehe Du auch an diesem Tage zur hl. Kommunion und halte Deine Ostern. Gehe oft in die Kirche und bitte den lieben Gott, daß er uns bald andere Zeiten schickt. Hast du das Versäumte schon nachgeholt? Es ist zu unangenehm für uns."[44]

44 NL J. Suwelack, Brief Katharina Suwelacks an Josef Suwelack vom 18.4.1915. Mit dem Versäumten ist wohl gemeint, dass Josef nicht regelmäßig beichtete und in die Kirche ging, was er seiner Mutter in einem Feldpostbrief vom 9. Januar 1915 mitteilte: „Deinen lieben Brief habe ich erhalten, und ich werde Deine guten Ratschläge getreulich soweit es geht befolgen. Du mußt Dir hier die Sache nicht so einfach vorstellen. Morgen, am 10ten haben wir alle Feldgottesdienst. Das ist die einzige Gelegenheit. Ich habe vor 8 Tagen einen kath[olischen] Feldgeistlichen kennengelernt und

Abb. 3: Die Kinder der Suwelacks, Billerbeck 1906: v. l. Josef, Richard, Aloysius, Mathilde, Helene, Wilhelm, Otto und Maria Suwelack (Foto: Familienarchiv Walter Suwelack, NL Josef Suwelack)

Billerbeck war während des gesamten 19. Jahrhunderts eine konfessionell homogene Stadt. Noch vor dem Ersten Weltkrieg standen lediglich 29 Juden und 35 Protestanten einer katholischen Mehrheit von 4.940 Gläubigen gegenüber.[45] In die Zeit nach dem Kulturkampf fiel der Bau des Billerbecker „Ludgerusdoms" (1892–1898). Für dessen Erbauung waren Spendengelder gesammelt worden, welche zu fast 74 Prozent aus der Billerbecker Bevölkerung stammten.[46]

Constance Sieger charakterisiert Billerbeck trotzdem nicht als katholisches Milieu, sondern spricht stattdessen von einer „traditionalen Lebenswelt",[47] die sich vornehmlich durch eine ländliche Struktur und eine geringe Bevölkerungszu-

den will ich nachher besuchen, um zu fragen, wo und wann ich beichten und kommunizieren kann. Du mußt bedenken, daß die Sachsen alle protestantisch sind, und es gibt hier wenige kath[olische] Geistliche." In einem Brief vom 2. März 1915 schrieb er seinen Eltern: „Jetzt gibt es hier in Lille eine Garnisionskirche. Aber da wir morgens jetzt schon um 1/2 8 Uhr auf dem Flugplatz sein müssen, komme ich wochentags nicht dazu, in die Kirche zu gehen."

45 Vgl. Sieger, Die Ludgerusstadt, hier S. 167.

46 Der Name „Ludgerusdom" bürgerte sich in Billerbeck ein, obwohl es sich nicht um den Sitz des Bischofs handelte. Trotzdem wurde bei der Einweihungsfeier der Propsteikirche darauf hingewiesen, dass das Bauwerk die Bezeichnung Dom durchaus tragen könne. Die Spenden für den Dombau beliefen sich auf insgesamt 317.745 Reichsmark, wovon aus Billerbeck 234.752 RM beigesteuert worden waren. Vgl. Sieger, Die Ludgerusstadt, S. 188 und S. 191.

47 Sieger, Die Ludgerusstadt, S. 200. Vgl. zum Begriff „Lebenswelt": Arbeitskreis für kirchliche Zeitgeschichte, Konfession und Cleavages. Ein Erklärungsmodell zur regionalen Entstehung des

nahme auszeichne. Außerdem sei das traditionell milieukonstituierende katholische Vereinswesen dort nur schwach ausgeprägt gewesen. In Billerbeck sei eher eine „stark ritualisierte Alltagsfrömmigkeit“ zu beobachten gewesen, die z. B. im dreimal täglichen Angelus-Läuten deutlich geworden sei, bei dem die Billerbecker sich zum Gebet zurückgezogen hätten.

An wichtigen kirchlichen Feiertagen fanden Prozessionen statt, außerdem regelmäßige Wallfahrten nach Telgte oder Kevelaer.[48] Doch auch Billerbeck hatte sich zum Wallfahrtsort entwickelt; schon kurz nach dem Tod des heiligen Ludger in Billerbeck im Jahr 809 erhielt das Städtchen den Status eines Wallfahrtsortes. So trägt auch die nach dem ersten Bischof von Münster benannte Basilika dessen Namen, die 1889 an der Stelle erbaut wurde, wo der Überlieferung nach der heilige Ludger gestorben sei.[49] Aufgrund des ausgeprägten Katholizismus seines Umfeldes ist es wahrscheinlich, dass Josef Suwelack ebenfalls rege am Gemeindeleben partizipierte.

Doch wäre es einseitig, Billerbeck ausschließlich unter religiösen Gesichtspunkten zu betrachten, da sich die Kleinstadt keineswegs der Moderne beziehungsweise Modernisierungsprozessen verschloss, wie am Beispiel des 1875 bereits geplanten und 1908 schließlich eingeweihten Bahnanschlusses gezeigt werden kann.[50] Zuvor hatten die BewohnerInnen zu den benachbarten Bahnhöfen nach Münster, Appelhülsen, Darfeld oder Coesfeld laufen und dabei Fußwege von ein bis fünf Stunden auf sich nehmen müssen.[51] Neben der Molkerei, die für den Anschluss Billerbecks an das Streckennetz vor allem mit dem Hinweis auf etwaige Einsparungen von (Transport-)Kosten argumentierte und den generellen wirtschaftliche Nutzen eines eigenen Bahnhofs für die gesamte Bevölkerung betonte, setzten sich auch die Bewohner für die Errichtung eines Bahnhofs ein. Dabei griffen sie zu unterschiedlichen demokratischen Mitteln: Sie unterstützten Petitionen, hielten Versammlungen ab und gründeten, als die Realisierung beschlossen war, ein Festkomitee, das die Einweihung des Bahnhofs vorbereiten sollte.[52] In einer kleinen Annonce elf Tage vor der Eröffnungsfeier der Bahnstrecke rief man zudem die StadtbewohnerInnen dazu auf, „durch Beflaggen ihrer Häuser ihr Interesse an der Feier zu bestätigen“ und sich damit gleichsam zum Fortschritt zu bekennen.[53] Immerhin begriff man den Bahnanschluss im wahrs-

katholischen Milieus in Deutschland, in: Historisches Jahrbuch 120 (2000), S.358–395, hier S.365.

48 Vgl. Sieger, Die Ludgerusstadt, S.200f., Zitat S.200.

49 Siehe http://www.baumberge-touristik.de/pages/posts/pfarrkirche-st.-ludgerus-in-billerbeck–33.php, online seit Oktober 2011 (3.10.2014).

50 Siehe zu diesem Thema Caroline Alice Brzoska, Ein Bahnhof für Billerbeck. Wirtschaftliche Bedeutung und kleinstädtisches Selbstverständnis um 1900, unveröffentlichte Masterarbeit, Münster 2010.

51 Vgl. Sieger, Die Ludgerusstadt, S.232.

52 Vgl. Broszka, Ein Bahnhof, S.79f.

ten Sinne des Wortes als eine Art Anschluss an die neue Zeit, denn, wie der „Billerbecker Anzeiger" betonte, sonst heiße es: „gute Nacht Billerbecker", was einem „Todesurtheil" gleich käme.[54] Ganz praktisch hoffte man, mit dem Bahnanschluss Billerbeck einerseits wirtschaftlich, andererseits aber auch als Pilger- und Erholungsort für Touristen attraktiver zu machen. Dabei sollte Billerbeck auf keinen Fall zu einer „Fabrikstadt" werden; stattdessen sollte sich der Fortschritt dem Stadtbild anpassen.[55]

Die zu beobachtenden „Häufigkeitsverdichtungen von Veränderungen",[56] als die Jürgen Osterhammel Epochenschwellen bezeichnet, lassen sich auch in der Kleinstadt Billerbeck nachweisen. Angefangen bei der Gründung einer florierenden Molkerei als Aktiengesellschaft, einer bis Anfang der 1830er Jahre in Preußen wenig verbreiteten Geschäftsform,[57] über die Anschaffung moderner Land- und Fortbewegungsmaschinen bis hin zur Ausrichtung eines eigenen Flugtages im Jahr 1912 erlebte die Billerbecker Bevölkerung ihren Aufbruch in die Moderne.

Die Informationen über Josef Suwelacks (Aus-)Bildung, vor allem was seine ersten Schuljahre betrifft, sind in der privaten Familienüberlieferung äußerst lückenhaft. Nachzuweisen ist, dass er von Ostern 1902 bis Weihnachten 1904 das Collegium Augustinianum in Gaesdonck[58] besuchte, ein am Humboldt'schen Bildungsideal ausgerichtetes katholisches Internat. Anschließend wechselte er zur Rektoratsschule in Billerbeck, danach – zusammen mit seinem Bruder Richard – von Ostern 1905 bis Ostern 1906 auf das Gymnasium Arnoldinum in Burgsteinfurt. Im Collegium Augustinianum in Gaesdonck standen Religion, Deutsch, Latein, Griechisch, Französisch, Geschichte, Erdkunde, Mathematik, Naturlehre und Zeichnen auf dem Stundenplan, außerdem „Technischer Unterricht", worunter Turnen, Zeichnen und Gesang subsumiert wurden.[59] Das Internat beher-

53 Billerbecker Anzeiger vom 16.2.1908.

54 Billerbecker Anzeiger vom 1. und 8.11.1891, zit. nach Sieger, Die Ludgerusstadt, S.233.

55 Vgl. Sieger, Die Ludgerusstadt, S.244.

56 Jürgen Osterhammel, Die Verwandlung der Welt, 3. Aufl., München 2009, S.114–115.

57 Hans Pohl zufolge waren vor 1843 in der Rheinprovinz insgesamt nur 41 Aktiengesellschaften – Eisenbahnen und Chausseen ausgenommen – beantragt worden. Vgl. Hans Pohl, Zur Entwicklung der Formen der Betriebs- und Unternehmensorganisation, insbesondere der Großorganisation im Verhältnis zum persönlich geführten Geschäft, in: ders., Wirtschaft, Unternehmen, Kreditwesen, soziale Probleme. Ausgewählte Aufsätze Teil 1, Stuttgart 2005, S.536–568, hier S.561.

58 Das Collegium Augustinianum, welches bis heute besteht, wurde von der Diözese Münster 1849 als kirchliches Privatgymnasium gegründet und war zuvor, seit 1400, ein Augustiner-Chorherrenstift, das 1802 aufgehoben wurde. Nach der Aufhebung wurde der verstaatlichte Besitz von Chorherren als persönliches Eigentum erworben; 1825 setzten diese den Bischof von Münster zum Erben ein. Das Collegium Augustinianum ist heute wegen seiner *Bibliotheca domus presbyterorum* bekannt, die sechzehn spätmittelalterliche Handschriften aus der ersten Hälfte des 16. Jahrhunderts besitzt. Siehe Universitäts- und Landesbibliothek Düsseldorf (Hg.), http://digital.ub.uni-duesseldorf.de/ihd/content/pageview /40419 83, online seit 2010 (12.5.2013).

59 Vgl. Jahresberichte des Collegium Augustinianum Gaesdonck, online verfügbar für den Zeitraum 1896–1910: http://digital.ub.uni-duesseldorf.de/ulbdsp/periodical/structure/4419969?query=Gaes donck, siehe für den Zeitraum 1902–1904 die entsprechenden Jahresberichte (12.5.13).

bergte insgesamt rund 150 Schüler für den Zeitraum von 1902 bis 1904.[60] Ostern 1906 verließen Richard und Josef das Gymnasium Arnoldinum mit dem Versetzungszeugnis für die Obersekunda.

Abb. 4: Josef Suwelack als „Einjähriger", 1907
(Foto: FamilienarchivWalter Suwelack, NL Josef Suwelack)

Nach dem „Einjährigen",[61] so ist zu vermuten, begann Josef in der Molkerei seines Vaters eine Ausbildung sowie dann auch ein Studium an der Technischen Hochschule Darmstadt. Im Wintersemester 1907/08 belegte er dort verschiedene Vorlesungen und Übungen, die sich nicht nur auf das Molkereiwesen beschränkten. Neben den Veranstaltungen „Grundzüge der Rechtswissenschaft" und „Volkswirtschaftslehre", „Kaufmännische Buchführung", „Wiesenbau und Drainage" sowie „Elemente der Landwirtschaftslehre" hörte er auch Vorlesungen zum Sozialismus und zur Kunstgeschichte.[62] Diese Fächerbreite verweist auf die viel-

60 Siehe ebd., „Chronik der Anstalt".

61 „Einjähriges freiwilliges Examen" bzw. kurz „Einjähriges" wurde die Reifeprüfung in deutschen Mittelschulen nach neun beziehungsweise zehn Schuljahren genannt. Es berechtigte den Absolventen im Kaiserreich, einen einjährigen Militärdienst statt des üblichen dreijährigen zu leisten.

62 Siehe NL J. Suwelack, Großherzogliche Technische Hochschule Darmstadt, WS 1907/08, belegte Vorlesungen und Übungen, beglaubigtes Zeugnis.

seitigen Interessen Josefs, der sich noch beruflich zu orientieren schien. Nach Verlassen der Technischen Hochschule Darmstadt wechselte er an die Düsseldorfer Kunstakademie, um seinen künstlerischen Neigungen weiter nachzugehen. Dies gestaltete sich jedoch nur als ein kurzes Intermezzo, bevor er sich 1909 vollständig der Fliegerei widmete.

Aufgewachsen in einem ökonomisch abgesicherten und weltoffenen Elternhaus, konnte Josef sich ausprobieren und war nicht gezwungen, in jungen Jahren eigenes Geld zu verdienen. Bevor er sein Hobby – die Fliegerei – zum Beruf machte, war er zunächst erst einmal längere Zeit „Sohn begüterter Eltern". Und auch als er gegen Ende des Jahres 1910 das Münsterland verließ und seinen Wohnsitz nach Berlin verlegte, finanzierte ihm sein Vater sein kostspieliges, ja luxuriöses Leben als Flieger.

1.3. Eine moderne Familie in einer beschleunigten Zeit?

Während seiner Jugend in Billerbeck verbrachte Josef Suwelack seine Freizeit unter anderem mit Tennisspielen; darüber hinaus interessierte er sich schon früh für technische Innovationen.[63] Durch die Tätigkeit seines Vaters in der Molkerei kam er bereits früh in Kontakt mit technischen Apparaten, mit Zentrifugen, Transportbändern wie auch mit dem firmeneigenen Automobil. Zwar konnten in Josef Suwelacks Nachlass keine technischen Zeitschriften aus den Jahren vor 1909 nachgewiesen werden wie „Prometheus", „Die Umschau" oder die „Technische Rundschau", die sich an interessierte Laien richteten und allgemeinverständlich technische Fragen behandelten,[64] doch zeigt eine Analyse der gesellschaftlichen Verhältnisse um die Jahrhundertwende, dass neuartige Mobilitätsmaschinen wie das Auto oder das Flugzeug sowie Technik allgemein große öffentliche Beachtung fanden. Zudem stieg das Angebot an Praxishandbüchern, die zum Nachbau von technischen Geräten anregten und kostengünstig die Reihe „Selbst ist der

63 Ein Foto, das Josef Suwelack auf dem Tennisplatz in Billerbeck zeigt, sowie sein 1912 vor Publikum gehaltener Vortrag, in dem er auf seine bisherige Fliegerkarriere zurückblickt, stützen diese These. Gleichwohl ist hinzuzufügen, dass viele Flieger in ihren Vorträgen und Biografien von einer seit Kindertagen bestehenden Technikfaszination und sportlichem Ehrgeiz berichten und scheinbar ein Bedürfnis nach Rechtfertigung bestand, das aus der Neugier des Publikums heraus geboren wurde. Vgl. Schüler-Springorum, Vom Fliegen und Töten, S.212.

64 Siehe zum Anstieg der naturwissenschaftlichen und technischen Zeitschriften: Georg Jäger, Das Zeitschriftenwesen, in: ders. (Hg.), Geschichte des deutschen Buchhandels, Frankfurt a.M. 2003, S.368–389, bes. S.368–380. Vgl. auch Rohkrämer, Eine andere Moderne?, S.48 sowie Möser, Fahren und Fliegen, S.77ff. Prometheus. Zeitschrift für Technik, Wissenschaft und Industrie, Berlin 1891–1920; Technische Rundschau, Leipzig 1887ff.; Die Umschau. Übersicht über die Fortschritte und Bewegungen auf dem Gebiete der Wissenschaft, Technik, Literatur und Kunst, Frankfurt a.M. 1897–1944.

Mann"[65] heraus, die, mit zahlreichen Abbildungen und Zeichnungen versehen, der Zielgruppe der männlichen Jugend den Nachbau von (Spielzeug-)Kameras, Dynamos und nicht zuletzt von Flugzeugen ermöglichte.

Einführungen in die Flugtechnik in der Reihe „Bibliothek für Luftschiffahrt und Flugtechnik",[66] Berichte aus dem seit 1908 vom Ingenieur Oskar Ursinius herausgegebenen „Flugsport" sowie Veröffentlichungen zum Handwerk der Fliegerei lieferten relevante Informationen,[67] um sich autodidaktisch den Flugzeugbau und das Fliegen anzueignen. Die genannten Bücher und Schriften verdeutlichen, dass zeitgleich mit der Entstehung der Fliegerei ein Absatzmarkt entstand, der zur Verbreitung von Innovationen und Konstruktionsplänen etc. beitrug. Bereits 1911 waren annähernd alle Flugzeugmodelle als Bauplan zu erwerben.[68]

Suwelacks Generation sah sich zudem mit neuen, geradezu revolutionären Erfindungen konfrontiert: der Glühlampe (1876), der elektrischen Straßenbahn (1881), dem Automobil (1885/86) oder der Telegrafie und Fotografie. Die möglich gewordene Serienproduktion von Elektromotoren führte zur Unabhängigkeit ganzer Industriezweige von der Dampfmaschine und „Technik wurde Gegenstand einer kulturell vermittelten öffentlichen Inszenierung und Popularisierung".[69] Schriftenreihen, die sich an die Zielgruppe der Jugend richteten, trugen zur Ausbildung der Technikbegeisterung bei und schufen schließlich auch Nachahmer.[70] Für manche Leute waren diese technikbegeisterten Tüftler – ebenso wie die Flugbegeisterten – seltsame Exoten, deren zumeist primitive erste (Flug-)Versuche zunächst belächelt wurden,[71] wie der Bremer Flugpionier Heinrich Focke zu berichten wusste:

65 W. Endig, Selbst ist der Mann, Leipzig 1911ff.

66 Hier zwei exemplarische Titel aus den Anfangsjahren der Reihe: Ferdinand Ferber, Die Kunst zu Fliegen, Berlin 1910; Paul Painievé/Émile Borel, Theorie und Praxis der Flugtechnik, Berlin 1911.

67 Siehe z.B. Otto Lilienthal, Der Vogelflug als Grundlage der Fliegekunst. Ein Beitrag zur Systematik der Flugtechnik, Berlin 1889; Dr. R. Wegner von Dallwitz, Hilfsbuch für den Luftschiff- und Flugmaschinenbau, Rostock 1909; Paul Hermuth (Hg.), Der junge Aviatiker. Eine Anleitung zum Bau von Flugmodellen, Stuttgart 1911 sowie Heinz Erblich, Fliegerschule. Was muß ich wissen, wenn ich Flieger werden will? Ein Lehr und Handbuch für den Flugschüler, 2. Aufl., Berlin 1915.

68 Vgl. Möser, Fahren und Fliegen, S.268f.

69 Kehrt, Moderne Krieger, S.87.

70 Die Entwicklung des Flugzeuges verlief weitgehend unbemerkt von der Öffentlichkeit. Siehe Michael Epkenhans/Andreas von Seggern, Leben im Kaiserreich. Deutschland um 1900, Stuttgart 2007, S.172.

71 So auch im Fall von Josef Suwelack, der in seinem Lichtbild-Vortrag von 1912 davon berichtete, dass ihn der Spott seiner Freunde dazu bewogen habe, sich als Konstrukteur und Flieger zu beweisen, vgl. NL J. Suwelack, Vortrag „Aus meinem Fliegerleben" gehalten am 24. Februar 1912 vor Mitgliedern des Luftschiffahrt-Verein Münster für Münster und das Münsterland E.V. Ein weiteres Beispiel ist August Euler, der in den Presseberichten zur Internationalen Luftfahrt Ausstellung (ILA) verhöhnt wurde und im Ausspruch: „August laß das Fliegen sein" öffentlich verspottet wurde. Zit. nach Christian Kehrt, August Euler – ein Luftfahrtpionier in Darmstadt-Griesheim, gekürzte Version des Beitrages für den August Euler-Jubiläumssammelband, hg. von Anne Holtmann-Mares und Andreas Göller, S.1–20, hier S.5, online unter https://www.yumpu.com/de/document/view/2401636/august-euler-ein-luftfahrtpionier-in-darmstadt-griesheim (20.3.2018).

„Wer in den Jahren 1908 bis 1912, zumal in einer konservativen Hansestadt wie Bremen, vom Fliegen sprach, sich darin betätigte und nun gar selbst die blaue Jacke anzog und eigenhändig Flugzeuge baute, der war in den Augen des braven Bürgers umwittert von einer Atmosphäre, gruselig gemischt aus Abenteuertum, Zirkus, Irrenhaus und Friedhof."[72]

Die Masse der Bevölkerung – sofern sie dafür überhaupt Interesse zeigte – informierte sich über Neuerungen in der Technik durch die Berichterstattung in den Zeitungen, die seit der Jahrhundertwende z. B. über die Fahrten des Grafen Zeppelin schrieben, oder schlicht durch Mund-zu-Mund-Propaganda. Noch bevor die Gebrüder Wright 1903 ihre ersten motorbetriebenen Flüge fernab der Öffentlichkeit erfolgreich ausführten, wurde vor allem den lenkbaren Luftschiffen um die Person Ferdinand von Zeppelins Beachtung geschenkt, der um 1900 zum gefeierten Helden und zur nationalen Identifikationsfigur avancierte.[73] Kaiser Wilhelm II. persönlich würdigte ihn 1908 bei der Verleihung des schwarzen Adlerordens als „den größten Deutschen des zwanzigsten Jahrhunderts",[74] der mit der Innovation des Luftschiffes die menschliche Kultur bereichert habe.

Auch in Billerbeck schien es eine Zeppelinbegeisterung gegeben zu haben, folgt man zumindest den Berichten der Heimatzeitung Josef Suwelacks, dem „Billerbecker Anzeiger". Während Meldungen über die Flugversuche der Wrights äußerst rar waren, wurde den Fahrten der lenkbaren Luftschiffe weit mehr Aufmerksamkeit geschenkt.[75] So meldete die Zeitung in der Ausgabe vom 16. August 1908 zu den Wrights lediglich: „Erfolgreicher Flugversuch. Le Mans, 11. August. Wright unternahm heute einen neuen Flugversuch, der 3.43 Minuten dauerte. Er beschrieb 3 Kreise in verschiedener Höhenlage. Die Geschwindigkeit betrug 56 Kilometer".[76]

Erst knapp ein Jahr später, am 1. August 1909, erschien eine halbspaltige Meldung zur Reaktion der Franzosen auf die Kanalüberquerung Louis Blériots am 25. Juli 1909 sowie zum Versuch der Kanalüberquerung per Flugzeug des Franzosen Hubert Latham:

„Dover, 27. Juli. Der Aviatiker Latham, der um 5,07 Uhr in Sangatta zu einem neuen Flug über den Kanal aufgestiegen und um 6,15 Uhr in Dover gesichtet wurde, ist zwei Seemeilen von Dover wiederum ins Meer gefallen und von einem ihm folgenden Torpedoboot aufgefischt worden. Später ist Latham in Dover eingetroffen und mit Begeisterung begrüßt worden."[77]

72 Zit. nach Peter Supf, Das Buch der deutschen Fluggeschichte. Vorzeit, Wendezeit, Werdezeit. Bd. 1, 2. Aufl., Stuttgart 1956, S.213.
73 Siehe Fritzsche, A Nation of Fliers, S.9–58.
74 Zit. nach Rohkrämer, Eine andere Moderne?, S.51.
75 Siehe z. B. Billerbecker Anzeiger vom 26.7.1908: „Grund des Zeppelin'schen Unfalls"; vom 16.8.1908: „Der Kronprinz bei Zeppelin", „Ein neuer Aufstieg in Sicht"; vom 23.8.1908: „Unfall des Parsevalluftschiffes", „Nachtfahrt des Militärluftschiffes" etc. Weitere Meldungen z. B. am 25.10.1908, 14.2.1909, 4.7.1909 etc.
76 „Erfolgreicher Flugversuch", in: Billerbecker Anzeiger vom 16.8.1908.
77 „Die glücklichen Franzosen", in: Billerbecker Anzeiger vom 1.8.1909, Zitat ebd. Die Kurzmeldungen über Blériots Flugvorführungen nahmen in der Folgezeit etwas zu, siehe z. B. Billerbecker

Schon in der Anfangszeit der Fliegerei wurden Verbesserungen in der Flugdauer oder Unterschiede zwischen den Flugzeug-Typen interessiert verfolgt. Seit Ende des Jahres 1909 berichtete daher nun auch der „Billerbecker Anzeiger" vermehrt über motorisierte Flugmaschinen.[78] In der Rubrik „Von der Luftschiffahrt" meldete die Zeitung am 19. September des gleichen Jahres:

> „Paris, 14. Sept. Der Aviatiker Santos Dumont unternahm infolge einer Wette mit seinem Einflächer Libelle einen Flug von St. Ehr nach Buc und legte die acht Kilometer lange Strecke in fünf Minuten zurück. Santos Dumont erklärte einem Berichterstatter, daß er behufs Förderung der Aviatik auf jeden Patentschutz an dem von ihm erfundenen Aeroplan verzichte und jedermann die diesbezüglichen Pläne zur Verfügung stelle."[79]

Obwohl das Luftschiff stets mehr Platz in der Berichterstattung beanspruchte, erfolgte eine Zunahme der Nachrichten über Flugzeuge, nicht zuletzt, da die Piloten mehr und mehr mit „Langstreckenflügen" auf sich und ihre Maschinen aufmerksam zu machen versuchten, um ihren Bekanntheitsgrad zu steigern.

Das neue Potenzial im Bereich der Mobilität rief einerseits Begeisterung, andererseits aber auch – meist aufgrund mangelnden Verständnisses für die Funktionsweise der Technik – Furcht und Misstrauen hervor. So galten die Luftfahrt oder auch das Automobil nicht selten als „moderne Wunder",[80] die sich dem technischen Verständnis der Bevölkerung entzogen. Im Zusammenhang mit diesen technischen Innovationen las man außerdem neben Erfolgen auch immer wieder von Automobilunfällen oder Schiffsunglücken, so dass Sensation und Katastrophe gleichermaßen die Neugier und Faszination der Leserschaft weckten.[81]

Die durch das Flugzeug neu erschlossene Dimension der Luft eröffnete darüber hinaus unbekannte visuelle und emotionale Erfahrungshorizonte, die von den handelnden AkteurInnen in Erlebnis- und Erfahrungsberichten verarbeitet und publiziert wurden.[82] Da die AutorInnen gerade in der Anfangszeit der Fliegerei nicht mit Vorwissen der RezipientInnen rechnen konnten, gestalteten sie durch

Anzeiger vom 10.10.1909. Der Münsterische Anzeiger brachte am 26. Juli 1909 eine Meldung zu diesem Ereginis: „Ein Flug über den Kanal!".

78 Vgl. „Die Konkurrenz der Flugmaschinen", in: Billerbecker Anzeiger vom 29.8.1909 oder auch die Ausgabe vom 10. Oktober 1909, „Von der Luftschiffahrt": „Das Publikum verlangte stürmisch nach Bleriot [sic!], der schließlich mit seiner Maschine unter lebhaftem Beifall aus dem Zelte erschien und zwei gelungene Aufstiege von 3 bezw. 4 Minuten Dauer machte."

79 Billerbecker Anzeiger vom 19.9.1909.

80 Vgl. Bernhard Rieger, Technology and the Culture of Modernity in Britain and Germany, 1890–1945, Cambridge 2005, S.20ff. Siehe zur ambivalenten Technikeinstellung im Deutschen Kaiserreich: Berghoff, „Dem Ziele der Menschheit entgegen", S.50f.

81 Auf den Aspekt des Unfalls und des Risikos wird an späterer Stelle genauer eingegangen. Vgl. Ausgaben des Billerbecker Anzeigers vom 26.7.1908: „Grund des Zeppelin'schen Unfalls", „Revolte auf der Radrennbahn"; vom 16.8.1908: „Automobil-Unfall"; vom 23.8.1908: „Unfall des Parsevalluftschiffes"; vom 14.2.1909: „Der Unfall des Ballons ‚Le Reve"' u.a.

82 Siehe z.B. Hirth, Meine Flugerlebnisse, S.173 sowie Josef Suwelack, Mein Flug zur Nordsee, in: Flugsport (1913), Nr.5, S.173f.

ihre Berichterstattung das über die FliegerInnen öffentlich kursierende Bild aktiv mit.

Neben der Tagespresse und dem Zeitschriftenwesen kam der Science-Fiction-Literatur, auch bekannt unter dem damaligen Label der Zukunftsliteratur, eine zentrale Bedeutung bei der diskursiven Aneignung der Technik zu.[83] Die Machbarkeitsstimmung der Zeit spiegelt sich besonders in diesem Genre, thematisierten die Romane doch sowohl die von der Technik ausgehenden Gefahren als auch ihr Fortschrittspotenzial. Während H. G. Wells in seinem Zukunftsroman „The War in the Air“ aus dem Jahr 1909,[84] der noch im gleichen Jahr in deutscher Übersetzung erschien, Englands Angst vor einer Invasion aus der Luft beschrieb, begriffen etwa Georg Korf und Bertha von Suttner die Fliegerei als Chance für grenzenlose Völkerverständigung und Frieden.[85] So war Korf der Meinung: „Die Eroberung des Luftreiches würde die Kriegsführung zu einer so ungehinderten Zerstörung der Städte und Zerfleischung der Bevölkerung gestalten, daß sie zur Unmöglichkeit geworden wäre.“[86] Zivilisationskritiker fürchteten einen Sittenverfall und eine Entmenschlichung durch die Zunahme technisierter Arbeitsprozesse.[87]Aller Skepsis und Furcht zum Trotz war die Zeit zwischen 1890 und 1914 jedoch geprägt von Technikbegeisterung und einem relativen Fortschrittsoptimismus.[88]

Zu einer Zeit, da die Fliegerei noch eine ganz junge Disziplin war und aufgrund der hohen Kosten, die sie verursachte, nur den wohlhabenden Schichten offen stand, war es etwas Exklusives, Flieger zu werden.[89] So empfahl Heinz Erb-

83 Felix Ingold, Literatur und Aviatik. Europäische Flugdichtung 1909–1927, Frankfurt a.M. 1980; Stephen Kern, The Culture of Time and Space. 1880–1918, Cambridge/Massachusetts 1983, S.90ff.; Michael Salewski/Ilona Stölken-Fitschen (Hg.), Moderne Zeiten. Technik und Zeitgeist im 19. und 20. Jahrhundert, Stuttgart 1994.

84 Herbert George Wells, The War in the Air, London 1909.

85 Siehe Georg Korf, So werden wir fliegen. Die natürliche Lösung des Flugproblems. Der Mensch als Segler der Lüfte in naher Zukunft. Mit einem 2. Teil: „Wenn wir fliegen“, Oranienburg 1909, S.71. Vgl. zu Bertha von Suttner: dies., Die Barbarisierung der Luft, Berlin 1912. Suttner bezog sich auf Wells Roman „The War in the Air“.

86 Korf, So werden wir fliegen, S.75.

87 Vgl. Eduard von Mayer, Technik und Kultur. Gedanken über die Verstaatlichung des Menschen, Berlin 1906: „Der Geist der Technik, das Wesen der Technik als selbstständige Macht betrachtet ist also: mittels der Arbeitsteilung den Menschen zu zerstückeln; durch die Arbeitseinheit ihn zu entpersönlichen; ... So wird er, entwürdigt zum willenlosen Gemächte der monotheistischen Idee, zu Händen ihrer politisch-sozialen Verkörperung, der militärisch-bureaukratischen Staatsgewalt.“, S.158, online unter: https://archive.org/stream/technikund kultuoomayegoog#page/n7/mode/2up (10.8.2014).

88 Vgl. Felix Holtschoppen, Rebellische Technik. Maschinenphantasien in der literarischen Phantastik um 1900, Berlin 2005, S.19–20 sowie Alexander Schmidt-Gernig, Zukunftsmodell Amerika? Das europäische Bürgertum und die amerikanische Herausforderung um 1900, in: Ute Frevert (Hg.), Das neue Jahrhundert. Europäische Zeitdiagnosen und Zukunftsentwürfe um 1900, Göttingen 2000, S.79–112, hier S.81.

89 Vgl. Kehrt, Moderne Krieger, S.72, siehe auch ders., „Das Fliegen ist immer noch ein gefährliches Spiel“ – Risiko und Kontrolle der Flugzeugtechnik von 1908 bis 1914, in: Gunter Gebauer/Stefan

lich, selbst Flieger und Herausgeber von Büchern zur Fliegerei, jungen Menschen, die beabsichtigten, sich in diesem Bereich zu betätigen, finanziell abgesichert zu sein, denn es sei „dringend erwünscht, daß der sich dem Flugsport Widmende über etwas Vermögen verfügt oder einen Beruf hat, zu dem er jeden Augenblick zurückkehren kann und der ihn anständig ernährt".[90]

Auf die Frage, was ihn bewogen habe, Flieger zu werden, soll Josef Suwelack während eines Vortrages über sein Fliegerleben 1912 geantwortet haben, dass „es vor 4 Jahren, als die ersten Nachrichten über die Erfolge der Gebrüder Wright, des Brasilianers Santos Dumont und des Franzosen Farman durch die Zeitungen gingen, ihn [sic!] keine Ruhe mehr gelassen habe, auch das Fliegen zu lernen, wenigstens zu versuchen."[91]

Bei den Gesprächen auf dem Tennisplatz, so die mündliche Überlieferung, sei es um das gleiche Thema gegangen.[92] Von seinen Landsleuten sei er verlacht worden, habe aber an seinem Vorhaben festgehalten. Möglich, dass die Neugierde Suwelacks und der Spott seiner Freunde allein ausschlaggebend waren, eine Fliegerlaufbahn einzuschlagen, jedoch scheint diese monokausale, aus der Rückschau entstandene Erklärung etwas dürftig. Die Aussicht auf Geldgewinne auf Flugwettbewerben gepaart mit „Ruhm" und Anerkennung durch das Publikum dürften ebenso wichtige Faktoren gewesen sein, die Fliegerei auszuüben. Jugendliche Technikbegeisterung und Abenteuerlust, Phänomene, auf die man in Fliegerberichten stößt oder von denen Fliegeralben Zeugnis geben,[93] haben Josef Suwelack sicherlich ebenfalls motiviert, sich dem neuen Berufszweig der Fliegerei zuzuwen-

Poser/Robert Schmidt/Martin Stern (Hg.), Kalkuliertes Risiko. Technik, Spiel und Sport an der Grenze, Frankfurt a. M./New York 2006, S.199–224, hier S.200.

90 Erblich, Fliegerschule, S.13.

91 Münsterischer Anzeiger, Rubrik „Aus der Stadt", vom 26.2.1912. Siehe auch NL J. Suwelack, stichwortartige Notizen Suwelacks zum Vortrag. Eine fast identische Begründung gibt Hellmuth Hirth in seiner Autobiografie: „Wie ich auf die Fliegerei kam? Mit großem Interesse las ich in England von den ersten Versuchen von Santos Dumont, Farman und Voisin und dann besonders auch von den Wrights. Ich studierte nun das Buch Otto Lilienthals, das allerdings damals für mich noch viel zu theoretisch gehalten war. In all meiner Zeit waren meine Gedanken ausschließlich mit dem Flugzeug beschäftigt". Hirth, Meine Flugerlebnisse, S.21.

92 Julia Koch-Suwelack hörte oft Geschichten über „Onkel Josef" von ihrem Vater Richard, dem jüngeren Bruder des Piloten. Zudem existieren neun Seiten handschriftliche Erinnerungen Richard Suwelacks an seinen älteren Bruder: NL J. Suwelack, Erinnerungen Richard Suwelacks an Josef Suwelack.

93 Siehe z. B. Hirth, Meine Flugerlebnisse: „Zu meinem sechsten Geburtstag bekam ich mein erstes Fahrrad, und damit ging mir ein heftiger Wunsch in Erfüllung, denn jede Maschine interessierte mich von klein auf, und ich konnte stundenlang meine Nase an Fensterscheiben plattdrücken, um in ein Maschinenhaus hineinsehen zu können", S.9, oder auch „Nach etwa weiteren acht Monaten packte mich ein unwiderstehlicher Abenteuerdrang und ich reiste ... nach Südamerika", S.18. Richard Dietrich erinnert sich in seinem Buch, dass er die Treppen zur Redaktion des „Flugsport" emporgestiegen sei und dort die Hefte durchgesehen habe: „Die Sache wurde von Heft zu Heft interessanter. Nach und nach hatte ich den ganzen Jahrgang 1910 vor mir liegen. Von keinem dieser Hefte wollte ich mich wieder trennen, denn ich fand nicht nur überzeugende Berichte und Abbildungen, sondern auch Bauzeichnungen. Schließlich kaufte ich den ganzen Posten und

den. Dafür sprechen die umfangreichen Fotosammlungen, die er seit seiner ersten Flugversuche in der Heimat anlegte und die seine fliegerischen Etappen dokumentieren.[94]

Da Josef seit seiner frühesten Kindheit technisch sozialisiert war, verfolgte er interessiert die Entwicklungen auf dem Gebiet der Flugtechnik, dieser seiner Meinung nach „welterschütternde[n] Neuerung".[95] Suwelacks erster selbst gebauter Gleiter bestand neben Tragflächen aus einem „primitive[n] Sitz, Höhen- und Seitensteuer und Gleitkufen".[96] Mit Unterstützung von Freunden und Leihgaben aus der Nachbarschaft versuchte Josef zunächst, dem Gleiter mittels eines Ambosses, der an einem Seil auf einen acht Meter hohen Holzturm „auf Möllerings Hügel" gezogen wurde, und zwei mit Seife eingeschmierten Gleitschienen den nötigen Auftrieb zu geben. Dieser erste Versuch misslang jedoch. Daraufhin baute Suwelack in einem Schuppen auf der Weide im Brock einen neuen Gleiter, der dieses Mal mit einem Automobil die nötige Antriebsgeschwindigkeit erhalten sollte. Er lieh sich ein Auto aus der Umgebung,[97] doch auch dieser zweite Versuch blieb erfolglos: „Dann kam er auf den Gedanken[,] durch ein paar Pferde den Start zu vollziehen. Mersmann wurde mit seinem mit zwei Pferden bespannten Landauer bestellt. Wesentlich war, die Pferde so anzutreiben, daß sie mit einem Sprung in vollen Galopp fielen."[98]

Suwelacks Bemühungen blieben nicht unbemerkt: Der „Billerbecker Anzeiger" berichtete am 31. Oktober 1909 erstmals über seine Flugversuche:

> „Schon seit einiger Zeit beschäftigt sich ein Mitbürger unserer Stadt, Herr Jos. Suwelack jr. mit dem Bau einer Flugmaschine, ähnlich wie solche schon in Berlin und Paris construiert sind. Wenngleich nun auch die erdachte Maschine in der Entwicklung begriffen ist, namentlich fehlt noch der Motor dazu, so sind doch schon einige Versuche damit angestellt, um die Maschine auf ihre Flugfähigkeit zu erproben. Diese Versuche waren zwar anfangs noch negativ, können aber jetzt schon mehr Erfolg aufweisen. Heute nun haben Versuche stattgefunden, wobei die Maschine tatsächlich auf einer Strecke von 60 m die Höhe von 10–12 m erreicht haben soll. Wir wünschen dem Erfinder, Herrn Suwelack, weiteren guten Erfolg."[99]

Mit dieser Leistung konnte Suwelack jedoch nicht mit den Flügen, die in Frankreich oder auf dem Flugplatz in Berlin-Johannisthal bereits ausgeführt wurden,

den Jahrgang 1909 noch dazu. Das war mein damals wertvollster Besitz". Richard Dietrich, Im Flug über ein halbes Jahrhundert, Gütersloh 1942, S.22.

94 Der Nachlass umfasst z.B. Fotoalben mit Aufnahmen der ersten selbstgebauten Gleitflieger aus Josefs Zeit bei Rumpler und den Kondor-Werken sowie Bilder von Flugwettbewerben, an denen er teilnahm etc.

95 NL J. Suwelack, Stichwortartige Notizen Suwelacks zum Vortrag am 24.2.1912.

96 NL J. Suwelack, Erinnerungen Richard Suwelacks an Josef Suwelack. Die weiteren Ausführungen basieren ebenfalls auf Richard Suwelacks schriftlichen Erinnerungen.

97 Ebd.: „In Rorup gab es eines der ersten Autos des Kreises Coesfeld. Der Besitzer erklärte sich bereit, den Wagen zur Verfügung zu stellen, um das Gleitflugzeug von den Schienen zu ziehen."

98 Ebd.

99 Billerbecker Anzeiger vom 29.10.1909.

Abb. 5: Josef Suwelack in seinem selbstgebauten Apparat mit 25-PS-Escher Motor, dahinter der Schuppen im Brock, Billerbeck 1909 (Foto: Familienarchiv Walter Suwelack, NL Josef Suwelack)

mithalten.[100] Aus diesem Grund nahm er im November 1909 schriftlich und telefonisch Kontakt zu Motorenfabriken in Frankreich und Deutschland auf und kaufte schließlich einen 25-PS-Motor der 1907 gegründeten Sächsischen Werkzeugmaschinen-Fabrik Bernhard Escher aus Chemnitz.[101] Damals reichte die Preisspanne von 2.600 Mark für einen Motor von 15 bis 18 PS bis hinauf zu 13.000 Mark für einen 175- bis 200-PS Motor bei entsprechender Zunahme der Tourenzahl, des Hubraums und des Gewichtes.[102]

Die Flugversuche des Billerbeckers schienen auch über die heimatlichen Grenzen hinaus Interesse geweckt zu haben, denn im „Billerbecker Anzeiger" war zu lesen, dass Ausschussmitglieder des Luftschiffahrtvereins für Münster und das

100 Nachdem im Jahr 1908 einige wenige Flugtage stattgefunden hatten, wurden 1909 die ersten Flugwochen abgehalten, z. B. vom 22.8.–28.8.1909 in Reims oder vom 26.9–2.10.1909 in Berlin-Johannisthal zur Eröffnung des Flugplatzes. Dabei wurden einige Rekorde erzielt.

101 Siehe NL J. Suwelack, Schreiben von „Établissements Robert Esnault-Pelterie" vom 3.11.1909 sowie Telegramm vom 3.11.1909, in welchem der Motoren-Preis mit „sieben tauend francs" angegeben wird; Schreiben der Sächsischen Werkzeug-Maschinenfabrik Bernhard Escher Aktiengesellschaft vom 3.11.1909 und Schreiben der Bielefelder Maschinen-Fabrik vorm. Dürkopp & Co. vom 3.11.1909.

102 Siehe NL J. Suwelack, Katalog der Firma Escher mit Informationen zum Motorenfabrikat und zu Preisen.

Münsterland E.V. sowie der Aviatiker Fritz Schlüter aus Frankfurt am Main Suwelacks Flugapparat besichtigt hätten.[103] Sogar der „Flugsport“ berichtete über die „Flugmaschine Suwelack“.[104]

Der Bau eines neuen Apparates nahm einige Monate in Anspruch, brachte Josef Suwelack aber dennoch nicht den erhofften Durchbruch, da sowohl das Fahrgestell den Belastungen durch die unebene Startfläche nicht standhielt, als auch beim Motor Probleme mit dem Vergaser und der Zündung auftraten. Nach einer durch winterliche Kälte bedingten Pause nahm Suwelack seine Flugversuche am 12. März 1910 wieder auf, und zwar mit einem Eindecker, der eine Spannweite von 8,80 m und eine Länge von 8,90 m hatte. Angetrieben von einem 30-PS-Motor, kam die Maschine zusammen mit dem Flugzeugführer auf ein Gewicht von 300 Kilogramm.[105] Da wesentliche Fortschritte sich dennoch nicht einstellten, wechselte der Billerbecker seinen Standort Mitte des Jahres 1910 und fuhr zur Loddenheide nach Münster.[106]

Es ist anzunehmen, dass Suwelack die Berichte in der Heimatzeitung über die Fortschritte in der Flugtechnik anspornten, sich weiter als „Konstrukteur“ zu beweisen.[107] Offenbar war er optimistisch eingestellt, denn schon im ersten Jahr seiner Konstruktions- und Flugversuche präsentierte er sich potenziellen Kunden mit einem Briefbogen, auf dem er für seine automatisch stabilen Eindecker warb. Zudem stellte er umfangreiche Überlegungen zu Vor- und Nachteilen der Ein- und Doppeldecker an, verglich Daten von verschiedenen Flugzeugtypen und stellte die Vorzüge seiner Konstruktion heraus: „Die Flugmaschine J. Suwelack Jr ist ein Eindecker von 14 1/2 mtr. Spannweite und 12 mtr. Länge, die die Vorteile der Ein- und Doppeldecker in sich vereinigt, ohne deren Nachteile zu besitzen.“[108] Aufzeichnungen dieser Art belegen, mit welcher amateurhaften Seriosität Josef Suwelack seine Idee verfolgte, Flieger zu werden.

103 Vgl. Billerbecker Anzeiger vom 28.11.1909.

104 Vgl. Flugsport (1910), Nr. 13, S. 415f.

105 Vgl. Billerbecker Anzeiger vom 13.3.1910.

106 Billerbecker Anzeiger vom 31.7.1910: „Herr Suwelack hat sich einen anderen Platz für seine Flugversuche ausgesucht und zwar die Loddenheide bei Münster, welche sich schon durch die allerdings mißglückten Flugversuche des Aviatikers Schlüter einigen Namen erworben hat.“

107 Am 7. November schrieb der Billerbecker Anzeiger über Hans Grade, einen Magdeburger Flieger: „Berlin, 30. Okt. Auf dem Flugplatz Johannisthal bewarb sich heute mit Erfolg der Ingenieur Grade ... um den Lanz-Preis der Lüfte in Höhe von 40.000 Mk. Der Lanz-Preis wurde im Frühjahr vorigen Jahres gestiftet und sollte dem deutschen Flugtechniker zufallen, der zuerst mit einem aus vollständig deutschem Material hergestellten Aeroplan über eine 100 Meter lange Startlinie zwei 1.000 Meter von einander entfernte Marken und zwar die zweite in entgegengesetzter Richtung wie die erste, umfliegt und wieder zum Startplatz zurückkehrt.“ Gleich darunter stand eine Meldung über einen Langstreckenflug.

108 NL J. Suwelack, Überlegungen zu Flugzeugkonstruktionen „Die Flugmaschine J. Suwelack Jr.“, 1909, 4 Seiten, und weitere „Studien“ über die Fliegerei ebd.

Bis zur Jahrhundertwende nutze das Militär die Loddenheide, ein großes, ebenes Feld, als Exerzierplatz.[109] Doch wenige Jahre später wurde das Gelände als Übungsplatz zu klein, so dass sich die Militärverwaltung 1906 zum Kauf eines neuen Grundstücks im Norden Münsters entschloss. Zu einem Verkauf der Loddenheide kam es nicht, sondern sie blieb weiterhin als Exerzierplatzbestehen, wurde jedoch teilweise der Öffentlichkeit zur Verfügung gestellt. Bereits 1836 war das Gelände als Pferderennbahn genutzt worden, seit 1908 fanden dann auf der Loddenheide die ersten Flugversuche und später auch Flugtage statt.

Am 23. Juli 1910 meldete der „Billerbecker Anzeiger" den Umzug von Josef Suwelack auf die Loddenheide:

> „Der Brock, bisher das Ziel so mancher Einwohner Billerbecks, liegt nun einsam und verlassen da ... Am Samstag nachmittag gegen 4 Uhr verließ Herr Suwelack den Brock, um über Darup, Nottuln, Appelhülsen, Albachten nach Münster zu fahren, zu fahren mit seinem Eindecker mit eigener Kraft, d.h. die gewaltige Zugkraft des Propellers als Fortbewegungsmittel gebrauchend. Diese Ueberlandfahrt ist dem kühnen Aviatiker gelungen, allerdings mit einigen Hindernissen ... Auf der Fahrt nach Münster [am Sonntag] passierte Herr Suwelack, wie der ‚Münst[erische] Anz[eiger]' zu berichten weiß, in rasanter Fahrt die Weseler Straße, Aegidistraße, Rothenburg und die Königstrasse. Trotz der kolossalen Menschenmengen, die in Folge dieser Fahrt in Münster und anderen Orten zusammenliefen, ist auch nicht das kleinste Unglück zu verzeichnen."[110]

Berichte über Schaulustige, die in großer Zahl Aufstiege von Ballons beobachteten, lassen sich für das Münsterland bereits für das Jahr 1909 nachweisen, zu einer Zeit, als Ballon- und Zeppelinfahrten dort noch Seltenheitswert besaßen. So bemerkte etwa der „Münsterische Anzeiger" Ende September 1909: „Der Aufstieg eines Ballons ist bislang in Münster immer noch ein seltenes Ereignis".[111] Ein Flugzeug war etwas noch weniger Alltägliches, weshalb die „kolossalen Menschenmengen" durchaus verständlich waren.

Trotz des Umzugs stellte sich für Suwelack weiterhin kein Erfolg ein.[112] Daraufhin, so erinnerte sich Richard Suwelack, reiste sein älterer Bruder in den Osten Deutschlands zur Motorenfabrik, um vor Ort ein aus Stahlrohren gearbeitetes Flugzeug zu bauen: „Er ist auch damit geflogen. Aber bei einem Versagen des Motors stürzte er damit ab, ohne selbst Schaden zu nehmen".[113] Dies scheint der

109 Vgl. für die folgenden Ausführungen die im StdA Münster vorliegende Schülerarbeit: StdA Münster, BU 9–0096, Betina Vogt, „Unsere Zukunft liegt in der Luft!". Die Entwicklung der zivilen Luftfahrt auf dem Flugplatz Loddenheide 1909–1933, unveröffentlichte Arbeit, hervorgegangen aus dem Geschichtswettbewerb des Bundespräsidenten 1990/91, Münster 1990/91, S.6–9.

110 Billerbecker Anzeiger vom 31.7.1910. Ganz unfallfrei verlief die Fahrt nicht, denn beim Anblick des fahrenden Flugzeuges wurden zweimal Pferde scheu und einmal sogar der Fahrer verletzt. Josefs Vater erhielt entsprechende Rechnungen. Vgl. Krewerth, Ein westfälisches Fliegerleben, S.32.

111 „Der Aufstieg eines Ballons", in: Münsterischer Anzeiger vom 27.9.1909.

112 Münsterischer Anzeiger vom 28.8.1910.

113 NL J. Suwelack, Erinnerungen Richard Suwelacks an Josef Suwelack. Vgl. auch Meldung im Billerbecker Anzeiger vom 28.8.1910.

letzte Versuch gewesen zu sein, mit „elterlichen" Mitteln in der Heimat einen flugfähigen Apparat zu bauen.

Im Herbst 1910 reiste Suwelack nach Berlin-Johannisthal zur „Wiege des deutschen Motorflugs in Deutschland", um Anschluss an die Entwicklung der Flugtechnik zu gewinnen:[114]

> „Der Aviatiker hat inzwischen den Flugplatz Loddenheide verlassen und sich nach dem großen Flugplatz in Johannisthal bei Berlin begeben. Seine Flugversuche seien erfolgreich fortgeschritten und er habe bereits mit einem Eindecker dreimal eine Runde von 15 bis 20 Minuten geflogen, wird von dort berichtet."[115]

In Johannisthal wohnte er fortan im Hotel Bürgergarten an der Parkstr. 9 zusammen mit anderen Fliegern, etwa Hellmuth Hirth oder Otto Linnekogel.[116] Und hier war es auch, wo die eigentliche „Fliegerkarriere" Josef Suwelacks begann, denn hier lag das deutsche Zentrum der Fliegerei, hier versammelten sich die damaligen Flugpioniere und solche, die es werden wollten.

Den Kontakt zu seiner Familie hielt Suwelack durch Briefe und Postkarten, wahrscheinlich auch durch Anrufe per Fernsprecher aufrecht. Sein Bruder Richard wurde nun zu seinem ständigen Begleiter, sei es, um „persönliche und geschäftliche Angelegenheiten für ihn zu ordnen", sei es als eine Art Leibwächter, da Josef „ständig umlagert [wurde] von Journalisten, Künstlern und vor allen Dingen sensationslüsternen Damen": „Wenn ich da war, musste ich ihn vor diesem Ansturm schützen und machte so meinerseits manch interessante Bekanntschaft."[117]

Der junge Unternehmersohn Suwelack war keineswegs der Einzige, der sich im Münsterland mit der Fliegerei beschäftigte. Auch Anton Knubel und Carl Rösner nutzten die Loddenheide als Test- und Konstruktionsgelände.[118] Ein Blick über

114 Werner Schwipps, Riesenzigarren und fliegende Kisten. Bilder aus der Frühzeit der Luftfahrt in Berlin, Berlin 1984, S.97.

115 Münsterischer Anzeiger vom 28.3.1911.

116 Siehe zu den Wohnanschriften der Johannisthaler Flieger: http://www.johflug.de/pdf/wohnanschri ften_neu.pdf (25.2.2014). Hellmuth Hirth (24.4.1886–1.7.1938) stammte aus Heilbronn. Nach dem Militärdienst als „Einjähriger" ging er mit einem Geschäftspartner seines Vaters in die USA, wo er in verschiedenen Motorenfabriken arbeitete. Während dieser Zeit reiste er viel, kehrte 1904 kurz nach Deutschland zurück und ging dann nach England. Erste Erfahrungen im Fliegen sammelte er bei August Euler, der in Darmstadt-Griesheim eine Flugzeugfabrik aufbaute und Hirth engagierte. Dort machte er seine ersten Flugversuche, wurde 1911 in Wien Flugschüler bei Karl Illner und erwarb im März 1911 sein Flugzeugführerzeugnis in Berlin-Johannisthal. Anschließend wurde Hirth Fluglehrer bei Rumpler, bestritt auf der Rumpler-Taube erfolgreich Flugwettbewerbe, stellte zahlreiche Rekorde auf und machte die Taube zum erfolgreichsten Flugzeug der Vorkriegszeit. Später stieg er auf den Albatros-Doppeldecker um. Nach dem Ersten Weltkrieg baute Hirth hochwertige Flugmotoren. Vgl. Schwipps, Schwerer als Luft, S.72–73.

117 NL J. Suwelack, Erinnerungen Richard Suwelacks an Josef Suwelack, alle Zitate ebd.

118 Der im Jahr 1857 geborene, radsportbegeisterte Anton Knubel war der Sohn eines Eisenbahnarbeiters und gelernter Drechsler. Zwei seiner acht Geschwister teilten die Leidenschaft für den Radsport. Knubel wurde als Radsportler überregional bekannt und gründete 1885 den ersten Fahrradhandel Münsters, den er elf Jahre später an seinem jüngeren Bruder Bernhard übertrug. An der

die Grenzen Westfalens hinaus verrät, dass sich die Mehrheit der Flieger aus dem Bürgertum rekrutierte und meist auf staatliche oder private Subventionen angewiesen war.[119] Ein Großteil der Piloten, die bis zum Ersten Weltkrieg die Flugprüfung erfolgreich ablegten – bis 1914 waren dies 871 Personen –, hatte entweder eine technische oder eine militärische Berufsausbildung vorzuweisen.[120]

Darüber hinaus versuchten auch verschiedene Städte wie Münster oder später Essen z. B. durch die Gründung von Luftfahrtvereinen an der Entwicklung der Luftfahrt teilzuhaben.[121] Das rege Interesse der Bevölkerung an Ballonaufstiegen nahm beispielsweise ein Journalist im „Münsterischen Anzeiger" zum Anlass zu fragen, „ob es nicht möglich ist, daß in Münster etwas mehr als bisher geschieht, um dem Verkehrsmittel der Zukunft hier eine Heimstatt zu gründen".[122] Daraufhin erfolgte bereits im nächsten Monat, am 18. Oktober 1909, auf einer Versammlung, bei der sich circa siebzig Herren zu einer Mitgliedschaft bereit erklärten, die Gründung eines solchen Vereins. Als vorläufiges Vereinsziel wurde festgehalten, „die Luftschiffahrt in Münster heimisch zu machen und zu fördern",[123] welches später noch um die „Pflege und Förderung der Luftschiffahrt in wissenschaftlicher, sportlicher und nationaler Hinsicht, sowie der Flugtechnik" ergänzt wurde.[124] Neben dem Interesse an der Luftschifffahrt waren es aber auch ganz handfeste Konkurrenzen, die diese Vereinsgründung begünstigten: Monate zuvor hatte sich nämlich bereits in der Nachbarstadt Osnabrück ein Verein gegründet.

Gründung des Luftschiffahrtvereins für Münster und das Münsterland E.V. 1909 war Anton Knubel beteiligt, seit 1910 versuchte er sich zusammen mit dem aus Hamm stammenden Carl Rösner im Fliegen. Auf der Loddenheide besaß Knubel eine eigene Konstruktionshalle, vertrieb dort Flugzeugteile und stellte seine Maschinen unter anderem in Berlin-Johannisthal vor. Später eröffnete er auf der Loddenheide eine „Flugzeugbauanstalt", in der er Ein- und Doppeldecker herstellte. Zur Zeit des Ersten Weltkrieges entwarf er eine transparente Flügelbespannung, damit die Flugzeuge für den Feind schwieriger zu erkennen wären, fand jedoch bei der Heeresverwaltung keinen Anklang. Am 8. September 1915 stürzte Knubel bei einem Testflug auf der Loddenheide tödlich ab. Carl Rösner hatte sich das Fliegen wie Josef Suwelack autodidaktisch beigebracht und Flugzeuge entworfen. 1911 konstruierte er mit Unterstützung einer Baugesellschaft einen Eindecker, ebenfalls mit transparenter Flügelbespannung, und beteiligte sich im Juni gleichen Jahres beim Deutschlandflug an einem Wettbewerb für Nichtpiloten, den er gewann. Über seinen weiteren Lebensweg ist nichts bekannt. Informationen zu Knubel und Rösner entnommen aus Bernd Haunfelder, Bernhard Knubel Kraftfahrtzeuge – Eine Chronik, Münster 1995, S.12; Langenfeld/ Prange, Münster. Die Stadt und ihr Sport, S.173 sowie Supf, Das Buch der deutschen Fluggeschichte, Bd. 2, S.468. Siehe auch Flugsport (1911), Nr. 18, S.641ff.

119 Vgl. Fritzsche, A Nation of Fliers, S.96.

120 Vgl. Kehrt, Moderne Krieger, S.72.

121 Für Essen siehe Rißmann-Ottow, Glück ab!, S. 63ff.; für Münster Ernst Kühl, Das schöne Münster. 50 Jahre Sportfliegen in Münster, Münster 1960; Langenfeld/ Prange, Münster. Die Stadt und ihr Sport, S.118ff. Der erste „Flug" – ein kurzer Luftsprung – mit einer Essener Flugmaschine des Niederrheinischen Vereins für Luftschiffahrt, Sektion Essen, fand am 1. Februar 1910 auf den „Ruhrwiesen" statt, vgl. Flugsport (1910), Nr. 4, S.112.

122 „Der Aufstieg eines Ballons", in: Münsterischer Anzeiger vom 27.9.1909, 1. Ausgabe.

123 Münsterischer Anzeiger vom 19.10.1909, 2. Ausgabe.

124 § 2 des Statutenentwurfs, zit. nach Willy-Bernd Bothe, Ballonfahren in Münster. 75 Jahre Freiballonsport, Münster 1984, S.15.

Dementsprechend bemerkte Anton Knubel anlässlich der Vereinsgründung in Münster in seiner Rede:

> „Sie wissen, daß Münster in technischen Fragen heute noch etwas hinternach folgt; das liegt an dem schwerfälligen Naturell der Westfalen, speziell des Münsterländers. Aber zähe, wie der Westfale ist, hat er auch, wenn er etwas erfaßt hat, zähe daran festgehalten. Hoffen wir, daß wir das auch von der Luftschiffahrt sagen können, der wir heute durch die Gründung eines Luftschiffahrtvereins dienen wollen."[125]

Überhaupt erhielt das Vereinswesen eine entscheidende Bedeutung bei der Entwicklung der organisierten Luftfahrt in Deutschland.[126] Zum wichtigsten Repräsentanten des deutschen Flugwesens entwickelte sich der 1902 als Dachverband aller Luftfahrtvereine gegründete Deutsche Luftschiffer-Verband, der 1910 in Deutscher Luftfahrer-Verband (DLV) umbenannt wurde.[127]

Ansbert Vorreiter schätzte die Mitgliederzahl flugtechnischer Vereine für 1912 auf zweitausend, weitere dreitausend Personen seien in Automobilvereinen im Bereich der Luftfahrt organisiert gewesen.[128] Bis 1907 stieg die Zahl der Flugvereine auf 16; in einer Gründungswelle von 1908 und 1909 nahm sie um ein Vielfaches zu, so dass im Jahr 1914 unter dem DLV 88 Vereine mit annähernd 79.000 Mitgliedern existierten.[129]

Die Aktivitäten im Bereich des Flugwesens kulminierten im Jahr 1909, dem „Schlüsseljahr der Entwicklung des Motorflugzeugs in Europa".[130] An mehreren Orten, zunächst in Frankreich, dann auch in Deutschland, wurde mit dem Bau stabiler Eindecker begonnen. Gleichzeitig hielt die Fliegerei Einzug in die Alltagskultur: Motive von Zeppelinen, Ballons und Flugzeugen zierten Teller, Vasen oder Postkarten und waren Gegenstand von Liedern.[131] Ferner überschlugen sich in den Anfangsjahren der Fliegerei die Meldungen über neue Rekordleistungen, zumeist aus Frankreich, da die Flugzeugtechnik ständig verbessert wurde und immer größere Höhen, höhere Belastungen oder weitere Strecken erreicht wurden. Betrug die Flugdauer z. B. 1906 nur 21 Sekunden, lag sie Ende des Jahres 1908 bereits bei zwei Stunden und 20 Minuten und steigerte sich ein Jahr darauf nochmals um weitere zwei Stunden (4 Std. 19 Min.).[132]

Joachim Radkau, der sich ausführlich mit dem um die Jahrhundertwende aufkommenden Phänomen der Nervosität befasst hat, ein Aspekt, der auch für den

125 Münsterischer Anzeiger vom 19.10.1909, 2. Ausgabe.
126 Christiane Eisenberg, Arbeiter, Bürger und der „bürgerliche Verein" 1820–1870. Deutschland und England im Vergleich, in: Jürgen Kocka (Hg.), Bürgertum im 19. Jahrhundert. Deutschland im europäischen Vergleich, Bd. 2, Göttingen 1995, S. 187–219, hier S. 189f.
127 Vgl. Jahrbuch des DLV 1911, S. 78.
128 Vgl. Ansbert Vorreiter (Hg.), Jahrbuch der Luftfahrt, München 1912, S. 601.
129 Jahrbuch des DLV 1914, S. 34.
130 Behringer, Der Traum vom Fliegen, S. 400.
131 Vgl. ebd., S. 412. Siehe auch Matthias Wenzel, Gotha auf alten Postkarten, Erfurt 2005, S. 101ff.
132 Siehe für eine Übersicht der Rekordleistungen: „Die Erfolge der Flugmaschinen", in: Flugsport (1910), Nr. 2, S. 42–45.

Fliegerdiskurs relevant ist, betont, dass die Menschen um die Jahrhundertwende in einen „Geschwindigkeitsrausch" verfallen und Sportler und Zuschauer gleichermaßen süchtig nach Rekorden gewesen seien.[133] Der Wunsch, alles möge schneller ablaufen, wird auch in den Erinnerungen Richard Suwelacks deutlich, der schreibt, dass es seinem Bruder nicht schnell genug gegangen sei, in der Fliegerei die nötigen Fortschritte zu machen.[134]

Vielleicht haben die Sensationsmeldungen und Berichte über Flugpioniere Josef Suwelack den Anstoß gegeben, nach Berlin-Johannisthal in das deutsche Zentrum der Fliegerei zu ziehen. Hierfür sprechen die einleitenden Worte Heinz Erblichs in seinem Handbuch für Flugschüler sowie die Erzählungen Anthony Fokkers,[135] der in seinen Memoiren „Der fliegende Holländer" von 1933 unter anderem auf seinen Aufenthalt in Johannisthal zurückblickte: „Johannisthal war eine lebhafte kleine Weltstadt. Die Fliegerei war ein neuer Sport, der tapfere Männer, aber auch Taugenichtse und Abenteurer aus der ganzen Welt dorthin gezogen hatte. Viele Amateurflieger waren Söhne reicher Eltern, die in Johannisthal Gelegenheit suchten und fanden, sich auszuleben."[136] Letzteres trifft auch auf Josef Suwelack zu, der durch die Unterstützung seiner Familie und die Hilfsbereitschaft von Nachbarn und Freunden diesen Schritt wagen konnte. Zeit seines Lebens blieb sein Elternhaus jedoch ein wichtiger Bezugspunkt für ihn. Mit seinen Eltern und Geschwistern stand er brieflich in Kontakt und nutzte verschiedene Anlässe, um seiner Familie und seinen Freunden aus Billerbeck und Umgebung einen Besuch abzustatten. Durch seinen Vater wurde er über aktuelle Ereignisse in seiner Heimatstadt informiert und die Billerbecker Bevölkerung konnte von Josef Suwelacks Aktivitäten fern der Heimat im „Billerbecker Anzeiger" und in anderen Zeitungen des Münsterlandes lesen.

Suwelacks Interesse an der Fliegerei war, so die These, insbesondere sportlich motiviert. Abgesehen von Skeptikern wurde den Flugpionieren vonseiten der Bevölkerung für ihre Flugshows Bewunderung entgegengebracht und auch die Flieger selbst waren sich ihrer „abgehobenen" Stellung durchaus bewusst.[137]

133 Vgl. Joachim Radkau, Das Zeitalter der Nervosität. Deutschland zwischen Bismarck und Hitler, Wien 1998, S.11ff., und ders., Technik in Deutschland. Vom 18. Jahrhundert bis heute, Frankfurt a. M./New York 2008, S.240.

134 Vgl. NL J. Suwelack, Erinnerungen Richard Suwelacks an Josef Suwelack.

135 Erblich, Fliegerschule, S.5: „Die Begeisterung für die Eroberung der Luft ist in alle Kreise der Bevölkerung eingedrungen, besonders die männliche Jugend nimmt an allem, was die moderne Technik auf dem Gebiete der Luftfahrt Neues zutage fördert, den regsten Anteil. In manchem Herzen ist wohl schon der Wunsch entbrannt, selbst mitzuwirken an dem großen Werk, und tausende von jungen Leuten sind gern bereit, ihre Kraft und ihr Leben für die Flugtechnik einzusetzen".

136 A.H.G. Fokker/Bruce Gould, Der fliegende Holländer. Das Leben des Fliegers und Flugzeugkonstrukteurs A.H.G. Fokker, Zürich/Leipzig/Stuttgart 1933, S.68.

137 Siehe z. B. Archiv-DTMB, NL Willy Stiasny 003, „Mit Bomben und Feuer gegen Flugpioniere", in: Der Rundblick, Berlin 1939, hier S.271–273 (1909).

Davon zeugen sowohl zahlreiche Berichte, in denen die Flieger als Männer mit der nötigen Ruhe und guten Nerven stilisiert wurden beziehungsweise sich selbst stilisierten, als auch Fotos, auf denen sie sich als todesmutige Dandys inszenierten. Sich mit anderen Flug(sport)begeisterten zu messen, die Herausforderung zu suchen, nach der individuellen Leistung bewertet und gegebenenfalls prämiert zu werden, machte die Flugzeuge zu „Dandyplattformen" für in der Hauptsache risikobejahenden jungen Männer.[138]

Geflogen werden konnte aus Sicherheitsgründen nur bei guten Wetterbedingungen, so dass sich die Flugpioniere bei starkem Wind, Regen oder nach absolvierten Flügen in den am Rande des Flugplatzes Johannisthal befindlichen Gaststätten und Cafés trafen. Zwei zentrale Treffpunkte waren das oben bereits erwähnte Hotel Bürgergarten und das Cafe Senftleben.

Josef Suwelack war in Johannisthal unter Gleichgesinnten und konnte mit einer „professionellen" Ausstattung und verbesserten Bedingungen seinem Wunsch, Berufsflieger zu werden, näher kommen. Seine Familie unterstützte ihn weiterhin in finanzieller und organisatorischer Hinsicht, doch seine Eltern mahnten ihn auch, nichts zu riskieren und „bodenständig" zu bleiben, wie dies aus einer Passage aus einem späteren Brief seines Vaters ersichtlich ist: „Aber große Angst haben wir auch gehabt wegen deiner Waghalsigkeit. Sei doch vorsichtig! Fliege nur bei gutem Wetter. Du mußt nicht denken, den Leuten etwas zeigen zu müssen."[139]

In der Tat erscheint die Bereitschaft zahlreicher junger Männer, ihr Leben aufs Spiel zu setzten, angesichts hoher Unfallzahlen und ungewisser finanzieller Zukunftsaussichten erklärungsbedürftig. Für den Verfasser des Lehrbuches „Fliegerschule", Heinz Erblich, musste „die Begeisterung für die Sache die Triebfeder sein", denn „die großen Preise, die sich gewiß mancher erträumt, sind nur vereinzelten beschieden".[140] Zudem schien sich eine „Kultur der Todesbereitschaft" herausbildet zu haben.[141] Die Gefahr des Fliegens war durch Zeitungsberichte, Fotografien oder Karikaturen medial präsent und vor allem bei Flugwettbewerben Teil der Attraktion.[142] In den von den Piloten angelegten Fotoalben finden sich immer wieder Aufnahmen von beschädigten, brennenden oder gänzlich zertrümmerten Flugzeugen, die Zeugnis von Unfällen und Katastrophen ablegen, die anscheinend einen besonderen Reiz sowohl auf die Flieger als auch auf die Zuschauer ausübten.[143] Darüber hinaus wurde das Unfallrisiko verharmlost und

138 Möser, Fahren und Fliegen, S.138.
139 NL J. Suwelack, Brief Joseph Suwelacks senior an Josef Suwelack vom 2.10.1911.
140 Erblich, Fliegerschule, S.12 (dort beide Zitate).
141 Möser, Fahren und Fliegen, S.464.
142 Zum Thema Unfälle als Attraktion siehe die Ausführungen zum Feld 3.
143 Siehe z. B. NL J. Suwelack, Fotoalbum aus der Zeit in Johannisthal sowie Fotos von Flugwettbewerben in der Fotosammlung ebd.

in den Fliegerdarstellungen zum Teil ausgeblendet.[144] Viele Luftfahrtinteressierte stimmten mit Otto Lilienthals Meinung überein, der vor seinem Tod zu seinem Bruder Gustav gesagt haben soll: „Opfer müssen gebracht werden."[145] Die Opferbereitschaft erfuhr in der Zeit des Ersten Weltkrieges sogar eine Transformation ins Religiöse, indem die (herunter-)gefallenen Krieger durch eine sprachliche und ikonografische Metaphorik sakralisiert wurden.[146]

Feld 2: Flugpionier und -lehrer, Konstrukteur und Unternehmer

2.1. Fliegermilieu und Risiko der Flugzeugtechnik

Im Alter von 21 Jahren verließ Josef Suwelack seine ländliche Heimat, in der das kirchliche Gemeindeleben, die Produktionsabläufe der väterlichen Molkerei und die Großfamilie sein Leben bestimmt und strukturiert hatten. Mit 1.564 Einwohnern für das Jahr 1910 war Billerbeck eine Stadt, in der den Bewohnern der Name Suwelack unter anderem aufgrund des Bekanntheitsgrades der Molkerei sowie der gesellschaftlichen Partizipation Joseph Suwelacks senior durchaus geläufig war.

In Berlin-Johannisthal hingegen, mit einer ungefähren Einwohnerzahl von 3.930 Personen,[147] war Suwelack junior eine noch gänzlich unbekannte Person und musste sich mit den Begebenheiten auf dem Flugplatz sowie mit den geltenden ex- und impliziten Regeln erst vertraut machen. Die rückblickende Beurteilung Anthony Fokkers zu seinem eigenen Umzug von Mainz nach Johannisthal trifft in ähnlicher Form wohl auch auf Josef Suwelack zu:

> „Mein Umzug nach Johannisthal brachte zunächst eine nicht unbeträchtliche Änderung in meinem Leben hervor. In Mainz war ich ein dicker Frosch in einem kleinen Teich gewesen, in Johannisthal war ich ein sehr kleiner Frosch in einem grossen Teich. Bis dahin hatte ich gar nicht gewusst, was für ein Provinzler ich war."[148]

Die unmittelbare Nachbarschaft zu anderen Fliegern, die entweder wie Josef Suwelack im Hotel Bürgergarten wohnten oder aus der Umgebung zum Flugplatz kamen, mag dem Westfalen jedoch schon bald zu neuen Bekanntschaften verholfen haben.

144 Siehe z. B. Flugsport (1912), Nr. 3, S. 102: „Die Schilderung von Unglücksfällen im Flugwesen, besonders von tödlichen Abstürzen, erfolgen manchmal in der Presse in einer Weise, die geeignet ist, die in Entwicklung stehende Industrie in höchstem Maße zu schädigen. Daß jedes neue Verkehrsmittel seine Opfer erfordert, ist selbstverständlich".
145 Zit. nach Behringer, Der Traum vom Fliegen, S. 391.
146 Vgl. Esposito, Mythische Moderne, S. 202ff.; Schnürer, Die Totenfeiern, S. 167–192.
147 Angaben zit. nach www.johflug.de (20.5.2014).
148 Fokker/Gould, Der fliegende Holländer, S. 68.

Am 17. Januar 1911 schloss Suwelack einen Vertrag mit der Deutschen Flugmaschinenbau Gesellschaft (DFG) mit Sitz in Berlin-Rummelsburg. Darin ging er die Verpflichtung ein, innerhalb von drei Monaten das Flugzeugführerzeugnis durch eine entsprechende Prüfung zu erwerben. Die im Voraus zu zahlenden Ausbildungskosten betrugen 1.500 Mark, bei erfolgreich bestandener Prüfung wurden weitere 1.500 Mark fällig; die Ausbildung galt spätestens nach drei Monaten als abgeschlossen, gleichgültig ob erfolgreich oder nicht.[149] Hierzu stellte die DFG ihm vertraglich eine Maschine vom Typ „Schultze-Herford" zur Verfügung. Der Flugzeugführer in spe verpflichtete sich seinerseits, nach Abschluss der Ausbildung in den Dienst der DFG gegen ein festes Monatsgehalt von 250 Mark und eine fünfzig-prozentige Gewinnbeteiligung an erzielten Geldpreisen auf Flugwettbewerben einzutreten. Zu seinen Aufgaben sollte das Testen von Maschinen und deren Vorführung auf Wettbewerben gehören. Für die erfolgreiche Ausbildung von Flugschülern wurde ihm zudem eine besondere Prämie zugesichert.[150]

Bis zum angesetzten Prüfungstermin am 17. April 1911 übte Suwelack sich im Fliegen. Von einem missglückten Flugversuch Suwelacks auf einem Fabrikat der DFG berichtete die „BZ am Mittag"[151] am 15. Februar, dass er tags zuvor auf dem Flugplatz in Johannisthal mit einem neuen Eindecker der DFG Rollversuche habe machen wollen: „Plötzlich erhob sich die Maschine vom Boden und schoß ca. 5 Meter fast kerzengerade in die Höhe. Suwelack erhob sich vom Sitze, offenbar, um abzuspringen, in diesem Moment kippte der Eindecker nach rechts über und stürzte nieder." Suwelack sei „mit dem Schrecken davon" gekommen, berichtete die Zeitung weiter, doch an der Maschine seien Fahrgestell und Propeller zu

149 Vgl. ebd. Eine Ausbildung zum Flugzeugführer kostete durchschnittlich 3.000 Mark, wobei von einer fundierten und geregelten Ausbildung keine Rede sein konnte, da der Ausbilder Fluglehrer sein konnte, ohne selbst das Pilotenzeugnis innezuhaben. Darüber hinaus hatte der Flugschüler, sofern er nicht beabsichtigte, einen Apparat zu kaufen, das volle Bruchrisiko zu tragen und musste oftmals lange warten, bis er zum Fliegen kam. Vgl. Schwipps, Schwerer als Luft, S.55. Siehe auch Erblich, Fliegerschule, S.127ff.: „Vertragsbedingungen des Verbandes Deutscher Motorfahrzeugindustrieller für die Ausbildung zum Flugzeugführer." Demnach mussten 1.500 Mark bei Vertragsunterzeichnung gezahlt werden und weitere 1.500 Mark bei erfolgreich bestandener Prüfung.

150 Vgl. NL J. Suwelack, Vertrag zwischen der DFG und Josef Suwelack vom 17.1.1911.

151 Die „Berliner Zeitung am Mittag", kurz „BZ am Mittag", 1904 gegründet und vom Ullstein Verlag herausgegeben, erhob für sich den Anspruch, die „schnellste Zeitung der Welt" zu sein. Ihre Aktualität zog eine heterogne Leserschaft an. Da es die Zeitung nicht im Abonnement gab, musste das Blatt täglich seine Kunden vom Kauf überzeugen. Die „BZ am Mittag" war zudem die erste Zeitung Deutschlands, die eine eigene Sportbeilage einführte. Ab 1912 erstattete sie durch einen Spezialkorrespondenten Bericht vom Flugplatz Johannisthal, sie sei das „Leib- und Magenblatt der damaligen Johannisthaler Fliegergemeinde" gewesen. Siehe zur „B.Z. an Mittag": Erich Wagner, BZ am Mittag und BZ, in: W. Joachim Freyburg/Hans Wallenberg (Hg.), Hundert Jahre Ullstein, Bd. 2, Frankfurt a. M./Berlin 1977, S.47–85, bes. S.50–62 sowie Fokker/Gould, Der fliegende Holländer, S.71.

Bruch gegangen.[152] Rund einen Monat später schrieb das gleiche Blatt, Suwelack habe „drei tadellose Runden in 10 bis 12 Meter Höhe“[153] ausgeführt.

Zwischen der DFG und Josef Suwelack bestanden offenbar jedoch Differenzen, was die Modalitäten der Ausbildung betraf, dies geht aus einem Einschreiben Suwelacks vom 12. April hervor. Darin forderte er die Gesellschaft auf, ihm „binnen 24 Stunden eine Schulmaschine zur Verfügung zu stellen, auf der ich die Bedingungen des Führerzeugnisses bis zum 17.ten [sic!] April erfüllen kann.“[154]

Der Besitz des Flugzeugführerzeugnisses war seit Anfang des Jahres 1910 Voraussetzung, um an Flugwettbewerben teilnehmen zu können, die von einem Verein des DLV ausgeschrieben waren. Die Prüfungsbedingungen sahen drei Rundflüge zu je mindestens fünf Kilometern Länge ohne zwischenzeitlichen Bodenkontakt bei einer Flughöhe von 50 Metern vor. Nach jedem Rundflug musste gelandet und der Motor abgestellt werden. Landung und Anhalten des Motors durften höchstens 150 Meter weit „von dem Punkte erfolgen, welcher dem Bewerber vorher hierfür bezeichnet wurde“.[155] Seit dem 1. Februar 1911 galten dann neue erschwerte Prüfungsbedingungen. So mussten nun bestimmte Marken in Form einer Acht umflogen werden, der Motor musste unmittelbar nach der Landung abgestellt werden und die Abweichung vom zuvor bezeichneten Landepunkt durfte nur noch 50 Meter betragen.[156]

Aus Suwelacks Notizen zu seinem 1912 gehaltenen Vortrag wird ersichtlich, dass er sich aufgrund des guten Rufes des Fabrikats für den Schulze-Herford-Eindecker entschied. Vermutlich als Referenz gab er den Namen „Behrend“,[157] einen Johannisthaler Flieger, an.[158]

152 „Fliegerunfall in Johannisthal“, in: B.Z. am Mittag vom 15.12.1911. In der Absturzstatistik über das Jahr 1911 von 1912 heißt es hingegen, „Suvaleck“ [sic!] sei „leichtverletzt“ gewesen. Vgl. Emil Neyen, Die Flugkunst ohne Schleier. Statistik von 1911 über die Flieger-Unfälle, Berlin 1912, S.20f.

153 „Auf dem Flugplatz Johannisthal“, in: B.Z. am Mittag vom 29.3.1911. Siehe auch „Suvellack [sic!]“, in: Flugsport (1911), Nr.8, S.271.

154 NL J. Suwelack, Einschreiben Josef Suwelacks an die DFG vom 12.4.1911.

155 Ansbert Vorreiter, Jahrbuch über die Fortschritte auf allen Gebieten der Luftschiffahrt 1911, München 1911, S.470.

156 Die neuen, verschärften Bedingungen sahen vor, dass der Prüfling „zweimal 5 Kilometer in geschlossenem Fluge zurück[legte] und dabei eine Höhe von mindestens 50 Metern“ erreichte. „Die Flugbahn wird durch zwei Marken festgesetzt, die höchstens 500 Meter voneinander entfernt sein dürfen. Diese Marken sind derart zu umfliegen, dass nach jeder Umkreisung der einen die Flugrichtung geändert, d.h. die zweite Marke im entgegengesetzten Sinne umflogen wird, der Flug also die Form einer Acht annimmt. In diesem Sinne muss so lange geflogen werden, bis 5 Kilometer zurückgelegt sind. ... Die Bedingungen für die Landung sind ebenfalls wesentlich verschärft worden. Vor allem muß der Motor spätestens in dem Moment abgestellt werden, in dem die Maschine auf dem Boden aufsetzt. Ferner muß das Flugzeug weniger als 50 Meter von einem vor dem Aufstiege bezeichneten Punkte zum Stillstand kommen. Das Flugzeug muß nach der Landung flugfähig sein.“ Geänderte Bestimmungen abgedruckt in: B.Z. am Mittag vom 15.2.1911.

157 Adolf Behrend (1869–1945) war ein deutscher Varietékünstler und um 1910 als Pilot der Deutschen Flugmaschinenbau-Gesellschaft in Berlin-Rummelsburg sowie als Fluglehrer auf dem Flugplatz Johannisthal tätig. Mit dem Pilotenschein Nr.7 gehörte er zu den ersten deutschen Flugpio-

Sein erster Versuch, den Piloten-Schein zu erwerben, misslang jedoch, wovon der „Münsterische Anzeiger" am 2. Mai 1911 berichtete. Der Zeitung zufolge habe er aufgrund eines Motordefekts einen Unfall gehabt und der Apparat sei bei der Landung außerhalb der vorgesehenen Flugbahn schwer beschädigt worden, der Flieger hingegen unverletzt geblieben.[159] Die DFG musste den Schaden übernehmen, trennte sich jedoch von Suwelack mit Wirkung vom 7. Juni. Der Flieger selbst stellte es in seinem späteren Vortrag jedoch so dar, als ob er sich von der DFG getrennt hätte, und auch der „Münsterische Anzeiger" teilte mit, Josef habe sich „auf den Trümmern eines dieser von ihm zu ‚Kleinholz' gefahrenen Apparate ... [geschworen,] diesem Typ Valet zu sagen und sich der ‚Taube' zuzuwenden."[160]

Abstürze, Bruch- und Notlandungen gehörten zu den Erfahrungen, die jeder Pilot in den ersten Jahren machte. Dies verleitete Erblich zu der Feststellung, der Fliegerberuf verlange buchstäblich „die ganze Aufopferung der Person".[161] Die noch unausgereifte Technik war störungs- und die Baumaterialien wetteranfällig, so dass ein erfolgreicher Flug wesentlich vom praktischen Können des Akteurs und regelmäßigen Kontrollen der Maschinen durch Monteure oder den Flieger selbst abhing.[162] Gleichwohl unterlag ein „sicherer" Flug nicht allein dem habitualisierten Wissen des Flugzeugführers, sondern beruhte auf weiteren Faktoren, deren Auswirkungen nicht immer vom Piloten abgeschätzt werden konnten. Häufig waren Windböen, Flügelbrüche oder Defekte am Motor, wie z. B. ein Vergaserbrand, für Abstürze verantwortlich.[163] Ansbert Vorreiter thematisierte eben diese Aspekte – individuelles, habitualisiertes Können und Wissen der Flieger auf der einen Seite, technische Defekte und Mängel der Maschinen auf der anderen Seite – in seinem Vorwort zum „Jahrbuch der Luftfahrt" von 1912:

> „Flugzeuge und Motoren sind in ihrer Konstruktion wesentlich verbessert worden, die Flugzeugführer vermehrten ihre Erfahrungen und ihr Können, und so wurden ganz bedeutende Flugleistungen erzielt. Allerdings ist dieser Fortschritt mit Opfern erkauft worden. Sind doch im vergangenen Jahre allein zehn deutsche Flugzeugführer tödlich verunglückt. Leider ist nicht immer die Ursache der Unfälle festzustellen, denn ... aus den Trümmern der Maschine kann man nicht einwandsfrei [sic!] erkennen, was erst bei der Landung, und was schon vorher beim Fluge defekt geworden ist."[164]

nieren und belegte Ende April 1910 den 1. Platz im mit 7.000 Mark dotierten 2. Lanz-Preis der Lüfte auf einem Schultze-Herfort Eindecker. Vgl. Sport im Bild (1911), Nr.7, S.184, online unter: http://anno.onb.ac.at/cgi-content/anno-plus?aid=sib&datum=1911&page=198&size=45 (22.5.2014); sowie Sonja Steiner-Welz, Schütte-Lanz-Luftfahrzeuge aus Mannheim, Bd. 1: Die Luftschiffe, Mannheim 2006, S.72.

158 Siehe NL J. Suwelack, Notizen zum Vortrag „Aus meinem Fliegerleben", S.2.

159 Vgl. Münsterischer Anzeiger vom 2.5.1911.

160 Münsterischer Anzeiger vom 26.2.1912.

161 Erblich, Fliegerschule, S.13.

162 Vgl. ebd., 3. Kapitel: „Das Flugzeug", S.32ff.

163 Siehe z. B. Erblich, Fliegerschule, S.10 sowie Neyen, Die Flugkunst.

164 Ansbert Vorreiter, Jahrbuch der Luftfahrt. II. Jahrgang 1912, München 1912, S.V.

Die Frage nach der Rolle des Fliegers als aktiver *airman* oder passiver Chauffeur war regelmäßig Gegenstand von Buchkapiteln, Presseartikeln und Fachzeitschriften.[165] Vielfach suchten die Autoren die Schuld an den Unfällen jedoch nicht bei der Technik, sondern bei den Fliegern, die z. B. nachlässige Kontrollen ausführen würden, sich selbst überschätzt hätten, oder denen es umgekehrt an den nötigen Nerven mangeln würde. Psychische und physische Ruhe sowie vorausschauendes Denken und Handeln waren Eigenschaften, die vor dem Hintergrund der bestehenden Konkurrenz, insbesondere zwischen Frankreich und Deutschland, zu nationalen Tugenden erhoben wurden. In den Jahren vor dem Ersten Weltkrieg und dann vor allem während des Krieges selbst war stets die Rede von den „guten Nerven", die man benötige, um als Flieger, als Soldat oder bei der Fliegertruppe bestehen zu können. Nervenstärke wurde gewissermaßen zu einem nationalen Charakteristikum funktionalisiert und von den Piloten auch öffentlich auf Flugschauen unter Beweis gestellt. Erblich schrieb in seiner „Fliegerschule" über die nationalen Charakterzüge:

> „Betreffs der Anpassung des Flugzeuges an die Individualität des Führers kann man sagen, daß nervöse Menschen für den französischen Typ geeigneter sind ... Für die Taube passt besser der Phlegmatiker. So haben sich die Flugzeuge in origineller Weise dem Volkscharakter entsprechend entwickelt."[166]

Während man einerseits explizit auf die Gefahren der Fliegerei verwies und von einem „Massenmord in schönster Friedenszeit"[167] sprach, berief sich andererseits der „Flugsport" 1913 auf eine Statistik, die, der Zeitschrift zufolge, den Rückgang der Todesstürze und damit die verbesserte Sicherheit der Apparate belegte.[168] Allerdings wurde ebenso vor Leichtsinn und mangelhafter Instandhaltung der Flugzeuge gewarnt. Merkverse[169] zu Arbeitsroutinen bei der Flugzeugwartung

165 Siehe z. B. Erblich, Fliegerschule, S.20f.: „Die präzise Wirkung der Steuer setzt ihn [den Flieger] in die Lage, den Apparat in jeder Stellung wieder aufzurichten, allerdings hängt viel von der persönlichen Geschicklichkeit des Piloten ab"; oder auch ebd. S.45: „Die Einstellung der Maschine ist in erster Linie Sache des Piloten, nur er kann die Maschine im Fluge beurteilen und Schlüsse ziehen."

166 Erblich, Fliegerschule, S.29f.

167 Neyen, Die Flugkunst, S.29.

168 Vgl. „Die Sicherheit des Maschinenfluges nimmt von Tag zu Tag zu", in: Flugsport (1913), H. 2, S.55.

169 „Flugregeln", in: Flugsport (1914), H. 23, S.890f. (Auszug): „Mensch willst du in die Lüfte segeln, alsdann befolge diese Regeln. 1. Zündkerzen sind nur dann verschmutzt, wenn man sie vorher nicht geputzt. ... 4. Ertrinkt der Motor in Benzin, mein junger Freund, dann drossle ihn. ... 10. Macht die Maschine eine Schwenkung entgegen der gewollten Lenkung, so sage nicht in diesem Fall, der gute Vogel habe Drall, nein gib es unumwunden zu, den Drall, o Flieger, den hast du! ... 13. Wenn man dir nichts beweisen kann, sag, der Motor sei schuld daran, er habe keine Tourenzahl, die Dämpfungsfläche sei zu schmal, das Höhensteuer wär verspannt, die Kolbenringe festgebrannt, dann habe auch zu guter Letzt schließlich die Zündung ausgesetzt, allein die eigne schwarze Seele, sei frei von jeder Schuld und Fehle."

oder auch Regeln wie die „Zwölf Gebote für Flugschüler“[170] waren Präventivmaßnahmen beziehungsweise Versuche zur Unfallverhütung und Disziplinierung.

Die Verletzungen, die die Flieger davontrugen, waren häufig schwerwiegend und setzten manch junger Karriere ein jähes Ende. Emil Neyen listete in seiner Statistik über Fliegerunfälle für das Jahr 1911 insgesamt 92 tödliche internationale Abstürze auf, wovon „7 [Personen] noch durch Benzinexplosion völlig verbrannt“[171] seien. Tabellarisch vermerkte er akribisch den Namen des jeweiligen Fliegers, den Unglücksort, die Art der Verletzung etc. und gab, sofern möglich, Ursachen für den Unfall an. Aus der Schwere der Verletzung, wenn der Flug keinen tödlichen Ausgang hatte, wurde zudem abgeleitet, ob der Flieger zum „Krüppel“ würde. Ein Auszug aus einer Tabellenzeile gibt einen ungefähren Eindruck von Neyens Buchführung:

> „Datum: 7. Mai, Name des Abgestürzten: Ingenieur Schendel und Meister Voss, Ort oder Stadt des Falles: Johannisthal, Todessturz oder Schwerverletzte: Todessturz, Art der Verletzung: beide schrecklich verstümmelt, Ob Krüppel?: †, Höhe des Absturzes: 1680 [m], Art des Absturzes: seitlich steil abgerutscht, Ursache des Sturzes: durch Luftstrudel unt. dunkl. Wolke, Typ des Apparates: Dorner E ..., Beschaffenheit der Reste des Flugzeuges: Trümmer.“[172]

Zahlreiche Flieger erlitten Frakturen der Arme oder Beine, des Schädels und der Wirbelsäule, innere Verletzungen sowie Verbrennungen, so dass sie zum Teil für den Rest ihres Lebens arbeitsunfähig waren. Speziell vor 1912 waren die Betroffenen und deren Familien nach einem solchen Absturz finanziell ruiniert, da sich Versicherungen bis dahin weigerten, den Fliegern eine Police auszustellen. Eine Änderung trat erst mit der National-Flugspende ein, deren Kuratorium eine Förderung und weitere Popularisierung der (Berufs-)Fliegerei anstrebte und Abhilfe versprach.[173] Darüber hinaus war angesichts des hohen Gefahrenpotenzials und

170 Erblich, Fliegerschule, S.181f.: Dort heißt es z. B.: „1. Nimm jede Gelegenheit wahr, etwas zu lernen, dennoch erziehe dich in Geduld und überstürze nichts, 2. Dränge nicht, allein zu fliegen, sondern überlasse es deinem Fluglehrer, zu urteilen, ob du reif genug bist ..., 8. Hast du eine Maschine zerbrochen, so suche dir deinen Fehler klar zu machen, und schiebe die Schuld nicht auf den Apparat, ..., 10. Gerätst du in Gefahr, so bemühe dich, bis zum letzten Moment derselben geistesgegenwärtig entgegenzuarbeiten, Unerschrockenheit ist die erste Zierde des Fliegers.“

171 Neyen, Die Flugkunst, S.28. Neyen fügt hinzu, dass die Dunkelziffer der schwerwiegenden Flugzeugabstürze höher liege (ebd.). Einer internationalen Statistik der Zeitschrift „Aeolus“ zufolge ereignete sich Mitte November 1911 der einhundertste Todessturz, vier Fünftel dieser 100 tödlichen Unfälle ereigneten sich auf Flugplätzen und ein Fünftel bei Überlandflügen. Unter den Opfern hätten sich 91 Flugzeugführer (75 mit Pilotenschein, 16 ohne abgelegte Prüfung) und neun Fluggäste befunden. Von den 100 Toten entfielen 34 auf Frankreich, je 15 auf Deutschland und Amerika, 12 auf England, 8 auf Italien, 5 auf Russland, 3 auf Spanien, 2 auf Belgien und jeweils ein Todesopfer auf die Niederlande, Luxemburg, Bulgarien, Brasilien, Peru und die Schweiz. Vgl. Aeolus: „Die Todesopfer des Flugsportes und ihre Lehren, in: Automobilwelt-Flugwelt vom 8.12.1911, S.3.

172 Ebd., S.22f.

173 Vgl. Kehrt, „Das Fliegen ist immer noch ein gefährliches Spiel“, S.202 sowie Rudolf Schmetzke, Luftfahrt und Versicherung. Eine historische Betrachtung über die Entstehung der deutschen Luftfahrtversicherung, Berlin 1954, bes. S.160–175.

der mentalen und körperlichen Belastung der Akteure die Entlastung der Piloten durch ein „automatisch-stabiles“[174] Flugzeug eine zentrale Forderung der Vorkriegszeit, besonders des Militärs. Medial hingegen wurden die Piloten zu Beherrschern der Technik und zu geschickten Akrobaten stilisiert, die das Publikum auf Flugschauen anzogen.[175] Anthony Fokker zufolge sei sogar ein Anstieg der Besucherzahlen nach schweren Abstürzen zu verzeichnen gewesen: „Wir Flieger wussten das, nahmen es hin, schlugen sogar Kapital daraus – aber trotzdem ... durfte man uns nicht für Gladiatoren halten“.[176]

Die Flugpioniere hatten Respekt vor der Maschine und ihnen war bewusst, dass sie einen risikoreichen Sport und Beruf ausübten. Hellmuth Hirth gab in seiner Autobiografie zu, dass er aufgrund seiner technischen Kenntnisse dem Flugzeug „eine beinahe ängstliche Vorsicht entgegen“ gebracht habe: „Die vielen Drähte, ... der ganze Aufbau aus Holz, die Verbindungen durch die Schrauben, das alles erschien mir als Maschinenbauer anfangs etwas unsicher“.[177] Diese Vorsicht, so Hirth weiter, habe ihn vor Unfällen bewahrt. Josef Suwelack, obwohl durchaus couragiert, war ebenfalls nicht bereit, um jeden Preis zu fliegen. In einem Feldpostbrief an seine Eltern schrieb er im August 1915: „Dieses Fokker Kampfflugzeug fliege ich nicht. Es sind mir zu viele Unfälle in letzter Zeit passiert damit“.[178]

Der öffentlich und von den Flugzeugfirmen propagierten Betriebssicherheit der Flugzeuge standen somit Defekte, Fliegerunfälle und Todesstürze gegenüber. Die neue Technik unterlag eben nicht nur, wie in Zeitschriften und Memoiren teilweise behauptet,[179] der Kontrolle und dem Können des Flugzeugführers, sondern war eine hoch komplexe und kontingente Erfahrung, mit der sich der Pilot jedes Mal aufs Neue auseinandersetzen musste.[180] Fokker meinte in seiner Auto-

174 Dies war die damals geläufige Formulierung für ein Flugzeug, das ohne große Einwirkung durch den Piloten eine stabile Flugbahn beibehielt. Siehe z. B. „Der Dunne-Doppeldecker. Stabilität bei schönem Wetter. – Ein Fortschritt auf dem Wege zur automatischen Stabilität?“, in: Allgemeine Automobil-Zeitung (1913), Nr. 34, S. 16ff., oder „Der automatisch-stabile Fokker-Eindecker“, in: Flugsport (1912), Nr. 3, S. 126.

175 Xaver Hafer, Die Anfa?nge der Stabilita?tsbetrachtung in der Flugmechanik, in: Zeitschrift für Flugwissenschaft und Weltraumforschung 12 (1988), S. 37–44 sowie Christian Kehrt, Chauffeur oder Airman? Pilotenbilder und Technikerfahrung 1908–1914, in: thema/forschung 3 (2008), S. 52–55.

176 Fokker/Gould, Der fliegende Holländer, S. 96. Ernst Heinkel äußerte sich in ähnlicher Form: „Bei den Zuschauern in Johannisthal wusste man noch tausendmal weniger als bei denjenigen auf dem Cannstatter Wasen, was sie eigentlich lieber sahen, den gelungenen Flug oder einen möglichst tödlichen Absturz. Jedenfalls zogen Todesstürze unzählige Menschen herbei, die dann kaum zu bändigen waren. Sie wollten die Toten sehen“; Jürgen Thorwald (Hg.), Ernst Heinkel, Stürmisches Leben, Stuttgart 1953, S. 33.

177 Hirth, Meine Flugerlebnisse, S. 27.

178 NL J. Suwelack, Feldpostbrief an seine Eltern und Geschwister vom 14.8.1915.

179 Hirth meinte, Fliegen sei „keine Kunst“ und verlange lediglich anfangs etwas Überwindung, auf sich gestellt zu sein. Vgl. Hirth, Meine Flugerlebnisse, S. 29. Diese Meinung vertrat auch Erblich, Fliegerschule, S. 106.

180 Vgl. Kehrt, Moderne Krieger, S. 24.

biografie, Abstürze hätten „fast zum täglichen Brot" gehört und er war sich sicher: „Dass ich heute noch lebe, kann ich nur einem ganz unverschämten Glück zuschreiben, denn bei jedem Unfall hat man von einem bestimmten Augenblick an die Beherrschung der Ereignisse nicht mehr in der Hand".[181] Vor diesem Hintergrund erscheint Josef Suwelacks fehlgeschlagene Flugprüfung beinahe alltäglich.

Anders als Suwelack brachen viele Flugschüler ihre Ausbildung letztlich auch wieder ab, denn immerhin war der Pilotenschein kostspielig, die Anstellungschancen nur mäßig und die Fliegerei körperlich und mental anstrengend.[182]

Eine dauerhafte Anstellung als Berufspilot oder Flugzeugkonstrukteur hing von mehreren Faktoren ab. Erstens musste man über ausreichend finanzielles und mentales Kapital verfügen, um „mithalten" zu können, denn ein ausschweifender Lebensstil und sicheres, selbstbewusstes Auftreten entschieden darüber, ob man in den exklusiven Fliegerkreis aufgenommen oder von diesem zurückgewiesen wurde. Zweitens war auch ein gewisses Durchhaltevermögen vonnöten. Erfolge auf Wettbewerben oder Beteiligungen an Einnahmen aus Verkäufen von Flugzeugen gaben hierbei den Ansporn, nicht aufzugeben und Rückschläge in Kauf zu nehmen. Durch ein gestaffeltes Preissystem wurden auch kleine Erfolge belohnt und schufen Anreize, sich an den zum Teil hoch dotierten, von Privatleuten, Unternehmen oder staatlich subventionierten Wettbewerben zu beteiligen. Diese boten somit die Möglichkeit, sich als Amateur gleichermaßen finanzielles wie auch symbolisches Kapital zu sichern.[183] Schließlich hielt der Fliegerberuf Chancen bereit, gesellschaftlich aufzusteigen und vor allem über Kontakte zum Militär Einfluss und Ansehen zu gewinnen. Die genannten Faktoren lassen sich im Folgenden exemplarisch an Josef Suwelacks Tätigkeit bei den Rumpler- und den Kondor-Werken nachvollziehen. Der Westfale hätte sich ohne Geldgewinne aus Flugwettbewerben und die Unterstützung durch seinen Vater seinen luxuriösen Lebensstil nicht leisten können. Seine Entscheidungen die Zukunft betreffend lassen darauf schließen, dass er eigenständig agieren und möglichst hohe Erträge aus Preisgeldern erwirtschaften wollte, ohne in einem abhängigen Arbeitnehmerverhältnis zu stehen.

Suwelack setzte im Juli seine Ausbildung bei den Rumpler-Werken fort, ließ sich bei dem dort tätigen Fluglehrer Hellmuth Hirth auf der „Rumpler-Taube" im Fliegen unterweisen und schloss einen Vertrag mit Edmund Rumpler[184] (1872–

181 Fokker/Gould, Der fliegende Holländer, S.100 (dort beide Zitate).

182 Vgl. Kranzhoff, Edmund Rumpler, S.62.

183 Vgl. Möser, Fahren und Fliegen, S.105.

184 Edmund Rumpler (1872–1940) war Jude und stammte aus Wien. Nach einem Maschinenbau- und Elektrotechnikstudium von 1890 bis 1895 an der TH Wien absolvierte er Praktika in Fabriken für Waggon- und Dampfmaschinenbau. 1896 erhielt er eine Stelle als Konstrukteur bei der Wagenbau-Fabriks-Gesellschaft in Nesselsdorf. Nach Tätigkeiten bei weiteren Motorenfabriken machte Rumpler sich selbstständig und gründete 1906 ein Konstruktionsbüro, dem er im selben

1940) ab, der am 2. August mit der dieses Mal erfolgreich bestandenen Pilotenprüfung in Kraft trat.[185] In einem Schreiben vom 3. August 1911 gratulierte Rumpler Josef Suwelack zur bestandenen Prüfung:

> „Wir beglückwünschen Sie herzlichst zu der gestern in tadelloser Weise ausgeführten, erfolgreich bestandenen Pilotenprüfung. Durch die erfolgreich abgelegte Pilotenprüfung tritt Ihr Anstellungsvertrag, wie vereinbart, in Kraft ... . Als kleines Zeichen unserer Anerkennung Ihres Verhaltens und Ihrer bisherigen erfolgreichen Tätigkeit in unserer Firma werden wir Ihnen als Extragratifikation auch Ihren [sic!] Gehalt für den Monat Juli zuwenden."[186]

Auch Hellmuth Hirth gratulierte Josefs Vater zu seinem erfolgreichen Sohn: „Zur glänzend bestandenen Pilotenprüfung Ihres Herrn Sohnes gratuliert herzlichst Hellmuth Hirth. Freund von Josef."[187] Am 30. August wurde Josef Suwelack dann die Fluglizenz (Nr. 102) ausgestellt.[188] Seine bestandene Prüfung feierten die Johannisthaler Flieger mit einem Fackelzug, um den Neuen in ihre Reihen aufzunehmen (Abb. 6).

Das Foto verrät einiges über den Habitus der Piloten, denn auf dem Bild inszenieren sie sich betont lässig und fortschrittsorientiert. Albert Colombo und Hellmuth Hirth sieht man Zigarette rauchend, Oberleutnant Reep steht breitbeinig mit den Händen in den Hosentaschen, Josef Suwelack „hängt" entspannt und lächelnd mit Schlapphut im Handwagen und Siegfried Hoffmann tritt mit Fliegermütze und hochgeschobener Fliegerbrille auf. Das Motorrad, auf dem drei der Männer sitzen, unterstreicht zusätzlich ihre Technikbegeisterung und den Hang zur Geschwindigkeit. Johannisthal wurde zu einem Treffpunkt für technik- und fluginteressierte junge Männer, die die öffentliche Aufmerksamkeit suchten, Kontakte knüpften und hofften, als Flieger reich und berühmt zu werden. Sie waren Stellvertreter einer neuen Generation der Moderne, die für den Augenblick lebte und eine turbulente und beschleunigte Zeit für sich nutzbar machte.

Jahr eine Gesellschaft zum Bau von Luftfahrzeugen angliederte. Seit 1900 war er ferner Redakteur der Zeitschrift „Der Motorwagen" und seit 1901 Herausgeber des „Automobiltechnischen Kalenders". Im November 1908 erfolgte der Eintrag der Firma Edmund Rumpler Luftfahrzeugbau als erstes deutsches Flugzeugunternehmen in das Berliner Handelsregister. 1910 erwarb er die Lizenz für den Bau und Vertrieb der Taube von Igo Etrich. Nach Streitigkeiten mit Etrich – Rumpler hatte die Lizenzzahlungen eingestellt – entfielen im Februar 1913 die Lizenzvereinbarungen, woraufhin die Taube von anderen Firmen lizenzfrei nachgebaut werden konnte und auch wurde. Rumplers zahlreiche Erfindungen, die er sich häufig patentieren ließ, hatten Einfluss auf die Flugzeugentwicklung bis in die Zeit des Ersten Weltkrieges. Er zählte zu den ersten, denen den Sprung von der handwerklichen Einzelfertigung zur rationellen Serienfertigung gelang. Siehe Hans Holzer, Rumpler, Edmund, in: Neue Deutsche Biographie 22 (2005), S. 256–257, online unter: http://www.deutsche-biographie.de/ppn117599026.html (20.5.2014).

185 Vgl. Flugsport (1911), Nr. 17, S. 612.

186 NL J. Suwelack, Schreiben der E. Rumpler Luftfahrzeugbau G.m.b.H an Josef Suwelack vom 3.8.1911.

187 NL. J. Suwelack, Postkarte Hellmuth Hirths an Joseph Suwelack senior, o.D. [wahrscheinlich Anfang August 1911].

188 Siehe Flugsport (1911), Nr. 19, S. 687.

Abb. 6: Fackelzug nach Suwelacks Pilotenprüfung am 2. August 1911 in Johannisthal: vorne v.l. Josef Suwelack (im Wagen), Hans Vollmöller, Albert Colombo, auf dem Motorrad: Benno König, Siegfried Hoffmann und Albert Rupp, hintere Reihe v.l. Heinrich Oelerich, Karl Krieger, Edmund Bohlig, Oberleutnant Reep und Hellmuth Hirth (Foto: Familienarchiv Walter Suwelack, NL Josef Suwelack)

2.2. Exkurs: Die Rumpler-Taube als Inbegriff des Flugzeuges der Vorkriegszeit

Hinter dem Namen „Taube“ verbarg sich das wohl bekannteste deutsche Flugzeug der Vorkriegszeit, das in zeitgenössischen Fachzeitschriften für seine guten Flugeigenschaften gelobt und durch die Erfolge des Piloten Hellmuth Hirth äußerst populär wurde: „Der Apparat zeichnet sich besonders durch seine beinahe automatische Stabilität, seine außerordentlich gefällige, vogelähnliche Form und seine leichte Bedienung aus“,[189] hob 1911 der „Flugsport“ die Rumpler-Taube positiv hervor. Und Jörg Armin Kranzhoff urteilte: „Niemals zuvor oder nachher hat ein speziell technisches Gebilde eine solche Volkstümlichkeit in Deutschland genossen.“[190] Noch nach dem Ersten Weltkrieg wurde die Meinung vertreten, dass dieses Flugzeug „den Ruhm deutschen Erfindergeistes ... weit über die Grenzen trug“.[191] Zwischen 1911 und 1914, aber auch später noch wurde der Tauben-Typ

189 Flugsport (1911), Nr. 25, S. 880.
190 Kranzhoff, Edmund Rumpler, S. 362.
191 Illustrierte Motor-Zeitung 5 (1919), S. 68, zit. nach Kranzhoff, Edmund Rumpler, S. 362.

von zahlreichen Flugzeugfirmen nachgebaut und auch Josef Suwelack orientierte sich bei seinen späteren Flugzeugkonstruktionen am Modell „Taube".[192]

Der Erfolg der Maschine war kein Zufallsprodukt, denn Edmund Rumpler, dem mit dem Patent an der Etrich-Taube[193] 1910 der Durchbruch als Flugzeugbauer gelang, ließ durch groß angelegte Werbekampagnen keine Zweifel daran aufkommen, dass die Rumpler-Taube an erster Stelle auf dem Flugzeugmarkt stehe.[194] Geworben wurde zum einen mit der Stabilität und Zuverlässigkeit des Fabrikats, zum anderen mit auf Wettbewerben erzielten Preisen und erreichten Rekorden (Abb. 7). Seit 1910 konnte mit der Serienproduktion von Etrich-Rumpler-Tauben die französische Doppeldecker-Hegemonie in Deutschland erstmals überwunden werden und Rumplers Firma entwickelte sich zu einem „Flugzeugbau-Imperium".[195]

Abb. 7: Werbung Rumplers für die mit der „Taube" gewonnenen Preise (Foto: Archiv-DTMB Ssg. III Firmenschriften Rumpler)

192 Vgl. ebd., S. 137–139 sowie Akten zur Gründung der Kondor-Werke 1912. Schätzungen zufolge hätten über 50 Firmen der Taube ähnliche Flugzeuge in circa 137 Grundbauarten konstruiert. Vgl. John A. de Vries, Taube, Dove of War. Historical aviation album publication. o. O. 1974, S. IV.

193 Der Österreichische Flugpionier und Fabrikant Igo (Ignaz) Etrich (1879–1967) hatte einen Eindecker nach dem Vorbild der Natur (Zanoniasamen) entwickelt, der eine bislang unerreichte Flugstabilität besaß.

194 Vgl. Kranzhoff, Edmund Rumpler, S. 9. Rumpler holte beispielsweise Beurteilungen zur Rumpler-Taube in Fachkreisen, etwa beim „Luftschiffahrtverein Münster. Für Münster und das Münsterland", ein: „Wir bestätigen Ihnen gern, daß Ihr Pilot, Herr Jos. Suwelack bei dem von uns veranstalteten Schaufliegen am 24. Oktober [1911] bei einem vom Observatorium Aachen gemeldeten 19 Sekunden-Meter Wind aufgestiegen ist und einen Flug von 65 Minuten Dauer ausgeführt hat. ... Wir bestätigen Ihnen ferner gern, daß die ‚Taube' elegante Kurven flog, allen Weisungen der Steuer sicher folgte und am Schluß sehr glatt landete"; Archiv-DTMB III. Ssg. 2 Firmenschriften: E. Rumpler Luftfahrzeugbau G.m.b.H. Lichtenberg Berlin, Johannistahl 1912, S. 32.

195 Vgl. Kranzhoff, Edmund Rumpler, S. 9 und Schmitt, Oldtimer, S. 82–84, Zitat S. 84.

Im öffentlichen Raum hielt die Taube Einzug in Lieder, konnte von Modellbauern nachgebaut werden und war ein beliebtes Motiv auf Alltagsgegenständen, wie z. B. auf Zigarettenpackungen der Firma Manoli.[196] Kranzhoff zufolge war ganz Deutschland in ein „Taubenfieber" verfallen.[197]

Beim Militär galt der Tauben-Typ, nachdem der Nutzen des Flugzeuges gegenüber dem lange präferierten Zeppelin-System erkannt worden war, als perfekte Verkörperung eines deutschen Militärflugzeuges.[198] Die Vorbehalte der Heeresverwaltung gegenüber Flugzeugen waren in den ersten Jahren durchaus berechtigt gewesen, denn die Feststellung, es handele sich bei den Flügen internationaler Flugpioniere nur um ein „Akrobaten-Kunststück"[199] ohne militärischen Nutzen, traf vielfach zu. Noch im Jahr 1910, nach der erfolgreichen Kanalüberquerung Louis Blériots und einigen abgehaltenen Flugwochen, war selbst der spätere Leiter der Deutschen Versuchsanstalt für Luftfahrt, Professor Bendemann, der Auffassung, dass das Flugzeug zwar als Transport- und Verkehrsmittel zukunftsfähig sei, gegenüber dem Zeppelin jedoch keine militärische Relevanz besäße.[200] Im Unterschied zu Frankreich, das die Flugzeugentwicklung durch Preise und Prämien auf Flugwettbewerben sowie durch Zuschüsse für Flugzeugfirmen förderte, wurde die gezielte staatliche Subventionierung des Flugwesens „schwerer als Luft" in Deutschland in den ersten Jahren vernachlässigt. Edmund Rumpler nannte in einem Vortrag am 24. Oktober 1911 im Kaiserlichen Aero-Club mangelnde finanzielle Zuschüsse als Grund für die geringere Anzahl an Flugzeugen gegenüber dem Nachbarn Frankreich. Dennoch hob er die Qualität der deutschen Erzeugnisse hervor und war der Meinung, sie stünden den französischen Modellen in nichts nach, überträfen sie sogar mitunter.[201]

Erst allmählich reifte in Militärkreisen die Einsicht, dass das deutsche Flugwesen gegenüber Frankreich in Rüstungsfragen stark im Rückstand sei, woraufhin an konkreten Bewertungskriterien für Militärflugzeuge gearbeitet und die Flugzeugtechnik stärker gefördert wurde.[202] Ziele waren z. B. die Tragfähigkeit für zwei Personen, technische Voraussetzungen für Start und Landung in jedem freien Gelände, eine Flugweite von mindestens 50 Kilometern, eine stabile und sichere Fluglage auch bei stärkerer Luftbewegung sowie eine sichere Landung.[203]

196 Vgl. Kranzhoff, Edmund Rumpler, S.362.
197 Vgl. ebd., S.9.
198 Vgl. Kehrt, Moderne Krieger, S.125.
199 Militärgeschichtliches Forschungsamt, Die Militärluftfahrt bis zum Beginn des Ersten Weltkrieges 1914: Anlagenband, Frankfurt a. M. 1966, Nr. 59, S.114.
200 Vgl. Friedrich Bendemann, Der heutige Stand und die Zukunftsaussichten der Fliegekunst, in: Flugsport (1910), Nr. 2, S.55–58, bes. S.58.
201 Vgl. „Die erfolgreichsten Flugzeuge des Jahres 1911", in: Flugsport (1911), Nr. 23, S.836–837.
202 Vgl. Kehrt, Moderne Krieger, S.60.
203 „Abnahmebedingungen für Flugapparate (Preußisches Kriegsministerium August 1907)", in: Fliegerrevue (2013), Nr. 39, S.30.

Obwohl die leichten und wendigen Eindecker bei Flugwettbewerben von den Piloten genutzt wurden, um schnelle Flugzeiten zu erreichen und dem Publikum Aufsehen erregende Flugshows zu bieten, war aus militärischer Sicht nicht die Wendigkeit der Maschinen, sondern vielmehr ihre Stabilität und Zuverlässigkeit, also die Betriebssicherheit für einen Kauf ausschlaggebend, denn die Flugzeuge sollten in erster Linie zu Aufklärungszwecken zum Einsatz kommen. Die Inspektion des Luft- und Kraftfahrwesens hatte 1912 diesbezüglich jedoch Bedenken, auch hinsichtlich des sportlichen Charakters der Fliegerei. Die öffentlichen Flugveranstaltungen hätten „ungesunde Züge" angenommen und der Wert der Fliegerei entspräche „nicht annährend [sic!] den hierfür aufzuwendenden großen Mitteln".[204] Anstelle der Flugzeuge sollten vielmehr Konstruktionen prämiert und staatlich gefördert werden, „die einen wesentlichen Fortschritt darstellen und dem *praktischen* Bedürfnis entsprechen".[205] Dennoch: Die Orientierung der Flugzeugfirmen an den Vorgaben der Heeresverwaltung lässt sich an Werbeplakaten nachvollziehen, auf denen etwa mit erfolgreichen Flügen, erfolgten Lieferungen an das Militär oder mit der Zuverlässigkeit des Motors oder Propellers geworben wurde.[206]

Da ein privater Absatzmarkt für Flugzeuge noch nicht bestand beziehungsweise nur wenige reiche Privatleute bei Flugzeugfirmen einen Apparat erstanden, wurde das Militär fast zum einzigen Abnehmer. Im Jahr 1912 arbeiteten 90 Prozent der Flugzeugindustrie im Dienst des Heeres.[207] Das Kriterium der Betriebssicherheit zwang die Flugzeugfirmen, sich ganz nach den Vorgaben und Wünschen der Heeresverwaltung zu richten: „Lacing a private market for airplanes, the industry concentrated on producing relatively large, slow, and stable airplanes to meet the military requirements".[208]

Die Konstruktionsmerkmale der Taube waren Vorbild für viele weitere Flugzeugbauer in den Vorkriegsjahren und trugen der Etrich-Konstruktion viel Lob ein: „Der Etrich-Flieger ist somit zur Zeit das formvollendetste und stabilste Flugzeug und so unbedingt sämtlichen anderen Systemen überlegen."[209] Doch bedeutete die Eigenstabilität der Rumpler-Taube zwar einerseits eine Entlastung des Flugzeugführers, der während des Fluges etwa für die Regulierung des Motors zuständig war und multitaskingfähig sein musste,[210] so besaß das Modell anderer-

204 BA-MA PH 9-XII/5, Schreiben der Inspektion des Luft- und Kraftfahrwesens vom 27.5.1912 betr. der Denkschrift des Deutschen Luftfahrer Verbandes von April 1912.

205 Ebd., Hervorhebung im Original.

206 Siehe dazu den Abschnitt „Werbung und Profit".

207 Vgl. Helmut Reinicke, Deutsche Auffahrten. Mentalitätsformen der Technikbegeisterung in der Entwicklung der Aviation, in: Wechselwirkung 54 (1992), S.22–28, hier S.27.

208 John Morrow, Building German Air Power. 1909–1914, Knoxville 1976, S.62.

209 „Über die Stabilisierung der Flugmaschine", in: Zeitschrift für Luftschiffahrt (1911), Nr.22, S.22.

210 Vgl. Möser, Fahren und Fliegen, S.132.

seits auch Nachteile, wie hohes Eigengewicht und geringe Transportlast. Von militärischer Seite wurde bemängelt, dass nennenswerte Verbesserungen in der Belastbarkeit seit 1911 ausgeblieben seien, was die Prüfungskommission von Serienflugzeugen zur Kritik veranlasste, die Flugzeugfirmen würden aus Profitgier nicht in die Qualität der Flugzeuge investieren: „Bei einer Anzahl von Firmen wird zweifellos das Geschäftsinteresse höher gestellt, als die notwendige Rücksicht auf die Sicherheit der Flieger."[211]

Gleichwohl etablierte sich die Taube in den Militärfliegerschulen und wurde zu Ausbildungszwecken verwendet. Ihr hoher Stellenwert zeigte sich unter anderem darin, dass sie als Symbol in das Flugzeugführerabzeichen aufgenommen wurde. Ferner äußerte man sich in der Schlussbemerkung zum Prinz-Heinrich-Flug[212] 1913 dahingehend, dass es in Deutschland keine wirkliche Alternative zur Taube gebe, die die geforderten Bedingungen, speziell das Kriterium der Betriebssicherheit, erfüllen würde:

> „Diese Bedingungen auf der einen Seite, das große Gewicht der Motoren auf der anderen, schränken die Konstruktionsmöglichkeiten derart ein, daß es bald bei uns keine anderen Flugzeuge mehr geben wird, als Tauben und Doppeldecker vom L.V.G.-Typ od. ähnl. Schon jetzt gehört die Hälfte der Apparate der Taubenart an."[213]

Am Tauben-Diskurs lässt sich das Dilemma der Flugzeugentwicklung und -prüfung aufzeigen, denn während zwischen 1907 und 1914 die Heeresverwaltung die Anforderungen entsprechend der fortschreitenden Entwicklung der Flugzeugtechnik modifizierte und „neue Vorschriften und Anregungen für die Flugzeugfabriken"[214] formulierte, sparte sie Details zu Konstruktionsfragen aus und war weiterhin auf den Vergleich unterschiedlicher Flugzeugmodelle in der Praxis, auf die Erfahrungen der Flieger und auf Unterredungen mit den Piloten und Konstrukteuren angewiesen. Die wechselseitige Abhängigkeit zwischen der Heeresverwaltung und den Flugzeugfabriken gestaltete sich jedoch vor allem für kleine bis mittlere Unternehmen als problematisch, da die Gesellschafter kontinuierlich in die Flugzeugverbesserung investieren mussten, um den wachsenden Ansprüchen gerecht zu werden. Eine Gewähr für die Abnahme von Apparaten durch das Militär bestand dabei zum Teil nicht.[215] Stattdessen hafteten die Flugzeugfirmen mit

211 BA-MA PH 24/165, Prüfung von Serienflugzeugen und Zubehör: Kommissionsverhandlung, Unfall Eckenbrecher, Döberitz 21.9. bis 9.10.1913.

212 Siehe den Abschnitt zum Prinz-Heinrich-Flug in Kapitel 3. Hierbei handelte es sich um militärische Flugwettbewerbe, bei denen „Aufklärungsflüge" absolviert werden mussten (Kameras testen, Karten zeichnen, den Feind „bombardieren" etc.). Diese Wettbewerbe waren gezielte Vorkriegsübungen mit hohen Preisgeldern.

213 Denkschrift über den Prinz Heinrich Flug 1913 <Dritten deutschen Zuverlässigkeitsflug am Oberrhein> Im Auftrage der Südwestgruppe herausgeben von Oberstleutnant b. St. d. Inf.-Regts. Nr. 105 Frhr. v, Oldershausen, Straßburg 1914, S.63.

214 BA-MA PH 24/165, Prüfung von Serienflugzeugen und Zubehör: Kommissionsverhandlung, Unfall Eckenbrecher, Döberitz 21.9. bis 9.10.1913.

215 Siehe dazu ausführlich das Kapitel zu den Kondor-Werken.

ihrem Betriebskapital für Flugzeugbrüche und mussten für hohe Materialqualität garantieren. Die Prüfungskommission des Heereswesens nahm sich hierbei unter anderem das Recht heraus, den Motor eingehend im Betriebszustand zu prüfen und ihn nach erfolgtem Flug auseinanderzunehmen. Darüber hinaus hieß es in den „Besondere[n] Bedingungen“ zur Flugzeugabnahme von 1913 zu Haftungsfragen:

> „Schäden, welche (innerhalb der ersten 8 Wochen) nach der Vorführung nachweislich durch minderwertiges Material, nachlässige Arbeit oder Fehler an der Konstruktion verursacht sind, erlegen der liefernden Firma die Verpflichtung zum kostenlosen Schadenersatz auf.“[216]

Zwar übernahm die Heeresverwaltung die Kosten für die Offiziersausbildung zum Flugzeugführer, doch waren die an die Firmen gestellten Anforderungen äußerst hoch, so dass die Flugzeugfabriken bis zuletzt das volle finanzielle Risiko zu tragen hatten und lange im Ungewissen blieben.

Nur durch Erfolge, die als relevant für kriegerische Zwecke bewertet wurden, hatten Flugzeugfirmen eine Chance, zu Lieferanten der Heeresverwaltung zu werden, weshalb Flugwettbewerbe spätestens seit 1911 – mit der Ausschreibung zum Ersten deutschen Zuverlässigkeitsflug am Oberrhein – nicht mehr rein sportlich zu bewerten sind, sondern zu militärischen Sichtungen für Maschinen und potenzielle Militärflieger wurden. Für die Flieger wiederum waren Wettbewerbe „technische Experimentierfelder, auf denen im sportlichen und spielerischen Umgang mit der Technik die neuesten Flugzeugtypen unter Praxisbedingungen getestet wurden“.[217] Außerdem konnten sich die Piloten und ihre Maschinen auf Flugwettbewerben einen Namen machen, was sich positiv auf den Ruf der Flugzeugfirma oder Fliegerschule auswirkte, die aus Erfolgen und Siegen Kapital schlagen und mit Werbung zusätzlich ihre Maschinen vermarkten konnte. Neben der „Öffentlichkeitsarbeit“ und den Marketingkampagnen war das Leben als Fluglehrer und Pilot in erster Linie strukturiert durch die Instandhaltung und Erprobung der Maschinen auf dem Flugplatz sowie durch die Ausbildung von Flugschülern. Es gilt daher zwischen einer internen und einer öffentlichen Seite der Fliegerei zu differenzieren, zwischen Arbeits- und Wettbewerbsfliegerei.

2.3. Flugplatzroutinen und Ausbildungsbetrieb I: Rumpler-Werke

Edmund Rumpler, der sich seit den erfolgreichen Flügen der Brüder Wright 1908 mit dem Flugzeugbau beschäftigte, gründete am 10. November 1908 die Firma

216 BA-MA PH 9-XV/2, Ausarbeitung einer Fliegervorschrift: „Besondere Bedingungen“ 1913 (Flugzeugabnahme).

217 Kehrt, Das Fliegen ist immer noch ein gefährliches Spiel, S.206.

Edmund Rumpler, Luftfahrzeugbau.[218] Die Gründung erfolge etwa zur gleichen Zeit, als konkrete Pläne zur Errichtung des Flugplatzes Johannisthal-Adlershof,[219] des ersten Motorflugplatzes Deutschlands, in die Tat umgesetzt wurden. Mit schon zu Beginn seiner unternehmerischen Tätigkeit groß angelegten Werbeversprechen[220] zog Rumpler so viele interessierte Flieger, Konstrukteure und Auftraggeber an, dass die Werkstatträume schon bald zu klein wurden und er Gesellschaftsform und Namen der Firma in Rumpler-Luftfahrzeugbau GmbH änderte, welche über ein beachtliches Stammkapital von 90.000 Mark verfügte. Anfang des Jahres 1910 zog er dann mit einem Teil seiner Werkstatt von Lichtenberg nach Johannisthal auf den „neuen Startplatz“ und mietete dort einen Schuppen an.[221]

Der Flugplatz Berlin-Johannisthal, der am 26. September 1909 mit einer Flugveranstaltung offiziell eröffnet wurde, war ursprünglich ein 300 Hektar großes Waldgebiet zwischen den Gemarkungen Adlershof und Johannisthal. Auf Initiative Eduard von Pustaus, Arthur Müllers und Georg von Tschudis wurde im Dezember 1908 auf Befehl des preußischen Kriegsministeriums mit der Rodung des Waldstücks und der Ebnung der Fläche begonnen.[222]

Pustau war Kapitän zur See, Luftfahrtpublizist und Schriftführer des Vereins Deutscher Flugtechniker, Müller Holzbauunternehmer mit guten Kontakten zur preußischen Forstbehörde und von Tschudi ehemaliger Luftschiffer, Offizier und Ingenieur sowie 1908 Leiter der ersten deutschen „Internationalen Luftfahrtausstellung“ (ILA) mit Kontakten zu Militärkreisen. Es handelte sich also um Männer in den „richtigen“ Positionen und mit dem nötigen Einfluss für das Vorhaben.

Sahen sich die Investoren zu Beginn mit finanziellen Problemen konfrontiert, so erfolgte im Jahr 1910 nach einem erfolgreichen Sanierungskonzept[223] der Aus- und Umbau des Flugplatzes, um Johannisthal für Zuschauer noch interessanter zu machen und Planungsfehler zu beheben. So wurden z. B. die Zuschauertribünen verlegt, um für die Besucher den Weg vom Bahnhof bis zu ihren Plätzen zu verkürzen.

Dies und die Umbenennung in „Flugplatz Johannisthal“ hatte weitere praktische Gründe, denn die Bahntarife bis zu den umliegenden beiden Bahnstationen Niederschönweide-Johannishtal und Adlershof-Alt-Glienicke betrugen einmal 10

218 Die folgenden Ausführungen basieren auf Schmitt, Oldtimer, S.82–84.

219 So lautete der Name bis 1910, bevor der Platz in „Flugplatz Johannisthal“ umbenannt wurde. Siehe Schmitt, Oldtimer, S.31.

220 Rumplers Konstruktionen erwiesen sich als fluguntauglich, bis ihm mit dem Nachbau der Etrich-Konstruktion der Durchbruch gelang. Siehe Kranzhoff, Edmund Rumpler.

221 Vgl. im Folgenden ebd., S.16ff.

222 Vgl. Schmitt, Oldtimer, S.18.

223 Das Sanierungskonzept umfasste die Gewinnung von Aktionären, Verhandlungen mit den Geldgebern und die Aussetzung der Zinsen für einen Zeitraum von zehn Jahren. Kapitalspekulationen und die gelungene Überzeugungsarbeit der Flugplatzgründer machten es schließlich möglich, den Flugplatz lebensfähig zu machen.

und einmal 20 Pfennig. Dementsprechend wurde der Haupteingang in Richtung der Haltestelle Johannisthal verlegt, das angelegte Restaurant, die Post und eine große Tribüne wieder abgerissen und auf die Johannisthaler Seite konzentriert, somit besucherfreundlich umgestaltet. Am ehemaligen Startplatz, der künftig den Namen „alter Startplatz" trug, verblieben lediglich 17 Flugzeugschuppen, die kurz darauf um sieben erweitert wurden, eine Fliegerkantine sowie zwei Flugzeughallen. Auch am „neuen Startplatz" wurden Schuppen errichtet.

Während von Tschudis Amtszeit als Flugplatzdirektor betrugen die Mieten für einen Schuppen (15x15 m) zwischen 200 Mark und 400 Mark monatlich. Weitere Einnahmen kamen durch Eintrittsgelder zustande, gestaffelt nach Steh-, Sitz- und Logenplätzen. Die Preise bewegten sich zwischen 30 Pfennig und zwei Mark an Tagen ohne Flugveranstaltungen und 50 Pfennig bis 10 Mark an Flugtagen. Abonnements und Jahreskarten wurden ebenfalls angeboten, deren Preise zwischen 75 und 100 Mark lagen. Ferner verdienten die Flugplatzbetreiber an der Vermietung von Reklameflächen, an den vierteljährlich erscheinenden Flugplatzheften und am Verkauf von Rechten an Interviews, Fotos und Ähnlichem.[224] Seit 1910 erhielt der Flugplatz außerdem jährlich staatliche Subventionen des preußischen Kriegsministeriums in Höhe von 20.000 Mark.

Derartig gut ausgestattet siedelten sich im Jahr 1910 zwölf Unternehmer und Unternehmen auf dem „neuen Startplatz" an, darunter auch die Rumpler-Werke, oder siedelten vom alten über. 21 Flugzeugfirmen beziehungsweise selbstständige Konstrukteure ließen sich zwischen 1910/12 und 1914 auf dem „alten Startplatz" nieder.

Der Auf- und Ausbau des Flugplatzes Johannisthal zeigt exemplarisch, wie kostenintensiv die frühe Luftfahrt sich gestaltete und dass sie wesentlich von Unternehmern und Industriellen getragen wurde. Kontakte zu einflussreichen Persönlichkeiten in der Wirtschaft und zu den führenden Namen der Luftfahrtvereine waren wichtige Bedingungen, um ein Großprojekt wie den Flugplatzbau in Johannisthal zu finanzieren. Intensiv betriebene Werbekampagnen und eine großflächige Berichterstattung in den Zeitungen trugen zum Gelingen des Flugplatz-Projekts bei, das nicht zuletzt auch durch große Versprechen und Spekulationen angeschoben wurde. Als Publikumsmagnete sollten Flugwettbewerbe dienen. Doch auch jetzt, wenige Jahre nach den ersten Motorflügen der Wrights, steckte die Fliegerei gewissermaßen noch in den Kinderschuhen. Um den wenigen fähigen Fliegern dennoch einen Anreiz zu geben, das Risiko von Schäden und Verletzungen einzugehen und an Flugwettbewerben teilzunehmen, mussten immense Summen als Preisgelder in Aussicht gestellt werden. Nichtsdestotrotz faszinierten

224 Von Tschudi verkaufte etwa an den Johannisthaler Fotografen Franz Fischer die alleinigen Rechte zum Fotografieren auf dem Flugplatz. Siehe Schmitt, Oldtimer, S.36.

Luftschiffe und Flugzeuge die Menschen länderübergreifend, in Deutschland etablierte sich sogar eine neue Rubrik in den Tages- und Wochenzeitungen, die unter dem Titel „Von der Luftschiffahrt“ oder im Sportteil Wissenswertes aus der „Fliegerwelt“ berichteten. Mit dem Flugplatz Johannisthal wurde die Fliegerei massentauglich und der Platz zum Sammelbecken für Flieger und Flugzeugfabriken in Deutschland.

Ab Juli 1911 war Josef Suwelack zusammen mit den bereits angestellten Piloten Hellmuth Hirth und Hans Vollmöller bei Edmund Rumpler unter Vertrag und unter anderem für die Ausbildung von Flugschülern zuständig. Da Hirth und Vollmöller hauptsächlich mit Flugvorführungen auf den für die Heeresverwaltung bestimmten Flugzeugen beschäftigt waren, fiel Josef Suwelack fast der gesamte Ausbildungsbetrieb zu.[225] Im Oktober betreute er 18 Flugschüler, wovon zwölf die Ausbildung nicht beendeten, das heißt, sie vorzeitig abbrachen oder bei einem anderen Piloten oder einer anderen Flugschule fortsetzten.[226]

Der Flugbetrieb in Johannisthal war in der zweiten Jahreshälfte 1911 sehr rege und stieg im August auf über tausend Flüge an, wovon 383 Aufstiege mit Etrich-Rumpler-Tauben erfolgten. Suwelack konnte mit 152 Aufstiegen die meisten Flüge verzeichnen und führte mehr als die Hälfte in Begleitung von Flugschülern aus.[227] Hier ist allerdings hinzuzufügen, dass viele Flüge nur wenige Minuten dauerten, so dass sich ihre hohe Anzahl relativiert.

Die Ausbildungskosten beliefen sich auf 3.000 Mark, die der Flugschüler zu bezahlen hatte. Suwelacks Aufgabe als Fluglehrer war es, die Schüler einerseits im Gebrauch des Flugzeuges zu unterweisen, sie andererseits aber auch mit der Wartung und Instandhaltung der Maschine vor, während und nach Flügen vertraut zu machen.[228]

Die Schulmaschinen Rumplers verfügten über zwei Sitze und zwei Steuer, so dass der Flugschüler zunächst an der Seite des Ausbilders ein Gefühl für das Steig-, Flug- und Landeverhalten der Maschine entwickeln und schließlich selbst die Kontrolle übernehmen konnte. Auf Schulflüge mit dem Fluglehrer folgten nach und nach Alleinflüge, bis der ausbildende Pilot der Meinung war, der Schüler sei prüfungsbereit. Für die Schüler-Lehrer-Beziehung war es wichtig, dass

225 Vgl. Kranzhoff, Edmund Rumpler, S.62.

226 NL J. Suwelack, Verzeichnis der Flugschüler, datiert auf den 10.10.1911. Nachnamen der Flugschüler: Linnekogel, Boutard, Temmel, Bümmler, Spandow, Langer, Kauders, Albers, Schwarz, Aug. Max Müller, Rosenstein, Schiedeck, Ompteda, Fremery, Lübbe, Oblt. Geerdtz, Hösli, Bundesmann, Hätschel.

227 Vgl. Kranzhoff, Edmund Rumpler, S.62. Hirth absolvierte 9, Vollmöller 57, Marchal 67, Beese 31 und Bertram 13 Flüge. Angaben aus dem Hauptflugbuch des Flugplatzes Johannisthal, zit. nach Kranzhoff, Edmund Rumpler, S.381 Anm. 178.

228 Da Josef Suwelack bei Hirth in die Lehre gegangen war, ist es wahrscheinlich, dass er seine Flugschüler in ähnlicher Form wie Hellmuth Hirth unterwies. Dieser berichtete über seine Ausbildungsweise in seiner Autobiografie. Siehe Hirth, Meine Flugerlebnisse, S.31–40.

gegenseitiges Vertrauen bestand: vonseiten der Schüler auf die Erfahrung des Piloten, vonseiten des Lehrers in die Eignung des Pilotenanwärters. War Letzteres nicht der Fall, konnte sich der Lehrer gegebenenfalls weigern, die Ausbildung zu übernehmen.

Unter den damaligen Fliegern herrschte weitgehend Einigkeit darüber, dass die Fliegerei eine reine Männerdomäne sei und Frauen weder körperlich noch mental den Anforderungen gewachsen seien. Den „Weibern" gab man zudem die Schuld, wenn die Flieger ihre Nerven verloren oder körperliche Schwäche zeigten. Willy Hahn, selbst Flieger, meinte 1913 über die Johannisthaler Frauen:

> „Wie verrückt waren die auf die Flieger ... Ja, diese Weiber, sie waren an vielem, vielem Unglück schuld. Aus dem Lindenkasino, dem Palais de danse, dem alten Ballhaus, überall wurden sie hergeschleppt; ... aber sie saugten manchem jugendfrohen jungen Blut den Verstand aus dem Hirn und das Mark aus den Knochen. Scharfe, sichere Augen wurden schlaff und wenn es dann galt, wieder zu stehen draußen in Gefahr, im Kampfe mit den Elementen, dann versagten die Nerven."[229]

Dennoch: Neben den zahlreichen männlichen Fliegern bewarb sich 1911 auch eine Frau, Amelie Hedwig Beese[230] (1886–1925), genannt Melli, um den Pilotenschein. Im Herbst 1910 hatte sie ihre Flugausbildung bei Robert Thelen auf einem Wright-Doppeldecker begonnen, wechselte jedoch nach einem erlittenen Nasenbeinbruch im Sommer 1911 auf die Etrich-Rumpler-Taube. In seiner Autobiografie erklärte Hellmuth Hirth, Beeses neuer Fluglehrer, seine Skepsis: „Der Unterricht von Damen hat seine Schattenseiten." Gemeint waren hier wohl die vermeintlich schwächeren Nerven der Fliegerinnen, ihre körperliche Konstitution oder auch ihre angenommene fehlende Kenntnis von der Technik. Nachdem Rumpler Hirth aber eröffnet hatte, dass er bereits einen Ausbildungsvertrag mit Melli Beese abgeschlossen habe, habe Hirth sich gezwungen gesehen, sie zu unterrichten. Trotz seiner Vorbehalte wollte er sein Können und die Qualität der Maschine unter Beweis stellen: „Ich wollte ihr [Beese] beweisen, wie leicht das

229 Willy Hahn, Aus meinem Fliegerleben, in: Die Flieger. Jahrbuch des Bund Deutscher Flugzeugführer e.V. 1913, Berlin 1914, S.58–59. Diese Aussage ist vor dem Hintergrund zu sehen, dass der Fliegerdiskurs männlich konnotiert war und Frauen als Gefährdung des Flieger-Images (Härte, Nervenstärke, Ausdauer) angeführt wurden.

230 Beese stammte aus reichem Elternhaus in Dresden und besuchte zunächst die königliche Akademie der freien Künste in Stockholm, um Bildhauerin zu werden. Sie hatte ein eigenes Atelier, zuerst in München, dann in Dresden, sammelte aber auch Berichte über die frühe Fliegerei. Beese versuchte sich im Bau von Flugzeugen und ging im Herbst 1910 nach Johannisthal, um sich zur Fliegerin ausbilden zu lassen. Einige männliche Fluglehrer weigerten sich, sie auszubilden, doch schließlich erwarb sie 1912 (Prüfung noch 1911) auf einer Etrich-Rumpler-Taube als erste deutsche Frau das Fliegerzeugnis Nr. 115. Sie nahm an Wettbewerben teil, stellte einige (Welt-)Rekorde auf und gründete 1912 mit dem Franzosen Charles Boutard, den sie heiratete, eine eigene Flugschule in Johannisthal, später auch eine eigene Flugzeugfabrik. Aufgrund der Nationalität ihres Mannes wurde sie von der Teilnahme an der Nationalen Flugwoche ausgeschlossen und bei Kriegsausbruch zusammen mit ihrem Ehemann interniert. Nach dem Krieg versuchte sie vergeblich, an ihre Erfolge der Vorkriegszeit anzuknüpfen, und nahm sich schließlich 1925 in Berlin das Leben. Siehe zum Leben Melli Beeses: Livia Käthe Wittmann/Barbara Zibler, Melli Beese und die „Flügel am Horizont". Die Geschichte der ersten deutschen Pilotin, Berlin 2009.

Fliegen auf einer Taube sei", sogar für Frauen.[231] Sein Urteil über Beeses Fähigkeiten fiel dennoch vernichtend aus:

> „Nach vieler Mühe und Not, erschwert durch den Oppositionsgeist, der den Frauen angeboren sein muß, gelang es der Schülerin endlich, die Pilotenprüfung abzulegen. Doch wie in allen Berufen, die gelegentlich die physischen Kräfte eines Mannes erfordern, glaube ich nicht, daß auf den heutigen Flugzeugen die Frauen etwas Großes leisten werden."[232]

Hirths Meinung verdeutlicht, dass der Einzug des „schwachen" Geschlechts in die Fliegerei für ihn einer Unterwanderung gleichkam. Der frauenfeindliche Fliegerdiskurs war eng verknüpft mit militärischen Überlegungen, denn ausschließlich Männer konnten in den Dienst bei der Fliegertruppe und beim Militär im Allgemeinen eintreten.[233]

Nach Unstimmigkeiten zwischen Beese und Hirth führten Vollmöller und Suwelack ihre Ausbildung fort, allerdings ebenfalls widerwillig und bewusst verzögernd. In ihren Aufzeichnungen schildert Beese, wie ihre männlichen Kollegen, die sie nicht in ihren Reihen willkommen hießen, sondern sie als „Rivalin" betrachteten, ihr das Leben schwer machten: „Die anderen Piloten der Firma, vor allem wohl die jüngeren, Vollmöller und Suwelak [sic!], sahen also im mir ihren Rivalen ... Grund genug für sie, mich, wenn irgend möglich, an der Erreichung meines Zieles zu hindern".[234] Die Pilotenscheinanwärterin sah sich mit Sabotagen an ihrer Schulmaschine konfrontiert und führte verrußte Zündkerzen oder abgelassenes Benzin als Beispiele an. Diese hätten sie auch zu Notlandungen gezwungen.[235]

In den Augen der Piloten mischten sich weibliche Flieger wie Beese in eine „Männersache" ein und sie wurden als störend empfunden. Man(n) traute ihnen darüber hinaus nicht zu, den technischen Anforderungen gewachsen zu sein, da der Fliegerberuf auch handwerkliche Fähigkeiten erforderte und der Pilot über umfangreiche Motorenkenntnisse verfügen musste. Bei der Fliegertruppe verlangte man z. B. von Flugzeugführer und Beobachter, Reparaturen an beschädigten Flugzeugen wie „Schreiner-, ... Schlosser- und Tapezierer-Arbeiten ausführen und auch berechnen oder schätzen [zu] können, welche Beanspruchung des Materials an der beschädigten Stelle eintritt".[236]

231 Hirth, Meine Flugerlebnisse, S.57.

232 Ebd.

233 Vgl. Evelyn Zegenhagen, Zwischen Herd und Himmel, zwischen Kochbuch und Flugbuch. Frauen in der deutschen Luftfahrt, in: Helmuth Trischler/Kai-Uwe Schrogl (Hg.), Ein Jahrhundert im Flug. Luft- und Raumfahrtforschung in Deutschland 1907–2007, Frankfurt a. M./New York 2007, S.70–87, hier S.74.

234 Erinnerungen Melli Besses zit. nach Wittmann/Zibler, Melli Beese, S.183.

235 Vgl. ebd. Aufgrund anderer Berichte von Fliegerinnen, die in ähnlicher Weise von Diskriminierungen durch männliche Kollegen sprechen, ist anzunehmen, dass Beese sich in ihren Erinnerungen nicht bewusst als „Opfer" inszeniert.

236 BA-MA PH 9-XV/2, Ausarbeitung einer Fliegervorschrift: Feldmässige Wiederherstellung nach Beschädigung bei der Landung, Straßburg, 21.1.1913.

Beeses Diskriminierung war kein Einzelfall, auch anderen Fliegerinnen und Luftschifferinnen wurden bewusst Steine in den Weg gelegt und sogar neue Vorschriften wurden erfunden oder eigene Frauen-Flugwettbewerbe ausgeschrieben, um eine Teilnahme an den übrigen Konkurrenzen zu verhindern.[237] Die gezielte Ausgrenzung der Fliegerinnen und die Boykotthaltung männlicher Flieger verweist auf den Fliegerhabitus, der nach den Vorstellungen der Zeitgenossen mit Eigenschaften konnotiert war, über die angeblich ausschließlich Männer verfügten. Dazu zählten körperliche und mentale Belastbarkeit, handwerkliche Fertigkeiten, geschicktes Agieren in Gefahrensituationen und vor allem die Fähigkeit, in jeder Situation die Nerven zu behalten sowie professionell, ruhig und routiniert zu agieren.

Angesichts der Vorurteile und Vorbehalte gegenüber weiblichen Flugschülerinnen stellten Männer also den weit überwiegenden Anteil an Pilotenanwärtern. Nicht anders verhielt es sich auch bei Josef Suwelacks Lehrtätigkeit. Für die erfolgreiche Ausbildung der Flugschüler Heinrich Lübbe, der am 17. November 1911 die Fluglizenz erwarb,[238] und Willy Rosenstein (Fluglizenz am 14. März 1912)[239] erhielt Suwelack jeweils 100 Mark und zusätzlich 200 Mark Gehalt, wovon 50 Mark auf ein Kautionskonto flossen.[240] Des Weiteren übernahm Suwelack die Flugausbildung der Herren Bohlig und Basenach.[241] Ab Mitte des Jahres 1911 bildeten Josef Suwelack und Hellmuth Hirth dann auch die ersten Offiziere in Johannisthal aus.[242]

Neben der Schulung von Fliegern hatte Josef Suwelack für die Instandhaltung der Maschinen Sorge zu tragen, musste die Rumpler-Tauben auf Flugwettbewerben vorstellen sowie durch Passagierflüge für die Taube und die Rumpler-Werke werben (Abb. 8). Ferner gehörte es zu seinen Aufgaben und denen der beiden anderen Werkspiloten Hirth und Vollmöller, Konstruktionsänderungen an den Maschinen vorzunehmen oder neue Motoren in der Praxis zu testen.

237 Vgl. Zegenhagen, Herd und Himmel, S. 72–74 sowie Bettina Gundler, „Nie wird die Fliegerei als Beruf überfüllt sein von Frauen." Geschlechterverhältnisse und berufliche Perspektiven in den Anfängen der Luftfahrt, in: Metis 4 (1995), H. 1, S. 29–40, bes. S. 33f.

238 Vgl. Flugsport (1911), Nr. 25, S. 901.

239 Vgl. Flugsport (1912), Nr. 7, S. 280.

240 NL J. Suwelack, Gehaltsabrechnung Josef Suwelacks bei den Rumpler-Werken, 4.10.1911 und Kautionskonto Suwelacks für die Monate Oktober bis Dezember 1911.

241 NL J. Suwelack, Schreiben Rumplers an Suwelack vom 13.1.1912. Drei Tage später erhielt Suwelack postalisch 218,75 Mark „konform Ihrer Aufstellung". Dabei handelte es sich wahrscheinlich um die Ausbildungsprämie für besagte Flugschüler. Siehe Schreiben Rumplers an Suwelack vom 16.1.1912.

242 Vgl. Erinnerungen von Hans-Joachim Siber, in: Willy Hackenberger, Die Alten Adler, München 1960, S. 92.

Abb. 8: Fluglehrer Suwelack mit seinen Schülern v.l. Heinrich Lübbe, Hermann Fremery, Willy Rosenstein, Charles Boutard, Erwin Schwarz und Otto Linnekogel, Johannisthal, Herbst 1911 (Foto: Familienarchiv Walter Suwelack, NL Josef Suwelack)

Von einem Motoren-Testflug Suwelacks berichtete die „Illustrierte Zeitschrift für das gesamte Ballon- und Flugwesen" in ihrer Ausgabe vom 20. August 1911:

> „Der Etrich-Rumpler-Flieger Joseph Suvelak [sic!] hat auf dem Flugplatz Johannisthal eine neue Type des Aeolus-Flugzeugmotors, 65 PS.8 Zylinder, der in eine Rumpler-Taube eingebaut ist, ausprobiert. Der Versuch hatte einen vollen Erfolg. Der neue Motor arbeitete tadellos. Suvelak flog etwa 3/4 Stunde über Land in der Umgebung des Flugplatzes herum und erreichte eine Höhe von 1200 m. Sonntag hat der Flieger nicht weniger als 27 Aufstiege mit Passagieren ausgeführt, die sämtlich sehr gut gelangen."[243]

Edmund Rumpler arbeitete konsequent an einer Verbesserung der Rumpler-Taube. Bis zum Herbst 1911 waren drei Prototypen fertiggestellt, die wichtige Voraussetzungen für den ab 1912 produzierten „Einheitstyp" schufen.[244] War die Taube im August 1911 noch mit einem 50-PS-Aeolus-Motor ausgestattet, so suchte Rumpler nach leistungsstärkeren Antrieben und ließ dafür seine Piloten neue Motoren erproben. Nach erfolgreich verlaufenen Versuchen wurde das neu ausgerüstete, von Suwelack getestete Flugzeug von der Heeresverwaltung als Modell A

243 Luftschiffahrt, Flugtechnik und -Sport. Illustrierte Zeitschrift für das gesamte Ballon- und Flugwesen, redigiert von Gustav Riefenstahl, Verlag E. Gundlach A.G. Bielefeld, Nr. 23, 20. August 1911, S. 15.

244 Vgl. Kranzhoff, Edmund Rumpler, S. 94.

7/11 angekauft. Anfang des Jahres 1911 hatte jedoch nur ein einziger Auftrag der Heeresverwaltung vorgelegen, so dass Rumpler auf zivile Abnehmer, das heißt auf Privatpersonen, angewiesen war.[245] Bis August wurden insgesamt neun Etrich-Rumpler-Eindecker der Militärtypen A 1/11 bis A 7/11 und dann weitere vier Eindecker neuerer Bauart mit kufenlosem Fahrgestell der Typen A 8/11 und A 9/11 an die Heeresverwaltung geliefert. Letztgenannte Typen wurden im September nach Metz transportiert und von Hirth vorgeführt. Sie waren die ersten für den operativen Einsatz gedachten „Kriegsflugzeuge", welche zu einem Stückpreis von 9.701 Mark verkauft wurden.[246]

Von den zwischen 1910 und 1911 gefertigten 21 Rumpler-Flugzeugen wurden, der Bestandsliste der Heeresverwaltung zufolge, zehn Tauben an die Fliegertruppe geliefert, die Ende 1911 insgesamt über 25 Apparate verschiedener Firmen verfügte.[247]

Rumpler war auf die Erfahrungen seiner Werkspiloten bei Flugwettbewerben oder auf dem Johannisthaler Flugplatz angewiesen. Die Flieger sollten ihm einerseits von Problemen berichten, andererseits aber auch konstruktive Verbesserungsvorschläge unterbreiten. Gewünscht waren vor allem Vereinfachungen an der Konstruktion, um die Maschine sowohl effizienter als auch ihre Herstellung kostengünstiger zu machen. An einem Schreiben Rumplers an Josef Suwelack vom 4. Januar 1912 wird dies exemplarisch deutlich:

> „Wir [die E. Rumpler Luftfahrzeugbau G.m.b.H.] wenden uns ... mit diesem Schreiben an Sie[,] um Sie zu bitten, uns bis spätestens 10. d.M. Vorschläge zu machen, welche sich auf unsere Taube beziehen und welche konstruktive Verbesserungen zum Gegenstand haben, die sich auf Vereinfachung, Verbilligung und Verringerung des Stirnwiderstandes beziehen."[248]

Die Zielsetzung für das Jahr 1912 lautete, eine „verkehrstaugliche Touren- [oder] Zuverlässigkeitsmaschine" zu entwickeln. Mithilfe mathematischer Formeln wurden aus den Flugergebnissen des Jahres 1911 technisch-wissenschaftliche Daten gewonnen, um genaue Auskünfte über den Zusammenhang verschiedener Faktoren wie Motorleistung, Flügelspannweite etc. zu erhalten.[249]

245 Franz Oster und Wilhelm Witterstätter waren beispielsweise private Käufer von Rumpler-Flugzeugen. Siehe Vorreiter, Jahrbuch der Luftfahrt, S.496. Insgesamt gingen während der Existenz der Rumpler-Werke neun Etrich-Rumpler-Tauben in den Privatbesitz von Piloten und Flugschulen über. Vgl. Kranzhoff, Edmund Rumpler, S.105.

246 BA-MA PH 9-XX/68, Beschaffungsbeschluss für Tauben-Typen A 7/11 bis A 9/11 vom 28.7.1911.

247 Vgl. BA-MA PH 9-XX/68, „Nachweisung der sämtlich bisher beschafften Flugzeuge". Die Heeresverwaltung verfügte der Bestandsliste zufolge Ende 1911 über neun Albatros-, zwei Aviatik-, je einen Wright-, Euler-, einen L.V.G- und zwei Farman-Doppeldecker sowie über neun Etrich-Rumpler-Eindecker. Der zehnte Rumpler-Apparat, Eigentum einer Familie Fritzsche, ging Kranzhoff zufolge durch eine Schenkung in den Besitz der Marine über. Vgl. Kranzhoff, Edmund Rumpler, S.105.

248 NL J. Suwelack, Schreiben Edmund Rumplers an Josef Suwelack vom 4.1.1912.

249 Vgl. Friedrich Rau, Bewertungen von Flugergebnissen, in: Zeitschrift für Flugtechnik und Motorluftschiffahrt (1911), Nr. 24, S.305ff.

Basierend auf den Erfahrungen und Rückmeldungen erfolgte schließlich eine Verringerung der Flügelspannweite der Taube unter Beibehaltung der Flügeltiefe. An Veränderungen nahmen die Rumpler-Werke weiterhin den Einbau bruchsicherer Stahlseile, eine Reduzierung der Spanndrähte sowie Vereinfachungen an Spannschlössern und Anschlussverbindungen vor: „Unterhalb der Flügel quer zur Flugrichtung läuft ein verspannter Träger, welcher mit der Hauptfläche durch Druckstützen und Drähte verbunden ist." Durch diese Maßnahme werde „selbst bei Reißen der Spanndrähte das Hochklappen eines Flügels unmöglich".[250] Bis zum Herbst 1911 wurden 17 Etrich-Rumpler-Tauben mit sieben verschiedenen Motoren (50- bis 95-PS) gebaut, wobei Rumpler sich stetig von der ursprünglichen Etrich-Konstruktion entfernte.[251]

Die Innovationsförderung im Flugzeugbau „von unten" durch Piloten und Konstrukteure, die neue Ideen an Gesellschafter oder an die Inspekteure der Fliegertruppe herantrugen, kennzeichnet sowohl die Flugzeugentwicklung der Vor- als auch der Kriegszeit: „Die praxis- und nutzungsdominierte technische Fortentwicklung war zwischen 1914 und 1918 kein Produkt militärischer Forderungen oder Initiativen der Industrie, sondern wurde vorwiegend von den Piloten angestoßen."[252] Mösers Feststellung lässt sich mit Einschränkung auf die Vorkriegszeit ausweiten, denn schon vor 1914 sahen sich Flugzeugbauer durch die Vielzahl anderer Firmen einem hohen Innovationsdruck ausgesetzt, um wettbewerbsfähig zu bleiben und Abnehmer für ihre Modelle zu finden. Der Ausbruch des Ersten Weltkrieges fungierte jedoch nochmals als Katalysator für neue Erfindungen, so entwickelten Frontflieger z. B. die Voraussetzungen für den Einbau eines mit dem Propeller synchronisierten Maschinengewehrs ab 1915.[253]

Zur Verbesserung der Flugzeuge trug neben den individuellen Praxiserfahrungen der Flieger auch der gegenseitige Ideenaustausch bei. Zwar standen die auf dem Flugplatz ansässigen Firmen in Konkurrenz miteinander, doch ließen sich Veränderungen an den firmeneigenen Konstruktionen kaum geheim halten. Erstens sahen die Piloten die Konkurrenz auf dem Flugplatz und auf Flugwettbewerben fliegen, zweitens wurde regelmäßig in Fachzeitschriften und in der Tagespresse von Neuerungen auf dem Gebiet der Flugtechnik berichtet. So stellte der „Flugsport" neue Apparate in allen wesentlichen Punkten vor, informierte über

250 Archiv DTMB III. Ssg. 2, Firmenschriften Rumpler, E. Rumpler Luftfahrzeugbau. Firmenprospekt, März 1912, S.7. Rumpler ließ sich die so genannte „Schellenbefestigung für Anschlußverbindungen für Flugzeuge" patentieren (Patent Nr. 253 479), abgedruckt in: Kranzhoff, Edmund Rumpler, S.421.

251 Vgl. Kranzhoff, Edmund Rumpler, S.92–93.

252 Möser, Fahren und Fliegen, S.316. Siehe auch den Abschnitt zu den Kondor-Werken und zu Feld 4.

253 Siehe zum Ersten Weltkrieg als Katalysator: Aribert Reimann, Der Erste Weltkrieg – Urkatastrophe oder Katalysator?, in: Aus Politik und Zeitgeschichte (2004), H. 29/30, S.30–38.

zum Einsatz gelangte Materialien und druckte detaillierte Konstruktionspläne der Modelle mit Maßangaben ab.[254] Angesichts der öffentlichen Verbreitung der Konstruktionen, die zwar zum Teil durch Patente vor Nachbau geschützt waren, überrascht es nicht, dass es viele Kopisten gab, die sich bei ihren Entwürfen an erprobten und bewährten Modellen wie der Rumpler-Taube orientierten und zum Teil den Namen „Taube“ in ihre Typenbezeichnungen aufnahmen.[255] Drittens warben die Flugzeugfirmen in Fachzeitschriften und auf Plakaten in eigener Sache, so dass die Konkurrenz zwangsläufig von Innovationen erfuhr.

Beese sprach in ihren Erinnerungen an den Flugplatz Johannisthal von „stiller, opfermutiger Arbeit hinter den Schuppenwänden“[256] und Kurt Zwickau berichtete von neu ausgearbeiteten Ideen im Fliegercafé: „Bei allem Erzählen lernte man Meinungsaustausch unbewusst. Täglich machte dieser oder jener neue Erfahrungen ... Dann wurde geknobelt, gemalt und aufgezeichnet, man verrenkte sich die Arme und Beine, um recht plastisch den anderen sich verständlich zu machen.“[257] Berichte wie der von Zwickau, die für ein interessiertes Laien- und Fachpublikum gleichermaßen bestimmt waren und in erster Linie unterhalten sollten, sind gefärbt von Melancholie und mit der Intention geschrieben, das „wilde und grenzenlose“ Fliegerleben par excellence zu skizzieren. Dennoch scheint nicht alles erfunden und überspitzt dargestellt zu sein, so dass auch Aussagen zu Arbeiten an Konstruktionen zutreffend sein mögen. Anzunehmen ist auch, dass viel Arbeitszeit darauf verwandt wurde, kleine und größere Änderungen an den Maschinen vorzunehmen und in regelmäßigen Abständen zu testen, um schnellstmöglich Problemen, den Motor, die Flügelbespannung oder das Fahrgestell betreffend, vorzubeugen und so langfristig Kosten zu sparen.

Ein Schreiben der Fokker-Aeroplanbau G.m.b.H. an die Versuchstechnische Prüfungskommission in Adlershof zeigt auf, dass sich die Flugzeugfirmen zwar nach den Wünschen und Vorgaben des Heeres richteten und „nach bestem Wissen und Gewissen einwandfreie Flugzeuge zur Ablieferung bringen“ wollten, jedoch eine Bevormundung und Einsicht in Geschäftsinterna ablehnten. Aufgrund kostenintensiver, umfassender „Prüfungen und Werkstattfahrten“ war die

254 Siehe entsprechende Berichte und Modellzeichnungen im „Flugsport“, z. B. Nr. 25 (1911): „Die Rumpler-Taube“, S. 880f.; oder Nr. 6 (1912): „Der Deperdussin-Renneindecker“, S. 203ff.

255 Der von Josef Suwelack 1912 entworfene Kondor glich der Rumpler-Taube in vielen Einzelheiten. Allerdings betonten die Gesellschafter der Kondor-Werke nicht die Gemeinsamkeiten mit der Taube, sondern vielmehr die Verbesserungen am Kondor. Man trug sich sogar mit dem Gedanken, sich die Neuerungen patentieren zu lassen, sollten sie sich in der Praxis als vorteilhaft erweisen. Vgl. Protokoll der Gesellschafter-Versammlung vom 4.6.1912. Vgl. auch Tabelle mit an der Taube orientierten Eindecker-Konstruktionen zwischen 1911–1914 bei Kranzhoff, Edmund Rumpler, S. 137–139.

256 Melli Beese, Unser Flugplatz – in Memoriam, in: Motor (1921), H. 5/6, S. 160.

257 Kurt Zwickau, „Auf eigene Rechnung und Gefahr“. Ein alter Adler erzählt aus den Kindertagen unserer Fliegerei, Berlin 1940, zit. nach Joachim Knappe, 100 Jahre „Alte Adler“: Pioniere der Luftfahrt 1900–2000, Essen 2002, S. 8.

Flugzeugfirma Fokkers nicht gewillt, „der Versuchsanstalt, Adlershof, einen Eingriff in ihren Betrieb und ihr Geschäft zu gestatten, der voraussichtlich zu Unzulänglichkeiten führen würde".[258] Auch bei den Rumpler-Werken war man auf die Wahrung von Autonomie bedacht. Fehler sollten intern behoben werden, denn es sei „naturgemäss [sic!] nicht im Interesse unserer Firma, dies [die Fehler] weiterzumelden, sondern vielmehr im eigenen Wirkungskreise entsprechende und energische Abhilfe zu treffen".[259]

Nach außen warb Rumpler mit der Betriebssicherheit seiner Flugzeuge und ließ seine angestellten Piloten zu diesem Zweck Werbeflüge mit prominenten Personen durchführen. In den Monaten Oktober und November 1911 unternahm Suwelack Passagierflüge mit dem Ballonführer Alfred Dierlamm und dem Flugzeugkonstrukteur Hans Alberti und hatte auch den Präsidenten von Mexiko, Porfirio Díaz als Fluggast an Bord.[260] Im darauffolgenden Monat begleiteten ihn Oberbaurat Dr. Hermann Zimmermann, Mitglied der Akademie der Wissenschaften, und der Kaiserliche Gesandte Scheller-Schwartz.[261]

Die Namen verdeutlichen, dass gezielt solche Passagiere ausgewählt und zu einem Flug eingeladen wurden, die gesellschaftlichen Rang und Namen hatten und als Werbefiguren für Rumplers Firma dienen konnten. Werbeflüge waren ein gängiges Mittel vieler Flugzeugfirmen, um auf ihre Modelle aufmerksam zu machen und möglichst positive und wohlwollende Urteile von Passagieren zu erhalten. Mit entsprechenden Dankesschreiben zufriedener Passagiere warb Rumpler in einer Firmenschrift. Darin abgedruckt findet sich ein Schreiben des Luftschiffers la Quiante, der am 9. Januar 1912 mit Suwelack einen Flug erlebte:

> „Von dem Passagierfluge ... unter Führung des Herrn Suwelack ... kann ich nur sagen, daß es ein wirkliches Vergnügen war. Ich habe bei demselben keinen Augenblick das Gefühl der Unsicherheit gehabt, sondern war über die erstaunliche Sicherheit überrascht. Sowohl das schnelle unmerkliche Verlassen des Bodens beim Aufstieg, der Flug mit seinen schönen Kurven selbst wie auch das sanfte völlig stoßlose Landen lösten bei mir ein Gefühl der absoluten Sicherheit aus, daß [sic!] ich an jenem Flugtage besonders nicht erwartet hatte, da der Wind so böig war, daß sich andere Apparate als die Taube nicht in der Luft befanden. Ein ‚Hoch' der Taube."[262]

Diese von Rumpler scheinbar gezielt inszenierte, plakative Marketingkampagne unterstreicht, dass die Bekanntheit einer Flugzeugfirma und ihrer Fabrikate wesentlich vom Erfolg der Öffentlichkeitsarbeit und Eigenvermarktung abhängig war. Die von Rumpler in Auftrag gegebenen Werksschriften sind höchst einseitig

258 BA-MA PH 24/99, Unfälle von Flugzeugen: Schreiben Fokker-Aeroplanbau m.b.H. an die versuchstechnische Prüfungskommission vom 21.2.1914 (Stellungnahme zu Unfall).

259 BA-MA PH 24/99, Unfälle von Flugzeugen: Schreiben der E. Rumpler Luftfahrzeug G.m.b.H an die verkehrstechnische Prüfungskommission vom 11.3.1914 (Bezugnahme auf Untersuchung an Flugzeugbrüchen).

260 Vgl. B.Z. am Mittag vom 4.11.1911.

261 Vgl. Kranzhoff, Edmund Rumpler, S.100.

262 Archiv DTMB III. Ssg. 2, Firmenschriften: E. Rumpler Luftfahrzeugbau G.m.b.H. Lichtenberg Berlin Johannistahl 1912, S.32.

und inszeniert, doch spiegeln sie implizit den Konkurrenzkampf unter den Flugzeugfirmen wider, die sich nur durch glaubhaft vermittelte beziehungsweise auf Wettbewerben nachgewiesene Erfolge auf dem Flugzeugmarkt etablieren konnten. Edmund Rumpler trat in einem Brief vom 8. Februar 1912 persönlich an Josef Suwelack heran, um ihn zu bitten, „einiges über das Schulen auf unseren Apparaten mitzuteilen, insbesondere über die Vorzüge, die hierbei gegen andere Systeme zu beobachten sind“,[263] da an einem Firmenprospekt – vermutlich an der oben zitierten Werbeschrift – gearbeitet wurde.

Mit dem verklärten Bild der romantischen Fliegerei hatte die Realität auf dem Flugplatz jedoch wenig zu tun. Edmund Rumpler verlangte von seinen Piloten Disziplin, Kreativität und Ausdauer. Eine Berufslaufbahn als Flieger implizierte Entbehrungen, Ehrgeiz, handwerkliche Fertigkeiten und hohe Belastbarkeit. Die Arbeitsfliegerei unterschied sich somit erheblich vom nach außen demonstrierten Dandytum und von einem Leben in Luxus. Gleichwohl profitierten Flieger wie Josef Suwelack von der Bekanntheit des Standorts Johannisthal und Edmund Rumplers als prominenter Arbeitgeber. Wenngleich sich Suwelack 1912 von Rumpler abwandte, konnte er nicht nur auf sein Erfahrungswissen zurückgreifen, sondern auch Rumpler als national bekannten Flugzeugbauer als Referenz angeben. Der eigene Bekanntheitsgrad hing maßgeblich von fliegerischen Erfolgen, persönlichen Kontakten und einem selbstbewussten Auftreten ab. Johannisthal als Sammelbecken für erfahrene und angehende Flieger war für Josef Suwelack eine wichtige Station, um sich überregional als Flieger einen Namen zu machen.

Werbung und Profit I: Rumpler-Werke

„Der Etrich-Rumpler-Eindecker ist zur Zeit zweifellos das populärste deutsche Flugzeug“, würdigte Ansbert Vorreiter 1912 die Konstruktion und führte ihre Beliebtheit auf ihr Aussehen und die Erfolge Hellmuth Hirths zurück.[264] Im Jahr 1911 nahmen Rumpler-Piloten an zahlreichen ausgeschriebenen Wettbewerben teil und erzielten, nach Angaben eines Rumpler'schen Werbeplakats aus dem gleichen Jahr, ca. 300.000 Mark Geldpreise und zusätzlich „zahllose Ehrenpreise“. Zahlen, das heißt Gewinnsummen und (Ehren-)Preise, dienten als Maßstab für Erfolg und Qualität der Flugzeuge. So druckte die „B.Z. am Mittag“ im Juli 1911 ein Rumpler'sches Inserat, das mit den Erfolgen der Rumpler-Piloten auf Wettbewerben die Qualität der Fabrikate unterstrich. Der Annonce zufolge hätten die Piloten innerhalb eines Vierteljahres eine Preissumme von 271.285 Mark erzielt. Außerdem hob man hervor, zu Lieferanten des Kriegsministeriums zu zählen.

263 NL J. Suwelack, Schreiben Edmund Rumplers an Josef Suwelack vom 8.2.1912.
264 Vorreiter, Jahrbuch der Luftfahrt, S.129, Zitat ebd.

Die unmissverständliche Botschaft lautete, dass der Name Rumpler für Leistung stehe und die Firma im wahrsten Sinne des Wortes im Geschäft sei.

Von den gewonnenen Preisgeldern standen den Fliegern bei Rumpler dreiunddreißig bis fünfzig Prozent zu, jedoch scheint es, als habe Rumpler sich nur unzureichend an diese Vereinbarung gehalten. Rumpler investierte statt dessen einen Großteil der Gelder in den Ausbau seiner Firma, doch angesichts gut gehender Geschäfte und hoher Gewinne forderten seine Mitarbeiter mehr Mitspracherecht und eine angemessenere Bezahlung. Da Rumpler seine Piloten nach Flugsaisonende regelmäßig entließ, organisierten sich die angestellten und selbstständigen Flieger im am 29. Februar 1912 gegründeten Bund Deutscher Flugzeugführer (BDF) mit Sitz in Berlin.[265] In einem Schreiben Ellery von Gorrissens, Fluglehrer der Ago-Fluggesellschaft mbH, Teil des Gründungsausschusses und Mitwirkender an einem Statutenentwurf, wurde Josef Suwelack gefragt, ob er der Organisation, die „die Flieger vor einer mehr und mehr um sich greifende[n] Ausbeutung zu schützen [und] ihre wirtschaftliche Lage zu verbessern" suchte, beitreten wolle.[266] Seine Antwort sollte er innerhalb von vier Tagen auf beiliegender Postkarte mitteilen. Suwelack bejahte und trat der Organisation bei.

Im Statutenentwurf vom 4. März wurde in § 2 als Vereinszweck „die Wahrung der beruflichen und wirtschaftlichen Interessen der in Deutschland tätigen Flieger" angegeben.[267] Ehrenmitglied konnte werden, wer sich „um den Vereine besondere Verdienste erworben" hatte oder wer durch seinen Beruf als Flieger erwerbsunfähig geworden war. Zu ordentlichen Mitgliedern wurden Flieger gezählt, die im Besitz des deutschen Flugzeugführerzeugnisses waren, sowie ausländische Piloten, die zu nationalen Wettbewerben zugelassen waren. Eine dritte Gruppe bildeten außerordentliche Mitglieder, wozu Flugschüler gezählt wurden. Während die ersten beiden Gruppen Sitz und Stimme in der Hauptversammlung hatten, besaß die letztgenannte nur eine beratende Stimme. Der Jahresbeitrag belief sich auf 24 Mark. Vom Verein ausgeschlossen werden konnte man durch fristgerechte Kündigung aufgrund ausbleibender Beitragszahlungen oder durch

265 Vgl. Kranzhoff, Edmund Rumpler, S.132. Vgl. auch Schmitt, Oldtimer, S.98. Der BDF forderte als monatliches Mindestgehalt 300 Mark, 250 Mark für die Ausbildung eines Flugschülers und 500 Mark für das Einfliegen eines neuen Apparates. Rumpler war allerdings nur bereit, zwischen 200 und 250 Mark für die erste, 100 bis 150 Mark für die zweite und 100 Mark für die dritte Forderung zu zahlen. Vgl. ebd.

266 Schreiben des Ausschussmitgliedes des Bundes Deutscher Flugzeugführer, Herrn von Gorrissen, an Josef Suwelack vom 19.2.1912. Ellery von Gorrissen wurde später Vorsitzender des BDF. Angefügt war eine Liste mit den Namen bisher beigetretener Flieger: Karl Krieger, Paul Schwandt, Albert Buchstätter, Edmund Bohlig, Wsewolod Abramowitsch, Benno König, Arthur Grünberg, Victor Stoeffler, Max Joachim Noelle, Bruno Jablonsky, Gordian Hösli, A. Stiploschek, Hans Golde, Felix Laitsch, Rubin Wechsler, Paul Wertheim, Nelly Bese [sic!], Ch[arles] Boutard, Willi Albers, Gerhard Sedlmayr, Dr. Lissauer, Krüger, Hoff, W. Rosenstein, Rupp, Bruno Hanuschke, Bodzéna Lagler, Blattmann, von Gorrissen.

267 NL J. Suwelack, Schreiben des Vorstandsmitglieds E. von Gorissen und des Schriftführers Hans Golde an Josef Suwelack vom 4.3.1912 mit einem Statutenentwurf des BDF in der Anlage.

Erwerbsunfähigkeit, die nicht auf den Fliegerberuf zurückzuführen war.[268] Satzungsänderungen oder eine Auflösung des Vereins waren nur bei einer Hauptversammlung durch Mehrheitsbeschluss zulässig und auch nur in Anwesenheit von mindestens drei Vierteln der Mitglieder.[269]

War das in § 2 des Statutenentwurfs formulierte Ziel noch recht allgemein gehalten, so äußerte man sich wenige Wochen später schon deutlicher zu Vereinsanliegen. Beklagt wurde die zunehmend militärische Ausrichtung der Flugveranstaltungen, die „ausschließlich die effektive Höchstleistung respektive Geschwindigkeit der Maschinen" belohnen würden und nur für die Heeresverwaltung von Interesse seien. Hingegen seien die Ausschreibungen „zum Nachteil des Publikums", das lediglich noch Zeuge des Starts und der Landung sein könne. Darüber hinaus würden die Wettbewerbe zum Bankrott selbstständiger Flieger führen, da diese die Kosten für den Transport und die Reparaturen der Maschinen selbst übernehmen mussten. Die Flieger könnten daher „aus pekuniären Gründen kaum mithalten" und würden bei Flugwettbewerben „vollständig ausgeschaltet". Eine Forderung des Bundes lautete deshalb, die Fliegerei „der Allgemeinheit" zuzuführen und Apparate zu unterstützen, die „bei geringster Motorenstärke und billigsten Anschaffungskosten trotzdem möglichst hohe Leistungen erzielen." Abschließend wurde festgehalten: „Wenn wir dieses errungen haben, wird das Fliegen sich erst als Sport und gleichfalls als Verkehrsmittel entwickeln, und dies soll unserer Ansicht nach unser Endziel sein."[270]

Dem Aufruf folgten jedoch keine Taten, denn weder konnten sich die Mitglieder auf ein einheitliches Programm verständigen, noch war Edmund Rumpler zu Zugeständnissen bereit. Infolgedessen setzte Anfang des Jahres 1912 eine Abwanderung vieler Werkspiloten und Flugschüler ein. Die meisten wechselten zu anderen Flugzeugfirmen wie Gotha oder Grade und waren dort z. B. als Einflieger oder Konstrukteure beschäftigt.[271] Andere, wie später Josef Suwelack, gründeten mithilfe von Kapitalgebern aus der Industrie eigene Flugzeugfirmen.

Als Hellmuth Hirth im Februar 1912 sein Arbeitsverhältnis formell aufgelöst

268 Siehe ebd., § 3, § 5 und § 6.

269 Vgl. ebd., § 15 und § 16.

270 NL J. Suwelack, Schreiben Georg Müraus an Josef Suwelack mit entsprechender Anlage vom 12.3.1912. Ähnlich argumentierte auch Hauptmann a.D. Dr. Hildebrandt in der Ausgabe des „Essener Generalanzeigers" vom 5. Juni 1912. Dort schreibt er unter anderem: „Man muss unbedingt dafür sorgen, daß auch Privatleute imstande sind, sich die teuren Drachen anzuschaffen und zu unterhalten. Die Unterhaltung eines Rennstalles ist schon sehr kostspielig, aber die Unterhaltung eines einzigen Flugzeuges erfordert weit höhere Mittel. Dabei setzen die Flieger andauernd im Kampf mit den Böen ... ihr Leben aufs Spiel! Wir haben genug junge Leute, die aus Liebe zum Sport, aus Wagemut, von Unternehmungsgeist getrieben und infolge technischer Veranlagung, die Laufbahn eines Fliegers ergreifen möchten."

271 Ein Einflieger testete die Maschinen und justierte gegebenenfalls das Fahrgestell neu, setzte einen anderen Motor ein oder verwendete eine andere Flügelbespannung. Er testete das Flugzeug auf die erreichbare Höhe, die Steiggeschwindigkeit, die Belastbarkeit etc. Hans Vollmöller wurde Einflieger bei Oertz und dann bei den Albatros-Werken, Hellmuth Hirth wurde dort 1912 zum techni-

hatte, war Suwelack nach eigenen Angaben zur Gänze beruflich ausgelastet, da er sowohl die Verantwortung für 28 Arbeiter als auch für 18 Maschinen trug. Darüber hinaus oblag ihm der gesamte Ausbildungsbetrieb, er musste regelmäßig an den Johannisthaler „Sonntagspreisen" teilnehmen, Passagier- und Schauflüge wahrnehmen und bei so genannten „Offizierstreffen" zugegen sein. Doch trotz dieser gestiegenen Belastungen erhielten die Piloten keine zusätzliche Entlohnung, ja zum Teil noch nicht einmal das vollständige Gehalt oder die vereinbarten Boni.[272] Betrachtet man die Diskrepanz zwischen vertraglich vereinbarter und tatsächlich gezahlter Gewinnbeteiligung (Tab. 1) bei den Rumpler-Werken, wird verständlich, warum auch Suwelack sein dortiges Arbeitsverhältnis 1912 kündigte.[273]

Suwelack war vertraglich eine Gewinnbeteiligung von 33 bis 50 Prozent zugesichert, doch weder, wie aus dem Kautionskonto ersichtlich wird, erhielt er die üblichen Spesen in Höhe eines Monatsgehalts von rund 500 Mark noch den entsprechenden Anteil aus den Einnahmen durch Flugwettbewerbe. Einzig Hellmuth Hirth, der eine Sonderstellung genossen hatte, äußerte sich entsprechend positiv über sein Beschäftigungsverhältnis:

> „Auf allen meinen Flügen, die ich als Angestellter der Firma E. Rumpler unternahm, stellte mir die Firma Flugzeug und Monteure auf ihre Rechnung zur Verfügung. Die Gewinne wurden so verteilt, daß ich vom Bruttogewinn ein Drittel und die Firma zwei Drittel erhielt. Ich fand das damals auch durchaus gerecht, denn es ist für ein Flugzeugunternehmen immer ein großes Risiko ... Die Unkosten, die Reparaturen, der Transport, der Unterhalt des Fliegers, alles das ist mit großen Ausgaben verknüpft".[274]

Das erste Jahr als Berufsflieger endete für Josef Suwelack mit einem persönlichen Erfolg, da er am 8. Dezember 1911 einen „Dauerweltrekord mit Passagier" (die Flugdauer mit Passagier betreffend) aufstellte, der ihn über Nacht international bekannt machte.[275]

Vor allem deutsche, österreichische und französische Zeitungen sowie Zeitschriften veröffentlichten in den folgenden Tagen Meldungen zu Suwelacks Welt-

schen Direktor ernannt. Zum Werdegang weiterer Rumpler-Flugschüler siehe Kranzhoff, Edmund Rumpler, S.133f.

272 NL J. Suwelack, Schreiben Suwelacks an Rumpler vom 13.2.1912 und Brief Suwelacks an Hubert Ahlers vom 13.2.1912.

273 In Suwelacks Aufstellung fehlt der Gewinn von 2.218 Mark aus dem Schauflug in Hannover am 8.10.1911 (Dauerpreis 1.068 M, Höhenpreis 800 M, 350 M Prämie). Angabe enthalten in B.Z. am Mittag vom 9.10.1911 und im Billerbecker Anzeiger vom 14.10.1911. Angaben zu den übrigen Gewinnen entnommen aus „Flugsport", „Sport im Bild" und „Westfälischer Merkur" (25. Oktober 1911). Die Angabe zur Ausbildung von Bohlig und Basenach geht aus einem Schreiben Rumplers vom 13.1.1912 im Nachlass Suwelacks hervor. Die Aufstellung erhebt allerdings keinen Anspruch auf Vollständigkeit. Ein Herr Kauders findet in einem Schreiben Rumplers an Suwelack vom 2.3.1912 Erwähnung.

274 Hirth, Meine Flugerlebnisse, S.104–105.

275 Im Nachlass von Josef Suwelack befinden sich 66 Berichte der nationalen und internationalen Presse zu seinem Weltrekord.

Tabelle 1: Kautionskonto Josef Suwelacks bei den Rumpler-Werken und erhaltene Preisgelder aus den Gewinnen bei Flugwettbewerben in den Monaten Oktober bis Dezember 1911 (in Mark)

Datum	Veranstaltung / Flugschüler	Gewinn / Prämie	Gewinnbeteiligung / Lohn	Kautionssumme, Anteil 25%
24. Sept. –1. Okt.	Berliner Herbstflugwoche WB A WB B	4609 3258	2622.90	655.70
3. Okt.	Ausb. Bertram + Sonntags-Preis	116.10	116.10	29.02
3. Okt.	Ausb. Bertram / Ausb. Beese + Sonntags-Preis	116.10	116.10	56.65
3. Okt.	Ausb. Lübbe / Ausb. Rosenstein	200	200	50
3. Okt.	Ausb. Albers / Ausb. Schwarz	250	250	62.50
8. Okt.	Schaufliegen Hannover	2218	nicht bekannt	nicht bekannt
15. Okt.	Schaufliegen Detmold	3150	139.87	34.97
24. Okt.	Schaufliegen Münster	1100 + 700	156	39
November	Ausb. Bohlig / Ausb. Basenach	218.75	218.75	54.69
20. Nov.?	Schaufliegen Dresden	–	144.73	36.15 [36.18]
	Nachtrag Detmold	–	133.33	33.33
?	Ausb. Kauders	nicht bekannt	nicht bekannt	nicht bekannt
				1052.04

rekord, doch auch englische, spanische, italienische und einige wenige portugiesische und schwedische Zeitungen nahmen Notiz von diesem Ereignis. Mehrheitlich enthielten die Artikel Informationen zur Flugzeit, technische Daten zum Flugzeug und zum Motor sowie die Angabe, wer den Rekord zuvor innehatte. Die meisten dieser Berichte aus dem Ausland waren allerdings nur wenige Zeilen lang.[276]

276 Siehe z. B. Republique de l'isbre vom 8.12.1911 und Notre Sportivo vom 9.12.1911.

Hingegen berichteten die deutschen Blätter nachvollziehbarerweise wesentlich ausführlicher über die neue Bestmarke im Passagierflug. Die „Polytechnische Rundschau" beispielsweise war eines jener Magazine, das sämtliche Details zum Rekordflug mitteilte:

> „Ein neuer Dauer-Weltrekord auf Rumpler-Taube. Am Freitag, den 8. Dez, 1911, 10 Uhr 32 Min. stieg der Rumpler-Pilot Josef Suwelack mit Passagier Lange[r] am Flugplatz Johannisthal zu einem Dauerrekord auf. Das Flugzeug, eine Rumpler Taube, stellte den neuesten Militärtyp dar, und zwar mit einem neuen 65 PS Achtzylinder-Aeolus-Motor ausgerüstet. Der Flug dauerte bis 3 Uhr 6 Min. Es wurde also eine reine Flugzeit von 4 Std. 34 Min. erzielt. Durch diese Leistung ist der Weltrekord, welchen bisher Leutnant Gerrard mit 4 Std. 13 Min. gehalten hat, nach Deutschland gekommen. Flugzeug wie Motor haben sich bis zum Schluß ganz ausgezeichnet gehalten. Der Pilot Suwelack mußte nur wegen der namentlich in größeren Höhen herrschenden Kälte landen, früher, als er beabsichtigt hatte. Das Tragvermögen der Flugmaschine war sehr gut, da außer Pilot und Passagier, welche zusammen 153 kg wogen, noch 160 l Benzin und 30 l Oel mitgenommen wurden. Als Sportzeugen traten Herr Oswald Struthmann und Herr Schiedeck auf. Das Rekordflugzeug ist eine Militärmaschine. Wir bemerken, daß nunmehr sämtliche in Deutschland bestehenden Weltrekorde von der Rumpler-Taube gehalten werden, denn außer dem Suwelackschen Rekord sind noch die von Melly [sic!] Beese geschaffenen Damenrekorde für Dauer und Höhe auf diesem System aufgestellt worden."[277]

Für Josef Suwelack persönlich war der Weltrekord ein Höhepunkt in seiner bisherigen Fliegerlaufbahn und durchaus besonders. Seinem Bruder Richard, der sich zu dieser Zeit in Ehrenbreitstein in Koblenz beim Trainbataillon aufhielt, teilte er auf einer Postkarte von seinem Weltrekord mit.

Nicht ohne Grund erscheint der Weltrekordflug Suwelacks in einem Kapitel über Werbung und Profit, denn mit Rekordleistungen konnte sich eine Flugzeugfirma von der Konkurrenz abheben und neue Kunden gewinnen. Der flugsportliche, durchaus freundschaftliche Wettkampf zwischen den Nationen, aber auch der Wettbewerb zwischen den Unternehmen war geprägt von Überbietungen, denn nachgewiesene Leistungen waren gleichsam Gütesiegel für die Maschinen, die von der Heeresverwaltung angekauft wurden. Der von Suwelack am 8. Dezember geflogene Apparat wurde bereits am folgenden Tag als Typ „A7" auf dem Luftweg nach Döberitz transportiert.[278] Auf dem dort 1895 eingerichteten Truppenübungsplatz war 1910 ein Provisorisches Fliegerkommando entstanden, das sich wenig später zur zentralen Ausbildungs- und Erprobungsstelle der preußischen Militärluftfahrt entwickelte.[279] Aufträge aus Döberitz zu erhalten, war ein erster wichtiger Schritt, um ein Flugzeugunternehmen rentabel zu machen, denn zählte man zu den Lieferanten der Heeresverwaltung, so konnte man seine Ressourcen besser ausschöpfen und gegebenenfalls in Serie produzieren. Dies wiederum kam

277 „Ein neuer Dauer-Weltrekord auf Rumpler-Taube", in: Polytechnische Rundschau (1912), Bd. 327, S. 13–15.

278 B.Z. am Mittag vom 9.12.1911. Dem Artikel zufolge kam ein Propeller der Firma Garuda zum Einsatz. Siehe auch Flugsport (1911), Nr. 27, S. 977f.

279 Kai Biermann/Erhard Cielewicz, Flugplatz Döberitz. Geburtsort der militärischen Luftfahrt in Deutschland, Berlin 2005.

dem Geschäft, dem Bekanntheitsgrad der Firma und den dort tätigen Piloten zugute.

Die Einzelleistung bewirkte zudem eine Steigerung der Popularität des Rekord-Fliegers, doch auch die am Erfolg beteiligten Unternehmen, wie Propeller- oder Motorenhersteller, profitierten von der Rekordleistung und warben damit in Produktanzeigen. So pries die Garuda-Flugzeug- u. Propeller-Bau GmbH aus dem Jahr 1912 ihren Propeller als unerreicht an und verwies auf durch ihn erzielte „zahlreiche Rekorde".[280] Rund zwei Monate nach seinem Weltrekord dann erhielt Suwelack beispielsweise eine Anfrage vom Inhaber der Wilhelm Morell Tachometerwerke aus Leipzig, der ihn bat, ihm eine Fotografie seiner Weltrekordmaschine zur Verfügung zu stellen, „auf der auch mein [Morells] Tachometer Phylax, den Sie bei Ihrem erfolgreichen Fluge geführt haben, deutlich sichtbar" sei.[281] Die beiden Beispiele verdeutlichen, dass über die Werbung mit einem prominenten Flieger und seinen Leistungen der Absatz gesteigert werden und allgemein Kapital aus der Bestmarke geschlagen werden konnte.

Suwelack selbst wird den Rekordversuch sehr wahrscheinlich gewagt haben, um auf sich aufmerksam zu machen und mit Alfred Pietschker gleichzuziehen. Pietschker hatte am 1. Oktober des gleichen Jahres auf einem Albatros-Doppeldecker mit zwei Passagieren und einer Flugdauer von 2 h und 19 min. in Johannisthal den ersten Weltrekord nach Deutschland gebracht.[282] Zwar hatte Melli Beese ebenfalls einen fast dreistündigen Flug mit einem Passagier absolviert, jedoch wurde ihre Leistung nur als Damenrekord verbucht.

In einer späteren Firmenschrift rekurrierten die Rumpler-Werke auf die Rekordleistungen ihrer Flugzeuge, die zu gesteigerter Nachfrage und zum Ausbau des Unternehmens geführt hätten: „Diese zahlreichen Erfolge haben uns nicht nur zur Haupt-Lieferantin für die deutsche Armee gemacht, sondern auch die Kauflust des Auslandes derart gesteigert, daß wir unsere Fabrikräume andauernd erheblich vergrößern mußten."[283] Nachfolgend waren die aufgestellten Erfolge und Rekorde der Jahre 1911 bis 1913 beziehungsweise 1914 tabellarisch aufgeführt.

Die Rekordversuche sind jedoch auch noch aus einer anderen Perspektive zu sehen, denn die Flieger und Firmen waren medial und vonseiten des Militärs einem hohen Erwartungsdruck ausgesetzt. Rekord- und Überlandflüge, wie die

280 Vgl. Anzeige der Garuda-Flugzeug u. Propeller-Bau-GmbH, in: Sport im Bild (1912), Nr. 14, S. 499.

281 NL J. Suwelack, Anfrageschreiben Wilhelm Morells (Wilhelm Morell Tachometerwerke) an Josef Suwelack vom 17.2.1912.

282 Übersicht der Weltrekorde bei Supf, Fluggeschichte, Bd. 1, S. 546f.

283 DTMB: III. Ssg. 2, Firmenschriften Rumpler, Erfolgsaufstellung der Rumpler-Eindecker. Hervorhebung im Original.

Kanalüberquerung Louis Blériots 1909,[284] dienten nicht ausschließlich einem Selbstzweck, sondern wurden durch die Presse mit ihren spannungsgeladenen Berichten und den von ihr ausgelobten Preisgeldern und Sponsorengeldern zu Sensationen gemacht. So stellt Möser fest: „Dies war ... die neue Konstruktion eines Typs der Kombination von Luftmobilität, Wettbewerb, Massenbegeisterung und Boulevardjournalismus, der die Diffusion der Fliegerei von da an begleitete."[285] Robert Wohl sprach in Bezug auf Blériots Kanalflug von einem „social drama, which would be played out time and again throughout the twentieth century by a long series of aviators".[286] Tatsächlich blieb Suwelacks Weltrekord nicht seine einzige „besondere" Flugleistung, von der (inter-)national Notiz genommen wurde. Um die Zuverlässigkeit seiner Kondor-Konstruktion zu demonstrieren, plante Josef 1913 nämlich einen Langstreckenflug von Essen nach London, der ihm, obwohl missglückt, viel Aufmerksamkeit für den Eindecker zuteilwerden ließ.[287]

Weiterhin wurden die Flieger durch den Wettstreit zwischen den Nationen, hier besonders zwischen Frankreich und Deutschland, dazu motiviert, die französische Dominanz in der Fliegerei zu schwächen und Frankreich an der Spitze abzulösen. Der „Flugsport" beklagte regelmäßig die Rückständigkeit des deutschen Flugwesens gegenüber der französischen Industrie[288] und appellierte an Zeitungen und an das „deutsche Volk", „in Deutschland eine Bewegung zu entfachen, die der französischen gleichkommt und in der das deutsche Volk zu einer vaterländischen Stiftung aufgefordert wird".[289] Die Zeitschrift plädierte für einen verantwortungsvolleren Umgang mit den zur Verfügung stehenden Mitteln und war der Meinung, dass „Schauflugveranstaltungen nicht in der Lage [seien], auf die Entwicklung der Flugmaschinen in gehörigem Maße veredelnd einzuwirken", sondern nur unnötige Ausgaben verursachten.[290]

Trotz hoher finanzieller Aufwendungen vonseiten des Militärs und der Industrie waren Flugwettbewerbe und Rekordleistungen schlussendlich jedoch unentbehrlich, weil sie einerseits eine wichtige Einnahmequelle für die Flugzeugfabriken und die Piloten darstellten und andererseits Vergleichsmöglichkeiten boten und den Austausch zwischen Militär und Flugzeugindustrie ankurbelten. Suwelacks Rekordflug am 8. Dezember, der den einzigen Weltrekord auf der Rumpler-Taube für das Jahr 1911 darstellte, setze den Schlusspunkt hinter das für das Unternehmen erfolgreiche Jahr.

284 Vgl. zur Berichterstattung über Blériots Kanalüberquerung Ingold, Literatur und Aviatik, S.86ff.
285 Möser, Fahren und Fliegen, S.438.
286 Robert Wohl, A Passion for Wings. Aviation and the western imagination. 1908–1918, Yale 1994, S.66.
287 Siehe Kapitel zu den Kondor-Werken.
288 Siehe z. B. „Michel schläft!", in: Flugsport (1912), Nr. 5, S.165ff.
289 Ebd., S.166.
290 Vgl. Flugsport (1911), Nr. 22, S.766f.

Transfer in die Heimat und (Um-)Brüche

Während der Weihnachtszeit 1911 war Suwelack zu Hause bei seinen Eltern und Geschwistern und bekam dafür von Rumpler Urlaub, der, nach einer ärztlichen Genehmigung, bis Mitte Januar 1912 währte.[291] In dieser Zeit entstand in Billerbeck die Idee zu einem eigenen Flugtag. Suwelack unterrichtete Rumpler von dem Vorhaben und gab an, dass er den Flugtag auf einer Rumpler-Taube bestreiten wolle.[292] Rumpler, der sich erfreut über den geplanten Flugtag zeigte, wies seinen Piloten darauf hin, „dass es empfehlenswert [sei], wenn Sie sich von den Herren, welche die Veranstaltung übernehmen, bzw. welche die von Ihnen erwähnte Summe von M. 1600,- garantieren, das schriftlich bestätigen lassen würden." Das Schreiben enthielt außerdem Vorschläge zu organisatorischen Fragen wie etwa zu der Errichtung von stabilen Flugzeugschuppen. Hier legte Rumpler vor allem Wert auf kostengünstige Lösungen: „Vielleicht baut den Schuppen am billigsten der Zimmermeister, welcher die Schuppen in Münster [für den dortigen Flugtag] gebaut hat".[293]

Die Organisation der Billerbecker Flugtage, wie sie auf späteren Plakaten angekündigt wurden, war ein Gemeinschaftsprojekt flugbegeisterter Männer, die die Flugtechnik und „ihren" Flieger Josef Suwelack lokal bekannt machen wollten. Der damalige Billerbecker Amtmann und Chronist Heinrich Fasse schrieb über die Vorbereitungen zu den Flugtagen:

> „Am heutigen Neujahrstag fand im Domhotel eine Versammlung von Freunden des Flugsports statt. Zweck war die Abhaltung eines Flugtages in Billerbeck durch den heimatlichen Flieger J. Suwelack. 8 Tage später wurde ein Komitee gebildet, das die Vorbereitungen zu dem am 2. Ostertag [1912] stattfindenden Flugtag eifrigst betrieb."[294]

Da auf die Billerbecker Flugtage im Kapitel zu Flugwettbewerben (Feld 3) gesondert eingegangen wird, um Veranstaltungsformen idealtypisch unterscheiden zu können und verschiedene dahinterstehende Motivationen aufzuzeigen, soll an dieser Stelle vorerst nur auf den Schriftwechsel zwischen dem Billerbecker Hubert Ahlers, der Schriftführer im Flugtag-Komitee und laut einer Broschüre der Gaststätte Ahlers „einer der eifrigsten Initiatoren"[295] in dieser Angelegenheit gewesen war, und Josef Suwelack eingegangen werden.

Suwelack hatte seine Teilnahme an den Flugtagen in seiner Heimatstadt im Dezember 1911 beziehungsweise Anfang Januar 1912 zwar bereits mündlich zuge-

291 Siehe NL J. Suwelack, Schreiben Rumplers vom 6., 13. und 17.1.1912 an Josef Suwelack.
292 NL J. Suwelack, Schreiben Suwelacks an Rumpler vom 15.1.1912.
293 NL J. Suwelack, Schreiben Rumplers an Suwelack vom 17.1.1912.
294 StdA Billerbeck, Heinrich Fasse, Geschichtliche Nachrichten aus dem Amte Billerbeck 1900–1924 (mit 60 Abbildungen), Bd. 1, S.66. Die Bände befinden sich in Privatbesitz.
295 StdA Billerbeck, Firmenchroniken – Handbibliothek, 250 Jahre 1734–1984 Handelshof, Gaststätte und Brennerei Franz-Josef Ahlers Billerbeck Langestraße 18.

sagt, jedoch erfolgte eine vertraglich abgesicherte Bestätigung seitens des Fliegers erst, als die Vorbereitungen schon weit fortgeschritten waren. Dieser Umstand bereitete Ahlers offensichtlich Kopfzerbrechen:

> „Da sitze ich armer Schriftführer eines hochwohllöblichen Flugkomitees nun schon 4 Wochen hier in meiner Angst, renne jeden Tag zur Post oder zu ‚Vater Suwelack', aber jedesmal muß ich mit langem Gesicht wieder abziehen. Dann höre ich ‚Brief unterwegs', aber weder Brief noch Kontrakt kommen an."[296]

Er warf Josef „Schreibfaulheit" vor, glaubte, er vergnüge sich in Berlin, und versuchte ihm deutlich zu machen, dass die Flugtage nur noch von seiner definitiven Zusage abhängig seien: „Die Flugsache macht gute Fortschritte. Freiherr von Twickel hat seine Weide zur Verfügung gestellt. Die Musik (Kürassiere von Münster) ist bestellt, mit der Bahn habe ich schon wegen der Extrazüge verhandelt."[297] Ahlers ermahnte Suwelack, sich zu beeilen, da für Sonntag, den 11. Februar, eine Generalversammlung angesetzt sei: „Wenn du bis dahin das Ding [den Vertrag] nicht schickst, gefährdest du das ganze Unternehmen." Zwar waren auch noch andere Piloten für die Flugtage gemeldet, doch war Suwelack als „Ur-Billerbecker" der wohl größte Publikumsmagnet, mit dem auch auf eigens angefertigten Plakaten für die Veranstaltung geworben wurde.[298]

Mit dem Hinweis auf seine weiter oben bereits angesprochene angespannte Arbeitssituation bei den Rumpler-Werken äußerte sich Josef gegenüber Ahlers, dass er „verdammt viel" zu tun habe. So sei er für zwei Dutzend Flugschüler verantwortlich, stehe mit Flugzeugfirmen in Verhandlung, sei in Hamburg und Frankreich gewesen und führe täglich etwa drei- bis vierstündige Konferenzen. Ferner müsse er Maschinen einfliegen, habe am Vortag eine sechsstündige Konferenz mit einem französischen Konstrukteur gehabt und arbeite unter Zeitdruck: „In 10 Minuten erwarte ich noch einen Herren."[299] Ungehalten über den Vorwurf, er gebe sich seiner Freizeit hin, erwiderte Suwelack Ahlers' Behauptung mit dem Kommentar: „Also lieber Hubert von Grossstadt [sic!] Vergnügungen merke ich sehr wenig."[300]

Es ist anzunehmen, dass Josef Suwelack die Aufgaben Hellmuth Hirths übernahm und für den Transport und das Einfliegen der für das Militär bestimmten Apparate zuständig war. Vermutlich, denn hier ist die Quellenlage äußerst spärlich, fuhr er im Auftrag Rumplers zur Fliegerstation nach Metz, die fünf Etrich-Rumpler-Eindecker ankaufen wollte.[301] In Hamburg war Suwelack möglicher-

296 NL J. Suwelack Brief von Hubert Ahlers an Josef Suwelack in Berlin-Johannisthal vom 12.2.1912.
297 Ebd.
298 Ebd.
299 NL J. Suwelack, Antwortschreiben Josef Suwelacks an Hubert Ahlers vom 13.2.1912.
300 Ebd.
301 Vgl. Militärgeschichtliches Forschungsamt, Die Militärluftfahrt. Technischer Band, 2. Aufl., o.O. 1966, S.39.

weise zwischen dem 8. und 10. Februar, um eine Rumpler-Taube, die von Oberleutnant Barends am 8. Februar bei einem militärischen Übungs-Erkundungsflug von Döberitz nach Hamburg geflogen werden sollte, instand zu halten. Dem „Flugsport" zufolge seien der Flieger und sein Beobachter auf dem Exerzierplatz Wandsbek gelandet.[302]

In Anbetracht der brieflichen Schilderungen Suwelacks und der folgenden (Um-)Brüche in seinem Leben ist davon auszugehen, dass er mit seiner beruflichen Situation nicht mehr zufrieden war und sich nach alternativen Arbeitgebern und Betätigungsfeldern in der Fliegerbranche umsah. Mehrere Aktivitäten Suwelacks, die im Folgenden näher beleuchtet werden, stützen diese These.

Im Januar 1912 schloss Suwelack einen privatrechtlichen Dienstvertrag mit dem Deutschen Reich, vertreten durch die Inspektion des Militär-Luft-und Kraftfahrwesens, den er sich am 2. Februar 1915 ausdrücklich bestätigen ließ.[303] Darin verpflichtete er sich, im Kriegsfall „vom 1. Mobilmachungstag bis zur Demobilmachung zur Kampfesleistung im Heer als Flugzeugführer" zur Verfügung zu stehen und darüber hinaus für das Militär Piloten auf dem Militärflugplatz in Döberitz oder aber an der Front auszubilden. Die benötigten Flugzeuge und Monteure musste er der Militärverwaltung zur Verfügung stellen sowie für die „Beschaffung von Bekleidung, Wäsche und Ausrüstung" selbst Sorge tragen. Hierfür leistete die zuständige Behörde einmalig eine Zahlung in Höhe von 500 Mark und übersandte ihm zudem eine Armbinde. Für „kriegstaugliche" Flugzeuge, sofern sie vom „Abteilungsführer" als solche eingestuft wurden, wurde, dem Vertrag zufolge, der „Normalwert" ersetzt, das waren rund 15.000 Mark.

Als Grundsold stand Josef Suwelack im Kriegseinsatz eine tägliche Vergütung von 25 Mark zu, die „bis zur Ermittelung des pensionsfähigen Einkommens immer zu Grunde [zu] legen" sei. Der Sold wurde zehn Tage im Voraus bezahlt und konnte sich um ein Vielfaches erhöhen, wenn folgende Kriterien erfüllt waren: Bei einem Dienstauftrag bekam Josef Suwelack 100 Mark bezahlt, dauerte der Auftrag, das heißt der Flug, mehr als zwei Stunden, erhöhte sich die Summe um weitere 100 Mark. Ein besonderer Anreiz wurde dadurch geschaffen, dass „für besonders erfolgreiche Flüge – durch den Chef des Generalstabes der Armee Prämien in Höhe von 5000,- Mark zuerkannt" werden konnten. Durch die Aussicht auf Belohnung sollte eine gewissenhafte Auftragserfüllung gesichert werden.

Weitere Vertragsabschnitte regelten gesondert die Versorgung und Bezahlung Suwelacks und seiner Hinterbliebenen. So kam das Militär für Unterkunft und

302 Vgl. „Eine Glanzleistung eines deutschen Offizierfliegers. Von Döberitz nach Hamburg", in: Flugsport (1912), Nr. 4, S. 130f.

303 NL J. Suwelack, Vertrag zwischen dem Deutschen Reich und Josef Suwelack vom 2.2.1915. Der dreiseitige Vertrag enthält mehrere verschiedene Datumsangaben, was vermutlich darauf zurückzuführen ist, dass er von mehreren Behörden bzw. zuständigen Personen gegengezeichnet wurde. Die folgenden Zitate finden sich im Vertrag wieder.

Verpflegung sowie für ärztliche Behandlungen während der Dienstzeit auf. Pensionsansprüche und Versorgung bei Invalidität wurden nach dem Offizierpensionsgesetz vom 31. Mai 1906 geregelt, wovon die Paragrafen 11, 12 und 35 für Suwelack Gültigkeit besaßen.[304] Eine „Verstümmelungszulage" stand ihm beim Verlust von Gliedmaßen, der Sprache oder des kompletten Gehörs zu und betrug jährlich 900 Mark, allerdings nur, wenn ihm zuvor Pensionszahlungen gewährt wurden und die Verletzungen im Dienst entstanden waren. Bei vollständiger Erblindung verdoppelte sich die Zulage. Gleiches galt bei schwerer körperlicher oder geistiger Schädigung.[305] An Pensionsleistungen standen Suwelack nach § 35 1/20 des Monatsgehaltes seiner Dienstzeit zu, die einmalig als monatliche Zulage für die Dauer von einem Jahr gezahlt wurden. Bei völliger Erwerbsunfähigkeit, entstanden durch den Einsatz im Heer, wurden 75 Prozent des Diensteinkommens gezahlt.[306]

Der Personen- und Gepäcktransport an die Front und zurück in die Heimat sollte mit der Bahn in der zweiten Klasse auf Kosten der Heeresverwaltung erfolgen, die ebenso die Ausbildungskosten der Flieger trug.

Der Vertrag konnte vonseiten der Militärbehörde vierzehn Tage vor der Mobilmachung nur noch formell gekündigt werden. Vorher galt jedoch: „In der Zeit zwischen dem Abschlusse des Vertrages und dem Abrücken ins Feld kann – die Behörde ohne Einhaltung der Kündigungsfrist vom Vertrage zurücktreten." Diese Regelung erstreckte sich auch auf Pflichtverletzungen des Flugzeugführers während der Dienstzeit.

Mit dem privatrechtlichen Dienstvertrag sorgte Suwelack für seine Zukunft vor. Durch die Arbeit bei den Rumpler-Werken war er immer wieder mit Militärbehörden in Kontakt gekommen und wird wahrscheinlich, zumindest indirekt, Einblick in Kriegsplanungen und Karrieremöglichkeiten beim Militär erhalten haben. Im Bewusstsein, dass die Heeresverwaltung der größte Abnehmer von Flugzeugen und einer der wichtigsten Arbeitgeber in der Fliegerbrache war, scheint sich Suwelack frühzeitig bemüht haben, ein für ihn günstiges und sicheres Abkommen zu schließen, das es ihm auch in Zukunft ermöglichen würde, als Flieger tätig zu sein. Da er seine militärische Ausbildung nach dem Einjährigen nicht weiter vertieft hatte, sorgte er auch deshalb vor, um im Kriegsfall nicht zum Landsturm eingezogen zu werden. Das hätte nicht nur eine weit geringere Besoldung zur Folge gehabt, sondern vor allem bedeutet, Privilegien einzubüßen und als „einfacher" Soldat die Grenzen des Deutschen Reiches zu verteidigen. Als

304 Vgl. A. Romen, Die Militärpensionsgesetze vom 31. Mai 1906 nebst den Ausführungsbestimmungen, 1. Teil: Das Offizierpensionsgesetz. Berlin 1907, hier S.77ff. und S.155ff., online unter: https://openlibrary.org/books/OL23414458M/Die_militärpensionsgesetze_vom_31._mai_1906_nebst_den_ausführungsbestimmungen.

305 Ebd., S.78, § 11.

306 Ebd., S.157, § 3, 5, 7.

Zivilflieger im Krieg hingegen bestand für Suwelack die Möglichkeit, Kontakte zu einflussreichen Militärs zu knüpfen und so Fürsprecher zu gewinnen, z. B. bei der Unterbreitung von Vorschlägen zu Konstruktionsverbesserungen an Flugzeugen. Wichtig erscheint nicht zuletzt auch der Umstand, dass Josef auf diesem Wege schlicht weiter seiner Tätigkeit als Flieger nachgehen und „Vergnügen" mit Arbeit verbinden konnte.

Während der Ausbildung der ersten Militärpiloten bei Rumpler hatte Josef Gelegenheit, bei Schulflügen und auf dem Flugplatz mit den bereits fertig ausgebildeten Militärpiloten ins Gespräch zu kommen, die ihm vermutlich von ihrer Tätigkeit berichteten und vielleicht auf Nachfrage detailliert Auskunft über Pläne und berufliche Perspektiven bei der Fliegertruppe geben konnten. Auch später, als Suwelack schon bei den Kondor-Werken arbeitete, stand er in Verbindung zu Mitarbeitern des Militärs. So bat ein Leutnant Meyer Suwelack in einem Brief vom Juli 1913, ihm eine Maschine zur Verfügung zu stellen, mit der er an einem Offiziersfliegen in Leipzig teilnehmen könne.[307] Solche und ähnliche persönliche Kontakte zu Militärs waren für alle Firmen von hoher Bedeutung, da die Inspekteure der Fliegertruppe die Firmen und Konstrukteure über aktuelle Entwicklungen im Flugzeugbau in Kenntnis setzten beziehungsweise Verbesserungsvorschläge machten. Die Flugzeugfirmen konnten aus den Unterredungen wiederum Rückschlüsse ziehen und gegebenenfalls Maßnahmen ergreifen, um den Forderungen gerecht zu werden. Der privatrechtliche Dienstvertrag Suwelacks war gewissermaßen ein zweites Standbein, eine weitere Option, sollte er als Konstrukteur oder Angestellter bei einer Flugzeugfabrik keinen Erfolg haben. Dass Josef Suwelack vor allem jedoch weiterhin als Flieger, insbesondere im sportlichen Bereich, aktiv sein wollte, zeigt sich deutlich an seinen weiteren Schritten.

Hatte er noch am 14. Februar 1912 mündlich gegenüber Rumpler erklärt, er wolle sich zwei Rumpler-Tauben, ausgerüstet mit einem 100-PS Argus Motor, kaufen,[308] so nahm er wenig später wieder Abstand davon. Vermutlich war seine berufliche Situation ein Grund dafür, dass er stattdessen in Kontakt mit der Flugzeugfirma Automobil & Aviatik AG trat, die ihren Sitz in Mühlhausen-Burzweiler

307 NL J. Suwelack, Schreiben Leutnant W. Meyers an Josef Suwelack vom 25.7.1913. Suwelacks Antwort vom 4.8.1913 fiel negativ aus: „Lieber Herr Meyer! Von einer Reise zurückgekehrt, komme ich erst heute dazu, Ihren Brief vom 25. Juli zu lesen und muss Ihnen zu meinem Bedauern leider mitteilen, dass ich Ihnen leider keinen Kondor-Eindecker für das Offiziersfliegen zur Verfügung stellen kann, da sowohl ich als auch unser Chefpilot in den nächsten Wochen durch Konkurrenzfliegen verhindert sein werde."

308 NL J. Suwelack, Schreiben der E. Rumpler Luftfahrzeugbau G.m.b.H. an Josef Suwelack vom 15.2.1912. Darin heißt es unter anderem: „Wir würden Ihnen dieselben [2 Rumpler-Tauben] aufgrund unserer beiliegenden Lieferungsbedingungen Nr. 167 zum Preise von je Mk: 18.500 ... liefern und zwar den einen Apparat Mitte März, den zweiten im Laufe des Monats April. Die Zahlungen hätten zu erfolgen ein Drittel bei Bestellung, zwei Drittel bei Uebergabe der Apparate. Die Uebergabe erfolgt nach einem viertelstündigen Flug."

im Elsass hatte und ebenfalls auf dem Flugplatz Johannisthal eine Filiale besaß.[309] Die Automobil & Aviatik AG war das Ergebnis einer Fusion zwischen Jules Spengler, der 1910 die Aviatik GmbH ins Leben gerufen hatte, und den Gesellschaftern Georges Châtel und Henri Jeannin, die beide bei der Argus-Motoren-GmbH arbeiteten. Die Aviatik-Flugzeugwerke waren wiederum aus der 1900 von Châtel gegründeten Fahrrad- und Automobilfabrik hervorgegangen.

„Aviatik", wie das Unternehmen fortan genannt wurde, gab den Fahrradbau auf und konzentrierte sich nur noch auf den Automobil- und Flugzeugbau. Die Unternehmer erwarben die Patentrechte am französischen Hanriot-Eindecker und kopierten ferner den französischen Farman-Doppeldecker. Ein Jahr später ließen sich Zweigstellen der Firma in Habsheim und Johannisthal nieder, die erfolgreich Flugschüler ausbildeten und Flugzeuge konstruierten. Aufgrund wachsender Konkurrenz zog sich Aviatik im Sommer 1913 aus Johannisthal zurück und gründete stattdessen 1914 eine Tochtergesellschaft in Österreich. Während des Ersten Weltkrieges wurde der Hauptsitz zunächst nach Freiburg im Breisgau und 1916 nach Leipzig-Heiterblick in dort neu erbaute Fabrikanlagen verlegt. Eine Fliegerschule errichtete man in den Luftschifferhallen in Leipzig-Mockau, die im Herbst 1916 jedoch ausbrannte. Nach dem Ersten Weltkrieg stellte Aviatik den Flugzeugbau ein. Das Unternehmen baute zwar überwiegend Doppeldecker, war aber auch im Eindeckerbau erfolgreich. In einem Inserat pries Aviatik seine Modelle als „schnellste deutsche Flugzeuge der Gegenwart" an und verwies ebenso wie die Rumpler-Werke darauf, zu Lieferanten der Heeresverwaltung zu zählen.[310]

Obwohl sein Vertrag bei Rumpler eine Konkurrenz-Klausel enthielt, die es ihm weder gestattete, in den Dienst einer anderen Firma zu treten, noch bei der Konkurrenz ein Flugzeug zu kaufen, setzte Suwelack sich darüber hinweg, schloss am 23. Februar 1912 einen Arbeitsvertrag mit Aviatik und erwarb zwei Renneindecker der Firma zu einem Stückpreis von 12.000 Mark, die er ratenweise mit 40 Prozent der auf Flugwettbewerben gewonnenen Preissummen abbezahlen wollte.[311]

Suwelacks weitere berufliche Schritte scheinen im Folgenden spontan und nicht überlegt. Er sah sich nach für ihn anscheinend besseren Arbeits- und finanziellen Bedingungen um, doch die Ereignisse überschlugen sich geradezu. Am 14. Februar 1912 reichte er seine Kündigung bei Rumpler ein und leitete rechtliche Schritte gegen ihn ein.[312] Rumpler wiederum erkundigte sich am 23. des Monats, weshalb Suwelack seinen „vertragsmässigen verpflichtungen [sic!] seit einigen

309 Die folgenden Ausführungen sind angelehnt an Schwipps, Als die Oldtimer flogen, S.69 sowie an ders., Schwerer als Luft, S.178f.
310 Siehe Inserat der Aviatik AG in Sport im Bild (1911), Nr. 5, S.127.
311 NL J. Suwelack, Kaufbestätigung der Automobil & Aviatik AG für zwei Renneindecker à 12.000 Mark und Vertrag mit der Aviatik AG vom 23.2.1912.
312 NL J. Suwelack, Schreiben Rumplers an Rechtsanwalt Dr. Kelk vom 23.2.1912.

Tagen nicht mehr" nachkomme, denn er habe „sechswöchige Kündigungsfrist und müsste vertragsmässig bis 31. März" bleiben. Zuvor hatte Suwelack allerdings bereits am 18. Februar ein neues Abkommen mit Rumpler geschlossen, wodurch er schon zum 15. März aus dem Unternehmen ausscheiden konnte, unter der Bedingung, dass er „sogleich eine Maschine fest zu dem ihm offerierten Preis kauft, jedoch in der Voraussetzung, dass er bis Ende März ... aushilfsweise Flugunterricht erteilt."[313]

Suwelack warf Rumpler „unsittliches Verhalten" vor, dieser wies die Anschuldigungen jedoch als „schroff" und „unmotiviert" zurück.[314] Ferner stellte Suwelack Zahlungsforderungen an Rumpler, denen dieser nur zum Teil nachkam. Die „Abrechung", persönlich wie finanziell, fiel entsprechend knapp aus. So schrieb Rumpler am 1. März 1912: „Wir bemerken, dass wir die Abrechnung erst nach Einigung mit Herrn Rosenstein zur Auszahlung bringen können ... . Wir senden Ihnen vorläufig Ihren Fixus."[315]

Suwelack forderte noch ausstehende Zahlungen von Schauflügen, verzichtete aber auf Honorare für Schülerausbildung und das Einfliegen der Flugzeuge für 1912. Daraufhin unterbreitete ihm sein vormaliger Arbeitgeber, dass nicht alle Abrechnungen vorlägen und eine exakte Bezahlung daher schwierig sei: „Da wir bis heute weder eine offizielle Abrechnung noch das Geld für den Flug [in Hannover] erhalten haben, könnten bei diesem Flug noch etwaige Verschiebungen stattfinden".[316] Offenbar bestanden auf beiden Seiten Differenzen, so dass Rumpler möglicherweise um Klärung bemüht war, denn am 9. März bat er Suwelack schriftlich: „Vor Ihrem Abgang aus Berlin möchte ich Sie gern persönlich sprechen".[317]

Das Verhältnis zwischen Suwelack und Rumpler entspannte sich später wieder, denn Ende November 1913 übersandten die Rumpler-Werke Lichtbilder von der Rumpler-Taube, von Motoren und Fertigungsprozessen etc., die Suwelack für einen Lichtbildvortrag benötigte.[318]

Hinsichtlich des Wechsels zu den Aviatik-Werken hatte Josef in seinem Vortrag „Aus meinem Fliegerleben", den er am 24. Februar 1912 auf der Hauptversammlung des Luftschiffahrtverein Münster hielt – das Versprechen hierzu hatte er nach dem Schaufliegen in Münster am 27. Juni 1911 gegeben[319] – berichtet,

313 Ebd.
314 Siehe NL J. Suwelack, Schreiben Suwelacks an Rumpler vom 5.3. und Antwort Rumplers vom 7.3.1912.
315 NL J. Suweiack, Schreiben Rumplers an Suwelack vom 1.3.1912.
316 Vgl. NL J. Suwelack, Schreiben Suwelacks an Rumpler vom 6.3. und Antwort Rumplers vom 7.3.1912.
317 NL J. Suwelack, Schreiben Rumplers an Suwelack vom 9.3.1912.
318 NL J. Suwelack: Schreiben der E-Rumpler Luftfahrzeug GmbH vom 14. und 22.11.1913.
319 Vgl. Billerbecker Anzeiger vom 28.6.1911.

dass er vom Erfolg der Aviatik-Eindecker gehört habe und beabsichtige, an künftigen Konkurrenzen auf dem Aviatik-Apparat teilzunehmen.[320] Möglich, dass die Werbung Aviatiks zusätzlich Ausschlag gab, sich für dieses Flugzeug zu entscheiden.

In den damaligen Fliegerkreisen waren der Name Suwelack und seine Leistungen mittlerweile bekannt, so dass sich die Nachricht von seiner Kündigung rasch in Johannisthal verbreitete. Am 13. Februar 1912 teilte Suwelack Hubert Ahlers mit: „Wo ich am I. III. von Rumpler fort gehe[,] fallen die Firmen wie die Raben über mich her. Ich habe Angebote von 2.000 Mk monatlich e.t.c.".[321]

Diese Aussage gewinnt zusätzlich an Gewicht, wenn man Abwerbungsschreiben anderer Flugzeugfirmen mit einbezieht. So wandten sich die Flugzeugfirmen Fokker-Aeroplanbau, Flugwerke Haefelin & Co. und weitere Hersteller an Suwelack, um ihn als Piloten zu gewinnen. Wie rasant die Meldung über seine Kündigung kursierte, zeigt das Datum des Schreibens der Haefelin-Werke, die bereits am 15. Februar 1912 Kontakt aufnahmen und ihr Schreiben per Eilboten zustellten:

> „Wie ich höre haben Sie bei der Firma Rumpler gekündigt. Wir brauchen, wie Sie ja wissen, einen tüchtigen Piloten, denn unser Herr Marschal ist für die Unterrichtserteilung ganz gut, jedoch ohne Befähigung für grössere Leistungen. Bitte teilen Sie mir postwendend mit, ob Sie im Prinzip bereit wären, zu uns zu kommen. Bejahendenfalls bitte ich um Ihren Besuch heute oder morgen zwischen 4–6 Uhr nachmittags."[322]

Nicht ganz zwei Wochen später versuchten die Haefelin-Flugzeugwerke erneut, mit Suwelack einen Vertrag zu schließen, der zwischenzeitlich scheinbar Bedingungen gestellt hatte, unter denen er eventuell bei der Firma arbeiten werde:

> „Ich habe Sie leider bei meinen wiederholten Besuchen in Johannisthal stets verfehlt. ... Wir [die Haefelin-Flugzeugwerke] wären selbstverständlich bereit, Ihnen die beiden gewünschten Apparate zu unseren Selbstkosten zu bauen, da wir einen außerordentlichen Wert darauf legen würden, dass Sie unsere Apparate fliegen."[323]

Bruno Jablonsky (1893–1978), der im Alter von 18 Jahren als jüngster deutscher Pilot seinen Pilotenschein machte, war 1911 Direktor der Garuda Propeller Fabrik in Johannisthal geworden.[324] Am Weltrekord-Flug Suwelacks war die Firma inso-

320 Vgl. „Aus der Stadt", in: Münsterischer Anzeiger vom 26.2.1912.

321 NL J. Suwelack, Brief Suwelacks an Hubert Ahlers vom 13.2.1912.

322 NL J. Suwelack, Einschreiben der Flugwerke Haefelin & Co. an Josef Suwelack vom 15.2.1912. Schreiben der Fokker-Aeroplanbau vom 23.2.1912: „Sehr geehrter Herr! Sie würden uns sehr zu Dank verpflichten durch gütige Mitteilung, ob wir Sie als ersten Reflektanten auf eine unserer neuen Maschinen vormerken dürften. Es ist uns sehr daran gelegen zu erfahren, wieviel Maschinen wir für uns in Auftrag geben müssen, um Ihnen evtl. mit schnellster Lieferung dienen zu können."

323 NL J. Suwelack, Schreiben der Flugwerke Haefelin & Co. an Josef Suwelack vom 27.2.1912. Josef hatte am 15. Februar auf das Schreiben der Haefelin Werke vom 15. Februar reagiert, welches jedoch nicht im Nachlass vorliegt.

324 Vgl. William D. Rubinstein (Hg.), The Palgrave Dictionary of Anglo-Jewish History, New York 2011, S.463.

fern beteiligt, als sie den Propeller hergestellt hatte. Garuda wollte 1912 nun selbst Flugzeuge bauen, daher versuchte auch Jablonsky im Mai, Suwelack dazu zu bewegen, in die Firma einzutreten. Er teilte ihm mit, dass ihr gemeinsam entwickelter Apparat „tadellos geflogen" sei.[325]

Offenbar genoss Josef also einen guten Ruf in Johannisthal und war ein gefragter Pilot. Die Frage, weshalb er die mitunter großzügigen Anstellungsangebote ablehnte, lässt sich nicht definitiv beantworten. Es entsteht der Eindruck, dass er nach mehr Unabhängigkeit strebte und von den erzielten Preisgeldern auf Wettbewerben einen Großteil selbst behalten wollte. Die Negativerfahrung bei den Rumpler-Werken wird ihn wahrscheinlich gelehrt haben, die Vertragsbedingungen sorgfältig abzuwägen. Eventuell trug er sich aber auch bereits nach seinem Fortgang von Rumpler mit dem Gedanken, sich selbstständig zu machen und eigene Flugzeuge zu entwerfen. Dies lässt die weitere Entwicklung Suwelacks in diesem Bereich vermuten.

Der Flieger lehnte alle Angebote ab und schloss einen Anstellungsvertrag mit der Automobil & Aviatik AG, der das Einfliegen und Vorstellen der Aviatik-Flugzeuge auf Wettbewerben umfasste. Hierzu hielt sich Suwelack in Mühlhausen im Elsass auf, um seinen neuen Aufgaben nachzukommen.[326] Die Zusammenarbeit war jedoch nur von kurzer Dauer, von Mitte März bis Ende Mai 1912, denn die Maschine erwies sich als wenig zuverlässig und ging mehrfach zu Bruch, so dass das Arbeitsverhältnis wieder aufgelöst wurde.[327]

Der Wechsel zu Aviatik scheint zudem nicht reibungslos funktioniert zu haben, denn gegenüber Hubert Ahlers äußerte sich Josef zu seiner Teilnahme an den Billerbecker Flugtagen: „Einen definitiven Bescheid[,] ob und mit welchem Apparat ich Ostern komme, kann ich dir erst in ca. 8 Tagen geben." Der folgende Absatz des Briefes wurde wahrscheinlich später angefügt, als zwischen Rumpler und Suwelack beziehungsweise zwischen Aviatik und Suwelack Klarheit herrschte: „Anbei lege [ich] den Contract von mir vollzogen vor und teile Dir mit, dass ich mit einem Renneindecker [der Firma Aviatik] nicht Taube nach dort komme. Weiteres später. Hoffentlich seid Ihr Bande jetzt zufrieden."[328]

Verzögerungen und möglicherweise weitere Ungereimtheiten vonseiten Josef Suwelacks führten zu Gerüchten über eine Erkrankung des Fliegers, was jedoch

325 NL J. Suwelack, Schreiben Bruno Jablonskys vom Mai 1912. Siehe auch Beschreibung des Garuda -Eindeckers im Flugsport (1912), Nr. 7, S. 241f.

326 Siehe NL J. Suwelack, Hotelrechnung des Hôtel Central ausgestellt auf Josef Suwelack mit Quittierung vom 22.3.1912.

327 Vgl. NL J. Suwelack, Schreiben der Automobil & Aviatik AG an Josef Suwelack vom 11.7.1912. Darin forderte das Unternehmen Suwelack auf, eine Summe von 3.870,57 Mark zu überweisen. Darunter waren tabellarisch vom 20. März bis zum 14. Mai Zahlungsleistungen aufgeführt.

328 NL J. Suwelack, Brief Suwelacks an Hubert Ahlers vom 13.2.1912. Der Vertrag ist leider nicht überliefert.

vom „Billerbecker Anzeiger" dementiert wurde: „Ein in den letzten Tagen umgehendes Gerücht, die Veranstaltung sei durch eine schwere Erkrankung des Herrn Suwelack in Frage gestellt, entspricht ganz und gar nicht den Tatsachen."[329]

Die Billerbecker Flugtage fanden trotz stürmischen Wetters zwar statt, viele Flüge fielen aber aufgrund eines Unwetters buchstäblich ins Wasser. Suwelack gelang ein nur vierminütiger Aufstieg, da der Motor aussetzte und nach einer eineinhalbstündigen Reparaturpause seinen Dienst erneut aufgab.

Mit dem Aviatik-Eindecker beteiligte sich Josef Suwelack am 23. Mai 1912 auch am Flugtag in Gelsenkirchen-Essen-Rotthausen, der zur Eröffnung des neuen Flugplatzes an Pfingsten abgehalten wurde.[330] Der „Essener Generalanzeiger" berichtete in seiner Ausgabe vom 29. Mai 1912: „Suwelack vollführte auf seinem Aviatik-Eindecker prachtvolle Rundflüge, hatte jedoch das Mißgeschick, daß sein Apparat beim Landen infolge zu scharfer Kurve umkippte. Die Tragflächen brachen ab. Suwelack blieb gänzlich unverletzt".[331]

Ende Mai gab Suwelack – angesichts mehrfacher Unzuverlässigkeiten der Maschinen durchaus verständlich – die beiden Renneindecker wieder zurück, allerdings unvollständig und beschädigt, so dass Aviatik einen Prozess gegen ihn anstrengte, der ihm einige Unkosten bescherte. Gleichzeitig kam die Rückgabe der Flugzeuge einer Kündigung gleich, hatte Suwelack sich doch vertraglich dazu verpflichtet, die Aviatik-Maschinen einzufliegen und vorzustellen. Das Verfahren zog sich über ein Jahr hin und endete mit einer Geldstrafe für den Flieger sowie den zu tragenden Gerichts- und Anwaltskosten. Der Vorwurf lautete auf versäumte Geldrückzahlungen sowie auf Flugzeugzubehör-Unterschlagung.[332]

Die Entscheidung, sich von Aviatik ab- und einem neuen Projekt zuzuwenden, wird durch das langwierige Verfahren nachvollziehbar. Trotz der nur wenige Monate andauernden Beschäftigung als Pilot bei Aviatik bleibt aber festzuhalten, dass Suwelack auch hier Erfahrungen als Konstrukteur und Flieger verschiedener Modelle sammelte, die er für seine weitere Zukunft als Berufsflieger und Flugzeugbauer nutzbar machen konnte.

Um die folgenden Ereignisse, das heißt den Wechsel Josef Suwelacks zu den Kondor-Werken in Essen, zu verstehen, ist es nötig, auf die Luftfahrt- und Vereinsentwicklung im Ruhrgebiet, mit dem Fokus auf Essen, einzugehen sowie einige Personen und Institutionen vorzustellen, die bei der Gründung der Kondor-Werke 1912 von Bedeutung waren. Denn dass Josef Suwelack mehr oder weni-

329 Billerbecker Anzeiger vom 23.3.1912.

330 Zur Gründung des Flugplatzes Gelsenkirchen-Essen-Rotthausen siehe den Abschnitt zur Luftfahrt in Essen.

331 „Vom Flugtag Gelsenkirchen-Essen-Rotthausen", in: Essener Generalanzeiger vom 29.5.1912.

332 Vgl. NL J. Suwelack, Klageschrift „Automobil & Aviatik Ges. gegen Suwelack", zugestellt im Auftrag des Juristen Dr. Bell am 21.5.1913; Abschrift der Urteilsverkündung vor dem königlichen Amtsgericht Essen am 22.10.1913 im Fall „Aviatik Ges. gegen Suwelack", S.2f.

ger nahtlos eine Anstellung bei den neu gegründeten Kondor-Werken fand, war keineswegs selbstverständlich. So rasch, wie die Anzahl der Flugzeugführer im Jahr 1912 gewachsen war, so rasant stieg auch der Anteil arbeitsuchender Flieger und Monteure, die nach ihrer Ausbildung beziehungsweise nach Erhalt ihres Flugscheins keine Stelle fanden.[333]

Die neu entstandene und noch im Aufbau begriffene Luftfahrtindustrie zog vor allem junge Männer aus dem Wirtschaftsbürgertum, aber auch Handwerker und Arbeiter an. Die rege landesweite Erfindertätigkeit ließ zudem den Wunsch einiger Städte wachsen, zu einem attraktiven Fliegerstützpunkt aufzusteigen und Subventionen vom Kriegsministerium zu erhalten. Die Städte versprachen sich von der Errichtung von Flugplätzen und der Ansiedelung von Flugzeugfirmen gute Pachterträge und Einnahmen aus Flugveranstaltungen. Vor allem in Städten mit einer stark ausgeprägten Industrie oder mit örtlichen Garnisonen konnten diese Pläne erfolgreich in die Tat umgesetzt werden. Dort existierten geeignete Gelände und die nötigen Baustoffe konnten ebenfalls bereitgestellt werden.

Für Josef Suwelack boten sich im Ruhrgebiet die Chance eines Neuanfangs und größere Gestaltungsspielräume, denn er konnte als angehender Konstrukteur der neu gegründeten Kondor-Werke sowohl auf seine Erfahrungen bei Rumpler und Aviatik zurückgreifen, als auch seine eigenen Ideen einbringen. Ferner verringerte sich die räumliche Distanz zu seinem Elternhaus, was ihn möglicherweise ebenfalls bewogen hatte, seinen Arbeitsplatz zu wechseln.

Rückblickend mutet Suwelacks Wechsel nach Essen äußerst gewagt an. Obwohl sich kapitalkräftige Industrielle und Unternehmer fanden, die bereit waren, in ein Flugzeugwerk zu investieren, war die Gründung eines solchen Unternehmens, gerade auch angesichts des relativ späten Entstehungszeitpunktes, risikoreich und kostenintensiv. Wie schon sein Vater trat Josef Suwelack junior als Visionär auf, der an seinen Prinzipien festhielt, auf sein Wissen vertraute und unter seinem Namen Flugzeuge vertreiben und vorstellen wollte. Seine beruflichen Erfahrungen mit Rumpler und Aviatik, die so oder ähnlich auch anderen Piloten widerfahren sein mögen, verdeutlichen zugleich, wie unberechenbar und kurzlebig die Fliegerei für manche Piloten gewesen sein wird. Sich langfristig in der Fliegerbranche zu etablieren, war nicht nur eine Frage des persönlichen Durchhaltewillens und des privaten Kapitals, sondern wurde wesentlich von persönlichen Kontakten und gegenseitigem Vertrauen bestimmt.

Suwelacks Vertrauen in Edmund Rumpler und in die Qualität des Aviatik-Eindeckers hatte im Laufe der Zeit Risse bekommen, ebenso wie seine finanzielle Situation sich verschlechtert hatte. Der Bruch mit Rumpler und Aviatik stellte ihn vor die Wahl, an Altem festzuhalten oder Berlin den Rücken zu kehren und mit

333 Siehe z. B. Flugsport (1911), Nr. 22, Anzeigenteil.

der Vergangenheit (vorerst) abzuschließen. So hoffte der Westfale vielleicht, im Ruhrgebiet eine zweite Chance nutzen zu können und sich größere Freiräume zu schaffen. In diesem Sinne eiferte er seinem Vater nach, der mit Ideenreichtum und einem zunächst kleinen Betrieb über mehrere Jahre in sein Unternehmen investiert hatte, bis er expandieren konnte und es schließlich zu Ansehen und Wohlstand gebracht hatte. Suwelack junior mag mit seiner Entscheidung gezeigt haben wollen, dass er ein „echter Suwelack" und in der Lage war, es seinem Vater – wenn auch auf einem anderen Gebiet – gleichzutun.

2.4. Luftfahrt in Essen: Motive und Interessengruppen

Wie in Münster bildeten sich auch in Essen zu Beginn des 20. Jahrhunderts Luftfahrtvereine heraus, die Ballonfahrten sowie Flugveranstaltungen organisierten und Versammlungen abhielten. Die Stadt Essen als Körperschaft war in verschiedenen Flugvereinen Mitglied, ebenso wie der damalige Essener Oberbürgermeister Wilhelm Holle und sein Stellvertreter Paul Brandi. Beide gehörten dem Niederrheinischen Verein für Luftschiffahrt sowie dem Vorstand der Sektion Essen an und waren damit am Aufbau einer Essener Luftfahrt beteiligt.[334] In den Vereinen saßen Großindustrielle aus dem Wirtschaftsbürgertum sowie Vertreter der Politik.

Der Niederrheinische Verein für Luftschiffahrt (NVfL) wurde am 15. Dezember 1902 durch Dr. Karl Bamler (1865–1926)[335] in Wuppertal-Barmen gegründet. 1895 war Bamler als Reservist während einer Übung zu einer Berliner Luftschifferabteilung gekommen, und auch zuvor hatte es bereits erste Berührungspunkte mit der Ballonfahrt gegeben. Bamler erkannte den Nutzen der Ballonfahrten für die Meteorologie und schaffte es, durch Vorträge und Lichtbilder 21 Leute zu gewinnen, die ihm bei der Gründung eines Luftfahrtvereins helfen sollten. Dieser hieß zunächst Niederrheinischer Verein zur Förderung der Luftschiffahrt, erhielt bald darauf aber seinen neuen Namen.[336] Bamler wurde Fahrtenwart des Vereins und der Barmer Oberbürgermeister Lentze zum ersten Vorsitzenden gewählt, 1907 wurde er dann vom neuen Oberbürgermeister Voigt abgelöst. Im Jahr 1908

334 Vgl. Jahrbuch des DLV (1910), S.192.

335 Die folgenden Ausführungen basieren auf Erwin Dickhoff, Essener Köpfe. Wer war was?, Essen 1985, S.11.

336 Vgl. Hugo Rieth, Der Wind bestimmt die Richtung, in: Jahrbuch Essen (1988), S.91–104, hier S.94. Zu den Gründungsmitgliedern gehörten unter anderem: der Fabrikant Hugo Toelle, der Kaufmann Hugo Eckert, Oberbürgermeister Lentze, Dr. Pistor und der Kommerzienrat Albert Molineus, allesamt also Leute aus den gehobenen Schichten, siehe Ernst Milach, Aus dem Reich der Lüfte, Bonn 1908, S.2 und ders., Die Geschichte des Niederrheinischen Vereins für Luftschiffahrt, in: Festschrift zum XXI. Deutschen Luftfahrer-Tag anläßlich des 25jährigen Jubiläums des Deutschen Luftfahrt-Verbandes und des Niederrheinischen Vereins für Luftschiffahrt: Sektionen Wuppertal und Bonn, Juni 1927, Barmen 1927, S.26–44, hier S.27.

ging das Amt an Hugo von Abercron über, auf den 1912 der Essener Rechtsanwalt Victor Niemeyer – dieser vertrat Josef Suwelack im Prozess gegen Aviatik – folgte.[337]

Niemeyer (1863–1949) stammte aus Berge in Hamm und kam 1876 nach Essen.[338] Wie sein Vater studierte er nach dem Abitur Rechtswissenschaften und ging für seine Studien nach Freiburg, Leipzig, Greifswald und Berlin, bevor er 1887 für das Referendariat nach Essen zurückkehrte. Er engagierte sich politisch, war von 1898 bis 1929 durchgängig Abgeordneter der Stadt Essen und gehörte verschiedenen kulturellen Vereinen an. Niemeyer schaffte den Anschluss an die wirtschaftliche Elite und pflegte beispielsweise Kontakte zum Hause Krupp. Gleichzeitig widmete er sich seit der Jahrhundertwende der Ballonfahrt. So verfasste er Artikel zur Luftfahrt und zum Luftfahrtrecht, war 1911 Vizepräsident des ersten internationalen Kongresses für Luftrecht in Paris,[339] Mitbegründer einiger Luftfahrtvereine, Vorstandsmitglied im DLV und Vorsitzender des NVfL.

Zweck des NVfL war gemäß § 3 der Satzung, „die Luftschiffahrt in den gebildeten Kreisen Deutschlands beliebt zu machen, einen gesunden Luftsport zu pflegen und die wissenschaftliche Erforschung der Atmosphäre nach Möglichkeit zu unterstützen." Diese Ziele sollten durch Vorträge, Veranstaltungen von Ballonfahrten, Förderung von Fachzeitschriften und durch die Beteiligung an internationalen Fahrten erreicht werden.[340] Der NVfL pflegte gute Kontakte zum DLV, wovon Bamler auch im Vereins-Jahrbuch von 1913 berichtete. Im DLV waren vier der sechs Mitglieder des NVfL aus der Essener Sektion vertreten, namentlich Justizrat Dr. Niemeyer, Professor Bamler, Dr. Peill und Ernst August Schröder.[341] Diese Herren waren bis auf Dr. Peill alle spätere Gesellschafter der Kondor-Werke.

Unter den Mitgliedern des NVfL waren zahlreiche – bis heute – bekannte Namen der Industrie, darunter Franz Haniel, die Goldschmidts, Fritz Thyssen oder Gustav Krupp von Bohlen und Halbach. Sie alle, mit Ausnahme der Goldschmidts, die selbst Ballonfahrer waren, waren passive Förderer der Luftfahrt. Gustav Krupp von Bohlen und Halbach unterstützte den NVfL mit Spenden. So investierte er z. B. 3.000 Mark in den Bau der Essener Flugmaschine und stiftete

337 Vgl. hierzu die entsprechenden Jahrbücher des DLV.

338 Die folgenden Ausführungen sind angelehnt an Dickhoff, Essener Köpfe, S.175; Reichshandbuch der deutschen Gesellschaft – Das Handbuch der Persönlichkeiten in Wort und Bild, Bd. 2, Berlin 1931, S.1329 und Viktor Niemeyer, Lebenserinnerungen eines Siebzigjährigen, Berlin 1937, bes. S.135ff.

339 Siehe dazu den Artikel in: Flugsport (1911), Nr. 14, S.496f.

340 Jahrbuch des DLV (1903), S.204, zit. nach http://www.ballon-bve.de/?page_id=31, online seit Februar 2012 (3.6.2014).

341 Vgl. Karl Bamler, Jahresbericht, in: Jahrbuch des NVfL (1912/13), S.20, Zitat ebd.

Preisgelder für Flugwochen. Die Goldschmidts wurden später Hauptgesellschafter der Kondor-Werke.[342]

Widmete sich der Verein zunächst hauptsächlich dem Ballonsport, so wurde seit 1908 auch verstärkt der Fliegerei schwerer als Luft Aufmerksamkeit zuteil. Hierzu richtete die Essener Sektion eine flugtechnische Kommission ein, die sich im Flugzeugbau versuchte. Ihr gehörten unter anderem Karl Bamler und Ernst-August Schröder an.[343] Durch Spenden aus der Industrie und freiwillige Zahlungen der Mitglieder standen Bamler zufolge 1909 mehrere zehntausend Mark zum Bau eines Zweideckers zur Verfügung. Ein Jahr später, am 1. Februar 1910, fanden auf den Ruhrwiesen Probeflüge statt, die allerdings nicht erfolgreich verliefen.[344] Da Fortschritte sich auch danach nicht einstellten, gab die Kommission 1912 die Arbeit an weiteren Flugzeugen auf, was auch den erschöpften finanziellen Ressourcen und einer regen Flugzeugproduktion in anderen Teilen Deutschlands geschuldet war.[345] Dennoch wurde die Etablierung einer Luftfahrt weiter verfolgt, so z. B. durch Ernst August Schröder, der zusammen mit Karl Bamler in der Geschäftsführung der 1908 gegründeten Rheinisch-Westfälischen Motorluftschiff-Gesellschaft saß und sich für die Durchführung von Flugveranstaltungen im rheinisch-westfälischen Industriegebiet stark machte.[346]

Schröder kann als einer der wichtigsten Förderer der Essener Luftfahrt gelten. Neben Vorstandstätigkeiten in Luftfahrtvereinen war er an der Gründung des Flugplatzes Gelsenkirchen-Essen-Rotthausen und der Kondor-Werke finanziell beteiligt.[347] Er sah sich dabei selbst anscheinend eher in der Rolle eines Mäzens, denn sowohl nach erfolgter Gründung des Flugplatzes als auch bei den Kondor-Werken zog er sich aus dem Tagesgeschäft zurück, obwohl er in beiden Gesellschaften im Vorstand saß.[348] Unter Schröders Leitung wurde nicht nur der Sitz des NVfL von Elberfeld nach Essen verlegt, gleichzeitig verlagerte sich auch die

342 Dazu HA-Krupp WA 41/73–130, Spendenliste Krupp; HA-Krupp WA 41/72–116, Schreiben Direktor Gillhausen, Krupp, an OB Machens, Gelsenkirchen, vom 29.6.1912.

343 Vgl. Bamler, Jahresbericht, in: Jahrbuch des DLV (1909), S.269f.

344 Berichte über die „Essener Flugmaschine" sind zu finden in: DLZ (1910), Nr. 3, S.28 und (1910), Nr. 6, S.21 oder auch in: Flugsport (1910), Nr. 4, S.112 und (1910), Nr. 6, S.179.

345 Vgl. ebd., S.7 und S.18.

346 Siehe zur Entwicklung dieses Vereinszweiges: Rißmann-Ottow, Glück ab!, S.99ff.

347 Darstellungen zu Schröders Leben finden sich in: Reichshandbuch der Deutschen Gesellschaft, S.1710; Ernst August Schröder, Die Sportluftfahrt im Ruhrland, in: Ruhrland Almanach der Lüfte, Essen 1929, S.19–23; Ernst August Schröder (jun.), Ernst August Schröder 1873–1949. Ein deutsches Bürgerleben vom Ausgang des 19. in das folgende schicksalhaft-wechselvolle Halbjahrhundert hinein aus eigener Kraft gestaltet, Essen/Nümbrecht-Benroth 1975–1977, bes. S.69ff. und ders., Von den Wehen in der Geburtsstunde des Luftverkehrs, in: Beiträge zur Geschichte von Stadt und Stift Essen (1991/92), H. 104, S.101–122 und Dickhoff, Essener Köpfe, S.209.

348 Siehe entsprechende Akten: StdAE Bestand 102 I, Akte 1059–1061 und Protokolle der Kondor-Werke.

Arbeit der Gesellschaft auf die Subvention und Durchführung von Flugveranstaltungen.

Nach der Organisation einer Fahrt des Parseval-Luftschiffes PL 6 nach Essen im Sommer 1911 leitete Schröder Verhandlungen mit den Kommunen und Behörden ein, um die Errichtung eines Flugplatzes in Rotthausen zu erwirken.[349] Um seine potenziellen kommunalen Geldgeber von der Notwendigkeit eines Flugplatzes zu überzeugen, unterstellte er eine französische Kriegsbedrohung für das Industriegebiet. In seinen in den 1940er Jahren entstandenen Tagebuchaufzeichnungen schrieb er: „Das Ruhrrevier ... liegt als offenes Weichteil an der Westflanke des Reiches, ungeschützt gegen eine überlegene französische Fliegerei."[350]

Noch deutlicher trat der von Schröder gesehene militärische Wert einer Flugzeugindustrie im Ruhrgebiet in einer Bittschrift um Subventionen an das Berliner Kriegsministerium vom November 1911 hervor. In dem Schreiben unterrichtete Schröder das Kriegsministerium von der Absicht der Essener Sektion, einen Flugplatz und entsprechende Anlagen zu errichten, um „im Falle eines Feldzuges im kriegsbrauchbaren Zustande der Heeresverwaltung zur Verfügung zu stehen." Weiter hieß es dort:

> „Sowohl Luftschiff- wie Flugzeugführung sollen herangebildet werden. Wir streben im Grundgedanken an, dazu beizutragen, dass auf der Strecke Berlin – Aachen und an der ganzen Westgrenze entlang Luftschiffe und Flieger weitmöglichst Platz mit den erforderlichen Einrichtungen für Unterkunft finden können."[351]

Für die Umsetzung dieser Ziele beantragte Schröder Subventionen zwischen 125.000 und 135.000 Mark. Am 12. Dezember sicherte ihm das Kriegsministerium eine einmalige Anzahlung in Höhe von 15.000 Mark zu, stellte jedoch noch weitere Zuschüsse in Aussicht, sollte sich der Flugplatz rentieren.[352] Das war durchaus bemerkenswert, denn das Kriegsministerium äußerte noch im Mai 1912 gegenüber der General-Inspektion des Militär- und Verkehrsflugwesens Zweifel an der Rentabilität von Flugplätzen. Es wusste aber ebenso, dass die Fluggesellschaften und -vereine sowie die Städte nur „lebensfähig" waren, „wenn ihnen durch Bewährung von Subventionen eine angemessene Verzinsung des Anlagekapitals zugesichert wird." Flugplätze, die „nach Lage und Beschaffenheit den militärischen Anforderungen" entsprachen, sollten daher unterstützt werden.[353]

349 Vgl. Rißmann-Ottow, Glück ab!, S.102.

350 Vgl. Schröder (jun.), Ernst August Schröder 1873–1949, S.109.

351 HA-Krupp WA 41/72–116, Schreiben der Rheinisch-Westfälischen Motorluftschiff-Gesellschaft verfasst von E. A. Schröder an das Kriegsministerium Berlin (Bitte um Subventionen des Kriegsministeriums für die Rheinisch-Westfälische Flug- und Sportplatzgesellschaft mbH), o.O./o.D. (wahrscheinlich Ende des Jahres 1911).

352 HA-Krupp WA 41/72–116, Antwortschreiben des Berliner Kriegsministeriums an E. A. Schröder vom 12.12.1911.

353 BA-MA PH 24/72, Schreiben des Kriegsministeriums an die General-Inspektion des Militär- und Verkehrsflugwesens vom 9.5.1912.

Das Kriegsministerium registrierte ein zunehmendes Interesse vieler Städte an der Schaffung solcher Anlagen und hatte Anträge aus Aachen, Duisburg, Dresden, Gelsenkirchen, Leipzig, Schwerin, Wanne und Wiesbaden erhalten.[354] Man rechnete mit Gesamtkosten zwischen 500.000 und 600.000 Mark, denen nur „geringe Rückeinnahmen (aus Schaustellungen usw.)" gegenüberstünden, so dass „daß [sic!] Kriegsministerium in absehbarer Zeit nicht in der Lage" sei, „diese Summe für eine größere Anzahl derartiger Anlagen aufzubringen." Allerdings dürfe man den Auf- und Ausbau von Flugplätzen nicht vernachlässigen, da Vorkehrungen „für spätere Fliegerstationen" getroffen werden müssten und eine genaue Abwägung von Investitionen nötig sei.[355]

Im Juni 1912 wurden Auswahlkriterien für Flugstützpunkte formuliert, die sich auf die Ausstattung und die Lage des Geländes konzentrierten: „Derartige Flugstützpunkte müssen nach Möglichkeit Schuppen für 4–6 Flugzeuge und eine Werkstatt erhalten, ein Fluggelände von 500x1000 m besitzen, schließlich noch Gelegenheit zur Unterbringung von einer Fliegerabteilung dem neuen verstärkten Etat bieten."[356]

Weiterhin mussten die Flugplätze die „Abhaltung von Wiederholungskursen" für bereits ausgebildete Flugzeugführer ermöglichen, sich als Ausgangspunkte für Truppenübungen eignen sowie als Stützpunkte für Zwischenlandungen bei Überlandflügen dienen können, was die Bereitstellung von Werkstatträumen, Öl- und Benzinlagern etc. umfasste. Entscheidend war zudem das Vorhandensein von Garnisonen. Das schloss wiederum einige Städte aus: „Auf Duisburg, Wanne und Gelsenkirchen dürfte, weil sie im Industriegebiet liegen und keine Garnision [sic!] besitzen, zu verzichten sein."[357] Die zuständige Behörde konzentrierte sich auf „leistungsfähige und kapitalkräftige Firmen", die den „verhältnismäßig geringen Bedarf" decken sollten, denn es sei nicht „die Aufgabe des Staates ..., eine große Zahl kleiner wenig leistungsfähiger Fabrikanten durch Subventionen künstlich über Wasser zu halten."[358] Einmal mehr zeigt sich an dieser Stelle, wie kritisch und vorsichtig das Kriegsministerium Neugründungen gegenüberstand.

Die Vorbereitungen zur Flugplatzgründung begannen nach der von Schröder

354 BA-MA PH 24/72, Schreiben des Kriegsministeriums an die General-Inspektion des Militär- und Verkehrsflugwesens vom 21.5.1912.

355 Ebd., Klammern im Original.

356 BA-MA PH 24/72, Betr. Subvention von Flugplätzen. Schreiben an die General-Inspektion des Militär- und Verkehrsflugwesens vom 7.6.1912. Als Kriterien für Flugplätze wurden aufgeführt: Tauglichkeit für militärische Zwecke, Schuppen (16x16 m) für mindestens sechs Flugzeuge, Vorhandensein einer Reparaturwerkstatt, Möglichkeit der Unterbringung von mindestens 3.000 kg Benzin und 1.000 kg Öl, dauerhaft guter Zustand für mindestens zehn Jahre und Nähe zu Garnisonen. Siehe ebd.

357 Ebd.

358 BA-MA PH 24/72, Antwortschreiben der Inspektion des Luft- und Kraftfahrwesens vom 27.5. betr. der Denkschrift des Deutschen Luftfahrer-Verbandes von April 1912.

organisierten, im Juli 1911 stattfindenden Fahrt des Parseval-Luftschiffes über Essen. Dabei sei, so konnte man einer Druckschrift von September des gleichen Jahres entnehmen, ein „hervorragend geeignetes" Gelände für einen Flugplatz im nordöstlichen Teil des Landkreises Essen entdeckt worden.[359] Durch die Gründung einer Flugplatzgesellschaft sollte erreicht werden, „die Luftschiffahrt im Westen unseres Vaterlandes im Interesse einer weiteren gesunden Entwicklung und vor allem unserer Landesverteidigung zu fördern."[360]

Bereits im Dezember 1911 lag ein Satzungsentwurf für die Gründung einer Rheinisch-Westfälischen Flug- und Sportplatz GmbH vor, mit geplantem Sitz in Gelsenkirchen und einer weiteren Niederlassung in Aachen. Auf den am jeweiligen Ort zu errichtenden Flugplätzen sollten Luftschiffe in Betrieb gehen sowie Flugzeugschuppen errichtet und dann vermietet werden, ferner Flugveranstaltungen stattfinden.[361]

Man kalkulierte mit einem Gründungskapital von 1,2 Millionen Mark, das von den Städten Aachen und Gelsenkirchen zu je 40.000 Mark, zu je 50.000 Mark von der Stadt Leichlingen und Herrn Herwarth zu Bittenfeld[362] sowie zu 100.000 Mark von Arthur Müller, dem Gründer des Johannisthaler Flugplatzes, aufgebracht werden sollte. Für den verbliebenen Betrag mussten weitere Geldgeber gefunden werden.[363]

Die Rentabilitätsberechnung für ein Jahr legte optimistisch einen Jahresumsatz von 732.000 Mark und einen Reingewinn von 200.000 Mark zugrunde sowie Subventionen des Kriegsministeriums von 40.000 Mark. Dies wurde, entgegen der üblichen Vorgehensweise, dem geplanten Gründungskapital zugerechnet.[364] Die kalkulierte Subvention des Kriegsministeriums blieb jedoch weit hinter den Erwartungen zurück, die Stadt Aachen stieg ganz aus dem Flugplatz-Projekt aus und die Stadt Gelsenkirchen war bereit, höchstens 300.000 Mark zu bewilligen, so dass das Gründungskapital auf 800.000 Mark nach unten korrigiert wurde.

Am 19. März 1912 wurde die Gründung der Gesellschaft unter dem Namen Westdeutsche Flugplatzgesellschaft mbH mit Sitz in Gelsenkirchen vollzogen.

359 HA-Krupp WA 41/72–116, Anonym (evtl. E. A. Schröder), Über die Vorzüge des geplanten Flugplatzes bei Gelsenkirchen. Ausgegeben September 1911, 2 Seiten, Zitat ebd.

360 Ebd.

361 HA-Krupp WA 41/72–116, Rheinisch-Westfälische Flug- und Sportplatz GmbH. Entwurf der Satzungen, o.O./o.D. Der Entwurf muss im Dezember entstanden sein, da der Landrat des Landkreises Essen in einem Schreiben vom 19. Dezember an Gustav Krupp von Bohlen und Halbach diesen zu einer finanziellen Beteiligung am Flugplatz-Projekt bewegen wollte, vgl. ebd.

362 Diese waren in die Durchführung der Luftschifffahrt des Parseval 6 im Juli 1911 über Essen involviert, ebenso wie auch die Stadt Aachen und Arthur Müller, der das Luftschiff bereitstellte. Siehe Rißmann-Ottow, Glück ab!, S.163ff.

363 HA-Krupp WA 41/72–116, Veranschlagte Kapital-Dispositionen für die Rheinisch-Westfälische Flug- und Sportplatz GmbH, o.D. (wahrscheinlich Dezember 1911).

364 HA-Krupp WA 41/72–116, Rentabilitätsberechnung der Rheinisch-Westfälischen Flug- und Sportplatz GmbH., o.D. (vermutlich Dezember 1911).

Die Ziele wurden im Unterschied zum Satzungsentwurf von 1911 noch um „die Errichtung und [den] ... Betrieb einer Automobil- und Radrenn-Lehr- und Versuchsbahn, ferner [um] die Abhaltung und Förderung sportlicher Veranstaltungen jeder Art“ erweitert.[365] Im zehnköpfigen Vorstand der neu gegründeten Flugplatzgesellschaft saßen, bis auf drei Ausnahmen, ausschließlich Mitglieder des NVfL, am 4. April wurde nachträglich noch Victor Niemeyer hinzugewählt.[366] Geschäftsführer wurden der Kaufmann Paul Giersberg sowie der Luftschiffführer und Oberleutnant a.D. Leo Forsbeck. Beide kamen aus Essen und zählten ebenfalls zu Mitgliedern des NVfL.[367] Das Entscheidungsmonopol lag damit in der Hand des Essener Luftfahrtvereins.

Für die Zeit vom 25. bis zum 28. Mai 1912 war die Eröffnung des Fluglatzes mit einer Flugwoche geplant, auf der sowohl Flieger als auch Luftschiffer ihr Können zeigen sollten (Abb. 9). Zwischen 1912 und dem Beginn des Ersten Weltkrieges wurden weitere Flugwochen veranstaltet, für die besonders zu Beginn erhebliche Preisgelder ausgesetzt waren.[368] Die Krupp-Flugwoche nahm aufgrund des hohen Preisgeldes und der Teilnehmerzahl einen vorderen Platz unter den insgesamt vierzehn großen Flugveranstaltungen des Jahres 1912 ein.

Nichtsdestotrotz schrieb die Westdeutsche Flugplatzgesellschaft rote Zahlen und stand mehrfach vor dem Aus, denn die Mietzahlungen der Pächter waren nicht gesichert, Pachterträge aus der Flugplatzgastronomie etc. gering und Subventionen blieben zusehends aus. Allein im Jahr 1914 betrug der Jahresverlust über 100.000 Mark und zwischen der Gründung der Flugplatzgesellschaft 1912

365 Vgl. StdAE Bestand 102 I, Akte 1059, Gesellschaftsvertrag der Westdeutschen Flugplatzgesellschaft vom 19.3.1912, § 2 und 3. Dem „Flugsport“ zufolge waren weitere Gemeinden an der Gründung finanziell beteiligt. Die Zeitschrift sprach von einem Gesamtkapital von 1,1 Millionen Mark. Allerdings erscheint der Gesellschaftsvertrag als Quelle glaubwürdiger, da es wenig Sinn gemacht hätte, weitere Geldgeber nicht aufzuführen. Siehe Flugsport (1912), Nr. 5, S. 191.

366 Vgl. StdAE Bestand 102 I, Akte 1059, Gesellschaftsvertrag der Westdeutschen Flugplatzgesellschaft vom 19.3.1912. Im Vorstand saßen: OB Theodor Machens und Bürgermeister Karl von Wedelstaedt für die Stadt Gelsenkirchen, Landrat Hans von Eynern für den Landkreis, Bürgermeister Heinrich Hohoff für Rotthausen, Beigeordneter Robert Schmidt für die Stadt Essen, für den NVfL zu Barmen Ernst August Schröder und Prof. Dr. Karl Bamler, für die Rheinisch-Westfälische Motorluftschiff-Gesellschaft Dr. jur. Karl Heimann aus Bochum und Bankdirektor Ewald Sachs sowie Bergwerksdirektor Ernst Linderhaus aus Katernberg für das Steinkohle-Bergwerk Zollverein. Hohoff, Schmidt und Linderhaus waren als einzige nicht Mitglied im NVfL. Zur Nachernennung Niemeyers siehe StdAE Bestand 102 I, Akte 1059, Abänderung des Gesellschaftsvertrages der Westdeutschen Flugplatzgesellschaft vom 4.4.1912.

367 Vgl. ebd.

368 Folgende Flugwochen fanden nach der Eröffnungs-Flugwoche auf dem Flugplatz statt: „Krupp-Flugwoche“ vom 4.–11.8.1912 (ges. Preisgeld 100.000 M, ausgezahlt: 75.100 M); „Jubiläums-Flugwoche“ vom 27.7.–3.8.1913 (ges. Preisgeld 45.000 M); „Flugzeugmodell-Ausstellung und -Wettfliegen“ vom 21.–28.9.1913; „Flugwoche Industrie-Gebiet“ vom 12.–19.7.1914. Informationen entnommen aus Flugsport (1913), Nr. 15, S. 549; (1913), Nr. 17, S. 636f.; (1913), Nr. 21, S. 816; (1914), Nr. 15, S. 664f. Vgl. auch Flugzeugwettbewerbe, in: DLV-Jahrbuch (1913), S. 155–166, und DLV-Jahrbuch (1914), S. 167–187, bes. S. 173; ebenso Karl Bamler, Die Jubiläums-Flugwoche in Gelsenkirchen, in: DLZ (1913), Nr. 17, S. 409–411.

Abb. 9: Werbeplakat für die Flugplatzeröffnung am 25. Mai 1912
(aus: Essener Generalanzeiger vom 23. Mai 1912)

und Kriegsende insgesamt über 480.000 Mark.[369] Es wurde sogar überlegt, die Gesellschaft ganz aufzulösen.[370]

Vonseiten der Vereine wurde für den Erhalt des Platzes plädiert. Schröder schrieb im März 1916 an den Gelsenkirchener Oberbürgermeister: „Zweifellos drängt die Entwickelung [sic!] des Flugzeugbaues dahin, dass auch in Zukunft grosse Plätze nötig sein werden. ... Wir hoffen und wünschen nur, dass auch in der zukünftigen Entwickelung unsere Westdeutsche Fluggesellschaft [sic!] und unser Platz lebhaften Anteil haben."[371] Alle Bemühungen zur Flugplatzrettung scheiterten jedoch letztendlich, denn Essen und die drei anderen Kommunen schieden aus der Westdeutschen Flugplatzgesellschaft freiwillig aus, wenngleich sich das Auflösungsverfahren lange hinzog und die Stadt Essen zwischenzeitlich wieder in die Gesellschaft eintrat.[372] Die Gesellschaft überdauerte zwar den Ersten Weltkrieg und auch einen Teil der Weimarer Zeit, doch schließlich wurde sie Ende der 1920er Jahre aufgelöst. Die Akten der Westdeutschen Flugplatzgesellschaft enden am 28. September 1928 mit einem am 15. Oktober 1930 hinzugefügten Vermerk, dass die Flugplatzangelegenheit mit der „Uebertragung der Essener Geschäftsanteile an die Stadt Gelsenkirchen erledigt"[373] sei.

Die zentralen Motive, Geld in einen Flugplatz zu investieren, waren sportlicher und wirtschaftlicher Natur. Essen sollte als Luftfahrtstandort attraktiv und das Kriegsministerium von der Notwendigkeit einer wehrhaften Luftfahrtindustrie überzeugt werden. Das Interesse von Kaufleuten wie Ernst August Schröder und Industriellen wie Krupp, Haniel oder den Goldschmidts an der Luftfahrt zeigt die grundsätzliche Akzeptanz der Fliegerei in den Städten und die Bereitschaft, in neue Technologien zu investieren.

Dass die Gesellschafter trotz hoher finanzieller Verluste bis Ende der 1920er Jahre am Flugplatz festhielten, unterstreicht um ein Weiteres, welche Bedeutung die Städte und Vereinsmitglieder der Westdeutschen Flugplatzgesellschaft und einer funktionierenden Flugzeugindustrie im Ruhrgebiet beimaßen. Die Förderer der Luftfahrt ließen nichts unversucht, um das Geschäft und den Betrieb aufrechtzuerhalten, ein Verhalten, das sich später auch in den Gesellschafterprotokollen der Kondor-Werke widerspiegelt.

369 StdAE Bestand 102 I, Akte 1059, Blatt 107–108, Bilanz und Gewinn-und-Verlust-Rechnung der Westdeutschen Fluggesellschaft mbH, Gelsenkirchen, den 28. Februar und den 31. Dezember 1918, Abschrift.

370 Vgl. StdAE Bestand 102 I, Akte 1059, Blatt 7, Schreiben der Stadt Essen an OB Machens, Gelsenkirchen, Essen, 3.11.1914.

371 StdAE Bestand 102 I, Akte 1059, Blatt 58–62, Rheinisch-Westfälische Motorluftschiff-Gesellschaft [Verfasser: E. A. Schröder], Über die Entwicklung der Flugtechnik, Essen, den 6. März 1916.

372 Vgl. StdAE Bestand 102 I, Akte 1059, Blatt 95, Schreiben OB Machens an den OB der Stadt Essen vom 27.2.1918. Zur Anteilsübertragung und zum Auflösungsprozess siehe ebd., Blatt 98–100.

373 StdAE Bestand 102 I, Akte 1061, Vermerk vom 15.10.1930 auf einem Schreiben an den OB der Stadt Gelsenkirchen vom 20.9.1928.

Für aufstrebende Flieger wie Josef Suwelack, die auf der Suche nach besseren Arbeitsbedingungen waren, stellte der Flugplatz Gelsenkirchen-Essen-Rotthausen mit neu entstehenden Flugzeugwerken eine Option dar, die Aussicht auf Erfolg, Einfluss und persönliche Entfaltung versprach. Prominente Persönlichkeiten wie Bamler, Ernst August Schröder oder die Goldschmidts verhalfen dem Unterfangen aufgrund ihrer Tätigkeiten als Bankiers oder Unternehmer zu einem seriösen und erfahrenen Ansehen. Nicht zuletzt das daraus resultierende gegenseitige Vertrauen war eine wesentliche Voraussetzung dafür, sich gemeinsam auf die Flugzeug-Branche einzulassen.

2.5. Flugplatzroutinen und Ausbildungsbetrieb II: Kondor-Werke

Nebenbetriebe auf dem Rotthausener Flugplatz

Um eine Einordnung der Kondor-Werke im Kontext der Entstehung einer „Flugzeugindustrie" im Ruhrgebiet vornehmen zu können, ist es erforderlich, die Entwicklung der Fliegerei schwerer als Luft in Essen abseits des Flugplatzes Gelsenkirchen-Essen-Rotthausen, zumindest ansatzweise, zu skizzieren. Die Rekonstruktion erwies sich jedoch zum Teil als schwierig, da nur wenige Informationen über die betreffenden Unternehmen an die Öffentlichkeit gedrungen sind. Hinweise auf Essener Flugzeugfirmen finden sich vor allem im „Jahrbuch der Luftfahrt" aus dem Jahr 1912 von Ansbert Vorreiter. Dort sind Firmen aufgelistet, die sich seit 1911 mit der Luftfahrt im weitesten Sinne befassten, darunter auch Firmen aus Essen. Dies waren: die Kreiselmotoren GmbH, Hansa-Haus, die Firma Weiland & Gildemeyer, die Niederrheinische Flugzeug-Bauanstalt aus Altenessen sowie die Friedrich Krupp AG, die Munition und Luftabwehrgeschütze gegen Luftschiffe anbot.[374]

Im Nachlass Josef Suwelacks sowie in wenigen Luftfahrtzeitschriften finden sich Hinweise auf zwei der drei genannten Firmen. So umfasst der Nachlass Suwelacks zwei Schreiben des Vereins Creditreform an die Molkerei Billerbeck aus dem Jahr 1911, in denen sich Informationen zur Niederrheinischen Flugzeug-Bauanstalt finden.[375] Gegründet im Sommer 1910 von Otto Hilsmann, Sohn eines Arztes aus Barmen, der sich „dem Flugsport widmete", vertrieb die Firma Motoren und baute Flugzeuge, eine genaue Stückzahl ist jedoch nicht bekannt. Nachweisen lassen sich zwei Flugzeugbauten, doch möglicherweise wurden noch mehr

374 Informationen aus Vorreiter, Jahrbuch der Luftfahrt 1912, S. 311ff., 623, 626, 634 und 646. Siehe auch die Annonce der Niederheinischen Flugzeugbauanstalt Hilsmann & Co, in: Flugsport (1911), Nr. 10, Anzeigenteil.

375 NL J. Suwelack, Verein Creditreform, Bochum an die Molkerei Billerbeck i. W. vom 15. und 17.1.1911. Vgl. für die folgenden Informationen im Fließtext ebd.

Typen produziert.[376] Im Februar 1911 wurde die Firma ins Handelsregister eingetragen. Der „Flugsport“ brachte dazu eine kurze Notiz mit folgendem Inhalt:

> „Die Gesellschaft hat, nachdem die Versuche [gemeint sind wohl Prüfverfahren für Motoren] beendet sind, mit dem Bau von Motoren in größerem Maßstabe begonnen. Weiter hat die Gesellschaft mit der Gemeinde Holten bei Oberhausen einen 8jährigen Pachtvertrag abgeschlossen, wonach das dortige Fluggelände von 800 x 200 m zur Verfügung steht. ... Im Mai dieses Jahres soll mit den aus der Fabrik hervorgegangenen Maschinen ein Schaufliegen veranstaltet werden.“[377]

Besagter Flugtag fand in Holten statt und wurde von der Bauanstalt organisiert; auch eine Maschine der Firma kam zum Einsatz.[378] Eine weitere Notiz findet sich ebenfalls im „Flugsport“, wonach „der Flieger Hechtfischer“ als Fluglehrer bei der Fliegerschule der Niederrheinischen Flugzeug-Bauanstalt tätig sei.[379] Darüber hinausgehende Meldungen fehlen; einzig eine Darstellung aus den 1940er Jahren nennt Hilsmann noch einmal im Kontext der Fliegerschule in Holten.[380] Bereits zwei Jahre später, am 23. September 1912, wurde die Firma aus dem Handelsregister wieder gelöscht. Nach der Aufgabe des Flugbetriebes übernahm der Duisburger Karl Strack den Platz, baute eine Fliegerschule und bildete in der Folgezeit Flugschüler aus.[381] Da über die anderen Essener Flugunternehmen noch weniger Hinweise zu finden sind, ist anzunehmen, dass sie ebenfalls nur für kurze Zeit existierten.

Die Firma Weiland & Gildemeyer hatte ihren Sitz in Bremen und errichtete 1909 eine weitere Niederlassung in Essen, die 1910 zum Hauptsitz wurde. In Essen wurde die Firma nur noch von Adolf Weiland unter Beibehaltung des Firmennamens geführt, da Heinrich Gildemeyer ausgeschieden war. Das Unternehmen vertrieb Flugmotoren, wurde am 17. Juli 1911 jedoch schon wieder aufgelöst.[382]

Die Kurzlebigkeit dieser Firmen ist kennzeichnend für den Beginn der kommerziellen Luftfahrt, denn viele Unternehmer und Konstrukteure konnten mit den größeren Flugzeugfabriken nicht konkurrieren und lösten sich wieder auf. Trotzdem zeigen die Firmengründungen, dass die Fliegerei in der zeitgenössischen Erwartung Aussicht auf ökonomischen Erfolg versprach.

Auf dem Flugplatz Gelsenkirchen-Essen-Rotthausen waren neben den Kondor-Werken noch drei weitere Unternehmen beheimatet: die Alstra-Flugzeugwerke,

376 Vgl. Luftschiffahrt, Flugtechnik und Sport (1911), Nr. 13, S. 13.
377 Vgl. Flugsport (1911), Nr. 4, S. 138, zitierte Passage ebd.
378 Vgl. Luftschiffahrt, Flugtechnik und Sport (1911), Nr. 13, S. 13 sowie Karl Bamler, Jahresbericht, in: Jahrbuch des NVfL (1911), S. 27 und S. 41.
379 Vgl. Flugsport (1911), Nr. 20, S. 721. Zitat ebd. Dort auch eine Abbildung mit vermutlich Flugschülern, -lehrer und Monteuren, S. 722.
380 Hubert Delvos, Die Entwicklung des Düsseldorfer Luftverkehrs, Düsseldorf 1941, S. 33.
381 Vgl. Flugsport (1912), Nr. 13, S. 518.
382 Bruno Lange, Das Buch der deutschen Luftfahrttechnik, 2 Bde., Mainz 1970, S. 385. Siehe auch die Anzeige von Weiland & Gildemeyer, in: Flugsport (1910), Nr. 2, Anzeigenteil.

die Flugschule Georg Mürau sowie der Unternehmer Josef Schlatter. Keiner der genannten Betriebe erreichte jedoch die Größe der Kondor-Werke, die als einzige den größeren Firmen wie Rumpler oder Fokker Konkurrenz machen konnten.[383]

Die Alstra-Flugzeugwerke erteilten bis 1914 Flugunterricht und bauten Flugzeuge, wenngleich Carl Strathmann schon kurz nach der Firmengründung wieder austrat und der Betrieb von Wilhelm Albers, einem ehemaligen Flugschüler Suwelacks, alleine fortgeführt wurde.[384] Albers selbst berichtete in Hackenbergers Buch „Die alten Adler" von 1960, dass er 1913 mit falschen Versprechungen nach Rotthausen gelockt worden sei und 1914 aufgrund des Baus der Trabrennbahn zum Flugplatz nach Wanne gewechselt habe.[385] Bei Bruno Lange und Peter Supf heißt es hingegen, Albers sei bereits Anfang 1913 nach Wanne umgezogen. Dort habe er beispielsweise Tragflächen und Rümpfe an andere Firmen wie die Kondor-Werke geliefert. Dass er auch am Flugzeugbau festhielt, legt ein Foto bei Supf nahe, das mehrere im Bau befindliche Flugzeuge vor der Parseval-Halle auf dem Wanner Flugplatz zeigt.[386]

Die Nachfrage durch Flugschüler war allem Anschein nach nicht besonders groß, denn bis 1914 lässt sich nur ein Flugschüler der Alstra-Flugzeugwerke, Leonard Dirks, nachweisen. Dieser kaufte im September 1912 eine Maschine, die der Taube nachempfunden war, und beabsichtigte laut dem „Flugsport" nach Ostfriesland in seine Heimat zurückzukehren, um dort eine eigene Flugschule zu eröffnen.[387] Dennoch ist von weiteren Flugschülern auszugehen, die möglicherweise deshalb in Listen nicht auftauchen, weil sie ihre Ausbildung abbrachen. Mit dem Ausbruch des Krieges beendete Albers seine unternehmerische Tätigkeit, wurde Soldat und wandte sich nach Kriegsende einer anderen Beschäftigung zu.[388]

Georg Mürau war ebenfalls nur ein Jahr auf dem Flugplatz in Rotthausen tätig. Er hatte im Februar 1912 seinen Piloten-Schein auf einem Grade-Apparat gemacht und schulte in seiner 1912 entstandenen Fliegerschule in Rotthausen ebenfalls auf dem Grade-System. Weiterhin hatte er die Generalvertretung der Hans Grade-Flugzeugwerke für Rheinland, Westfalen, Hessen, Belgien und Luxemburg.[389]

Mürau bildete auch Flugschüler aus – sechs lassen sich nachweisen – und nahm, allerdings ohne Erfolg, an Flugwettbewerben mit einem Grade-Eindecker teil. Als ihm die Stadt Krefeld ein Angebot zum Aufbau einer Fliegerschule machte, verlegte er sein Werk im April 1914 zum Exerzierplatz Engelsberg bei

383 So das Urteil Kranzhoffs; vgl. Kranzhoff, Edmund Rumpler, S.134.
384 Vgl. Lange, Luftfahrttechnik, S.11.
385 Vgl. Hackenberger, Die Alten Adler, S.98–101.
386 Vgl. Lange, Luftfahrttechnik, S.15 und Supf, Fluggeschichte, Bd. 2, S.148.
387 Vgl. Flugsport (1912), Nr. 21, S.800.
388 Vgl. Supf, Fluggeschichte Bd. 2, S.148.
389 Die folgenden Informationen sind entnommen aus Vogt, Seidene Kugel, S.106–112 sowie aus Flugsport (1914), Nr. 1, S.38; (1912), Nr. 15, S.577 und Jahrbuch des DLV 1914, S.126 und S.130.

Traar und kündigte in Rotthausen die Verträge. Im Mai begann der Aufbau der Crefelder Fliegerschule Georg Mürau, die Zahl der Schulflugzeuge wurde von vier auf sechs erhöht und Mürau hoffte, wie viele andere auch, Militärpiloten ausbilden zu dürfen. Bei Kriegsausbruch wurde Mürau eingezogen, die Flugschule kurzzeitig geschlossen und kurz darauf der Betrieb wieder aufgenommen, bevor im Dezember 1914 die endgültige Schließung erfolgte. Über den weiteren Lebensweg Müraus ist nichts bekannt.

Der dritte Betrieb, der neben den Kondor-Werken in Rotthausen existierte, war Josef Schlatters Flugzeugbauanstalt. Schlatter (Jg. 1871)[390] hatte am 25. Mai 1912 auf einem Otto-Doppeldecker seine Fluglizenz erworben und nahm, wie übrigens Mürau und Albers auch, an der Krupp-Flugwoche teil.[391] Wie die beiden anderen schnitt er nicht sonderlich gut ab.[392] Im September 1912 begann er, selbst Flugzeuge zu bauen, und bildete bis 1914 Schüler aus. Die letzten Meldungen über sein fliegerisches Wirken stammen aus den Jahren 1913 und 1926. Bei der Jubiläums-Flugwoche 1913 auf dem Flugplatz Gelsenkirchen-Essen-Rotthausen platzierte er sich im Mittelfeld und erhielt über 2.500 Mark Preisgeld. Der letzte Hinweis ist im Krupp-Archiv enthalten, denn Schlatter wandte sich 1926 an die Westdeutsche Flugplatzgesellschaft mit der Bitte um Neugründung seiner Fliegerschule in Rotthausen, wozu es nicht kam.[393]

Nennenswerten Erträge konnte vermutlich keiner dieser kleinen Betrieben erwirtschaften, denn weder scheinen die drei Firmen sonderlich erfolgreich bei Wettbewerben abgeschnitten zu haben, noch dürften sie über viel Kapital verfügt haben. Für diese Annahme spricht, dass die Westdeutsche Flugplatzgesellschaft den Unternehmern Albers, Mürau und Schlatter, obwohl keiner der drei besondere Leistungen bei der Krupp-Flugwoche vorweisen konnte und Müraus Apparat infolge eines Überschlags schwer beschädigt wurde, je 1.500 Mark stiftete.[394] Womöglich diente das Geld als Finanzspritze und Innovationsförderung. Nimmt man durchschnittlich Einnahmen von 3.000 Mark pro Schüler[395] als zu entrichtende Ausbildungskosten an und wiegt diese mit dem Bau und der Instandhaltung der Maschinen sowie mit Mietzahlungen, Angestellten- und Materialkosten etc. auf, so waren die Firmen ohne größere (Heeres-)Aufträge oder ein hohes Stammkapital nicht überlebensfähig.

390 Sein Todesjahr ist unbekannt.
391 Vgl. Flugsport (1912), Nr. 13, S. 517; (1912), Nr. 16, S. 620 und (1912), Nr. 17, S. 649f.
392 Vgl. Bamler, Krupp-Flugwoche, 4.–11. August 1912, in: DLZ (1912), Nr. 17, S. 420 sowie Flugsport (1913), Nr. 21, S. 800. Zwei Aufnahmen, die Schlatter in einem Otto-Doppeldecker zeigen, sind enthalten in: Radzicki, Der Traum vom Fliegen im Ruhrgebiet, S. 45–46. Zu den Ergebnissen der Jubiläums-Flugwoche siehe Jahrbuch des DLV 1914, S. 173.
393 HA-Krupp WA 41/72–116, Schreiben Schlatters an Krupp-Direktor Gillhausen, 20.5.1913.
394 Vgl. Bamler, Krupp-Flugwoche, 4.–11. August 1912, in: DLZ (1912), Nr. 17, S. 420.
395 Diese Summe nennt Albers bei Hackenberger, Die alten Adler, S. 99.

Gemein ist allen Konstrukteuren, Fabrikanten oder Unternehmern der Glaube an eine Karriere in der Flugzeugbranche und eine grundsätzlich optimistische Einstellung. Bei Kriegsausbruch waren die Kondor-Werke die einzig verbliebene Fabrik am Rotthausener Platz, wenngleich es möglich ist, dass Georg Schlatter ebenfalls noch bis 1914 ansässig war.

Gründung, Struktur und Gesellschafter der Kondor-Werke

Etwa zeitgleich mit der Westdeutschen Fluggesellschaft und dem Flugplatz Gelsenkirchen-Essen-Rotthausen entstanden die Kondor-Werke.[396] Eine Woche nach der Eröffnungsflugwoche des Flugplatzes trafen sich Ernst August Schröder, Otto May, Hauptbuchhalter bei der Th. Goldschmidt AG, und Josef Suwelack im Juni 1912 zu finalen Vorbereitungen einer Firmengründung.[397] Geregelt wurden Personal- und Räumlichkeitsfragen.

Ursprünglich war wohl geplant, Emil Jeannin, der auch an der Eröffnungsflugwoche teilgenommen hatte, als Zulieferer für Flugzeuge zu gewinnen. In einem Schreiben vom 2. Juli teilte Suwelack den anderen Gesellschaftern mit, dass er bei seinem Besuch in Berlin auch Jeannin aufgesucht und ihn wegen einer Kooperation befragt habe. Jeannin habe den Kondor-Werken das Recht eingeräumt, seine Apparate mit einer Lizenzzahlung von jeweils 1.500 Mark pro Apparat nachzubauen. Suwelack schlug jedoch vor, sich zunächst vom Kriegsministerium Aufträge zusichern zu lassen, bevor eine endgültige Zusage gemacht werde.[398] Da

396 Obwohl das Unternehmen durchaus bekannt geworden war, liegt keine Monografie über die Kondor-Werke vor. Es existieren lediglich einige kurze, zum Teil fehlerhafte oder überzeichnete knappe Darstellungen. Zu nennen sind: Lange, Luftfahrttechnik 1970 und überarbeitete Neuauflage 1986; Wolfgang Wagner, Die Flugzeuge der Kondor-Werke, in: Aerokurier 1 (1974), S.67–69, der besonders die spätere Entwicklung der Jahre 1917–1918 skizziert, allerdings fehlerhaft. Ferner: Hugo Rieth, Kondor-Flugzeugwerke 1912–1918, in: Heimatstadt Essen (1975), Nr.76, S.121–130 (basierend auf Wagners Darstellung); Schmitt/Schwipps, 20 Kapitel frühe Luftfahrt, S.70. Weiterhin: Günter Rhiemeyer, Nichts dauert ewig. Vom Aufstieg und Niedergang des Kondor-Werkes, in: Lemgoer Hefte (1995), Nr.3, S.22–28 (Darstellung über die Niederlassung des Werkes in Lemgo. Zu diesem Zeitpunkt war der Flugzeugbau bereits eingestellt worden.); Krewerth, Josef Suwelack, S.62ff. (quellenbasierte, aber einseitige Darstellung); Ralf Peters/Annette Zehnter, Grenzen überwinden. 150 Jahre Th. Goldschmidt, Bottrop 1997 (dort nur eine Randnotiz über die „Goldschmidt-Flieger“, S.31) und schließlich als fundierte und quellenkritische Darstellung, die sich auf den sozialökonomischen Aspekt der Kondor-Werke konzentriert: Rißmann-Ottow, Glück ab!, S.238ff.

397 Die Akten über die Kondor-Werke liegen in Kopie im Nachlass Josef Suwelacks vor und werden im Folgenden zitiert als: Archiv Th. Goldschmidt AG (Archiv-TG). Vgl. Archiv-TG, Akte Kondor I, Nr.0406/1912, Protokoll über die Unterredung vom 4.6.1912 zwischen den Herren E. A. Schröder, J. Suwelack und O. May betreffend Gründung der Kondor-Flugzeugwerke.

398 Vgl. Archiv-TG, Akte Kondor I, Nr.0207/1912, Niederschrift über Verhandlungen mit der Westdeutschen Fluggesellschaft und Emil Jeannin, Urheber ist sehr wahrscheinlich Suwelack.

Jeannin bei weiteren Flügen seine Apparate durch Abstürze etc. verlor, wurde von einer Zusammenarbeit letztlich jedoch wieder Abstand genommen.

Als Flieger sollte neben Suwelack Karl Krieger[399] gewonnen werden, der aufgrund seines Engagements im Flugzeugbau geeignet schien, ferner sollten ein Zeichner, ein Schreiner und ein Hilfsarbeiter eingestellt werden. Suwelack wurde darüber hinaus beauftragt, einen Monteur, für den ein monatliches Gehalt von 220 Mark vorgesehen war, zu kontaktieren. Mit der Kontaktaufnahme zu den übrigen Arbeitern wurde das Büro der Rheinisch-Westfälischen Motorluftschiff-Gesellschaft betraut, das bereit war, diese Aufgabe „gegen eine geringe Vergütung" zu übernehmen. Ein Buchhalter sollte bei Bedarf „nebenamtlich beschäftigt" werden.

Hinsichtlich der erforderlichen Örtlichkeiten planten die Initiatoren, Räume zu mieten beziehungsweise bauen zu lassen. Ernst August Schröder erklärte sich bereit, Büroräume und eine Angestellte der Rheinisch-Westfälischen Motorluftschiff-Gesellschaft zur Verfügung zu stellen, wozu er als Vorsitzender dieser Gesellschaft die Kompetenz gehabt haben dürfte. Ferner bemühte er sich, Krupp davon zu überzeugen, die zwei gestifteten Flugzeuge den Kondor-Werken für die Ausbildung von Militärpiloten zur Verfügung zu stellen. Suwelacks und sein Interesse hatte er bereits im April bekundet.[400] Auf den Apparaten sollten „5 Offiziersflieger" ausgebildet werden, außerdem war die Teilnahme an „einer Anzahl von Schauflügen im Laufe des Sommers oder Herbstes" vorgesehen, so z. B. an der Krupp-Flugwoche im August. Infolge weiterer Verzögerungen und eines noch nicht fertiggestellten Flugzeuges musste dieser Plan jedoch verworfen werden.

Für die Unterbringung der Maschinen sollte ein Werkstattschuppen durch die Westdeutsche Flugplatzgesellschaft auf einer Grundfläche von 80x18 Meter gebaut werden, so dass acht Apparate darin Platz finden konnten. Weiterhin waren ein „Zeichen- und Kontorraum, ein Privat-Kontor und ein Konferenzzim-

399 Karl Krieger (Jg. 1885) war Chauffeur des Kaisers und hatte erstmals 1909 auf der ILA ein Flugmodell ausgestellt. Kurz darauf ließ er sich von seinen Chauffeurdiensten freistellen, um sich ganz dem Flugzeugbau widmen zu können. Als der Flugplatz Johannisthal entstanden war, zog Krieger zum „alten Startplatz", richtete eine Werkstatt ein und baute einen Eindecker. Seine Haupteinkünfte bezog er jedoch aus dem Einfliegen von Flugzeugen für andere Firmen, so auch für die Kondor-Werke. Am 12. September 1911 erhielt er den Pilotenschein, baute weitere Eindecker und zog 1913 auf den „neuen Startplatz" um. Ferner bildete er Flugschüler aus und beteiligte sich an Flugwettbewerben. Zusammen mit seinem Bruder Oskar baute er 1914 ein Torpedoflugzeug, welches er sich patentieren lassen wollte. Alarmiert durch diesen Schritt, beschlagnahmte das Reichsmarineamt den Dreidecker, doch Krieger prozessierte dagegen und gewann. Das Reichsmarineamt musste daraufhin 50.000 Mark Schadenersatz zahlen, eine Auszahlung blieb aufgrund des Kriegsausbruchs jedoch aus. Im Jahr 1915 oder 1916 wurde die Firma stillgelegt. Informationen entnommen aus Schmitt, Als die Oldtimer flogen, S. 64–65. Eine Notiz über Kriegers ersten Eindecker ist enthalten in: Flugsport (1911), Nr. 17, S. 612.

400 HA-Krupp WA 41/72–116, Rheinisch-Westfälische Motorluftschiff-Gesellschaft [Verfasser E.A. Schröder] an Direktor Gillhausen, 27.4.1912.

mer“[401] vorgesehen. Man rechnete beim Schuppen-Bau mit Kosten in Höhe von 25.000 Mark, die jährlichen Mietzahlungen der Kondor-Werke sollten 2.500 Mark betragen. Für die Installation von Wasser- und Stromleitungen hatte die Westdeutsche Flugplatzgesellschaft Sorge zu tragen. Da sich die Verlegung entsprechender Leitungen allerdings erheblich verzögerte, waren die Kondor-Werke nicht bereit, „eher Miete ... [zu bezahlen], bis die Errichtung fertig ist“.[402]

Zu den bisherigen Kosten kamen weitere Material- und Angestelltenausgaben hinzu, ebenso Werbe-, Versicherungs-, Reise- und Betriebskosten sowie Mieten für Büroräume etc. Zusätzlich wollte man zwei Schweißapparate zu je 200 Mark und Werkzeuge im Wert von 1.000 Mark beschaffen. Als Provisorium wurden für die Montage des geplanten Flugzeuges Maschinenfabriken oder eine Schreinerei in Essen-Caternberg beziehungsweise Stoppenberg ins Auge gefasst.[403]

Aufnahmen, die Josef Suwelack in einem „Kondor“-Album aufbewahrte, geben einen Eindruck von den Dimensionen der Werkstatträume und Einblick in den Entstehungs- und Bauprozess eines Flugzeuges (Abb. 10). Die Fotos rücken zudem eine Personengruppe ins Licht, die in den Protokollen und sonstigen Fliegerberichten allenfalls marginal vorkommt, obwohl sie den Hauptanteil an der Flugzeugproduktion trug: Dies waren Arbeiter, Monteure und Techniker. Doch die Männer an der Basis, wie man sie nennen könnte, die dafür Sorge trugen, dass Flugzeuge die Werkstatthallen flugfähig verließen, und für die Sicherheit des Motors und des Apparates garantieren mussten, bleiben im Unterschied zu den Gesellschaftern, zu Militärs und anderen Flugzeugfabrikanten gesichtslos. Nur selten werden sie namentlich erwähnt und dann auch nur, wenn ein besonderer Anlass vorlag. Das war z. B. der Fall, als ein „Bierabend“ ausgerichtet wurde, mit dem den Arbeitern für ihren Beitrag zum Gelingen des erfolgreichen Erstflugs des Kondor gedankt wurde:

> „Aus Anlass des guten ersten Erfolges des Flugzeugs soll jedem, an demselben beschäftigten Arbeiter M: 20,– den beiden Vorarbeitern M: 25,– dem Meister Riemann M: 150,– dem Techniker Küchen M: 100,– Gratifikation gegeben werden. Ausserdem ist an einem Samstag ein kleines Fest für das Personal zu veranstalten.“[404]

Insgesamt bewilligten die Gesellschafter für „Gratifikation an Meister und Arbeiter sowie für einen Bierabend“ 610 Mark.

401 Archiv-TG, Akte Kondor I, Nr. 0406/1912, Protokoll über die Unterredung vom 4.6.1912 zwischen den Herren E. A. Schröder, J. Suwelack und Otto May betreffend Gründung der Kondor-Flugzeugwerke.

402 Archiv-TG, Akte Kondor I, Nr. 2709/1912, Protokoll der Gesellschafter-Versammlung vom 27.9.1912.

403 Archiv-TG, Akte Kondor I, Nr. 0406/1912, Protokoll über Unterredung vom 4.6.1912 zwischen den Herren E. A. Schröder, J. Suwelack und Otto May betreffend Gründung der Kondor-Flugzeugwerke.

404 Archiv-TG, Akte Kondor I, Nr. 1210/1912, Protokoll der Gesellschafter-Versammlung vom 12.10.1912.

Josef Suwelack wurde mit der Aufgabe betraut, „mit Hülfe des zu engagierenden Zeichners unverzüglich anhand seiner Skizzen mit der Anfertigung der Pläne und Werkstattzeichnungen zu beginnen." Ebenso sollten Angebote bei nicht näher benannten Firmen für die nötigen Baumaterialien und bei den Argus-Motorenwerken für einen „100 PS-Motor" eingeholt werden. Gebaut werden sollte ein der Rumpler-Taube nachempfundener Typ „mit Verbesserungen im Fahrgestell, der Bootskonstruktion und den Flügeln, die Herr Suwelack in seiner Praxis als zweckmäßig erkannt hat".[405]

Abb. 10: Schreinerei der Kondor-Werke, um 1912/13
(Foto: Familienarchiv Walter Suwelack, NL Josef Suwelack)

Am 15. Juni 1912 erfolgte die Unterzeichnung des Gründungsvertrages der Kondor-Werke durch die Gesellschafter Ernst August Schröder, Otto May und Bernhard Goldschmidt, jüngerer Sohn des Teilhabers der Th. Goldschmidt AG, Karl Goldschmidt.[406] Das Stammkapital belief sich auf insgesamt 50.000 Mark und wurde zu 25.000 Mark von Schröder, zu 2.000 Mark von May und zu 23.000 Mark von Bernhard Goldschmidt geleistet. Gegenstand des Unternehmens war gemäß § 2 „die Herstellung und der Verkauf von Flugzeugen, die Ausbildung von Flugzeugführern sowie die mit diesen Aufgaben zusammenhängenden Geschäfte

405 Ebd.
406 Information zu den Goldschmidts bei Dickhoff, Essener Köpfe, S. 74f.

und Unternehmungen." Josef Suwelack wurde zum Geschäftsführer bestellt und erhielt einen Anstellungsvertrag.[407]

Tabelle 2: Gesellschafter, Einlagen und Stammkapital der Kondor-Werke 1912/13

	Juni 1912 (Gründung)	Januar 1913	Mai 1913 (Kostendeckung)	August 1913	Oktober 1913**
Stammkapital (in Mark) / Gesellschafter	50.000	100.000*	122.000	205.000	300.000
Karl Goldschmidt[1)]	–	5.000	15.000	75.000	
Bernhard Goldschmidt[2)]	23.000	25.000	25.000	25.000	
Ernst August Schröder[3)]	25.000	29.000	29.000	30.000	
Otto May[4)]	2.000	4.000	4.000	4.000	
Josef Suwelack[5)]	–	Sacheinl./ 10.000	Sacheinl. / 10.000	Sacheinl. / 10.000	
Alfred von Back-Begavár[6)]	–	5.000	8.250	17.000	
Wilhelm Schäfer[7)]	–	3.000	3.000	5.000	
Heinrich Hillebrand[8)]	–	1.000	1.000	1.000	
Victor von Back[9)]	–	5.000	5.000	5.000	
Hans Goldschmidt[10)]	–	5.000	13.750	10.000	
Eduard Staßfurth[11)]	–	–	–	3.000	
Wilhelm Marx[12)]	–	3.000	3.000	3.000	
Richard Passavant-Gontard[13)]	–	–	–	10.000	
Hugo Racine[14)]	–	3.000	3.000	5.000	
Dr. Georg Hirschland[15)]	–	1.000	1.000	1.000	
Kurt Hirschland[16)]	–	1.000	1.000	1.000	

Im Jahr 1913 stießen weitere dreizehn Gesellschafter zu den Kondor-Werken hinzu, was zu erheblichen Kapitalerhöhungen führte. Zunächst wurde das Stammkapital im Januar 1913 um 50.000 Mark aufgestockt[408] und im Mai „zur

407 Vgl. Archiv-TG, Akte Kondor I, Nr. 1506/1912, Gesellschaftsvertrag der Kondor-Flugzeugwerke, Essen, 15.6.1912.

408 Der Beschluss hierzu wurde bereits im November 1912 gefasst, doch mussten weitere Kapitalgeber gewonnen werden. Siehe Archiv-TG, Akte Kondor I, Nr. 1211/1912, Protokoll der Gesellschafter-Versammlung vom 12.11.1912.

Quellen zu Tab.2: Protokolle der Gesellschafter-Versammlungen etc. der Kondor-Werke; Peters/Zehnter, Grenzen überschreiten; Manfred Rasch, Adelige Unternehmer am Ende der wilhelminischen Epoche, in: Hartmut Berghoff/Jürgen Kocka/Dieter Ziegler (Hg.), Wirtschaft im Zeitalter der Extreme. Beiträge zur Unternehmensgeschichte Österreichs und Deutschlands, München 2010, S.21–46; Detlef Krause, Zur Geschichte der Mitteldeutschen Creditbank, in: Bankhistorisches Archiv. Zeitschrift für Bankengeschichte (1991), H. 17, S.17–41; Walther Däbritz/Wendt Paulick, Th. Goldschmidt AG Essen. Neun Jahrzehnte Geschichte einer deutschen Fabrik, Essen 1937; Hermann Schröter, Die Familie Hirschland in Essen, in: Stadt Essen (Hg.), Geschichte und Schicksal der Essener Juden. Gedenkbuch für die jüdischen Mitbürger der Stadt, Essen 1980, S.167–179, bes. S.174ff.
Anm.: 1) Inhaber der Th. Goldschmidt AG, Mitglied im NVfL, Vorsitzender des Aufsichtsrates der KW, Aufsichtsratsmitglied bei der Mitteldeutschen Creditbank in Meinigen 1911–1926; 2) Sohn von K. Goldschmidt, Marineoffizier, Prokurist der Th. Goldschmidt AG, Sitz im Aufsichtsrat der KW; 3) Inhaber der Firma Schröder & Baum (Linoleumhandel), Vorsitzender der Rheinisch-Westfälischen Motorluftschiff-Gesellschaft, Mitglied im NVfL, Gesellschafter der Westdeutschen Fluggesellschaft, Mitglied im Aufsichtsrat der KW; 4) Direktor der Th. Goldschmidt AG, Mitglied des Aufsichtsrates der KW; 5) Technischer Direktor der KW, Mitglied des NVfL; 6) Direktor der Th. Goldschmidt AG mit Prokura, Mitglied des Aufsichtsrates der KW, ein Onkel Ernst Füßmanns; 7) Direktor der Th. Goldschmidt AG; 8) Jurist, leitender Angestellter der Th. Goldschmidt AG; 9) ?; 10) Bruder von K. Goldschmidt, Inhaber der Th. Goldschmidt AG, Mitglied des NVfL; 11) Direktor der Th. Goldschmidt AG; 12) ehemaliger OB der Stadt Düsseldorf, Gesellschafter und Vorsitzender des Aufsichtsrates der Th. Goldschmidt AG, Vorsitzender des Düsseldorfer Luftfahrtvereins; 13) Aufsichtsratsmitglied der Th. Goldschmidt AG, Aufsichtsratsmitglied der Mitteldeutschen Creditbank von 1898–1923; 14) Medizinalrat, Aufsichtsratsmitglied der KW; 15) Jurastudium und Promotion zum Dr. jur. 1907, 1912 Eintritt als Teilhaber in das Familienunternehmen Bankhaus Simon Hirschland; 16) Bankier, seit 1905 Teilhaber des Bankhauses Simon Hirschland, Bruder von Georg Hirschland, Mitglied des NVfL, Aufsichtsratsmitglied der Th. Goldschmidt AG.
* Am 12. November 1912 war noch die Rede davon, dass sich Alfred von Back-Begavár mit 10.000, Bernhard Goldschmidt mit 7.000, Otto May mit 3.000 und Ernst August Schröder mit 5.000 Mark an der Kapitalerhöhung beteiligen sollten. Für die noch fehlenden 26.000 Mark sollten weitere kapitalkräftige Leute gewonnen werden. Vgl. Archiv-TG, Akte Kondor I, Nr.1211/1912, Protokoll der Gesellschafter-Versammlung vom 12.11.1912.
** Die Verteilung des Gesellschaftskapitals für Oktober 1913 ist nicht überliefert, da entsprechende Protokolle fehlen. Es existiert lediglich eine notarielle Abschrift vom 9. Oktober 1913, worin die Kapitalerhöhung um weitere 95.000 Mark bekannt gegeben wird. Archiv-TG, Akte Kondor I, Nr.0910/1913.

Bezahlung der dringendsten Schulden" um weitere 22.000 Mark erhöht. Im August wurden nochmals weitere 105.000 Mark investiert, bis das Stammkapital im Oktober 1913 schließlich insgesamt 300.000 Mark betrug.[409]

Der Vertrag galt für zwei Jahre und verlängerte sich ohne eingehaltene sechsmonatige Kündigungsfrist automatisch jeweils um ein weiteres Jahr. Im Falle von Krankheit oder eines Unfalls mit einer Genesungsdauer von mehr als zwei Monaten konnte das Arbeitsverhältnis jedoch vorzeitig beendet werden. Suwelack hatte Anspruch auf einen Jahresurlaub von vier Wochen, den Zeitpunkt bestimmten allerdings die Gesellschafter.[410] Sein Jahresgehalt betrug 4.800 Mark, außerdem

409 Siehe dazu die entsprechenden Gesellschafterprotokolle: Archiv-TG, Akte Kondor I, Nr.2001/1913, Notarielles Schreiben über Kapitalerhöhungen vom 20.1.1913; Nr.2205/1913, Gesellschafter-Versammlung vom 22.5.1913; Nr.0708/1913, Protokoll über die Verhandlung vom 7.8.1913 sowie Nr.0910/1913, Notarieller Beschluss vom 9.10.1913.

410 NL J. Suwelack, Anstellungsvertrag zwischen den Kondor Flugzeugwerken und Josef Suwelack vom 15.7.1913, 4 Seiten. Weitere Ausführungen ebd.

stand ihm ein Gewinnanteil aus dem durch die Geschäftsfelder erzielten Reingewinn zu. So erhielt er vertragsgemäß ein Zehntel der Gewinne aus Flugwettbewerben, ein Drittel der Gewinne aus von ihm persönlich bestrittenen Flugwettbewerben, ein Fünftel der Einnahmen durch die Flugschüler- und Offiziersausbildung sowie schließlich ein Zehntel des Gewinns aus dem Flugzeugbau. Im Gegenzug hatte er folgenden Verpflichtungen nachzukommen: In § 3 wurde ihm ein einjähriges Berufsverbot als Konstrukteur erteilt, sollte er aus der Firma ausscheiden. So durfte er weder bei einer anderen Firma eintreten, noch sich als Alleinunternehmer betätigen. Bei Zuwiderhandlung drohte ihm eine Vertragsstrafe von 15.000 Mark.[411] Weiterhin musste er den Gesellschaftern auf Verlangen jederzeit Rechenschaft über seine Arbeit und die Situation der Firma ablegen und war an Weisungen des Aufsichtsrates gebunden. Seine Erfahrung als Konstrukteur hatte er der Gesellschaft zur Verfügung zu stellen und an seinen Konstruktionen durfte er „weder für sich, noch für andere Schutzrechte ... im In- und Auslande ... erwerben".[412] Für die erfolgreiche, das heißt gewinnbringende Umsetzung seiner Ideen stellte ihm die Gesellschaft nicht näher benannte Prämienzahlungen in Aussicht. Schließlich wurde Suwelack auf eigenen Wunsch als Gesellschafter mit einer „Bar- oder Sacheinlage von M. 10–30.000,–" aufgenommen, welche er in Form seiner Konstruktionspläne einbrachte, deren Wert auf 10.000 Mark festgelegt wurde. De facto blieb er jedoch auch weiterhin Angestellter.[413]

Abgesehen von Josef Suwelack, der zusätzlich durch seinen Vater finanzielle Unterstützung erhielt, musste keiner der übrigen Gesellschafter seinen Lebensunterhalt durch den Bau und das Vorstellen von Flugzeugen bestreiten, denn alle waren entweder bei einer anderen Firma angestellt oder sogar selbst als Unternehmer tätig.[414] Dies versetzte die Beteiligten in die Lage, flexibel zu agieren, denn es drohte einzig der Verlust des Betriebskapitals, nicht aber der persönliche Ruin.

Die hohen finanziellen Ausgaben und die Beziehungen zu gesellschaftlich hochgestellten Kreisen hoben die Kondor-Werke von den übrigen, in der näheren Umgebung ansässigen Flugzeugfirmen deutlich ab.[415] Nicht nur, wie im Folgenden dargelegt wird, unternahmen sie mehrfach Versuche zu Fusionen mit anderen Flugzeugfirmen, sondern sie richteten ihren Betrieb nach ökonomischen

411 Ebd. Der entsprechende Abschnitt lautet: „Herr Suwelack verpflichtet sich, innerhalb eines Jahres nach Beendigung des Vertrages im Gebiete des Deutschen Reiches weder eine Stellung in einem gleichartigen oder ähnlichen Unternehmen anzunehmen, noch selbst ein solches Geschäft zu errichten, oder auch unmittelbar oder mittelbar an einem solchen Geschäft sich zu beteiligen. Für den Fall der Zuwiderhandlung hat er eine Vertragsstrafe von M. 15.000,- zu zahlen. Durch die Zahlung dieser Vertragsstrafe wird der Anspruch auf Erfüllung nicht ausgeschlossen.", S.1f.

412 Ebd., S.2, § 5.

413 Vgl. Archiv-TG, Akte Kondor I, Nr.1312/1912, Bericht der Gesellschafter-Versammlung vom 13.12.1912.

414 Siehe Zusatzinformationen zur Tabelle „Gesellschafter der Kondor-Werke".

415 Vgl. Rißmann-Ottow, Glück ab!, S.245.

Prinzipien aus, waren bereit, ins Ausland zu gehen und dem nationalen Fluggedanken den Rücken zu kehren.

Die Aufstellung der Gesellschafter zeigt, dass die Mehrheit der Investoren aus dem gut situierten Kreis des Wirtschaftsbürgertums kam. Vertreten waren Firmeninhaber, Bankiers und Leute mit Positionen in Vorständen. Bis auf Josef Suwelack war jedoch keiner der Gesellschafter ein ausgebildeter Motorflieger.

Die immensen finanziellen Aufwendungen und Ausgaben unterstreichen einerseits, dass die Investoren an die Zukunftsfähigkeit einer Flugzeugindustrie im Ruhrgebiet und an die Kondor-Werke glaubten; sie zeigen andererseits aber auch die Notwendigkeit von hohen finanziellen Opfern, um den Kontakt zur Militärbehörde und zu Fusionspartnern herzustellen. Die Beteiligungen hatten spekulativen Charakter, denn trotz ausbleibender Renditen wurde weiter in den Auf- und Ausbau der Firma investiert. Man hoffte langfristig, als Zulieferer des Militärs herangezogen zu werden und so das Unternehmen rentabel zu machen.[416]

Wider Erwarten zog sich der Bau des ersten Flugzeuges länger hin als geplant, so dass der Kondor erst Anfang Oktober 1912 fertiggestellt war und am 12. Oktober seinen Erstflug hatte. Zur Entlastung Suwelacks war am 27. September 1912 Ernst Füßmann als kaufmännischer Geschäftsführer mit einem Jahresgehalt von 1.150 Mark eingestellt worden,[417] wodurch Suwelack sich ganz auf den technischen Teil konzentrieren konnte. Füßmanns Aufgabe umfasste die Bearbeitung des „gesamten kaufmännischen Betrieb[s] einschließlich Lohnwesen" und war zunächst als Halbtagsstelle konzipiert. Zum Jahreswechsel 1913 erhielt er eine Ganztagsstelle bei einem Gehalt von 3.600 Mark und mit der Aussicht auf 4 Prozent Tantieme aus dem Reingewinn des Betriebes, die 3.600 Mark jedoch nicht übersteigen sollte.[418] Am 21. Dezember war schließlich die zweite Maschine abflugbereit, und am 3. Januar 1913 erhob sich ein Kondor mit Suwelack am Steuer zu einem Flug über die Stadt Essen. Für diese Leistung erhielt er einen Ehren- und Erinnerungsbecher vom Oberbürgermeister, was die einzige bekannt gewordene Aktivität der Stadt in Verbindung mit den Kondor-Werken darstellt.[419]

In einer Kalkulationsaufstellung Ernst Füßmanns vom 14. Dezember 1912 rechnete dieser durch, welchen Gewinn die Kondor-Werke beim Verkauf von zehn Apparaten à 20.000 Mark zu erwarten hätten. Füßmann bezifferte die jährlichen

416 Vgl. hierzu auch ebd.

417 Vgl. Archiv-TG, Akte Kondor I, Nr. 2709/1912, Protokoll der Gesellschafter-Versammlung vom 27.9.1912. Zu Ernst Füßmann liegen keine biografischen Informationen vor. Bis 1906 betrieb Füßmann in Essen-Altendorf eine Schmiedewerkstatt, die von Paul Füßmann weitergeführt wurde. Vgl. Rißmann-Ottow, Glück ab!, S. 257 und Der Motorwagen, Bd. 1 (1906), S. 237.

418 Vgl. Archiv-TG, Akte Kondor I, Nr. 1312/1912, Protokoll der Gesellschafter-Versammlung vom 13.12.1912.

419 Der Becher befindet sich ebenfalls im Familienbesitz der Suwelacks. Siehe auch NL J. Suwelack, Schreiben des OB der Stadt Essen an Josef Suwelack vom 28.1.1913.

Unkosten (Löhne, Reise-, Strom-, Versicherungskosten etc.) auf insgesamt 83.000 Mark, dazu kamen Baukosten für zehn Flugzeuge von insgesamt 90.000 Mark. Dem stellte er den möglichen Verkaufserlös gegenüber. Aus Füßmanns Kalkulation wird ersichtlich, dass er bereits große Aufträge und damit die Serienfertigung einplante und mit hohen Erträgen rechnete. Für das Jahr 1913 prognostizierte er einen Reingewinn von 10.000 Mark.[420] Hierbei war nur der Gewinn aus Flugzeugverkäufen berücksichtigt, mögliche Einnahmen aus Flugwettbewerben und Schülerausbildung waren noch nicht enthalten.

Bereits im November 1912, ein halbes Jahr nach Gründung der Kondor-Werke, legte Josef Suwelack den anderen Gesellschaftern seine Ansichten über die Notwendigkeit zum Bau einer Flugschule dar, für die er vier Schulmaschinen vorsah. Sein Vorschlag stieß auf Zustimmung: „Die Versammlung beschloss die Errichtung einer derartigen Flugschule zur Ausbildung von Offizierspiloten." Der Aufbau wurde 1913 energisch betrieben und dazu auch Kontakt zur Stadt Düsseldorf aufgenommen.[421] Darüber hinaus trat Suwelack mit dem Wunsch an das Unternehmen heran, in den Bau von Wasserflugzeugen zu investieren. Hierfür wurden einmalig 3.000 Mark bewilligt, mit denen sämtliche Kosten bestritten werden sollten.[422] Da die Kondor-Werke schon früh in eigener Sache warben und mit der „B. Z. am Mittag" sowie mit dem „Flugsport" Abkommen bezüglich der Anzeigenaufgabe etc. geschlossen hatten, liegt es nahe, dass auch Suwelack selbst die mediale Berichterstattung über das Flugwesen verfolgte.[423] Möglicherweise fand er dort Inspirationen für seine Pläne, denn im „Flugsport" wurde bereits seit 1911 und dann verstärkt 1912 über Neuerungen und Aktivitäten im Bereich der Marinefliegerei und des Wasserflugsports berichtet.[424]

Über Zeitungen oder Zeitschriften wurde aber auch weiteres Personal gesucht, so z. B. für den geplanten Serienbau der Kondor-Flugzeuge. Hierfür wurden entsprechende Anzeigen, beispielsweise im „Flugsport", geschaltet, die Anfang Dezember 1912 sowie in der ersten Januarausgabe 1913 erschienen.[425] Es wurde

420 Vgl. Archiv-TG, Akte Kondor I, Nr. 1412/1912, Bericht über die Gesellschafter-Versammlung vom 14.12.1912.

421 Siehe hierzu die Protokolle der Besprechungen und Gesellschafter-Versammlungen ab August 1913ff.: Archiv-TG. Akte Kondor I, Nr. 0708/1913; Nr. 1108/1913; Nr. 1610/1913; Nr. 0201/1914 und Nr. 0000/1914. Außerdem: Delvos, Düsseldorfer Luftverkehr, S. 117f.

422 Vgl. Archiv-TG, Akte Kondor I, Nr. 1211/1912, Zitat ebd.

423 Vgl. Archiv-TG, Akte Kondor I, Nr. 1312/1912, Bericht über die Gesellschafter-Versammlung.

424 Siehe „Neues über den Stand des Flugwesens in Amerika", in: Flugsport (1911), Nr. 18, S 625ff.; „Ausland. Wasserflugmaschine in England", ebd. S. 646ff.; (1911), Nr. 26, S. 946 (Werbung für das im Januar 1912 erscheinende Buch Oskar Ursinius, Wasserflugmaschinen. Praktischer Leitfaden für Flugmaschinen-Konstrukteure, Berlin 1911); „Die Wasserflugmaschine ist die Sportmaschine der Zukunft", in: Flugsport (1912), Nr. 27, S. 905f.; 1912 wurde sogar eigens ein „Fisch"-Symbol eingeführt, das Artikel zu Wasser-Flugmaschinen kennzeichnete. Siehe z. B. Flugsport (1912), Nr. 10, S. 373ff.

425 Siehe Flugsport (1912), Nr. 25, S. 965 sowie (1913), Nr. 1, Anzeigenteil, 2. Seite.

Wert auf ausführliche schriftliche Bewerbungen und Zeugnisse gelegt, aufgrund derer die Bewerber zu einem Vorstellungsgespräch eingeladen wurden. Die Firma lockte potenzielle Mitarbeiter mit der Aussicht auf eine „Lebensstellung“.[426]

Die Protokolle der Gesellschafter-Versammlungen der Kondor-Werke und einige Kostenaufstellungen über Löhne, Flugzeugteile, Werbekosten etc. geben Einsicht in den Geschäftsbetrieb und belegen die hohe Relevanz von guten Beziehungen zu den „richtigen“ Personen, das heißt zu einflussreichen Militärs, Firmeninhabern, Fliegern und (Geschäfts-)Freunden. Immer wieder ist in den Protokollen die Rede davon, Kontakte zu reichen Personen in anderen Städten oder zu persönlich bekannten Militärs zu knüpfen, um sie für das Flugzeugunternehmen zu interessieren. Als über den Bau einer Wasserflugmaschine diskutiert wurde, versprach Bernhard Goldschmidt, „sich persönlich mit massgebenden Personen des Reichs-Marine-Amtes in Verbindung zu setzen[,] um so evtl. zu erfahren, auf welche Punkte die Reichs-Marine-Verwaltung bei Wasserflugzeugmaschinen den größten Wert legt“.[427] In seiner Funktion als Marine-Offizier war es naheliegend, dass er diese Erkundigungen einholte. Gleichzeitig standen die Kondor-Werke in Verbindung mit einem Freund Goldschmidts, dem Oberstleutnant zur See Otto Bertram,[428] der über seine Kontakte zum Militär für die Gesellschaft Erkundigungen einholen, Auskünfte erteilen und an Flugzeugfirmen herantreten sollte, da die Kondor-Werke Fusionen mehrfach in Erwägung zogen.[429] Allein zu Suwelacks Zeiten bei den Kondor-Werken – ihm wurde am 30. Juni 1914 als erstem Angestellten gekündigt[430] – kontaktierten die dafür beauftragten Gesellschafter folgende Flugzeugfirmen: die Gothaer Waggonfabrik, die Albatros-, Euler- und die Harlan-Werke, die AEG, die Continental-Flugzeugwerke und auf Initiative Josef Suwelacks die Aviatik-Werke. Aber auch ausländische Firmen wurden seit 1914 zwecks einer Zusammenarbeit aufgesucht, so z. B. die englische Sopwith Aviation Company, ferner wurden Anfang des Jahres 1913 Verbindungen zur spanischen Militärverwaltung hergestellt, um neue Absatzmärkte zu schaffen.[431] Selbst Süd-

426 Flugsport (1913), Nr. 1, Anzeigenteil 2. Seite.

427 Archiv-TG, Akte Kondor I, Nr. 1211/1912, Protokoll der Gesellschafter-Versammlung vom 12.11.1912.

428 Bertram (geb. 9. Oktober 1886) war Kapitänleutnant bei der Marine, erhielt am 9. Oktober 1911 seinen Flugschein (Nr. 123) und veröffentlichte 1912 zusammen mit Hans Steffen das Buch: Orientierung auf Überlandflügen auf Grund praktischer Erfahrungen, Berlin 1912. Vgl. Jahrbuch des DLV 1914, S. 115.

429 Vgl. Archiv-TG, Akte Kondor I, Nr. 1312/1912, Bericht über die Gesellschafter-Versammlung vom 13.12.1912.

430 Vgl. Archiv-TG, Akte Kondor I, Nr. 3006/1914, Protokoll der Gesellschafter-Versammlung vom 30.6.1914. Nachfolger wurde ein Herr Anton Ludwig, Angestellter bei der Th. Goldschmidt AG. Als Konstrukteur wurde Paul Westphal eingestellt. In späteren Protokollen war beschönigend die Rede davon, dass Suwelack die Kondor-Werke verlassen habe.

431 Die Allgemeine Elektrizitäts-Gesellschaft (AEG) begann 1911 unter der Leitung von Emil Rathenau (1838–1915), dessen Aufgaben in Folge von Krankheit seit 1912 durch seinen Sohn Walther all-

amerika und Russland kamen, zumindest als Abnehmer für Flugzeuge, jedoch scheinbar als letzte Option, für Suwelack in Betracht. Gegenüber Karl Goldschmidt erklärte er im Oktober 1914 schriftlich: „In Süd-Amerika hat die Firma Rumpler eine Serie von 6 Apparaten abgesetzt und kommt dieses Gebiet für spätere Zeiten evtl. auch in Betracht." Allerdings präferierte der Flieger das Deutsche Reich: „Immerhin ist es notwendig, dass wir zuerst unser Augenmerk auf die deutsche Heeresverwaltung richten und uns bemühen, zunächst Apparate hier zu verkaufen und Flugschüler zu erwerben".[432] Ebenfalls brachte Suwelack Geschäfte mit Russland als Möglichkeit zur Profitsteigerung ins Gespräch, fügte jedoch hinzu, dass die Kondor-Werke erst bei der deutschen Militärverwaltung Erfolg haben müssten, bevor das Geschäft vergrößert werden könne.[433] Möglicherweise begriff Suwelack einen direkten Gang ins Ausland als Niederlage, denn immerhin hatte er in den Vorjahren bereits gute Kontakte zu inländischen Firmen geknüpft und die Hoffnung auf deutsche Aufträge noch nicht aufgegeben.

Die Verhandlungen mit anderen Flugzeugfirmen scheiterten jedoch aus unterschiedlichen Gründen. Zum einen stießen die Kondor-Werke auf Widerstand, wie z.B. im Fall von Euler, da mit der Firma „sehr schwer zu verhandeln" sei.[434] Im Geschäftsbericht für das erste Halbjahr 1914 hieß es diesbezüglich: „Von Herrn Schröder mit Euler angeknüpfte Verhandlungen führten ebenfalls zu keinem

mählich übernommen wurden, unter dem Namen „AEG Abteilung Flugzeugbau" in Henningsdorf sich auch auf diesem Gebiet zu betätigen. Das Unternehmen baute nach französischem Vorbild Ein- und Doppeldecker, wovon die Militärverwaltung 1913 zwei Doppeldecker abnahm. Seit 1913 baute das Unternehmen auch Wasserflugzeuge und nahm im gleichen Jahr an einem Wasserflugzeug-Wettbewerb in Warnemünde teil. Im Ersten Weltkrieg baute das Unternehmen über 650 Beobachtungs- und Tiefangriffsflugzeuge, ferner 523 Bomber. Nach Kriegsende wurde der Flugzeugbau eingestellt.
Die Gothaer Waggonfabrik gliederte ihrem Unternehmen 1913 ebenfalls eine Versuchsabteilung für Flugzeuge an und baute bis Kriegsbeginn der Taube nachempfundene Ein- und Doppeldecker. Im Ersten Weltkrieg stellte die Firma Bomben- und Marineflugzeuge her, kam aber nach 1918 durch den Versailler Vertrag zum Erliegen. 1933 nahm sie die Flugzeugproduktion wieder auf und führte diese bis 1944 fort, bis sie infolge von massiver Kriegszerstörung erneut zum Erliegen kam.
Die Continental-Flugzeugwerke bauten unter diesem Namen keine Flugzeuge, sondern aus ihnen gingen 1913 die Union-Flugzeugwerke hervor, die den Bomhard-Pfeildoppeldecker produzieren. Informationen entnommen aus Schwipps, Schwerer als Luft, S.223ff. und S.235f.; Supf, Fluggeschichte, Bd.2, S.156 und Lange, Luftfahrttechnik, S.18 und S.46.
Vgl. Protokolle der Kondor-Werke vom 20.1.1913 (Archiv-TG, Akte Kondor I, Nr.2001/1913); 5.2.1913 (Archiv-TG, Akte Kondor I, Nr.0502/1913); 21.8.1913 (Archiv-TG, Akte Kondor I, Nr. 2108/1913); 14.10.1913 (Archiv-TG, Akte Kondor I, Nr.1410/1913); 13.12.1912 (Archiv-TG, Akte Kondor I, Nr.1312/1913); 2.1.1914 (Archiv-TG, Akte Kondor I, Nr.0201/1914); 9.1.1914 (Archiv-TG, Akte Kondor I, Nr.0901/1914); 13.1.1914 (Archiv-TG, Akte Kondor I, Nr.1301/1914); 29.1.1914 (Archiv-TG, Akte Kondor I, Nr.2901/1914); 4.3.1914 (Archiv-TG, Akte Kondor I, Nr.0403/1914); 19.3.1914 (Archiv-TG, Akte Kondor I, Nr.1903/1914); ff. und Geschäftsbericht über den Gang der Verhandlungen im ersten Halbjahr 1914 (Archiv-TG, Akte Kondor I, Nr.0000/1914).

432 NL J. Suwelack, Schreiben J. Suwelacks an Dr. Karl Goldschmidt vom 14.10.1913.

433 Vgl. ebd.

434 Archiv-TG, Akte Kondor I, Nr.1312/1912, Bericht der Gesellschafter-Versammlung.

Resultat.“[435] Zum anderen waren die Kondor-Werke und ihre potenziellen Geschäftspartner uneins über Gewinnbeteiligungen, denn entweder wiesen die Kondor-Werke die Forderungen anderer Hersteller zurück, oder die Firmen waren ihrerseits nicht bereit, auf die Wünsche des Essener Unternehmens einzugehen.

Mit Otto Wiener, dem Direktor der Albatros-Werke, traf sich Füßmann am 5. Januar 1914 zu Sondierungsgesprächen, nachdem er bereits im August 1913 schriftlich Kontakt aufgenommen hatte. Doch auch dieses Gespräch brachte nicht den erhofften Zusammenschluss, denn Wiener wies den Vertragsentwurf der Kondor-Werke zurück, der in ihm „absolut kein Interesse“ weckte.[436] Stattdessen regte er eine mindestens fünfjährige vertragliche Zusammenarbeit und die Übernahme einer Generalvertretung durch die Kondor-Werke an, da Lizenzverträge laut Füßmann „jedem der Kontrahenten Hintertüren offen lasse[n]“. Offenbar befanden sich die Albatros-Werke Anfang des Jahres 1914 in einer komfortablen Ausgangsposition, denn dem Bericht Füßmanns zufolge hatten sie gegenüber den Kondor-Werken einen Vorsprung bei der Militärverwaltung, des Weiteren war man bereits wegen einer zusätzlichen Niederlassung der Albatros-Werke in Süddeutschland an Wiener herangetreten. Darüber hinaus eröffnete Wiener Füßmann, dass die Kondor-Werke mit der Festlegung auf den Eindecker-Bau ins Hintertreffen gerieten, da „der letzte Pariser Salon jedem Fachmann klar und deutlich vor Augen geführt“ habe, dass „Tauben und Eindecker sich ... überholt hätten“. Zwar war Wiener für weitere Gespräche offen, doch führten diese zu keinem Ergebnis. Im Geschäftsbericht für das erste Halbjahr 1914 wurde schließlich festgestellt, dass die Verhandlungen „an der Gegenforderung und dem ganzen Verhalten der Albatros Werke“ gescheitert seien.[437]

Auch Gespräche mit den Harlan-Werken waren wenig erfolgreich: Zu einer Fusion mit den Harlan-Werken waren die Gesellschafter der Kondor-Werke nicht bereit, da Harlan fünfzig Prozent des Reingewinns beansprucht und zusätzlich für die verbliebenen fünfzig Prozent „eine gleichmäßige Verteilung“ gefordert

435 Archiv-TG, Akte Kondor I, Nr. 0000/1914, Geschäftsbericht über den Gang der Verhandlungen im ersten Halbjahr 1914.

436 Vgl. Archiv-TG, Akte Kondor I, Nr. 0808/1913, Mitteilung Ernst Füßmanns an Dr. Karl Goldschmidt vom 8.8.1913 sowie Archiv-TG, Akte Kondor I, Nr. 0501/1914, Unterredung mit Herrn Direktor Wiener am 5.1.1914 (Ernst Füßmann). Alle zitierten Passagen ebd.

437 Archiv-TG, Akte Kondor I, Nr. 0000/1914, Geschäftsbericht über den Gang der Verhandlungen im ersten Halbjahr 1914, Zitat ebd. Der Pariser Salon, auch bekannt unter dem Namen „Pariser Aérosalon“, fand zum ersten Mal 1908 im Rahmen der Pariser Automobilausstellung statt und entwickelte sich in den folgenden Jahren zu einer Luftfahrtmesse. Bis 1924 fand die Veranstaltung jährlich statt, nur unterbrochen durch den Ersten Weltkrieg. Ab 1924 ging man zu einem Zwei-Jahresrhythmus über, der bis heute gilt. Auf dem Pariser Salon präsentierten alle größeren Flugzeugfirmen wie Rumpler, Albatros, Aviatik etc., aber auch einzelne Konstrukteure ihre Maschinen. Ferner stellten Motoren-, Propeller-, Tachometerhersteller etc. ihre neuesten Modelle aus. Siehe z. B. „Pariser Salon“, in: Flugsport (1912), Nr. 24, S. 854ff.

habe. Die gestellten Ansprüche wiesen die Gesellschafter als „unannehmbar" zurück.[438]

Beziehungen zu einflussreichen Kontaktmännern bei Flugzeugfirmen oder dem Militär zu unterhalten, zu taktieren und mit mehreren Parteien parallel Verhandlungen zu führen, waren Voraussetzungen, um im Flugzeuggeschäft erfolgreich zu sein, das machen die Protokolle der Kondor-Werke deutlich. Sondierungsgespräche wurden von persönlicher Zu- oder Abneigung ebenso wie von Glück und dem richtigen Zeitpunkt geleitet. Die Flugzeugbranche war deshalb in einem ständigen Wandel begriffen, Entwicklungen waren sprunghaft und oftmals unberechenbar. In Erwägung gezogene und tatsächlich erfolgte Verhandlungen hatten zumeist vorläufigen Charakter und waren gekennzeichnet durch gegenseitiges Ab- und vorsichtiges Herantasten. Zudem spielten die finanzielle Situation der anderen Firmen sowie gute Auftragsbestände eine wichtige Rolle bei der Wahl der Geschäftspartner: „In Betracht kommen in erster Linie Gotha und Albatros; Gotha infolge seiner guten finanziellen Position, Albatros infolge seiner starken Auftragsbestände, die es allein auszuführen Schwierigkeiten hat."[439]

Die Kondor-Werke spielten den anderen Unternehmen gegenüber nicht mit offenen Karten, sondern holten im Vorfeld bei Militärs oder Banken Erkundigungen über den Geschäftsbetrieb ein und gingen äußerst bedacht vor, bevor sie einer Fusion zustimmten. Dieses Verhalten war jedoch keineswegs überraschend, denn auch die Konkurrenzfirmen ließen Vorsicht walten. Als Ende August 1913 Gespräche mit der Gothaer Waggonfabrik und den Albatros-Werken geführt werden sollten, beschlossen die Gesellschafter, dass Suwelack und Schröder mit Albatros in Verbindung treten sollten, ohne die Firma über die Absichten der Kondor-Werke ins Bild zu setzen:

> „Die Herren werden ihren Schritt als vorläufig rein persönlichen hinstellen und sich von Albatros das nötige Zahlenmaterial erbitten, um auf Grund dessen den Gesellschaftern einen Vorschlag unterbreiten zu können. Die Herren werden bei der Gelegenheit auch auf die von uns zu gründende Flugschule hinweisen, ohne Nennung des Namens der Stadt Düsseldorf, die bei Zustandekommen einer Verständigung auf gemeinsame Rechnung betrieben werden könnte."[440]

Zu Gotha hieß es: „Herr Oberleutnant Goldschmidt wird über den Umweg einer Berliner Bank vorsichtig in Erfahrung zu bringen suchen, ob bei Gotha prinzipiell ein Interesse vorliege, gegebenenfalls mit Kondor Hand in Hand zu gehen."[441]

Das Essener Unternehmen war darauf bedacht, stets den richtigen Mann zu beauftragen, um spezielle Auskünfte zu erhalten. Verhandlungen mit der AEG

438 Archiv-TG, Akte Kondor I, Nr. 1312/1912, Bericht von der Gesellschafter-Versammlung am 13.12.1912.
439 Archiv-TG, Akte Kondor I, Nr. 2108/1913, Treffen zwischen B. Goldschmidt, J. Suwelack, E. Füßmann und V. von Back am 21.8.1913 im Büro der Th. Goldschmidt AG, Essen.
440 Ebd.
441 Ebd.

schob man z. B. so lange auf, bis Karl Goldschmidt von einer Reise nach Berlin zurückgekehrt war, da er Kontakt zu Walther Rathenau, dem Inhaber der AEG, pflegte und ein gutes Wort für die Kondor-Werke bei ihm einlegen sollte.[442] Mitarbeiter bei der Militärbehörde beziehungsweise Männer mit Kontakten in die obersten Kreise des Militärluftwesens suchte vor allem Bernhard Goldschmidt auf, um Fingerzeige zu erhalten, welche Forderungen die Generalinspektion an die Fabriken stellte. So wandte er sich mehrfach an Felix Wagenführ,[443] der als Adjutant beim Generalinspekteur des Verkehrswesens von Entwicklungen der Militärfliegerei zuerst erfuhr. In einer privaten Unterredung am 5. Januar 1914 setzte Wagenführ seinen Gesprächspartner darüber in Kenntnis, dass die Kondor-Werke mit ihren Flugzeugen nicht mehr zeitgemäß seien, beziehungsweise sie diese zu spät auf den Markt gebracht hätten: „Er sagte, dass die Sache für uns insofern ungünstig stände, als wir ein Jahr zu spät gekommen seien. Die Taube habe heute einen Stand der Vervollkommnung erreicht, der einschneidende Neuerungen so ziemlich illusorisch mache."[444] Das gleiche gelte für Flugschulen, von denen mehr zur Verfügung stünden als nötig. Dennoch gab Wagenführ Ratschläge, wie das Unternehmen zum Lieferanten der Heeresverwaltung werden könne, diese waren jedoch sehr vage und allgemein gehalten. So erklärte er, dass nur „wirklich auffällige Neuerungen" das Interesse wecken würden, bezweifelte gleichzeitig jedoch, dass die Kondor-Werke kurzfristig in die engere Auswahl kämen. Ferner verwies er auf seine fehlende Entscheidungskompetenz. Trotzdem versprach er, alles „an ihn gelangende [sic!] wohlwollend" einzubringen. Wagenführ riet den Kondor-Werken zudem, ihre finanzielle Situation gegenüber der General-Inspektion offen darzulegen und um eine eindeutige Auskunft zu bitten, „ob wir nach Beseitigung der vorhandenen kleinen technischen Mängel Aussicht auf Bestellungen hätten oder nicht, damit weitere unnötige Geldopfer vermieden würden."

Bereits drei Tage später, am 8. Januar 1914, fand eine diesbezügliche Unterredung mit Hauptmann Wilhelm Grade[445] und Leutnant Waldemar Walrab von Buttlar[446] (1885–1952) in Döberitz statt.[447] Grade war Kompaniechef im Flieger-

442 Ebd.

443 Felix Wagenführ (1873–1947) war seit 1911 Leiter der Lehr- und Versuchsanstalt für das Militärflugwesen in Döberitz und wurde später Fliegeroffizier. In den Jahren 1913/14 war er Adjutant beim Generalinspekteur des Verkehrswesens und damit in der Position, von Entwicklungen der Militärfliegerei zuerst zu erfahren. Im Krieg war er Leiter der FfA 30. Vgl. Wagenführ, Felix, in: Rudolf Vierhaus (Hg.), Deutsche biographische Enzyklopädie, 2. überarb. u. erw. Ausgabe, München 2008, S. 338.

444 Archiv-TG, Akte Kondor I, Nr. 0501/1914, Unterredung mit Herrn Hauptmann Wagenführ am 5.1.1914, Zitate ebd.

445 Grade war der Bruder des Fliegers Hans Grade, er erwarb am 29. September 1910 den Pilotenschein (Nr. 29) und wurde bei Kriegsbeginn Leiter der FfA 7.

446 Freiherr von Buttlar (1885–1952) war seit 1912 Leutnant, er machte im März 1912 seinen Flugschein (Nr. 167), war im Krieg in der Forschung tätig und bei Kriegsende Hauptmann. Nach dem

bataillon Nr. 1 und von Buttlar unter anderem für die Flugzeuginspektion und -abnahme in Döberitz zuständig. Das Treffen ergab, dass der vorgeführte Kondor-Eindecker „einen sehr guten Eindruck" gemacht habe, doch Tauben nicht mehr in größerer Zahl angeschafft würden. Gewünscht werde nun ein schneller Eindecker „für strategische Zwecke", den die Kondor-Werke nach vorgenommenen Änderungen erneut vorfliegen sollten. Auf die konkrete Frage nach Bestellungen antworteten die beiden Militärs diplomatisch und unverbindlich, dass die General-Inspektion „nicht abgeneigt" sei, „einen Probeauftrag zu erteilen", wenn entsprechende Veränderungen an der Maschine vorgenommen würden. Erst auf weitere Nachfragen nach neu ausgearbeiteten Vorschriften wurden die Informationen detaillierter. Neue Vorschriften seien noch nicht im Druck, doch müsse ein Militärflugzeug die folgenden Anforderungen erfüllen: „Höchste Spannweite 12 m, Gewicht des Apparates 530 kg. Betriebsstoff für 2 Stunden, Nutzlast 175 kg, kurzer An- und Auslauf, 1000 m [Steigfähigkeit] in 8 Minuten."

Wenngleich solche Unterredungen für Flugzeugfirmen von immenser Wichtigkeit waren, weil sie auf diese Weise gewissermaßen aus erster Hand von Neuerungen erfuhren, so verdeutlichen sie ebenso, dass die Firmen unter hohem Innovationsdruck standen und bei ausbleibenden Aufträgen der Bankrott drohte. Die Kondor-Werke trugen stets das volle finanzielle Risiko sowie das Bruchrisiko für die Maschinen. Auffällig ist in allen Protokollen, die über Treffen mit Militärs und Geschäftsleuten Auskunft geben, die Unverbindlichkeit der Verhandlungen, wobei den Kondor-Werken nie gänzlich die Hoffnung genommen wurde: „Obgleich man im allgemeinen [sic!] mit neuen Fabriken nicht mehr [Verträge] abschliessen wolle, so sei es doch nicht gesagt, dass bei guter Ausführung auch wir bei der M[ilitär-]Verw[altung] ins Geschäft kämen."[448]

Josef Suwelack legte ebenfalls Wert auf das Urteil der Militärverwaltung und berief sich in seinem technischen Bericht über das Geschäftsjahr 1912 mehrfach auf Oberstleutnant zur See Otto Bertram, der ihn an seinem Wissen über Versuche mit Wasserflugzeugen teilhaben ließ:

> „Herr Oberltn. Bertram hatte die Freundlichkeit, mir einen großen Teil seiner Erfahrungen, die er mit Wasserflugmaschinen bei der Marine-Versuchsstation in Putzig angestellt hat[,] zur Verfügung zu stellen[,] namentlich auch Fingerzeige zu geben, welche Maschinen von der Marineverwaltung bevorzugt werden."[449]

Krieg arbeitete er bei den Bayrischen Motoren Werken (BMW), Informationen unter: http://www.frontflieger.de/3-b-f.html#buttlarwaldemar (11.10.2014)

447 Archiv-TG, Akte Kondor I, Nr. 0801/1914, Unterredung mit Hauptmann Grade und Leutnant von Buttlar in Döberitz vom 8.1.1914. Alle weiteren Ausführungen ebd.

448 Ebd.

449 Archiv-TG, Akte Kondor I, Nr. 2001/1913, Technischer Bericht Josef Suwelacks für das Geschäftsjahr 1912, vorgelegt auf der Gesellschafter-Versammlung am 20.1.1913.

Fürsprecher und Kontaktmänner bei der Militärverwaltung hatten dementsprechend indirekt Einfluss auf die interne Firmenarbeit, da sie Kriterien vorgaben oder anregten, nach denen Militärflugzeuge entwickelt werden sollten. Das Verhältnis zwischen Militär und Flugzeugfirmen war dabei einerseits hierarchisch gegliedert, weil die Weisungen „von oben“ kamen. Andererseits brachten die bei den Firmen tätigen Werkspiloten selbst Verbesserungsvorschläge ein, um sich zu profilieren und die Absatzchancen zu steigern. Insbesondere diese „Innovationsschriften“ von der Basis zeigen die Mischung aus Unterwürfigkeit und Standhaftigkeit, mit der die Firmen beziehungsweise die Konstrukteure ihre Ideen vorbrachten.

In den Anträgen achtete man penibel auf demütige Formulierungen und überlegte gemeinsam, wie eine Eingabe zu formulieren sei.[450] Im Mai 1915, nachdem Josef Suwelack bereits länger an der Front stand, richtete er z. B. ein Schreiben an einen Oberleutnant,[451] um diesem seine Idee bezüglich des Einbaus eines Maschinengewehrs in ein Kampfflugzeug vorzustellen:

> „Der Zweck meines Schreibens ist[,] Herrn Oberleutnant meine Meinung über Kampfflugzeuge mitzuteilen und Herrn Oberleutnant zu bitten, meine Bitte, die ich zum Schluß ausspreche, im Sinn des Herrn Oberleutnant unterstützen zu wollen.“

Nach Erläuterung seines Konzeptes schloss Suwelack: „Ich bitte nun Herrn Oberleutnant ganz gehorsamst um seine Meinung fragen zu dürfen, ob ein Gesuch der Kondor-Werke Aussicht auf Genehmigung hat“.[452] Die erhofften Aufträge durch das Militär blieben jedoch aus und auch Josef Suwelacks Vorschlag zum Einbau eines Maschinengewehrs verlief ins Leere.[453]

Die Gesellschafter-Protokolle spiegeln, wie sich Hoffnung und Enttäuschung abwechselten und zeitweise in Resignation endeten. In der Wahrnehmung der Gesellschafter und besonders auch in der Josef Suwelacks war der Kondor eine „technisch erstklassig durchgearbeitete Maschine“, mit der er in der Lage sei, „jeder Zeit mit jeder Konkurrenz in den Kampf zu treten“.[454] Solche Aussagen zeugen von großem Selbstbewusstsein, doch standen der Westfale und die übrigen Gesellschafter mit dieser Meinung keineswegs alleine da. Die zeitgenössische

450 Vgl. Archiv-TG, Akte Kondor I, Nr. 2108/1913: Treffen zwischen B. Goldschmidt, J. Suwelack, E. Füßmann und V. von Back am 21.8.1913 im Büro der Th. Goldschmidt AG, Essen: „Herr Oberleutnant Goldschmidt übernimmt zusammen mit Herrn Suwelack die Stilisierung der an die General-Inspektion zu richtende Eingabe.“

451 Der Name ist nicht bekannt, es ist stets die Rede von „Herr Oberleutnant“.

452 NL J. Suwelack, Feldpostbrief an einen Leutnant betr. des Einbaus eines Maschinengewehrs in ein Kampfflugzeug, Lille, 23.5.1915 [unvollständiges Manuskript mit vielen Korrekturen].

453 Vgl. NL J. Suwelack, Feldpostbrief an Richard Suwelack vom 30.7.1915 sowie Brief an seine Eltern und Geschwister vom 8.8.1915. Siehe auch NL J. Suwelack, Rechnung des Patentanwalts Max Wohlgemuth an J. Suwelack betr. „Neuanmeldung Anordnung für Maschinengewehre in Flugzeugen“ in Höhe von 95,50 Mark vom 22.6.1915.

454 Archiv-TG, Akte Kondor I, Nr. 2001/1913, Technischer Bericht Josef Suwelacks für das Geschäftsjahr 1912, vorgelegt auf der Gesellschafter-Versammlung am 20.1.1913, beide Zitate ebd.

Presse bewertete den Kondor durchaus wohlwollend und positiv. Von militärischer Seite jedoch wurden Mängel festgestellt und das Flugzeug nicht als etwas grundsätzlich Neues bewertet. Diese Diskrepanz zwischen Selbst- und Fremdwahrnehmung soll im Folgenden exemplarisch aufgezeigt werden.

Josef Suwelack und der Kondor-Eindecker zwischen Selbst- und Fremdwahrnehmung

In seinem technischen Bericht für das Geschäftsjahr 1912 beurteilte Josef Suwelack das zurückliegende Halbjahr sehr positiv und sah der Zukunft optimistisch entgegen: „Wenn ich am Ende des Jahres 1912 einen Rückblick werfe auf das Bestehen der ,Kondor-Werke', so kann man in der Tat behaupten, dass wir in der letzten Zeit sehr viel erreicht haben."[455] Während schon der erste Apparat bei seinem Erstflug am 12. Oktober 1912 „sofort flug- und gebrauchsfähig" gewesen sei, stellten die folgenden zwei Maschinen nach Suwelacks Meinung nochmals Steigerungen dar: „Der 2. Apparat bewährte sich in ganz hervorragender Weise. Wie bekannt ist[,] flog der Apparat sofort und erwies sich als vollständig automatisch stabil." „Während des Fliegens konnte ich in aller Ruhe die Hände vom Steuer loslassen, ohne dass der Apparat dadurch beeinflusst wurde[,] und was noch mehr bedeutet, der Apparat geht bei Ausschalten des Motors selbsttätig in den Gleitflug über und gleitet zur Erde ohne jegliche Steuerbetätigung." Ferner überträfe er die Taube in der Stabilität, was ihm mitfliegende Passagiere bestätigt hätten. Als weitere Referenz nannte Suwelack Oberstleutnant zur See Otto Bertram, der auf Einladung der Kondor-Werke mit dem Apparat geflogen und „aufs Höchste befriedigt" gewesen sei. Sichtlich stolz bemerkte der Konstrukteur, dass sich das Unternehmen innerhalb kürzester Zeit „einen guten Namen geschaffen" habe, die Kondor-Werke „in ganz Deutschland bekannt" seien und „in der Flugzeugindustrie ... schon einen guten Klang" hätten. Er kam am Ende des Berichtes zu dem Schluss, dass Grund zur Hoffnung bestehe, Aufträge der Militärverwaltung zu erhalten. Er sei in seiner Ansicht von Bertram bestärkt worden und führte die Vergrößerung des Betriebes als Beleg für den Erfolg des Flugzeugwerkes an. Die Fertigstellung einer dritten und vierten Maschine erfolge voraussichtlich Anfang beziehungsweise Mitte Februar 1913.[456] Aufgrund seiner Besuche im Kriegsministerium, bei denen man ihm „sehr günstige Bescheide" gegeben habe, war Suwelack zudem voller Hoffnung, bald „zu größeren Lieferungen herangezogen zu werden". Auch mit Blick auf einen Zusammenschluss mit den Continental-Flugzeugwerken[457]

455 Ebd. Dort auch alle weiteren Zitate.
456 Vgl. ebd.
457 Die Continental-Flugzeugwerke stellten den sogenannten Bomhard-Pfeildoppeldecker her.

vertraute er auf Otto Bertram, der sich für eine Fusion einsetzte. Außerdem wollte Suwelack ebenfalls persönlich in Berlin mit der Firma sprechen.

Insgesamt zeugen die Protokolle davon, dass Fürsprecher und Vertrauenspersonen in die Geschäftsstrukturen der Kondor-Werke eingeweiht waren und auf ihre Weisungen und Rückmeldungen großer Wert gelegt wurde. De facto waren die Kondor-Werke Ende des Jahres 1912 jedoch zahlungsunfähig, denn es war bislang nicht gelungen, irgendwelche Einnahmen zu erzielen. Daher wurde eine Kapitalerhöhung in Höhe von 14.000 Mark beschlossen, für eine weitere Beteiligung wollte Karl Goldschmidt „maßgebende Persönlichkeiten interessieren".[458]

Die zeitgenössische Presse berichtete positiv über die Flugzeuge der Kondor-Werke, der „Flugsport" stellte den Kondor sogar mit einer Maßzeichnung ausführlich vor, wenngleich die hier gezeigte dreisitzige Variante nie gebaut wurde (Abb. 11).[459] Hervorgehoben wurde vor allem die sorgfältige bauliche Ausführung. So hieß es etwa zum Fahrgestell: „Das Fahrgestell ist aus Eschenholz-Balken, die sehr leicht auswechselbar sind, hergestellt. Eine durchgehende 4 Räder tragende Achse gestattet bei sehr schlechtem Terrain ohne Gefahr zu landen." Ferner lobte der „Flugsport", dass „auf die konstruktive Durchbildung der Maschine ... sehr großes Gewicht gelegt" worden sei.[460] In der nächsten Ausgabe berichtete die Zeitschrift von mehreren längeren Flügen Suwelacks auf seinem Eindecker, die „die große Steigfähigkeit und Schnelligkeit glänzend bewiesen" hätten.

Weiterhin informierte der Artikel über die Pläne der Kondor-Werke: „In allernächster Zeit wird Suwelack den Apparat einer Militärkommission in Döberitz vorführen und wird derselbe hier beweisen, daß er selbst den höchsten Anforderungen der Militärbehörde entspricht." Das Flugzeugwerk habe zudem eine Wasserflugmaschine im Bau, die der Vollendung entgegenstehe und am Rhein ausprobiert werden solle, und beabsichtige außerdem, eine Flugschule „mit vorläufig vier Apparaten" zu errichten.[461]

Über den zweiten Apparat hieß es im Januar 1913, dass er „durch sein elegantes Aeußere [sic!] noch mehr [anspreche] wie der erste" und mit „verschiedenen Verbesserungen versehen" worden sei. Berichtet wurde von erfolgreichen Flügen Suwelacks über Rotthausen, Gelsenkirchen und Heßler, bei denen sich der Apparat als „vollständig automatisch stabil" erwiesen habe.[462] In ähnlicher Weise setzte sich die Berichterstattung in den beiden folgenden Ausgaben fort:

458 Archiv-TG, Akte Kondor I, Nr. 1811/1912, Protokoll der Gesellschafter-Versammlung vom 18.11.1912.
459 Siehe zur Produktion der Kondor-Werke: Protokolle der Gesellschafter-Versammlungen; Archiv P.M. Grosz/Archiv DTMB; Lange, Luftfahrttechnik, S. 86 und Rißmann-Ottow, Glück ab!, S. 259 (Tabelle mit zwischen 1912 und 1920 gebauten Flugzeugen der Kondor-Werke).
460 Ebd., S. 904.
461 Flugsport (1912), Nr. 25, S. 949, dort alle Zitate.
462 Flugsport (1913), Nr. 1, S. 29.

> „Neuerdings hat Suwelack seinen zweiten ‚Kondor'-Apparat versucht. Die Maschine flog sofort, ohne daß ein Draht nachgespannt werden mußte. … Gleich darauf [am 7. Januar 1913] bestieg Oberl[eu]t[nant] zur See Bertram die Maschine und führte sofort einen wunderbaren Flug in 300–400 m Höhe über Rotthausen und Schalke aus und landete in elegantem Kurvengleitflug auf dem Flugplatz. Er war mit der Maschine aufs höchste [sic!] zufrieden und bezeichnete sie als sehr stabil."[463]

Damit noch nicht genug, habe Suwelack noch einen Passagierflug absolviert, bei dem er die Hände vom Steuer gelassen habe. Dieser Flugvorführung habe „Se[ine] Exzellenz Generalleutnant von der Goltz" beigewohnt, anschließend habe er die Kondor-Werke besichtigt.[464]

Die Formulierungen im „Flugsport" decken sich mit beziehungsweise ähneln sehr stark den Ausführungen in den entsprechenden Protokollen und Berichten der Kondor-Werke, was die Vermutung nahelegt, dass das Blatt angehalten wurde, werbewirksam und geschäftsfördernd zu schreiben. Es soll nicht unterstellt werden, dass die Fachzeitschrift die Unwahrheit schrieb, doch ist in den Artikeln eine euphemistische und positive Berichterstattung zu erkennen, der die Gesellschafter erfreut begegneten.[465]

In einem von ihm selbst verfassten Artikel über „Flugtechnische Fragen", 1912 erschienen im Jahrbuch des NVfL, versuchte Suwelack seine Expertise unter Beweis zu stellen, indem er seine Meinung über die Sicherheit und Gefahr der Flugtechnik sowie über die Verwendung von Ein- oder Zweideckern darlegte.[466] Er bedauerte zwar, dass nach wie vor tödliche Abstürze zu verzeichnen seien, doch sah er insgesamt einen Rückgang von Todesstürzen und damit eine erhöhte Sicherheit in der Fliegerei: „Daß die Zahl der Todesstürze relativ erheblich abgenommen hat, geht auch aus dem Vergleich der tödlichen verunglückten Flieger mit der Zahl der ausgebildeten Flugzeug-Führer hervor." Er verglich die Daten der Jahre 1909 bis 1912 und appellierte an den Flugzeugbauer, „sein Hauptaugenmerk auf die Sicherheit und Stabilität seiner Maschine zu richten".[467] Die Unfallursachen führte er zum einen auf nachlässig verarbeitetes Material zurück, zum anderen auf das fehlende Können beziehungsweise die mangelnde Erfahrung des Fliegers. Selbst ein guter Flieger könne während eines Fluges nicht allen Proble-

463 Flugsport (1913), Nr. 2, S. 63–64.

464 Von einer Besichtigung der Kondor-Werke berichtete auch die Zeitschrift „Der Motorwagen", Bd. 16 (1913), S. 120. Der „Flugsport" berichtete von einer Besichtigung der Kondor-Werke durch Gustav Krupp von Bohlen und Halbach und von begeistertem Publikum, das den freihändigen Flügen Suwelacks zugeschaut hätte: „Vom Flugplatz der Kondor-Werke", in: Flugsport (1913), Nr. 3, S. 106.

465 NL J. Suwelack, Schreiben der Kondor-Werke an die Gesellschafter vom 28.1.1913: „Ferner überreichen wir Ihnen einliegend Nr. 2 des ‚Flugsportes' mit einem beachtenswerten Artikel über die letzten Flüge von Suvelack."

466 Josef Suwelack, Flugtechnische Fragen, in: Jahrbuch des NVfL 1912, S. 126–141.

467 Ebd., S. 127.

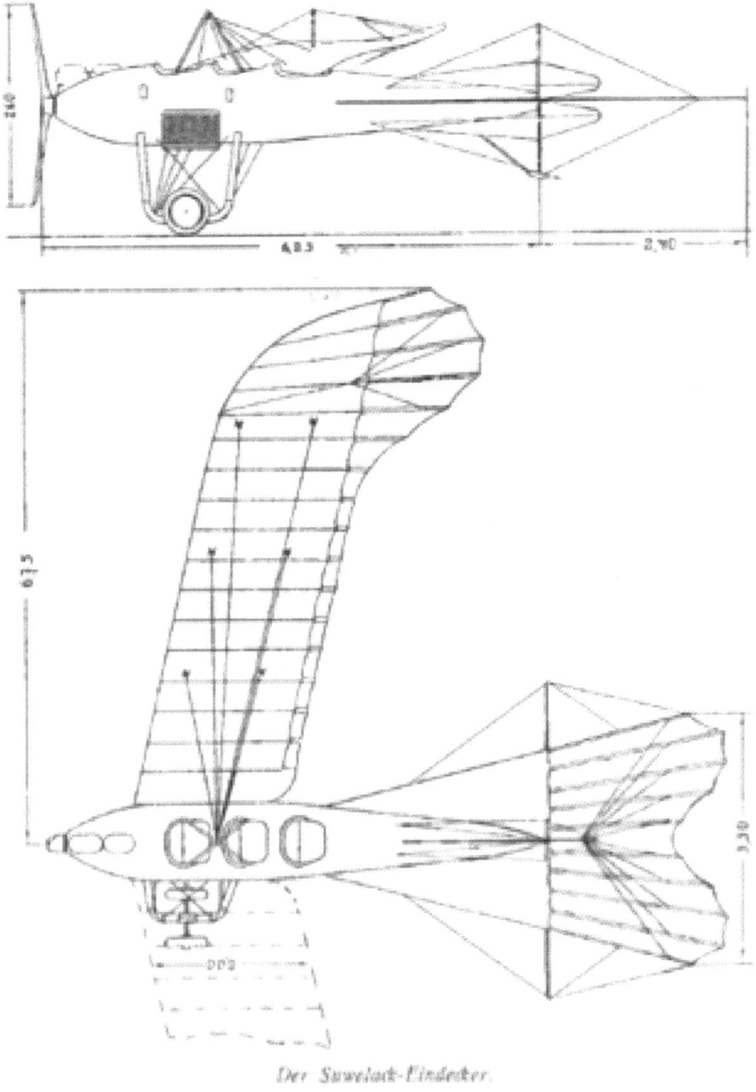

Abb. 11: Der Kondor-Eindecker Josef Suwelacks (aus: Flugsport Nr. 24/1912, S. 903)

men begegnen, sei manchmal „absolut machtlos und seinem Geschick preisgegeben".[468]

An anderer Stelle warb Suwelack demonstrativ für „seinen" Eindecker und hob dessen Eigenstabilität und Solidität hervor: „Der von mir konstruierte Eindecker nimmt jede Kurve ohne Zuhilfenahme der Verwindung absolut korrekt und nimmt selbsttätig genau die Schräglage ein, die der Kürze oder Länge der Kurve

468 Ebd., S. 128.

entspricht." Außerdem erlaube Suwelacks Maschine dem Piloten, sich während des Fluges zu entspannen, da sie so konstruiert sei, dass „der Führer jederzeit während des Fluges, selbst bei böigem Wetter, das Höhensteuer loslassen" könne.[469] Nach weiteren Lobpreisungen des Kondors und damit seiner eigenen Fähigkeiten sowie nach der „Beweisführung", dass 56 von insgesamt 82 Weltrekorden in der Fliegerei auf Eindeckern erzielt worden seien, kam Suwelack zu dem Resultat, dass die Frage, ob „Eindecker oder Zweidecker" vorzuziehen seien, eindeutig zugunsten des Eindeckers beantwortet werden könne.[470]

Wie groß sein Vertrauen in seine eigenen Fähigkeiten und in den Kondor gewesen sein muss, unterstreicht Suwelacks Versuch, im Februar 1913 mit dem Flugzeug von Essen nach London zu fliegen, um die Zuverlässigkeit der Maschine unter Beweis zu stellen und seinen Bekanntheitsgrad sowie den der Kondor-Werke zu steigern. Die scheinbar eigenmächtig und länger geplante Ärmelkanal-Überquerung, bei der er aufgrund schlechter Sicht die Orientierung verlor, einen falschen Kurs einschlug, schließlich wendete, über der Zuidersee tiefer ging und bei Deventer notlandete, war eine waghalsige und naive Unternehmung, denn der Apparat befand sich noch in der Testphase und hatte somit seine Sicherheit noch gar nicht bewiesen. Denkbar ungünstig war auch der gewählte Zeitpunkt, denn Suwelack startete am 23. Februar, einem Wintertag, von dem er annehmen musste, dass die Wetterbedingungen suboptimal für einen Langstreckenflug waren.

Während Suwelack seine Sinneseindrücke und Empfindungen beim Fliegen schriftlich in „Mein Flug zur Nordsee"[471] festhielt und von der Qualität des „Kondors" begeistert schien, stieß seine Flugreise bei den Gesellschaftern auf Unverständnis. Man hielt derlei riskante Flüge für wenig sinnvoll und empfahl, sie künftig einem angestellten Piloten zu überlassen.[472]

Der Bericht Suwelacks über seinen Langstreckenflug steht ganz in der Tradition der Blériot'schen Kanalüberquerung und bedient sich des Narrativs der Einsamkeit und des Naturerlebnisses. Suwelack beschreibt in seinem Text ein ekstatisches Glücks- und Ergriffenheitsgefühl sowie überwältigende Landschaftsimpressionen:

> „Dort, wo ich das Häusermeer von Essen glaubte, erhoben sich mächtige Wolkenberge, die sich in riesigen Ballen auftürmten. Es war ein derartig herrlicher Anblick über diese endlose, glatte weiße

469 Ebd., S.131.

470 Vgl. ebd., S.134ff.

471 NL J. Suwelack, Maschinenschrift „Mein Flug zur Nordsee", leicht verändert und gekürzt abgedruckt in: Flugsport (1913), Nr. 5, S.173f.

472 Vgl. Archiv-TG, Akte Kondor I, Nr. 2502/1913, Protokoll der Gesellschafter-Versammlung vom 25.2.1913. Die Aussagen der Gesellschafter verdeutlichen eine unterschiedliche Wertschätzung der Piloten: Es gab entbehrliche Flieger und unentbehrliche, wobei Suwelack zu den unentbehrlichen gehörte.

Wolkendecke hinweg zu fliegen, daß ich mich beinahe immer veranlaßt fühlte, aus meinem Apparat auszusteigen, um hier oben auf dieser unendlichen wunderschönen Fläche, wo es kein lebendes Wesen außer mir gab, ganz abgeschieden von allen einherzugehen."[473]

An anderer Stelle ist die Rede von Wolkenbergen, die „eine wunderbare, rosarote Färbung und eine eigenartige blaue Schattenbildung" gezeigt hätten. Der Kondor wird in Suwelacks Bericht sakralisiert und als „Wesen" behandelt, das ihm gehorcht: „Der Schatten des Kondor wurde in seiner ganzen schönen Form auf der Wolkendecke wiedergegeben und um ihn herum legten sich in den herrlichsten Regenbogenfarben Kreise, die ihn wie ein Glorienschein umgaben."

Eine Kontrastierung zwischen Erde und Himmel, welcher bisher unerschlossene Räume bereithielt und eine „Idealnatur"[474] schuf, findet sich auch in anderen zeitgenössischen und später erschienenen Fliegerberichten in ähnlicher Form, so z. B. bei Arthur Pfleger und Hellmuth Hirth. Pfleger schrieb 1930: „Hoch türmten sich Wolkenberge auf, bildeten Gassen und winkelige Ecken, anzusehen wie eine Stadt aus dem Mittelalter. Rosig angehaucht von der schon tiefstehenden Sonne waren die einen, in Schatten getaucht, wie vertra?umte Pla? tzchen, die anderen."[475] Hellmuth Hirth beschrieb seinen ersten Wolkenflug rückblickend mit den Worten: „Es überkam mich eine Art Stimmung, als wa?re alles nur ein Ma?rchen. ... Die Spitzen der Türme waren noch weiß von der sinkenden Sonne, die Wolkenta?ler aber zeigten alle Schattierungen eines Violett, das die tiefsten Stellen wie mit einer dunklen Flüssigkeit angefüllt hatte."[476] Weitere Fliegerberichte ließen sich ergänzen, die das romantische Flugmotiv ebenfalls aufgreifen und den Piloten während des Fluges in eine Naturidylle eintauchen, ja geradezu in sie versinken lassen.[477] Das entworfene Pilotenbild lebt von der Einsamkeit des Einzelnen, der eins mit seiner Maschine und zum „Bezwinger" der Lüfte wird.

Ein weiteres gängiges Erzählmotiv ist der in der Fliegersprache als „verfranzen"[478] bezeichnete Orientierungsverlust bei Störung der Kompassnadel. Josef Suwelack, der sich bei seinem Flug verirrte, führte dies auf die „Schwungmassen des Motors" zurück, durch die sein Kompass gestört worden sei. Auch durch dichte Wolken zu fliegen galt als gefährlich, weil der Flugzeugführer kein Gefühl mehr für oben und unten hatte und aufgrund eingeschränkter Sicht Bordinstrumente wie den Höhenmesser nicht mehr zur Hilfe nehmen konnte.

473 Flugsport (1913), Nr. 5, S. 173, alle weiteren Zitate ebd.
474 Möser, Fliegerblick, S. 102.
475 Arthur Pfleger, Franz im Feuer. Vier Jahre Flugzeugbeobachter, Regensburg 1930, S. 67.
476 Hirth, Flugerlebnisse, S. 189.
477 Vgl. Möser, Fliegerblick, S. 99–106.
478 Der Begriff geht zurück auf die als „Fliegerehe" bezeichnete Beziehung zwischen „Emil" (dem Flugzeugführer) und „Franz" (dem Beobachter), der die Richtung vorgab. Verlor man die Orientierung, sprach man von „verfranzen". Siehe Hackenberger, Die Alten Adler, S. 69 und Supf, Fluggeschichte, Bd. 2, S. 308.

Tatsächlich war die Navigation ein viel diskutiertes Problem in Fachzeitschriften, denn die Instrumente waren empfindlich gegenüber Umwelteinflüssen und das Kartenlesen in den Anfangsjahren der Fliegerei eine militärische Fähigkeit, die erst allmählich auch Zivilflieger beherrschen lernten.[479] Das „Verfranzen" bot manch einem Flieger Stoff für Anekdoten. So dichtete Willy Hackenberger im Ersten Weltkrieg: „Und Franz döst dahin / mit geschwundenem Sinn / und lenkt dann die Kiste, / die Richtung erwägend, / völlig ‚verfranzt' / in befremdliche Gegend."[480]

Um Abhilfe zu schaffen wurden verschiedene Verfahren entwickelt, die die Orientierung aus der Luft erleichtern sollten. Man riet den Fliegern, sich an markanten Landschaftspunkten wie Eisenbahnlinien oder Flussläufen zu orientieren. Auch überdimensional große Zahlen oder spezielle Markierungspunkte am Boden sollten dazu beitragen, ohne Umwege ans Ziel zu kommen. Für die deutschen Flugplätze und -felder gab es Schattenumriss-Karten, die es ermöglichen sollten, einen Flugplatz an seiner Form von oben wiederzuerkennen. Die Orientierung an markanten Landschaftsformationen wurde auch noch während des Ersten Weltkrieges von Beobachtungsflugzeugen genutzt. Die große Anzahl von Luftaufnahmen im Nachlass Josef Suwelacks von der französischen und belgischen Westfront, die er verbotenerweise nach Hause sandte, belegt die Bedeutung der Landschaft und der Infrastruktur als Wegweiser. Auf viele Flieger wirkte der Boden jedoch wie ein Mosaik aus Linien und Rechtecken, das nur von speziell geschulten Militärs gedeutet und gegebenenfalls für strategische Planungen verwendet werden konnte. Allerdings war auch die Ausbildung der Offiziere zum Beobachter bei den dafür eingerichteten Stellen zeitlich zu knapp bemessen, um sie gänzlich in die Lage zu versetzten, Luftbilder problemlos dechiffrieren zu können.[481]

Suwelack ging in seinem Bericht über den Langstreckenflug auch auf Turbulenzen während des Fluges ein, die er jedoch nicht auf seinen Apparat, sondern auf die schlechte, nebelige Sicht zurückführte. Der Motor habe die ganze Zeit mit einer „wunderbaren Gleichmäßigkeit" gearbeitet, und am Ende seiner Schilderung schloss er: „Glänzend bewährt hat sich auf dem ganzen Fluge mein Kondor, der, dem geringsten Steuerdruck gehorchend, mir es ... ermöglichte, alle Hindernisse noch im letzten Augenblicke zu überfliegen."[482]

479 Vgl. Möser, Fahren und Fliegen, S.180f. Siehe zur Diskussion über die Navigation im Luftraum z.B. Vorreiter, Jahrbuch der Luftfahrt 1912, S.412ff.; „Die Orientierung der Flieger", in: Flugsport (1912), Nr.3, S.90ff.; „Zur Orientierungsfrage", in: Flugsport (1912), Nr.5, S.176ff. Hans Steffen/ Otto Bertram, Orientierung auf Überlandflügen, Berlin 1912, S.54; „Orientierung aus Luftfahrzeugen", in: Sport im Bild (1912), Nr.14, S.474ff.

480 Hackenberger, Die Alten Adler, S.69.

481 Vgl. BA-MA PH 17-I/25, Die Bildmeldung der Flieger (Druck) – Teil I: Beobachter, Teil II: Bildoffizier. Darin: A: Ausbildung des Beobachters (Entwurf): Die Weiterbildung bei der Fliegerabteilung 1917–18. Siehe auch die Ausführungen zum Feld 4.

482 Flugsport (1913), Nr.5, S.174.

Die Miniaturisierung des Bodens, wie sie auch Hirth und Suwelack beschrieben, erlaubte es, sich „in der Rolle des Regisseurs [zu] wähnen",[483] der den Überblick, die (vermeintliche) Kontrolle und die Macht über das Geschehen am Boden hatte. Während Suwelack die Halle des Rotthausener Flugplatzes aus der Luft als „winziges Pünktchen"[484] beschrieb, verglich Hirth Häuser und hohe Türme mit „Kinderspielzeugschachteln" und einer „Puppenstadt", da das Auge aus großer Höhe keinen „geeigneten Maßstab" fände.[485]

Die Piloten bedienten einerseits mit ihren Erzählungen die zeitgenössischen Lesererwartungen und verarbeiteten andererseits in ihnen ihre Sinneseindrücke und Empfindungen. Sie überspielten damit zugleich auftretende Probleme während des Fluges und verharmlosten das Risiko der Flugzeugtechnik, indem sie auf die Zuverlässigkeit des Motors und die Stabilität der Maschine verwiesen. Unerwähnt blieben hingegen stets die anfallenden Arbeiten wie z. B. die notwendigen Regulierungen der Geschwindigkeit und des Motors durch den Flugzeugführer, der Motorengeräusche und das Flugverhalten des Apparates richtig deuten musste.

Suwelacks kontraproduktiver Langstreckenflug wurde in den deutschen Zeitungen registriert und brachte eine Vielzahl von zumeist kurzen Meldungen hervor. Zu mehr Beachtung bei der Militärverwaltung führte er jedoch nicht.[486] Der überwiegende Teil der Zeitungen wählte neutrale beziehungsweise dem Verlauf des Fluges gerecht werdende Überschriften wie: „Im Aeroplan über dem Meere", „Der Überlandflug Suvelacks", „Mißglückter Flug über den Kanal" oder „Irrfahrt im Aeroplan über der Nordsee". Lediglich die mit der Heimat Suwelacks in Verbindung stehenden Blätter sowie Essener Zeitungen betonten den Versuch in euphemistischen Überschriften. Die „Essener Volkszeitung" titelte: „Eine Glanzleistung Suwelacks" und die „Rheinisch Westfälische Zeitung" druckte den bereits weiter oben zitierten Bericht „Mein Flug zur Nordsee" in voller Länge.[487] Angesichts der Verbundenheit zu Josef Suwelack war diese wohlwollende Berichterstat-

483 Köppen, Das Entsetzen, S.228.

484 Flugsport (1913), Nr. 5, S.173.

485 Hirth, Meine Flugerlebnisse, S.223.

486 Siehe die Zeitungsberichte im NL J. Suwelack, z. B. „Mißglückter Flug über den Kanal", in: Freiberger Anzeiger und Tageblatt vom 26.2.1913; „Ein gefährlicher Flug", in: Pester Lloyd, Budapest, vom 24.2.1913; „Ein Flugabenteuer", in: Frankfurter Nachrichten vom 25.2.1913; „Im Aeroplan über dem Meere", in: Kösliner Zeitung vom 26.2.1913; „Irrfahrt im Aeroplan über der Nordsee", in: Das deutsche Blatt vom 25.2.1913; „Der Überlandflug Suvelacks", in: Die Zeit, Wien, vom 25.2.1913; „Flieger Suvelack im Nebel über der Nordsee", in: Hamburger Nachrichten vom 24.2.1913; „Die Fahrt des Fliegers Suwelack", in: Hamburger Fremdenblatt vom 25.2.1913; „Vom Flugwesen", in: Waldenburger Wochenblatt vom 25.2.1913; „Der Flieger Suvelack in schwerer Gefahr", in: Dresdner Nachrichten vom 25.2.1913.

487 NL J. Suwelack, „Eine Glanzleistung Suwelacks", in: Essener Volkszeitung vom 25. 2.1913 und „Mein Flug zur Nordsee am 23. Februar", in: Rheinisch Westfälische Zeitung vom 25.2.1913.

tung durchaus nachvollziehbar. Auch einige französische Zeitungen nahmen zumindest Notiz von seinem Überlandflug.[488]

Wenn Suwelack insgesamt die Berichterstattung in den Zeitungen auch akzeptieren musste, so zeigt ein Brief des Fliegers vom 15. März 1913 an Ernst Nienhausen, der der Westdeutschen Flugplatzgesellschaft das Flugplatzgelände verpachtete,[489] doch, dass er auf seinen Ruf bedacht war. Er, Suwelack, habe von seinem Bruder gehört, dass Nienhausen „gelegentlich einer Versammlung meinem Vater gegenüber Aeusserungen über meinen Flug nach Holland gemacht habe ..., die nicht den Tatsachen entsprechen." Nienhausen habe das Gerücht verbreitet, dass an besagtem Tag kein Nebel geherrscht habe und Suwelacks Bericht daher nicht auf Tatsachen beruhe: „Ich erblicke in diesen Aeusserungen eine Beleidigung, da mir vorgeworfen wird, dass ich nicht die Wahrheit berichtet habe." Suwelack erwartete von Nienhausen, sich bis zum 18. März bei ihm zu entschuldigen oder ihm die Namen der Männer zu nennen, die ihm diese Nachricht mitgeteilt hätten. Andernfalls habe er die Konsequenzen zu tragen.[490]

Wäre der Flug gelungen, so hätte diese Leistung den Kondor-Werken möglicherweise zu mehr Aufträgen, insbesondere von militärischer Seite, verholfen. Doch so blieben Militäraufträge aus, Fusionen scheiterten und auch auf Flugwettbewerben rangierte das Unternehmen im Mittelfeld. Bis die Gesellschafter zu dieser Einsicht gelangten, dauerte es jedoch bis Anfang des Jahres 1914. Nachdem weitere Unterredungen ohne Resultat geblieben waren, hielt man es „nicht für zweckmäßig, [w]eitere Geldopfer in den Bau von Tauben hineinzustecken",[491] und plante die Liquidation der Gesellschaft zum 1. April 1914. Zur Geschäftsaufgabe kam es dann allerdings nicht, weil über einen Hamburger Exporteur, der Verbindungen zum spanischen Militär hatte, ein neuer Absatzmarkt in Spanien geschaffen werden konnte.[492]

Die Verhandlungen mit Firmen, Militärs und Prüfungskommissionen für Flugzeuge zeugen davon, wie sehr die Gesellschafter und besonders auch Josef Suwelack an den Erfolg des Unternehmens und des Kondors glaubten. Obwohl

488 Vgl. NL J. Suwelack, „Une Descente intemperstive", in: Le Progres Lyon vom 25. 2.1913; „Un atterrissage émouvant", in: Le Petit Bleu Bruxelles vom 25. 2.1913; „Aviation", in: Dernière Heure Bruxelles vom 26.2.1913; „Un Aviateur Allemand au Zuyderzée", in: Journal de Caen vom 25.2.1913 und weitere Berichte.

489 Siehe StdAE Bestand 102 I, Akte 1059, Blatt 136–138, Vertrag zwischen Herrn Gutsbesitzer Nienhausen und der Westdeutschen Fluggesellschaft m.b.H., Gelsenkirchen, den 19.3.1912.

490 NL J. Suwelack, Beschwerdebrief J. Suwelacks an Herrn Ernst Nienhausen vom 15.3.1913. Ein Antwortschreiben ist nicht überliefert.

491 Archiv-TG, Akte Kondor I, Nr. 0000/1914, Geschäftsbericht über den Gang der Verhandlungen im ersten Halbjahr 1914.

492 Siehe Archiv-TG, Akte Kondor I, Nr. 1203/1913, Protokoll der Gesellschafter-Versammlung vom 12.3.1913; Nr. 3003/1914, Protokoll der Gesellschafter-Versammlung vom 30.3.1914 und Archiv-TG, Akte Kondor I, Nr. 0000/1914, Geschäftsbericht über den Gang der Verhandlungen im ersten Halbjahr 1914.

die Flugzeuge der Kondor-Werke bei Flugvorführungen und Wettbewerben im Mittel- oder Schlussfeld rangierten, war die Zuversicht der Gesellschafter ungebrochen.[493]

Tabelle 3: Teilnahme der Kondor-Flieger Beck und Suwelack an Flugwettbewerben im Jahr 1913

Ausgeschriebener Wettbewerb	Flieger	Ergebnis	gewonnener Anteil aus dem Gesamtpreis (in Mark)
Prinz-Heinrich-Flug 10. – 17. Mai 1913	Suwe-lack	ohne Platzierung wegen Bruch und Aufgabe am 14. Mai	1.500 / 54.259,70 + 2 Ehrenpreise
Jubiläumsflugwoche Flugplatz GE-E-Rotthausen 27. Juli – 3. August	Beck	Platz 6 von 11	3.062,34 / 43.449
Ostpreußischer Rundflug 9. – 14. August	Suwe-lack	Platz 6 von 6 Absturz mit Armbruch	1.100 oder 1.500 / 39.185
Aeroplanturnier Gotha 16. – 18. August	Beck	Platz 3 von 6	3.200 / 13.900
Rund um Berlin 30. – 31. August	Beck	Platz 8 von 20	3.145,14 / 101.771,92

In der Gesellschafter-Versammlung wenige Tage nach Suwelacks Teilnahme am Prinz-Heinrich-Flug dankte Karl Goldschmidt dem Flieger für seine „glänzende Leistung", während er gleichzeitig sein „Bedauern über das Herrn Suwelack bei Heidelberg ereilte Missgeschick" äußerte.[494] Suwelack hatte sich während des Fluges verirrt, musste am dritten Wettbewerbstag aufgrund eines Motordefekts bei Heidelberg notlanden und kam beim erneuten Start nicht rechtzeitig vom Boden los, so dass der Apparat in eine Baumgruppe rollte und schwer beschädigt wurde. Angesichts des Unfalls und der Beschädigung am Apparat wäre eine Schelte vonseiten der Gesellschafter zu erwarten gewesen, doch offenbar wurde das vorzeitige Ausscheiden Suwelacks aus dem Wettbewerb als nicht dramatisch angesehen. Betont wurden stattdessen Erfolge und die Qualität der Maschine, wohingegen technische Fehler und Motorversagen nur als vorübergehende Pro-

493 Immerhin kamen die beiden Flieger Suwelack und Beck im Jahr 1913 bei Flugwettbewerben auf eine Gesamtgewinnsumme von rund 12.000 Mark (siehe Tab. 3), worin jedoch die Prämienzahlungen für mehrstündige Flüge nicht inbegriffen waren. Zusammenstellung mithilfe der Informationen aus „Flugzeug-Wettbewerbe", in: Jahrbuch des DLV 1914, S.167–178 und „Der Ostpreußische Rundflug", in: Allgemeine Automobilzeitung (1913), Nr. 35, S.17–18.

494 Archiv-TG, Akte Kondor I, Nr. 2205/1913, Protokoll der Gesellschafter-Versammlung vom 22.5.1913. Vgl. Denkschrift über den Prinz-Heinrich-Flug, S.26 und S.59f.

bleme, als „kleine ... technische ... Mängel“ bewertet wurden, die sich nach Meinung der Gesellschafter leicht beheben ließen.[495]

Im August 1914 erhielten die Kondor-Werke durch die Inspektion der Fliegertruppen den Auftrag zum Bau eines Doppeldeckers, der kurze Zeit später fertiggestellt war und einer Abnahmekommission auf dem Rotthausener Flugplatz vorgeführt wurde. Während der Apparat zu diesem Zeitpunkt die Bedingungen „glänzend“ erfüllt habe, fiel die zweite Prüfung durch die Kommission in Berlin-Adlershof negativ aus. Die Prüfungskommission, bestehend aus den Herren von Buttlar, Ingenieur Dorner und Dr. Heller, äußerte an achtzehn Konstruktionsmerkmalen Kritik, welche die Gesellschafter jedoch als unerheblich einschätzten. Dem Protokoll der Gesellschafter-Versammlung zufolge beruhten die Beurteilungen „teilweise auf Irrtum der Kommission“, teils seien die Fehler „so geringfügiger Natur ..., dass wir sofort Versuche unternahmen, den Beschluss der Inspektion der Fliegertruppen rückgängig zu machen“. Dieser Versuch blieb jedoch ohne Erfolg.[496]

Nach außen warben die Kondor-Werke, wie die übrigen deutschen Flugzeugfirmen auch, in Zeitungen und Fachmagazinen sowie auf Reklamemarken für ihre Fabrik und ihre Flugzeuge. Zugleich wurde der Name „Suwelack“ manchmal mit abgedruckt, was den Bekanntheitsgrad des Konstrukteurs wiederum erhöhte und aus Suwelack eine „Marke“ machte (Abb. 12).

Überzeugt von der Konstruktion ihrer Flugzeuge meldeten die Kondor-Werke zwischen 1913 und 1918 insgesamt vier Bauvorrichtungen zum Patent an, darunter ein „Fahrgestell für Flugzeuge mit hochziehbaren Rädern“ im Jahr 1913. Ferner ließen sie zwischen 1912 und 1920 – so lange wurden unter dem Namen der Gesellschaft Flugzeuge produziert – 46 Gebrauchsmuster eintragen.[497]

Die Kondor-Werke standen, was den betriebenen Aufwand, die Professionalisierung und die Ausdifferenzierung des Betriebes anbelangte, Firmen wie Albatros und Rumpler in nichts nach. Allerdings hatten die Unternehmer zu einer Zeit in den Aufbau einer Flugzeugfirma investiert, als der Markt gesättigt und große, leistungsstarke Flugzeugfirmen im Deutschen Reich bereits vorhanden waren. Hinzu kam, dass das Ruhrgebiet unter fliegerischen Gesichtspunkten für die Militärverwaltung wenig attraktiv war, weil einerseits Garnisonen fehlten, andererseits ausreichend Flugplätze vorhanden waren. Bis 1914 gab es im Deutschen Reich 62 Flugplätze beziehungsweise -felder, von denen zahlreiche Städte wie Augsburg, Bremen, Kassel, Köln, Darmstadt, Freiburg, Mainz, Münster oder Wiesbaden, um nur einige zu nennen, als Exerzierplätze genutzt wurden und zum Teil Garniso-

495 Siehe Archiv-TG, Akte Kondor I, Nr. 0501/1914: Unterredung mit Herrn Hauptmann Wagenführ am 5.1.1914; und Archiv-TG, Akte Kondor I, Nr. 0000/1914: Geschäftsbericht über den Gang der Verhandlungen im ersten Halbjahr 1914.

496 Archiv-TG, Akte Kondor I, Nr. 0000/1914, Geschäftsbericht für das zweite Halbjahr 1914.

497 Siehe Deutsches Patentamt München, Patente und Gebrauchsmustereinträge Kondor Flugzeug-Werke.

Abb. 12: Werbung der Kondor-Werke in einer Luftfahrerzeitschrift von 1913
(Foto: Archiv-DTMB NL Stiasny)

nen vorzuweisen hatten. Flugzeugfirmen waren ebenfalls über das Land verteilt, so z. B. in Köln, Berlin, Bonn, Danzig, Erfurt, Flensburg oder im Ruhrgebiet.[498] Bedingt durch die große Anzahl und Streuung der Flugplätze und -fabriken waren Neugründungen daher durchaus entbehrlich.

Ernst August Schröder hatte gegenüber Hugo Stinnes im Oktober 1913 auf die Gefahr für das rheinisch-westfälische Industriegebiet durch französische Flieger hingewiesen und für den Aufbau einer leistungsfähigen Flugzeugindustrie im Ruhrgebiet plädiert. Bei dem geheimen Vier-Augen-Gespräch brachte er die Kondor-Werke ins Spiel und versuchte, Stinnes als Geldgeber für das ehrgeizige Projekt einer abwehrbereiten Flugzeug- und Rüstungsindustrie zu gewinnen: „Ich schlug Herrn Stinnes vor, mit etwa 1–2 Millionen, einer Summe, die für die Industrie wenig bedeute, für die Flugtechnik aber eine große Summe darstelle, an die Sache heranzugehen, etwa in der Weise, dass sich die verschiedenen großen Industrie-Verbände beteiligten."[499] Abgesehen von der Niederschrift dieser geheimen Unterredung mit dem Militär finden sich in den Protokollen der Kondor-Werke jedoch keine weiteren Äußerungen, die eine Fixierung des Unternehmens auf Rüstungsfragen nahelegen. Vielmehr standen Fragen nach Absatzmöglichkeiten und lukrativ erscheinenden Geschäftsbereichen im Vordergrund, deren Auf- und Ausbau 1913 betrieben wurde.

Die divergierende Sicht der Gesellschafter intern und der Militärverwaltung bei der Bewertung der Flugzeuge kennzeichnet allgemein die Beziehung zwischen der Flugzeuginspektion und den Firmen. Während die Fabriken stetig neues Kapital generieren mussten, um den wachsenden Ansprüchen der Militärfliegerei gerecht zu werden, konnte die Prüfungskommission neue Vorschriften verkünden, Kritik an der Konstruktion üben und die Flugzeugfirmen auf ein nächstes Mal vertrösten, sie hinhalten. Trotz aller vorgegebenen, vermeintlich objektiven Bewertungskriterien hing ein Ankauf wesentlich vom Wohlwollen der Prüfungskommission und dem Ruf der Flugzeugfirma ab. Eine beständige Auftragslage konnten nur wenige Firmen wie Albatros, Fokker oder Rumpler vorweisen und selbst diese Firmen, wie bereits gezeigt, mussten sich Eingriffe und Einblicke in ihren Geschäftsbetrieb gefallen lassen. Neue Namen in der Fliegerei mussten sich zunächst einen Kundenstamm aufbauen und mit sicheren, innovativen und leistungsstarken Maschinen auf sich aufmerksam machen. Der Flugzeugmarkt war geprägt von einem gleichsam darwinistischen Kampf um Anerkennung, abhängig von Empfehlungen und zu einem gewissen Grad vom Zufall begünstigt. Auch die Kondor-Werke als unbekannte Größe waren auf Werbung, positive Kritik und Für-

498 Siehe die Übersicht über Flugplätze, Flugfelder und Fluggelände bei Erblich, Fliegerschule, S. 174ff.

499 Archiv-TG, Akte Kondor I, Nr. 0310/1913, Niederschrift der geheimen Unterredung zwischen Ernst August Schröder und Hugo Stinnes am 3.10.1913.

sprecher angewiesen. Langfristige, auf Sicherheit setzende Planungen waren in diesem Berufs- und Geschäftszweig schwierig.

In der Heimatzeitung Josef Suwelacks wurden die Gründung der Kondor-Werke und Suwelacks Flüge im Kondor mit Interesse verfolgt. Am 22. Juni 1912 berichtete der „Billerbecker Anzeiger“, dass der „Plan des Fliegers Suwelack, auf dem Flugplatz [in Rotthausen] eine Flugzeugfabrik zu errichten“,[500] in Kürze verwirklicht werde:

> „Herr Suwelack hat bereits eine ganze Reihe von Ingenieuren, Monteuren usw. für sein Unternehmen verpflichtet. Die finanzielle Seite ist durchaus gesichert, denn ein Konsortium von Großkapitalisten aus Essen und anderen Städten des Industriebezirkes hat Herrn Suwelack eine Summe bis zu einer Million Mark zur Verfügung gestellt.“[501]

Suwelack erscheint hier als Initiator, und auch in späteren Artikeln der Lokalzeitung wurde die Arbeit des Westfalen besonders herausgestellt: „Allgemein bekannt sein dürfte es, daß Herr Jos. Suwelack jr. schon seit einiger Zeit mit dem Bau eines Flugzeuges in seiner Condor-Flugzeugwerkstatt auf dem Flugplatz ... beschäftigt ist.“[502] Über den ersten herausgebrachten Apparat hieß es anerkennend, dass er sich bei einem Probeflug in einer Weise präsentiert habe, „wie es in der Geschichte der Flugtechnik einzig dasteht.“[503] Suwelacks Konstruktion ähnele zwar äußerlich der Rumpler-Taube, überträfe sie jedoch an „Eleganz und Formschönheit“.[504] Insgesamt beurteilte die Zeitung das Flugzeug als ebenbürtig, wenn nicht sogar überlegen im Vergleich mit vorhandenen Systemen, wobei die positive Bewertung vermutlich eher auf Stolz und Lokalpatriotismus fußte denn auf der fachlichen Kompetenz der Redakteure. Ausführlich ging auch der „Münsterische Anzeiger“ im Februar 1913 auf die Verarbeitung des Kondors ein und hob, wie der „Flugsport“, die Besonderheiten und Neuerungen des Apparates hervor.[505] Die Beschreibung des Kondor als elegantes oder „rassiges“[506] Flugzeug war typisch für die zeitgenössische Charakterisierung der Maschinen.

In Billerbeck sah man den jungen Westfalen als erfolgreichen Unternehmer, der sich mit „seinem“ Flugzeugwerk verwirklicht und es zu Ansehen gebracht hatte. Auch Suwelack selbst präsentierte sich nach außen als erfolgreicher

500 „Luftschiffahrt“, in: Billerbecker Anzeiger vom 22.6.1912.

501 Ebd. Damit war wohl die Entstehung der Westdeutschen Fluggesellschaft gemeint, denn der „Flugsport“ schrieb, dass das Projekt mit einer Million Mark gefördert werde. Vgl. Flugsport (1912), Nr. 4, S. 158.

502 „Luftschiffahrt“, in: Billerbecker Anzeiger vom 17.8.1912.

503 „Luftschiffahrt. Ein neuer Eindecker“, in: Billerbecker Anzeiger vom 19.10.1912.

504 Ebd.

505 Vgl. „Luftschiffahrt. Von den Kondor-Flugzeugwerken, G.m.b.H.“, in: Münsterischer Anzeiger vom 16.2.1913. Die Beschreibung ähnelt sehr den Ausführungen im „Flugsport“, was die Vermutung nahelegt, dass die Kondor-Werke den berichtenden Blättern eine Beschreibung der Maschine zur Verfügung stellten.

506 Die Formulierung findet Verwendung bei Vogelsang, Die deutschen Flugzeuge in Wort und Bild, Teil 1, S. 21.

Geschäftsmann. Denn einerseits war er zwar ein auf die Kontakte der Gesellschafter angewiesener Pilot, der ohne sie und das generierte Kapital vermutlich weiterhin als Arbeitnehmer sein Geld durch die Ausbildung von Flugschülern und die Teilnahme an Wettbewerben verdient hätte. Andererseits aber hatte er bei den Kondor-Werken als Mitgesellschafter und Konstrukteur mit einem passablen Jahresgehalt Gestaltungsmöglichkeiten, konnte seine Ideen in die Besprechungen einbringen und in Gegenwart von Militärs als erfolgreicher Unternehmer auftreten: „Schon im November vorigen Jahres habe ich versucht, einen kurzen Urlaub zu erhalten und zuhause bei den Kondor-Werken ein Kampfflugzeug zu bauen. Einerseits wollte mich die Abteilung nicht gern entbehren ..., andernteils stand meine Firma technisch nicht auf der Höhe wie heute.“[507]

Möglicherweise war es diese Form der gesellschaftlichen Anerkennung, die Suwelack 1912 bewogen hatte, die durchaus großzügigen Angebote anderer Flugzeugfirmen abzulehnen. Hohe Selbstwirksamkeitserwartungen waren vielleicht ebenfalls Triebfedern, auf dem Gebiet des Flugzeugbaus seinen Weg „nach oben“ zu gehen. Der Glaube an den eigenen Erfolg hielt auch noch während des Ersten Weltkrieges an und lässt Josef Suwelack insgesamt als jungen Mann erscheinen, der vor allem an sich und seine Ideen glaubte. Er war, wie noch zu zeigen sein wird, überzeugt, als Unternehmer dauerhaft bestehen zu können. Suwelack hielt sich die Kondor-Werke „warm“,[508] wie er im Dezember 1914 an seinen Bruder Richard schrieb, war aber beispielsweise auch offen für eine Kooperation mit Aviatik oder dem Militär in Spanien. Maßgeblich für seine Planungen waren für ihn vermutlich eine langfristige berufliche Perspektive und die Chance, seine Konstruktionsentwürfe umzusetzen sowie sich als Flieger betätigen und beweisen zu können.

Projekte und Planungen des Jahres 1913

Im Laufe des Jahres 1913 bauten die Kondor-Werke ihre Geschäftszweige weiter aus, nahmen an Flugwettbewerben teil und bemühten sich um die Errichtung einer Flugschule. Im November 1912 wurde erstmals erwogen, einen Apparat zur Vorführung nach Döberitz zu schicken mit der Hoffnung auf einen Serienauftrag.[509] Im Februar erhielten die Kondor-Werke einen Brief von der Fliegerstation in Döberitz mit der Aufforderung, den Apparat „möglichst bald“ dort vorzustellen. Suwelack setzte sich, seinem technischen Bericht zufolge, daraufhin telefonisch

507 NL J. Suwelack, Schreiben Suwelacks an einen Oberleutnant, Lille, den 23.5.1915.
508 NL J. Suwelack, Feldpostbrief an Richard Suwelack, Lille, 3.12.1914.
509 Archiv-TG, Akte Kondor I, Nr. 1811/1912, Protokoll der Gesellschafter-Versammlung vom 18.11.1912.

mit der Fliegerstation in Verbindung und vereinbarte für die erste Februarwoche einen Termin.[510]

Am 5. Februar 1913 führte Suwelack mit dem für die Militärverwaltung bestimmten Apparat einen zwanzigminütigen Flug bei einer Flughöhe von 600 Metern aus: „Von diesem wichtigen Ereignis sind die Tagesblätter sofort benachrichtigt worden."[511] Am 12. März teilte er seine Absicht mit, „in ca. 14 Tagen nach Döberitz"[512] zu gehen. Doch das Vorfliegen wurde in den folgenden Monaten immer wieder verschoben und fand schließlich erst am 15. Dezember 1913 statt. Zwischenzeitlich hatte Josef Suwelack unter anderem am Prinz-Heinrich-Flug (siehe Feld 3) teilgenommen und mithilfe der dort gesammelten Erfahrungen „eine ganz neue Type" mit Verbesserungen gleich an mehreren Flugzeugteilen entwickelt. Statt vier kamen künftig nur zwei Räder zum Einsatz, das Gewicht der Maschine wurde um 150 Kilo reduziert und größere Schnelligkeit sollte durch den Einbau eines Stirnkühlers zur Reduzierung des Luftwiderstandes erreicht werden.[513]

Nichtsdestotrotz führten die Konstruktionsänderungen nicht zu den erwarteten Aufträgen. Der Bericht der Inspektion der Fliegertruppen, erschienen am 22. Dezember 1913, beurteilte den Kondor als „etwas über dem Durchschnitt", was für die Erteilung eines Bauauftrages nicht ausreichend war: „Das Flugzeug ist eine Taube mit normalen Tragflächen, rundem Holzfournierrumpf und einem Fahrgestell, das dem Albatrosfahrgestell ähnelt. Die Anwendung von Kabeln statt Drähten und die sonstigen Forderungen aufgrund der Belastungsproben sind nur teilweise erfüllt."[514]

Die Anforderungen an Flugzeuge für die Teilnahme an Flugwettbewerben für das Jahr 1913 legte ein 32-Punkte-Katalog fest. Achtundzwanzig Kriterien waren vorgeschrieben, die übrigen vier „erwünscht".[515] Neben schneller Montage und Demontage des Flugzeuges sowie Unempfindlichkeit und leichter Auswechselbarkeit des Materials waren hohe Stabilität, Tragkraft und rasche Steigfähigkeit der Maschine zentrale Forderung der Militärverwaltung. Zudem musste das Flugzeug in allen Teilen aus deutschem Material hergestellt sein und mit Waffen und Munition ausgerüstet werden können. In einem weiteren Regelwerk, den zwölf Punkte

510 Archiv-TG, Akte Kondor I, Nr. 2001/1913, Technischer Bericht Josef Suwelacks auf der Gesellschafter-Versammlung am 20.1.1913.

511 Archiv-TG, Akte Kondor I, Nr. 0502/1913, Protokoll der Gesellschafter-Versammlung vom 5.2.1913.

512 Archiv-TG, Akte Kondor I, Nr. 1203/1913, Protokoll der Gesellschafter-Versammlung vom12.3.1913.

513 NL J. Suwelack, Brief J. Suwelacks an Karl Goldschmidt vom 21.5.1913.

514 Archiv P.M. Grosz/Archiv DTMB, Inspektion der Fliegertruppen. Vorführung eines Kondor-Flugzeuges, Berlin Schöneberg, 22.12.1913.

515 Bedingungen für den Flugwettbewerb 1913 (Stand Juli/August 1912), enthalten in: MGFA, Die Militärluftfahrt bis zum Beginn des Ersten Weltkrieges 1914. Technischer Band, 2. Aufl., Frankfurt a. M. 1966, Anlagen, S. 212.

umfassenden „Vorschriften und Anregungen für die Flugzeugfabriken", hieß es unter Punkt drei, dass an folgenden Stellen Kabel anzubringen seien: „a) An den Hauptaufhängungspunkten der Unterverspannung der Tragflächen, b) an der Verspannung der Tragflächen gegen Stirndruck, c) an sämtlichen Steuerorganen." Stahldrähte durften hingegen nur dort Verwendung finden, „wo konstruktiv die Anbringung ... unerläßlich scheint".[516] Diese letzte Formulierung ließ wiederum Raum für Interpretation und war somit unverbindlich gewählt. Insgesamt zielten die Richtlinien und Vorschläge auf die Betriebssicherheit des Flugzeuges und die Sicherheit der Insassen ab. Gefordert wurden qualitativ hochwertiges Material und eine sorgfältige Verarbeitung der Drähte, Schrauben und Spannschlösser etc. Unter Punkt fünf wurde z. B. gefordert, dass „[a]lle Holme, Streben, Träger, Stützen und Zapfen ... da, wo sie in Schuhen, Hülsen oder Vertiefungen gelagert sind, gegen Herausspringen zu sichern"[517] seien.

In einer Gesellschafter-Versammlung vom 12. März 1913 hatte man darüber diskutiert, ob der Kondor die gestellten Anforderungen erfüllen könne. Suwelack war zu diesem Zeitpunkt zuversichtlich, denn auch etablierte Firmen hätten dem Essener Unternehmen gegenüber keinen Vorteil, da „noch keiner der bestehenden Apparate die neuen Bedingungen habe erfüllen können." Ernst August Schröder schloss sich dieser Meinung an und prognostizierte Gewinne, da die Konjunkturaussichten gut seien.[518]

Doch auch Skeptiker waren unter den Gesellschaftern vertreten. Otto May, der Josef Suwelack offensichtlich misstraute, regte eine Beteiligung Suwelacks in Höhe von 5.000 Mark an, damit dieser so seine Loyalität unter Beweis stellte. Zugleich sprach er sich gegen weitere Investitionen aus. Suwelack wollte „die Sache mit seinem Vater verhandeln" und der Gesellschaft dann seine Entscheidung mitteilen.[519] Die Rückversicherung seines Handelns bei seinem Vater und die finanzielle Abhängigkeit von ihm lassen den Unternehmer Josef Suwelack mehr denn je als Unternehmersohn erscheinen, der auf die Hilfe seiner Familie, besonders auf die seines Vaters, angewiesen war und auch blieb.[520]

Die Gesellschafter beriefen sich außerdem auf Auskünfte ihre Kontaktmänner bei der Militärverwaltung, die ihnen versichert hätten, dass der Kondor die Bedingungen „in einer weit über dem Durchschnitt stehenden Weise erfüllt" habe.[521] Als das Reichsmarineamt im Februar 1914 jedoch auch die Bestellung von Kon-

516 „Vorschriften und Anregungen für die Flugzeugfirmen, Beilage 1", zit. nach MGFA, Militärluftfahrt. Technischer Band, Anlagen, S.228.

517 Ebd.

518 Archiv-TG, Akte Kondor I, Nr.1203/1913, Protokoll der Gesellschafter-Versammlung vom 12.3.1913.

519 Ebd.

520 Siehe Ausführungen zum Feld 4.

521 Archiv-TG, Akte Kondor I, Nr.0000/1914, Geschäftsbericht über den Gang der Verhandlungen im ersten Halbjahr 1914.

dor-Wasserflugzeugen definitiv ablehnte, stellte man die ohnehin spärlichen Versuche mit Wasserflugzeugen ein und zeigte sich resigniert:

„Herr Dr. Karl Goldschmidt ... kommt zum Schluss zu der Einsicht, dass alle uns gemachten Versprechungen trotzdem das Flugzeug die Bedingungen erfüllt hat nicht eingehalten werden und uns auch jetzt keine Aufträge direkt zuerteilt werden. Es sei daher schwierig, weitere Kreise für das Unternehmen zu gewinnen und muss man daher sehen, durch Anschluss an eine andere Fabrik unsere Position zu stärken."[522]

Umso dringlicher erschien nun eine Verlagerung der Kondor-Werke ins Ausland oder eine Fusion mit einer ausländischen Flugzeugfirma. Geplant war eine Vertretung der Sopwith Aviation Company durch die Kondor-Werke in Deutschland oder Österreich, die dann den Bomhard-Pfeildoppeldecker als Wasserflugzeug in Lizenz nachgebaut und vertrieben hätten. Die Kondor-Werke wären damit zur deutschen Filiale des englischen Unternehmens geworden. Auch mit den Avro-Werken sollten Gespräche geführt werden. Alfred von Back-Begavár reiste für Verhandlungen nach Paris, um dort bei der Firma vorzusprechen.[523] Die Verhandlungen entwickelten sich so weit, dass mit Sopwith Anfang 1914 ein Vorvertrag geschlossen wurde und im Februar ein Musterexemplar vorgeflogen werden sollte, ein Zusammenschluss blieb dann aber aus.[524]

Auch jetzt noch fuhren die Kondor-Werke bei ihren Überlegungen zweigleisig. Zum einen verhandelten sie mit dem Ausland, vor allem mit Spanien, zum anderen gaben sie trotz der ablehnenden Haltung des Militärs die Hoffnung nicht auf, Aufträge von der Heeresverwaltung zu erhalten. Sie bewarben sich um den Preis der Nationalflugspende und versuchten auch weiterhin, ihre Maschinen in Döberitz vorzuführen.[525]

Der Aufbau einer Flugschule verlief ebenfalls ungünstig für die Kondor-Werke. Nachdem Josef Suwelack die Gründung einer solchen Einrichtung im November 1912 angeregt hatte, nahmen die Pläne im Sommer 1913 konkrete Züge an. Hatte das Unternehmen den regelmäßigen Schulbetrieb auf dem Rotthausener Flugplatz zwar bereits begonnen und bis Ende 1913 einen Flugschüler fertig ausgebildet,[526] so hielt Josef Suwelack den Flugplatz dennoch ungeeignet für einen Ausbildungsbetrieb. Er brachte Köln und Düsseldorf als Standorte einer Flugschule ins Gespräch, denn das „Wetter in Düsseldorf [sei] überaus viel günstiger ... wie

522 Archiv-TG, Akte Kondor I, Nr. 1301/1914, Protokoll der Aufsichtsratssitzung vom 13.1.1914.

523 Archiv-TG, Akte Kondor I, Nr. 02011914, Besprechung zwischen A. von Back-Begavár, B. Goldschmidt, E. Füßmann und J. Suwelack am 2.1.1914.

524 Vgl. Archiv-TG, Akte Kondor I, Nr. 2901/1914, Besprechung zwischen K. Goldschmidt, B. Goldschmidt, A. von Back-Begavár, E. A. Schröder, W. Schäfer. H. Racine, J. Suwelack und E. Füßmann am 29.1.1914; NL J. Suwelack, Brief Suwelacks an A. von Back-Begavár vom 9.1.1914 und Schreiben E. Füßmanns an Suwelack vom 18.2.1914.

525 Vgl. NL J. Suwelack, Nachricht der Kondor-Werke an J. Suwelack vom 12.2.1914.

526 Walter Hoefig (geb. 18.01.1889, Hamburg) erhielt am 8. November 1913 seine Fluglizenz und wurde als Militärflieger ausgebildet. Siehe Flugsport (1913), Nr. 25, S. 973. Vgl. auch Liste der Flugzeugführer, in: Jahrbuch des DLV 1914, S. 114–143.

hier in Rotthausen", ferner die „Nebelbildungen" geringer, so dass der Flugbetrieb schon morgens beginnen könne. In Köln könne man außerdem „Kölner vermögende Kreise" für die Kondor-Werke interessieren.[527]

Suwelack kalkulierte mit fünf Apparaten und Baukosten von jeweils 15.000 Mark. Dazu kämen Werkzeuge, Büroeinrichtung, Betriebsstoffe, Miete, Strom-, Licht-, und Heizkosten sowie Angestellten- und Bruchkosten eines Flugzeuges mit einer Gesamthöhe von rund 138.000 Mark, denen Einnahmen in Höhe von 164.000 Mark durch Schülerausbildung gegenüberstünden.

Ausgestattet mit zwei Fluglehrern könnten Suwelack zufolge „10 Schüler in 3 Monaten gut ausgebildet sein, wenn der Flugbetrieb stetig aufrecht erhalten wird." Die Betriebsunkosten habe er „ziemlich genau nach unseren bisherigen Erfahrungen festgelegt" und die angesetzte Summe für Flugzeugbruch sei „eher zu hoch als zu niedrig" angesetzt (Tab. 4).[528]

Tabelle 4: Rentabilitätsberechnung einer Flugschule nach Josef Suwelack

Bedarf und Betriebskosten	Kosten pro Jahr
Anschaffung / Bau von 5 Apparaten à 15.000 M.	75.000 M. (einmalig)
Einrichtungen, Werkzeuge, Maschinen, Büroeinrichtung etc.	5.000 M. (einmalig)
2 Fluglehrer à 400 M. Monatsgegehalt	9.600 M.
Prämien für die Fluglehrer	2.500 M.
4 Monteure à 45 M. wöchentlich	8.600 M. [8.640 M.]
Betriebsstoffe Oel und Benzin	12.000 M.
Kraft, Licht und Heizung	3.000 M.
Miete	2.000 M.
Apparatbruch	20.000 M.
Gesamt	ca. 138.000 M.

Suwelack rechnete mit zwanzig Schülern pro Jahr, durch die jeweils 7.000 Mark Einnahmen zu verzeichnen wären. Hinzu kämen drei aus der Nationalflugspende zugeteilte Flugschüler, namentlich Offiziere, für die die Heeresverwaltung bei erfolgreich bestandener „Feldpilotenprüfung" jeweils 8.000 Mark an Ausbil-

527 Archiv-TG, Akte Kondor I, Nr. 2205/1913, Protokoll der Gesellschafter-Versammlung vom 22.5.1913 und NL J. Suwelack, Schreiben J. Suwelacks an Kommerzienrat Dr. Karl Goldschmidt vom 14.10.1913. Alle weiteren Angaben ebd.

528 Kostenaufstellung nach J. Suwelack, ebd.

dungskosten an die zuständige Flugschule überweisen würde.[529] Alle genannten Zahlen und Aufträge basierten auf den Wunschvorstellungen, das heißt den Hoffnungen Suwelacks, denn die Kondor-Werke erhielten, zumindest bis Mitte des Jahres 1914, keine Militäraufträge.

Die Stadt Düsseldorf sowie der Düsseldorfer Luftfahrer-Club waren an einer örtlichen Fliegerschule interessiert und unterstützten die Kondor-Werke bei ihrem Vorhaben. Nicht zuletzt die weitreichenden Beziehungen des ehemaligen Oberbürgermeisters der Stadt Düsseldorf, Wilhelm Marx, selbst Gesellschafter der Kondor-Werke, erleichterten die Vertragsverhandlungen. Mit Direktor Paul Probst, dem Vorsitzenden des Düsseldorfer Luftfahrer-Clubs, und dem Düsseldorfer Oberbürgermeister Oehler einigte man sich auf Folgendes:[530] Der Luftfahrer-Verein und die Stadt erklärten sich bereit, den Aufbau einer Kondor-Fliegerschule auf dem Luftschifflandeplatz Düsseldorf-Lohausen mit 75.000 Mark zu subventionieren. Den Kondor-Werken wurden vom Düsseldorfer Luftfahrer-Club zwei Schuppen zur Verfügung gestellt; des Weiteren hatte der städtische Oberbürgermeister den Bau eines dritten Schuppens mündlich zugesichert, eine Bestätigung sollte im Herbst folgen. Der Flugbetrieb, der am 22. August 1913 begann, sollte durch Walter Hoefig, einen ehemaligen Flugschüler des Unternehmens, durchgeführt werden, da Suwelack und Otto Beck, der Anfang des Jahres 1913 bei den Rumpler-Werken gekündigt hatte und Pilot bei den Kondor-Werken geworden war,[531] mit dem Vorstellen der Maschinen in Döberitz und Spanien beschäftigt waren.[532] In einem Entwurf, gerichtet an die Inspektion des Militärflugwesens, kündigten die Kondor-Werke die Entstehung einer umfangreich ausgestatteten

529 Vgl. ebd.

530 Vgl. Hubert Delvos, Die Entwicklung des Düsseldorfer Luftverkehrs, Düsseldorf 1941, zu Probst S.41ff. und zu Kondor S.117f.

531 Vgl. Kranzhoff, Edmund Rumpler, S.134, und Rieth, Die Kondor-Flugzeugwerke, S.126. Otto Beck (1890–1954) war der Sohn eines Bahnbeamten und stammte aus Kornwestheim bei Stuttgart. Nach einer Ausbildung im Motorenbau arbeitete er bei mehreren Automobilfabriken und ging 1911 zu Rumpler nach Berlin-Johannisthal. Dort engagierte ihn Hellmuth Hirth als Motorenfachmann, zu seinen Aufgaben gehörte die Kontrolle des Motors und die Begleitung auf Flugwettbewerbe. Ein Jahr später ließ er sich bei Hirth zum Flugzeugführer ausbilden und erhielt am 20. Januar 1913 die Lizenz Nr. 352. Ende des Jahres 1913, am 25. Dezember, stellte er einen Rekord im Dauerflug auf und erhielt für seine achtstündige Flugzeit durch das Kuratorium der Nationalflugspende eine Prämie von 8.000 Mark. Anfang des Jahres hatte er bei Rumpler gekündigt und war zu den Kondor-Werken als Einflieger gegangen. Nach Suwelacks Weggang wurde er der dortige Chefpilot und bildete während des Krieges an der in Großenhain aufgebauten Kondor-Fliegerschule, die 1917 nach Nordhausen verlegt wurde, Militärpiloten aus. Nach dem Ende des Ersten Weltkrieges, das, zumindest was den Flugbetrieb anbelangte, auch das Ende der Kondor-Werke darstellte, war Beck arbeitslos. Er machte sich mit einer Autoreparaturwerkstatt selbstständig, bis ihm Hellmuth Hirth eine Stelle bei den Elektron-Werken in Cannstatt verschaffte. Von dort schaffte Beck den Sprung zu den Bayrischen Leichtmetallwerken in München, wo er Betriebsleiter, später Direktor wurde. Vgl. auch Reinhold Kienzle, Otto Beck erwarb 1913 den Pilotenschein, in: Kornwestheimer Geschichtsblätter 18 (2008), S.53–58.

532 Vgl. Archiv-TG, Akte Kondor I, Nr. 2108/1913.

Flugschule an, die so ausgerüstet werden sollte, dass qualitativ hochwertiger Flug- und Theorieunterricht erteilt werden könnte. Die Fertigstellung der Anlagen war bis zum 1. Februar 1914 geplant.[533]

Die Entscheidung über den Betrieb einer Flugschule konnte jedoch nicht ohne die örtliche Garnisonsverwaltung getroffen werden, die den Platz als Exerzierfeld nutzte. Sie genehmigte zwar die Aufnahme eines provisorischen Schulbetriebes mit bis zu fünf Flugzeugen, verweigerte aber ihre Zustimmung zu einer weiteren Ausdehnung des Betriebes. Die Duldung währte bis zum Frühjahr 1914, dann weigerte sich die Garnisonsverwaltung, die Genehmigung zu verlängern.[534] Der Schulbetrieb wurde daher im Frühjahr 1914 von den Kondor-Werken eingestellt, danach wurden scheinbar auch nur noch wenige Piloten ausgebildet.[535] Auch die Errichtung eines durch Oberbürgermeister Oehler angeregten Flugstützpunktes scheiterte am Beschluss des Kriegsministeriums, das das Gelände als ungeeignet für eine derartige Anlage ansah.[536]

Angesichts der schlechten wirtschaftlichen Situation des Unternehmens „entspann sich eine längere Diskussion“[537] über Sanierungsmaßnahmen, die ergebnislos blieb. Man trug sich mit dem Gedanken einer Liquidation, wollte jedoch parallel dazu an einem anderen Standort eine neue Gesellschaft ins Leben rufen. Über einen Hamburger Exporteur, eine Firma Möller, bestand Kontakt nach Spanien, wo vonseiten der Regierung und eines Privatinvestors Interesse an einer Flugzeugfirma vorlag. Die Kondor-Werke baten ihren Fürsprecher, mit seinen „spanischen Freunden“ nach Essen zu kommen, um ihnen den Kondor vor Ort vorzustellen.[538]

Vermutlich wäre es den Gesellschaftern lieber gewesen, sich bei der Militärverwaltung zu etablieren und im Inland zu arbeiten. Da es sich aber um eine Existenzfrage handelte, erscheint der Plan, das Werk nach Spanien zu verlagern, nach-

533 Vgl. NL J. Suwelack, Entwurf Maschinenschrift betreffend Antrag der Kondor Flugzeugwerke an die Inspektion der Militär-Fliegertruppen, Berlin-Schöneberg, undatiert (verm. Ende 1913/Anfang 1914), 4 Seiten mit handschriftlichen Korrekturen.

534 Die Ausführungen basieren auf den Informationen aus den Gesellschafter-Protokollen der Kondor-Werke vom 7., 11. und 21.8.1913 (Archiv-TG, Akte Kondor I, Nr. 0708/1913; Nr. 1108/1913; Nr. 2108/1913), vom 16.10.1913 (Archiv-TG, Akte Kondor I, Nr. 1610/1913), vom 2.1.1914 (Archiv-TG, Akte Kondor I, Nr. 0201/1914), dem Geschäftsbericht für das erste Halbjahr 1914 (Archiv-TG, Akte Kondor I, Nr. 0000/1914) sowie auf Delvos, Entwicklung, S. 117ff.

535 Nachweisen lassen sich die Pilotenprüfungen von Albert Riemann (geb. 1886, Johannisburg), der im Februar 1914 seine Lizenz mit der Nr. 653 erhielt, vgl. Deutsche Luftfahrer-Zeitschrift (1914), Nr. 4, S. 75; Erich Beerman (geb. 28.7.1893, Pegestorf, Kreis Hameln), Nr. 783 (30.5.1914); Bernhard Nierhoff (geb. 20.2.1895, Gelsenkirchen), Nr. 788 (9.6.1914). Angaben nach Flugsport (1914), Nr. 14, S. 598.

536 Vgl. Delvos, Entwicklung, S. 119.

537 Archiv-TG, Akte Kondor I, Nr. 1403/1914, Protokoll der Gesellschafter-Versammlung vom 14.3.1914.

538 Vgl. Archiv-TG, Akte Kondor I, Nr. 1203/1914, Protokoll der Gesellschafter-Versammlung vom 12.3.1914.

vollziehbar. Das Unternehmen sah einen möglichen spanischen Auftrag „als günstiges Moment für eine Liquidation", als eine zum richtigen Zeitpunkt erschienene Rettung. Es ging darum, „die bisherigen hineingesteckten Kapitalien wenigstens zum teil [sic!] zu retten".[539] Das Spaniengeschäft war jedoch ebenfalls kein von langer Hand geplantes Unterfangen, sondern eine von weiteren in Erwägung gezogenen Optionen.

Im September 1913 informierte Suwelack Karl Goldschmidt, dass die Maschine, die in Spanien vorgestellt werden sollte, pünktlich geliefert werden könne; im November war sie dann zeitgleich mit einem für Döberitz bestimmten Apparat fertiggestellt.[540] Nach weiteren vier Monaten wurde das Flugzeug mit der Bahn zum Flugplatz Cuatro Vientos nach Madrid transportiert und eine Delegation der Kondor-Werke, bestehend aus Suwelack, Bernhard Goldschmidt, Heinrich Hillebrand und zwei Monteuren, dorthin entsandt. Am 12. Februar notierte Suwelack:

> „Gestern Abend kamen wir in Madrid an. Unser Apparat war leider auf einer Station unterwegs stehen geblieben. Herr Pujol veranlasste aber, dass derselbe am selben Tag weiter spediert wurde und morgen eintreffen soll. Sodann besuchten wir die hiesige Mercedes-Niederlassung. Herr Direktor Frommel sagte mir seine volle Unterstützung zu. ... Wir vereinbarten, dass bei den Vorführungen ein Motorenspezialist kostenlos zugegen sein werde. Heute Abend erhält Herr Pujol technischen Unterricht, damit er die Konstruktion erklären kann. Herr Pujol ist der bestimmten Meinung, dass wir hier Erfolg haben werden."[541]

Während Goldschmidt und Hillebrand die Verhandlungen mit spanischen Regierungsvertretern und dem Kaufmann Eduardo Pujol übernahmen, begann Suwelack am 17. Februar 1914 mit den Flugvorführungen – mit Erfolg. Im folgenden Monat kaufte das spanische Militär die präsentierte Maschine und erteilte einen Auftrag über vier weitere. Die vier für Spanien bestimmten Flugzeuge wurden im Juni gebaut und sollten Mitte Juli vertragsgemäß geliefert werden. Für die Baukosten und die Anschaffung von Motoren traten Karl Goldschmidt und Alfred von Back-Begavár in Vorleistung, denn sie hofften, die Bestände der Kondor-Werke an Spanien verkaufen zu können.[542] Suwelack sollte eine diesbezügliche „Offerte" zum Kauf eines Flugzeugs gemacht werden und Ernst Füßmann wurde beauftragt, an andere inländlische Firmen heranzutreten, um sie für das Spaniengeschäft zu interessieren.[543] Wegen des Kriegsausbruchs kam es jedoch nicht mehr

539 Archiv-TG, Akte Kondor I, Nr. 0000/1914, Geschäftsbericht über den Gang der Verhandlungen im ersten Halbjahr 1914, alle Zitate ebd.

540 NL J. Suwelack, Schreiben Suwelacks an Dr. Karl Goldschmidt vom 13.9.1913, und Archiv-TG, Akte Kondor I, Nr. 0611/1913, Protokoll der Aufsichtsratssitzung vom 6.11.1913.

541 NL J. Suwelack, Handschriftliche Notizen Josef Suwelacks zum Besuch in Spanien vom 12.2.1914.

542 Vgl. Archiv-TG, Akte Kondor I, Nr. 3006/1914, Protokoll der Gesellschafter-Versammlung vom 30.6.1914.

543 Vgl. ebd.

zu einer Kooperation mit deutschen Firmen und auch die Zusammenarbeit mit Spanien war „in Folge des Krieges vollständig eingeschlafen", wenngleich wenige Monate nach Kriegsbeginn wieder Kontakt mit Spanien aufgenommen und 1915 der Vertrag erneuert wurde.[544]

Das wiederholte Scheitern bei der deutschen Heeresverwaltung und die bestehende Möglichkeit, in Spanien neu anzufangen beziehungsweise einen Schlussstrich zu ziehen, führte am 30. März 1914 zum Beschluss über die Liquidation der Gesellschaft. Sämtliche Arbeiter, mit Ausnahme von zwei Leuten, die das für Döberitz bestimmte Flugzeug fertigstellen und instand halten sollten, sollten entlassen, für die gebauten oder bereits in Betrieb stehenden Flugzeuge, Flugzeugteile und Werkzeuge Käufer gefunden werden.[545] Worten folgten jedoch nicht sofort Taten, denn man hatte die Hoffnung noch immer nicht ganz aufgegeben, Bauaufträge vom Militär zu erhalten. Suwelack wurde allerdings gekündigt, was vermuten lässt, dass vor allem Ernst Füßmann nach wie vor an seiner Integrität zweifelte. Die Abneigung beruhte offenbar auf Gegenseitigkeit, denn Suwelack äußerte sich gegenüber seinem Bruder Richard mehrfach abwertend über Füßmann und unterstellte ihm „Stockdummheit".[546] Auch die übrigen Gesellschafter beurteilte er als „wüste Schieber", die keine Ahnung vom Flugzeugbau oder von der Fliegerei hätten.[547] Seine Kündigung stand höchst wahrscheinlich im Zusammenhang mit seinem 1912 mit dem Deutschen Reich geschlossenen privatrechtlichen Vertrag und einem Schreiben der Inspektion der Fliegertruppen vom 30. März 1914. Darin wurde er aufgefordert, „sich im Laufe des 2. Mobilmachungstages bei der 3. Komp[agnie] Flieger B[a]t[ai]l[lon] 1 in Grossenhain zu melden",[548] so dass er möglicherweise von seiner Arbeit bei den Kondor-Werken freigestellt war.

Suwelacks Fortgang macht deutlich, dass seine Vorstellungen und die der übrigen Gesellschaften divergierten. Ging es Suwelack eher um die sportliche Fliegerei und die Entwicklung eigener neuer Modelle, so waren für die anderen Gesellschafter vor allem der Umsatz und die Frage nach kapitalkräftigen Partnern von Bedeutung. Das gegensätzliche Verständnis vom Nutzen der Fliegerei trat beispielsweise deutlich bei Suwelacks missglücktem Flug nach London hervor. Für ihn war es ein Abenteuer, der Versuch, den Kondor und nicht zuletzt seine Person

544 Ebd. Siehe auch spätere Protokolle der Kondor-Werke und Feldpostbriefe Josef Suwelacks.
545 Vgl. Archiv-TG, Akte Kondor I, Nr. 3003/1914, Protokoll der Gesellschafter-Versammlung vom 30.3.1914.
546 NL J. Suwelack, Feldpostbrief J. Suwelacks an Richard Suwelack vom 3.12.1914. Siehe auch NL J. Suwelack, Feldpostbriefe vom 1. und 9.5.1915.
547 Vgl. ebd., dort das Zitat und Feldpostbrief an Richard Suwelack vom 30.7.1915.
548 NL J. Suwelack, Schreiben der Inspektion der Fliegertruppen an Josef Suwelack, Berlin-Schöneberg, 30.3.1914.

durch Werbung berühmt zu machen, für die Gesellschafter hingegen ein leichtsinniges, risikoreiches und kostspieliges Unterfangen, das verlustreich und tödlich hätte enden können. Zwar stand Suwelack auch noch während seines Kriegseinsatzes mit den Kondor-Werken in Kontakt und traf sich mit einigen Gesellschaftern in Essen auf Heimaturlaub, doch rechnete er wohl nicht damit, nach seinem Militäreinsatz in Essen eine dauerhafte Anstellung zu finden. In den Briefen von der Front an seinen Bruder waren die Kondor-Werke des Öfteren Thema und Josef der Meinung, dass er sein Glück nach dem Krieg dort nicht machen könne.[549]

Im Mai 1914 knüpften die Kondor-Werke erneut Gespräche mit der Militärverwaltung an, „die zu irgendwelchen Resultaten nicht führten".[550] Sie wandten sich sogar an den Gelsenkirchener Bürgermeister und den Aufsichtsrat der Westdeutschen Flugplatzgesellschaft mit der Bitte, sich bei der Heeresverwaltung, „unter Darlegung der Verhältnisse",[551] für den Erhalt der Kondor-Werke einzusetzen, was ein schwerer Schritt gewesen sein muss, denn das Verhältnis zwischen der Flugplatzgesellschaft und den Kondor-Werken war von Spannungen begleitet.[552] Am 25. Juni legte das Unternehmen der Verkehrstechnischen Prüfungskommission (VPK) in Döberitz Konstruktionszeichnungen eines Doppeldeckers vor, dessen Bau durch den noch in Arbeit befindlichen spanischen Auftrag vorerst hintangestellt wurde. Eine dem Generalinspekteur des Flugwesens Anfang Juli vorgelegte ausführliche Denkschrift mit der Bitte um Flugzeugbestellungen sowie eine von Suwelack unter der Hand an Aviatik gerichtete Anfrage eine Fusion betreffend[553] blieben ebenfalls vergeblich. Sichtlich frustriert und ernüchtert schloss der Halbjahresbericht Füßmanns von 1914:

549 Siehe Abschnitt „Die Wege trennen sich".

550 Archiv-TG, Akte Kondor I, Nr. 0000/1914, Geschäftsbericht über den Gang der Verhandlungen im ersten Halbjahr 1914.

551 Archiv-TG, Akte Kondor I, Nr. 3003/1914, Protokoll der Gesellschafter-Versammlung vom 30.3.1914.

552 Das Verhältnis zwischen den Kondor-Werken und der Westdeutschen Flugplatzgesellschaft wurde schon seit 1912 von Spannungen begleitet, denn der Verpächter hatte die nötigen Installationen (Wasser, Gas etc.) erst nach und nach veranlasst, so dass die Kondor-Werke sich weigerten, für nicht erbrachte Leistungen Miete zu bezahlen. Im August 1913 strengten die Flugplatzdirektoren Giersberg und Forsbeck eine „Beleidigungs"-Klage gegen Suwelack an, so dass angenommen werden kann, dass die Kommunikation zwischen den Unternehmen gestört war. Vgl. diverse Schriftwechsel, Anwaltsschreiben etc. im Nachlass Josef Suwelacks für das Jahr 1913.

553 Vgl. NL J. Suwelack, Schreiben der Automobil- & Aviatik-AG an Josef Suwelack vom 31.7.1914. Aufgrund des Kriegsausbruchs kamen die im Juli begonnenen Verhandlungen zum Erliegen. Vgl. Archiv-TG, Akte Kondor I, Nr. 0000/1914, Geschäftsbericht über den Gang der Verhandlungen im ersten Halbjahr 1914. In einem Feldpostbrief an seinen Bruder Richard schrieb Suwelack: „Chatel [Henri Châtel, Direktor der Aviatik-Gesellschaft] von der Aviatik Ges. soll tot sein. Beim Gefecht bei Mühlhausen. Eine Granate ist in seine Villa geschlagen". Feldpostbrief vom 27.10.1914.

„Es ergibt sich aus vorstehendem also, dass die Verhandlungen mit der Heeresverwaltung zu keinem Resultat geführt haben. Die Verwaltung steht vielmehr auf dem Standpunkt, dass die Firmen auf ihr eigenes Risiko hin immer Versuche machen sollten, dass aber andererseits die Verwaltung gar keine Zusage wegen Auftragserteilung machen könne. Es war daher für ein Unternehmen wie das unsrige ausgeschlossen, auf das Geradewohl [sic!] hin weiter zu arbeiten."[554]

Die Schulden waren mittlerweile auf über 200.000 Mark angewachsen und neue Investoren nicht zu gewinnen.[555] Sämtliche Bemühungen unterstreichen, wie verzweifelt das Unternehmen versuchte, einer Geschäftsaufgabe zu entgehen.

Mittelständische Unternehmen wie die Kondor-Werke, die gewissermaßen zu spät ins Flugzeuggeschäft eingestiegen waren und dadurch der Entwicklung hinterherliefen,[556] konnten sich gegenüber länger etablierten Firmen nicht behaupten. Dies war auch dem Umstand geschuldet, dass die Kondor-Werke nicht in den Genuss von Fördermitteln aus der Nationalflugspende kamen, die, 1912 erstmalig durchgeführt, das deutsche Flugwesen stärken und damit Frankreich in seiner Führungsrolle in der Flugzeugindustrie ablösen sollte.

Nach dem Vorbild der Zeppelinspende[557] hatte Prinz Heinrich von Preußen auf Anregung August Eulers die Nationalflugspende[558] ins Leben gerufen, um „unaufhaltsam weiterzuarbeiten an der Vervollkommnung der Flugapparate, an der Ausbildung der Flieger, auf daß die Gefahren vermindert, die Leistungen erhöht werden". Außerdem sollte die Nationalflugspende dazu beitragen, durch „ausgedehnte Wettkämpfe ... Erfindungsgeist, Mut und Tatkraft an[zu]spornen und immer stolzere Leistungen von Mann und Maschine" zu erzielen.[559] Militärisch relevante Leistungen auf Wettbewerben und eine leistungsstarke Flugzeugindustrie wurden zu einem „nationale[n] Prestigeprojekt"[560] und zu einer Aufholjagd mit Frankreich.[561]

Im Jahr 1912 kamen durch Spenden der Bevölkerung über sieben Millionen Mark zusammen, die im folgenden Jahr auf die genannten Bereiche verteilt wur-

554 Archiv-TG, Akte Kondor I, Nr. 0000/1914, Geschäftsbericht über den Gang der Verhandlungen im ersten Halbjahr 1914.

555 Vgl. Archiv-TG, Akte Kondor I, Nr. 1903/1914, Protokoll der Besprechung am 19.3.1914, sowie NL J. Suwelack, Feldpostbrief J. Suwelacks an seinen Vater, Köln, 30.7.1915.

556 Die Rückständigkeit der Kondor-Werke seit seinem Weggang beklagte auch Suwelack in einem Brief an seinen Bruder: NL J. Suwelack, Feldpostbrief J. Suwelacks an Richard vom 26.12.1914.

557 Als Ferdinand Graf von Zeppelin am 5. August 1908 mit seinem Luftschiff in Echterdingen notlanden musste und das Luftschiff wenig später durch einen Sturm fortgerissen wurde und explodierte, löste dieses Ereignis eine Welle der Empathie und Hilfsbereitschaft des deutschen Volkes aus. Ein Unbekannter sprach daraufhin bei Zeppelin vor und forderte das deutsche Volk auf, Geld zu sammeln, um den Bau eines neuen Zeppelin zu ermöglichen. Die „Zeppelinspende" brachte über sechs Millionen Mark ein und war die bis dahin größte freiwillige Spenden-Sammelaktion im Deutschen Kaiserreich. Siehe Frank, Zeppelin-Ereignisse und de Syon, Zeppelin!.

558 Karl-Dieter Seifert, Mit Sammelgroschen zur deutschen Luftmacht – Die Nationalflugspende 1912–1914, Berlin 2014.

559 Zit. nach MGFA, Militärluftfahrt. Technischer Band, S. 53.

560 Kehrt, Moderne Krieger, S. 68.

561 Vgl. Reinicke, Deutsche Auffahrten, S. 26.

den.[562] Auf militärisch ausgerichteten Flugwettbewerben wie dem Prinz-Heinrich-Flug 1913 sichtete und prämierte man die erfolgreichsten Piloten, Flugzeuge und Flugzeugfirmen. Außerdem wurden exklusiv für Zivilflieger Geldprämien auf Stundenflüge ausgesetzt,[563] die 1.000 Mark pro geflogener Stunde betrugen und für die Zeit vom 1. März bis zum 31. Dezember 1913 ausgelobt wurden. Führte ein Pilot einen solchen Flug zusätzlich mit einem Passagier oder mit 65 Kilo Ballast aus, erhöhte sich die Prämie um weitere 500 Mark. Maximal konnte man so auf eine Prämienzahlung von 10.000 Mark kommen. Josef Suwelack hatte z. B. am 20. Oktober 1913 einen zweistündigen Flug und am 28. Dezember einen neunstündigen Flug absolviert, wonach ihm zweitausend beziehungsweise siebentausend Mark durch das Kuratorium der National-Flugspende überwiesen wurden.[564] Wenngleich diese Summen für die damalige Zeit erheblich und Ansporn für weitere Innovationen waren, deckten sie in der Fliegerszene nur vorübergehend die laufenden Kosten. Mit der Annahme der Prämie verpflichteten sich die Piloten, gleichgültig ob Zivil- oder Militärflieger, im Kriegsfall der Fliegertruppe zur Verfügung zu stehen und im Frieden innerhalb eines Jahres an einer dreiwöchigen Übung teilzunehmen. Bis zum Ende des Jahres 1913 erhielten 236 Zivilpiloten Prämien aus der Geldsammlung, weitere 424 Personen bei Kriegsbeginn.[565]

Ferner wurden auf Kosten der Nationalflugspende 1913 insgesamt neunzehn Firmen[566] zur Pilotenausbildung zugelassen, darunter Albatros, Aviatik, Fokker, Harlan oder die Ago-Flugzeugwerke – Unternehmen, mit denen die Kondor-Werke vergeblich versucht hatten zu fusionieren. Die mit der Spende verbundene Zuteilung von Flugschülern sicherte einerseits stabile und hohe Einnahmen und führte verstärkt zu Flugzeugkäufen durch die Heeresverwaltung, führte aber andererseits bei Firmen wie Kondor, die nicht berücksichtigt worden waren, zu Absatzschwierigkeiten.

Im November 1912 verfügte die Heeresverwaltung über 43 Ein- und 65 Doppel-

562 Vgl. MGFA, Militärluftfahrt. Technischer Band, S.53–54. Zur Verwendung der Gesamtsumme konnte hier ermittelt werden: Von den 7.234.506 Mark standen 5.212.691,41 Mark frei im Interesse des deutschen Flugwesens zur Verfügung, 1.060.018,81 Mark wurden für die Beschaffung von Heeresflugzeugen verwandt, 577.000 Mark für die Gründung der deutschen Versuchsanstalt für Luftfahrt in Berlin-Adlershof und 338.287,97 Mark für verschiedene Zwecke wie Prämien- und Rentenzahlungen.

563 Vgl. Schwipps, Schwerer als Luft, S.103.

564 Vgl. NL J. Suwelack, Schreiben des Kuratoriums der National-Flugspende an J. Suwelack vom 9.1.1914. Die Summen, die Suwelack erhielt, sind dem Schreiben entnommen, wobei dort nicht erläutert wird, warum er für den zweiten Flug lediglich 7.000 Mark bekam. Piloten, die den weitesten Überlandflug geschafft hatten, erhielten solange eine Monatsrente von 2.000 beziehungsweise 4.000 Mark, bis ihre Leistung durch einen anderen überboten wurde. Für Prämien- und Rentenflüge wendete man 1913 annähernd 480.000 Mark auf, weitere 435.000 Mark wurden 1914 hierfür bereitgestellt, doch „nur" 150.000 Mark ausgezahlt, weil die Mittel bereits im Februar erschöpft waren. Vgl. Schwipps, Schwerer als Luft, S.104.

565 Vgl. ebd., S.103.

566 Siehe die Übersicht bei Schwipps, Schwerer als Luft, S.101 sowie im Flugsport (1913), Nr. 8, S.300.

decker; Ende des Jahres waren es dann insgesamt 153 Flugzeuge.[567] Weitere Maschinen wurden 1913 aus Mitteln der Nationalflugspende beschafft, sie sollten die mittlerweile veralteten Schulflugzeuge ersetzten. Zwar hatten vor Ausbruch des Ersten Weltkrieges 817 „Alte Adler" ihren Pilotenschein gemacht, doch widmeten sich nicht alle aktiv der Fliegerei, sondern waren stattdessen als Flugzeugbauer, Monteure oder Ausbilder tätig. Bei Kriegsausbruch im August 1914 standen damit schätzungsweise rund 240 Flugzeuge, 254 ausgebildete Piloten und 271 Beobachter der Fliegertruppe zur Verfügung, ohne dass jedoch sämtliche Maschinen einsatzbereit waren.[568] Österreich-Ungarn stellte weitere 39 Flugzeuge und 85 Feldpiloten.[569] Frankreich hatte zum Vergleich zu diesem Zeitpunkt eine Flugzeugstärke von 141 Maschinen, allerdings mehr als doppelt so viele – zwischen 500 und 650 – ausgebildete Militärpiloten.[570]

Die Überalterung der Maschinen sowie die Kriegsuntauglichkeit der überwiegenden Zahl der Flugzeuge hatte Ernst August Schröder gegenüber dem Industriellen Hugo Stinnes bereits 1913 beklagt:

> „[V]on den 345 [deutschen Zivil-Fliegern fliegt] ungefähr die Hälfte *Grade*apparate ..., d.h. Apparate, mit denen man bestenfalls an kleineren Plätzen den Leuten mal eine halbe Stunde etwas vorfliegen kann, die aber als kriegsbrauchbar gar nicht in Betracht kommen und mit denen irgendwelche nennenswerten Ueberlandflüge niemals ausgeführt sind und auch niemals gemacht werden können."[571]

Neben der Innovationsförderung und Unterstützung in der Flugzeugindustrie führte die Nationalflugspende auch zum Versicherungsschutz der ersten Piloten. Gegen einen vierteljährlich zu bezahlenden Beitrag von 45 Mark[572] konnte ein Flieger bei „vorübergehender Arbeitsunfähigkeit" einen Tagessatz von acht Mark, bei „völliger Arbeitsunfähigkeit" eine einmalige Zahlung in Höhe von 24.000 Mark beziehen.[573] Josef Suwelack war seit 1913 bei der Victoria zu Berlin Allgemeine Versicherungs-Actien-Gesellschaft unfallversichert und bekam nach einem Armbruch, den er sich beim Ostpreußischen Rundflug am 11. August 1913 bei einem Flugzeugabsturz zugezogen hatte,[574] im Jahr 1914 rückwirkend für den

567 Vgl. MGFA, Militärluftfahrt. Technischer Band, S.53 und S.59.

568 Je nach Autor schwankt die Zahl der 1914 zur Verfügung stehenden Flugzeuge zwischen 220 und 450 Maschinen. Siehe Peter Kilduff, Germany's First Air Force 1914–1918, London 1991, S.10; Morrow, Building German Air Power, S.87. Sebastian Rosenboom kam jüngst zu den im Text genannten Zahlen: Rosenboom, Im Einsatz, S.20. Siehe zur Einsatzfähigkeit der Flugzeuge: Morrow, Building German Air Power, S.60 und S.75f.

569 Reinhard Desoye, Die k.u.k Luftfahrtruppe. Die Entstehung, der Aufbau und die Organisation der österreichisch-ungarischen Heeresluftwaffe 1912–1918, Bd.1., 2. Aufl., Wien 1999, S.81.

570 Kehrt, Moderne Krieger, S.71.

571 Archiv-TG, Akte Kondor I, Nr.0310/1913, Niederschrift der geheimen Unterredung zwischen Ernst August Schröder und Hugo Stinnes am 3.10.1913.

572 Der Betrag schwankte zwischen 45 und 50 Mark je nach Art des Versicherungsschutzes.

573 Vgl. Angaben bei Erblich, Fliegerschule, S.13 und S.173.

574 Siehe Ausführungen zum Feld 3 sowie Allgemeine Automobil-Zeitung (1913), Nr.35, S.17f.

Zeitraum vom 11. August bis einschließlich 14. Oktober 1913 325 Mark Krankengeld ausgezahlt.[575]

Die ökonomischen Impulse der Nationalflugspende kamen jedoch nur einem geringen Teil der Flugzeugindustrie zugute, denn die Hoffnung vieler Aviatiker, für ihre Mühen Entschädigungen aus dem Fonds zu erhalten, erfüllte sich größtenteils nicht. Firmen mit wenig Kapital und dementsprechend günstiger produzierten Maschinen mit geringer Motorenstärke von maximal 50 PS wurden gar nicht berücksichtigt, denn bei der Vergabe der Gelder war die militärische Brauchbarkeit der Flugzeuge ausschlaggebend. Über die Verteilung der Fördermittel entschied ein Kuratorium von 50 Mitgliedern, bestehend aus Vertretern aller Bundesstaaten, der Reichsregierung, der größten Luftfahrtvereine[576] sowie drei aktiven Fliegern: Hellmuth Hirth, Ellery von Gorrissen und August Euler.[577] Die Kehrseite der Nationalflugspende wird auch am Beispiel der Kondor-Werke deutlich, die trotz wiederholter, unermüdlicher Unterredungen mit Militärs und schriftlichen Gesuchen bei der Heeresverwaltung keine Aufträge erhielten. Des Weiteren wurde auf einer Versammlung des Vereins Deutscher Motorfahrzeug-Industrieller (VDMI) am 2. Februar 1914, an der Ernst Füßmann für die Kondor-Werke teilnahm, neben anderen Kritikpunkten beklagt, dass am Prinz-Heinrich-Flug „Offiziersflieger gleichfalls wie Zivilflieger um die ausgesetzten Geldpreise konkurrieren". Und während den Privatfirmen „ganz bedeutende pekuniäre Opfer" abverlangt würden, wären die Offiziersflieger durch die Heeresverwaltung und die „Fliegerzulage"[578] in Höhe von 150 Mark abgesichert.[579]

Sämtliche Planungen, wie z. B. Fusionen mit anderen Flugzeugproduzenten oder die Zusage von Heeresaufträgen, waren 1913 gescheitert. Der Aufbau einer Flugschule in Düsseldorf-Lohausen wurde zwar betrieben und von der Stadt Düs-

575 Vgl. NL J. Suwelack, Bestätigungschreiben der Victoria zu Berlin Allgemeine Versicherungs-Actien-Gesellschaft an Josef Suwelack (Versicherungsschein No. 53) vom 13.1.1914. Siehe auch Schreiben der Victoria vom 17.3.1914 betreffs Verlängerung des Vertrages für ein weiteres Jahr mit Auskünften über den durch die Nationalflugspende herbeigeführten zusätzlichen Versicherungsschutz. Demnach war jetzt auch der Tod des Fliegers mit maximal 4.000 Mark versichert, wofür ein jährlicher Versicherungsbeitrag von 200 Mark (50 Mark pro Quartal) verlangt wurde. Für die Vertragsbedingungen (z. B. Zahlung bei Invalidität) siehe NL J. Suwelack, „Allgemeine Bedingungen für die Lebenslängliche [sic!] Eisenbahn- und Dampfschiff-Unglück-Versicherung"/ Versicherungsschein No. 331781. Vereinbarte Versicherungssummen: a) im Todesfall: 30.000 Mark, b) bei Invalidität: 22.500 Mark, c) bei vorübergehender Arbeitsunfähigkeit: 7,50 Mark pro Tag. Die wöchentlich zu zahlende Prämie betrug 3 Mark, Vertrag ausgestellt am 4.9.1913.

576 Hierzu zählten der Deutsche Luftfahrer-Verband (DLV) als Gesamtorganisation aller Luftfahrtvereine, der Verein Deutscher Motorfahrzeug-Industrieller als Repräsentant der Luftfahrtindustrie und der Kaiserliche Automobil-Club.

577 Vgl. Schwipps, Schwerer als Luft, S. 83.

578 Die Offizierspiloten erhielten neben ihrem Gehalt eine monatliche Fliegerzulage, eine Art Gefahrenzulage, in Höhe von 150 Mark.

579 NL J. Suwelack, Protokoll der am 2.2.1914 im Generalsekretariat des V.D.M.I. stattgefundenen Sitzung der Flugzeug-Konvention, 12 Seiten, hier S. 2. Diese Zulage sollte die Militärflieger für die bei Flugveranstaltungen eingegangene (Lebens-)Gefahr entschädigen.

seldorf sowie dem dortigen Luftfahrtverein unterstützt, kam dann aber aufgrund der Weigerung der Garnisonsverwaltung, die Flugschule auf dem Exerzierplatz noch länger zu tolerieren, nach nur einem Jahr bereits wieder zum Erliegen. Umso bedeutender und weitsichtig erscheinen in diesem Zusammenhang Josef Suwelacks 1912 getroffene Vorkehrungen für seine Zukunft. Durch den privatrechtlichen Vertrag mit dem Deutschen Reich hatte er die Möglichkeit, weiter im Dienst der Fliegerei zu stehen und zu handeln. Unmittelbar nach Kriegsausbruch nahm er seine Tätigkeit bei der Fliegerabteilung auf. Ernst Füßmann, der ebenfalls zum Krieg einberufen wurde, schied vorübergehend als Geschäftsführer der Kondor-Werke aus. Für ihn wurden am 3. August 1914 Anton Ludwig[580] sowie am 24. August Caspar Maria Grod, beide aus Essen, als stellvertretende Geschäftsführer eingestellt. Am ersten September 1915 wurde Grod außerdem Prokura erteilt.[581]

Die Wahrnehmung der Rolle Josef Suwelacks bei den Kondor-Werken in der Rezeptionsgeschichte ist erheblich zu relativieren. Aus dem heimatlichen „Bastler“, der bei Rumpler seine autodidaktischen Fertigkeiten durch die Arbeit auf dem Flugplatz Johannisthal erweitert hatte, war ein privilegierter technischer Angestellter und wenig später ein Unternehmensteilhaber geworden. Ein Autodidakt, wenngleich recht begabt, war Suwelack auch noch während seiner Zeit bei Kondor. In Anschreiben, Teilnehmerlisten, Artikeln und auch in den Protokollen der Gesellschafter-Versammlungen wurde Josef Suwelack zwar regelmäßig als „Ingenieur“[582] ausgewiesen, doch erscheint die Bezeichnung eines technischen Angestellten seiner Tätigkeit angemessener.

Von Suwelacks Konstruktionen wurde in der zeitgenössischen Presse Notiz genommen und sein Name mit dem Ruf des Unternehmens in Verbindung gebracht. Allerdings waren es andere Gesellschafter, vor allem die Goldschmidts, die durch die Bewilligung von neuem Betriebskapital das Unternehmen am Leben hielten. Ohne die Beziehungen Bernhard Goldschmidts zur Marine und zu ande-

580 In den Protokollen der Kondor-Werke taucht Ludwigs Vorname stets nur in abgekürzter Form auf. In einer Firmenschrift der Goldschmidt AG werden sein vollständiger Name und seine Position beim Chemiewerk genannt. Oberingenieur Anton Ludwig war Angestellter der Th. Goldschmidt AG und seit Anfang 1920 Leiter der technischen Abteilung. Vgl. Däbritz/Paulick, Th. Goldschmidt AG Essen, S.96.

581 Vgl. zum Ausscheiden Füßmanns und zur Einstellung der stellvertretenden Geschäftsführer: Archiv-TG, Akte Kondor I, Nr. 0000/1914, Geschäftsbericht für das zweite Halbjahr 1914. Siehe auch Automobil- und flugtechnische Gesellschaft (Hg.), Der Motorwagen, Bd. 19, Berlin 1916, S.458. Zu Spanien vgl. NL J. Suwelack, Feldpostbrief an Richard Suwelack vom 30.7.1915. Ernst Füßmann schied im Oktober 1916 endgültig aus den Kondor-Werken aus. Vgl. Archiv-TG, Akte Kondor I, Nr. 0109/1915, Protokoll der Aufsichtsrats-Sitzung am 1.9.1915, und Archiv-TG, Akte Kondor I, Nr. 1610/1916, Protokoll der außerordentlichen Gesellschafter-Versammlung am 16.10.1916.

582 Siehe z. B. diverse Protokolle der Kondor-Werke; Der Motorwagen Bd. 19 (1916), S.458; Jahrbuch des DLV 1913, S.130.

ren, ihm persönlich bekannten Militärs sowie das finanzielle Gewicht weiterer Gesellschafter hätten die Kondor-Werke wahrscheinlich nicht lange existieren können. Flexibilität, der Kontakt zu den „richtigen“ Leuten, das Ausschöpfen unterschiedlicher Möglichkeiten und zu einem gewissen Grad auch eine beständige Hoffnung auf Erfolg trugen dazu bei, dass das Werk als Flugzeugunternehmen noch bis 1920 bestand.

Josef Suwelack selbst war flexibel und keineswegs nur auf die Kondor-Werke fixiert, wie aus seinen späteren Feldpostbriefen ersichtlich wird. Oberste Priorität hatte für ihn die Fliegerei, doch über das Wie und Wo dachte er während seiner Kriegszeit oft und ausführlich nach. Nicht nur für das Unternehmen, sondern auch für den Flieger selbst war der Ausbruch des Ersten Weltkrieges eine kontingente, ungewisse Situation, in der die Weichen neu gestellt wurden. Der Krieg bildete einen tiefen Einschnitt in den Erfahrungen Suwelacks und ließ Vergangenheit und Zukunft, die sich schon in der Vorkriegszeit immer weiter voneinander entfernt hatten, nun endgültig auseinanderdriften.[583]

Die Wege trennen sich: Der Rüstungsbetrieb der Kondor-Werke während des Ersten Weltkrieges und Suwelacks Zukunftsgedanken

Mit dem Beginn des Ersten Weltkrieges kam der von den Kondor-Werken lang ersehnte Umschwung. Die von der spanischen Regierung in Auftrag gegebenen vier Flugzeuge wurden zwar von der deutschen Militärverwaltung beschlagnahmt; durch die gezahlte Entschädigung von vermutlich jeweils 25.000 Mark[584] kam dies jedoch einem Kauf gleich.[585]

Bereits Ende Juli hatte das Unternehmen von der Inspektion der Fliegertruppen die Zusage erhalten, eine Kondor-Taube abzunehmen, sofern diese die militärischen Bedingungen erfülle. Der Bau wurde unverzüglich aufgenommen und am 31. August das Flugzeug auf dem Luftweg nach Köln befördert. Zwei Tage später wurde es „in Gegenwart des Herrn Oberst von Eberhardt“ – dem Inspekteur

583 Siehe zum Bruch durch den Ersten Weltkrieg: Stefan Breuer, Bürokratie und Charisma. Zur politischen Soziologie Max Webers, Darmstadt 1994, bes. S.139ff.; Lucian Hölscher, Der Geschichtsbruch von 1914. Vortrag, gehalten im Ruhr-Museum Essen am 1. Juli 2014, online unter: http://www.ruhr-uni-bochum.de/lehrstuhl-ng3/publikationen/hoelscher/DerGeschichtsbruchVon1914.pdf (12.2.2014); Inka Mülder-Bach (Hg.), Modernität und Trauma. Beiträge zum Zeitenbruch des Ersten Weltkrieges, Wien 2000 sowie Bernd Hüppauf, Was ist Krieg? Zur Grundlegung einer Kulturgeschichte des Kriegs, Bielefeld 2013, bes. S.237ff.

584 Dieser Betrag wird im Protokoll einer Unterredung vom 4. September 1914 genannt: Archiv-TG, Akte Kondor I, Nr.0409/1914. In einem späteren Protokoll heißt es lediglich: „Die für die beschlagnahmten Tauben erzielten Preise waren zufriedenstellend“. Archiv-TG, Akte Kondor I, Nr.0000/1914, Geschäftsbericht für das zweite Halbjahr 1914.

585 Vgl. Archiv-TG, Akte Kondor I, Nr.0000/1914, Geschäftsbericht für das zweite Halbjahr 1914.

der Fliegertruppen – vorgeflogen und von der Heeresverwaltung übernommen. Es folgte der lang erwartete Serienauftrag: „Gleichzeitig wurden wir von Herrn Oberst von Eberhardt beauftragt, nun mehr mindestens 1 von diesen Tauben wöchentlich zur Ablieferung zu bringen." Bis Ende des Jahres 1914 lagen Aufträge für 18 Maschinen vor.[586]

Angesichts der bis dahin ablehnenden Haltung der Militärverwaltung gegenüber den Kondor-Werken und der negativen Beurteilung der Maschinen erscheint dieser Sinneswandel erklärungsbedürftig. Die neue Kaufbereitschaft des Militärs resultierte vor allem aus dem Umstand, dass die in der Vorkriegszeit beschafften Flugzeuge überaltert waren und schlicht nicht ausreichend einsatzfähige Flugzeuge zur Verfügung standen. Dieser Mangel an kriegstauglichen Maschinen bewirkte, dass nun auch mittlere Unternehmen verstärkt Aufträge erhielten und sich finanziell sanieren konnten. So bauten die Kondor-Werke ihre Anlage aus, vergrößerten den „Waschraum der Arbeiter", bauten eine Schmiede und eine „Klosettanlage" an und beschafften neues Werkzeug. Ohne die akute Bedrohung durch den bevorstehenden Krieg hätte sich die wirtschaftliche Situation der Essener Firma wahrscheinlich weiter verschärft.

Trotz einiger Schwierigkeiten, die durch den Kriegsausbruch entstanden waren, wie z. B. die Sperrung des Eisenbahnverkehrs oder Forderungen der Lieferanten nach Zahlungsvorschüssen, hatten die Kondor-Werke in den Monaten November und Dezember acht Apparate geliefert, weitere sechs waren im Bau, so dass die Schulden kurzzeitig auf 130.000 Mark abgebaut werden konnten.[587] Zeitgleich entstand nach Konstruktionsplänen Josef Suwelacks ein Doppeldecker, der sogenannte „Typ A", dessen Vorführung im November stattfand. Die Beurteilung fiel vernichtend aus, denn im „Ganzen [wurden] 18 Bemängelungen seitens der Kommission aufgestellt",[588] das Flugzeug war nach dem Urteil des Inspekteurs der Fliegertruppen „absolut nicht sicher und zuverlässig in der Ausführung."[589]

Bis Ende des Jahres 1914 wuchsen die Schulden erneut wieder an auf 268.095,17 Mark,[590] doch der vorliegende Militärauftrag gab Anlass zur Hoffnung, dass das Unternehmen aus der Absatzkrise ausbrechen könnte. Diese Hoffnung erwies sich jedoch als trügerisch: Ende des Jahres 1914 beziehungsweise Anfang 1915 distanzierte sich die Militärverwaltung, die die Abnahme der Kondor-Tauben vertraglich zugesichert hatte, von dem Versprechen und lehnte sechs bestellte Apparate ab, da sie den neuen Anforderungen nicht entsprächen und

586 Vgl. ebd.
587 Vgl. NL J. Suwelack, Feldpostbrief J. Suwelacks an Richard S.vom 3.12.1914.
588 Archiv-TG, Akte Kondor I, Nr. 0000/1914, Geschäftsbericht für das zweite Halbjahr 1914.
589 NL J. Suwelack, Feldpostbrief Josefs an Richard Suwelack vom 26.12.1914.
590 Vgl. Archiv-TG, Akte Kondor I, Nr. 2807/1915, Protokoll der Gesellschafter-Versammlung vom 28.7.1915.

damit als veraltet galten.[591] Die Gesellschafter wichen daraufhin in den Folgemonaten von ihrem verständnisvollen und unterwürfigen Verhandlungsverhalten ab und wollten für geregelte Verhältnisse und garantierte Bestellungen sorgen, notfalls gerichtlich: „Wenn die Verhandlungen erfolglos sein sollten, soll die Klageanordnung wegen Nichtabrufs der Tauben erfolgen."[592] Sie stellten konkrete Bedingungen an die Militärverwaltung und bestanden auf einer verbindlichen Zusage:

> „Es soll im Angebot vermerkt werden, dass wir bereit sind, nach den vorgelegten Entwürfen ein Flugzeug zu bauen zu den noch festzustellenden Preisen, dass es als abgenommen gilt, wenn es die vereinbarten Bedingungen erfüllt, und dass die Inspektion sich verpflichtet, in diesem Falle eine Serie von 20 oder 50 Flugzeugen dieser Bauart in Bestellung zu geben zu noch zu vereinbarenden Preisen."[593]

Auch im weiteren Verlauf waren die Kondor-Werke nun vor allem um Wirtschaftlichkeit bemüht. Die Abhängigkeit von der Heeresverwaltung und der Konkurrenzdruck durch Großunternehmen erschwerten jedoch die Planung des Geschäftsbetriebs und machten ökonomische Prognosen schwierig. Gefordert war ein hohes Maß an Flexibilität und Anpassungsfähigkeit, um einer Insolvenz, wie sie viele kleinere Firmen erlebten, zu entgehen. Nicht nur die Fliegerei war eine kontingente Erfahrung, sondern auch der Erfolg beziehungsweise Misserfolg eines Flugzeugunternehmens und damit auch die Zukunft der Investoren wurde ebenfalls von verschiedenen Variablen beeinflusst und war riskant. Die Entwicklungen ab 1915 sollen daher hier in knapper Form nachgezeichnet werden.

Als Vertretung für Josef Suwelack wurde Paul Westphal[594] am 1. Februar 1915 als Konstrukteur mit einem Monatsgehalt von 600 Mark eingestellt und erhielt später auch Prokura. Außerdem wurde ihm eine Gewinnbeteiligung in Aussicht gestellt, sofern die Heeresverwaltung die von ihm konstruierten Doppeldecker abnahm.[595] Westphals erster Apparat, ein Doppeldecker („Typ W 1"), wurde im Juli 1915 der bayerischen Militärverwaltung zum Kauf angeboten, eine Vorführung blieb jedoch aus, da die „Leistungen des Apparates" als „nicht hervorragend und von besonderem Interesse erschienen".[596] Auch die im Spätherbst 1915 von sechzehn Arbeitern gebaute W 2 erfüllte die Bedingungen bei der Vorführung

591 Vgl. Archiv-TG, Akte Kondor I, Nr. 2011/1914, Protokoll der Gesellschafter-Versammlung vom 20.11.1914.

592 Archiv-TG, Akte Kondor I, Nr. 0109/1915, Protokoll der Aufsichtsratssitzung vom 1.9.1915.

593 Archiv-TG, Akte Kondor I, Nr. 2809/1915, Protokoll der Aufsichtsratssitzung vom 28.9.1915.

594 Paul Westphal (Lebensdaten nicht bekannt) hatte von 1911 bis 1914 einen Schuppen auf dem Flugplatz Johannisthal gemietet, wo er in einer Werkstatt Eindecker baute. Seine Konstruktionen erwiesen sich als flugfähig und ein Eindecker wurde sogar von der Militärverwaltung abgenommen; Bestellungen blieben jedoch aus. Von 1914 bis Anfang 1916 war er bei den Kondor-Werken beschäftigt, über seinen weiteren Lebensweg ist nichts bekannt. Siehe Schmitt, Oldtimer, S. 50.

595 Vgl. Archiv-TG, Akte Kondor I, Nr. 2201/1915, Protokoll der Aufsichtsrats-Sitzung am 22.1.1915 sowie Archiv-TG, Akte Kondor I, Nr. 0907/1915, Protokoll der Aufsichtsrats-Sitzung am 9.7.1915.

596 Archiv P.M. Grosz/Archiv DTMB, Schriftwechsel der K. B. Insp. d. Mil. Luft- u. Kraftf-Wesens und der K. B. Flieger-Ersatz-Abteilung vom 17., 21. und 23.7.1915.

Ende Februar beziehungsweise Anfang März 1916 nicht und stürzte durch Motorversagen sogar ab. Ein Weltrekord im Passagierflug mit einer Flughöhe von 3.280 Metern änderte nichts an der Ablehnung des Flugzeuges durch die Heeresverwaltung[597] und dies bedeutete das Ende der Karriere Westphals bei den Kondor-Werken.

Im Frühjahr 1915 hatten die Kondor-Werke Josefs Suwelacks Doppeldecker „Typ A" den gestellten Anforderungen angepasst, doch konnte der Apparat die mittlerweile neu geforderte Geschwindigkeit nicht erfüllen, was schließlich zur Aufgabe der Entwicklung eigener Flugzeuge führte. Stattdessen ging man 1916, nach erfolgreichen Verhandlungen mit den Albatros-Werken, dazu über, Flugzeuge in Lizenz nachzubauen.[598] Des Weiteren konzentrierte sich das Unternehmen auf die Errichtung einer Fliegerschule in Großenhain, Sachsen, die im April anlief.[599] Ein Vertrag mit der Militärverwaltung über drei Jahre wurde geschlossen, wobei jährlich 75 Schüler zugewiesen werden sollten. Der Schulbetrieb lief gut, denn bereits im gleichen Jahr wurden 63 Schüler, 1916 weitere 221 zu Militärpiloten ausgebildet. Alfred von Back-Begavár zufolge, der Suwelack im April 1915 bei der Feldfliegerabteilung besuchte, belief sich der Umsatz durch den Schulbetrieb auf 500.000 Mark.[600]

Die Kondor-Werke bildeten Militärpiloten mit bis zu fünf Fluglehrern aus, mit der Leitung der Flugschule wurde Paul Füßmann betraut, der im September 1915 in das Unternehmen eingetreten war.[601] Die Instandhaltung der circa zwanzig Maschinen vor Ort in Großenhain übernahmen 30 Arbeiter, weitere 150 Arbeiter kümmerten sich auf dem Flugplatz in Rotthausen um den Bau neuer Apparate für die Flugschule. Zusätzliche Einnahmen erzielte der Betrieb durch die Reparatur von Flugzeugen von der Front beziehungsweise von der Fliegerstation in Köln.[602]

Das Unternehmen verlegte seine Flugschule im September 1917 erneut, dies-

597 Vgl. Der Motorwagen, Bd. 18 (1915), S. 50: „Die neue Welthöchstleistung im Passagierflug".

598 Ein Auftrag über 50 Albatros- BII- Schulflugzeuge lag im Oktober 1916 vor.

599 Ernst Füßmann, der seinen Kriegsdienst bei der Militärverwaltung in Sachsen ableistete, hatte die erneute Gründung einer Flugschule angeregt und wurde beauftragt, die Verhältnisse in Großenhain und Altenburg zu prüfen. Vgl. Archiv-TG, Akte Kondor I, Nr. 2201/1915, Protokoll der Aufsichtsrats-Sitzung am 22.1.1915 Obwohl keinerlei biografische Informationen zu Ernst und Paul Füßmann vorliegen, ist ein Verwandtschaftsverhältnis wahrscheinlich, denn die beiden betrieben zwischen 1897 und 1906 gemeinsam eine Schmiedewerkstatt in Essen-Altendorf. Siehe Rißmann-Ottow, Glück ab!, S. 257. Vgl. auch Der Motorwagen, Bd. 1 (1906), S. 237.

600 Vgl. NL J. Suwelack, Brief an Richard Suwelack, Petit Ronchain [sic!] [Renchain], 1.5.1915. Ein Foto der Kondor-Fliegerschule in Großenhain findet sich auf einer Internetseite zum Flugplatz in Großenhain, Redaktion: Dr. Hannes Täger, Dresden, online seit September 2013 unter: http://www.flugplatzgeschichte-grossenhain.de/1913-bis–1919.htm (2.6.2014)

601 Zur Einstellung Paul Füßmanns siehe Archiv-TG, Akte Kondor I, Nr. 0109/1915, Protokoll der Aufsichtsrats-Sitzung am 1.1915.

602 Vgl. NL J. Suwelack, Feldpostbrief J. Suwelacks an seinen Vater, Köln, 30.7.1915.

mal nach Nordhausen im Harz (Abb. 13), da die Kapazitäten nicht mehr ausreichten. Das Stammkapital war im Mai 1917 auf 900.000 Mark erhöht worden, was deutlich macht, dass die Gesellschafter neuen Mut gefasst hatten und weiterhin an den ökonomischen Erfolg glaubten.[603]

Abb. 13: Anzeige der Kondor-Werke aus dem Jahr 1918 (aus: „Motor" Mai/Juni 1918, S.184)

Zugleich hatte sich die Aufstellung der Gesellschafter zwischenzeitlich erheblich verändert: Wilhelm Schäfer, Heinrich Hillebrand, Eduard Staßfurth und Viktor von Back hatten im Juli 1915 ihre Anteile an Alfred von Back-Begávar verkauft.[604]

603 Vgl. die Angaben bei Rißmann-Ottow: Glück ab!, S.257ff.
604 Vgl. NL J. Suwelack, Feldpostbrief J. Suwelacks an seinen Vater, Köln, 30.7.1915.

Dieser kaufte 1917 auch noch die Anteile der Herren May, Marx, Passavant-Gontard und den Geschäftsanteil Hans Goldschmidts. Geblieben waren, abgesehen von Josef Suwelack, der trotz seiner Kündigung noch immer Gesellschafter war, die Gründungsmitglieder, von Back-Begávar, die Hirschlands und Hugo Racine. Suwelacks Anteil wurde nach dessen Tod von den Gesellschaftern übernommen und 1917 unter ihnen versteigert.[605]

Der Lizenzbau von Albatros-Doppeldeckern wurde seit 1916 betrieben, zwischen 1917 und 1918 bauten die Kondor-Werke rund 400 Albatros-B2-Schulflugzeuge. Zudem entstanden unter der Leitung von Walter Rethel[606] und Paul Georg Erhardt, letzterer blieb nur für kurze Zeit bei Kondor, mehrere Prototypen neuer Doppeldecker, aber auch Entwürfe für Schul- und Kampfflugzeuge. Ein Hochdecker, gebaut im Sommer 1918, bewährte sich sogar bei einem militärischen Wettbewerb im Oktober 1918 und eröffnete die Aussicht auf einen Auftrag zur Serienfertigung von 100 Exemplaren. Da wenig später der Waffenstillstand folgte und die Militärfliegerei nach Kriegsende verboten war, war die Bestellung jedoch hinfällig.[607] Trotzdem zeigen die Produktionszahlen der Kondor-Werke, dass die bis 1914 beschafften Flugzeuge der Heeresverwaltung bei Weitem nicht ausreichten, um gegen Frankreich und England im Kampf bestehen zu können.[608] Das Unternehmen profitierte damit von der großen Flugzeugnachfrage, die regelmäßige (Nach-)Bestellungen zur Folge hatte. Die Hauptproduktion von Schulflugzeugen fand vor allem in den letzten beiden Kriegsjahren statt (vgl. Tab. 5).[609]

Doch obwohl die Auftragslage der Kondor-Werke durchaus gut war und die Zahl der Beschäftigten im Sommer 1918 auf einen Spitzenwert von 1.200 Mann stieg, konnten die Schulden weiterhin langfristig nicht abgebaut werden.[610] Im Gegenteil: Bis zur Aufgabe des Flugzeugbaus und der Umwandlung beziehungsweise der Änderung des Unternehmensgegenstandes in ein Säge- und Holzbear-

605 Vgl. ebd., S.244.

606 Walter Rethel (1892–1977). Eine Artikelserie über ihn erschien in: Der Flieger (1977), H. 9, S.399–402; (1977), H. 10, S.427–430; (1977), H. 11, S.488–489 und (1977), H. 12, S.527–529. Rethel arbeitete als Konstrukteur bei den Kondor-Werken, ging 1919 dann zu den Fokker-Flugzeugwerken, bevor er bei den Arado-Flugzeugwerken als Chefkonstrukteur engagiert wurde. Ungeachtet des Flugzeugbau-Verbotes nach dem Krieg entwickelte Rethel weiterhin militärische Flugzeuge und ging 1936 zu Willy Messerschmidt. Informationen zu Rethel auch bei: Jörg Armin Kranzhoff, Die Arado-Flugzeuge. Vom Doppeldecker zum Strahlflugzeug, Bonn 2000, S.23ff.

607 Siehe zum Kondor E 3: „Kondor E 3", in: Air International, Dezember 1983, S.305f.

608 Zwischen 1914 und 1918 wurden im Deutschen Reich rund 48.000 Flugzeuge mobilisiert, von denen 3.128 durch Feindeinwirkung verloren gingen. Siehe BA-MA PH 9 XV/7, Verluste der deutschen Fliegertruppen (inkl. bayerische Fliegerverbände) vom 2.8.1914 bis 11.11.1918. Vgl. auch Neumann, Die deutschen Luftstreitkräfte, S.81.

609 Angaben übernommen aus Rißmann-Ottow, Glück ab!, S.259.

610 Vgl. Wolfgang Wagner, Die Flugzeuge der Kondor Werke, in: Aerokurier (1974), H. 1, S.67–69, hier S.67.

Tabelle 5: Die Flugzeugproduktion der Kondor-Werke zwischen 1912 und 1920

Bezeichnung	Typ	Bauzeit	Anzahl	Anmerkungen
Taube	Eindecker	1912–1915	mind. 26	mind. drei versch, Ausführungen mit Änderungen
Typ A	Doppeldecker	1914–1917	ca. 40	nur für Flugschule
W 1	Doppeldecker	Frühjahr 1915	1	Prototyp, Verwendung in Flugschule
W 2	Doppeldecker	Herbst 1915	1	Prototyp, Absturz Anfang 1916
Albatros B II	Doppeldecker	1917–1918	ca. 400	Lizenzbau, Schulfugzeug
	Dreidecker	Sommer 1917	1	Prototyp
D 7	Doppeldecker	Herbst 1917	1	Prototyp
D 1	Doppeldecker	1917/18	1	Prototyp Jagdflugzeug
D 2	Doppeldecker	Frühjahr 1918	2	Weiterentwicklung des D 2, Prototypen
B I (K 100)	Doppeldecker	Ende 1917	1	Prototyp nur aus Holz
D 6	Doppeldecker	Frühjahr/ Sommer 1917	1	Prototyp Jagdflugzeug
E 3 (D I)	Hochdecker	Sommer 1918	mind. 4	Prototypen, Jagdfluzeug, für Serienbau vorbereitet
Luftverkehrs-Limousine	Hochdecker	1919/20	1	Prototyp von Kondor, Bau in Holland 1920 bei Navo
Gesamt			ca. 479	

beitungswerk im Jahr 1920 belief sich das Minus auf schätzungsweise 3,2 Millionen Mark.[611]

Bereits während des Krieges, zeitgleich mit der Erhöhung des Stammkapitals am 14. Mai 1917, war der Gegenstand des Unternehmens um Geschäfte erweitert worden, „die nicht unmittelbar dem Bau von Flugzeugen dienen, die sich aber technisch oder geschäftlich einem Flugzeugunternehmen angliedern lassen." An diesem Tag wurde auch der Beschluss zur Auflösung des Unternehmens von Ende März 1914 aufgehoben.[612] Da die Kondor-Werke seit 1917 verstärkt Holzflug-

611 Vgl. die Berechnung des Gewinns und des Verlustes der Kondor-Werke bei Rißmann-Ottow, Glück ab!, S.265.

612 Archiv-TG, Akte Kondor II, Nr.1405/1917, Protokoll der Gesellschafter-Versammlung am 14.5.1917. Siehe auch Der Motorwagen, Bd.24 (1921), S.295.

zeuge bauten, suchten sie einen Zulieferer für den Werkstoff, den sie in einem Sägewerk in Lemgo fanden. Das Sägewerk wurde von Wilhelm Schnakenberg betrieben und als dieser im Oktober 1915 starb, verkaufte seine Frau den Betrieb drei Jahre später an die Kondor-Werke.[613] Schließlich verlegten diese ihren Sitz im Juni 1918 nach Lemgo, wo sie das dort gekaufte Sägewerk in „Kondor Flugzeugwerke GmbH – Abt. Sägewerk" umbenannten. Im gleichen Jahr entstand in Nordhausen ein weiterer Betrieb, das „Kondor-Werk – Gesellschaft für Holzbearbeitung". Der Umstieg auf die Holzverarbeitung versetzte das Unternehmen nach Kriegsende in die Lage, auf eine Friedensproduktion umzustellen.[614]

Über die Zeit der Kondor-Werke nach Kriegsende existieren nur wenige Unterlagen.[615] Bekannt geworden ist, dass 1919 das Stammwerk auf dem Flugplatz Gelsenkirchen-Essen-Rotthausen als Autowerkstatt und Möbelschreinerei genutzt wurde, dann Zug um Zug die Anlagen demontiert und die Gebäude verkauft wurden. Anderthalb Jahre nach Kriegsende, im Mai 1920, war die Räumung vollzogen. Die Aufgabe des Flugzeugbaus wurde wahrscheinlich Ende des Jahres 1919 beschlossen, möglicherweise aber auch erst 1920, denn im Jahresabschlussbericht findet eine Auflösung des Geschäftsfeldes keine Erwähnung. In einem Programmheft aus dem Jahr 1919 warben die Kondor-Werke zudem noch mit ihrer „Kondor-Luftverkehrs-Limousine".[616]

Am 28. Oktober 1920 wurde ein neuer Gesellschaftsvertrag aufgesetzt, der die Neuausrichtung des Unternehmens nach sich zog. Die Firma nannte sich nun „Kondor-Werke Gesellschaft für Holzbearbeitung mit beschränkter Haftung" und die Zweigstelle in Lemgo wurde zur Tochtergesellschaft. Zwei Jahre später – der Flugplatz in Nordhausen war mittlerweile umgepflügt und somit für fliegerische Aktivitäten unbrauchbar gemacht worden – wurden die Kondor-Werke in Lemgo zur Lippischen Kondorwerke AG umgewandelt. Caspar Maria Grod und alle anderen Gesellschafter bis auf die Goldschmidts schieden aus dem Unternehmen aus. Mit der Produktion von Fenstern, Türen und Treppen war das Sägewerk in den folgenden Jahrzehnten schließlich äußerst erfolgreich. Die Nachfahren der Goldschmidts waren bis zum Jahr 1989 als Teilhaber an der Möbelproduktion des Werkes beteiligt, das 1993 aufgelöst wurde.[617]

613 Vgl. Günter Rhiemeier, Nichts dauert ewig. Vom Aufstieg und Niedergang des Kondor-Werkes, in: Lemgoer Hefte (1995), Nr. 3, S. 22–28, hier S. 22.

614 Vgl. zur Nachkriegsentwicklung die Ausführungen bei Rißmann-Ottow, Glück ab!, S. 260 und S. 264ff.

615 Archiv-TG, Akte Kondor II, Nr. 2810/1920, Protokoll der Generalversammlung am 28.10.1920. Dort war die Rede davon, dass während der März-Unruhen die Geschäftsräume der Kondor-Werke geplündert und ein Teil der Geschäftsakten vernichtet worden seien. Der Geschäftsbericht für das Jahr 1919 erschien erst im September 1920: Archiv-TG, Akte Kondor I, Nr. 1709/1920, Geschäftsbericht für das Jahr 1919.

616 „Erster deutscher Fliegertag auf dem Flugplatz Gelsenkirchen-Essen-Rotthausen am 21. und 22.9.1919", in: Der Deutsche Flieger (1919), Nr. 1, S. 25–27.

Die Entwicklung der Kondor-Werke und damit auch die Entwicklung Josef Suwelacks stehen beispielhaft für die Ambivalenz der Moderne: Alles wurde immer weniger plan- und berechenbar, der Innovations- und Fortschrittsdruck, mit dem sich die Flugzeugindustrie konfrontiert sah, machte es zunehmend schwieriger, den Vorgaben zeitnah zu entsprechen. Der Fortschritt und die mit ihm verbundene Beschleunigung überholten sich gleichsam selbst. Was wenige Monate, manchmal sogar nur wenige Wochen zuvor noch als innovativ und „neu“ bewertet worden war, galt kurz darauf bereits wieder als rückständig und „alt“. Wer in der Flugzeugindustrie mithalten und dauerhaft Fuß fassen wollte, musste deshalb ein hohes Maß an Flexibilität besitzen, über finanziellen Spielraum verfügen und über das Einholen von Informationen bei persönlich bekannten Kontaktmännern in der Militärverwaltung in der Lage sein, Zukunft zu antizipieren.

Der Bruch mit der Vergangenheit, der im Ausbruch des Ersten Weltkrieges kulminierte, kündigte sich bereits in der Entwicklung der Flugtechnik an. Noch war allerdings, wie Josef Suwelacks „technischer Bericht“ von 1913 zeigte, eine Rückbesinnung auf das bislang Erreichte, ein Rückbezug zur Vergangenheit möglich. Doch die Kluft zwischen Vergangenheit und Zukunft vergrößerte sich stetig und führte zu Zukunftsängsten und Perspektivlosigkeit. Die Gegenwart wurde zu einer „leeren Zeit“, zu einer „uferlosen Gleichzeitigkeit“.[618]

Rund drei Monate nach Kriegsbeginn machte sich Josef Suwelack noch Hoffnung, bei den Kondor-Werken nach Kriegsende – er rechnete mit einem Sieg Deutschlands bis Juni/Juli 1915 – eine feste und sichere Anstellung zu bekommen. Seinem Bruder Richard schrieb er am 20. November 1914, als er zu Besuch in Essen war: „Condor baut jede Woche einen Apparat und soll die Fabrikation auf 2 Maschinen gebracht werden. Das Militär hat auch für Friedenszeiten Aufträge garantiert.“[619] Hinsichtlich des spanischen Auftrages war er ebenfalls zuversichtlich: „Die sp. Sache steht auch sehr gut und ich arbeite sehr für Condor. Meine Doppeldecker-Construction ist ausgeführt und hat der Apparat in 26 Minuten 2100 mtr. Höhe erreicht[,] verlangt wird 2000 mtr. in 40 Minuten.“ Er wolle mit Kondor in stetiger Verbindung bleiben und war an diesem Punkt seiner Überlegungen noch überzeugt, eine Möglichkeit zu finden, beim Essener Flugzeugbauer erneut als Konstrukteur engagiert zu werden.

Mitte November war Suwelack für wenige Tage auf Heimaturlaub, den er für Kurzbesuche in Billerbeck und Essen nutzte, wo er vom Geschäftsbetrieb des Unternehmens erfuhr.[620] Alfred von Back-Begavár, mit dem Suwelack während des Krieges in regem brieflichen und persönlichen Kontakt stand, lud ihn zweimal

617 Vgl. Rhiemeier, Nichts dauert ewig, S.23ff. und Rißmann-Ottow, Glück ab!, S.265f.
618 Assmann, Ist die Zeit aus den Fugen?, S.42.
619 NL J. Suwelack, Feldpostbrief an Richard Suwelack, Essen, 20.11.1914.
620 Vgl. ebd. und NL J. Suwelack, Feldpostbrief an seine Familie, Lille, 27.11.1914.

in den „Kaiserhof" zum Essen ein und prognostizierte ihm, dass er in naher Zukunft mit seinem Firmenanteil gut verdienen werde. Auch hierüber schrieb Suwelack in einem Brief an seinen Bruder.[621] Er berichtete von seinem Plan, sich bei Kondor ein Kampfflugzeug bauen zu lassen, fügte jedoch im nächsten Satz hinzu: „Ich will Dir [Richard] ja sagen, dass ich im Allgemeinen die Nase ziemlich voll habe von der ganzen Flugzeugfabrikation. Es sind alles zu wüste Schieber und ich glaube doch, dass nach dem Krieg ein gewisser Rückschlag kommt. Jedenfalls das ist sicher."

Der Wunsch nach einem geregelten und sicheren Leben, nach Sicherheit, taucht immer wieder in den Briefen an seinen Bruder oder seine Eltern auf. Geleichzeitig bekannte Josef freimütig, dass er nicht recht wisse, was er nach dem Krieg beruflich machen solle, er war hin- und hergerissen in seinen Überlegungen. Die Kondor-Werke wären im Unterschied zu ihm selbst nicht auf dem aktuellen Stand, denn das Unternehmen habe beispielsweise nicht mitbekommen, dass „seit Wochen ständig Fliegeroffiziere in Essen" seien, „um bei Krupp Panzerungen zu holen." Außerdem laufe der Betrieb der Entwicklung hinterher. Josef fächerte seine berufliche Zukunft in drei Optionen auf: eine Zukunft als Teilhaber bei Kondor, als selbstständiger Unternehmer oder eine militärische Laufbahn bei der Fliegertruppe. Er wog jeweils ab, inwieweit sich seine Situation verbessern oder verschlechtern würde.

Das Spaniengeschäft wollte Josef nutzen, um möglichst schnell zu viel Geld zu kommen: „Ich würde dann vielleicht auf fünf Jahre gehen mit gutem Vertrag[,] um schnell Geld zu verdienen und mich dann in Deutschland in einer anderen Sache niederzulassen, das wird man ja sehen." Richard, der seinen Bruder in geschäftlichen Fragen beriet, empfahl ihm in seinem Antwortbrief, an Eduardo Pujol zu schreiben, um einerseits die Verbindung zu ihm aufrecht zu erhalten und andererseits „irgend welche[n] Intrigen" vorzubeugen.[622]

Als Reserveoption wollte Josef sich die Kondor-Werke „warm halten", sperrte sich innerlich jedoch gegen diese Möglichkeit aus oben genannten Gründen.[623] Eine Position beim Militär schien ihm aus finanzieller Sicht verlockend, doch widerstrebte ihm die dortige Abhängigkeit von Offizieren: „Beim Militär als Ingenieur. Man bekommt dann irgend eine Werft in Köln, Hannover oder dergleichen. Das Gehalt beträgt 6.000 Mark. ... Aber das will auch überlegt sein und man ist doch viel abhängig von Offizieren." Selbstbewusst und von sich und seinen Fähigkeiten überzeugt gab er sich optimistisch: „Ich glaube ja bestimmt, dass ich eine solche Stellung bekommen kann."

Bei allen gefassten oder wieder verworfenen Plänen war nie die Rede davon,

621 Vgl. im Folgenden NL J. Suwelack, Feldpostbrief an Richard Suwelack, Lille, 3.12.1914.
622 NL J. Suwelack, Feldpostbrief Richard Suwelacks an Josef Suwelack vom 15.12.1914.
623 NL J. Suwelack, Feldpostbrief an Richard Suwelack, Lille, 3.12.1914.

ins elterliche Molkereigeschäft einzusteigen. Stattdessen zeugen die Briefe Suwelacks davon, dass er seine Zukunft in der Fliegerei sah, sich nur nicht im Klaren über die genaue Form dieser Tätigkeit war.

Am zweiten Weihnachtstag 1914 äußerte er sich gegenüber Richard erneut zu Kondor.[624] Er habe unlängst mit dem Inspekteur der Fliegertruppen von Eberhardt gesprochen, der zu Besuch bei der Feldfliegerabteilung 24 gewesen sei. Der Oberst habe die Ansicht vertreten, dass die Kondor-Werke keinen fähigen Konstrukteur mehr hätten und daher die Maschinen nicht abgenommen würden: „Der Adjutant sagte dann zum Oberst ‚Ja, Herr Suwelack ist seit Beginn des Krieges im Felde und daher ist solches wohl erklärlich'."

Auch über sein Vorhaben, sich bei den Kondor-Werken ein Kampfflugzeug mit schwenkbarem Maschinengewehr[625] bauen zu lassen, unterrichtete Suwelack Major Wilhelm Siegert – seit 1915 neuer Inspekteur der Fliegertruppen – bei dessen Besuch auf dem Flugplatz der Fliegerabteilung. Siegert habe sich „sehr lobend über die Idee ausgesprochen",[626] zumal Suwelack weitere Referenzen anführte, die ihm bestätigt hätten, dass „die Zielvorrichtung absolut einwandfrei" sei.[627] Am 15. Juni 1915 leitete Josef Suwelack die Anmeldung seiner MG-Vorrichtung zum Patent ein.[628] Der zuständige Patentanwalt Max Wohlgemuth aus Berlin, bei dem die Kondor-Werke bereits mehrfach Gebrauchsmuster hatten eintragen lassen, riet Suwelack, zusätzlich zum Patent einen „Eventual-Gebrauchsmusterschutz" eintragen zu lassen, falls seine Vorrichtung z. B. in Frankreich oder England doch schon Verwendung gefunden haben sollte.

Wie rasch sich Suwelacks Einstellung zu den Kondor-Werken erneut änderte, zeigen zwei Briefe, die im Abstand von etwa zwei Monaten verfasst wurden. Ende Mai oder Anfang Juni 1915 schrieb Suwelack in einem Brief an seinen Vater: „Von Herrn von Back erhielt ich heute einen sehr liebenswürdigen Brief. Ich stehe jetzt in stetem Briefwechsel mit Condor. Beim Patentanwalt habe ich auch alles fertig."[629] Am 30. Juli sandte Josef einen Brief aus Köln an seinen Bruder, in dem er ihm abrissartig die Ergebnisse der Gesellschafter-Versammlung darlegte.[630]

624 NL J. Suwelack, Feldpostbrief an Richard S., Lille, 26.12.1914. Alle weiteren Angaben ebd.

625 Im Unterschied zu den bisher verwendeten MG sollte Suwelacks Vorrichtung ein Schießen nach vorne, nach unten und nach hinten, also gewissermaßen ein 360°-Feuer ermöglichen. Vgl. NL J. Suwelack, Schreiben J. Suwelacks an einen namentlich nicht bekannten Oberleutnant, Lille, 23.5.1915.

626 NL J. Suwelack, Feldpostbrief an Richard Suwelack vom 9.5.1915.

627 NL J. Suwelack, Schreiben J. Suwelacks an einen namentlich nicht bekannten Oberleutnant, Lille, 23.5.1915.

628 NL J. Suwelack, Antwortschreiben des Patentanwalts Max Wohlgemuth, Berlin, 22.6.1915. Das Schreiben vom 15. des Monats ist nicht im Nachlass erhalten, findet aber in der Antwort Wohlgemuths Erwähnung.

629 NL J. Suwelack, Feldpostbrief Suwelacks an seinen Vater, undatiert, Poststempel vom 8.6.1915. Suwelack wollte sich seine Maschinengewehr-Vorrichtung patentieren lassen.

630 NL J. Suwelack, Feldpostbrief an Richard Suwelack, Köln, 30.7.1915. Alle weiteren Zitate ebd.

Einige Auszüge verdeutlichen die ungewisse Zukunft des Unternehmens und damit auch Suwelacks wenig aussichtsreiche Perspektive:

„18 Apparate sind verkauft. Davon 13 abgenommen, fünf stehen unabgenommen fertig im Schuppen. Flugschule geht gut. – Apparat in Adlershof ist beim hereinschieben [sic!] in Schuppen zur Hälfte zerbrochen (ganze untere Tragdecken). – Wie gesagt, wenn auch die Lage sich im Allgemeinen gebessert [hat,] habe ich nicht den geringsten Glauben an Erfolg."

Suwelack schloss:

„Jedenfalls ist eines sicher, dass ich bei Condor nach [dem] Krieg hier in Deutschland keine Anstellung finde und wenn auch[,] so ist meiner bestimmten Überzeugung nach doch [an eine] dauernde Stellung nicht zu denken. – Nachdem ich jetzt ein Jahr lang die Fliegerei auf der anderen Seite mitgemacht habe, habe ich die feste Überzeugung, dass hier in Deutschland so leicht nichts zu holen ist. Denke dir mal das günstige in Spanien bei meiner Kriegserfahrung etc. Schreibe mir bitte deinen Rat umgehend –. [P.S.:] Maschinengewehrsache ist nichts geworden, schreibe dir darüber noch ausführlich."

Seinem Vater teilte er am gleichen Tag schriftlich mit:

„Meine Meinung, dass ich hier bei Condor keine dauernde oder auch nur vorübergehende Stellung haben kann[,] hat sich jetzt nur noch verstärkt. Es sind noch mehrere dazwischen getreten[,] die ich nur dann verdrängen kann[,] wenn ich viel Glück habe. Andernfalls ist es natürlich nicht zu verachten[,] wenn ich als Gesellschafter einen kleinen Einfluss bei Condor behalte. Aber Dr. Carl und von Back machen ja doch was sie wollen."[631]

Sein Bruder erteilte ihm den Rat, seine Anteile zu behalten, und Josef vertrat die gleiche Ansicht.[632] Der letzte überlieferte Brief, in dem Josef sich über seine fliegerische Zukunft äußerte, ist datiert auf den 1. September 1915. Offensichtlich war er zwischenzeitlich weiter in seinem Entschluss bestärkt worden, nach dem Krieg in Spanien neu anzufangen, denn er schrieb, dass Eduardo Pujol ihm gegenüber versichert habe „ohne mich nichts [zu] machen". Von Back-Begávar habe auch bekundet, nach Spanien zu gehen: „Er sagt, das Land sei ja momentan sehr am aufblühen [sic!], so daß sich die Industrie in einigen Jahren gut entwickeln würde. Man wird ja sehen."[633]

Die abwartende Haltung Josef Suwelacks kennzeichnet durchgängig seine beruflichen Überlegungen während des Krieges. Regelmäßig schloss er einen Gedankengang mit der Formulierung: „Man wird ja sehen" oder auf Französisch mit „on verra".[634] Insgesamt dachte er „sehr viel" über seine Zukunft und seine „Existenz nach dem Kriege" nach.[635] Beispielhaft für die Hin-und-Hergerissenheit des Fliegers steht ein Brief vom April 1915 an Richard: „Nach vielen Überlegungen komme ich immer wieder darauf, dass Spanien das Beste ist. ... Die K[ondor] W [erke] werden glaube ich nie ins Geschäft kommen. Eine Möglichkeit war die

631 NL J. Suwelack, Feldpostbrief an Joseph Suwelack senior, Köln, 30.7. 1915.
632 Vgl. NL J. Suwelack, Feldpostbrief an Richard Suwelack, Petit Ronchain, 14.8.1915.
633 NL J. Suwelack, Feldpostbrief an Richard Suwelack, Petit Ronchain, 1.9.1915.
634 Siehe z. B. NL J. Suwelack, Feldpostbriefe vom 1.11.1914, 3.12.1914, 26.12.1914, 22.5.1915 oder vom 1.9.1915.
635 NL J. Suwelack, Feldpostbrief an „Natz" [Richard] Suwelack, Lille, 8.1.1915.

Kampfflugmaschine. Aber wenn die Leute gar nichts unternehmen[,] ist nicht zu helfen."

Suwelack bereitete seine ungewisse Zukunft große Sorgen und ließ ihn inständig auf ein baldiges Ende des Krieges hoffen: „Hoffen wir, daß der Brei bald vorüber ist",[636] schloss er in einem Brief an Richard Anfang Februar 1915, und in einem Schreiben an seine Familie bat er: „Gott gebe uns bald den endlichen Sieg und Frieden".[637] Solche Äußerungen sind häufig in der Feldpost von Soldaten zu finden, sie können trotz ihrer Floskelhaftigkeit als Ausdruck der Hoffnung auf ein baldiges Ende von Krieg und Gewalt gelesen werden. Suwelack selbst hoffte wohl vor allem auf das Kriegsende, um wieder ein „normales", sicheres (Berufs-)Leben führen zu können.

Mit zunehmender Dauer des Krieges wandelte sich Suwelacks anfänglicher Optimismus, dass „in zwei Wochen ... Schluss" sei,[638] in Resignation oder vielmehr in die Erkenntnis, dass der propagierte „kurze", auf den Erfahrungen des letzten großen Krieges gegen Frankreich von 1870/71 basierende Feldzug[639] sich länger als erwartet gestalten werde. Bereits im Oktober 1914 gab Richard Suwelack zu bedenken: „In der Hoffnung Weihnachten wieder zu Hause zu sein, werden sich wohl viele getäuscht haben, denn jetzt fängt die Sache für uns an etwas schwieriger zu werden. Aber kriegen werden wir die Luders [sic!] ja doch."[640] Am 27. Oktober 1914 schrieb Josef seinem Bruder: „Bald haben wir die ganze Küste und dann dauert die Chose hoffentlich nicht mehr lange. Du wirst auch die Nase voll haben."[641]

In seiner Stellung als Aufklärungsflieger bei einer Feldfliegerabteilung war Suwelack im Unterschied zur Artillerie oder zu den Soldaten in den Schützengräben in der komfortablen Situation, (Frei-)Zeit, Erholung und Zugang zu ausreichend Lebensmitteln und Schreibutensilien zu haben.[642] Bedingt durch die Wetterabhängigkeit der Flieger konnte er „in Ruhe" über seine Zukunft nachdenken.

636 NL J. Suwelack, Feldpostbrief an Richard Suwelack, Lille, 1.2.1915.

637 NL J. Suwelack Feldpostbrief an die Eltern und Geschwister, Lille, 3.1.1915.

638 NL J. Suwelack, Feldpostkarte an die Eltern und Geschwister, Reims, o.D. [Poststempel vom 5.9.1914].

639 Siehe zu den durchaus ambivalenten Kriegsvorstellungen des Volkes und der Militärführung: Frank Becker, Strammstehen vor der Obrigkeit? Bürgerliche Wahrnehmung der Einigungskriege und Militarismus im Deutschen Kaiserreich, in: Historische Zeitschrift (2003), Bd. 277, S.87–113, bes. S.110ff.; Stig Förster, Der deutsche Generalstab und die Illusion des kurzen Krieges, 1871–1914. Metakritik eines Mythos, in: Militärgeschichtliche Mitteilungen 54 (1995), S.61–95; Dirk Schumann, Der brüchige Frieden. Kriegserinnerungen, Kriegsszenarien und Kriegsbereitschaft, in: Ute Frevert (Hg.), Das neue Jahrhundert. Europäische Zeitdiagnosen und Zukunftsentwürfe um 1900, Göttingen 2000, S.113–145 sowie Oliver Stein, Das Kriegsministerium und der Ausbau des deutschen Heeres 1871–1914, in: Müller/Rogg, Das ist Militärgeschichte!, S.48–62. Siehe auch die Ausführungen zum Feld 4.

640 NL J. Suwelack, Feldpostbrief Richard Suwelacks an Josef, Vorges, 12.10.1914.

641 NL J. Suwelack, Feldpostbrief an Richard Suwelack, Lille, 27.10.1914.

642 Siehe die Ausführungen zum Feld 4.

Bereits zweieinhalb Monate nach Kriegsbeginn war die Begeisterung für die Militärfliegerei in Ernüchterung umgeschlagen: „Ob ich nach dem Krieg bei der Fliegerei bleibe, weiß ich nicht. Zur Militärfliegerei könnte ich so übertreten, aber ich habe hier schon davon die Nase voll. Und dann weiß ich nicht wie die Chancen nach dem Krieg sind. Jedenfalls hat man mir sehr davon abgeraten."[643]

Möglich, dass er diese Äußerung aus einer Laune heraus aufschrieb, doch scheint er tatsächlich nicht ernsthaft eine militärische Berufslaufbahn in Erwägung gezogen zu haben. Auch hätte ihm als Bürgerlichem seit 1890 der Weg in die prestigeträchtige Offizierslaufbahn offen gestanden,[644] doch hielt der Beruf wenig Aussicht auf Beförderung und hohes Einkommen bereit, was zahlreiche potenzielle Bewerber abschreckte.[645] Vermutlich hatten Suwelack die Militärs bei der Fliegertruppe, befreundete Flieger und nicht zuletzt seine vor Ort gemachten Erfahrungen für die Realität sensibilisiert und ihm die Hoffnung auf eine Militärkarriere mit großer Entscheidungsfreiheit genommen.

Zwischen 1910 und 1914 war aus dem technikbegeisterten Billerbecker ein Flieger geworden, dem es gelungen war, mit der Hilfe reicher Förderer und Mitstreiter ein neues Flugzeugunternehmen aufzubauen. Während Suwelack bei den Rumpler-Werken weisungsgebunden gewesen und vertraglich stark in Anspruch genommen worden war, konnte er seine Arbeitszeit bei den Kondor-Werken flexibler gestalten. Als Gesellschafter hatte er Mitbestimmungsrechte und konnte Flugzeuge nach seinen Vorstellungen entwickeln. Somit war aus dem kleinen Angestellten Suwelack, einem von Edmund Rumpler abhängigen Arbeitnehmer, ein privilegierter technischer Leiter geworden, der im Laufe von etwa zwei Jahren Kontakte zu Militärs, Großindustriellen und anderen Flugzeugfirmen geknüpft hatte.

Doch nicht erst der Ausbruch des Krieges stellte den Westfalen vor neue Herausforderungen. Bereits der aus einem Zerwürfnis mit Edmund Rumpler resultierende Fortgang aus Berlin-Johannisthal hatte Suwelack vor die Wahl gestellt, wie er seine weitere Zukunft gestalten wollte. Mehrere lukrative Angebote

643 NL J. Suwelack, Feldpostbrief an Richard Suwelack, Lille, 27.10.1914.

644 Siehe zum „synthetischen Militarismus", der Zusammenfassung von „Königsheer" und „Volksheer": Frank Becker, Synthetischer Militarismus. Die Einigungskriege und der Stellenwert des Militärischen in der deutschen Gesellschaft, in: Michael Epkenhans (Hg.), Das Militär und der Aufbruch in die Moderne 1860 bis 1890. Armeen und Marinen und der Wandel von Politik, Gesellschaft und Wirtschaft in Europa, den USA sowie Japan, München 2003, S.125–141. Zur Integration des Bürgertums in die Armee siehe Bernhard R. Kroener, Integrationsmilitarismus. Zur Rolle des Militärs als Instrument bürgerlicher Partizipationsbemühungen im Deutschen Reich und in Preußen im 19. Jahrhundert bis zum Ausbruch des Ersten Weltkrieges, in: ders./Ralf Pröve/Bruno Thoß (Hg.), Kriegerische Gewalt und militärische Präsenz in der Neuzeit. Ausgewählte Schriften, Paderborn/München/Wien/Zürich 2008, S.83–107, hier S.98ff.

645 Vgl. Oliver Stein, Das Kriegsministerium und der Ausbau des deutschen Heeres 1871–1914, in: Christian Th. Mülller/Matthias Rogg (Hg.), Das ist Militärgeschichte!, Paderborn/München/Wien/Zürich 2013, S.48–62, hier S.54.

anderer Flugzeugfirmen, die ihm sofort eine Stelle angeboten hätten, lehnte er ab. Auf die Frage nach dem Warum gibt es jedoch aufgrund der Quellenlage keine eindeutige Antwort, es muss daher bei Vermutungen bleiben. Vielleicht trug er sich schon damals mit dem Gedanken, etwas Eigenes aufzubauen, vielleicht überschätzte er auch seine Fähigkeiten oder aber die Arbeitsbedingungen erschienen ihm nicht seinem Stand als Weltrekordhalter gemäß. Rein pragmatisch gesehen war Essen zudem räumlich näher an seiner Heimatstadt Billerbeck gelegen und damit der direkte Kontakt zu seinem Elternhaus leichter herzustellen als aus Berlin. Möglicherweise wollte er aber auch nicht erneut riskieren, durch eine Flugzeugfirma ausgebeutet und am Gewinn nicht ausreichend beteiligt zu werden. Dafür sprechen die eindeutigen Vertragsverhältnisse bei den Kondor-Werken, die ihm ausdrücklich Gewinnbeteiligungen garantierten.

Allerdings fügte sich Suwelack nicht in die Reihen der übrigen Gesellschafter ein. Ihm war durchaus bewusst, dass zum Fliegen und zum Flugzeugbau viel Geld erforderlich war, doch scheint er nicht ernsthaft besorgt gewesen zu sein, dass das Unternehmen bankrott gehen könnte. Er ging davon aus, dass der ökonomische Erfolg sich langfristig einstellen werde, und hatte zudem seine Konstruktionspläne als Sacheinlage eingebracht, jedoch kein Vermögen.[646]

Er war kein Selfmademan oder erfahrener Unternehmer, der immer und überall ein sicheres Gespür für verschiedene Sachverhalte hatte. Vielmehr war er ein ambitionierter, passionierter Flieger und Konstrukteur, der das Glück hatte, auf Unternehmer und Luftfahrtinteressierte zu stoßen, die gewillt waren, in eine Flugzeugfirma zu investieren. Gleichwohl brachte Suwelack sich aktiv ins Geschäft ein, regte z. B. die Gründung einer Flugschule an oder empfahl, in den Bau von Wasserflugzeugen zu investieren. Ihn jedoch als Firmeninhaber und Gründer der Kondor-Werke zu bezeichnen, wie es auch in zeitgenössischen Publikationen häufiger geschehen ist, kann mit der Forschungslage nicht gestützt werden. Seine eigene Sicht auf seine Stellung bei dem Essener Flugzeugunternehmen war ambivalent, denn während er einerseits in Briefen an das Militär selbstbewusst auftrat, äußerte er sich gegenüber Richard Suwelack vor allem abfällig über Ernst Füßmann, bei dem er Intrigen gegen seine Person vermutete.

Während seines zweijährigen Kriegseinsatzes tendierte Suwelack jedenfalls auf lange Sicht gesehen dazu, sich von den Kondor-Werken ab- und einer anderen Firma zuzuwenden. Da es sich bei den Feldpostbriefen und -karten jedoch nur um Momentaufnahmen handelt, lässt sich nicht definitiv bestimmen, welche Zukunft er für sich in dem Unternehmen sah. Da ihm die unsichere finanzielle

646 Für die geforderten 5.000 Mark, die Suwelack zusätzlich leisten sollte, gibt es keinen Beleg in den Bilanzen des Unternehmens.

Situation der Firma bewusst war, entwickelte er Pläne, einen Neuanfang zu versuchen.

Der Erste Weltkrieg als Zäsur von außen zwang Josef Suwelack, seine fliegerische Vergangenheit Revue passieren zu lassen, seinen Standpunkt immer wieder neu zu bestimmen. Der Krieg konfrontierte ihn mit einer kontingenten Situation, in der alles und nichts zugleich möglich schien. Wie tief dieser Bruch sein würde, war in den ersten Kriegsmonaten noch nicht abzusehen. Erst im Laufe des Jahres 1915 zeichnete sich ab, dass Vergangenheit und Zukunft gleichermaßen ihre Bedeutung verloren.

Feld 3: Flugwettbewerbe: Kontakt- und Heldenschmiede, Sichtungsplattform

Lag der Fokus im zurückliegenden Kapitel auf der internen Konstrukteur- und Flugzeugfirmenarbeit, den Personal- und Firmenstrukturen sowie den wechselseitigen Abhängigkeiten zwischen den Flugzeugfirmen und innerhalb der einzelnen Unternehmen, so geht es im Folgenden um die öffentliche Seite der Fliegerei. Anhand unterschiedlicher, von Josef Suwelack wahrgenommener Flugwettbewerbe sollen idealtypisch verschiedene Formen und Ausprägungen von Wettbewerben unterschieden beziehungsweise miteinander verglichen werden. Diese Vorgehensweise gestattet es, die inhaltlichen Schwerpunktthemen der Flugveranstaltungen herauszustellen und nach der Organisation, den Funktionen sowie der Wahrnehmung zu fragen, aber auch nach Veränderung von Flugwettbewerben durch den beziehungsweise die Piloten, das Publikum und das Militär. Damit liegt dieses Feld gewissermaßen quer zu den anderen Handlungsfeldern Josef Suwelacks, weist es doch Überschneidungen einerseits mit seiner unternehmerischen Tätigkeit und andererseits mit seinem späteren Kriegseinsatz im Ersten Weltkrieg auf. Flugwettbewerbe als öffentliche Seite der Fliegerei belegen das Eindringen der Flugtechnik in die Lebenswelt der Bevölkerung, das Flugzeug wurde zu einer Chiffre für eine neue Zeit- und Raumerfahrung und für den Aufbruch in die Moderne.

Bei der Auswahl der zu beleuchtenden Flugtage und -wochen wurde versucht, sowohl Groß- als auch regionale sowie militärisch ausgerichtete Veranstaltungen zu berücksichtigen, um die dahinterstehenden Intentionen und Mechanismen der Konstruktion von Fliegerhelden sowie des nationalen und internationalen Konkurrenzkampfes herauszuarbeiten.

Auch die Presse übte einen nicht zu unterschätzenden Einfluss auf den Ruf eines Fliegers und damit häufig zugleich auch auf den Ruf einer Flugzeugfirma aus und emotionalisierte die Fliegerei.[647] Dabei variierte die Berichterstattung in

647 Vgl. Borscheid, Tempo-Virus, S.193f.

Abhängigkeit vom lokalen Bekanntheitsgrad der Flieger. Der „Held des Tages", als der Josef Suwelack beim Flugtag in Münster 1911 gefeiert worden war, war zwei Jahre später beim Prinz-Heinrich-Flug angesichts der Dominanz von Militärpiloten durch das Reglement bedeutungslos geworden, bedeutungslos deshalb, da sich die Fliegerei von einer „Sportisierung" zu einer „Entsportlichung" verlagerte.[648] Die sich ändernde Wettbewerbsausrichtung gestaltete das Pilotenbild immer auch mit und konnte über Ruhm und Ehre oder über Bedeutungslosigkeit entscheiden.

Zwischen 1911 und 1914 nahm Suwelack für Rumpler, Aviatik und die Kondor-Werke an Wettbewerben teil. Doch während zwar sein Name in der Literatur in Teilnehmerlisten abgedruckt wurde, ist eine kritische Analyse der Berichterstattung, die von heldenhaft-topischen bis hin zu kritischen Beschreibungen reicht, bislang nur selten erfolgt.[649]

Idealtypisch lassen sich die Wettbewerbe in nationale, regionale und militärische Flugwettbewerbe unterscheiden. Internationale Konkurrenzen gab es zwar ebenfalls, allerdings dominierten die drei erstgenannten Typen, die vor allem für die eigenen Landsleute veranstaltet wurden. Am Beispiel der Nationalen Flugwoche Berlin-Johannisthal 1911 aus nationaler Perspektive, drei Flugveranstaltungen auf lokaler Ebene und schließlich am Prinz-Heinrich-Flug aus militärischer Sicht sollen entlang der Wettbewerbsteilnahmen Josef Suwelacks bestimmte Aspekte auf einer Mikroebene thematisiert und abschließend eine Beurteilung über die Bedeutung von Flugwettbewerben bei der Ausbildung von Identitäten – national, regional und individuell – und unternehmerischem Erfolg vorgenommen werden. Methodisch bietet diese Vorgehensweise die Chance, Suwelack als Teil der Werbebeziehungsweise Rüstungsindustrie zu begreifen und sein Wirken im Wechselspiel mit den örtlichen, aber auch gesellschaftlichen und politischen Voraussetzungen und den Mentalitäten der beteiligten Personen zu beschreiben. Noch deutlicher als im Kapitel zur unternehmerischen Tätigkeit Josef Suwelacks treten hier die Aspekte des Risikos, der (Selbst-)Vermarktung als wichtiger Erfolgsfaktor sowie der Bedeutung von Industriespionage hervor, denn Flugwettbewerbe lieferten Erkenntnisse hinsichtlich neuer Technologien und Bauweisen und boten eine Plattform des öffentlichen und kommerziellen Austausches und der Konkurrenz.

648 Die Termini werden im Folgenden erläutert und sind entnommen aus Norbert Elias/Eric Dunning, Quest for excitement. Sport and leisure in the civilizing process, Oxford 1986, hier wird in den weiteren Ausführungen die deutsche Ausgabe zitiert: Dies., Sport und Spannung im Prozess der Zivilisation. Frankfurt a. M. 2003; Eric Dunning, Sport Matters. Sociological studies of sport, violence and civilization, London/New York 1999; Möser, Fahren und Fliegen, S. 107ff.

649 Zwar werden in der Familienbiografie den Wettbewerbsteilnahmen Suwelacks mehrere Seiten gewidmet und auch einige Zeitungsartikel zitiert, doch übernimmt die Darstellung das Bild der heldischen Aura des Piloten, das die Zeitungen konstruiert hatten. Anders jedoch eine unveröffentlichte Masterarbeit aus dem Jahr 2012, die sich eingehend mit der Technikbegeisterung im Münsterland befasst hat: Martina Glanemann, Technikbegeisterung und kleinstädtische Strukturen. Flugtage im Münsterland 1908–1914, unveröffentlichte Masterarbeit Münster 2012.

3.1. Sportisierung und Sensationslust – Die Nationale Flugwoche Berlin-Johannisthal

Organisation

Die Nationale Flugwoche Berlin-Johannisthal vom 24. September bis zum 1. Oktober 1911 war der erste Wettbewerb, an dem Josef Suwelack teilnahm, seit er im Besitz des Flugscheins war. Unter dem Protektorat des Berliner Vereins für Luftschiffahrt, des Kaiserlichen Aero-Clubs, des Kaiserlichen Automobil-Clubs und des Vereins Deutscher Flugtechniker waren Preise in einer Gesamthöhe von 40.000 Mark ausgesetzt. Gestiftet wurden die Preisgelder von Luftfahrtvereinen, Unternehmen und vom Preußischen Kriegsministerium.

Die Flugwoche wurde veranstaltet von der Flug- und Sportplatz Berlin-Johannisthal GmbH, für die Durchführung war ein 26-köpfiger Organisations-Ausschuss zuständig. Darin vertreten waren unter anderem der Direktor des Flugplatzes Berlin-Johannisthal, Major a. D. Georg von Tschudi, als Vorsitzender des Ausschusses, ferner Vereinsvorsitzende, Militärs und weitere einflussreiche Unternehmer wie Arthur Müller oder Dr. Fritz Huth, letzterer Gründer der 1909 entstandenen Flugmaschinen- und Motorengesellschaft mbH.[650] Durch ihre Tätigkeit in Organisationsausschüssen und Luftfahrtvereinen bekannten sie sich zur Fliegerei und konnten durch ihre Kontrollfunktion Einfluss auf die Flugzeugentwicklung nehmen. Auch in den weiteren Veranstaltungskomitees saßen Luftfahrtaktivisten und -lobbyisten, die über die ausgesetzten Preisgelder und die Wettbewerbsbestimmungen die Ausrichtung der Flugzeugtechnik mittel- bis langfristig steuerten.

Auf die zwei ausgeschriebenen Wettbewerbe, einen Flugzeug- (Wettbewerb A) und einen Flieger-Wettbewerb (Wettbewerb B) entfielen Preisgelder von 10.000 beziehungsweise 30.000 Mark. Eine Ehrenplakette des Kaiserlichen Aero-Clubs erhielt der Pilot, der die größte Flughöhe erreicht hatte.[651]

Während es sich bei den Flugwochen der Jahre 1909 und 1910 noch um internationale Ausschreibungen gehandelt hatte, war die Nationale Flugwoche deutschen Fliegern und Apparaten vorbehalten und eine genuin deutsche Leistungsschau. Von den ursprünglich 41 gemeldeten Fliegern bekamen 37 Piloten eine Teilnahmegenehmigung. Letztlich beteiligten sich 24 Flugzeugführer mit 29 Flugzeugen,[652] darunter mit Melli Beese erstmals auch eine Frau, an den Wett-

650 Vgl. Schwipps, Schwerer als Luft, S.45.

651 Vgl. Ausschreibungen für die Nationale Flugwoche Berlin-Johannisthal vom 24. September bis 1. Oktober 1911, S.10f.

652 Vermutlich hatten einige Flieger Reserveflugzeuge im Schuppen untergebracht, was die Anzahl der in der Tabelle aufgelisteten Flugzeuge erklären würde. Die Angaben in Zeitungen und Zeitschriften sind zudem zum Teil verschieden.

kämpfen.[653] Teilnahmeberechtigt waren gemäß den „Allgemeinen Bestimmungen" der Ausschreibung „alle Flieger deutscher Nationalität ..., die vor dem Meldeschluß im Besitze eines Führerzeugnisses sind, das von einer der F.A.I. angehörigen Sportmacht ausgestellt ist."[654] An den Wettbewerb B waren darüber hinaus zusätzliche Bedingungen geknüpft. So musste das Flugzeug, abgesehen vom Motor, ein rein deutsches Erzeugnis sein und der Pilot durfte „noch keinen Geldpreis von 5000 M. oder mehr gewonnen haben".[655]

Bei der Anmeldung zur Flugwoche musste pro Flugzeug ein Nenngeld von 100 Mark entrichtet werden, das jedoch zurückgezahlt wurde, sofern mindestens fünf Minuten am Stück geflogen wurde. Diese Maßnahme sollte verhindern, dass auch „Bastler" am Wettbewerb teilnahmen. Flieger, die nicht zugelassen wurden, erhielten das Nenngeld „spätestens 8 Tage nach Mitteilung der Nichtzulassung" zurück.[656]

Da die Preisgelder nach der gemessenen Flugdauer verteilt wurden, wurden auch kleinere Leistungen honoriert und damit Anreize zur Teilnahme geschaffen. Somit war der Wettbewerb auch für Piloten interessant, die nicht bei einer Flugzeugfirma angestellt waren, sondern in Eigenverantwortung Flugmaschinen bauten. Sie spekulierten oftmals darauf, sich ein Leben als Flieger durch den Gewinn der zum Teil sehr hohen Preisgelder leisten zu können.[657] Darüber hinaus stellte das Kriegsministerium in Aussicht, „von denjenigen Stellen, die den bestqualifizierten Eindecker und den bestqualifizierten Zweidecker deutschen Ursprungs gemeldet haben, ein Flugzeug ... zu kaufen." Allerdings sollte nicht eine Firma beide Sieger-Flugzeuge stellen dürfen, sondern in diesem Fall die zweitbeste Fabrik ebenfalls Berücksichtigung finden. Zugleich mussten die Apparate „gute Beobachtungsmöglichkeit und militärische Brauchbarkeit" unter Beweis stellen.[658]

Neben den jeweils speziellen Regeln der einzelnen Wettbewerbe, so auch der Flugwoche in Berlin-Johannisthal, existierten auch allgemeine Regeln für den Flugsport, die in den „Vorschriften des Deutschen Luftschifferverbandes für den

653 Angaben zusammengestellt mithilfe der Informationen aus Flugsport (1911), Nr. 20, S.726; (1911), Nr. 22, S.793; Jahrbuch des DLV (1912), S.143; Sport im Bild (1911), Nr. 40, S.1150ff.; NL J. Suwelack, „Berliner Flugwoche. Schlechtes Wetter – schöne Flüge", in: [unbekannter Zeitungstitel] vom 30.9.1911.

654 Ausschreibungen für die Nationale Flugwoche Berlin-Johannisthal vom 24. September bis 1. Oktober 1911, S.3. F.A.I. ist die Abkürzung für die 1905 in Paris gegründete und bis heute bestehende Fédération Aéronautique Internationale, eine Organisation, die die Luftfahrtaktivitäten koordiniert und organisiert.

655 Ebd.

656 Ebd., S.4, dort auch das Zitat.

657 Vgl. Möser, Fahren und Fliegen, S.105; Sport im Bild (1911), Nr. 40, S.1150ff., hier S.1150.

658 Vgl. Ausschreibungen für die Nationale Flugwoche Berlin-Johannisthal vom 24. September bis 1. Oktober 1911, S.10, dort auch die zitierte Passage.

Flugsport"[659] aufgeführt waren. Diese befassten sich nicht nur mit dem Erwerb des Flugscheins, sondern auch mit der Ausrichtung von Flugwettbewerben und der Erteilung von Strafen, sie informierten über mögliche Flugrekorde und über bei der Veranstaltung zu tragende Abzeichen. Das Reglement war umfassend, streng und besonders die Strafmaßnahmen als Abschreckung gedacht. Sie reichten von einem offiziellen und öffentlichen Tadel über eine „Geldstrafe über 100 Mark" bis hin zum „dauernden Ausschluß von allen sportlichen Veranstaltungen". Unter Punkt 51 hieß es hierzu deutlich:

> „Der Ausschluß von einem Wettbewerb kann wegen Täuschung oder wegen Versuchs dazu ausgesprochen werden – Der Bewerber, den diese Strafbestimmung trifft, ist der Flugsportkommission anzuzeigen, welche gegebenenfalls seinen zeitweiligen oder dauernden Ausschluß auszusprechen hat. Der Ausschluß zieht für den davon Betroffenen die Nichtigkeitserklärung seiner Anmeldung und den Verlust seines Nenngeldes nach sich."

Die Verhängung einer Strafe erfolgte jedoch erst nach vorheriger Rücksprache der Flugsportkommission mit dem Veranstalter. Die Sanktionen erstreckten sich dabei nicht nur auf den Flieger selbst, sondern konnten auch über Helfer oder Flugpassagiere, die regelwidrig gehandelt hatten, sowie über Leute – betroffen waren hier wohl vor allem Journalisten und Informanten der Presse – verhängt werden, die bewusst Falschmeldungen veröffentlicht hatten. Der Juvenal'schen Frage: „Quis custodiet ipsos custodes?" trug das Regelwerk unter Punkt 55 Rechnung: „Auch Mitglieder des Organisationsausschusses und sportliche Leiter können in Strafe genommen werden. Für diesen Fall gelten die gleichen Bestimmungen."

Wenngleich die Wettbewerbe zwischen 1909 und 1911 noch in einer „Spielphase"[660] begriffen waren und der sportliche Wettkampf überwog, wurde die Wettbewerbsfliegerei schon früh von nationalistisch-kriegerischen Überlegungen begleitet und über Teilnahmebedingungen versucht, zum einen qualitativ hochwertigen Sport zu bieten, zum anderen aber auch über den sportlichen Nutzen hinausreichende Flugzeuge und erprobte Piloten zuzulassen.

Wettbewerbsfunktionen: Testfeld, Prestigegewinn und Sportisierung

Flugwettbewerbe waren, besonders in der Anfangszeit der Fliegerei, „technische Experimentierfelder, auf denen im sportlichen und spielerischen Umgang mit der Technik die neuesten Flugzeugtypen unter Praxisbedingungen getestet wur-

659 „Vorschriften des deutschen Luftschifferverbandes für den Flugsport", abgedruckt in: Flugsport (1911), Nr. 4, S. 123–130.
660 Kehrt, Das Fliegen, S. 200.

den",[661] und zugleich „Bühnen des technischen und ökonomischen Kräftemessens"[662] der Nationen.

Die Inspektion des Militär-, Luft- und Kraftfahrwesens stellte allerdings bereits 1912 den Sinn und Nutzen von Schauflügen infrage. Auf eine Denkschrift des Deutschen Luftfahrer-Verbandes vom April 1912 reagierend gab man zu verstehen, dass die aufgewendeten Preisgelder in keinem Verhältnis zu den fliegerischen Leistungen stünden, denn ihr Wert entspreche „nicht annährend [sic!] den hierfür aufzuwendenden großen Mitteln. [Die Wettbewerbe] tragen einen reinen sportlichen Charakter und dienen in erster Linie der Befriedigung der Schaulust der Menge und dem Sensationsbedürfnis der Presse".[663]

Tatsächlich glichen die ersten Flugwochen einem Volksfest,[664] sie waren Rekordveranstaltungen und spannende Wettkämpfe. Die Zeitungen berichteten immer wieder von mehreren tausend Besuchern, die zum Flugplatz geströmt seien, obwohl die Wetterlage durchaus Auswirkungen auf das Kommen oder Ausbleiben der zahlenden Besucher hatte. Die „Berliner Volks-Zeitung" berichtete in ihrer Ausgabe vom 25. September über die Eröffnung der Flugwoche tags zuvor:

> „Das Wetter war nicht gerade vielversprechend, und so blieb denn der erwartete Massenbesuch auf dem Startplatz aus. Auf den Stehplätzen aber drängte sich eine nach vielen Tausenden zählende Menge, die trotz des gegen 5 Uhr nachmittags ausbrechenden Regens bis zum Schluß aushielt."[665]

Der beschriebene Andrang auf den Stehplätzen erklärt sich vermutlich daraus, dass die Eintrittspreise in Preiskategorien gestuft waren und ein Stehplatz am günstigsten war.[666] Besonders in der zweiten Wettbewerbshälfte, so berichteten die Zeitungen, sei die Wetterlage unbeständig gewesen: „Trüb und grau und regnerisch sah der gestrige Tag drein. ... Trotzdem war dem Tage ein voller Erfolg beschieden. Die Flieger wollten zeigen, daß ihnen das Wetter nichts anhaben könne und flogen sogar in strömendem Regen, der gegen 5 Uhr einsetzte, ruhig weiter."[667]

Die schlechten Wetterbedingungen nutzte Karl-Heinz Bernius,[668] der für den „Berliner Lokalanzeiger" über die Flugwoche schrieb, als Aufhänger für seinen

661 Ebd., S.206. Vgl. auch Asendorf, Super Constellation, S.188.
662 Fuhs, Fliegende Helden, S.709.
663 BA-MA PH 24/72, Luftschiffer- und Fliegervereine. Subventionierung von Fliegervereinen, Fliegerschulen und Flugplätzen zur Förderung der Luftfahrt, darin: Schreiben der Inspektion des Luft- und Kraftfahrwesens vom 27.5.1912, betr. Denkschrift des Deutschen Luftfahrer Verbandes von April 1912.
664 Vgl. Reinicke, Deutsche Auffahrten, S.26.
665 „Beginn der Nationalen Flugwoche in Berlin-Johannisthal", in: Berliner Volks-Zeitung vom 25.9.1911, Abendausgabe.
666 Der Eintritt betrug 5 Mark in der höchsten und 50 Pfennig in der niedrigsten Preiskategorie.
667 NL Willy Stiasny D 006 (1911), „Berliner Flugwoche. Der erste Tag des Johannisthaler Meetings", in: [unbekannter Zeitungstitel] vom 25.9.1911, S.411.
668 Karl-Heinz Bernius (1887–1917), geboren in Niedernhausen/Darmstadt, war Sportredakteur beim

Artikel, um den mutigen Flügen der männlichen Flieger Respekt zu zollen. Unter der Überschrift „Der Sturmflugtag der Berliner Herbst-Flugwoche" sprach er sich anerkennend über die Leistungen der Piloten aus, die ihre Prüfungen „glänzend gemeistert"[669] hätten. Er vertrat die Ansicht, dass dazu „Manneskraft und Mannesmut" nötig seien, „um bei einem solchen Wetter Flüge auszuführen". Der Artikel steht einmal mehr beispielhaft für das durch die Medien konstruierte und von den Fliegern reproduzierte Pilotenbild der Vorkriegszeit, das sie zu Beherrschern ihrer Maschinen, zu todesmutigen Männern und Bezwingern der Lüfte stilisierte.

Diesen abgehärteten, explizit männlichen Flieger-Typus kontrastierte Bernius mit einem weiblichen Gegenentwurf in der Person Melli Beeses. Beeses Entschluss, an diesem Tag nur als Passagier bei Hellmuth Hirth mitzufliegen, kommentierte er süffisant: „[M]an darf es deshalb Fräulein M. Beese nicht zu sehr übel nehmen, wenn sie gestern die Steuerung ihrer Rumpler-Taube den bewährten Händen ihres Lehrmeisters Hirth überließ." Empört zeigte sich der Journalist über den – allerdings erfolglosen – Versuch der Fliegerin, die Wettbewerbe zu verschieben, bis eine Besserung der Flugbedingungen eintrete: „Es mag dies eine Mahnung sein für Fräulein Beese, die sich in den gefährlichsten männlichen Beruf hineinwagte, in dem sie aber, wenn sie wirklich als ernsthafte Konkurrentin gelten will, nicht die Ritterlichkeit ihrer männlichen Kollegen in Anspruch nehmen darf."

Als ein Musterbeispiel für die genannten „ritterlichen" Fliegertugenden führte Bernius Josef Suwelack an, bei dem er als Passagier an Bord war. Der Rumpler-Flieger sei ein „junger, vielversprechender Pilot" und Bernius zeigte sich positiv überrascht „von der Sicherheit und Ruhe, mit der mein Führer die Rumpler-Taube 57 Minuten lang, trotz der heftigen Böen, in der Luft zu halten wußte." Josef Suwelack besäße „ein bewundernswertes Gefühl für die Stärke der anstürmenden Windstöße" und wisse der Maschine stets „die vorteilhafteste Richtung zu geben, so daß er sie, ohne die Verwindung zu betätigen, tadellos im Gleichgewicht halten kann".

Die zitierte Aussage greift wesentliche zeitgenössische Merkmale des Flieger-Habitus auf: Ruhe, fliegerisches Gefühl und Risikobereitschaft – Eigenschaften, die in den Berichten den Männern vorbehalten waren. Josef Suwelack ging bewusst das Risiko ein, bei turbulenten Wetterverhältnissen zu fliegen, weil von den Fliegern erwartet wurde, den hohen Ansprüchen des Publikums sowie der berichtenden Presse gerecht zu werden. Erst in einer Gefahrensituation bewies ein Pilot in der zeitgenössischen Wahrnehmung sein Können und unterstrich

„Berliner Lokalanzeiger". Im Ersten Weltkrieg war er Leutnant der Reserve bei der Fliegertruppe und verunglückte am 24. Januar 1917 in Döberitz bei einem Flugzeugabsturz tödlich.

669 Karl Heinz Bernius, „Der Sturmflugtag der Berliner Herbst-Flugwoche", in: Berliner Lokalanzeiger vom 29.9.1911. Alle folgenden Zitate ebd.

seine außergewöhnlichen Fähigkeiten. Bei drohender Gefahr, z. B. dem Aussetzen des Motors oder Reißen eines Spanndrahtes, durften die Flieger keine Nerven zeigen, sondern mussten einen kühlen Kopf bewahren. Als Gustav Witte, ein weiterer Teilnehmer der Flugwoche, am gleichen Flugtag abzustürzen drohte, habe er „mit großer Geistesgegenwart und Kaltblütigkeit" seine Maschine vor einer Bruchlandung bewahrt und „unter dem tosenden Beifall der Stehplatzbesucher" seinen Flug fortgesetzt.[670] Die Sporthelden waren „Inkarnationen der Außeralltäglichkeit"[671] und erst das eingegangene Risiko garantierte Bewunderung und Ehre.

Standen einerseits die Flugleistungen – größte Höhe, längste Flugdauer etc. – im Vordergrund, so galt es andererseits, das Interesse des Publikums durch eindrucksvolle Flugvorführungen zu wecken. Das dabei eingegangene Risiko hatte durchaus handlungsrelevanten Nutzen, denn freihändig zu fliegen bewies nicht nur die Eigenstabilität des Flugzeuges, sondern erlaubte es dem Piloten, notwendige Regulierungen, z. B. am Vergaser, vorzunehmen. Allerdings waren die erhobenen Hände zum Teil auch eine demonstrative Geste, den eigenen Schneid zu beweisen.[672]

Die Ein- und Doppeldecker dienten zwischen 1909 und 1911/12 nicht primär als zweckrationale Transportmittel, sondern wurden zum Vergnügen, aus der Lust an Mobilität, zu sportlichen Zwecken und als Mittel genutzt, sich von der Masse abzuheben. Der Sport bot beiden Seiten, den fliegenden Akteuren und den Zuschauern, die Chance, die gesellschaftlich auferlegten Zwänge für die Dauer der Veranstaltung zu überwinden, dem Alltag kurzzeitig zu entfliehen und sich als Teil einer Gemeinschaft zu fühlen.

Die Soziologen Norbert Elias und Eric Dunning haben in diesem Zusammenhang sportliche Veranstaltungen als „quest for excitement"[673] beschrieben, als eine Suche nach Spannung und Vergnügen gleichermaßen. Zugleich stellen Wettkämpfe für sie eine reglementierte Form des Kampfes dar, die bei Sportlern und Publikum eine kathartische Wirkung erzeugen würden.[674] Sport fungiere als Kompensation zur Gefühlsunterdrückung und -kontrolle in der Arbeitswelt. Dunning und Elias sehen in der wachsenden Verbreitung und Popularität des Sports im 19. und 20. Jahrhundert eine Parallelgeschichte der Industrialisierung

670 Ebd.

671 Karl-Heinrich Bette, Sportsoziologie, Bielefeld 2010, S.109.

672 Vgl. Möser, Fahren und Fliegen, S.132.

673 Elias/Dunning, Sport und Spannung; Dunning, Sport Matters.

674 Zur kathartischen Wirkung von Unfällen bemerkt Kurt Möser kritisch, dass die Zuschauer Katastrophen nicht immer als reinigend und befreiend empfunden und sie als „notwendiges Opfer" in der Technikentwicklung begriffen hätten, sondern besonders bei „kleineren" Unfällen im Flug- oder Automobilsport betroffen und schockiert gewesen seien. Ferner widerspricht Möser Bernhard Rieger, der von einer zunehmenden Abhärtung der Zuschauer, einer Gewöhnung an Katastrophen ausgeht. Vgl. Rieger, Technology, S.85; Möser, Fahren und Fliegen, S.450f.

und des Kapitalismus. Nach Elias' Modell der Zivilisationstheorie vollzieht sich im Laufe der Sozialisation des Individuums eine fortschreitende Affekt- und Triebregulierung sowie eine Verinnerlichung des sozialen Zwangs.[675] Der Sport fungiert als Ventil, als ein zumeist sozial kontrolliertes Ausgleichsfeld, in dem Affekte, zumindest innerhalb der (Sport-)Regeln, ausgelebt werden können, sowohl durch den aktiven Sportler als auch durch den passiven Zuschauer. Denn einerseits darf hemmungslos applaudiert, gejubelt und gestarrt werden und eine Identifikation mit den Sportlern stattfinden, andererseits reguliert der aktive Sport die Affekte durch Fairplay und Gemeinschaft.[676]

Die vom Publikum gezeigte und durch die Medien verstärkte Sensationslust war jedoch doppelgesichtig: Während die Leistung der Piloten mehrheitlich[677] lobend hervorgehoben und den deutschen Flugzeugfirmen erstklassige Arbeit bescheinigt wurde, nahm gleichzeitig auch die Berichterstattung über Unfälle und Abstürze viel Raum ein; die Grenze zwischen Schaulust und Voyeurismus war fließend. Hauptmann a.D. Alfred Hildebrandt, der für die Zeitschrift „Sport im Bild" einen mehrseitigen, reich illustrierten Artikel über die Berliner Flugwoche verfasste, schrieb z. B.: „Die deutsche Flugtechnik hat zweifellos einen gewaltigen Sprung gemacht. Die hinter uns liegende Flugwoche war technisch und sportlich die interessanteste, die wir in Deutschland bis jetzt gehabt haben."[678] Gleichzeitig bedauerte er den Todessturz des Militärfliegers Paul Engelhard,[679] der am 29. September ums Leben gekommen war. Engelhard träfe keine Schuld, denn er sei schon „häufig gefahren [= geflogen] bei viel stärkerem und böigerem Wind" und habe seinen Wright-Apparat stets „meisterlich durch alle Fährnisse hindurchgebracht". Hildebrandt würdigte den 43-Jährigen als jemanden, der alle Fliegertugenden besessen habe: „Der ruhige und unerschrockene, edle und vornehme

675 Siehe hierzu die knappe und verständliche Einführung in Elias' Soziologie bei Ralf Baumgart/Volker Eichener, Norbert Elias zur Einführung, Hamburg, bes. S.76–95.

676 Vgl. Elias/Dunning, Sport und Spannung, S.3 sowie Bette, Sportsoziologie, S.103ff.

677 Eine Ausnahme bildete der „Flugsport", der die Ergebnisse der Nationalen Flugwoche für „normal" und damit für durchschnittlich hielt: „Man ist es gewohnt, große Loblieder über die veranstalteten Flugwochen zu hören. Gewiß, wo es die Leistungen gebieten, da soll auch nichts verschwiegen werden. Aber normale Leistungen sind eben keine, die ein großes Lob verdienen. Wenn hier nur die Tatsachen erwähnt werden, so geschieht dies, um in der Berichterstattung ... keine Lücke zu lassen"; „Von der Berliner Flugwoche", in: Flugsport (1911), Nr. 21, S.751f.

678 Hauptmann a. D. Hildebrandt, Die sechste Berliner Flugwoche, in: Sport im Bild (1911) Nr. 40, S.1150–1153, hier S.1150, online seit 2011 unter: http://anno.onb.ac.at/cgi-content/anno-plus?aid=sib&datum=1911&size=45&page =116 4 (12.05.2014).

679 Paul Emil Engelhard (27.7.1868, Münster–29.11.1911, Berlin-Johannisthal) trat 1886 in die Marine ein und wurde Korvettenkapitän. Am 15. März 1910 erwarb er den Flugschein in Johannisthal. Anschließend war er als Fluglehrer bei der Wright-Gesellschaft tätig und nahm an Flugwettbewerben teil. Nach seinem Todessturz 1911 wurde er in Bremen beigesetzt. Im Jahr 1915 wurde eine Straße in Johannisthal und 1936 eine Straße in Münster nach ihm benannt. Informationen zu Engelhard entnommen aus Bruno Lange Luftfahrttechnik Textteil, S.24; Luftschiffahrt, Flugtechnik und -Sport. Illustrierte Zeitschrift für das gesamte Ballon- und Flugwesen, redigiert von Gustav Riefenstahl, Verlag E. Gundlach A.G. Bielefeld, (1911), Nr. 28, S.16.

Mann, dessen Geschicklichkeit und Geistesgegenwart auch ich öfter bewundern konnte, ... wird uns Luftfahrern unvergeßlich bleiben".[680] Der Durchhaltewillen bis zum Schluss, der auch das Ende des Lebens bedeuten konnte, machte Engelhard zu einem Helden des Wagnissports.[681]

Schaulustige versammelten sich beispielsweise sowohl um die Flugzeugwracks Engelhards und Ellery von Gorrissens[682] als auch um unversehrt gelandete Flieger.

Die Zeitungen veröffentlichten in den nächsten Tagen Meldungen und Nachrufe zu Engelhards Tod, in denen sie an einen tüchtigen und erfolgreichen Mann erinnerten. Die Aviatik habe „einen ihrer ältesten und besten Vertreter"[683] verloren, urteilte ein Journalist, und Alfred Hildebrandt zufolge sei Engelhard „[e]iner der kühnsten Flieger" gewesen. Weiter hieß es: „Ein Mann von vornehmer Gesinnung ist mit ihm dahingerafft. Sein Tod bedeutet für alle, die ihn näher kannten, einen Verlust; der deutsche Flugsport hat mit ihm einen seiner Besten verloren."[684] Die „Berliner Volks-Zeitung" informierte ausführlich über die bekannten Hintergründe des Absturzes und über den Gesundheitszustand von Engelhards Passagier und Flugschüler Gerhard Sedlmayr.[685] Hier zeigt sich auch, wie schnell Gerüchte um sich griffen. Hatte es in der Morgenausgabe des gleichen Blattes noch geheißen, der Passagier Sedlmayr sei „lebensgefährlich verletzt", gab man in der Abendausgabe schon wieder Entwarnung.[686]

Über die Berichte und besonders auch über Bilder wurden die Unfälle und Katastrophen verarbeitet und als Sensation angeeignet.[687] Fotoalben, die sich um den Flugalltag in Johannisthal drehen, enthalten immer wieder Aufnahmen von Bruch- oder Baumlandungen sowie von Flugzeugwracks.[688] Bemerkenswert ist

680 Ebd., S.1151.

681 Der Wagnissportler zeichnet sich Martin Stern zufolge aus durch „Wagemut, herausragende Willensstärke, Durchhaltevermögen, asketische Härte gegen sich selbst, Bereitschaft zum totalen Engagement und ... [die] Fähigkeit, absolute Entscheidungen zu treffen, gepaart mit außergewöhnlichen Selbstkontrollkapazitäten." Stern, Heldenfiguren im Wagnissport, S.49.

682 Eine Abbildung des zertrümmerten Flugzeuges von Gorrissens ist abgedruckt bei: Hauptmann a. D. Hildebrandt, Die sechste Berliner Flugwoche, in: Sport im Bild (1911), Nr.40, S.1152.

683 NL Willy Stiasny D 006, „Kapitän Engelhard †. Todessturz beim Johannisthaler Flugmeeting", in: [unbekannter Zeitungstitel] o.D., S.415.

684 Alfred Hildebrandt, Flugzeug-Industrie, in: Jahrbuch der Motorluftschiff-Studiengesellschaft (1912), Bd.5, S.39–52, hier S.44.

685 Vgl. „Der Todessturz in Johannisthal", in: Berliner Volks-Zeitung vom 30.9.1911, Abendausgabe. Engelhard wurde vom Heckmotor erschlagen. Vgl. Kranzhoff, Edmund Rumpler, S.98.

686 Vgl. die Ausgabe der „B.Z. am Mittag" vom 30.9.1911 sowie die „Berliner Volks-Zeitung" vom 30.9.1911, Morgen- und Abendausgabe, sowie 1.10.1911, Morgenausgabe.

687 Im Nachlass Josef Suwelacks sind dutzende Fotos von Unfällen, Not-, Bruch- oder Baumlandungen enthalten, die die Sensationslust ebenfalls belegen. Vgl. hierzu auch Kehrt, Moderne Krieger, S.89; Rieger, Technology, S.51ff. sowie Frank, Zeppelin-Ereignisse, S.27ff. und S.86ff.

688 Siehe z. B. DTMB Fotosammlung VI.2-D019 [ohne Titel 1914–1918], darin einige Aufnahmen von abgestürzten beziehungsweise abgeschossenen Flugzeugen, online unter: http://www.sdtb.findbuch.net/php/view2.php?ar_id=3750&be_id=55&ve_id=99791&count=0 sowie ebd. VI.2-D063

hier die Perspektive der Fotografen zum Unfallobjekt. Die Bilder sind aus nächster Nähe aufgenommen und belegen das Bedürfnis, das Ereignis unmittelbar festzuhalten. Zum *excitement* gehörte sowohl der Erfolg als auch die (tödliche) Niederlage. Doch beschränkte sich die Sensationsaneignung nicht nur auf Fotos. In den Erinnerungen der Augenzeugen gibt es Aussagen, dass ganze Menschengruppen zur Unglücksstelle geeilt seien, um ein persönliches Andenken mitzunehmen. Gerhard Sedlmayr bemerkte später:

> „Es scheint bei Flugplatzkatastrophen eine gewisse Anarchie unter den Zuschauern auszubrechen. ... Uns armen Fliegern wird da nämlich ... gestohlen, was nicht niet- und nagelfest ist. ... Kapitän Engelhardt fehlte seine wertvolle Krawattennadel. Ich meine, das ist doch gemeiner Leichenraub, sonst nichts."[689]

Am 2. Oktober 1911 fand auf dem Flugplatz Johannisthal eine Trauerfeier für Engelhard statt, bei der „die Witwe des Verstorbenen, sämtliche in Berlin weilende Flieger, die Direktion der Flugplatzgesellschaft, die Direktion der Wright-Gesellschaft sowie zahlreiche Armee- und Marineoffiziere"[690] ihm die letzte Ehre erwiesen:

> „Der schwarze Eichensarg war von einer Kriegsflagge bedeckt, vor dem Flughaus aufgebahrt. General [Stephan] von Nieber, der Vorsitzende des Deutschen Luftschifferverbandes, widmete dem Verstorbenen einen warmen Nachruf. Am Sarge legten die Flieger Johannisthals sowie mit dem Flugsport in Beziehung stehende Gesellschaften, Vereine und Fabriken Kränze nieder. Nach der Trauerfeier wurde der Sarg zum Leichenwagen getragen, den zwei Propeller schmückten."[691]

Eine Bestattung mit militärischen Ehren und eine imposante Trauerfeier mit zahlreichen geladenen Gästen wurde aber nicht nur Militärpiloten zuteil, sondern erstreckte sich auch auf Zivilflieger; diese Praxis setzte sich im Ersten Weltkrieg bei der Fliegertruppe fort, sie galt transnational und garantierte den Fliegern noch im Tod Ehre und Würde.[692]

Der betriebene Aufwand stand dabei sowohl in Friedenszeiten als auch während des Krieges zum Teil in keinem Verhältnis zum militärischen Rang. Paul Engelhard war Korvettenkapitän, doch war diese Position der niedrigste Offiziersrang bei den Stabsoffizieren der Marine. Doch gaben die pompösen Totenfeiern für die „Helden der Lüfte" ihrem Sterben einen Sinn und wiesen ihnen einen Platz im kollektiven Gedächtnis der Nation zu.

[Kriegserinnerungen; Luftfahrt 1915–1918] http://www.sdtb.findbuch.net/php/view 2.php?ar_id=3750& be_id=55&ve_id=100000&count=0 (6.7.2014). Vgl. auch NL J. Suwelack, Aufnahmen aus Johannisthal und von Flugwettbewerben, verschiedene Alben.

689 Zit. nach Georg von Tschudi, Aus 34 Jahren Luftfahrt. Persönliche Erinnerungen, Berlin 1928, S.112.

690 Luftschiffahrt, Flugtechnik und -Sport (1911), Nr.16, S.16.

691 Ebd.; vgl. auch den Artikel im NL Willy Stiasny D 006 (1911), „Die Trauerfeier für Kapitän Engelhard. Ueberführung nach Bremen.", in: [unbekannter Zeitungstitel] o.D., S.377.

692 Florian Schnürer, „Nun hat der große Flieger dich geholt ..." – Die Totenfeiern für die „Ritter der Lüfte" als transnationales Medienereignis, in: Kulturwissenschaften exemplarisch. Gießener Forschungsbeiträge zu acht Kernkompetenzen, Trier 2011, S.167–192, bes. S.173f. und S.184ff.

Die Opfermetaphorik des Fliegertodes, die ebenfalls in der Zeit vor dem Ersten Weltkrieg ihren Anfang nahm, ließ die gestürzten und (hinab-)gefallenen Krieger als tapfere Männer erscheinen, die ihr Leben für die Fliegerei bereitwillig hingegeben hatten. Das erfahrene Trauma konnte durch die Totenfeier zum Triumph umgedeutet werden. Die von Bernhard Giessen beschriebene Dialektik von „Triumph und Trauma", zwischen deren Gegenpolen sich nationale Identität konstituiere, lässt sich auch am Opfergedenken der Flieger zeigen.[693] Bei der Trauerfeier Engelhards stand nicht die Katastrophe, das Trauma im Vordergrund, sondern der Triumph über die „dritte Dimension", den Luftraum.[694] Erst das bewusst eingegangene Risiko und „Abenteuer" verlieh den Piloten ihr Heldentum. Der Tod der modernen „Ikariden" konnte, indem er als Opfertod gedeutet wurde, mit Sinn versehen und über den Rückgriff auf den Ikarus-Mythos in eine mythische Traditionslinie gesetzt werden.[695] Noch in der ersten Kriegshälfte griff man auf das klassische Motiv des Ikarus zurück, um beispielsweise den Tod Oswald Boelckes im Jahr 1916 als Heldentod zu werten. Boelcke war nach Manfred von Richthofen der populärste deutsche Jagdflieger.[696]

Als Alfred Pietschker, der Sieger der Nationalen Flugwoche in Berlin-Johannisthal, am 15. November 1911 auf dem gleichen Flugplatz tödlich verunglückte, berichteten die Zeitungen in ähnlicher Weise über ihn. Pietschker blieb als „Sieger der Berliner Flugwoche" in Erinnerung und sein Tod wurde als „schwerer Verlust der deutschen Fliegerwelt" betrauert.[697] Zugleich wehrten sich die Albatros-Werke, bei denen der Pilot tätig war, vehement gegen den Vorwurf, der Apparat sei nachlässig verarbeitet gewesen. In einer Stellungnahme des Unternehmens, eingesandt an die Redaktion des „Flugsports", verwies man auf Pietschkers eigenmächtige Entscheidung – und damit auf das Risiko –, den neuen, nach seinen Wünschen umgestalteten Apparat auszuprobieren. Die Risiko-Argumentation erlaubte es den Albatros-Werken, jede Schuld am Absturz von sich zu weisen.[698]

Aber nicht nur um die Toten entstand ein Kult, sondern bereits zu Lebzeiten galt den Fliegern die Bewunderung des Publikums. Doch die kontaktfreudige Menge, die dem „Lufthelden" persönlich begegnen, ihn anfassen, ein Autogramm oder eine andere „Trophäe" erhalten wollte, beunruhigte die Piloten. Hellmuth

693 Vgl. Aleida Assmann, Der lange Schatten der Vergangenheit. Erinnerungskultur und Geschichtspolitik, München 2006, S.14.

694 Vgl. Helmuth Trischler, Die neue Räumlichkeit des Krieges. Wissenschaft und Technik im Ersten Weltkrieg, in: Berichte zur Wissenschaftsgeschichte 19 (1996), S.95–103.

695 Vgl. Rieger, Technology, S.33f.

696 Vgl. Fritzsche, Nation of Fliers, S.74ff., Abb. S.81, online unter: http://www. simplicissimus.info/index.php?id=6&tx_lombkswjournaldb_pi1%5BissueId%5D=1002&tx_lombkswjournaldb_pi1%5Baction%5D=showIssuePages&tx_lombkswjournaldb_pi1%5Bcontroller%5D=YearRegister&cHash=a6261e9e8fff3569a8ac53c7d7e122f0, Stand April 2012 (7.7.2014).

697 „Todessturz in Johannisthal", in: B.Z. am Mittag vom 15.11.1911.

698 Vgl. Flugsport (1911), Nr. 25, S.908.

Hirth berichtete von 20.000 Menschen, die ihn 1911 nach der Landung während eines Flugwettbewerbs empfangen hätten: „Der Anblick der Menge war unheimlich, ich konnte nur Köpfe sehen und über allem schwebte eine schwarze Staubwolke." Erst der Einsatz einer „200 Mann [starken] Infanterie" habe die Leute beruhigen können.[699] Auch Josef Suwelacks Bruder Richard erinnerte sich, dass er seinen Bruder vor dem Andrang neugieriger Damen habe beschützen müssen.[700]

Wenngleich solche Erinnerungen übertrieben scheinen mögen, gibt es dennoch Belege dafür, dass die Flieger von Menschen umringt waren und bejubelt wurden. Am Boden waren sie volksnah und Sportler zum Anfassen, in der Luft hingegen unnahbare „Bezwinger der Lüfte", einzigartig in ihren Fähigkeiten.[701] Im zunächst sportlichen Wettstreit mit Frankreich, der mehr und mehr nationalistische Züge annahm, galten sie als Beweis für die Überlegenheit der deutschen Technik: „Wenn wir zum Schluß unser Urteil zusammenfassen, so können wir feststellen, daß unsere deutschen Flieger den ausländischen, auch den Franzosen, vollkommen ebenbürtig sind. Ueberlegen [sic!] sind sie diesen sicher im Fliegen bei böigem Wetter."[702] Aus zeitgenössischer Sicht waren die Flieger „Auserwählte",[703] eine nationale Elite, die eine führende Position im internationalen Wettstreit erstritten hatte und die die französische Hegemonie in der Flugzeugtechnik wirksam schwächen würde.[704]

Suwelacks Eltern in Billerbeck hatten die Leistungen ihres Sohnes während des Wettbewerbs „mit großer Spannung tägl[ich] verfolgt" und waren hoch erfreut über seine Platzierung. Der Vater teilte Josef mit, dass man auf seine Berichte „wie versessen" gewesen sei: „Selbst die Kinder auf der Straße rissen sich um die Zeitungen. Also herzlichen Glückwunsch!"[705] Gleichzeitig brachte er seine Bestürzung über den Todessturz Paul Engelhards zum Ausdruck und bat seinen Sohn, „bodenständig" zu bleiben und vor einem Aufstieg zu beten: „Aber große Angst haben wir auch gehabt wegen Deiner Waghalsigkeit. Sei doch vorsichtig. ... Denke doch auch jetzt an Deine religiösen Pflichten. Offenbar hat Dich Gott beschützt. Viele sind, die tägl[ich] für Dich beten."

Neun Tage später folgte ein zweiter Brief, in dem der Molkereidirektor seinem ältesten Sohn erneut einschärfte, Risiken zu meiden: „Nimm Dich doch in Acht und fliege nur bei günstigem Wetter und stets mit der äußersten *Vorsicht*. Du mußt nicht denken[,] den Leuten mal etwas zu zeigen und dabei in Gefahr zu

699 Hirth, Meine Flugerlebnisse, S.102.
700 NL J. Suwelack, Erinnerungen Richard Suwelacks an Josef Suwelack.
701 Vgl. Kehrt, Moderne Krieger, S.82.
702 Sport im Bild (1911), Nr.40, S.1153.
703 Vgl. Rieger, Technology, S.64
704 Vgl. Fuhs, Fliegende Helden, S.712.
705 NL J. Suwelack, Brief Joseph Suwelacks senior an Josef, Billerbeck, 2.10.1911.

kommen."[706] Die elterliche Sorge ist angesichts der reellen (Lebens-)Gefahr des Flugsports verständlich und zudem nicht unbegründet. Der Appell an Josefs Vernunft, Sicherheit einer spektakulären Flugshow vorzuziehen, widersprach jedoch den Publikumserwartungen und dem standesgemäßen Fliegerhabitus. Die Gleichzeitigkeit von Waghalsigkeit und Beherrschung war kein Paradoxon, sondern Voraussetzung, um als Flieger vor der Masse bestehen zu können. Anders formuliert: Hätte Suwelack sich wie Beese geweigert, bei Sturm zu fliegen, hätte er die „symbolische Gewalt", das männliche Ideal von Mut und Stärke, unterlaufen, seinem Flieger-Image geschadet und die konstruierte „natürliche" Ordnung gestört, denn Angst und Unsicherheit hatte es im Fliegerdiskurs nicht zu geben.[707] Ein erfolgreicher Auftritt auf einem Flugwettbewerb war nur möglich, solange nach den im- und expliziten Regeln des Feldes gespielt und Männlichkeit als soziale Praxis im Sport reproduziert wurde.

Aufgeladen mit nationalem Pathos war die Nationale Flugwoche auch eine politische Veranstaltung. Persönlichkeiten aus der Industrie oder der Politik wohnten ihr bei und verliehen ihr durch ihre Anwesenheit zusätzlich Bedeutung. So waren beispielsweise Prinz Sigismund von Preußen sowie hochrangige Militärs in Johannisthal auf dem Flugplatz vor Ort.[708] Aufgrund ihrer Erfolge lud der Prinz Hellmuth Hirth und Josef Suwelack, möglicherweise auch noch weitere Flieger, am 9. September 1911 zu einer Tafel ein.[709] Diese Form der persönlichen Wertschätzung und Auszeichnung sicherte den Piloten zusätzliche Ehre und steigerte ihr Prestige. In der Gunst des Kaiserhauses zu stehen, verhalf zu sozialem Ansehen und Josef Suwelack war ebenso wie andere Piloten darauf bedacht,[710] sich stets mit der gesellschaftlichen Elite zu präsentieren und sich vertraut mit Militärs oder dem Adel zu inszenieren.

Zugleich profitierten beide Seiten von einem erfolgreichen Flug: Das Kaiserhaus konnte sich als großzügiger Förderer und „Freund" der Piloten präsentieren, die geladenen Flugzeugführer wiederum erhielten gewissermaßen die staatliche Legitimation für ihren Beruf. Die Personen, mit denen hier Josef Suwelack zusammenkam, waren adeliger Herkunft oder in gesellschaftlichen und militärischen Führungspositionen aktiv. Rittmeister a.D. von Frankenberg und Ludwigsdorf war beispielsweise gleich in mehreren Ausschüssen des Kaiserlichen Aero-

706 NL J. Suwelack, Brief Joseph Suwelacks senior an Josef, Billerbeck, 11.10.1911. Hervorhebung im Original.

707 Vgl. dazu Krais, Die männliche Herrschaft, S.37ff.

708 Vgl. NL J. Suwelack, „Berliner Flugwoche. Schlechtes Wetter – schöne Flüge", in: [unbekannter Zeitungstitel] vom 30.9.1911.

709 Vgl. „Luftschiffahrt", in: Billerbecker Anzeiger vom 14.10.1911.

710 Vgl. Hirth, Meine Flugerlebnisse, S.121. Dort abgedruckt ist eine Aufnahme vom September 1911, auf der Hirth sich mit dem württembergischen König und der Königin ablichten ließ, im Hintergrund zahlreiche Zuschauer. Siehe auch ebd., S.137.

Clubs[711] vertreten und der flugbegeisterte Prinz Sigismund von Preußen (1891–1927) baute 1911 sogar einen Eindecker.[712] Der Kaiserliche Aero-Club, dem Josef Suwelack als ordentliches Mitglied 1913 beitrat, bot den Fliegern die Möglichkeit, über die Teilnahme am Vereinsleben Kontakte zu politisch und ökonomisch einflussreichen Personen zu knüpfen. Beitreten konnte jedoch nur, wer über Großkapital verfügte oder wer durch ein Vereinsmitglied vorgeschlagen wurde, so dass die Mitgliedschaft exklusiv war.[713]

Die in einem privaten Fotoalbum Josef Suwelacks sorgfältig aufbewahrten und beschrifteten Aufnahmen belegen den Wunsch nach der Dokumentation der eigenen fliegerischen Laufbahn und des sozialen Aufstiegs. Sie spiegeln damit zugleich die Selbstreferentialität Suwelacks und sein Selbstverständnis wider, sich in der Fliegerszene etabliert zu haben, eine Sicht, die durch die Medien auch nach außen gelangte. Erst die Presse und die dort veröffentlichten und dominierenden männlichen[714] (Sieger-)Bilder machten Flieger wie Josef Suwelack zu Volkshelden und entschieden über ihr öffentliches Profil.[715]

Das Sich-Abheben und Abgrenzen von der Masse drückte sich nicht zuletzt auch im Kleidungsstil und in der Gestik der Akteure aus.[716] In den Worten Georg Simmels:

711 Der Kaiserliche Aero-Club, gegründet im Jahr 1907, stand unter dem Protektorat des Kaisers und war eine exklusive Organisation, in der sich Großbürgerliche und Vertreter der preußischen-deutschen Hocharistokratie zusammenschlossen, um nationale und internationale Flugsportveranstaltungen zu fördern und den deutschen Flugsport im In- und Ausland zu repräsentieren. Der Kaiserliche Aero-Club übernahm verstärkt eine soziale Vermittlungsfunktion zwischen den Flugsportlern, der Luftfahrtindustrie und der Politik. Eine Ausschuss- und Mitgliederliste ist abgedruckt in: Jahrbuch des DLV (1910), S.601ff.

712 Meldungen über den Eindecker enthalten in: Flugsport (1911), Nr. 2, S.59; (1912), Nr. 12, S.479 und (1912), Nr. 17, S.646, sowie in: Der Motorwagen XVI (1913), H. IX, Einlage nach S.216. Zu den Mitgliedern des Kaiserlichen Aero-Clubs siehe Jahrbuch des DLV 1910, S.601ff.

713 NL J. Suwelack, Schreiben des Kaiserlichen Aero-Clubs an Josef Suwelack mit angefügtem Aufnahmeformular vom 22.8.1913.

714 Christine Walther, die sich in ihrer Studie mit der Darstellung von „Siegertypen" in der Sportpresse beschäftigt hat, beziffert den Anteil von weiblichen Siegerinnen auf rund 10 Prozent für den Zeitraum zwischen 1900 und 1920. Ihre Untersuchung stützt die These einer Männerdomäne des Sports. Vgl. Christine Walther, Siegertypen. Zur fotografischen Vermittlung eines gesellschaftlichen Selbstbildes um 1900, Würzburg 2007, S.116ff.

715 Das gleiche Bild befindet sich auch in einem Fotoalbum Suwelacks mit der Beschriftung „In der Siegermaschine Hirths vom Oberrheinflug und dem Flug München – Berlin, die ich bei der Herbstflugwoche benutzte." Diese Aussage zeigt, dass Suwelack sich in eine Reihe mit Hirth stellte, an Hirths Leistungen anknüpfen wollte und auch konnte. Bezeichnend ist, dass auch auf anderen Fotos stets nur die Flugschüler mit Namen gekennzeichnet werden, nicht aber die Monteure.

716 Eric Hobsbawm betont die politische und soziale Rolle des Sports einerseits für die Formierung einer sozialen Elite, die die älteren aristokratischen Schichten abzulösen bestrebt war, andererseits als Abgrenzung nach unten gegen die Massen Er spricht von einem „specific new bourgeois pattern of leisure-activity and a life-style". Hobsbawm, Mass Producing Tradtitions, S.300.

> „Wo die quantitative Steigerung von Bedeutung und Energie an ihre Grenze kommen, greift man zu qualitativer Besonderung, um so, durch Erregung der Unterschiedsempfindlichkeit, das Bewußtsein des sozialen Kreises irgendwie für sich zu gewinnen."[717]

Suwelack sticht auf Gruppenfotos regelmäßig hervor. Entweder hat er die Arme in die Hüften gestemmt, während alle anderen eher steif wirken, hat die Beine überkreuzt, wo die anderen parallel stehen, oder trägt legere Kleidung, während der Rest der Truppe Uniform trägt.[718] Ebenso war das Zigarettenrauchen charakteristisch für viele Flieger, es findet sich auch auf mehreren Bildern mit Josef Suwelack. Rauchen gehörte zum standesgemäßen Fliegerdasein und wurde von den Piloten in ihren Selbstdarstellungen des Öfteren thematisiert. In Suwelacks Notizen zu seinem 1912 gehaltenen Vortrag findet sich unter dem Abschnitt „Episoden" auch der Punkt „Rauchen" und Hellmuth Hirth schrieb in seiner Autobiografie: „Da ich Zigarettenraucher bin, habe ich mein Punkfeuerzeug stets bei mir .... Bei den vielen Stundenflügen, die der Abnahme eines Militärflugzeuges voraufgehen [sic!], habe ich stets, um mir die Zeit zu vertreiben, Zigaretten geraucht."[719] Bei seiner missglückten Kanalüberquerung zündete sich Hubert Latham 1909 auf seinem Flugzeug im Moment des Scheiterns eine Zigarette an und ließ sich dabei fotografieren. Nur wenige journalistische Berichte verzichteten auf einen Kommentar zu dieser Geste, denn Latham hatte im Augenblick höchster Gefahr Haltung bewahrt und seine Extravaganz demonstriert.[720]

Die Flugwettbewerbe boten den Beteiligten die Aussicht auf soziales Kapital, gleichzeitig stellten die Konkurrenzen aber auch eine Geldquelle für Firmen und Piloten dar: Sie waren häufig „die einzige Möglichkeit, mit dem Flugzeug Geld zu verdienen", denn ein privater Absatzmarkt für Flugzeuge existierte nur in sehr geringem Umfang.[721] „Zeit" stand synonym für „Geld", denn die Preisgelder verteilten sich auf die längsten Flugzeiten. Insbesondere in den Anfangsjahren der Fliegerei kam es darauf an, die Maschinen möglichst lange ohne Bodenkontakt in der Luft zu halten und ihre Zuverlässigkeit unter Beweis zu stellen. Das gewonnene Geld investierten die Konstrukteure in den Bau neuer Flugzeuge oder in Verbesserungen an ihren Maschinen, so dass der Gewinn zugleich einen Teil des Betriebskapitals bildete. Gemessen am Gesamtpreisgeld entfielen allerdings jeweils nur relativ kleine Beträge auf die einzelnen Piloten, die jedes Mal das Risiko von Beschädigungen am Flugzeug einkalkulieren mussten. Und während angestellte Werkspiloten auf Kosten der jeweiligen Flugzeugfirma starteten, finan-

717 Georg Simmel, Die Großstädte und das Geistesleben, in: Klaus Lichtblau (Hg.), Georg Simmel. Soziologische Ästhetik, Darmstadt 1998, S.103–114, hier S.112.

718 Siehe einige Fotos im Feld 4.

719 Hirth, Meine Flugerlebnisse, S.229.

720 Vgl. Möser, Fahren und Fliegen, S.140.

721 Karl-Dieter Seifert, Hans Grade. Ingenieur – Flugpionier – Automobilbauer. Ein Leben in stürmischen Zeiten, Magdeburg 2008, S.238.

zierten Privatflieger ihren Start komplett aus eigenen Mitteln. Zwar trug der Veranstalter anteilig die Kosten, doch bargen die Flugveranstaltungen damit insgesamt viele Risiken. Josef Suwelack erhielt von den gewonnenen 7.867 Mark aus der Nationalen Flugwoche einen Anteil von 33 Prozent, doch musste er davon wiederum einen 25-prozentigen Anteil auf ein Kautionskonto einzahlen. Letztlich verblieben ihm noch immer rund 2.000 Mark.

Dass die Problematik – hohes Risiko, geringer Verdienst – auch in den folgenden Jahren bestehen blieb, wird besonders am Prinz-Heinrich-Flug im Jahr 1913 ersichtlich. Dieser Wettbewerb bevorzugte Militärflieger, die zivile Flugzeugindustrie hingegen hatte das Nachsehen.[722]

3.2. Flieger „auf Tour" – regionale Flugbegeisterung am Beispiel der Flugtage in Detmold, Münster und Billerbeck

Wenn im Folgenden die Konstruktion regionaler Identität am Beispiel von Flugveranstaltungen analysiert wird, so geschieht dies auf der lokalen Ebene. Im Zentrum stehen die Fragen danach, wie über diese Ereignisse und die Flugsportler berichtet wurde, welche Narrationsmuster dominierten und welche Bedeutung die Fliegerei und ihre Akteure für das Selbst- und Fremdbild der Region hatten. Ebenso soll untersucht werden, inwiefern diese Faktoren auch Einfluss auf das Bild von Josef Suwelack genommen haben.

Detmold: „Da hat Detmold wieder einen großen Tag gehabt"[723] – Flugzeuge und ihre Lenker als ‚gegenwärtige Zukunft'

Der Detmolder Flugtag fand am 15. Oktober 1911 statt und sorgte für Begeisterung unter den Zuschauern. So berichtete der für die „Lippische Landeszeitung" schreibende Professor August Weweler:[724] „Für die allermeisten ist es Sensation, für viele ein interessantes Schauspiel, für einige ein Objekt des Studiums. Für alle aber ein Nervenkitzel".[725] Der Musiker und Komponist zeigte sich fasziniert von

722 Siehe den Abschnitt zum Prinz-Heinrich-Flug.

723 „Der Detmolder Flugtag", in: Lippische Landeszeitung vom 16.10.1911 (Beilagenblatt).

724 August Weweler (1868–1952) war ein aus Westfalen stammender Musiker und Komponist, der 1898 Leiter des Sommertheaters in Detmold wurde. Zwischen 1908 und 1920 war Weweler Leiter des Detmolder Oratorienvereins und 1916 Mitbegründer des Lippischen Landeskonservatoriums, das bis 1945 bestand. Seit 1932 war er Mitglied der NSDAP. Weweler unterrichtete von 1935 bis 1940 an der Essener Folkwangschule Musik und kehrte 1943 nach Detmold zurück, an den Erfolg der Vorkriegszeit konnte er jedoch nicht mehr anknüpfen. Neun Jahre später starb er. Für ausführliche Informationen zu Weweler siehe Lukas Speckmann, August Weweler, „Zurück zu Mozart!" Ein konservativer Komponist des 20. Jahrhunderts und sein ästhetisches Konzept, Münster 1997.

725 Prof. August Weweler, Der Detmolder Flugtag, in: Lippische Landeszeitung vom 16.10.1911.

der Flugtechnik und dem Teilnehmerfeld, besonders auch von Melli Beese, der einzigen weiblichen Teilnehmerin.

Um die Zuschauer über die Teilnehmer des Wettbewerbs und die vorgeführte Techik genauer in Kenntnis zu setzen, war zuvor ein begleitendes Programmheft[726] erstellt worden. Es umfasste biografische Hintergrundinformationen zu den drei Teilnehmern Gustav Witte, Hans Vollmöller – an dessen Stelle erneut Josef Suwelack flog – und Melli Beese, ferner zur Geschichte und technischen Entwicklung der Rumpler-Taube. Solche Programmhefte richteten sich gleichermaßen an Experten und Laien, die darüber detaillierte Auskünfte zu den Flieger-Lebensläufen erhielten und (Fach-)Wissen über die Maschinen und ihre Lenker erwarben. Auch die Lokalzeitungen erkannten zunehmend das Interesse des Bürgertums an der Fliegerei und versorgten ihre Leser mit aktuellen biografischen und technischen Hintergründen. Auf diese Weise „diffundierte langsam die spezifische Wissensstruktur der wettbewerbsorientierten Mobilitätskultur in die Mainstream-Kultur“[727] und es bestand für jeden die Möglichkeit, sich zeitnah über aktuelle Entwicklungen in der Fliegerei zu informieren.

Der Flugtag sollte „Tausenden von Zuschauern Gelegenheit geben, sich selbst ein Urteil darüber zu bilden, ob der Ruf, den die ‚Rumpler-Taube' genießt, gerechtfertigt“ sei, so ließ das Programmheft wissen.[728] Am Ende des Heftes folgten Bitten an das Publikum, geduldig zu sein, Verständnis für den Ausfall von Flügen aufgrund schlechter Wetterbedingungen aufzubringen sowie „unter allen Umständen die Absperrung zu respektieren und keinesfalls das Flugfeld zu betreten.“[729] Ausdrücklich wies man darauf hin, dass der Veranstalter keine Haftung für Schäden bei Zuwiderhandlung übernehme.

Es regnete und gegen Abend zog Nebel auf, so dass, wie ein Zuschauer in einem Leserbrief an die „Lippische Landeszeitung“ mitteilte, „[u]nsere Geduld auf eine harte Probe“[730] gestellt worden sei. Die Zuschrift an die Zeitung zeugt dennoch von der Flugbegeisterung des Zuschauers und vom Stolz der Bewohner, in Detmold bei einem Flugtag dabei gewesen zu sein. Der folgende Auszug belegt die Massenwirksamkeit und Anziehungskraft der Fliegerei, wenngleich sich die genannten Zuschauerzahlen nicht verifizieren lassen:

> „Und als ich dann am Sonntag zur festgesetzten Stunde meine Schritte zum Flugplatz lenkte, da überkam mich doch ein Gefühl des Stolzes, als ich in der Lageschen Straße ganze Trupps von Männlein und Weiblein in Bewegung sah, die mit mir dasselbe Ziel hatten. ... [U]nd oben auf der

726 Das Programmheft, bereitgestellt durch das Landesarchiv NRW auf der Seite des Heimatvereins Jerxen-Orbke, online unter: http://www.heimatverein-jerxen-orbke.de/Flugtag.htm (5.7.2014).

727 Möser Fahren und Fliegen, S.110.

728 Programmheft „Detmolder Flugtag: Sonntag 15. Oktober 1911, auf der Rennbahn (Jerxerheide)“, S.12.

729 Ebd., S.14f.

730 Leserbrief eines Zuschauers [Name unbekannt]. „Der Detmolder Flugtag“, in: Lippische Landeszeitung vom 16.10.1911, Beilage.

> Klüter Chaussee und der Jerxer Heide auf allen Seiten des gewaltigen, mit einer niedrigen Umzäunung umgebenen Startplatzes, da standen die Menschen wie die Mauern, Tausende und aber Tausende [sic!]. Wirklich, wenn auch nur die Hälfte ihr Eintrittsgeld bezahlt hat, dann ist die Stadt glänzend auf ihre Kosten gekommen. Stolz kann Detmold auf alle Fälle sein."[731]

Interessanterweise hatte der Zuschauer nicht registriert, dass nicht wie ursprünglich geplant Vollmöller, sondern Suwelack flog. Dies verweist auf ein allgemeines Phänomen der Flugschauen: Hielten sich mehrere Flieger gleichzeitig in der Luft auf, steigerte es zwar einerseits die Spannung, doch konnten anderseits die Flugzeuge mit ihren Piloten, vor allem wenn es sich um den gleichen Maschinen-Typ handelte, mitunter nur schwer auseinander gehalten werden. Möglichweise hatte der Zuschauer auch die Informationen über den Pilotentausch nicht erhalten, und so schrieb er:

> „Lange dauerte es dann auch nicht, dann fliegt Vollmöller mit seiner Rumpler-Taube und so fahren sie [Vollmöller, d.h. Suwelack, und Witte] beide in den Lüften umher, bald ganz niedrig, so daß endloser Jubel ertönt, wenn sie wieder über unseren Häuptern erscheinen, bald so hoch, daß sie kaum mehr sichtbar sind."[732]

Der Stolz des Verfassers, ein Detmolder zu sein und in seiner Heimat im Kollektiv die moderne Technik mit eigenen Augen zu sehen, bringt die Verbundenheit zur eigenen Stadt zum Ausdruck, ein Loyalitätsbekenntnis, wie es sich auch für andere Städte beziehungsweise Regionen als größere Einheiten nachweisen lässt.

Deutlich wird auch, dass der Beobachter überzeugt war von der „große[n] Sicherheit" der Maschinen und der raschen Entwicklung der Flugtechnik, die nicht mehr in den „Kinderschuhen" stecke. Der Augenzeuge war zuversichtlich, dass der „Flugtechnik die Zukunft" gehöre und noch vorhandene Schwächen behoben würden, nachdem die Fliegerei schwerer als Luft zur Tatsache geworden sei.[733]

August Weweler, der Suwelacks (alias Vollmöllers) Flug ebenfalls verfolgt hatte, lobte Josefs „Mut, seine Sicherheit, seine Ruhe", die er bei der Witterung gezeigt habe: „Wundervoll war seine Landung, bei der er eine scharfe Drehung des Apparates zeitigte, die ein enormes Geschick voraussetzte." Gleiches gelte für Gustav Witte, das „Wunderkind" der Fliegerszene.[734] Fliegerisches Geschick und Gefühl, Ausgeglichenheit, Entschlossenheit und Mut, das waren wesentliche Bestandteile des zeitgenössischen Fliegernarrativs, das in ähnlicher Form stetig reproduziert wurde, so auch hier, und nationale Bedeutung erlangte. Piloten stellten eine scheinbar geradezu ideale Verkörperung von Männlichkeit dar.[735]

731 Ebd.
732 Ebd.
733 Ebd.
734 Prof. August Weweler, Der Detmolder Flugtag, in: Lippische Landeszeitung vom 16.10.1911, Zitate ebd.
735 Zu ähnlichen Befunden kamen auch Möser, Fahren und Fliegen; Kehrt, Moderne Krieger, S.33ff.; Haude, Grenzflüge, S.293ff.

Allerdings greife Alf Lüdtke zufolge das „Klischee ... einer eindeutig-polaren geschlechtsspezifischen Unterscheidung der Adressaten wie der Nutzer modernster Technik“[736] zu kurz, eine These, die in der Forschung kontrovers diskutiert wird. Denn während sich Frauen wie Melli Beese vor allem über die Kleidung zwar bewusst männlich inszeniert hätten,[737] habe dieses Verhalten nicht zur „Überwindung der Geschlechterdifferenz“ geführt.[738] Nichtsdestotrotz deuten die Berichterstattung und die visuelle Präsentation der Fliegerinnen darauf hin, dass sie sich dem Flieger-Habitus anpassten, um sich im Feld der Männer zu behaupten. Im Unterschied zu ihren männlichen Kollegen durften Frauen jedoch Schwäche zeigen und sich weigern, bei schlechtem Wetter zu starten, weil Schwäche und Angst im zeitgenössischen Diskurs nach wie vor als „typisch weiblich“ galten. Die Teilnahme von Fliegerinnen blieb so eine beliebte „Attraktion“ auf Flugtagen.

Flugtage versetzten Groß- und Kleinstädte immer wieder in Aufregung. Städte wie Detmold und, wie noch zu zeigen sein wird, auch die Provinz Westfalen versprachen sich von den Veranstaltungen hohe Besucherzahlen und damit Einnahmen für die (Stadt-)Kassen. Der organisatorische Aufwand für wenige, dafür aber landesweit bekannte Flieger, der sich anschaulich am Billerbecker Beispiel verdeutlichen lässt, unterstreicht die Bedeutung der Fliegerei in der Gesellschaft. Die veranstaltenden Regionen zeigten sich technikbegeistert und offen für eine „bessere“ Zukunft. Die Reaktion des Zuschauers, der in seinem Leserbrief in der „Lippischen Landeszeitung“ seinen Stolz zum Ausdruck brachte, Teil dieses Spektakels „seiner“ Stadt zu sein, macht die Wirkung der Fliegerei auf den „Durchschnittsbürger“ deutlich. Die mediale Vorbereitung auf und die Berichterstattung über einen Flugwettbewerb mobilisierte die örtliche Bevölkerung und erhöhte langfristig ihre Technikakzeptanz. Das „moderne Wunder“[739] der motorisierten Fliegerei war eines der bestimmenden Themen in den Lokalzeitungen, in denen sich das Gefühl, in einer Zeit des Auf- und Umbruchs zu leben, des scheinbar unaufhaltbaren Fortschritts,[740] auf der Mikroebene im Leserbrief widerspiegelte:

> „Da haben wir hier also gesehen die Flugmaschinen, die man noch vor einem halben Menschenalter für eine Unmöglichkeit hielt und an die wir selbst noch nicht glauben wollten, als uns die

736 Alf Lüdtke, Ikonen des Fortschritts. Eine Skizze zu Bild-Symbolen und politischen Orientierungen in den 1920er und 1930er Jahren in Deutschland, in: ders./Inge Marßolek/Adelheid von Saldern (Hg.), Amerikanisierung. Traum und Alptraum im Deutschland des 20. Jahrhunderts, Stuttgart 1996, S.199–210, hier S.206.

737 Vgl. Gundula Wolter, Sensationelle Sportkostüme. Die Kleidung der Flugpionierinnen, in: Wolfgang Meighörner (Hg.), Frau und Flug. Die Schwestern des Ikarus, Marburg 2004, S.211–237, hier S.233.

738 Maria Osiezki, „Der Flug ist das Leben wert“. Entgrenzung weiblicher Lebensentwürfe in der Luftfahrt des 20. Jahrhunderts, in: Meighörner, Frau und Flug, S.8–25, hier S.23.

739 Dazu Rieger, Technology, S.21ff.

740 Siehe zum Fortschrittsnarrativ Angela Schwarz, Der Schlüssel zur modernen Welt. Wissenschaftspopularisierung in Großbritannien und Deutschland im Übergang zur Moderne (ca. 1870–1914), Stuttgart 1999, S.164–172 sowie Loewenstein, Der Fortschrittsglaube, bes. S.339–344 und S.356f.

> Kunde von ihnen vor 8 Jahren aus Paris kam, wo sie den Eiffelturm umflogen hatten. ... Nach solchen Leistungen [auf der Flugschau in Detmold], bei denen der Flieger sein Fahrzeug wohl eine halbe Stunde und länger vollständig beherrscht hat, ... gehört der Flugtechnik die Zukunft."[741]

Zeppeline und Flugzeuge waren ein seltener Anblick in den Städten und es überrascht daher nicht, dass zahlreiche Menschen Zeuge des Besonderen sein und mit eigenen Augen den „realen Mythos" sehen wollten. Dass sich Flugveranstaltungen nicht nur auf Großstädte beschränkten, sondern auch in agrarisch geprägten Kleinstädten großen Zuspruch erfuhren, kann am Beispiel der Flugtage in Münster und Billerbeck, mit dem Schwerpunkt auf der Heimatstadt Suwelacks, exemplarisch aufgezeigt werden.

Münster: Westfälisches Selbstbewusstsein

Das Münstersche Schaufliegen auf der Rennbahn des Westfälischen Reitervereins sollte an einem Sonntag, dem 22. Oktober 1911, stattfinden. Hierbei handelte es sich nicht, wie beim Deutschen Rundflug vom Juni 1911, um einen groß angelegten Wettbewerb – Münster hatte ein Etappenziel gebildet –, sondern um ein Ereignis, das vom örtlichen flugtechnischen Ausschuss[742] initiiert worden war und nachhaltigen Eindruck auf die Umgebung hinterließ. Der Kartenvorverkauf löste einen solchen Andrang aus, dass die Plätze knapp wurden und die Sitzplatzeinteilung am Vortag der Flüge kurzfristig noch geändert werden musste.[743] Erneut wurden die Leser in den Zeitungen mit Hintergrundinformation über die Flieger versorgt und es wurde angekündigt, dass Josef Suwelack versuchen werde, einen Höhenrekord aufzustellen.[744]

Aufgrund eines „starken Sturmes"[745] musste das Schaufliegen jedoch abgesagt werden. Entgegen der Ankündigung auf dem Flugplakat, es werde „bei jeder Witterung"[746] geflogen, wollte man offenbar kein zu hohes Risiko eingehen. Der Ausfall der Veranstaltung war allerdings erst kurzfristig am Sonntag bekanntgegeben worden, so dass die aus der näheren Umgebung angereisten Zuschauer, darunter auch zahlreiche Billerbecker, leer ausgingen.[747] Der „Westfälische Merkur" berichtete, dass sich am „Sonntag ... halb, wenn nicht gar ganz Billerbeck auf den

741 Leserbrief eines Zuschauers [Name unbekannt], „Der Detmolder Flugtag", in: Lippische Landeszeitung vom 16.10.1911, Beilage.

742 Dieser Ausschluss war am 1. Oktober 1910 vom Verein für Münster und das Münsterland E.V. gegründet worden, um die Motorfliegerei aktiv zu fördern. Vgl. Klaus Prange, Tollkühn am Himmel über Münster, in: Westfälischer Heimatbund (Hg.), Jahrbuch Westfalen, Münster 2000, S.26–44, hier S.34.

743 Vgl. Münsterischer Anzeiger vom 21.10.1911, 2. Ausgabe.

744 Vgl. Münsterischer Anzeiger vom 20.10.1911, 1. Ausgabe.

745 Westfälischer Merkur vom 25.10.1911.

746 Vgl. Billerbecker Anzeiger vom 29.10.1911.

747 Vgl. Westfälischer Merkur vom 23.10.1911, Mittagsausgabe.

Weg nach Münster gemacht“ habe, „um seinen Suwelack als kühnen Flieger anzustaunen und zu bewundern.“[748] Da die Aachener Wetterwarte für die folgende Woche bessere Wetterbedingungen prognostiziert hatte, verschob der Luftschiffverein Münster die Flugveranstaltung auf Dienstag, den 24. Oktober, und teilte der wartenden Menge mit, dass das Spektakel „Nachmittag 4 Uhr“[749] stattfinden werde.

Doch auch am Dienstag war keine Besserung des Wetters eingetreten, woraufhin die geplanten Passagierflüge abgesagt wurden und zwei der vier angekündigten Flieger sich weigerten, zu starten.[750] Gleichwohl war der Zuschauerstrom sowie das Interesse der Bevölkerung am Flugtag ungebrochen und damit vor allem die Einnahmen des Verkehrsbetriebes hoch, wie ein Zeitungsbericht verdeutlicht:

> „Alle möglichen Vehikel, vom feinsten Auto bis zum Metzgerfuhrwerk, vom Kinderwagen bis zum Fahrrad, waren in den Dienst der Flugsache gestellt. Die ‚Elektrische' mit zwei oder drei Anhängewagen kroch schwer beladen durch die Straßen zum Schützenhof. Sie wird wieder ein gutes Geschäft gemacht haben, hatte sie doch, wie uns erzählt wird, am Sonntag die höchste Tageseinnahme seit ihrem Bestehen zu verzeichnen.“[751]

Die Flüge Gustav Wittes und Josef Suwelacks wurden mit Spannung erwartet und vor allem Suwelacks Überlandflug von Münster ins rund fünfundzwanzig Kilometer westlich gelegene Billerbeck hinterließ damals und auch noch heute einen bleibenden Eindruck. Die westfälischen Zeitungen berichteten ausführlich über diesen „Aus-Flug“ Suwelacks; der „Münsterische Anzeiger“ oder auch der „Westfälische Merkur“ brachten detailreiche Artikel über das Ereignis, schilderten die Reaktionen des Publikums und die in Freude umschlagende Sorge von Josefs Eltern, die ihren Sohn in Münster erstmals fliegen sahen. Im „Westfälischen Merkur“ konnte man am 25. Oktober lesen:

> „Inzwischen zog die Taube immer weiter ihre Riesenkreise in den Lüften, immer kleiner und kleiner ward sie, bis sie kurz nach 5 Uhr am Horizont in der Richtung nach Mecklenbeck verschwand. Einige bange Augenblicke vergingen, bis man des Rätsels Lösung fand: Suwelack stattete seiner lieben Vaterstadt eine Luftvisite ab! Dieser Versuch überraschte in Billerbeck allgemein.“[752]

Bei seiner Rückkehr sei er von einer jubelnden Zuschauermenge empfangen und von „seinen Eltern und Geschwistern, die ihren Sohn und Bruder zum ersten Male in der Luft sahen“, herzlich gegrüßt worden. Mit freudig „leuchtenden Augen“ sei er gelandet und von den „übrigen Fliegern und den Herren des Luftschiffahrtvereins“ gefeiert worden.

748 Westfälischer Merkur vom 25.10.1911.
749 Westfälischer Merkur vom 23.10.1911, Mittagsausgabe.
750 Vgl. Münsterischer Anzeiger vom 25.10.1911, 1. Ausgabe.
751 Westfälischer Merkur vom 25.10.1911.
752 Münsterischer Anzeiger vom 25.10.1911, 1. Ausgabe. Weitere Zitate ebd.

Suwelack selbst gab in einem anschließenden Interview seine Eindrücke von seinem ersten Überlandflug wie folgt wieder:

> „Ich hatte zwar Öl und Benzin für einen Stundenflug füllen lassen, wußte aber nicht, als ich aufstieg, ob ich einen Flug nach Billerbeck werde wagen können, da ich diesen Motor erst zum dritten Male benutzte und nicht wußte, welche Windverhältnisse ich oben antraf. Als ich aber merkte, daß – mein Motor so tadellos arbeitete, da steuerte ich kurz entschlossen auf Billerbeck zu."

Zweimal sei er von starken Windböen erfasst worden und habe nur im letzten Augenblick nahende Hindernisse überfliegen können: „Dann war ich über der Heimat. Ein stolzes Gefühl beseelte mich, als ich dicht an den Türmen [des Ludgerus-Doms] vorbei und über dem Hause meiner Eltern und dem Bahnhof dahinfuhr und den auf den Straßen versammelten Landsleuten zuwinken konnte."

Das Überraschungsmoment war auf Suwelacks Seite und verlieh seinem Flug in seine Heimatstadt zusätzlich Bedeutung. Das anschließende Interview ließ die Zuschauer nachträglich am Erlebnis teilhaben und erlaubte es Josef, die Berichterstattung über sich und seinen Flug gezielt zu lenken. Die Deutungshoheit über solche Ereignisse lag damit hier wie auch sonst oftmals nicht allein bei der Presse, sondern die Piloten nahmen durch ihre Leistungen, Statements und ihr (körperliches) Auftreten Einfluss auf ihr öffentliches Profil.

Die Karriere des Fliegers Suwelack konnte man im „Münsterischen Anzeiger" nachlesen, der an die ersten heimatlichen Flugversuche erinnerte und einen „Schwur" Suwelacks zitierte, den er gegenüber seinen Landsleuten nach seinen erfolglosen anfänglichen Flugversuchen geäußert haben soll: „Ik krieg et doch noch fäddig, ik kiek de Billerbeksen doch noch upp'n Kopp und fleig iähr auk no üm de Ludgerus-Täön." Jetzt habe er sein Wort wahr gemacht und beide Zeitungen gratulierten ihm zu seinem Erfolg.

Weiter berichteten die Zeitungen, das Publikum sei durch die Flüge Wittes und Suwelacks, „die Helden des Detmolder Flugtags", für ihr langes Warten „reichlich entschädigt" worden,[753] und auch die Billerbecker, die laut „Westfälischem Merkur" am 22. Oktober noch verärgert über die „Suvelackierung" gewesen seien, hätten sich am Dienstag „vollauf zufrieden" über seinen Flug geäußert.[754] Der Flug in die Heimat hatte Suwelack den Titel „Sturmflieger"[755] eingebracht, außerdem wurde er zusammen mit Gustav Witte in einer Vereinssitzung „in engerem Kreise im Hotel Moormann" geehrt. Die beiden Flieger hätten „durch ihre Entschlossenheit die Ehre des Tages gerettet" und wären von Landrat Fels[756] und dem Direktor der Rumpler-Werke, Herrn Zeyssig, mit anerkennenden

753 Vgl. Münsterischer Anzeiger und Westfälischer Merkur vom 25.10.1911, Zitate im Münsterischen Anzeiger, ebd.

754 Beide Zitate ebd..

755 Billerbecker Anzeiger vom 6.4.1912.

756 Fels war neben seiner politischen Tätigkeit auch im Münsterischen Luftfahrtverein als Beisitzer

Worten bedacht worden.[757] Im Gegenzug bedankte sich Suwelack beim Veranstalter für dessen Engagement in der Fliegerei und versprach, auch künftig dem Münsterischen Verein ein treues Mitglied zu sein und das verliehene Abzeichen zu tragen.[758]

Auch der „Billerbecker Anzeiger" ging in seiner Begeisterung noch weiter und historisierte den Flug des Piloten über Billerbeck. Mit seiner Aktion sei Suwelack dem Wunsch der Zeitung nachgekommen, sich am Zeppelin-Luftschiff „Schwaben", das am 19. Oktober 1911 über Billerbeck geflogen war, ein Beispiel zu nehmen und „seine Landsleute ebenfalls mit einem Besuch mit der ‚Taube' [zu] beehren." Nun sei „der Wunsch zur Tatsache" geworden und man wolle sich beeilen, „dieses Ereignis der Geschichte der Stadt Billerbeck einzuverleiben".[759] Die Nachricht von Suwelacks Flug-Besuch habe sich wie ein „Lauffeuer ... von Haus zu Haus, von Straße zu Straße [verbreitet], alles eilte auf die Straßen, um den verehrten Mitbürger fliegen zu sehen." Josef habe „die Herzen aller Billerbecker erobert" und keiner werde je „diesen denkwürdigen Tag, den 24. Oktober, vergessen":

> „Eine Generation wirds [sic!] der andern übermitteln, dass ein Sohn Billerbecks seine Vaterstadt überflogen und in stolzer Fahrt die hohen Türme von St. Ludger umkreiste! Unauslöschlich werden diese Worte in die Geschichte Billerbecks verzeichnet stehen allen nachkommenden Geschlechtern diese kühne Tat eines jungen Menschen, der sich dem Flieger-Beruf gewidmet hat [,] ins Gedächtnis rufend."[760]

Den Flugtag hatte sich die Stadt Münster 10.000 Mark kosten lassen, doch sei die Summe nach Zeitungsangaben trotz der vielen zahlenden Besucher letztlich „kaum zur Hälfte gedeckt" gewesen. Einen Eindruck von der Zusammensetzung der Kosten vermittelt der „Westfälische Merkur". Ihm zufolge zahlte man für „zwei der Apparate für ihre Anwesenheit in Münster je 1100 und zwei andere 600 bzw. 700 M." als Ausstellungsgeld. Zudem würden bei einer zwanzigminütigen Flugzeit weitere 700 Mark gezahlt.[761]

Nicht nur der „Billerbecker Anzeiger" (ver)ehrte Suwelack, der als „Billerbecker Flieger" bis heute (!) im regionalen Gedächtnis verankert blieb, sondern seine

aktiv. Vgl. Jahrbuch des DLV 1913, S.184. Weitere Mitglieder waren Bürgermeister Dieckmann aus Münster, Militärs, Fabrikanten, Kaufleute und Akademiker, ebd.

757 Münsterischer Anzeiger vom 25.10.1911.

758 Vgl. ebd.

759 Billerbecker Anzeiger vom 29.10.1911. Den vollständigen Bericht aus dem „Billerbecker Anzeiger" sowie die Meldung über die Ehrung im Hotel Moormann im „Münsterischen Anzeiger" liegen im Nachlass Suwelacks als Maschinenschrift neben dem Artikel selbst auch auf Briefbögen der Kondor-Werke vor. Da Suwelack sich die Mühe machte, den Bericht abzuschreiben, und auch sonst Meldungen über seiner Flieger-Aktivitäten sammelte, ist davon auszugehen, dass er auf diese Dokumente für Vorträge etc. zurückgriff und möglicherweise plante, seine Fliegererlebnisse zu veröffentlichen. Auch angesichts späterer Aussagen, Fotos für ein Album aufzubewahren, und weiterer Berichte im „Flugsport" oder im „Essenener Generalanzeiger" aus der Feder Josef Suwelacks mutet es wahrscheinlich an, dass er sein Flieger-Leben partiell öffentlich machen wollte.

760 Ebd.

761 Westfälischer Merkur vom 25.10.1911.

Leistung wurde sogar zum „Stoff" eines Schulaufsatzes. So ließ ein Mädchen namens Martha Drerup Suwelack in ihrer Phantasie bis in die Milchstraße fliegen: „Aber wie sah der Kerl aus! Er und sein Apparat saßen von oben bis unten voll Buttermilch! Er war nämlich so hoch geflogen,daß er mit dem Propeller in die Milchstraße geraten war."[762] Und die Billerbeckerin Johanna Schwering informierte ihren Sohn Walther (1885–1915),[763] der sich 1911 als wissenschaftlicher Assistent an der Deutschen Akademie der Wissenschaften in München aufhielt:

> „Prosper [ein Bruder] hat ein Gedicht auf Suwelack gemacht. Er hat es dir aufgeschrieben. Du kannst dir denken, wie aufregend das Schauspiel war. Ein großer Vogel da oben in der Luft und der Gedanke, ein Mensch ist da in Lebensgefahr. ... Martha [eine jüngere Schwester] sagte, als er am Turme der neuen Kirche hergeflogen, habe ich laut gerufen: ‚Nicht so nah am Turm vorbei.' ... Vater hat den Münst[erischen] Anz[eiger] mit dem Bericht an C[arl]-A[nton] [einen Bruder] geschickt mit dem Auftrage, ihn an dich weiterzusenden. Suwelack hat bei weitem besser geflogen (in Münster) als sein College."[764]

Der Schulaufsatz und der private Brief belegen das Eindringen der Fliegerei in die Alltagswelt der Bevölkerung sowie die Faszination und die gleichzeitige Furcht vor der Flugtechnik. Ein zwei Monate später erschienener Artikel, der über die Taufe eines zweiten Vereinsballons des Münsterischen Luftfahrtvereins berichtete, sowie ein im Januar 1914 an Josef Suwelack adressierter Brief Winfried Schwerings zeigen ebenso die Nachwirkung der Kirchturmumrundung Suwelacks. Anlässlich der Ballontaufe hielt Münsters Regierungspräsident Jaroslaw von Jarotzky eine Rede, in der er die Bedeutung der Fliegerei für das Münsterland hervorhob und die Flugschau Josef Suwelacks ins Gedächtnis rief, mit der dieser die Zukunftsfähigkeit Westfalens bewiesen und der Fliegerei zu noch mehr Ansehen verholfen habe: „Damals stand es für mich fest: Das Münsterland hat den Geist der Zeit erfaßt, der auf die Eroberung des Luftmeeres für den Flug des Menschen gerichtet ist."[765] Suwelack wurde als „Münsterländer" ausgewiesen und sei Teil eines Landes, „das da bewohnt wird von einem kernigen, einem aufrechten, einem heimattreuen Geschlechte, das immer und überall mit dem Namen seiner Heimat Ehre eingelegt" habe.[766] Winfried Schwering, der Josef Suwelack 1912 auf dem Flugtag hatte fliegen sehen, wollte sich daraufhin bei Suwelack zum Flieger

762 Martha Drerup, Schulaufsatz, o. D., abgedruckt in Krewerth, Ein westfälisches Fliegerleben, S.16–17, hier S.17.

763 Informationen zu Schwering finden sich bei Ursula König-Heuer/Julia Paulus (Hg.), „In der Ferne das Donnern der Kanonen". Briefwechsel eines Billerbecker Soldaten mit seiner Mutter im Ersten Weltkrieg. Münster 2014.

764 Familienarchiv Walter Suwelack, Nachlass Familie Schwering und Verwandte. Walther Schwering, Brief aus Billerbeck vom 27.10.1911.

765 NL J. Suwelack, „Der Taufakt", in: [unbekannter Zeitungstitel, verm. Münsterischer Anzeiger] vom Dezember 1911.

766 Ebd.

ausbilden lassen. Da er allerdings 1914 in München studierte, erteilte ihm der Flieger den Rat, sich besser dort ausbilden zu lassen.[767]

Suwelack war damit nicht nur ein Flieger, sondern ein stereotyper Repräsentant Westfalens und ein Vorbild, das in den Augen der Zeitgenossen seine Heimat auf Wettbewerben würdig vertrat. Gleichzeitig wurde er auch für politische Zwecke vereinnahmt. Der Flugsport und die Ausrichtung von Flugwettbewerben wirkten fördernd auf die regionale Identität, denn Suwelack machte seine Heimat über die Grenzen hinaus bekannt. Obschon der Flugtag überregional wenig[768] Beachtung fand, was auf die lokale Besonderheit zurückzuführen ist, war der Flugtag für die Billerbecker und die Lokalzeitungen ein Großereignis und Witte und Suwelack lokale Helden, die das Image der Region, der Provinz Westfalen, stärkten.

Billerbeck: Versöhung mit der Moderne

Nachdem Josef Suwelack am 8. Dezember 1911 seinen Weltrekord aufgestellt hatte, hielt er sich während der Weihnachtsfeiertage bei seiner Familie in Billerbeck auf. In dieser Zeit entstand die Idee zu einem eigenen Flugtag in der Ludgerus-Stadt.

Anfang des neuen Jahres trafen sich in Billerbeck an der Luftfahrt interessierte Männer, um darüber zu beraten, wie sie einen eigenen Flugtag mit Suwelack als Mittelpunkt der Veranstaltung organisieren könnten. Der örtliche Amtmann und Chronist Heinrich Fasse hielt bereits die Vorbereitungen für bedeutend und notierte:

> „Am heutigen Neujahrstage fand im Domhotel eine Versammlung von Freunden des Flugsports statt. Zweck der Veranstaltung war die Abhaltung eines Flugtages in Billerbeck durch den heimatlichen Flieger J. Suwelack. 8 Tage später wurde ein Komitee gebildet, das die Vorbereitungen zu dem am 2. Ostertage stattfindenden Flugtag eifrigst betrieb."[769]

Im Bewusstsein der Einmaligkeit des Flugtages gründete man keinen Luftfahrtverein, sondern lediglich ein Komitee. Damit war die Finanzierung nicht wie in Münster durch eine Mitglieds- und die Stadtkasse gewährleistet, sondern die Gel-

767 Brief Winfried Schwerings an Josef Suwelack vom Januar 1914 und Antwortschreiben Suwelacks vom Februar 1914.

768 Der „Flugsport" veröffentlichte eine kurze Meldung zum Münsterischen Flugtag unter der Rubrik „Flugveranstaltungen" und erwähnte sogar Suwelacks Flug nach Billerbeck. Vgl. Flugsport (1911), Nr. 23, S. 834.

769 Geschichtliche Nachrichten aus dem Amte Billerbeck 1900–1924 (mit 60 Abbildungen), I. Band, S. 66, 1.1.1912. Auf den folgenden Seiten wurden ausschließlich die Billerbecker Flugtage dokumentiert, ein Hinweis darauf, für wie bedeutsam das Ereignis in der Lokalgeschichte erachtet wurde.

der mussten privat aufgebracht werden. Aus diesem Grund sollten möglichst viele Einwohner vom Vorhaben überzeugt werden und Bürgschaften übernehmen.

Offenbar gelang die Überzeugungsarbeit, denn bereits auf der Versammlung am Neujahrstag 1912 wurden von „fünfzig bis sechzig" Personen Bürgschaften unterzeichnet. Im Vergleich mit den rund siebzig Gründungsmitgliedern des Münsterischen Luftfahrtvereins war dies eine bemerkenswert große Gruppe.[770] Da im Laufe der Woche noch viele weitere Bewohner unterschrieben hätten, sei „der Flugtag in finanzieller Beziehung als gesichert zu betrachten", so der „Billerbecker Anzeiger".[771] Die Zeitung lobte besonders das Engagement der weniger wohlhabenden Bürger: „Für die große Begeisterung spricht am besten, daß mancher ... gutsagte, der seine Groschen durch harte Arbeit sauer verdienen muß. (Bravo!)".[772] Sie wies ausdrücklich darauf hin, dass nach wie vor eine finanzielle Beteiligung möglich sei und Listen in mehreren Wirtschaften auslägen.[773] Als Termin für die Veranstaltung visierte man die Osterfeiertage 1912[774] an, so dass allen Billerbeckern ein Besuch der Flugtage möglich wäre.

Alle Geldgeber waren am nächsten Tag zu einer Versammlung im Domhotel zur Wahl eines Komitees eingeladen – das Ergebnis ist nicht überliefert. Aus späteren Schreiben geht allerdings hervor, dass Hubert Ahlers zum Schriftführer des Komitees bestellt worden war. Ahlers, der Volkswirtschaft und Fremdsprachen studiert und auf Bitten seines Vaters die elterliche Gaststätte übernommen hatte, war auch gesellschaftlich engagiert.[775] Seit 1908 bekleidete er mehrere öffentliche Ämter und war, so heißt es in einer späteren Broschüre über die Gaststätte, „einer der eifrigsten Initiatoren der Billerbecker Flugtage" gewesen.[776] Er führte im Vorfeld der Flugtage einen regen Schriftwechsel mit Josef Suwelack, informierte den Flieger über die neuesten Fortschritte, drängte ihn aber auch, ihm eine vertragliche Zusage seines Arbeitgebers zuzusenden, um Gewissheit an seiner Teilnahme zu haben.[777] In den folgenden Wochen traf sich das Komitee in diversen Gaststätten Billerbecks, um mit der Organisation fortzufahren, was der „Billerbecker Anzeiger" jeweils am Vortag unter der Überschrift „Flugtag Billerbeck" ankündigte.[778]

770 Vgl. Prange, Tollkühn am Himmel, S.26 sowie Billerbecker Anzeiger vom 6.1.1912.
771 Billerbecker Anzeiger, ebd.
772 Ebd.
773 Ebd.
774 Vgl. ebd.
775 StdA Billerbeck, 250 Jahre 1734–1984. Handelshof, Gaststätte und Brennerei Franz-Josef Ahlers Billerbeck Langstraße 18.
776 Ebd.
777 Vgl. hierzu den Abschnitt zu den Rumpler-Werken in Feld 2.
778 Siehe z. B. Billerbecker Anzeiger vom 13.1.,17.2. oder 24.2.1912.

Der Flugtag, wie Martina Glanemann in ihrer Darstellung treffend schreibt, sollte eine Veranstaltung „von Billerbeckern und für Billerbecker“[779] sein, ein Gemeinschaftsprojekt aller Bewohner und ein außergewöhnliches Ereignis, getragen von einem hohen Maß an Freiwilligkeit und Großzügigkeit. Freiherr von Twickel stellte beispielsweise seine „dem Hause Hameren gegenüberliegenden Weiden ... als Flugplatz unentgeltlich zur Verfügung“[780] und die Eintrittspreise waren niedrig gehalten, so dass den Eintritt „wohl ein jeder, der für den Flugtag Interesse besitzt, erschwingen“ könne.[781] Alle, die sich am Zustandekommen der Flugtage finanziell beteiligt hatten, erhielten überdies freien Eintritt und durften Suwelacks Apparat kostenlos besichtigen: „Diese Mitgliedskarten sind beim Schriftführer, Herrn Hubert Ahlers, Langestraße, in Empfang zu nehmen.“[782] Um auch dem Nachwuchs Gelegenheit zu geben, der Veranstaltung beizuwohnen, erklärte sich das Flugkomitee zu einem Preisnachlass für Schüler und Schülerinnen bereit, so dass die Kinder in Begleitung ihrer Lehrer nur 25 Pfennig zahlen mussten:

> „Diejenigen Schulkinder, die an dieser Vergünstigung teilzunehmen wünschen, haben sich an den beiden Ostertagen, unmittelbar nach Schluß des Nachmittags-Gottesdienstes auf dem Spielplatze der Knabenschule ... einzufinden und werden unter der Aufsicht der Herren Lehrer auf einen besonders hergerichteten Platz geführt, um dort den Schaufliegen beiwohnen zu können.“[783]

Die Sonderkondition erstreckte sich dabei jedoch nicht auf Kinder in Begleitung ihrer Eltern, die den regulären Preis zu zahlen hatten.[784]

Um zu gewährleisten, dass auch wirklich alle Billerbecker eine Eintrittskarte erstehen konnten, wurde der Vorverkauf bis Karsamstag verlängert und der „Billerbecker Anzeiger“ verwies erneut auf die Einmaligkeit der Flugveranstaltung: „Wohl niemals wird uns wieder eine solch schöne Gelegenheit, derartiges ganz aus der Nähe zu sehen, geboten, deshalb Billerbecker muß für Euch an den beiden Ostertagen die Parole lauten: ‚Auf, zum Flugplatz!‘“[785]

Die Ausrichtung als exklusiver *Billerbecker* Flugtag, als der er in der Zeitung und auf späteren Plakaten ausgewiesen wurde, zeigte sich auch daran, dass die Einnahmen aus der Veranstaltung größtenteils der Stadt zugutekommen sollten. So war es ausdrücklich nur örtlichen Händlern gestattet, ihre Waren an den Flugtagen zu verkaufen, wofür sie sich sogar eine Genehmigung von Hubert Ahlers erteilen lassen mussten.[786]

779 Glanemann, Technikbegeisterung, S.37.
780 Billerbecker Anzeiger vom 20.1.1912.
781 Die Karten kosteten laut „Billerbecker Anzeiger“ zwischen 3 Mark in der teuersten und 50 Pfennig in der günstigsten Kategorie im Vorverkauf und zwischen 4 Mark und 80 Pfennig an der Tageskasse. Billerbecker Anzeiger vom 30.3.1912.
782 Ebd. Alle anderen mussten 30 Pfennig bezahlen.
783 Billerbecker Anzeiger vom 6.4.1912.
784 Ebd.
785 Ebd.
786 Vgl. Billerbecker Anzeiger vom 23.3.1912.

Trotz aller Zukunfts- und Technikorientierung Billerbecks hatte das letzte Wort in der Organisation die Kirche. Da der Flugtag an den Osterfeiertagen stattfinden sollte, musste das Flugtag-Komitee seine Pläne nicht nur mit dem Landrat als weltlicher Regierung abstimmen, sondern auch mit der Geistlichkeit, war die Religion doch zentraler Bezugspunkt und Taktgeber der Stadt. Wenige Tage vor Ostern erhielt Hubert Ahlers als Vertreter des Komitees einen Brief Propst Brinkmanns, in dem dieser eine Überschneidung der Veranstaltung mit dem Nachmittagsgottesdienst und damit eine Ruhestörung befürchtete: „Ich möchte daher die Herrn dringend ersuchen[,] in der Billerbecker Zeitung ... oder auf andere geeignete Weise das hiesige Publikum zu benachrichtigen, daß erst nach 4 Uhr die Schauflüge stattfinden werden."[787] Eine Antwort ist nicht überliefert – angesichts des Datums nicht überraschend –, jedoch veranlasste Freiherr von Twickel, dass in der Zeit des Gottesdienstes zumindest kein Verkauf stattfinden dürfe.[788]

Hinsichtlich notwendiger Absperrungen und Parkmöglichkeiten etc. hatte man frühzeitig Absprachen mit dem Landrat getroffen. Am 14. März erhielt Ahlers die Zusage, dass „gegen die Sperrung der Chaussee-Billerbeck-Alstätte für die Dauer des 7. und 8. April d. Jrs. nichts einzuwenden"[789] sei. Zwei Wochen später folgte die Bestätigung, „Gendarmen an diesen Tagen nach Billerbeck zu kommandieren".[790] Der örtliche Turnverein erklärte sich bereit, sich unentgeltlich an der Straßenabsperrung zu beteiligen, was im „Billerbecker Anzeiger" lobend erwähnt wurde.[791] Als nach Abschluss der Flugtage über die Verwendung der Einnahmen beraten wurde, zeigte man sich gegenüber dem Turnverein erkenntlich und spendete 100 Mark. Weitere 200 Mark sollten für die Nationalflugspende verwandt und der verbliebene Überschluss „dem hiesigen St. Ludgerus-Hospital ... [zur] Anlage einer neuen Zentralheizung" gestiftet werden.[792]

Zusätzlich zu den Schau- und Passagierflügen war ein Rahmenprogramm angekündigt, das die Stimmung akustisch und optisch steigern sollte. Bereits im Februar hatte Hubert Ahlers gegenüber Josef Suwelack erklärt, dass die „Musik (Kürassiere von Münster)"[793] bestellt sei. Darüber hinaus war neben dem traditionellen Osterfeuer am Ostersonntag ein großes Feuerwerk geplant, wozu sich der Schriftführer eigens an einen Pyrotechniker aus Münster gewandt hatte. Dessen

787 NL J. Suwelack, Brief von Propst Brinkmann an Hubert Ahlers, Billerbeck, 2.4.1912.
788 Vgl. Billerbecker Anzeiger vom 6.4.1912.
789 NL J. Suwelack, Brief von Kreisassistent Freiherr von Fürstenberg, Coesfeld, an Hubert Ahlers vom 14.3.1912.
790 NL J. Suwelack, Brief von Kreisassistent Freiherr von Fürstenberg an Hubert Ahlers vom 30.3.1912.
791 Billerbecker Anzeiger vom 6.4.1912.
792 Billerbecker Anzeiger vom 11.5.1912.
793 NL J. Suwelack, Brief von Hubert Ahlers an Josef Suwelack in Berlin-Johannisthal vom 12.2.1912.

Empfehlung, das Feuerwerk auf dem Veranstaltungsplakat mit anzukündigen, wurde Folge geleistet.[794]

Das zum Flugtag gestaltete Plakat (Abb. 14), das die Silhouette Billerbecks und die Taube zeigte, repräsentierte das städtische Selbstverständnis einerseits als bedeutender katholischer Wallfahrtsort und andererseits als Wiege des Fortschritts, in der die Karriere des Josef Suwelack ihren Anfang genommen hatte. Die Silhouette der Stadt und die Überschrift „Billerbecker Flugtage 1912“ nahmen etwa die Hälfte des Plakates ein, darunter kündigte eine von Weitem deutlich lesbare Schrift die Veranstaltungstage und „Grosse Schau- und Passagier-Flüge“ an. In fett gedruckten Buchstaben und ebenfalls sehr gut lesbar stand Josef Suwelacks Name direkt darunter, sichtlich stolz ergänzt durch den Zusatz, dass er „der bedeutendste Flieger der Gegenwart“ und „Inhaber des Dauerweltrekords mit einem Passagier“ sei. Deutlich zum Ausdruck kommt hier die Prominenz Suwelacks in seiner Heimatstadt, aber auch die persönliche Note der Veranstaltung: Die Billerbecker Flugtage waren nicht nur für die Stadt konzipiert, sondern vor allem als Hommage an den „Tüftler“ von einst gedacht, der sich durch Tatendrang und Durchhaltevermögen seinen Platz in der Fliegerei erkämpft hatte. Den verbliebenen Platz auf dem Plakat füllten Informationen zu Eintrittspreisen, Speise- und Parkmöglichkeiten sowie zum Rahmenprogramm.

Komplettiert wurde die Vorbereitung auf den „großen Tag“ durch einen langen Bericht im „Billerbecker Anzeiger“, der die Bedeutung der Luftfahrt im Allgemeinen und Josef Suwelacks im Besonderen herausstellte. Erinnert wurde an den Besuch des Zeppelins „Schwaben“ sowie an Suwelacks Flug in die Heimat auf dem Münsterischen Flugtag, so dass die jetzige Veranstaltung gleichsam aus einer Reihe flugsportlicher Ereignisse hervorging. Das „Thema Luftsport“ sei deshalb aus Billerbeck nicht mehr wegzudenken.[795] Der Artikel verdeutlichte allen Billerbeckern, dass der Flugtag ihren gemeinschaftlichen Anstrengungen zu verdanken war, und stimmte sie auf den 7. und 8. April ein. Zugleich war den Veranstaltern bewusst, dass die endgültige Entscheidung über den Erfolg oder Misserfolg des Flugtages sozusagen in den Händen Gottes lag: „[V]on der Gunst des Wettergottes hängt bei unserem Unternehmen fast alles ab, namentlich in finanzieller Beziehung“.[796]

Am Freitag hatte Dauerregen die traditionelle Karfreitagsprozession unmöglich gemacht, so dass sie in die Kirche verlegt worden war, und am Karsamstag, so berichtete der „Billerbecker Anzeiger“, habe ein „furchtbarer Sturm ... die Hoffnung aller Billerbecker“ zunichte gemacht.[797] Ähnliche Verhältnisse wie in Müns-

794 NL J. Suwelack, Brief von L. Gillen aus Münster an Hubert Ahlers vom 14.3.1912.
795 Billerbecker Anzeiger vom 6.4.1912.
796 Ebd.
797 Billerbecker Anzeiger vom 13.4.1912.

Abb. 14: Plakat zum Billerbecker Flugtag, Billerbeck, Ostern 1912 (Foto: Familienarchiv Walter Suwelack, NL Josef Suwelack)

ter waren zu befürchten. Doch der Ostersonntag begann mit Sonnenschein und erleichtert konnte die Heimatzeitung verkünden: „Unser Flugtag gelingt uns, das Wetter ist günstig."[798] Anschließend hieß es, dass „abertausende von Menschen aus allen Himmelsgegenden" nach Billerbeck gekommen seien.[799]

Die für drei Uhr angekündigte erste Flugveranstaltung begann jedoch erst anderthalb Stunden später und es ist daher anzunehmen, dass auf die Bitte des Propstes Rücksicht genommen wurde. Möglicherweise verzögerte sich der Start aber auch aufgrund von Motorproblemen am Aviatik-Eindecker Suwelacks, denn es wurde berichtet, dass schließlich nach einem rund vierminütigen Flug der Motor ausgesetzt habe und Suwelack gezwungen gewesen sei, im Brock zu landen. Es folgte eine anderthalbstündige Pause, während der „die Probleme am Vergaser" behoben wurden.[800] Allerdings endete auch ein zweiter Aufstieg nach nur anderthalb Minuten und eine erneute einstündige Reparaturpause zögerte den Schauflug weiter hinaus. Als dann der Motor zum dritten Mal aussetzte habe das Publikum, „das bis dahin geduldig ausgehalten hatte, ... jetzt ziemlich enttäuscht den Flugplatz verlassen": „Enttäuscht und auch betrübt über das Mißgeschick, das den Flieger betroffen hatte, eilten die Zuschauer zur Stadt und von da an zum Hahnenkamp, um wenigstens am zweiten Teil des Programms, dem prachtvollen Riesenosterfeuer ... sich zu ergötzen."[801]

Alle Hoffnungen ruhten nun auf dem zweiten Flugtag, zu dem trotz eines aufgezogenen Sturmes erneut viele Zuschauer gekommen waren. Doch der Sturm vereitelte alle weiteren Flugversuche und auch der von Suwelack in Aussicht gestellte Flug am Dienstag konnte aufgrund des sich noch weiter verschlechternden Wetters nicht stattfinden. Aus den Billerbecker Flugtagen wurden „begrabene Hoffnungen".[802]

Josef Suwelack hatte jedoch durch den Ausgang der Veranstaltung nicht an Popularität eingebüßt. Seine Leistungen und Aktivitäten, auch als späterer technischer Direktor der Kondor-Werke, wurden weiterhin interessiert verfolgt und der Stolz auf den Flieger blieb bestehen. Ferner hatte die Stadt trotz aller Widrigkeiten noch immer vom Flugtag profitiert. Billerbeck war bekannter geworden und konnte sich gegenüber größeren Städten behaupten: „Das kleine Landstädtchen Billerbeck schwingt sich hierdurch hinauf in die Reihe größerer Städte, die ähnliches [sic!] veranstalten".[803] Die für gemeinnützige Vereine und die Nationalflugspende gestifteten Gelder lassen zudem erahnen, dass die Kartennachfrage hoch gewesen sein muss.

798 Ebd.
799 Ebd.
800 Ebd.
801 Ebd.
802 Ebd.
803 Billerbecker Anzeiger vom 6.4.1912.

Betrachtet man die Veranstaltung weniger unter dem Aspekt des fliegerischen „Erfolges" als vielmehr unter dem Aspekt der Stärkung der regionalen Identität, dann stehen die Billerbecker Flugtage beispielhaft für kleinstädtische Technikbegeisterung und kollektive Partizipation. Die Einbindung der Bevölkerung in die Vorbereitungen, die Möglichkeit, einen persönlichen Obolus beizusteuern, und das stimmige Rahmenprogramm an den Osterfeiertagen zeigen den organisatorischen Aufwand und die Begeisterung der Bürger, die Berichterstattung über das Ereignis wiederum betont die Wichtigkeit der Zeitung als Hauptinformationsquelle für die Bevölkerung. Die Billerbecker fieberten der Veranstaltung entgegen und taten alles in ihrer Macht stehende, um für reibungslose und in Erinnerung bleibende Flugtage zu sorgen. Josef Suwelack als Lokalhelden gebührend zu feiern, ihm einen angemessen Empfang zu bereiten, die Kleinstadt Billerbeck als Stadt des Flugsports überregional bekannt zu machen und sich gegenüber den benachbarten Städten Münster und Osnabrück zu behaupten, waren Motive, die bei der Initiierung der Veranstaltung im Vordergrund standen. Neben dem finanziellen Erfolg für die Stadt konnten sich die Angehörigen des Flugtag-Komitees, und hier besonders Hubert Ahlers, darüber hinaus sicher sein, dass ihre Namen und ihre Rolle bei der Vorbereitung nicht in Vergessenheit geraten würden. Trotz oder gerade wegen der Einmaligkeit des Ereignisses fanden die Billerbecker Flugtage Eingang in die Stadtchronik und werden bis heute in der Heimatzeitung, meistens anlässlich von Jubiläen, und in Publikationen zur Landesgeschichte Westfalens rezipiert.[804]

Die Flugtage in Detmold, Münster und Billerbeck führten zur Identifikation der Bevölkerung mit ihrer Stadt und verbreiteten sich über die Presse in der jeweiligen Region. Der von Bourdieu als *illusio* bezeichnete Glaube an gemeinsame Ziele, Wertvorstellungen und regionale Besonderheit zeigt sich in der Vorbereitung auf und Berichterstattung über die Flugtage. Die Städte gaben sich über die Flugzeugtechnik und „ihre" Helden der Lüfte als zukunftsgewandt, sportbegeistert und technikinteressiert. Hier „erfand" und konstruierte sich eine Region über das Zusammenwirken von Medien und Bevölkerung.[805] Der damit verbundene Integrationsdruck zeigt sich deutlich am Billerbecker Beispiel. Es wurde erwartet, dass sich wirklich alle Bewohner Billerbecks in irgendeiner Form am Flugtag beteiligten und die Lokalzeitung wies mehrfach auf die geringen Eintrittspreise und insbesondere auf die historische Dimension der Veranstaltung hin. Der „Billerbecker Anzeiger" machte deutlich, dass wer zu Billerbeck (auch noch nach dem

804 Siehe z. B. Sieger, Die Ludgerusstadt, S.243f.; Glanemann, Technikbegeisterung, S.36ff.; „Ein Pionier der Luftfahrt", in: Billerbecker Anzeiger von 1959; „Aus ‚Billerbecker Großflugtagen' wurden 1912 begrabene Hoffnungen", in: Billerbecker Anzeiger von 1982. Weitere Titel ließen sich ergänzen. Hierbei handelt es sich stets um westfälische Veröffentlichungen.

805 Vgl. Weichlein, Regionalism, S.95ff.

Flugtag) gehören wollte, sich zu Suwelack bekennen musste, zum Westfalen – zu Westfalen.

Ferner erlaubte die Fliegerei eine Versöhnung von Alt und Neu und machte die durch die Moderne erlebte Beschleunigung und Brucherfahrung erträglicher. Lokal und durch größere Flugtage zugleich auch national bekannte Flieger wie Josef Suwelack demonstrierten über ihre Flugvorführungen, dass die Technik beherrsch- und kontrollierbar war. Die durch die Medien diskursiv transportierten Werte, Pilotenbilder, Technikvorstellungen und Erwartungen wirkten aber auch auf den Habitus der individuellen Flugsportler zurück. Über ihre Teilnahme an Wettbewerben, ihre (Flug-)Vorstellung in der Luft und am Boden sowie über die aktive Gestaltung des Pilotenbildes durch Interviews, Fotos oder Postkartenmotive erlangten sie Ruhm, Anerkennung und Erfolg.[806] Die Inszenierung der Fliegerei als Abenteuer und Erlebnis, wie sie exemplarisch in Suwelacks Interview auf dem Flugtag in Münster oder im Leserbrief der „Lippischen Landeszeitung" zum Vorschein kam, spielte eine wichtige Rolle bei der öffentlichen Aneignung, Verbreitung und Akzeptanz der Flugtechnik.

Die öffentliche Vermittlung und Vermarktung der Fliegerei kennzeichnete dabei ein ambivalentes Nebeneinander von Faszination und Furcht. Vor allem über Bilder – etwa Postkarten oder Fotos – wurden Erfolge und Unfälle gleichermaßen als Sensation angeeignet. Rückschläge, so wurde es in der zeitgenössischen Wahrnehmung kommuniziert, waren notwendige „Opfer", um weiter am Fortschritt zu arbeiten.

Die „tourenden" Flieger lösten in Groß- und Kleinstädten immer wieder ähnliche Reaktionen aus. Stets war die Rede vom Stolz der Stadt, Ausrichter eines Flugtages zu sein, und immer wieder berichteten die Zeitungen von Menschenmassen, die dem Spektakel als Zuschauer beigewohnt hätten. Die Organisation war mit hohen finanziellen Ausgaben und logistischem Aufwand verbunden: Plakate und Broschüren mussten gedruckt, Informationen über die Presse verbreitet und die Piloten durch attraktive Prämien zum Fliegen animiert werden. Darüber hinaus erwartete das zahlende Publikum buchstäblich, auf seine Kosten zu kommen, so dass in der Regel zusätzlich zur eigentlichen Flugschau ein Rahmenprogramm geboten wurde. Schauplatz waren Rennbahnen, Weiden oder Flugplätze, die gute Start- und Landebedingungen sowie ausreichend Platz für das Publikum boten.

Die Billerbecker Flugtage nehmen unter den hier vorgestellten regionalen Flugveranstaltungen eine Sonderrolle ein, da sie zum einen von einer extremen Form der Flugbegeisterung zeugen und zum anderen von einem eigens gegrün-

806 Vgl. Kehrt, Moderne Krieger, S.33f.

deten lokalen Komitee statt von einem Luftfahrtverein organisiert wurden. Die Flugtage in Münster und Billerbeck erhielten ihre regional-historische Dimension durch die Berichterstattung, die das Bild des eifrigen, zielstrebigen und unerschrockenen Westfalen mit konstruierte, gleichsam ein „westfälisches Geschlecht", dem kommende Generationen gedenken und nacheifern sollten. Dass Josef Suwelack den Bericht über seinen Flug um die Türme des Ludgerus-Doms während seiner Zeit bei den Kondor-Werken auf der Schreibmaschine sowie auf einem Briefbogen des Unternehmens festhielt, macht deutlich, dass er sich der Tragweite dieser Umrundung für seine Bekanntheit in Westfalen bewusst und, wie er selbst im Interview 1912 geäußert hatte, stolz auf seinen Flug war. Das Plakat zum Billerbecker Flugtag mit Josef Suwelack als lokaler Identifikationsfigur im Mittelpunkt symbolisierte die Gleichberechtigung von Tradition und Innovation, von ländlicher Idylle und Modernität. Billerbeck, so kam es in den Berichten zum Ausdruck, war in der Moderne angekommen.

Im Vergleich des Detmolder Flugtages mit den westfälischen Veranstaltungen offenbart sich die Subjektivität der Zuschreibungen, denn jeder Flugtag schuf sich seine individuellen Helden: Mal stand Hellmuth Hirth im Vordergrund des Interesses, mal war Melli Beese die Hauptattraktion und schließlich galt die Bewunderung einzig Josef Suwelack. Angesichts der Kostenintensität der Wettbewerbe wollte jede Stadt sicherstellen, die „richtigen", d. h. die aktuell erfolgreichsten und reichsweit bekanntesten Flieger und Fliegerinnen zu engagieren. Daher ist es nachvollziehbar, dass Hellmuth Hirth aufgrund seiner zahlreichen Erfolge in Detmold zum Publikumsliebling avancierte und Josef Suwelack in Münster und Billerbeck der „Held des Tages" war. Über die Werbung mit ihren Namen versprachen sich die Veranstalter viele zahlende Besucher aus der Umgebung und damit gute Einnahmen für die Stadtkasse sowie ein positives Feedback von den Medien.

Parallel zu diesen friedlich-sportlichen Wettbewerben bildete sich spätestens seit 1911 ein neuer Flugwettbewerbs-Typ heraus, der unter militärisch-kriegerischen Vorzeichen stand und der Sichtung und Rekrutierung kriegstauglicher Flugzeuge und Piloten diente. Bombenabwurf- und Luftbild-Fotografie-Wettbewerbe bereiteten die Flugzeugführer auf ihre künftige Aufgabe als Aufklärungs- und Beobachtungsflieger an der Front vor; Langstrecken- und Etappenflüge sollten die Leistungsfähigkeit und Ausdauer der Flugzeuge, insbesondere der Motoren, beweisen und Innovationen bewirken. Für Zivilflieger wie Josef Suwelack wurde es zunehmend schwierig, eine Teilnahmeerlaubnis für die Wettbewerbe zu erhalten, denn die Militärflieger dominierten diese und verdrängten die Privatflieger mehr und mehr. Auch die Zuschauerorientierung und die Berichterstattung änderten sich. Zwar büßten die Flieger nichts von ihrer Popularität ein und Flugveranstaltungen wurden auch weiterhin auf Plakaten angekündigt, doch bekamen Schaulustige die Piloten häufig nur noch bei Start und Landung zu sehen und hatten nicht mehr die Möglichkeit, ihren Flug unmittelbar zu verfolgen. Stattdessen bestand seit dem Sommer 1913 die Möglichkeit, ähnlich wie im Pferderennsport auf Platz oder Sieg eines Flugzeuges zu setzen.[807] Die jahrmarktähnlichen Flugveranstaltungen, wie sie in Billerbeck oder Münster stattgefunden hatten, gehörten allmählich der Vergangenheit an, wenngleich der Publikumsandrang auf den Wettbewerben nach wie vor groß war. Trotzdem verlagerte sich die Fliegerei von einem „Risikosport" hin zu einem „Ausdauersport", ausgerichtet auf militärische Einsatzfähigkeit und Belastbarkeit.

3.3. Die Fliegerei im Zeichen der Kriegsvorbereitung – Der Prinz-Heinrich-Flug von 1913

Ausrichtung

Die ersten Überlandwettbewerbe hatten bereits 1910 stattgefunden und waren von Flugvereinen oder Unternehmen wie z. B. der Kaffeefirma Kathreiner oder dem Ullstein Verlag finanziert worden. So hatte die Kaffeefirma 1911 den „Kathreinerpreis" von München in Höhe von 50.000 Mark und der Ullstein Verlag den mit 100.000 Mark dotierten „B.Z. Preis der Lüfte" gestiftet. Die Frage, wie das deutsche Flugwesen durch Wettbewerbe sinnvoll gefördert und auf den praktischen Nutzen hin verbessert werden könnte, hatte schon zu diesem Zeitpunkt die Überlegungen maßgeblich bestimmt.[808]

Beim Kaisermanöver im September 1911 kamen neben Luftschiffen erstmals auch Flugzeuge zum Einsatz, die ihre Verwendbarkeit für strategische Erkundungszwecke, die Nahaufklärung, vorführen sollten. Im Bericht über das Kaisermanöver hieß es, dass das Flugzeug „volle Berechtigung besitzt, in die Aufklärungsmittel der Heeresverwaltung aufgenommen zu werden."[809] Der „Flugsport" lobte ebenfalls die fortgeschrittene Flugtechnik:

> „[Man] hatte ... endlich auch bei uns den Versuch angestellt, einige Flugmaschinen ... einzelnen Truppenteilen im Kaisermanöver beizugeben. Geleistet wurde Hervorragendes. Die Flugmaschine ist für eine moderne Kriegsführung ein gerade so unentbehrliches Mittel geworden, wie es das Automobil schon lange ist. Ja noch mehr, die Flugmaschine wird nicht allein als Erkundungs- und Verkehrsmittel, sondern als wirkliche Waffe, als sogenannte vierte Waffe, ausgebildet werden, und die Zukunft wird vielleicht noch die Wirkungen dieses furchtbaren Angriffsorgans der Menschheit vorführen.[810]

Einen Ersatz für die Kavallerie sah man hingegen im Flugzeug noch nicht, bislang hielt man es „nur bedingt" für ein „wertvolles Nachrichten- und Erkundungsmit-

807 Vgl. hierzu z. B. „Der Flugtoto", in: Berliner Tage- und Handelsblatt vom 14.8.1913, 1. Beiblatt.

808 Siehe „Wie können wir durch Wettbewerbe das deutsche Flugwesen fördern?", in: Flugsport (1911), Nr. 22, S. 767.

809 BA-MA PH 9-V/67, Bericht über die Verwendung des Fliegerkommandos Blau im Kaisermanöver 1911.

810 „Die Flugmaschinen bei dem diesjährigen Kaisermanöver", in: Flugsport (1911), Nr. 20, S. 702.

tel".[811] Den ersten militärischen Einsatz hatte ein Flugzeug im Italienisch-Türkischen Krieg von 1911/12, bei dem die Italiener diese neue Waffengattung einsetzten und Ende Februar 1912 auch die ersten Luftbilder eines Aufklärungsfluges zurückbrachten.[812] Beim Kaisermanöver im darauffolgenden Jahr wurde die Bedeutung der Fliegertruppe im Deutschen Reich symbolisch mit der Übergabe eines Luftbildes an Kaiser Wilhelm II. inszeniert.

Der sogenannte Deutsche Zuverlässigkeitsflug am Oberrhein, der im Mai 1911 zum ersten Mal durchgeführt wurde und bis einschließlich 1914 jährlich stattfand, betonte in besonderem Maße die Bedeutung der Luftfahrt. Ziel der Veranstaltung war es, wie es im Namen anklang, die Zuverlässigkeit der Flugzeuge zu fördern und die Entwicklung des deutschen Flugwesens herauszustellen. Paul Béjeuhr hob im „Flugsport" die Leistungsorientierung des Wettbewerbs hervor:

> „Die Zuverlässigkeit der Flugzeuge soll gefördert werden, der Begriff Zuverlässigkeit ist also nicht nur ein Aushängeschild, sondern ihm ordnen sich alle sonst üblichen Wettbewerbsmöglichkeiten unter. Keine Höhen- und Geschwindigkeitsflüge werden verlangt, nicht die Sensationslust soll durch verblüffende Rekorde angeregt werden, sondern ehrlich und ohne irgendwelche Verschleierung sollen die Teilnehmer den Beweis erbringen, daß unsere Flugtechnik im Stande ist, ... mehrtägige Flüge hintereinander mit demselben Material auszuführen."[813]

Prinz Heinrich von Preußen, der sich in der Automobilindustrie als Förderer engagierte, wandte sich nun auch dem Flugwesen zu und ließ sich bei August Euler zum Flieger ausbilden. 1913 übernahm er die Schirmherrschaft des Zuverlässigkeitswettbewerbes, der jetzt auch unter dem Namen „Prinz-Heinrich-Flug" veranstaltet wurde und erstmals am Mittelrhein entlangführte. Im Automobilkorps begleitete der Prinz die Piloten zu den Etappenzielen des Wettbewerbs.[814]

Das Symbol und zugleich ein Ehrenpreis des Fluges war eine nackte Ikarusgestalt,[815] die, auf einer Weltkugel stehend, auf einen Adler herabsah. Die von Bildhauer Carl Korschann entworfene Skulptur sollte „die Herausforderung des

811 Ebd.

812 John W. R. Taylor/Davin Mondey, Spione am Himmel über uns. Die Geschichte der Luftaufklärung, Stuttgart 1974, S.28–30; Die ersten Luftaufnahmen aus einem Ballon reichen zurück in die Mitte des 19. Jahrhunderts, seit 1911 experimentierte das Militär verstärkt auch mit der Fotografie aus Flugzeugen. Siehe zur Entwicklung der Militärfotografie: Helmut Jäger, Erkundung mit der Kamera. Die Entwicklung der Photographie zur Waffe und ihr Einsatz im 1. Weltkrieg, 2. Aufl., München 2007, S.23ff.

813 Paul Béjeuhr, Konstruktive Fragen der Flugtechnik und ihre Lösung durch Wettbewerbe, in: Flugsport (1911), Nr. 9, S.289–293, hier S.289.

814 „Der Prinz-Heinrich-Flug 1913", in: Sport im Bild (1913), Nr. 20, S.542–543, hier S.542.

815 „Ehrenpreis Seiner Königlichen Hoheit des Prinzen Heinrich von Preußen", entworfen von Bildhauer Carl Korschann, Höhe 83 cm, 1913. Abgedruckt in: Denkschrift über den Prinz Heinrich Flug 1913. Dritter deutscher Zuverlässigkeitsflug am Oberrhein, veranstaltet von der Südwestgruppe des deutschen Luftfahrerverbandes, hg. von Obstl. Frhr. v. Oldershausen, Straßburg 1914, S.9, online unter: http://anno.onb.ac.at/cgi-content/anno-plus?aid=sib&datum=1913&page=576&size=45 (25.5.2014).

Abb. 15: Reklamemarke zum Prinz-Heinrich-Flug 1913, gestaltet von Ernst Riess (Foto: Familienarchiv Walter Suwelack, NL Josef Suwelack)

Königs der Lüfte" darstellen und „das Trotzige und Herausfordernde des Menschen gegenüber der Natur" zum Ausdruck bringen.[816]

Der Prinz-Heinrich-Flug wurde als „Friedensübung" propagiert, der die Fliegerei „volkstümlich" machen sollte, beruhe die Militärfliegerei doch auf den Anstrengungen der Industrie und Zivilflieger.[817]

Das konnte jedoch nicht darüber hinwegtäuschen, dass die Veranstaltung eine nationalistisch-militärische Ausrichtung hatte, denn zugelassen waren ausschließlich deutsche Maschinen und deutsche Flugzeugführer, vornehmlich Militärflieger. Lediglich bei der Motorenwahl hatte man Freiheit gelassen, jedoch durfte der Motor im Unterschied zu den anderen Flugzeugteilen im Verlauf des Fluges nicht ausgetauscht werden. Außerdem kam erstmals eine neue Bewertungspraxis zum Einsatz, die die Flugleistung in Abhängigkeit von der zum Einsatz gelangten PS-Zahl wertete und Rückschlüsse auf die Schnelligkeit und Zuverlässigkeit erlauben sollte.[818]

Die Veranstaltung war als „Herrenfliegerwettbewerb" ausgeschrieben, das heißt, dass neben angestellten Firmen- und Militärpiloten auch selbstständige Privatpersonen teilnehmen durften, sofern sie von einem Luftfahrtverein vorgeschlagenen worden waren. Frauen waren von dem Wettbewerb ausgeschlossen.[819] Alle Teilnehmer mussten Flugerfahrung vorweisen können, die sie durch den Nach-

816 „Prinz Heinrich-Flug 1913", in: Flugsport (1913), Nr. 10, S. 341–356, Zitat S. 347–348.
817 „Zum Prinz Heinrich-Flug", in: Flugsport (1914), Nr. 11, S. 438–439.
818 Siehe Denkschrift über den Prinz-Heinrich-Flug, S. 57ff.
819 Vgl. Ausschreibung des Prinz-Heinrich-Fluges 1913, in: Flugsport (1913), Nr. 5, S. 180ff., § 2 und § 3b.

weis eines einstündigen Fluges, der Erfahrung im Gleitflug und der Ausführung von mindestens zehn Passagierflügen schriftlich garantieren mussten. Flugzeuge, die nicht dem Bestand der Heeresverwaltung angehörten, mussten eine Vorprüfung am Abend vor Beginn des Wettbewerbs in Wiesbaden bestehen.

Gefordert wurde eine Steigfähigkeit von 500 Metern in höchstens 15 Minuten mit einer durchschnittlichen Belastung von 200 kg, des Weiteren ein kurzer An- und Auslauf sowie eine Landung auf einem vorher bestimmten Gelände (§ 12 der Ausschreibung). Die Maschinen konnten ab dem 7. Mai auf dem Flugplatz Wiesbaden untergebracht werden und mussten am 9. Mai der Flugplatzleitung vorgeführt und von dieser abgenommen werden. An den Etappenorten erfolgte die Unterbringung der Maschinen in Zelten, für deren Bewachung Gewähr übernommen wurde (§ 19).

Zehn Flugzeuge in Privatbesitz von Militärpiloten waren zum Wettbewerb eingetroffen, außer Konkurrenz flog zudem Hans Grade auf einem Eindecker mit. Doch bereits bei der Vorprüfung am 10. Mai schieden drei Apparate aus, darunter auch Grades. Schließlich gingen acht Privatmaschinen an den Start; Josef Suwelack und Leutnant Victor Carganico (1887–1945) flogen als Reserveflieger mit. Von der Heeresverwaltung beteiligten sich elf der ursprünglich zwölf gemeldeten Militärflieger, so dass insgesamt 19 Piloten an den Start gingen.[820]

Der Wettbewerb gliederte sich in drei Tagesflüge, sogenannte Zuverlässigkeitsetappen, bei denen die Gesamtflugdauer bewertet wurde, und einen zweiten Wettbewerb, eine zweitägige Aufklärungsübung mit dem Ziel Straßburg.

Die erste sogenannte Zuverlässigkeitsetappe führte von Wiesbaden nach Kassel mit einer Zwischenlandung in Gießen und umfasste circa 170 Kilometer. Die zweite Etappe am 12. Mai führte von Kassel nach Koblenz über eine Flugstrecke von 187 Kilometern und die letzte Etappe, nach einem Ruhetag am 13. Mai, von Koblenz nach Karlsruhe mit Kontrollpunkten in Kreuznach, Mainz, Frankfurt am Main, Neustadt und Heidelberg bei einer zurückzulegenden Flugstrecke von weiteren 306 Kilometern.[821] An Preisgeldern standen 70.000 Mark zur Verfügung – ein Großteil finanziert mit Mitteln aus der Nationalflugspende–, von denen jedoch nur knapp 55.000 Mark ausgezahlt wurden.[822]

Für jede erfolgreich, das heißt bis zum Etappenziel geflogene Stecke erhielten die Piloten ein Preisgeld von je 500 Mark sowie das Nenngeld in gleicher Höhe zurück. Sofern zwei Etappen zurückgelegt worden waren, standen ihnen weitere 1.000 Mark zu, wer alle drei Etappen gemeistert hatte, erhielt 2.500 Mark. Für die

820 Vgl. Hans Zeyssig, Prinz-Heinrich-Flug, in: Der Motorwagen, H. 18 vom 30.6.1913, S.458ff., hier S.458.

821 Vgl. Denkschrift über den Prinz-Heinrich-Flug, Falteinlagen mit Teilnehmerlisten und Flugdaten etc. Ursprünglich betrug die dritte Etappe nur rund 200 km, war dann aber kurzfristig von der Flugleitung um weitere 100 km verlängert worden. Vgl. Sport im Bild (1913), Nr. 20, S.542.

822 Vgl. Zeyssig, Prinz-Heinrich-Flug, S.459, sowie Flugsport (1913), Nr. 5, S.183 (§ 13).

ersten drei Flieger mit der kürzesten Gesamtflugdauer warteten obendrein Preisgelder in Höhe von 5.000, 3.000 und 2.000 Mark (§ 13). Außerdem hatten unter anderem[823] der deutsche Kaiser, der Großherzog von Baden und Prinz Heinrich von Preußen Ehrenpreise gestiftet, die ebenfalls den drei erfolgreichsten Flugzeugführern zufallen sollten.[824]

Hatten die Firmen für den Transport ihrer Flugzeuge zum Bestimmungsort selbst die Kosten zu tragen, so übernahm der Veranstalter alle übrigen Aufwendungen, mit Ausnahme der Kosten für Personen- und Bruchschäden (§ 30). Er stellte Karten, Benzin und Öl gegen Wertbons sowie Ersatz- und Reparaturteile unentgeltlich zur Verfügung (§§ 26, 28 u. 29): „Zur Unterstützung der Flieger [wird] jedem Flugzeug ein Begleitwagen zur Verfügung gestellt ..., auf den [sic!] der Unparteiische mit dem Chauffeur und zwei Monteure mit dem leichten Reparaturmaterial Platz nehmen und ständig dem Flugzeug so schnell wie möglich zu folgen haben."[825]

Da der gesamte Wettbewerb als Aufklärungsübung ausgeschrieben war, mussten bei allen Flügen die Passagiere das Bordbuch führen, die Flugzeuge mit Barografen ausgestattet sein und beides „nach jeder Landung sofort an die sportliche Etappenleitung" zur Auswertung und Kontrolle übergeben werden (§ 21). Passagiere durften ausschließlich Offiziere sein, da der Flug „vor allem in militärischer Beziehung brauchbare Resultate ergeben"[826] sollte und nur Offiziere im Kartenlesen und Bordbuch-Führen geübt waren. Die Piloten hatten bestimmte Kontrollpunkte zu überfliegen und dazu auf 150 Meter zu sinken. Simulierte Kriegslagen sollten die Aufklärungsübungen am 17. und 18. Mai so real wie möglich gestalten und die Flieger erst kurz vorher über den Vormarsch des „Feindes" informiert werden:

> „Allgemeine Kriegslage: Während blaue Kräfte am Oberrhein sich versammeln, sind rote Kräfte im Vormarsch in östlicher Richtung. Besondere Kriegslage: Bei dem Armee-Oberkommando der dritten blauen Armee (Karlsruhe), welcher eine Fliegerabteilung unterstellt ist, geht am Abend des 15. Mai folgende Meldung der dritten Kavalleriedivision ein: ‚Am Katharinenthaler Hof, 4 km nördlich von Pforzheim, Pforzheim, 15. Mai, 8 Uhr abends, an 8:45 abends. Starke, feindliche Kolonnen, anscheinend rechter Flügel des Feindes, erreichte heute mit Anfängen Stuttgart und Ludwigsburg.' Die Flieger müssen bei diesen Uebungen eine kriegsmäßige Höhe von 800 m erreichen. Den Fliegern wird am Morgen mitgeteilt, wo die Kavalleriedivision sich befindet, damit sie über diese hinausfliegen können, um ihre Erkundigungen anzustellen."[827]

Vom Ausgang des Fluges erhoffte man sich nicht nur von militärischer Seite ein gutes Abschneiden, auch in Sport- und Fachzeitschriften hatte man große Erwartungen an die Piloten und ihre Flugzeuge: „Da aber nur unsere besten Offiziers-

823 Siehe die vollständige Liste der Preisstiftungen in Flugsport (1913), Nr. 8, S. 305.
824 Vgl. „Der Prinz-Heinrich-Flug 1913", in: Sport im Bild (1913), Nr. 20, S. 542–543, hier S. 542.
825 Ebd.
826 Ebd., S. 543.
827 „Prinz Heinrich-Flug 1913 (Fortsetzung und Schluß)", in: Flugsport (1913), Nr. 11, S. 394–395.

flieger und auch die Elite unserer Herrenflieger an dem Kampf am Mittelrhein teilnehmen, darf man dem Ergebnis des Prinz-Heinrich-Fluges 1913 in aller Ruhe entgegensehen."[828]

Verlauf und Bewertung des Fluges

Die Formierung einer nationalen Fliegerelite zeigte sich nicht zuletzt auch in der Teilnehmeraufstellung. Zugelassen waren fast ausschließlich Militärpiloten und nur wenige Zivilflieger wie Josef Suwelack, der vom Vereinsvorsitzenden des Oberrheinischen Vereins für Luftfahrt vorgeschlagen worden war[829] und zusammen mit Leutnant Victor Carganico als Ersatz für die am 10. Mai ausgeschiedenen Flieger antreten durfte.[830]

Im „Motorwagen" kritisierte Hans Zeyssig die Zurückdrängung der privaten Flugzeugindustrie sowie den hohen Kostenaufwand für den Wettbewerb, des Weiteren beklagte er das Missverhältnis von Kosten und Gewinn. Zeyssig bezifferte die entstandenen Unkosten für die zivile Flugzeugindustrie auf ein Vielfaches des in Aussicht gestellten Preisgeldes. Der Prinz-Heinrich-Flug bedeute ein „im Interesse der Sache gebrachtes freiwilliges Opfer von vielen Tausenden [Mark]"[831] und die Zivil- und Privatflieger kämen nicht annähernd auf ihre Kosten. Aufgrund zahlreicher Reparaturen und wetterbedingter Beschädigungen an den Flugzeugen hätten darüber hinaus viele Flieger vorzeitig aufgeben müssen, darunter Leutnant Zwickau, Leutnant Engwer und Josef Suwelack: „Engwer beschädigte seine Maschine schwer in Kassel und mußte dort aufgeben. Suwelack mußte das gleiche wegen einer Havarie kurz hinter Karlsruhe tun."[832] „Unsummen" seien so in die Organisation und Durchführung des Wettbewerbs geflossen, die nach Meinung Zeyssigs besser für die Flugzeugindustrie hätten eingesetzt werden sollen.[833]

828 Ebd.

829 NL J. Suwelack, Schreiben des Vereinsvorsitzenden des Oberrheinischen Vereins für Luftfahrt an Josef Suwelack betreffs der Aufnahme in den Oberrheinischen Verein für Luftfahrt und der Einladung zur Teilnahme am Prinz-Heinrich-Flug, Straßburg, 22.2.1913.

830 Vgl. Denkschrift über den Prinz-Heinrich-Flug, S.16. Aus Gründen des Zeitmangels wurde es Suwelack und Leutnant Carganico gestattet, die Vorprüfung auf einem anderen Flugplatz als in Wiesbaden zu absolvieren, so dass sie ihre Probeflüge in Koblenz bzw. Karlsruhe ausführten. Vgl. ebd. Siehe auch die Berichterstattung in Flugsport (1913), Nr. 10, S.352: „Nachdem bis abends der Goedecker-Eindecker [am 10. Mai] nicht eingetroffen war und Hirth durch Kleinholz ausschied, traten die Reserve-Flieger Suwelack und Carganico offiziell in den Wettbewerb ein."

831 Zeyssig, Prinz Heinrich Flug, S.372.

832 Ebd., S.458.

833 Zeyssig, Prinz-Heinrich-Flug (Fortsetzung und Schluß), S.459.

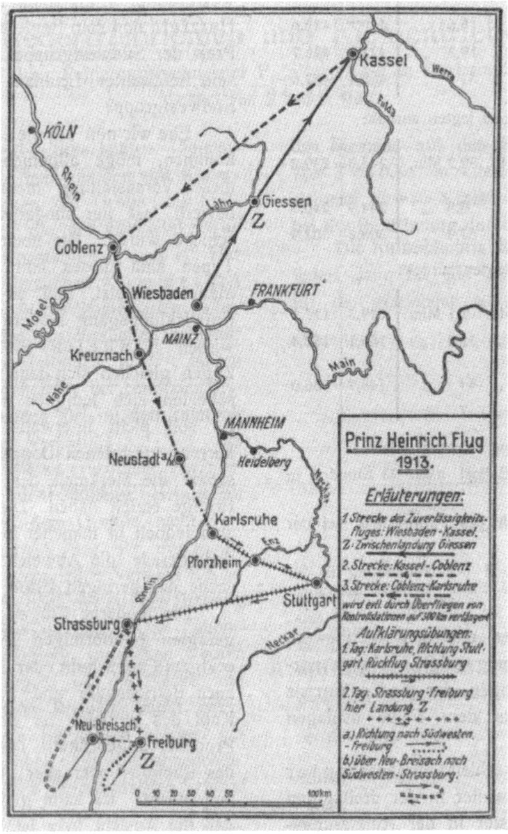

Abb. 16: Flugstrecke des Prinz-Heinrich-Flugs 1913 (aus: Paul Béjeuhr, „Technischer Rückblick auf den Prinz-Heinrich-Flug“, in: Dinglers Polytechnisches Journal 94 (1913), H. 26, Bd. 328, S. 401)

Der für 3:30 Uhr morgens angesetzte erste Flugtag am 11. Mai begann mit windigem und regnerischem Wetter, so dass der Start verschoben wurde. Schließlich gab Prinz Heinrich den Start für 7:30 Uhr frei, denn trotz der nach wie vor schlechten Witterung habe er „Vertrauen zu der im letzten Jahr wesentlich geförderten Ausbildung“ der Flugzeugführer und sei der Ansicht, dass „bei derartigen Wind- und Regentagen im Kriege unbedingt geflogen werden“ müsse.[834] Auch später in der 1914 erschienenen „Denkschrift über den Prinz-Heinrich-Flug“ formulierte man als langfristiges Ziel „die vollkommene Unabhängigkeit vom Wetter“, räumte aber ein, dass man sich noch im „Entwicklungsstadium“ befinde.[835]

Der Flug stellte Josef Suwelack, der sich mit der Startnummer 23 und Oberleutnant von Ascheberg als Passagier als Reserveflieger beteiligte, sowie einige andere

834 Denkschrift über den Prinz-Heinrich-Flug, S. 16. Siehe auch die Berichterstattung im „Berliner-Tage- und Handelsblatt“ vom 13. bis 16.5.1913.

835 Denkschrift über den Prinz-Heinrich-Flug, S. 68.

Flieger vor Herausforderungen, denn die zurückzulegenden Etappenstrecken waren lang und das Wetter zunächst unbeständig, so dass mit technischen Problemen zu rechnen war. Die Zeitschrift „Der Motorwagen" berichtete z. B.:

> „Schon die erste Etappe Wiesbaden-Gießen stellte an beide [Flugzeug und -führer] große Anforderungen. Die Apparate mußten Hagelböen durchfliegen, die auf den Propellern sichtbare Zeichen hinterließen. Alle Propeller waren an dem Angriffspunkt vollständig von Hagelkörnern zerkratzt, so daß auf der ganzen Länge und einer Breite von 1–2 cm die Politur verschwunden war und das Holz in der Naturfarbe aufgerauht zum Vorschein kam."[836]

Mit Ausnahme eines einzigen Apparates hätten alle Piloten „aus den verschiedensten Gründen kleinere und größere Reparaturen vornehmen" müssen, die sie entweder zur Aufgabe gezwungen oder schlechte Flugzeiten zur Folge gehabt hätten, was bei der „schweren Konkurrenz" jedoch nachvollziehbar gewesen sei.[837] Lediglich der Flieger Schlegel habe seinen Apparat ohne Beschädigungen und Reparaturen bis ins Ziel gebracht: „Daß der Apparat nicht besser placiert wurde, hatte seinen Grund darin, daß der Pilot in Straßburg in folge Nervosität aufgab."[838] Insgesamt aber hätten die Piloten ihre Maschinen ruhig gelenkt und beherrscht. Zeyssig war überzeugt, dass „[w]enn Schlegel nicht mehrfach vom Pech verfolgt worden wäre, – einmal versagte sein Beobachter, einmal lief ihm ein Platzarbeiter [in Neustadt] in den laufenden Propeller und zuletzt verlor er infolge der Aufregung selbst die Ruhe" – seine Platzierung besser ausgefallen wäre.[839]

Es ist auffällig, dass auch hier häufig ohne Empathie über Unfälle und Abstürze berichtet wurde und stattdessen eine hohe Risikoakzeptanz überwog. Über den „Platzarbeiter", der in den Propeller geraten war, zeigte man sich weder im „Motorwagen" noch in der „Denkschrift über den Prinz Heinrich Flug" schockiert, sondern betonte stattdessen, dass keine schweren Verletzungen der Piloten und kein Todessturz zu verzeichnen gewesen wären.[840] Zu Leutnant Geyer, der bei einem Probeflug in Wiesbaden abstürzte, aus dem zerstörten Apparat „herausgeschleudert" wurde und sich „den linken Unterschenkel und den linken Unterarm" gebrochen hatte, äußerte man sich im „Motorwagen" zuversichtlich: „Die Heilung verläuft sehr normal, sodaß genannter Herr schon mit Sicherheit hofft, in 4 Wochen wieder fliegen zu können." Ein weiterer Flieger hatte sich beim Sturz ein „scharfkantiges Zigaretten-Etuis [sic!] ... in den Unterleib" geklemmt und sich die Nieren verletzt, auch hierüber ließ man nichts weiter verlauten.[841]

In einer Denkschrift wurden die gesammelten Erfahrungen aus dem Prinz-Heinrich-Flug detailliert behandelt und über die künftig anzustrebende Entwick-

836 Zeyssig, Prinz Heinrich Flug, S.372.
837 Ebd., S.373.
838 Ebd.
839 Ebd.
840 Denkschrift über den Prinz-Heinrich-Flug, S.25.
841 Beide Zitate ebd.

lung der Flugtechnik spekuliert. Man verglich die teilnehmenden Flugzeuge miteinander und ging auf Gemeinsamkeiten und Unterschiede der Konstruktionen ein:

> „Von Grade abgesehen und den älteren Typen ... wiesen alle Apparate ... folgende charakteristische Merkmale auf: den vorne liegenden Motor, den geschlossenen bootsförmigen Rumpf, die hinten am Rumpf befestigten Steuerflächen. Für die Verminderung des Luftwiderstandes scheint diese Anordnung wohl die beste zu sein. Sie war am vollendetsten durchgebildet beim Hirthschen Eindecker ... . Auch der Torpedorumpf des Kondor-Eindeckers hat diesem viel an Geschwindigkeit eingebracht."[842]

In einer technischen Zusammenstellung wurden die Gesamtzahl an Notlandungen sowie die Ursachen und die Häufigkeit der Motordefekte aufgeführt. So habe es beim Zuverlässigkeitsflug „36 Notlandungen" gegeben, „davon 16 wegen Motordefekts und 20 aus anderen Ursachen (Benzineinnahme, Orientierung, starken [sic!] Böen usw.)." Die Motordefekte hätten sich „prozentual ganz gleichmäßig" auf die Marken Mercedes, Argus und Gnôme verteilt.[843]

Der Zuverlässigkeitsflug war die wichtigste fliegerische Großveranstaltung des Jahres 1913 und lieferte dem Militär, der zivilen Flugzeugindustrie und den Zulieferern neue Erkenntnisse hinsichtlich notwendiger Verbesserungen an den Motoren, der Organisation der Abläufe (Wartungen im Vorfeld, Abstimmungen etc.), der Nachschubversorgung und der Wetter(un)abhängigkeit der Flugzeuge. Auch Josef Suwelack leitete aus seinen Erfahrungen beim Prinz-Heinrich-Flug Maßnahmen zur Verbesserung seines Kondors ein. Er nahm Veränderungen am Fahrgestell vor, reduzierte das Gewicht und wollte eine größere Schnelligkeit mit der Anbringung eines Stirnkühlers zur Verringerung des Luftwiderstandes erreichen.[844]

Suwelack hatte die erste Etappe erfolgreich beendet, jedoch während des Fluges Orientierungsschwierigkeiten gehabt und war von der direkten Fluglinie abgewichen: „Die Flugzeiten der *Kondor*-Taube von Ingenieur Suwelack (Nr. 23) sind wohl nicht mit in Vergleich zu ziehen, da Herr Suwelack, wie aus dem Bordbuch hervorgeht, große Umwege gemacht hat."[845] Dem „Flugsport" zufolge habe er sich „in die Gegend von Worms verflogen" und sei „infolge des großen Zeitverlustes nach Wiesbaden zurück[gekehrt], um den Flug nachmittags nochmals auszuführen."[846] Die Zeitungen berichteten erneut von Zuschauermassen – das „Berliner Tage- und Handelsblatt" sprach z. B. von „30.000 Zuschauern", die die Flieger bei ihrer Ankunft in Kassel jubelnd empfangen und dazu lange ausgeharrt hätten: „Als erster traf dann um 12 Uhr 24 Minuten Leutnant v. Hiddessen unter großem

842 Denkschrift über den Prinz-Heinrich-Flug, S.49.
843 Ebd., S.46.
844 NL J. Suwelack, Brief J. Suwelacks an Karl Goldschmidt vom 21.5.1913.
845 Denkschrift über den Prinz-Heinrich-Flug, S.59. Hervorhebung im Original.
846 „Der Verlauf des Prinz Heinrich-Fluges", in: Flugsport (1913), Nr. 10, S.352ff., hier S.354.

Jubel der Zuschauer ein."[847] Mit Josef Suwelack, der als letzter Starter gegen 20 Uhr in Kassel landete, endete der erste Flugtag.[848]

Am 12. Mai musste Suwelack kurz hinter Karlsruhe notlanden und traf gegen 19:20 Uhr in Koblenz ein, denn aufziehende Gewitter hatten den Start für einige Flieger erst in den frühen Nachmittagsstunden möglich gemacht. Während der dritten Etappe am folgenden Tag führte Suwelack die geforderte Zwischenlandung in Heidelberg durch, rollte dann aber beim anschließenden Start in eine Baumgruppe[849] (Abb. 17) und brachte „einen schwer havarierten Apparat nach Essen zurück".[850] Sein Beobachter und er selbst blieben Berichten zufolge unverletzt.

Die Aufnahme vom „verunglückten Start" zeigt die hier gezielt inszenierte Massenwirksamkeit der Fliegerei einerseits und die militärische Ausrichtung anderseits, denn die Nummerierung am unteren linken Bildrand legt nahe, dass es sich um eine professionelle Aufnahme handelt, wie sie in ähnlicher Form auch in Zeitschriften zu finden waren.[851] Aufgrund der entstandenen Schäden am Kondor gab Josef Suwelack, der sich nicht auf der Aufnahme befindet, vorzeitig auf.

Wenngleich die gewonnenen Etappengelder durch die Reparaturen am Apparat rasch wieder aufgebraucht waren, erhielt Suwelack als bleibende Erinnerung den „Ehrenpreis der Stadt Heidelberg für die beste Flugleistung zwischen Coblenz und Heidelberg" in Form eines Pokals sowie einen „Ehrenpreis der Südwestgruppe", einen versilberten Teller mit Gravur, gestiftet von „Wiesbadener Freunden des Flugsports" (Abb. 18).[852] Außerdem erhielten die Kondor-Werke als eines von fünf weiteren Flugzeugwerken eine Geldentschädigung in Höhe von 1.500 Mark[853]

Nachdem am 17. Mai die letzten Aufklärungsübungen beendet waren, fand in Straßburg am Abend ein Schlussbankett im Hotel Stadt Paris unter dem Vorsitz Prinz Heinrichs mit 230 geladenen Gästen statt:[854]

> „Ferner waren außer den meisten Teilnehmern am Fluge erschienen Prinz Waldemar von Preußen, Prinz Georg von Bayern, der kaiserliche Statthalter von Elsaß-Lothringen Graf Wedel, Graf Zeppelin, der kommandierende General des 15. Armeekorps v. Deimling, Gouverneur Generalleut-

847 Berliner Tage- und Handelsblatt vom 13.5.1913, 1. Beilagenblatt, dort beide Zitate.

848 Vgl. ebd.

849 Vgl. „Die dritte Etappe des Prinz-Heinrich-Fluges", in: Berliner Tage- und Handelsblatt vom 14.5.1913 sowie Denkschrift über den Prinz-Heinrich-Flug, S.26.

850 Zeyssig, Prinz Heinrich Flug, S.373.

851 Siehe z.B. „Momentbilder von Zuverlässigkeits-Etappen des Prinz-Heinrich-Fluges 1913", in: Sport im Bild (1913), Nr.21, S.598–599, online unter: http://anno.onb.ac.at/cgi-content/anno-plus?aid= sib&datum=1913&page=606&size=45 (9.7.2014).

852 Denkschrift über den Prinz-Heinrich-Flug, S.74–75.

853 Auf die Deutschen Flugzeugwerke Leipzig-Lindenthal entfielen 9.500 M, auf die Automobil- und Aviatik AG 7.000 M; die Gothaer Waggonfabrik erhielt 8.000 M und die Albatros-Werke sowie die Luftverkehrsgesellschaft je 6.000 M, Angaben nach Jahrbuch des DLV 1914, S.166.

854 Vgl. „Prinz Heinrich-Flug 1913 (Fortsetzung und Schluß)", in: Flugsport (1913), Nr.11, S.393ff., hier S.400 sowie den Bericht im „Berliner Tage- und Handelsblatt" vom 19.5.1913, 1. Beilagenblatt: „Der Schluß des Prinz-Heinrich-Fluges".

Abb. 17: „Verunglückter Start“, Prinz-Heinrich-Flug, Heidelberg, 14. Mai 1913 (Foto: Familienarchiv Walter Suwelack, NL Josef Suwelack)

Abb. 18: Ehrenpreis der „Südwestgruppe“ für Suwelacks erfolgreiche Teilnahme am Prinz-Heinrich-Flug im Mai 1913 (Foto: Privatbesitz)

nant v. Eberhardt, Staatssekretär Zorn v. Bulach, Geheimrat Professor D. Hergesell, der Polizeipräsident und der Bürgermeister sowie zahlreiche hervorragende Angehörige flugtechnischer und industrieller Kreise.[855]

Vor diesem erlesenen Kreis hielt Prinz Heinrich eine Begrüßungsrede, in der er die Fähigkeiten der Piloten pries und sie als charakter- und willensstarke Männer präsentierte, die der Armee wertvolle Dienste leisteten und noch leisten würden: „Mit anderen Worten, die Herren Flieger sind vorzüglich durch- und ausgebildet. Sie haben ihre Maschinen vorzüglich in der Hand, sie riskieren bedeutend mehr als in den vergangenen Jahren, sie sind sicherer und ihrem Motor überlegen."[856] Auf die militärische Relevanz des Flugzeuges verwies der Prinz mit einem Vergleich zur Marine, dem außenpolitischen Prestigeprojekt des Kaiserreiches im Wettrüsten mit England um Weltgeltung.[857] Der Schutzherr des 1898 gegründeten Deutschen Flottenvereins und seit 1909 Generalinspekteur der Marine hatte den „Flottengedanken" unterstützt, verglich die Flugtechnik mit dem militärischen Seewesen und unterstrich den hohen Wert der Technik in der Bildung und Erziehung:

> „Das Torpedoboot für die Armee ist das Flugzeug. Der Charakter wird gestählt, hohe Anforderungen an den einzelnen Menschen und die Willenskraft werden gestellt. Er kommt unwillkürlich mit der Technik in Berührung. Die Technik ist auch für den Menschen bildend und erziehend. Die Armee sollte dankbar sein, daß sie einen großen und später ... noch größere Kreise in dieses Mittel hineinziehen und erzieherisch ausbilden kann."[858]

Im Anschluss an das Bankett, so berichtete der „Flugsport", habe der Prinz den Teilnehmern „eigenhändig die vom Preisgericht zuerkannten Preise" im Sachsencasino überreicht.[859]

Leutnant Ernst Canter, der sowohl die beste Gesamtflugzeit als auch die beste Leistung bei den Aufklärungsflügen vorzuweisen hatte, erhielt den „Kaiserpreis", den „Prinz Heinrich-Preis der Lüfte" sowie den zweiten Zuverlässigkeitspreis, den „Ehrenpreis des bayerischen Kriegsministers". Den ersten Platz im Zuverlässigkeitsflug belegte Leutnant Ferdinand von Hiddessen, er erhielt den mit 5.000 Mark dotierten Preis aus der Nationalflugspende.[860]

Über den Ausgang des Fluges äußerte sich die berichtende Presse zufrieden und bewertete die Veranstaltung als fest etablierte Leistungsschau, die auch künftig fortgeführt werden müsse: „Die Zuverlässigkeitsflüge der Südwestgruppe des D.L.V. ... haben sich mittlerweile zu einem klassischen Flugwettbewerb ausgebil-

855 Berliner Tage- und Handelsblatt vom 19.5.1913, 1. Beilagenblatt.
856 Vgl. „Prinz Heinrich-Flug 1913 (Fortsetzung und Schluß)", in: Flugsport (1913), Nr. 11, S. 393ff., hier S. 400.
857 Siehe zur Flottenpolitik im Deutschen Kaiserreich: Rolf Hobson, Maritimer Imperialismus. Seemachtideologie, seestrategisches Denken und der Tirpitzplan 1875 bis 1914, München 2004.
858 „Prinz Heinrich-Flug 1913 (Fortsetzung und Schluß)", in: Flugsport (1913), Nr. 11, S. 400.
859 Ebd.
860 Vgl. ebd., S. 400f.

det, auf den die Augen der ganzen Welt gerichtet sind."[861] Prinz Heinrich hatte in seiner Rede zugesichert, dass er auch künftig die Schirmherrschaft des Wettbewerbs übernehmen werde, und die Teilnehmer eingeladen, sich erneut zu beteiligen.[862]

Vom Lokal- zum Militärhelden: Die Etablierung einer männlichen Fliegerelite bis zum Ersten Weltkrieg

Die Piloten und die Flugzeugtechnik waren seit dem Beginn der motorisierten Fliegerei in eine mediale Öffentlichkeit eingebunden. Die Fliegerei symbolisierte Fortschrittlichkeit, Macht und nationale Größe.[863] Flugwettbewerbe als Felder des Kräftemessens hatten damit häufig nicht nur eine sportliche, sondern auch eine ideologisch-politische Funktion. Mit der zunehmenden Förderung des Militärflugwesens und der zivilen Luftfahrt durch Spendensammlungen wie die Nationalflugspende oder mit der Schaffung spezieller Aufklärungs- und Überlandflug-Veranstaltungen wie dem Prinz-Heinrich-Flug avancierte die Fliegerei zu einem ambitionierten nationalen Projekt. Ziel war es, die Entwicklung leistungsfähiger und zuverlässiger Flugzeuge staatlich zu subventionieren, Nachwuchspiloten zu rekrutieren und eine technische Vorreiterrolle einzunehmen.

Über ein gestaffeltes Preissystem, vom Königshaus oder zentralen militärischen Stellen gestiftete Ehrenpreise und mit der Aussicht, Zugang zum elitären Feld der Militärflieger zu erhalten, wurden Anreize für viele junge Männer geschaffen, sich in den Dienst der Fliegerei oder der Fliegertruppe zu stellen. Für einen Fernflug-Wettbewerb, bei dem mindestens 1.000 Kilometer als Tagesleistung zurückgelegt werden mussten, waren z. B. 300.000 Mark an Preisgeld aus der Nationalflugspende für die erfolgreichsten Teilnehmer ausgelobt.[864] Die hohe Motivation zahlreicher junger Männer – die häufig wie Josef Suwelack in den 1880er Jahre geboren worden waren –, sich als Flieger ausbilden zu lassen beziehungsweise sich freiwillig zur Fliegertruppe zu melden, erklärt sich auch aus diesen Kampagnen und der öffentlichen Präsenz der Fliegerei auf Wettbewerben.

Preisausschreiben oder Prämienzahlungen aus dem Fonds der Nationalflugspende bewirkten einen enormen Innovationsschub. Somit wurde die Schaffung von Prüfstellen und Organisationen notwendig, die durch Reglementierungen und Konstruktionsvorgaben gezielt Einfluss auf die Flugzeugentwicklung nah-

861 Ebd., S.401.
862 Vgl. Berliner Tage- und Handelsblatt vom 19.5.1913, 1. Beilagenblatt.
863 Vgl. Kehrt, Moderne Krieger, S.33f.
864 Vgl. „Die Preisträger der Nationalflugspende", in: Sport im Bild (1913), Nr.50, S.1544, online unter: http://anno.onb.ac.at/cgi-content/anno-plus?aid=sib&datum=1913&page=1552&size=45 (17.7.2014).

men. Wenngleich sich das Flugwesen im zeitgenössischen Fortschrittsdenken weiterentwickelte, so blieb die Militärverwaltung auch noch während des Krieges auf die Rückmeldungen ihrer Offiziers- und Zivilflieger angewiesen. Innovationen wurden deshalb nicht nur oktroyiert, sondern auch von den Fliegern selbst eingebracht, so dass sich die Militärfliegerei im wechselseitigen Erfahrungsaustausch zwischen der Industrie als Basis und der Militärbehörde als Kontrollinstanz entwickelte.

Bis Ende des Jahres 1913 fanden noch weitere zehn Flugwettbewerbe statt,[865] von denen der Ostpreußische Rundflug im August sowie die Herbstflugwoche in Johannisthal von Ende September bis Anfang Oktober mit Preisen des Kriegsministeriums dotiert waren. Der Ostpreußische Rundflug, an dem sich Josef Suwelack ebenfalls beteiligte, war offen für deutsche Offiziere auf Flugzeugen der Heeresverwaltung sowie für „andere Flugzeugführer deutscher und österreichischer Reichsangehörigkeit, sofern dieselben gegen Haftpflicht versichert sind (500.000 Mk. bei Katastrophen, 100.000 Mk. im Einzelfalle)".[866] Gefordert wurden Zuverlässigkeitsflüge und Aufklärungsübungen in Verbindung mit einem Fotografie- und Bombenabwurf-Wettbewerb sowie einem Wettbewerb für den kürzesten Auslauf. Etappenziele waren dieses Mal Pillau, Insterburg und Alleinstein, Start- und Zielpunkt war Königsberg.[867] Für Josef Suwelack endete die Veranstaltung mit einem Armbruch, denn seine Maschine war von einer Windböe erfasst worden und daraufhin auf dem Boden aufgeschlagen.[868]

Statistisch gesehen waren die Unfallzahlen mittlerweile jedoch rückläufig, denn die Anzahl der Flugzeugbeschädigungen ging – bezogen auf die Gesamtzahl der Flüge – von 0,84 auf 0,31 Prozent zurück, und tödliche Unfälle verringerten

865 Wasserflugzeugwettbewerbe ausgenommen: Schlesische Flugwoche (8.–15. Juni); Kieler Flugwoche (9.–16. Juli); Rundflug „Essen-Gelsenkirchen" (26. Juli–3. August); Ostpreußischer Rundflug (9.–14. August); Aeroplanturnier Gotha (16.–18. August); Leipziger Flugveranstaltung (23.–25. August); „Rund um Berlin" (30. u. 31. August); Dresdener Herbstflugwoche (14–21. September); Herbstflugwoche Johannisthal (28. September–5. Oktober); Herbstveranstaltung in München (11. u. 12. Oktober). Angaben nach Flugsport (1913), Nr. 7, S. 263, und (1913), Nr. 8, S. 304–306. Siehe auch die Übersicht bei Schwipps, Schwerer als Luft, S. 99f.

866 „Wettbewerbe", in: Flugsport (1913), Nr. 14, S. 521–522, hier S. 521. Mit Geldern aus der Nationalflugspende wurden bis zum Ende des Jahres 1913 361 deutsche Piloten gegen Unfall versichert, was auf lange Verhandlungen des Kuratoriums der Nationalflugspende mit den Versicherern zurückzuführen war, die sich bis dahin geweigert hatten, Zivilflieger zu versichern. Die Versicherung kostete 400 Mark, von denen die Nationalflugspende für Flugschüler die Kosten vollständig, und für Zivilflieger zur Hälfte übernahm. Vgl. Schwipps, Schwerer als Luft, S. 102.

867 Vgl. ebd.

868 Vgl. Allgemeine Automobil-Zeitung Nr. 35, Berlin 1913, S. 17. Ob er trotzdem einen Preis erhielt, ist nicht bekannt. In der „Berliner-Volkszeitung" wurde vermerkt: „Von den Offiziersfliegern erhielt den ersten Preis (Kaiserpreis) für die beste Zeit in allen Etappenflügen Leutnant Pretzell ... Von den Zivilfliegern erhielt für die beste Gesamtflugzeit den ersten Preis (Preis der Nationalflugspende 12.000 Mark) der Flieger Alfred Friedrich ... An die Offizier- und Zivilflieger und die Beobachtungsoffiziere wurden außerdem zahlreiche Ehrenpreise verteilt.", „Die Preisverteilung im ostpreußischen Rundflug", in: Berliner-Volkszeitung vom 15.8.1913.

sich um die Hälfte.[869] Allerdings bestand nach wie vor die tägliche Gefahr von zum Teil schweren Unfällen, denn die Ansprüche an Militärflugzeuge stiegen durch eine geforderte kürzere Steig- und Landedauer, größere Reichweite oder höhere Tragkraft. Beim Prinz-Heinrich-Flug im Mai 1914, der dieses Mal in fünf Etappen aufgeteilt war und über eine Strecke von mehr als 2.600 Kilometern führte – fast die fünffache Distanz im Vergleich zum Vorjahr –, starben vier Offiziere.[870]

Zur stetigen Verbesserung der Sicherheit und Leistungsfähigkeit der Flugzeuge waren seit 1912 neue wissenschaftliche Prüfstellen und Organisationen gegründet worden, deren Aufgabe es war, Prüfmethoden und Flugzeugstandards zu entwickeln, die zuverlässig Auskunft über den Stand und die weitere Entwicklung der Flugtechnik geben sollten, obschon von einer „Flugwissenschaft" noch nicht gesprochen werden kann.[871] So war am 3. April 1912 die Gründung der Wissenschaftlichen Gesellschaft für Flugtechnik in Berlin erfolgt, die Anfang 1913 vom Kuratorium der Nationalflugspende beauftragt wurde, die von zahlreichen Konstrukteuren eingereichten Vorschläge zu sichten. Hatte die Gesellschaft eine Erfindung für brauchbar befunden, reichte sie ihre Empfehlung bei der Versuchsabteilung des Militärverkehrswesens ein, die wiederum Gutachter damit beauftragte, die militärische Tauglichkeit zu prüfen.[872]

Basierend auf den Erkenntnissen und gesammelten Erfahrungen der Prüfstellen, den Wettbewerbsergebnissen und gehaltenen Vorträgen etc. kam es zur Gründung der Verkehrstechnischen Prüfungskommission (VPK), die durch Kabinettsorder des Kaisers vom 22. November 1913 geschaffen worden war und seit dem 1. Dezember dem Allgemeinen Kriegsdepartement des Kriegsministeriums unmittelbar unterstellt war.[873] Als direkte Verbindung zwischen Heeresverwaltung und Industrie war sie eine wichtige Anlaufstelle für beide Seiten und

869 Vgl. MGFA, Militärluftfahrt Technischer Band 1966, S.81.

870 Vgl. „Zum Prinz Heinrich-Flug", in: Flugsport (1914), Nr. 11, S.438. Siehe auch die Kritik zu den ungleichen Wettbewerbsvoraussetzungen für Militär- und Zivilflieger: „Johannisthaler Brief", in: ebd., S.454–456. Zur Bedeutung der Prinz-Heinrich-Flüge siehe z. B. Hauptmann a. D. Dr. Alfred Hildebrandt, Die Bedeutung des Prinz-Heinrich-Fluges, in: Sport im Bild (1914), Nr. 20, S.602–606, online unter: http://anno.onb.ac.at/cgi-content/anno-plus?aid=sib&datum=1914&size=45&page=610 (11.5.2014).

871 Vgl. MGFA, Militärluftfahrt Technischer Band 1966, S.113.

872 Vgl. ebd., S.711f. Die VPK wurde zur zentralen Abnahmestelle für Flugzeuge. Die Abnahme erfolgte in zwei Schritten: Eine provisorische Abnahme testete die Sicherheit und Stabilität der Flugzeuge, anschließend musste die Flugzeugfirma die Maschine zu einer Einheit der Fliegertruppe überführen und dort erneut montieren, um sie dann von einem Firmenpiloten einfliegen zu lassen. Anschließend musste sie von einem Militärpiloten geflogen werden und galt erst dann als tatsächlich abgenommen. Allerdings behielt sich das Militär vor, bei Schäden, die innerhalb der ersten acht Wochen auftraten, den Apparat zurückzugeben. Vgl. Schwipps, Schwerer als Luft, S.114, und BA-MA PH 9-XV/2, Ausarbeitung einer Fliegervorschrift: „Besondere Bedingungen" 1913 (Flugzeugabnahme).

873 Vgl. ebd., S.80.

befasste sich in der „Abteilung V Fliegerwesen" in verschiedenen Sektionen mit der Flugzeugkonstruktion, der Motorenfrage, der Fotografie sowie dem Waffengebrauch, kurz: mit kriegsrelevanten Fragen der Fliegerei.[874]

Den Stand des Militärflugwesens zum Jahresende 1913 bewertete der Bericht der General-Inspektion des Militär-Verkehrswesens vom 1. November des gleichen Jahres wie folgt:

> „In der Konstruktion der Flugzeuge macht sich eine Annäherung an Einheitstypen geltend, beim Eindecker die Taubenform, beim Doppeldecker zum Rumpf-Doppeldecker mit vorn liegendem Motor. Die Geschwindigkeit und Steigfähigkeit der Flugzeuge und die Zuverlässigkeit der Motoren sind bedeutend verbessert. Der größeren Sicherheit der Flugzeuge wird weiter große Beachtung beigemessen. ... Es kann jetzt mit einer Durchschnittsleistung von 80 bis 90 km/h und einem Aktionsradius von 300 bis 350 km gerechnet werden."[875]

Ferner hätten die Flugzeuge ihre Manöverfähigkeit sowie ihre Verwendbarkeit zum Artillerieschießen und zum gezielten Bombenabwurf bewiesen, wenngleich die Trefferquote noch verbesserungsfähig sei.[876]

Das Bestreben der Militärverwaltung, durch Vorschriften größere Einheitlichkeit und höhere Betriebssicherheit zu erzielen, bestand bereits seit 1912. Damit Flugzeuge abgenommen wurden, mussten sie fortan über eine sogenannte „Militärsteuerung" verfügen, einen Mechanismus, bei dem der Steuerhebel zur Betätigung des Höhensteuers mit den Verwindungsklappen gekoppelt war und das Seitensteuer mit den Füßen bewegt wurde.[877] Durch die Interaktion zwischen Mensch und Maschine sollte die Unfallgefahr verringert, intentionales Handeln geschult und eine Entlastung des Flugzeugführers erreicht werden.[878] Die Bedeutung von implizitem Wissen und Automatismen hatte man in der wissenschaftlichen Physiologie frühzeitig erkannt und auf die hohen Anforderungen an den Piloten sowie auf die Auswirkungen der Fliegerei auf das Kreislaufsystem verwiesen.[879] Die Instrumente sollten daher bedienerfreundlich angeordnet und intuitiv zu bedienen sein – eine Meinung, die nicht nur die Wissenschaft, sondern auch

874 Vgl. MGFA, Militärluftfahrt Technischer Band 1966, S.97f., sowie Supf, Fluggeschichte Bd. 2, S.120.

875 Bericht der General-Inspektion des Militär-Verkehrswesen vom 1.11.1913 zit. nach MGFA, Militärluftfahrt Technischer Band 1966, S.88.

876 Vgl. ebd.

877 Erblich, Fliegerschule, S.62f.

878 Vgl. Kehrt, Moderne Krieger, S.120.

879 Nathan Nuntz, Zur Physiologie und Hygiene der Luftfahrt, Berlin 1912, S. 9: „Unsere Sinne lenken unser Verhalten in doppelter Weise, einmal, indem sie bewußte Vorstellungen vermitteln, die dann unser Handeln bestimmen, und andererseits, indem auf die Sinneseindrücke *unbewußt* Reaktionen unseres Bewegungsapparates erfolgen, die wir als Sinnesreflexe bezeichnen. ... Von höchster praktischer Bedeutung werden diese Dinge für den Flieger, der in jedem Moment des Fluges auf die feinste Verschiebung der Lage seines Flugzeuges reagieren muss, zum Teil so schnell reagieren muss, dass jede bewußte Überlegung ausgeschlossen ist. ... Darum kann die Physiologie durch exaktes Studium der auf Gleichgewichtsstörungen erfolgenden unbewußten Bewegungen wichtige Anhaltspunkte für die Konstruktion der Aeroplane resp. ihrer Steuereinrichtungen liefern". Siehe auch ebd., S.35ff.

Flieger und Techniker vertraten. So ging beispielsweise Oskar Ursinius, der Herausgeber des „Flugsport", davon aus, dass solche Steuerungen am leichtesten zu bedienen seien, „die sich an die instinktive Gleichgewichtserhaltung des Menschen anschließen."[880] Langfristig sollte eine größere Einheitlichkeit im Flugmaschinenbau geschaffen werden, um standardisierte Flugzeugtypen zu produzieren, ohne sich stets auf neue Systeme umstellen zu müssen.

Mit der militärischen Ausrichtung der Flugwettbewerbe, der Förderung der Flugzeugindustrie durch das Kriegsministerium und Geldmitteln aus der Nationalflugspende wurde die Fliegerei allmählich offiziell „entsportlicht" und eine „militärische Funktionalisierung" vollzogen.[881] Das Leistungsprinzip drängte kleinere Flugzeugfirmen, aber auch weibliche Fliegerinnen, letztere waren zu den Militärwettbewerben ohnehin nicht mehr zugelassen, aus den nationalen Wettbewerben heraus und führte langfristig zur Dominanz von Militärpiloten. Disziplin, Ausdauer sowie mentale und körperliche Stärke waren Anforderungen an Flugzeugführer, wie sie bereits in der Vorkriegszeit systematisch auf Wettbewerben und Überlandflügen trainiert und schließlich in einer Denkschrift Leutnant Müllers vom Fliegerbataillon I in Döberitz aus dem Jahr 1914 expliziert wurden:

> „Männer fordert der Dienst, aber Männer von gefestigter Gesinnung und einwandfreiem Charakter fordert der Dienst als Flieger. ... Besonnenheit, aber doch rasche Entschlußkraft unter den schwierigsten Verhältnissen muß man von ihm fordern. ... Er muß gewandt und kräftig sein; sich bewußt der großen Verantwortung, die er bei Mitnahme eines Beobachters hat, muß er sein ganzes Können einsetzen, um der großen Sache zu dienen."[882]

Josef Suwelack passte sich diesem „gestählten", „abgehärteten" militärischen Pilotenhabitus an und optimierte auch seine Konstruktionspläne entsprechend den gestellten Anforderungen. Die Berührungen mit der Militärverwaltung und die Partizipation an militärischen Wettbewerben wie dem Prinz-Heinrich-Flug 1913 oder dem Ostpreußischen Rundflug im August des Jahres hatten ihn z. B. dazu veranlasst, Veränderungen am Kondor-Eindecker vorzunehmen. Der Erfahrungsaustausch zwischen den Firmen und Suwelacks fotografisch-bildliche „Inspektion" der Konkurrenz-Maschinen gaben Anstoß für Innovationen und offenbarten zudem, ob die Kondor-Werke den gestiegenen Ansprüchen gerecht wurden.

An der Wettbewerbsteilnahme Josef Suwelacks lässt sich die Entwicklung des Flieger-Sports exemplarisch aufzeigen. Waren die ersten Wettbewerbe prinzipiell für jeden kapitalkräftigen Amateur (gleichgültig ob Mann oder Frau) und jedes Flugzeug offen (typen-, motoren- und „marken"-unabhängig), so stiegen die Anforderungen an Flugzeuge und Piloten bedingt durch höhere Erwartungen,

880 Oskar Ursinius, Wissenswertes über die Konstruktionen der Betätigungseinrichtungen für die Steuerorgane, in: Flugsport (1910), Nr. 7, S. 194.

881 Möser, Fahren und Fliegen, S. 107.

882 BA-MA PH 9-XV/2, Ausarbeitung einer Fliegervorschrift: „Der Flugzeugführer. A. Anforderungen" von Leutnant Müller vom Fliegerbataillon I in Döberitz, 7.1.1914.

strengere Vorschriften und eine stärkere Auslese. Bis zum Ausbruch des Ersten Weltkrieges hatten sich einige Flugzeugfirmen wie Albatros, Aviatik, Rumpler, LVG oder Gotha[883] zu „Marktführern“ in der Flugzeugbranche entwickelt, die es mittelständischen Unternehmen wie den Kondor-Werken durch eine gute und beständige Auftragslage erschwerten, in die Spitzenpositionen vorzustoßen. Gleichwohl war der Militärverwaltung daran gelegen, stets neue Impulse, vor allem finanzieller Art, zu geben und den Bau sowie die Verbesserung von Flugzeugen zu fördern. Trotz neuer Vorschriften und „Anregungen für die Flugzeugfirmen“ blieb die Flugzeugindustrie ein Experimentierfeld und Flugwettbewerbe fungierten bis zuletzt als praxisnahe Leistungsschauen.

Obschon die Zuschauerzahlen konstant hoch blieben, verloren Flugwettbewerbe mit zunehmender „Entsportlichung“ auch den Kontakt zum Boden und die Nähe zu den Flugsportbegeisterten. Während lokale Veranstaltungen wie in Detmold, Münster oder Billerbeck individuelle Heldenfiguren hervorgebracht hatten und die Flieger des Öfteren persönlich bekannt waren, distanzierte sich die Militärfliegerei von Folklore und Artistik und war stärker auf Sicherheit und Stabilität fokussiert. Mit der Konzentration auf militärische Funktionalität und geprägt von nationalistischen Vorstellungen trat neben die Kunstflieger ein neuer militärischer Fliegertyp, der Teil einer neuen nationalen Elite war und nicht mehr länger nur als Repräsentant einer Stadt oder einer Region diente. Eine neue Fliegerelite, zu der nun auch Josef Suwelack zählte, wurde gezielt staatlich ausgebildet und durch reichsweit bekannte Förderer wie Prinz Heinrich von Preußen ideologisch vereinnahmt.

Flugwettbewerbe und hier vor allem die militärischen, ermöglichten den sozialen Aufstieg „nach oben“ einerseits und die Abgrenzung „nach unten“ andererseits. Der Flieger-Sport brachte ein „specific new pattern of leisure-activity and a lifestyle“[884] hervor, das besonders deutlich in den Posen der Flieger zum Ausdruck kam. Nur wer es verstand, selbstsicher und zugleich „locker“ aufzutreten, verschaffte sich Ansehen und Anerkennung in der Fliegerei: „Beides, Regelsetzung und Regeldurchbrechung, war in den Mobilitätskulturen inhärent. Dem Druck zur Integration ... und zur Selbstdisziplinierung entsprach auf der anderen Seite die demonstrative Befreiung von einengend empfundenen und systemischen Rahmenbedingungen.“[885] Die soziale Durchlässigkeit des Flieger-Sports verschob mitunter auch bestehende Hierarchien. Als Flugzeugführer übernahm Suwelack die Kontrolle über die Maschine und die Verantwortung für den mitfliegenden Oberstleutnant. Für die Dauer des Fluges hatte sich der ranghöhere

883 Vgl. MGFA, Militärluftfahrt Technischer Band 1966, S.103f.
884 Hobsbawm, Mass producing Traditions, S.300.
885 Möser, Fahren und Fliegen, S.115.

Beobachter dem Piloten unterzuordnen, wenngleich er arbeitsteilig für die Orientierung und die Eintragungen ins Bordbuches zuständig war.[886]

Mit simulierten Kriegsszenarien, Preisen für den besten Motor, den besten Flieger oder die zuverlässigste Maschine sollten das deutsche Flugwesen und seine Akteure kriegstauglich gemacht werden. Die Wettbewerbe beschränkten sich dabei nicht nur auf die Maschinen und ihre Piloten, sondern auch handwerkliche und logistische Aufgaben flossen in die Überlegungen mit ein. So wurden z. B. auch die erfolgreichsten Monteure prämiert und Preise für die schnellste Flugzelt-Montage sowie die größte Zelt-Stabilität verliehen.[887] Durch staatliche Förderprogramme und -kampagnen hatte das deutsche Flugwesen bis zum Sommer 1914 den Rückstand gegenüber Frankreich aufgeholt. Alle wichtigen Flugrekorde, ausgenommen der Geschwindigkeitsrekord, waren bis zum Kriegsausbruch in der Hand deutscher Piloten.[888]

Der Flieger-Sport hatte zwischen 1909 und 1914 individuelle Heldenfiguren geschaffen, die gesellschaftlich akzeptiert und hoch angesehen waren und die der Fliegerei zu einer enormen Popularität verhalfen. Die mediale Darstellung der Fliegerei und der Piloten in Zeitschriften und Zeitungen, auf Postkarten und Reklamemarken etc. verknüpfte traditionelle mit modernen Elementen und erlaubte es über den Rückgriff auf Mythen wie den sagenhaften Ikarus oder über Narrative des Fortschritts, der Eroberung und der nationalen Größe eine hochkomplexe Technikerfahrung mit Sinn zu versehen. Wettbewerbe als Kommunikationsorte zwischen Industrie, Militär und Öffentlichkeit waren Momentaufnahmen des Status Quo und ein Feld, in dem in einem reglementierten Rahmen sowohl über die Zukunft des Flugwesens als auch über Karrierechancen von Fliegern wie Josef Suwelack entschieden wurde. Piloten, die sich bereits zu Friedenszeiten einen Namen durch messbare Erfolge auf Wettbewerben gemacht hatten, meldeten sich bei Kriegsbeginn freiwillig zur Fliegertruppe, bestrebt, ihr geübtes fliegerisches Können im Krieg zu beweisen, an die „sportlichen" Siege anknüpfen zu können, und überzeugt, in einen gerechten, kurzen Verteidigungskrieg zu ziehen.[889]

886 Vgl. ebd., S.118ff.; Kehrt, Moderne Krieger, S.172ff. und Dr. H. Elias, Der Passagier im Flugzeug, in: Sport im Bild (1914), Nr. 20, S.612 und S.614, online unter: http://anno.onb.ac.at/cgi-content/anno-plus?aid=sib&datum=1914&size=45&page=620 (26.7.2014).

887 „Ausschreibung eines Wettbewerbes für transportable Flugzeug-Zelte", in: Flugsport (1912), Nr. 9, S.365. Veranstaltet wurde der Wettbewerb wie der Prinz-Heinrich-Flug von der Südwestgruppe des DLV in Zusammenarbeit mit dem preußischen Kriegsministerium und der Inspektion des Militär-, Luft- und Kraftfahrwesens im Anschluss an den 2. Deutschen Zuverlässigkeitsflug am Oberrhein.

888 Vgl. Schwipps, Riesenzigarren, S.97.

889 Vgl. z. B. „Deutsche Flieger!", in: Flugsport (1914), Nr. 16, S.677, sowie die Ausführungen zum Feld 4.

Feld 4: Die Fliegerei im Ersten Weltkrieg: Suwelacks Kriegserfahrung

4.1. Kriegsvorstellungen und Legitimationen

Szenarien und Prognosen für einen drohenden Krieg hatte es in Europa bereits seit den 1880er Jahren gegeben.[890] Politiker wie Friedrich Engels, Helmuth von Moltke oder Generalleutnant Colmar Freiherr von der Goltz, aber auch Publizisten wie Friedrich von Bernhardi, Bertha von Suttner, Herbert George Wells, Norman Angell oder der russische Geschäftsmann Ivan Bloch entwickelten Vorstellungen von möglichen kriegerischen Auseinandersetzungen.[891] Während Bernhardi in „Deutschland und der nächste Krieg" einen großflächigen Konflikt als „biologische Notwendigkeit"[892] ansah, führten Bloch und Angell rational-wissenschaftliche Argumente gegen einen Krieg an. Sie argumentierten, der technologische und wirtschaftliche Fortschritt würde alle bisherigen Konflikte in den Schatten stellen und in einem wirtschaftlichen, technologischen und menschlichen Verfall enden. Bloch sah voraus, dass die Fronten rasch erstarren würden und das Defensivsystem zu einem langen Abnutzungskrieg führen würde.[893] Auch Wilhelm Lamszus sprach in seinem Roman „Das Menschenschlachthaus" aus dem Jahr 1912 von einem Kampf Mensch gegen Maschine, bei dem „vom Feind ... nichts zu sehen", dafür aber die Luft mit „Eisendonner" erfüllt sein würde.[894]

Von einer Entgrenzung der Gewalt, einem maschinisierten Krieg[895] und einem

890 Vgl. die Ausführungen bei Jörn Leonhard, Die Büchse der Pandora. Geschichte des Ersten Weltkriegs. München 2014, S.67–74; Köppen, Das Entsetzen, S.201–218. Siehe zur ideellen und realen Aufrüstung auch die Ausführungen bei Möser, Fahren und Fliegen, S.487–497; ders., Technische Zukunftskriege vor dem Großen Krieg, in: Simone Finkele/Burkhardt Krause (Hg.), Technikfiktionen und Technikdiskurse. Ringvorlesung des Instituts für Literaturwissenschaft im Sommersemester 2009, Karlsruhe 2012, S.209–244.

891 Vgl. Leonhard, Die Büchse, ebd.

892 Friedrich von Bernhardi, Deutschland und der nächste Krieg, Berlin 1912, S.11, zit. nach Bernd Ulrich/Jakob Vogel/Benjamin Ziemann (Hg.), Untertan in Uniform. Militär und Militarismus im Deutschen Kaiserreich. Quellen und Dokumente, Frankfurt a.M. 2001, S.206.

893 Ulrich/Vogel/Ziemann, Untertan in Uniform, hier Bd. 1, S.XV: „Der Krieg ist jetzt in Folge der außerordentlichen Fortschritte der Waffentechnik, der hochgesteigerten Präzision der Feuerwaffen und ihres enormen Vernichtungsvermögens furchtbarer geworden. Vom nächsten großen Kriege kann man als ein Rendez-vous des Todes sprechen!". Zu den wirtschaftlichen Folgen siehe ebd., S.XVII-XVIII.

894 Wilhelm Lamszus, Das Menschenschlachthaus. Visionen vom Krieg, o.O. 1912, S.64. Lamszus war jedoch kein Pazifist, sondern der Auffassung, dass ein Krieg erst die Daseinsberechtigung eines Volkes sei: „Ein Geschlecht, das einem Krieg nicht mehr gewachsen ist, ist dem Untergang geweiht". Wilhelm Lamszus, Vom Anarchismus zum Gesetz, in: Die Friedenwarte 15 (1913), S.58–62, zit. nach Jost Dülffer, Kriegserwartung und Kriegsbild in Deutschland vor 1914, in: ders., Im Zeichen der Gewalt. Frieden und Krieg im 19. und 20. Jahrhundert, Köln/Weimar/Wien 2003, S.107–123, hier S.120.

895 Vgl. Köppen, Das Entsetzen, S.212ff.; Felix Holtschoppen, Rebellische Technik. Maschinenphantasien in der literarischen Phantastik um 1900, Berlin 2005; Günther A. Höfler, Die Invasionen der Kriegsmaschinen in die Literatur. Der Erste Weltkrieg als literarische Technikerfahrung, in: Theo Elm/Hans H. Hiebel (Hg.), Medien und Maschinen, Freiburg 1991, S.225–244.

unabsehbaren Ende gingen grundsätzlich alle Kriegsbefürworter oder -gegner aus: „Wenn dieser Krieg zum Ausbruch kommt, so ist seine Dauer und sein Ende nicht abzusehen ... [,] es kann ein siebenjähriger, es kann ein dreißigjähriger Krieg werden“,[896] warnte Moltke in seiner Reichstagsrede von 1890 vor einem „Weltenbrand“.[897] Eine ähnliche Sicht nahm auch Generalmajor Köpke, Oberquartiermeister im Generalstab, ein: „Jedenfalls sprechen die Anzeichen genug dafür, daß der Krieg der Zukunft ein anderes Ansehen, als der von 1870/71 haben wird. Schnelle Siege von entscheidender Bedeutung haben wir nicht zu erwarten.“[898] Die Unfähigkeit der handelnden Akteure, trotz alledem das tatsächliche Ausmaß eines Krieges abzuschätzen, hat Christopher Clark auf die Formel der „Schlafwandler“ gebracht. Die Protagonisten seien „wachsam, aber blind“ gewesen, „von Alpträumen geplagt, aber unfähig, die Realität der Gräuel zu erkennen, die sie in Kürze in die Welt setzen sollten.“[899] Die Sozialdemokraten gingen hierin sogar soweit, den wilhelminischen Militarismus als rückständigen „Dekorationsmilitarismus“ zu bewerten, der ohnehin nicht auf die Herausforderungen eines modernen Krieges vorbereitet sei.[900]

Einige militärische Zukunftsromane, darunter auch solche, die die Flugtechnik berücksichtigten,[901] verharmlosten hingegen die Auswirkungen eines großen Krieges und knüpften an die glorifizierende Erinnerungsliteratur des Deutsch-Französischen Krieges von 1870/71 an.[902] So machte Rudolf Martin die Fliegerei in zwei seiner zahlreichen fiktionalen Publikationen zum Thema. Während er in „Stehen wir vor einem neuen Weltkriege?“ die Bedeutung des Luftschiffes und des Flugzeugs in einem kommenden Krieg hervorhob,[903] beschrieb er in „Berlin-Bagdad. Das Deutsche Weltreich im Zeitalter der Luftschiffahrt 1910–1931“ aus dem Jahr 1907 die Bombardierung Berlins durch russische Luftschiffe, der er

896 Helmuth von Moltke, Rede im Reichstag, 14. Mai 1890, in: Reinhard Stumpf (Hg.), Kriegstheorie und Kriegsgeschichte. Carl von Clausewitz und Helmuth von Moltke, Frankfurt a.M. 1993, S.504–507, hier S.504.

897 Vgl. Leonhard, Die Büchse, S.74 und S.80; Jost Dülffer, Die zivile Reichsleitung und der Krieg: Erwartungen und Bilder 1890–1914, in: ders., Im Zeichen der Gewalt. Frieden und Krieg im 19. und 20. Jahrhundert, hg. von Martin Kröger, Ulrich S.Soénius und Stefan Wunsch, Köln/Weimar/Wien 2003, S.124–140, hier S.131f.

898 Zit. nach Stig Förster, Der deutsche Generalstab und die Illusion des kurzen Krieges 1871–1914. Metakritik eines Mythos, in: Militärgeschichtliche Mitteilungen 54 (1995), S.61–95, hier S.75.

899 Christopher Clark, Die Schlafwandler. Wie Europa in den Ersten Weltkrieg zog, München 2015, S.718.

900 Vgl. Bernhard Neff, „Dekorationsmilitarismus“. Die sozialdemokratische Kritik eines vermeintlich nicht kriegsmäßigen Militärwesens (1890–1911), in: Wolfram Wette (Hg.), Schule der Gewalt. Militarismus in Deutschland 1871–1945, Berlin 2005, S.91–110.

901 Beispielhaft: H. G. Wells, The War in the Air.

902 Vgl. Schumann, Der brüchige Frieden, S.119f.; Thomas Rohkrämer, Der Militarismus der „kleinen Leute“. Die Kriegervereine im deutschen Kaiserreich 1871–1914, München 1990, S.276–283.

903 Rudolf Martin, Stehen wir vor einem neuen Weltkriege?, Leipzig 1908.

jedoch einen deutsch-türkischen Sieg folgen ließ. In seinen Vorstellungen erstreckte sich das Deutsche Reich von Berlin bis Bagdad.[904]

Jenseits dieser militärisch-politischen, zum Teil sehr realistischen Einschätzungen[905] eines künftigen Krieges erwarteten große Teile der Bevölkerung einen Krieg, wie er sich 1870/71 zugetragen hatte. Man hoffte auf kurze, rasche Entscheidungsschlachten, die den Sieg für Deutschland bringen würden, und erwartete einen romantisch-verklärten Kampf „Mann gegen Mann“.[906] Einen folkloristischen Kult um die „Nation in Waffen“ hatten die zahlreichen Krieger-, Veteranen- und Flugvereine bereits vor 1914 zelebriert.[907] Militärische Strukturen prägten die Gesellschaft des Kaiserreiches ebenso wie sozialdarwinistische Vorstellungen der eigenen Überlegenheit, die vor allem aus dem Wunsch resultierten, sich mit anderen Nationen messen zu können.[908] Dies drückte sich beispielhaft in den Flugwettbewerben der Vorkriegszeit aus, die als fairer „Kampf“ zwischen den Nationen ausgetragen und propagiert wurden. Der Rüstungswettlauf, der auch an der vermehrten Beschaffung von Flugzeugen Anfang 1914 sichtbar wird, zeugt von einem „verbreiteten Bellizismus und ... einem Klima zunehmender Kriegserwartung“.[909] Das interessierte und beiderseitig misstrauische Beäugen der (rüstungs-)technischen Aktivitäten der anderen Nationen spiegelte sich aber auch in der medialen Berichterstattung wider. So enthielt der „Flugsport“ seit 1908 eine Rubrik „Pariser Brief“, welche über den Stand der französischen Flugzeugindustrie, militärische Flugübungen und Aufrüstungsambitionen informierte.

904 Vgl. Rudolf Martin, Berlin – Bagdad. Das Deutsche Weltreich im Zeitalter der Luftschiffahrt 1910–1931, Stuttgart/Leipzig 1907, S.85 und S.118.

905 Vgl. Förster, Illusion des kurzen Krieges, S.65ff.

906 Vgl. Jost Dülffer, Kriegserwartung und Kriegsbild, in: ders., Im Zeichen der Gewalt, S.107–123, hier S.116ff.

907 Jakob Vogel, Nationen im Gleichschritt. Der Kult der „Nation in Waffen“ in Deutschland und Frankreich. 1871–1914, Göttingen 1997, S.279.

908 Siehe die Darstellung der „Meistererzählungen“ zur Unvermeidbarkeit des Ersten Weltkrieges bei: Leonhard, Die Büchse, S.74ff.; Peter Walkenhorst, Nation – Volk – Rasse. Radikaler Nationalismus im Deutschen Kaiserreich 1890–1914, Göttingen 2007, S.119ff.; ders., Der „Daseinskampf des deutschen Volkes“: Nationalismus, Sozialdarwinismus und Imperialismus im wilhelminischen Deutschland, in: Jörg Echternkamp/Sven Oliver Müller (Hg.), Die Politik der Nation. Deutscher Nationalismus in Krieg und Krisen 1760–1960, München 2002, S.131–148.

909 Leonhard, Die Büchse, S.77. Ähnlich argumentiert auch Schumann, Der brüchige Frieden, S.121: „Nicht Kriegsbegeisterung, sondern eine fatalistische Hinnahme des Kriegsrisikos bestimmte ... [die Haltung der Veteranen] zum Krieg“. Köppen argumentiert mit dem Begriff des „Vitalismus“ als dominierende Stimmung der Vorkriegszeit: „Die Erneuerung des Geistes durch das Barbarische, die Katharsis durch den Rausch der Sinne und die Erfahrung des Unabdingbaren“ seien „Zielwerte vitalistischer Befindlichkeiten, von denen gleichermaßen die Generalstäbe, die kulturellen Eliten und die Bohemiens in den europäischen Hauptstädten affiziert“ gewesen seien. Köppen, Das Entsetzen, S.212ff., Zitat S.214. Vgl. auch Sönke Neitzel, Blut und Eisen. Deutschland und der Erste Weltkrieg, Zürich 2003, S.40.

Die massenhafte Verbreitung von literarischen Zukunftsromanen in Kombination mit der tatsächlichen Entwicklung der Technik bewirkte eine beschleunigte Aufrüstung im Deutschen Reich, aber auch in anderen Staaten. So war in England mit der sogenannten *Zeppelin Scare*, der Furcht vor einer Bedrohung Englands durch deutsche Luftschiffe, seit 1912 die Luftrüstung energisch betrieben und durch Propaganda in der Öffentlichkeit verbreitet worden.[910] Kriegsfantasien und die sich verfestigenden Bündnissysteme verstärkten Feindbilder und forcierten sowohl eine materielle wie auch eine „geistige Rüstung".[911]

Trotz der Gegenwärtigkeit eines drohenden Krieges darf jedoch nicht unerwähnt bleiben, dass die Zeit zwischen 1870/71 und 1914 eine „der längsten Friedensphasen in der europäischen Geschichte"[912] bildete, in der Konflikte an der Peripherie und lokal begrenzt ausgetragen wurden. Wirtschaftliche Prosperität, Diplomatie und Kooperation kennzeichneten diese Zeit vor dem Krieg ebenso wie die Entstehung einer staatlichen Sozialfürsorge im Kaiserreich und die Entwicklung neuer Technologien.[913] Die Ambivalenz zwischen Konflikten und Krisen auf der einen und Kooperation und Fortschrittsgeist auf der anderen Seite bestimmte die „Moderne".[914]

Josef Suwelack war als Schüler vermittelt worden, das Deutsche Reich sei ein „Hort des europäischen Friedens"[915] und bestrebt, seine Friedenspolitik fortzuführen. Im Juni 1913 hatte sich Kaiser Wilhelm II. anlässlich seines 25-jährigen Thronjubiläums gar als „Friedenskaiser" feiern lassen.[916] Das Denken in militärischen Kategorien hatte Suwelack dann über den Kontakt zur Militärverwaltung, aus der Berichterstattung über das Flugwesen in Zeitungen und Zeitschriften sowie über seine Teilnahme an militärischen – als „Friedensübungen" propagier-

910 Nigel Watson, The Scareship Mystery. A survey of worldwide phantom airship scare (1909–1918), Corby 2000; Guillaume de Syon, Zeppelin! Germany and the airship. 1900–1939, Baltimore 2002, S.72ff.

911 Möser, Technische Zukunftskriege, S.235.

912 Leonhard, Die Büchse, S.79.

913 Wehler, Deutsche Gesellschaftsgeschichte, Bd.3, S.533f.

914 Vgl. Schumann, Der brüchige Frieden, S.113f.; Ute Schneider, Spurensuche. Reinhart Koselleck und die „Moderne", in: dies./Lutz Raphael (Hg.), Dimensionen der Moderne. Festschrift für Christof Dipper, Frankfurt a.M./Berlin/Bern u.a. 2008, S.61–72; Christof Dipper, Italien und Deutschland seit 1800. Zwei Gesellschaften auf dem Weg in die Moderne, in: ders./Lutz Kinkhammer/Alexander Nützenadel (Hg.), Europäische Sozialgeschichte. Festschrift für Wolfgang Schieder, Berlin 2000, S.485–503.

915 W. Pfeiffer, Lehrbuch für den Geschichtsunterricht an höheren Lehranstalten, 6. Teil: Lehraufgabe der Oberprima, Breslau (Hirt) 1907, S.183, zit. nach Rainer Bendick, Kriegserwartung und Kriegserfahrung. Der Erste Weltkrieg in deutschen und französischen Schulgeschichtsbüchern (1900–1939/45), 2. Aufl., Herbolzheim 2003, S.66.

916 Jesko von Hoegen, Der Held von Tannenberg. Genese und Funktion des Hindenburg-Mythos, Köln/Weimar/Wien 2007, S.182 und Christopher Clark, Wilhelm II.: Die Herrschaft des letzten deutschen Kaisers, München 2008, S.199.

ten – Wettbewerben erfahren und die sich zuspitzende nationalistische Stimmungslage wahrgenommen. Mit dem 1912 geschlossenen Dienstvertrag mit dem Deutschen Reich hatte er sich auf einen möglichen Krieg eingestellt.

Diese Vorsorgemaßnahme jedoch mit Voraussicht oder Vorahnung gleichzusetzen wäre verfehlt, denn Suwelack war vor dem Kriegsausbruch fest in die Arbeitsstrukturen bei den Kondor-Werken eingebunden und damit beschäftigt, potenzielle Geschäftspartner zu akquirieren sowie einen neuen Absatzmarkt in Spanien zu erschließen. Allerdings ist es wahrscheinlich, dass ihn die Unterredungen mit Militärs wie Waldemar Walrab von Buttlar oder Felix Wagenführ und die Verbindung Bernhard Goldschmidts zur Marine für eine mögliche Mobilmachung sensibilisierten. Zudem war sich der Flieger bewusst, dass das Militär der wichtigste und größte Förderer der Luftfahrt war und es daher von Nutzen sein konnte, sich in den Dienst des Heeres zu stellen. Da er bereits am 14. Oktober 1911[917] offiziell ausgemustert worden und daher ausdrücklich nicht landsturmpflichtig war, bestand für Suwelack die Möglichkeit, sich freiwillig zur Fliegertruppe zu melden. Seit dem beginnenden 19. Jahrhunderts war das „Opfer für das Vaterland" positiv konnotiert,[918] es galt als erstrebenswert und nationale Pflicht, sich für das Vaterland als Soldat zur Verfügung zu stellen. Dass das Heldengedenken zwischen 1813 und 1870/71 auch am Anfang des 20. Jahrhunderts noch immer durch zahlreiche Denkmäler und Nationalfeiertage[919] gesellschaftlich präsent war, verdeutlicht seinen hohen Stellenwert in der Erinnerungskultur.

Ende März 1914 war Suwelack aufgrund des 1912 geschlossenen Vertrages von der Inspektion der Fliegertruppen aufgefordert worden, „sich im Laufe des 2. Mobilmachungstages bei der 3. Komp[agnie] Flieger B[a]t[ai]l[lon] 1 in Grossenhain zu melden".[920] Es ist daher anzunehmen, dass die Anstrengungen zur Rekrutierung von Fliegerpersonal Anfang des Jahres 1914 verstärkt wurden, um für den Ernstfall vorbereitet zu sein.

In seiner Rede an das deutsche Volk vom 31. Juli 1914 suggerierte der Kaiser einen aus Neid auf die gesellschaftlichen und wirtschaftlichen Errungenschaften Deutschlands von außen kommenden Angriff, der die Deutschen zu einer

917 NL J. Suwelack, Ausmusterungs-Schein, ausgestellt auf „Josef Heinrich Maria Suwelack" von der Königlichen Ober-Ersatzkommission am 14.10.1911. Gründe für die Dienstuntauglichkeit werden nicht genannt.

918 Manfred Hettling/Jörg Echternkamp, Heroisierung und Opferstilisierung. Grundelemente des Gefallenengedenkens von 1813 bis heute, in: dies. (Hg.), Gefallenengedenken im globalen Vergleich. Nationale Tradition, politische Legitimation und Individualisierung der Erinnerung, München 2013, S. 123–158, bes. S. 128–136, Zitat S. 136.

919 Siehe zur Feier- und Erinnerungskultur im Deutschen Kaiserreich und zu ihrem allmählichen Bedeutungsverlust während des Ersten Weltkrieges: Fritz Schellack, Nationalfeiertage in Deutschland von 1871 bis 1945, Frankfurt a. M./Bern/New York/Paris 1990, S. 107ff.

920 NL J. Suwelack, Schreiben der Inspektion der Fliegertruppen an Josef Suwelack, Berlin-Schöneberg, 30.3.1914.

„gerechten Verteidigung“[921] zwinge.[922] Diese Deutung aufgreifend schrieb der „Flugsport“ in seiner Ausgabe vom 5. August: „Russische Tücke und Französischer Uebermut haben uns diesen Kampf aufgezwungen, diesen Kampf, den wir nun um unsere Existenz, um unsere Zukunft führen und siegreich beenden werden!“[923] Von diesem Glauben an einen gerechten und siegreichen Krieg ließen sich Josef und seine Eltern, so geht es aus den folgenden Feldpostkarten und -briefen hervor, anfänglich ebenfalls leiten. Relativ schnell sollten Abenteuerlust und Pflichtgefühl jedoch in Ernüchterung und Desillusionierung umschlagen.[924]

4.2. Der Erste Weltkrieg als ‚touristischer Ausflug‘? Der Kriegsteilnehmer Josef Suwelack: Sinnbildung durch Zäsurerfahrung

Im Juli 1914 hatte Josef Suwelack noch mit Aviatik persönliche Verhandlungen wegen des Fortbestandes und der Auftragslage der Kondor-Werke geführt, als das Deutsche Reich mit der Mobilmachung begann und auch der Pilot auf der Grundlage seines Dienstvertrages zum Kriegsteilnehmer wurde. Ende des Monats besuchte Suwelack seine Eltern und Geschwister zu Hause in Billerbeck und konnte sich noch persönlich von ihnen verabschieden.

Seine Fahrt zur Front begann am 1. August um Mitternacht vom Bahnhof in Hamm nach Berlin. Nach einer elfstündigen Zugfahrt, die er auf seinem Koffer schlafend verbrachte, kam er in Berlin an, von dort sollte er dann weiter zu seinem Bestimmungsort, zum Flugplatz nach Großenhain in Sachsen, reisen.[925] In Berlin kaufte er sich bei einem Reise- und Herrenausstatter seine notwendige Ausrüstung und einige Nährmittel im Wert von fast 600 Mark,[926] denn als Einjährig-Freiwilliger hatte er für Verpflegung und Ausrüstung selbst aufzukommen.[927]

921 Deutscher Reichsanzeiger und Königlich Preußischer Staatsanzeiger, Berlin, 1.8.1914, Nichtamtliches, S.2.

922 Die lange Zeit präsente These von der „Einkreisung“ Deutschlands ist mittlerweile mehrfach widerlegt worden: Winfried Baumgart, Deutsch-russische Beziehungen 1890–1914/18, in: ders. (Hg.), Preußen-Deutschland und Rußland vom 18. bis zum 20. Jahrhundert, Berlin 1991, S.105–118, hier S.114; Wolfgang J. Mommsen, Großmachtstellung und Weltpolitik. Die Außenpolitik des Deutschen Reiches 1870–1914, Frankfurt a. M./Berlin 1993, S.162ff.; über die deutsche Perspektive hinausweisend: Andreas Rose, Zwischen Empire und Kontinent. Britische Außenpolitik vor dem Ersten Weltkrieg, München 2011, S.460f.

923 „Deutsche Flieger!“, in: Flugsport (1914), Nr.16, S.677.

924 Bernd Ulrich, Die Desillusionierung der Kriegsfreiwilligen von 1914, in: Wolfram Wette (Hg.), Der Krieg des kleinen Mannes. Eine Militärgeschichte von unten, München/Zürich 1992, S.110–127; Eva Horn, Erlebnis und Trauma. Die narrative Konstruktion des Ereignisses in Psychiatrie und Kriegsroman, in: Inka Mülder-Bach (Hg.), Modernität und Trauma. Beiträge zum Zeitenbruch des Ersten Weltkrieges, Wien 2000, S.131–162.

925 Vgl. NL J. Suwelack, Karte an seine Eltern vom 3. und 4.8.1914.

926 NL J. Suwelack, Kaufbeleg für Josef Suwelack des Herrenausstatters „Johannes Steinberg“, Berlin, 17.11.1914.

927 Vgl. Lothar Mertens, Das Einjährig-Freiwilligen Privileg. Der Militärdienst im Zeitgeist des deut-

Großenhain erreichte er mit der Bahn am 3. August. In den folgenden Wochen und Monaten stand er in regem brieflichen Kontakt mit seinen Eltern, ab Mitte Oktober dann auch mit seinem Bruder Richard.[928]

Richard Suwelack, der als Einjährig-Freiwilliger vor dem Ersten Weltkrieg in einem Kürassierregiment in Koblenz, Ehrenbreitstein, gedient hatte, war ebenfalls eingezogen worden und leistete seinen Dienst als Feldmagazin-Inspektor bei der 13. Reserve-Division des 7. Armeekorps.[929] Von Billerbeck über das neutrale Belgien kam Richard über Lüttich und Maubeuge in Nordfrankreich in Vorges, rund zwanzig Kilometer hinter Laon gelegen, an.[930]

Als Feldmagazin-Inspektor war er für die Bestellung, Lagerung und Verteilung von Gütern, Waffen und Lebensmitteln an Truppenteile zuständig sowie mit der Buchführung betraut.[931] Obschon er nicht ständig im Feuer stand, habe er seine „Feuertaufe" bereits kurz nach Kriegsausbruch erhalten. Am 12. Oktober 1914 teilte er seinem Bruder mit:

> „Gestern Abend erhielt ich zu meiner größten Freude den ersten Brief von Dir. Der Postverkehr war ... in der ersten Zeit sehr unregelmäßig, so daß auch anscheinend meine Briefe, welche ich an Dich geschrieben habe, nicht übergekommen sind. Mir geht es recht gut. An die Strapazen hat man sich jetzt gewöhnt, so daß man selbst die größten Märsche mit Leichtigkeit macht. Zudem habe ich ein tadelloses Reitpferd, das eine unverwüstliche Gesundheit hat. Ich habe bei den Eilmärschen, die wir z. B. von Lüttich nach Maubeuge machten, 80 km am Tage geritten. Meine Tätigkeit ist manchmal sehr anstrengend aber auch andererseits angenehm, weil wir sehr selbständige Posten haben. Die Feuertaufe habe ich schon in Belgien erhalten und bin auch hier schon mehrere Male im Feuer gewesen. Im allgemeinen sind wir aber immer einige km hinter der Feuerlinie."[932]

Josef Suwelack erlebte den Ausbruch des Krieges als einen befreienden Aufbruch, als spannendes Abenteuer, und schrieb seinen Eltern von einem „Riesentrubel", der auf den Straßen geherrscht habe. Von einem Offizier habe er erfahren, dass die deutschen Truppen bereits fünfzig Kilometer in Russland vorgerückt seien.[933] Josefs Eltern, die den Krieg als „gerecht" empfanden, erfuhren vor allem die Auswirkungen des Krieges abseits der Front: die rasche Mobilmachung, die grundlegende Umwälzung des Alltags und die Auswirkungen auf die Molkerei durch den Kriegszustand. Sein Vater schrieb am 10. August:

schen Kaiserreiches, in: Zeitschrift für Religions- und Geistesgeschichte 42 (1990), H. 4, S.316–329, hier S.319ff.

928 Vgl. NL J. Suwelack, Undatierter Feldpostbrief Josef Suwelacks an seine Eltern, wahrscheinlich geschrieben nach dem 3.10., und Feldpostbrief Richards an Josef vom 12.10.1914.

929 NL J. Suwelack, Brief Joseph Suwelacks senior an Josef Suwelack, Billerbeck, 8.11.1914. Hermann Klüting, Soldaten in Westfalen und am Niederrhein. Das Königlich Preußische VII. Armeekorps, Beckum 1982.

930 NL J. Suwelack, Brief Richard Suwelacks an Josef, Vorges, 12.10.1914.

931 A. Kunz, Der Verpflegungsbeamte im Kriege. Eine Sammlung von Bestimmungen über das Feld-Verpflegungswesen und die persönlichen Verhältnisse des Verpflegungsbeamten, nach amtlichen Quellen, 2. Aufl., Berlin 1913.

932 NL J. Suwelack, Feldpostbrief Richard Suwelacks an Josef, Vorges, 12.10.1914.

933 NL J. Suwelack, Karte an seine Eltern vom 3.8.1914.

„Der Krieg ist für uns äußerst schwierig geworden, nachdem England uns auch den Krieg erklärt hat und Italien uns im Stiche läßt. Hoffentlich wird Gott unserer gerechten Sache seinen Segen geben. Gebetet wird in ganz Deutschland Tag und Nacht. Jedermann ist bereit, mit im Feuer zu stehen. Der Verkehr ist seit 10 Tagen vollständig aufgehoben. Nur Militärzüge fahren selbst auf unserer Nebenbahn. In Münster werden täglich 176 Militärzüge abgelassen resp. durchgelassen. Nach Frankreichs Grenzen werden ungeheure Truppenmassen befördert. Pferde und Wagen werden genommen, sodaß die Landwirte ihre Äcker nicht mehr besorgen können. Gott Dank, daß wir durch Belgien den Feind angreifen. Mit der Molkerei habe ich fürchterliche Schwierigkeiten. Ich lasse täglich 8 bis 10 Fuhrwerke direkt von hier nach Essen und Bochum fahren.“[934]

Die individuelle Verarbeitung der Ereignisse offenbart ein Stimmungsgemisch zwischen emotionaler, medial verstärkter, zum Teil inszenierter Euphorie,[935] die zum Mythos des „Augusterlebnisses“[936] wurde, und Anspannung und Kriegsfurcht.[937] So sei es in Münster zu „patriotischen Kundgebungen“, „Hamsterkäufen“, einer „verstärkten Abhebung von Spargeldern“ gekommen und ein Klima von Endzeiterwartungen, Xenophobie und Gerüchten sei entstanden.[938]

Nach seiner Ankunft in Großenhain informierte Josef seine Eltern umgehend über seinen Verbleib und sein Wohlbefinden. Er habe ein „wunderbares Quartier“ und „immer die feinsten Kuchen“ zu essen. Den „furchtbaren Ernst der Sache“ sehe er zwar gelegentlich, doch insgesamt habe er ein „sehr gemütliches Leben“.[939] Äußerungen dieser Art lassen sich in ähnlicher Form auch in Feldpostbriefen anderer Soldaten[940] aus dem Ersten Weltkrieg finden. Die Häufung

934 NL J. Suwelack, Brief von Joseph Suwelack senior an Josef Suwelack vom 10.8.1914.
935 Leonhard, Die Büchse, S.129.
936 Siehe die zum Standardwerk gewordene Betrachtung des „Augusterlebnis“-Mythos beziehungsweise der Kriegswahrnehmung der bildungsbürgerlichen Schicht bei: Jeffrey Verhey, Der „Geist von 1914“ und die Erfindung der Volksgemeinschaft. Hamburg 2000. Siehe auch Wolfgang Kruse, Die Kriegsbegeisterung in Deutschland zu Beginn des Ersten Weltkriegs. Entstehungszusammenhänge, Grenzen und ideologische Strukturen, in: Marcel van der Linden/Gottfried Mergner (Hg.), Kriegsbegeisterung und mentale Kriegsvorbereitung, Berlin 1991, 73–87; Benjamin Ziemann, Zum ländlichen Augusterlebnis 1914 in Deutschland, in: Bedrich Loewenstein (Hg.), Geschichte und Psychologie. Annäherungsversuche, Pfaffenweiler 1992, S.193–203; Christian Geinitz/Uta Hinz, Das Augusterlebnis in Südbaden. Ambivalente Reaktionen der deutschen Öffentlichkeit auf den Kriegsbeginn 1914, in: Gerhard Hirschfeld/Gerd Krumeich/Dieter Langewiesche/Hans-Peter Ullmann (Hg.), Kriegserfahrungen. Studien zur Sozial- und Mentalitätsgeschichte des Ersten Weltkriegs, Essen 1997, S.20–35; Christian Geinitz, Kriegsfurcht und Kampfbereitschaft. Das Augusterlebnis in Freiburg. Eine Studie zum Kriegsbeginn 1914, Essen 1998, und speziell zu Münster: Kristina Thies, Das Augusterlebnis in Münster im Spiegel der Kriegschronik Eduard Schultes, in: Tobias Arand (Hg.), Die „Urkatastrophe“ als Erinnerung – Geschichtskultur des Ersten Weltkriegs, Münster 2006, S.99–131.
937 Vgl. Ziemann, Zum ländlichen Augusterlebnis, S.40f.
938 Thies, Das Augusterlebnis in Münster, S.101.
939 NL J. Suwelack, Karte an seine Eltern vom 4.8.1914.
940 In Auswahl seien folgende Titel stellvertretend für weitere Quellensammlungen genannt: Daniel Schmidt (Hg.), „Bin noch gesund und munter“. Gelsenkirchener Feldpost aus dem großen Krieg 1914–1918, Essen 2014; Ursula König-Heuer/Julia Paulus (Hg.), „In der Ferne das Donnern der Kanonen“. Briefwechsel eines Billerbecker Soldaten mit seiner Mutter im Ersten Weltkrieg, Münster 2014; Gundula Gahlen (Hg.), Feldpost 1916/17. Die Briefe und Tagebücher des Oberleutnants August Dänzer aus Siebenbürgen und Rumänien (2012), online unter: http://www.uni-cms.net/feldpostsammlung (12.1.2015); Feldpostsammlung des Imperial War Museum: http://www.iwm.

bestimmter narrativer Elemente und die Schilderung „alltäglicher Banalitäten“[941] – Thematisierung des eigenen Wohlbefindens, der Verpflegung, der „Arbeit“ an beziehungsweise hinter der Front, der Kameradschaft sowie die Übermittlung „neuer“ und scheinbar „verlässlicher“ Informationen – zeigen Chancen und Grenzen der Feldpost Suwelacks für die Rekonstruktion seines Kriegserlebnisses auf.

Der Quellenwert der Feldpost liegt auf den ersten Blick in ihrer unmittelbaren Nähe zum Geschehen, der kurzen zeitlichen Verarbeitung und der Wiedergabe der individuellen Eindrücke. Dies suggeriert einen hohen Grad an Authentizität,[942] der sich bei kritischer Auseinandersetzung mit der Gattung „Feldpost“ jedoch als „Authentizitätsfalle“[943] erweist.

In der Forschung ist mehrfach betont worden, dass Feldpostbriefe kulturellen Deutungsmustern und Wahrnehmungsfiltern unterliegen, geprägt durch die erfahrene familiäre und gesellschaftliche Sozialisation, die Ausbildung und den bevorzugten beziehungsweise zur Verfügung stehenden Lesestoff.[944] In diesem Zusammenhang darf auch die Bedeutung der offiziellen und der Selbstzensur nicht unterschätzt werden.[945] Erwartungshaltungen, Dispositionen und gesellschaftliche Konventionen bestimmten die Korrespondenz zwischen Sender und Empfänger. Aus der Fronterfahrung entstanden „Konversationsmaximen der Kriegsbriefe“, von denen Isa Schikorsky folgende fünf „emotive Sprachhandlungs-

org.uk/history/letters-to-loved-ones (12.1.2015). Siehe auch die Auflistung von Quelleneditionen bei: Bernd Ulrich, Die Augenzeugen. Deutsche Feldpostbriefe in Kriegs- und Nachkriegszeit 1914–1933, Essen 1997, S.12.

941 Aribert Reimann, Die heile Welt im Stahlgewitter. Deutsche und englische Feldpost aus dem Ersten Weltkrieg, in: Hirschfeld/Krumeich/Langewiesche/Ullmann, Kriegserfahrungen, S.129–145, hier S.130f.

942 Vgl. Elke Scherstjanoi, Als Quelle nicht überfordern! Zu Besonderheiten und Grenzen der wissenschaftlichen Nutzung von Feldpostbriefen in der (Zeit-)Geschichte, in: Veit Didczuneit/Jens Ebert/Thomas Jander (Hg.), Schreiben im Krieg – Schreiben vom Krieg. Feldpost im Zeitalter der Weltkriege, Essen 2011, S.117–125, hier S. 122. Sebastian Haak, „And Feel the Tingle of the Greatest Game of Them All“. Über eine (verdrängte) Facette der Wahrnehmung von Krieg und Gewalt in den Briefen US-amerikanischer Soldaten, in: ebd., S.175–184, hier S.182; Ulrich, Die Augenzeugen, S.11ff.

943 Jörg Echternkamp, Militärgeschichte, in: Frank Bösch/Jürgen Danyel (Hg.), Zeitgeschichte – Konzepte und Methoden, Göttingen 2012, S.293–312, hier S.305. Feldpostbriefe als Ego-Dokumente würden den „Eindruck des Unverfälschten“ erwecken, doch werde häufig übersehen, dass die Dokumente nicht mehr, aber auch nicht weniger als eine subjektiv-spezifische, kulturell beeinflusste Wahrnehmung eines kommunizierten Teils des Erlebten darstellten: „Was [die Soldaten] ... für nicht erwähnenswert hielten, was ihnen untersagt schien, was sie verschweigen wollten: davon erfährt man nichts.“, S.305f.

944 Bernd Ulrich, „Militärgeschichte von unten“. Anmerkungen zu ihren Ursprüngen, Quellen und Perspektiven im 20. Jahrhundert, in: Geschichte und Gesellschaft 22 (1996), S.473–503, hier S.499.

945 Vgl. Ulrich, „Militärgeschichte von unten“, S.493; Wilhelm Deist, Zensur und Propaganda in Deutschland während des Ersten Weltkrieges, in: ders. (Hg.), Militär, Staat und Gesellschaft. Studien zur preußisch-deutschen Militärgeschichte, München 1991, S.153–163; Thilo Eisermann, Pressephotographie und Informationskontrolle im Ersten Weltkrieg. Deutschland und Frankreich im Vergleich, Hamburg 2000.

kategorien" als zentrale, wiederkehrende Elemente benannt hat: „Verschweigen, Verharmlosung, Poetisierung, Phraseologisierung und Imagepflege".[946] Auch Paul Fussell hat das Schreiben von Feldpostbriefen als zum Teil belangloses Seitenfüllen bewertet: „The trick was to fill the pages by saying nothing and to offer the maximum number of clichés."[947]

Trotz aller gebotenen Vorsicht und Skepsis kann die Feldpost Josef Suwelacks dennoch als Schriftzeugnis der kulturell und mentalitätsgeschichtlich gefärbten individuellen Sinnbildung analysiert werden, wiederum verbunden mit einigen Einschränkungen. Die Briefe Suwelacks zeigen weder, wie „der" Krieg war, noch zeigen sie, was der Krieg für Suwelack bedeutete. Die Schreiben aus dem Krieg sind vielmehr Konstruktionen, eine „curious literariness of real life",[948] dies sollte nachdrücklich betont werden: Als Feldflieger erlebte Suwelack nur einen Ausschnitt des Krieges, der seine Wahrnehmung determinierte, „zugleich [griff] ... er Kolportiertes und Propagiertes auf, manches eher zufällig, vieles ungenau, einiges ist regelrecht falsch".[949] Diese räumlich gefilterten Eindrücke erfahren in der schriftlichen Verarbeitung zusätzlich eine weitere Reduktion: „Aus dem objektiv Sichtbaren gerinnt ... nur ein Teil, nämlich das selektiv und mit persönlich bevorratetem Deutungsvermögen Wahrgenommene zu potentiell brieftauglichen Informationen."[950] Mit der fortschreitenden Kriegsdauer eintretende Gewöhnungseffekte, auf den Adressaten zugeschnittene Informationen, die Verwendung von „Codes" und das Fehlen bestimmter Zusammenhänge, die sich nur für den Sender und den Empfänger erschließen, gilt es bei der Analyse der Feldpost zu beachten.[951]

Der fragmentarische und „gebrochene"[952] Zugang zu Suwelacks Feldpost sowie die Analyse einiger Kriegsfotografien aus seinem Nachlass lassen Aussagen über seine Tätigkeit bei der Fliegerabteilung, über Beziehungen, Aufenthaltsorte, Verpflegung und Logistik zu. Rückschlüsse von individuellen auf kollektive Struk-

946 Isa Schikorsky, Kommunikation über das Unbeschreibbare. Beobachtungen zum Sprachstil von Kriegsbriefen, in: Wirkendes Wort 42 (1992), H. 2, S.295–315, hier S.301. Solche „Standardthemen" der Feldpost hat Martin Humburg für den Zweiten Weltkrieg ebenfalls beschrieben: Martin Humburg, Das Gesicht des Krieges. Feldpostbriefe von Wehrmachtsoldaten aus der Sowjetunion 1941–1944, Wiesbaden 1998, S.93ff.

947 Paul Fussell, The Great War and Modern Memory, ND Oxford 2013, S.198.

948 Ebd., S.IX. Ähnlich äußerte sich auch Bernd Ulrich: Wer Feldpostbriefe als Quelle benutze, dürfe „nicht vergessen, dass der Akt des Briefeschreibens selbst schon Konstruktion" sei; Ulrich, „Militärgeschichte von unten", S.493. Siehe zur quellenkritischen Auseinandersetzung mit Feldpostbriefen auch Martin Humburg, „Jedes Wort ist falsch und wahr – das ist das Wesen des Wortes". Vom Schreiben und Schweigen in der Feldpost, in: Didczuneit/Ebert/Jander, Schreiben im Krieg – Schreiben vom Krieg, S.75–85, hier S.82f.

949 Scherstanoi, Als Quelle nicht überfordern!, S.122.

950 Ebd., S.123.

951 Vgl. ebd.

952 Ulrich, „Militärgeschichte von unten", S.493.

turen und Empfindungen, das heißt auf einen fliegerischen Kriegsdiskurs,[953] erlaubt die singuläre Analyse des privaten Briefwechsels hingegen nicht. Stattdessen ermöglicht der biografische Ansatz der Untersuchung, Josefs Kriegszeit an sein Vorkriegsleben anzubinden sowie Brüche und Kontinuitäten, aber auch die Bedeutung der Herkunft[954] aufzuzeigen – eine Möglichkeit, die angesichts häufig fehlender biografischer Informationen zu vielen Fliegeralben und losen Fotosammlungen nur selten gegeben ist.[955] Auch seinen Umgang mit der andauernden Gegenwart[956] wird zum Thema gemacht. In diesem Sinne will die folgende Darstellung einen Beitrag zur Sozialgeschichte der Luftstreitkräfte für die ersten beiden Kriegsjahre leisten.

Die Briefe von der Front in die Heimat und umgekehrt dienten primär der gegenseitigen Beruhigung, sollten Ängste und Sorgen zerstreuen helfen, aber auch Erwartungen bestätigen.[957] Die den Briefen zum Teil beigefügten Fotos entsprangen dem „Wunsch nach Dokumentation für Verwandte und Bekannte sowie für die spätere Erinnerung".[958] Sie waren aber ebenso der Versuch, durch eine Konservierung der Zeit im Bild[959] den Orientierungsverlust durch die überwältigende Flut an Sinneseindrücken zu kompensieren, Komplexität zu reduzieren und Sprachlosigkeit zu überwinden.

953 Achim Landwehr gibt als Voraussetzung für die Analyse eines Diskurses die Notwendigkeit der Existenz eines ausreichend großen Quellenbestandes und den konsequenten Einbezug der Kontexte an. Er unterscheidet zwischen einem situativen, medialen, institutionellen und historischen Kontext. Einen Diskurs versteht er als die Regelung des Sag-, Denk- und Machbaren. Vgl. Achim Landwehr, Historische Diskursanalyse, 3. überarb. Aufl., Frankfurt a. M. 2008, S.21 und 91ff. Eine wirkliche Diskursanalyse, denn der Begriff erfreut sich in der Geschichtswissenschaft großer Beliebtheit, kann aber selbst unter Einbezug weiterer Quellen im Rahmen dieses Kapitels nicht vorgenommen werden.

954 Herkunft ist hier verstanden als Rückhalt und ein Rest erhalten gebliebener Tradition. Siehe die Ausführungen zur „Kompensationstheorie" bei Assmann, Zeit aus den Fugen, S.239ff.

955 Vgl. Hüppauf, Kriegsfotografie und die Erfahrung des Ersten Weltkriegs, S.44. Unter „andauernder Gegenwart" soll hier verstanden werden, dass Suwelack sich gewissermaßen in einer Dauerschleife befand, er war handlungsunfähig, was die Planung seiner Zukunft anbelangte, gleichzeitig war der biografische Bruch durch den Kriegsbeginn so groß, dass auch ein Blick auf die Vergangenheit kaum Sinnstiftung bot.

956 Vgl. ebd., S.31ff.

957 Vgl. Ulrich, „Militärgeschichte von unten", S.494.

958 Bernd Hüppauf, Fotografie, in: Gerhard Hirschfeld/Gerd Krumeich/Irina Renz (Hg.), Enzyklopädie Erster Weltkrieg, Paderborn/München/Wien/Zürich 2003, S.499f. Vgl. auch Tobias Arand, Der Nachlass des Leutnants Bernhard Mangels aus Münster. Exemplarische Überlegungen zur inhaltlichen und formalen Erschließung von Kriegsfotoalben deutscher Veteranen des Ersten Weltkrieges, in: Westfälische Forschungen 58 (2008), S.421–439, hier S.422.

959 Siehe zum Gedanken des Zeit-Festhaltens: Paul Hugger, Die Photographie als Dokument des privaten Erinnerns, in: Brigitte Bönisch-Brednich/Rolf W. Brednich/Helge Gerndt (Hg.), Erinnern und Vergessen. Vorträge des 27. Deutschen Volkskundekongresses Göttingen 1989, Göttingen 1991, S.235–242; Martina Krauss, Kleine Welten. Alltagsfotografie – die Anschaulichkeit einer „privaten Praxis", in: Gerhard Paul (Hg.), Visual History. Ein Studienbuch, Göttingen 2006, S.57–75, hier S.57.

Zu den Motiven der Bilder gehörten sich wiederholende Elemente – Mannschaftsaufnahmen, Kriegsgerät, Zerstörung, Landschaftsaufnahmen etc.[960] Die Vorüberlegungen zur Analyse der Feldpost als Ego-Dokument gelten deshalb gleichermaßen auch für das fotografische Quellenmaterial. Die privaten Aufnahmen sind kein Abbild der Wirklichkeit, sondern Konstruktion, ein selektiv, durch die Perspektive Suwelacks und die Linse gefilterter Ausschnitt aus dem Feld des Krieges, oder, wie es Bourdieu Mitte der 1960er Jahre formuliert hat: „In Wirklichkeit hält die Photographie einen Aspekt der Realität fest, d. h. das Ergebnis einer willkürlichen Wahl und somit einer Bearbeitung: Von den Eigenschaften des Gegenstandes werden nur jene erfasst, die in einem besonderen Augenblick und unter einem besonderen Blickwinkel hervortreten."[961] Hinzu tritt, dass der vermeintlich eindeutige Bildinhalt, den der Betrachter entnehmen zu können glaubt, sich nicht immer mit der Bildbeschriftung durch den Fotografen deckt.

Im Unterschied zu den für die Öffentlichkeit bestimmten Fotografien der staatlichen Kriegsberichterstattung zeigen die Privataufnahmen als „private Praxis"[962] „manche Aspekte des Todes unverschämter"[963] und im Fall von Josef Suwelack aus dem „Hinterland" der Front sowie aus einer neuen Perspektive: der Luft. Bei der Entscheidung, etwas zu fotografieren oder nicht, spielte dementsprechend eher die Selbstzensur eine Rolle, d. h. die eigene Gewissensentscheidung darüber, was fotografiert, verewigt und erinnert werden sollte.

Schließlich ist noch auf die heutige Archivierung der Briefe und Fotos hinzuweisen. Der Nachlass Josef Suwelacks wurde von Walter Suwelack nach dem Tod des Fliegers in einem Schrank in seiner Essener Wohnung gefunden, von dort nach Billerbeck transportiert und dann systematisch in Kartons archiviert. Unabhängig davon, ob die Fotos von Josef Suwelack oder seinen Eltern ursprünglich einmal anders geordnet worden waren oder nicht, sind sie heute zum Teil von den Briefen getrennt und nach Themen sortiert – Luft-, Personenaufnahmen, Aufnahmen von Kirchen etc. –, so dass sich die Bilder häufig nicht mehr exakt kontextualisieren lassen. Sie wurden Teil eines neuen Diskurses, desjenigen des „disziplinierten und disziplinierenden Wissens der Archivare".[964]

960 Vgl. Hüppauf, Kriegsfotografie, S. 31.

961 Pierre Bourdieu, Die gesellschaftliche Definition der Photographie, in: ders. (Hg.), Eine illegitime Kunst. Die sozialen Gebrauchsweisen der Photographie, ND Hamburg 2006, S. 85–110, hier S. 85.

962 Den Begriff der „privaten Praxis" hat Jens Jäger als Sammelbegriff für die Amateur- und als „Knipser"-Fotografien bezeichneten Aufnahmen vorgeschlagen. Vgl. Jens Jäger, Fotografie und Geschichte, Frankfurt/New York 2009, S. 183ff.

963 Anton Holzer, Das fotografische Gesicht des Krieges, in: ders. (Hg.), Mit der Kamera bewaffnet. Krieg und Fotografie, Marburg 2003, S. 7–20, hier S. 15.

964 Hüppauf, Kriegsfotografie, S. 39.

Aus diesem Grund und im Wissen um den Skandal, den die „Wehrmachtsausstellung“[965] auslöste, beschränkt sich die Analyse des fotografischen Materials vorwiegend auf kontextualisierbare Personenaufnahmen, wenn möglich werden aber auch andere Aufnahmen in die Narration eingebunden, doch soll das Kapitel nicht zu einer „Bilder-“Geschichte von Josef Suwelacks Kriegserlebnis werden. Vielmehr werden die Fotos als „Ausdruck von Empfindungen, im Sinne von Selbstverwirklichung, unter dem Aspekt des gesellschaftlichen Prestiges“[966] sowie der individuellen Deutung der Kriegserfahrung verstanden.

Die von Schikorsky beschriebene Systematik der „Konversationsmaximen“ gibt im Folgenden die thematische Analyse-Struktur der Feldpost vor, bei der aspektorientiert vorgegangen wird. So lassen sich die Briefe unter verschiedenen Fragestellungen thematisch betrachten und Entwicklungen aufzeigen. Bei der Analyse der Fotos stehen vor allem die Inszenierung und der in Haltung, Gestik und Mimik zum Ausdruck gebrachte männlich-kameradschaftliche Fliegerhabitus im Fokus. Darüber hinaus soll an ausgewählten Aufnahmen der Alltag in einer Fliegerabteilung skizziert und unter Einbezug weiterer Quellen, etwa dem Flugbuch Josef Suwelacks sowie Zeitungsberichten und propagandistischen Bild- und Schriftquellen, die private mit der öffentlichen, medialen Darstellung des Luftkrieges in Ansätzen verglichen werden.

Bevor sich die Darstellung dem Kriegseinsatz Josef Suwelacks zuwendet, ist es geboten, einen kurzen Abriss der Entwicklung und Organisation der Luftstreitkräfte im Deutschen Reich voranzustellen, ferner auf die Bedeutung der Luftaufklärung und der Luftbildfotografie einzugehen. Nur vor diesem Hintergrund lassen sich der Weg Josef Suwelacks vom Etappenflugzeugpark zur Front und die an die Flieger gestellten Aufgaben verstehen. Die in den Abschnitten über die Kondor-Werke und den Prinz-Heinrich-Flug bereits beschriebene Entwicklung knüpft nahtlos an die folgenden Ausführungen an.

965 Hans-Ulrich Thamer, Eine Ausstellung und ihre Folgen. Impulse der „Wehrmachtsausstellung“ für die historische Forschung, in: Ulrich Bielefeld/Heinz Bude/Bernd Greiner (Hg.), Gesellschaft – Gewalt – Vertrauen. Jan Philipp Reemtsma zum 60. Geburtstag, Hamburg 2012, S.489–503.

966 Pierre Bourdieu, Kult der Einheit und kultivierte Unterschiede, in: ders., Eine illegitime Kunst, S.25–84, hier S.26f.

4.3. Exkurs: Zur Entwicklung und Organisation der Luftstreitkräfte und der Luftaufklärung im Deutschen Reich

Die Luftstreitkräfte

„1914 noch ein unbekannter Begriff", sollte aus einer skeptisch beäugten Fliegertruppe, die aus den „kümmerlichen Ansätzen eines deutschen Flugsportes" im Jahr 1909 hervorgegangen war,[967] eine bedeutsame Streitmacht werden,[968] „deren Wichtigkeit noch vor fünf Jahren kaum jemand ahnte. ... Der Krieg hat uns gezeigt, daß wir im Flugzeug ein außerordentliches Mittel besitzen, ... um Gelände auszukundschaften."[969]

Die zitierten Aussagen resümieren in Kurzform die enorme Entwicklung, die das deutsche Flugwesen zwischen 1909 und 1918 nahm. Hatte das Hauptaugenmerk des Militärs im späten 19. Jahrhundert und bis ins erste Jahrzehnt des 20. Jahrhunderts hinein noch auf den Luftschiffen und Fesselballons gelegen, beschloss der Große Generalstab 1909 nun auch die Beschaffung von Flugzeugen für militärische Zwecke. Die auf nur rudimentären Kenntnissen basierende „Ausbildung" der ersten Militärpiloten erfolgte parallel an der „Provisorischen Militärfliegerschule" in Döberitz.[970] Mit den seit 1911 jährlich stattfindenden militärischen Zuverlässigkeitsflügen sowie der im Zuge der Nationalflugspende erfolgten finanziellen Förderung der Flugzeugindustrie und der Fliegerausbildung wurde der Auf- und Ausbau einer Fliegertruppe vorangetrieben. Frankreichs führende Stellung in der Luftrüstung wirkte als Katalysator, denn Deutschland wollte den technologischen Rückstand aufholen.

Obschon sich viele junge Männer freiwillig zu einer Fliegerausbildung meldeten, fehlte es anfänglich an technisch-versiertem Bodenpersonal und die Unfallzahlen waren hoch.[971] Dessen ungeachtet nahm beim Kaisermanöver 1911 zum ersten Mal auch ein militärisches Flugzeug teil und der Ausbau des „Fliegerkommandos der Luftschiffer-Abteilung der Verkehrstruppen" zu einer Fliegertruppe wurde verstärkt.[972] Am 1. Oktober 1912 ging aus der provisorischen die Königlich-Preußische Fliegertruppe hervor, in der ein Jahr später sächsische und württembergi-

967 Georg Paul Neumann, Die deutschen Luftsreitkräfte im Weltkriege, ND Bremen 2011, S.3 und S.58.

968 Sönke Neitzel, Zum strategischen Mißerfolg verdammt? Die deutschen Luftstreitkräfte in beiden Weltkriegen, in: Bruno Thoß/Hans-Erich Volkmann (Hg.), Erster Weltkrieg – Zweiter Weltkrieg. Ein Vergleich. Krieg, Kriegserlebnis, Kriegserfahrung in Deutschland, Paderborn 2002, S.167–192, hier S.167 und S.191.

969 „Über unsere Luftflotte", in: Die Luftflotte VII (1915), Nr. 2, S.10–11, zit. nach Esposito, Mythische Moderne, S.208.

970 Vgl. Neumann, Die deutschen Luftstreitkräfte, S.61; Supf, Fluggeschichte, Bd. 2, S.99.

971 Vgl. Supf, Fluggeschichte Bd. 2, S.105; Rosenboom, Im Einsatz, S.20.

972 Vgl. Ernst von Hoeppner, Deutschlands Krieg in der Luft, Leipzig 1921 [ND Barsinghausen 2013], S.2.

sche Formationen aufgingen. Ferner entstand ein eigenständiges bayrisches Kontingent.[973] Im gleichen Jahr traten noch die Marineflieger- und Marineluftschifferabteilungen hinzu.

Bis zum August 1914 verfügte das preußische Heer über 33 Feldfliegerabteilungen. Jedem der acht Armeeoberkommandos und 25 aktiven Armeekorps war eine Feldfliegerabteilung zu je sechs Flugzeugen zugeteilt. Außerdem wurden sieben Festungsfliegerabteilungen zu je vier Flugzeugen, acht Etappenflugzeugparks und vier Fliegerersatzabteilungen[974] eingerichtet, welche der personellen und materiellen Nachschubversorgung der Abteilungen dienten.[975] Neben den anderen Luftfahrzeugen standen damit rund 240 Flugzeuge – sämtliche Privat- und Werksflugzeuge der Industrie wurden bei Kriegsausbruch beschlagnahmt –,[976] 254 ausgebildete Piloten und 271 Beobachter zur Verfügung, die, und das sollte betont werden, sich alle freiwillig zur Fliegertruppe gemeldet hatten.[977] Der Personalbedarf war damit jedoch bei Weitem nicht gedeckt, so dass das Kriegsministerium im August dazu aufrief, sich zur Fliegertruppe zu melden.[978] Zwischen 1914 und 1918 verzehnfachte sich das Personal der Fliegertruppen auf schätzungsweise 5.000 Mann; in der Heimat waren Harald Potempa zufolge etwa 80.000 Mann in den Flugzeugfirmen für die Pilotenausbildung und den Bau von Maschinen zuständig.[979] Für die Ausbildung neuer Piloten veranschlagte man einen Zeitraum von gerade einmal vier bis acht Wochen.[980]

Da die Fliegerabteilungen von den Bundesstaaten aufgestellt worden waren, besaßen sie keine zentralisierte Führung und gehörten organisatorisch zu den Verkehrstruppen. Innerhalb dieser Einheit hatten die Luftstreitkräfte 1913 eine eigene Inspektion der Fliegertruppen (IdFlieg) erhalten, doch weder bestand durch die Zuteilung zu den Armeekorps eine Kooperation unter den Fliegerabtei-

973 Harald Potempa, Die Königlich-Bayerische Fliegertruppe. 1914–1918, Frankfurt a. M. 1997, S. 2.

974 Am 18. August wurde eine zusätzliche Fliegerersatzabteilung eingerichtet. Vgl. Major a. D. Helmut Förster, Die Entwicklung der Fliegerei im Weltkriege, o. O. 1930, S. 39–66, hier S. 40.

975 Vgl. Hoeppner, Deutschlands Krieg, S. 7.

976 Siehe dazu auch den entsprechenden Abschnitt bei den Kondor-Werken.

977 Je nach Autor schwankt die Zahl der 1914 zur Verfügung stehenden Flugzeuge zwischen 220 und 450 Maschinen. Siehe Peter Kilduff, Germany's First Air Force 1914–1918, London 1991, S. 10; Morrow, Building German Air Power, S. 87. Sebastian Rosenboom kam jüngst zu den im Text genannten Zahlen: Rosenboom, Im Einsatz, S. 20. Siehe zur Einsatzfähigkeit der Flugzeuge: Morrow, Building German Air Power, S. 60 und S. 75f. Zur Freiwilligkeit der Flieger vgl. Neitzel, Zum strategischen Mißerfolg verdammt?, S. 189.

978 Förster zufolge hatten sich 15.000 Freiwillige als Flieger beziehungsweise Monteure gemeldet. Vgl. Förster, Die Entwicklung der Fliegerei, S. 40. Die Ausbildung der Flieger erfolgte bei den Flugzeugfirmen, während die Beobachter ihre Ausbildung bei der Fliegerersatzabteilung erhielten, ebd. Siehe auch den Aufruf im Flugsport: „Bekanntmachung für Flugzeugführer", in: Flugsport (1914), Nr. 16, S. 677.

979 Potempa, Fliegertruppe als Teil der Deutschen Luftstreitkräfte bis 1918, S. 15.

980 BA-MA PH 9-XV/2, Ausarbeitung einer Fliegervorschrift, „Die Ausbildung der Flieger", Straßburg, 25.1.1913.

lungen, noch gab es eine geregelte Kommunikationsbasis zwischen der Industrie und dem Militär.[981] Zu diesen Missständen notierte der damalige Inspekteur der Fliegertruppen, Oberst Walther von Eberhardt, im August 1914 in seinem Tagebuch, dass er sich mit einem „Haufen Telegramme von der Front" konfrontiert sehe, die Etappenflugzeugparks nur „ungenügend ausgestattet" seien und die „Tauben" nicht steigen würden.[982] Diesen Problemen wirkte die Oberste Heeresleitung im März 1915 mit der Ernennung des Major Hermann von der Lieth-Thomsen (1867–1942) zum Feldflugchef entgegen, der am 26. April 1915 ins Amt eingeführt wurde. Ihm unterstanden künftig alle Fliegereinheiten und das Ersatzwesen; seine vorrangige Aufgabe war es, die Abteilungen mit Flugzeugen auszustatten.[983]

Im Laufe des Krieges wurde die Flugzeugproduktion aller Großmächte erheblich gesteigert; bis zum Ende des Ersten Weltkrieges standen den deutschen Fliegertruppen rund 48.000 Flugzeuge zur Verfügung, von denen nur rund sieben Prozent durch Feindeinwirkung verloren gingen. Insgesamt starben 6.840 Angehörige der Fliegertruppen an der Front und in der Heimat: In der Heimat kamen 1.800 Männer bei der Arbeit in der Fliegerei zu Tode, davon 98 Prozent bei der Ausbildung, im Feld fielen 1.450 Flieger ohne, 1.420 mit Feindeinwirkung im Luftkampf, davon 710 Beobachter. Weitere 1.400 Personen wurden als vermisst gemeldet.[984]

Im Zuge einer Ausweitung der Einsatzgebiete der Flugzeuge, bedingt durch verbesserte Distanzwaffen, veränderte Frontverläufe und die Einführung des mit dem Propeller synchronisierten Maschinengewehrs,[985] wurde am 8. Oktober 1916 die Stelle des Kommandierenden Generals der Luftstreitkräfte (KoGenLuft) geschaffen, die Ernst von Hoeppner bekleidete. Ihm zur Seite gestellt wurden Major von der Lieth-Thomsen und der Inspekteur der Fliegertruppen Major Wil-

981 Vgl. Rosenboom, Im Einsatz, S.21.

982 Tagebucheintrag Oberst Walther von Eberhardt vom 15.8.1914, zit. nach Supf, Fluggeschichte, Bd. 2, S.270.

983 Vgl. Neitzel, Zum strategischen Mißerfolg verdammt?, S.168. Siehe für eine ausführliche Darstellung der Luftrüstung im Krieg: John H. Morrow, German Air Power in World War I, Lincoln (Nebraska)/London 1982. Siehe auch Förster, Die Entwicklung der Fliegerei im Weltkriege, S.44.

984 Vgl. Martin Kutz, Das Flugzeug als Waffe – der Erste Weltkrieg als Experimentierfeld des Luftkriegs, in: Museum für Verkehr und Technik (Hg.), Hundert Jahre deutsche Luftfahrt. Lilienthal und seine Erben, Gütersloh/München 1991, S.39–59, hier S.53. BA-MA PH 9-XV/7, Verluste der deutschen Fliegertruppen (einschl. bayer. Fliegerverbände) vom 2.8.1914 bis 11.11.1918.

985 Es wird in der Luftfahrtgeschichte bisweilen übersehen, dass die Versuche mit Maschinengewehren, Bombenabwurf-Vorrichtungen und anderen Distanzwaffen bereits in die Vorkriegszeit zurückreichen und Vorschläge eingereicht wurden. Siehe dazu David Edgerton, England and the Aeroplane. An essay on a militant and technological nation, Houndsmills/London 1992, S.17 und z. B. „Pariser Brief", in: Flugsport (1910), Nr. 20, S.338–649, hier S.642. Auch August Euler wollte sich bereits 1912 ein Maschinengewehr patentieren lassen: BA-MA PH 19/131, Lehr- und Versuchsanstalt Döberitz, 2.7.1912, Betreff: Euler Maschinen Gewehrflugzeug.

helm Siegert, der Anfang 1915 die Nachfolge von Oberst von Eberhardt angetreten hatte.[986]

Hauptaufgabe der Flieger war die Nah- und Fernaufklärung, das heißt das Fotografieren der feindlichen Stellungen, Unterstände und Munitionslager etc.[987] Feindliche Flugzeuge störten jedoch vermehrt die Aufklärung, brachten die Maschinen dazu abzudrehen oder versuchten, durch Bombenabwürfe den Gegner zum Absturz zu bringen. Von einem „richtigen Luftkampf" konnte allerdings noch nicht die Rede sein, denn die zur Verfügung stehenden Waffen – Karabiner und mitgeführte Bomben[988] – verfehlten meist ihr Ziel. Gleichsam als Vorhut und Schutz für die Aufklärer bildeten sich ab 1916 neue Einheiten wie die Jagdstaffeln heraus, 1917 erfolgte die Aufstellung der ersten Jagd- und Kampfgeschwader.[989] Leutnant Falkenstein hatte bereits im März 1914 eine Störung der Aufklärungsflugzeuge durch den Gegner und die daraus folgende Notwendigkeit zur Aufrüstung der Maschinen antizipiert: „Diese Sperre kann naturgemäß nur durch Kampf durchbrochen werden, also muß das Aufklärungsflugzeug auch ein Kampfflugzeug sein, oder sollen wir unsere Aufklärer kampf- und schutzlos jenen Torpedojägern der Luft ausliefern, nur um sie um ein geringeres vielleicht leichter und schneller zu machen?"[990]

Bis Kriegsende blieben die Fliegerabteilungen, die am 1. Januar 1917 in Fliegerabteilungen (FA) und Fliegerabteilungen (Artillerie) umstrukturiert wurden und deren Aufgabe die Aufklärung und das Einschießen der eigenen Artillerie umfasste, das „Rückgrat" der Luftstreitkräfte.[991]

Die fotografische Luftaufklärung

Die Luftaufklärung war die zentrale Voraussetzung für die Erstellung von detaillierten und den jeweils aktuellen Frontverlauf zeigenden Kriegskarten. Der Einsatz der Fotografie für militärische Zwecke war bereits seit den 1880er Jahren dis-

986 Vgl. Neitzel, Zum strategischen Mißerfolg verdammt?, S.169

987 Vgl. Kehrt, Schneid, S.177.

988 Vgl. Potempa, Fliegertruppe als Teil der Deutschen Luftstreitkräfte bis 1918, S.21. Erst im Frühjahr 1915 wurden die ersten Flugzeuge mit Maschinengewehren ausgerüstet. Generalleutnant Förster zufolge seien bis Dezember 1915 1.138 Maschinengewehre an die Fliegertruppe ausgegeben worden, davon 439 fest eingebaut. Ab April 1916 habe sich die Zahl um monatlich 300 bis 400 Stück erhöht. Vgl. Förster, Die Entwicklung der Fliegerei im Weltkriege, S.45f.

989 Vgl. Neitzel, Zum strategischen Mißerfolg verdammt?, S.176.

990 BA-MA PH 24/91, Lt. Falkenstein, Denkschrift über den weiteren Ausbau von Flugzeugen als Kampfmittel, Jüterbog, 10.3.1914.

991 Rosenboom, Im Einsatz, S.21. Mit dem Einschießen der Artillerie war gemeint, die Einschläge der Geschosse zu registrieren und die Artillerie gegebenenfalls anzuweisen, ihre Position zu verändern, um eine höhere Trefferquote zu erzielen. Dazu kommunizierten die Flieger über Handzeichen und Fahnen, 1915 wurde auch die Funktelegrafie praktisch erprobt. Vgl. die Einträge in Suwelacks Flugbuch von April und Mai 1915 zur Telegrafie.

kutiert worden. Durch die Erfindung der Trockenplatte und 1902 die Entwicklung der panchromatischen Platte konnten Farben in verschiedenen Graustufen wiedergeben werden. Damit wurde es möglich, mittels der Fotografie detailgetreuere Bilder als die aus Fesselballonen gemachten Handzeichnungen zu liefern.[992] Angesichts der dringenden Notwendigkeit von geeigneten Karten, die eine Orientierung aus der Luft in ständiger Bewegung sowie am Boden ermöglichen sollten, kam auch der Fotogrammetrie eine entscheidende Bedeutung zu. Sie war eine Visualisierungstechnik, die optische Verfremdungseffekte durch die Höhenperspektive und Verzerrungen durch die Eigengeschwindigkeit des Flugzeuges korrigieren und ein umfassendes Raumpanorama erzeugen sollte: „An die Stelle imaginierter Höhenblicke der klassischen Kartographie trat die Auseinandersetzung mit den photographisch fixierten Höhenblicken und einem daraus entstehenden neuen Raumbild."[993] Damit verbunden war zugleich eine neue, abstrakte Raumerfahrung, die auch in der Kunst in den Stilrichtungen des Kubismus, Expressionismus und Futurismus ihren Ausdruck fand.[994]

Die Flieger gehörten zu einer der wenigen militärischen Einheiten, die mit Kameras[995] ausgestattet waren. Bei Ausbruch des Krieges mangelte es allerdings sowohl an qualifiziertem Personal, das fähig gewesen wäre, die Luftbilder in entsprechende Karten zu übertragen, als auch an den nötigen Fotogrammetrie-Geräten und Kriegskarten. Siegfried Boelcke, im Krieg Chef des Feldvermessungswesens, schrieb diesbezüglich 1920:

> „Der monate-, oft jahrelang an ein und derselben Stelle im Boden festgewurzelte Stellungskampf erforderte eine Fülle bis ins kleinste genauer Karten. ... Vor allem waren es die mit rascher Vervollkommnung entstehenden Luftbilder der Flieger, die in Karten großen Maßstabs übertragen wer-

992 Helmut Jäger, Erkundung mit der Kamera. Die Entwicklung der Photographie zur Waffe und ihr Einsatz im 1. Weltkrieg, 2. Aufl., München 2007, S.25–41.

993 Stefan Siemer, Bildgelehrte und Geotechniker. Luftbild und Kartographie um 1900, in: Alexander Gall (Hg.), Konstruieren, kommunizieren, präsentieren. Bilder von Wissenschaft und Technik, Göttingen 2007, S.69–108, hier S.71.

994 Möser, Fliegerblick, S.99–106; Joseph Farrell, Das Fliegen. D'Annunzio und die Futuristen, in: Bodo-Michael Baumunk (Hg.), Die Kunst des Fliegens (anlässlich der Ausstellung „Die Kunst des Fliegens" vom 2. Juli bis 27. Oktober 1996 im Zeppelin-Museum Friedrichshafen), Friedrichshafen 1996, S.72–83.

995 Die Fliegerabteilungen waren 1914 mit einer zusammenrollbaren Dunkelkammer, einem Gerätekoffer und zwei Kameras von je 25 cm Brennweite ausgestattet. Die Kameras bestanden aus einem Holz- bzw. Metall-Hohlkörper, einem Objektiv und dem Verschluss. Die Form war so gewählt, dass sie einen möglichst geringen Luftwiderstand bot. Vgl. Carl Fink Generalmajor a.D., Die Geschichte des Luftbildwesens unter besonderer Berücksichtigung der organisatorischen Arbeiten, die für den Ausbau des Luftbildwesens bis Dezember 1917 geleistet worden sind, [1918] ND Hildesheim 1958, S.9. Im Juli 1915 besaß die Fliegerabteilung drei Handkameras mit einer Brennweite von 25 cm, je eine Handkamera (50 cm Brennweite), eine im Flugzeug fest montierte Kamera (70 cm Brennweite), einen Grundrissbildner, ein Vergrößerungsgerät (13x18 cm), ein Vervielfältigungsgerät (Opalograf) und einen Dunkelkammerwagen. An Kameras standen verschiedene Modelle zur Verfügung, z.B. von den Firmen Lechner, Goldmann oder Zeiss. Siehe auch BA-MA PH 9-XIII/126, Inspektion des Luftbildwesens, darin: Photographieren aus Flugzeugen.

den mußten. Dazu fehlte nahezu alles: der ‚Vorgang' ... und, nicht zuletzt, die Karte 1:25.000 selbst, in der man die Ergebnisse der Bildaufklärung hätte eintragen können."[996]

Aufgrund des erwarteten kurzen Bewegungskrieges war jede Einheit mit nur einem Satz Kriegskarten für die Bodentruppen ausgestattet worden, Nachschub war nicht vorgesehen.[997] Über Fliegerkarten verfügte bei Ausbruch des Krieges nur Frankreich.[998]

Rückblickend bemerkte Boelcke, man sei 1914 mit einer Kriegskarte ausgestattet gewesen, „um einen frisch-fröhlichen Feldkrieg von einigen Wochen zu führen", doch hätten die Feldvermesser der sich rasch ausdehnenden und wandelnden Kriegslandschaft gar nicht schnell genug begegnen können und so hätten die Aufklärungsflieger ihnen den Rang abgelaufen.[999] Gleichwohl sei eine enge Kooperation zwischen den Fliegern und den Bildauswertern des Vermessungswesens unverzichtbar gewesen:

„Die Flieger nahmen die Bilder auf, meldeten die Neuerscheinungen taktischer Art und stellten den rasch anwachsenden Bildstoff übersichtlich zusammen. Die Bildauswerter ... übertrugen die Bildergebnisse in die Karte und regten die Aufnahme etwa noch fehlender Bilder an. Die Ballonbilder hatten im Vergleiche zu denen der Flieger keine nennenswerte Bedeutung."[1000]

Während des Krieges arbeitete man an einer stetigen Verbesserung der Kameras, der Bildauswertung sowie der Schulung des Personals.[1001] Man wollte den Unterricht auf ein Minimum an Theorie beschränken, da der „Grundsatz, daß die Erfahrung der beste Lehrmeister der Photographie ist, ... die Richtschnur für die Ausbildung sein"[1002] müsse: „Es ist nur zu lehren, was unbedingt notwendig und mit den feldmäßigen Mitteln zu erreichen ist. Alle Künsteleien sind untersagt."[1003]

Die Anforderungen an den mitfliegenden Beobachtungs-Offizier sollten allmählich und seinen Fähigkeiten entsprechend gesteigert werden. Betont wurde vor allem auch die kommunikative Feinabstimmung zwischen Flugzeugführer und Beobachter im Vorfeld eines Aufklärungsfluges:

996 Oberleutnant z. D. Siegfried Boelcke, Kriegsvermessung und ihre Lehren, Berlin, S.5.

997 Vgl. Rudolf Hafeneder, Überblick über das militärische Geowesen Deutschlands im 19. und 20. Jahrhundert, Euskirchen 2004, S.19. Bei den Kriegskarten handelte es sich um Pläne mit Markierungen, Wegenetzen etc.

998 Vgl. ebd., S.69.

999 Vgl. ebd.

1000 Ebd., S.20.

1001 Vgl. Siemer, Luftbild, S.93.

1002 BA-MA PH 9-XV/2, Ausarbeitung einer Fliegervorschrift, T.D.V.F. Nr.59, Nur für den Dienstgebrauch. Photographieren aus Flugzeugen, Berlin 1913.

1003 Ebd. Siehe auch zum Aspekt des Lernens durch ständiges Üben: BA-MA PH 17-I/25, Die Bildmeldung der Flieger (Druck) – Teil I: Beobachter, Teil II: Bildoffizier: A: Ausbildung des Beobachters (Entwurf) 1917/18, S.21a: „Das Lesen und Auswerten der Fliegerbilder lernt der Beobachter nur durch dauernde Uebung". Es folgen Vorschläge für Übungsaufgaben.

> „Flugzeugführer und Beobachter besprechen die Aufgabe und machen sich über den einzuschlagenden Weg klar, wobei sich der Flugzeugführer besonders das Gelände in der Nähe des aufzunehmenden Gegenstandes einprägt. Der gewählte Weg wird mit Buntstift in die Karte eingetragen und in Abschnitte (z. B. 10 zu 10 km) eingeteilt."[1004]

Nur als Team, das sich auch ohne Worte – in der Luft war aufgrund der Motorengeräusche eine mündliche Kommunikation ohnehin schwierig – verstand und arbeitsteilig agierte, seien gute Aufklärungsresultate zu erzielen.[1005]

Einen wesentlichen Vorteil des Flugzeuges gegenüber der Kavallerie, die ebenfalls für die Aufklärung zuständig war und bei den Eröffnungsschlachten zum Einsatz kam,[1006] sah die Militärverwaltung vor allem in der Bewegungsfreiheit und schnellen Nachrichtenübermittlung, doch sollten die Flieger auch operativ eingesetzt werden. Die Flugzeuge seien dort von Nutzen, „wo Gefechts- oder Geländeverhältnisse die Kavallerieaufklärung beschränken oder verzögern, z. B. bei Flußübergängen".[1007] Außerdem könne die Luftaufklärung beim Übergang von der Fern- zur Nahaufklärung nahtlos fortgesetzt werden, während die Aufgabe der Kavallerie zumeist bei der Fernaufklärung ende. Die Luftaufklärung sollte genaue Informationen zur Belegung des Straßennetzes durch feindliche Kolonnen und zu deren Stärke und Länge liefern. Bei der Gefechtsaufklärung sei „neben der Feststellung der Flügel, Aufstellung der Reserven oder Abmarsch von Verstärkungen etc. die Erkundung verdeckt stehender Artillerie von größter Bedeutung."[1008] Mit dem Erstarren der Fronten im September 1914 gewann die Aufklärungsarbeit der Flieger zusätzlich an Bedeutung, war es doch die einzige Einheit, die sich zunächst relativ gefahrlos an der Front bewegen konnte.[1009]

Hinsichtlich der Aktualität der Informationen war man der Auffassung, dass innerhalb einer Stunde Nachrichten von Truppen übermittelt werden könnten, die zwei Tagesmärsche voneinander entfernt waren. Das Zerstörungspotenzial der Flugzeuge lag vor allem im Bombenabwurf über logistischen und versorgungsrelevanten Schlüsselpunkten wie „Pulvermagazinen, Kasernen, Munitionslagern, Elektrizitäts- und Gaswerken". Den militärischen Entscheidungsträgern war bewusst, dass die „Überlegenheit in der Luftaufklärung" den Ausgang einer Schlacht wesentlich beeinflussen konnte.[1010]

1004 Ebd., Der photographische Erkundungsflug.

1005 Vgl. BA-MA PH 17-I/25, Die Bildmeldung der Flieger (Druck) – Teil I: Beobachter, Teil II: Bildoffizier: A: Ausbildung des Beobachters (Entwurf) 1917/18, S.21a.

1006 Vgl. Stefan Kaufmann, Kriegführung im Zeitalter technischer Systeme – Zur Maschinisierung militärischer Operationen im Ersten Weltkrieg, in: Militärgeschichtliche Zeitschrift 61 (2002), S.337–367, hier S.355.

1007 BA-MA PH 9-XV/2, Ausarbeitung einer Fliegervorschrift, Taktische Verwendung [durchgestrichen].

1008 Ebd.

1009 Vgl. Carl Fink Generalmajor a.D., Die Geschichte des Luftbildwesens, S.9.

1010 Ebd.

Als „Auge des Heeres"[1011] waren die Flieger deshalb von großer Bedeutung für die strategische Planung. Sie hatten als Einzige einen Überblick über das Kampfgeschehen, zugleich aber auch einen distanzierten Blick, der sie das Ausmaß der Zerstörung und des Schreckens nur ahnen ließ. Aufgrund ihrer begrenzten technischen Leistungsfähigkeit hatten sie allerdings, wie Sönke Neitzel unterstrichen hat, angesichts der größeren Ressourcen, d. h. der zahlenmäßigen Überlegenheit der Alliierten keine strategische Wirkung, das heißt sie beeinflussten den Schlacht-, nicht aber den Kriegsverlauf.[1012] Während des Krieges wurden mehr als eine Million Luftbilder fotogrammetrisch ausgemessen und rund sieben Millionen Quadratkilometer Gelände aufgenommen.[1013] Ohne die fotografische Luftaufklärung lassen sich insbesondere die großen Schlachten nach 1915 in Verdun und an der Somme nicht in diesem Ausmaß vorstellen.[1014] Von einer organisierten und auf die zu bewältigenden Aufgaben vorbereiteten Fliegertruppe konnte in der ersten Kriegshälfte jedoch noch nicht die Rede sein; der Erwartungshorizont der Heeresleitung, aber auch der Soldaten entsprach nicht dem tatsächlichen Kriegsgeschehen.

4.4. Schreiben und Fotografieren im Krieg: Josef Suwelack als Aufklärungsflieger bei der Feldfliegerabteilung 24

Logistik: Von der Etappe an die Front: Mobilmachung und Mobilität

In Großenhain, das Josef Suwelack am 3. August mit der Bahn erreichte, befand sich seit 1913 eine noch im Aufbau befindliche Militärfliegerstation, die über Flugzeughallen, Unterkünfte, Kommandantur, Schulgebäude, eine Werft und Tankanlagen verfügte. Ihre Funktion bestand darin, Fliegereinheiten aufnehmen, mobilisieren und demobilisieren zu können.[1015] In den ersten Augusttagen rückten von der in Großenhain stationierten königlich-sächsischen 3. Kompanie etwa 30 Flugzeuge mit 479 Mann, d. h. Offiziere und Unteroffiziere samt Mannschaften, an die Front aus. Sie gehörten den Feldfliegerabteilungen 23, 24 und 29 sowie dem Etappenflugzeugpark 3 an. Hauptmann der Feldfliegerabteilung 24 war Horst von Minckwitz (1877–1956),[1016] die Leitung des Etappenflugzeugparks 3 war Major d. R. Georg Mardersteig übertragen worden.[1017]

1011 Hafeneder, Überblick über das militärische Geowesen, S.17.
1012 Vgl. Neitzel, Zum strategischen Mißerfolg verdammt?, S.176 und S.191.
1013 Vgl. Hafeneder, Überblick über das militärische Geowesen, S.18.
1014 Vgl. Jens Jäger, Fotografiegeschichte(n). Stand und Tendenzen der historischen Forschung, in: Archiv für Sozialgeschichte 48 (2008), S.511–537, hier S.527.
1015 Siehe für eine knappen Abriss über die Geschichte des Flugplatzes Großenhain: Polo Palmen (Hg.), Flugplatz Großenhain. Historischer Abriss, o. O. 2007, S.7ff.
1016 Informationen zu Minckwitz: Hannes Täger, Minckwitz, Horst Benno Adolf von, in: Sächsische Biografie, hg. vom Institut für Sächsische Geschichte und Volkskunde e.V., bearb. von Martina Schattkowsky, online unter: http://www.isgv.de/saebi/ (12.2.2015).
1017 Ebd., S.21.

Josef Suwelack war dem Etappenflugzeugpark 3 zugeteilt, der für den Transport, das Einfliegen der Maschinen und den personellen Ersatz für die drei ausgerückten Feldfliegerabteilungen zuständig war. Erst nachdem die Versorgung gewährleistet war, durfte er als Fliegerersatz Erkundungsflüge machen.[1018] Als „technischer Leiter",[1019] wie er seine Position in einem Brief an seine Eltern angab, war er für die Abnahme und das Einfliegen der von den Flugzeugfirmen gelieferten Flugzeuge zuständig. Am 9. August berichtete er seiner Familie:

> „Ich habe in den 8 Tagen nun schon 14 verschiedene Apparate geflogen. L.V.G., Euler, Albatros u.s.w. Doppeldecker und Eindecker. Heute Morgen habe ich einen Gleitflug von 13 Minuten aus 900 mtr. Höhe gemacht. Man kann also sehr weit kommen, wenn der Motor versagt und gerade dieses übe ich ein. Vorläufig werden wir zu Feldflügen nicht verwandt, erst zuallerletzt."[1020]

In Großenhain blieb Suwelack bis zum 15. August, dann fuhr er nachts um halb drei[1021] mit der Fliegerersatzabteilung den Fliegerabteilungen nach Niedermendig Richtung Belgien hinterher, sein Ziel erreichte er wahrscheinlich am 18. August. Die Fahrt im Durchgangszug im Zweierabteil[1022] führte über Naumburg an der Saale, Weimar, Hanau, Biebrich und Sankt Vith unmittelbar vor der belgischen Grenze.[1023]

Aus Niedermendig erreichte Josefs Eltern eine Karte vom 24. August, in der er von der Errichtung eines Flugplatzes und von einem Bombentransport für eine Fliegerabteilung nach Namur berichtete:

> „Bis jetzt waren wir in Niedermendig am Laacher See und haben hier in der Eifel einen Flugplatz eingerichtet. Unsere Abteilungen sind aber schon zu weit vorgerückt und wir müssen nach Belgien hinein. Vorige Woche hatte ich einen Bombentransport per Auto nach Namur für unsere Flieger. Ich hatte einen 80 PS Opel Wagen mit 4 Mann Besatzung. Jeder von uns hatte Karabiner. Wir fuhren über Prüm durch Luxemburg nach Marche. ... Wir fuhren über notdürftige reparierte gesprengte Brücken. Alle Chausseebäume lagen abgeschlagen über den Wegen und waren von den Pionieren fortgeräumt. Die Bewohner ängstlich. Kein Kopf ließ sich sehen. Die meisten Häuser verlassen. ... Das Schwierigste war, daß ich gar keine Karte mithatte und mich durch Befragen der Bevölkerung zurechtfinden mußte."[1024]

Anhand weiterer schriftlicher Darstellungen gewinnt man einen ungefähren Eindruck von der anfänglichen Ausstattung der Fliegerabteilung. Josef hatte seinen Eltern am 9. August berichtet, dass die Etappe mit „15 Apparaten, 10 Automobillastzügen und 5 Personenautos per Bahn" ausrücken werde,[1025] Die Fliegertruppe

1018 Vgl. NL J. Suwelack, Postkarte an seine Familie, Großenhain, den 4. August 1914 und Brief vom 11.8.1914.

1019 NL J. Suwelack, Brief Josef Suwelacks an seine Familie vom 9.8.1914.

1020 Ebd.

1021 NL J. Suwelack, Brief Josef Suwelacks an seine Familie vom 15.8.1914, Hotel de Saxe, Großenhain.

1022 Vgl. ebd.

1023 Vgl. NL J. Suwelack, Postkarten Josef Suwelacks aus Naumburg an der Saale (16.8.14), Weimar (ebd.), Hanau (17.8.14), Biebrich (ebd.). Der rekonstruierte Weg orientiert sich an den Postkartendatierungen.

1024 NL J. Suwelack, Feldpostbrief, Niedermendig, 24.8.1914.

kann somit als die „ersten vollmotorisierte Einheit der Militärgeschichte gelten“.[1026]

Ende August verlegte der Etappenflugzeugpark sein Quartier nach Marche in Belgien, wo die Fliegerersatzabteilung das Château des Bords-Champs bewohnte. Anschließend führte der Weg die Abteilung nach Rocroi in Frankreich, wo sie sich nur wenige Tage auf einem Bauerngut aufhielt. Am 4. September flog sie nach Reims und wohnte dort in einem Grand Hôtel. Weiter ging es nach Mourmelon und Chalons; von dort aus fuhr Suwelack mit dem Auto nach Doische, „um einen Transport Flugzeuge abzuholen“,[1027] und übernachtete in einer herrenlosen Villa. Am 13. September wurde er dann zur Feldfliegerabteilung 24, die dem XIX. Armeekorps unterstand, nach St. Etienne versetzt.[1028] Bei dieser Abteilung, die am 8. Oktober[1029] ihr Quartier zunächst nach Vendegies-au-Bois in die Nähe von Lille verlegte und am 18. Oktober 1914[1030] ganz nach Lille wechselte, blieb er bis zu seinem Tod am 13. September 1915.

Josef Suwelacks Arbeit als Aufklärungsflieger über den feindlichen Stellungen begann am 20. September 1914 mit einem Flug in 2.000 Metern Höhe über Suippes nordwestlich von Châlons-en-Champagne in Begleitung seines Beobachters Leutnant Lothar Wilisch (geb. 1888).[1031] Etwa jeden zweiten Tag war er im September und Oktober mit dem Flugzeug und wechselnden Beobachtern – Leutnant Wilisch, Oberleutnant Kurt Müller (1886–1915) und Oberleutnant Meinhard Rosenmüller (geb. 1883), dem Abteilungsführer der FfA 24 – in der Luft, um die Stellungen zu erkunden. Besonders in der letzten Oktoberwoche flogen sein Beobachter und er täglich.[1032]

War diese erste Phase für Josef Suwelack noch von einem hohen Grad an Luftmobilität gekennzeichnet, so stagnierte die Anzahl der Flüge in den Monaten November bis Februar, denn die Abteilung hatte die Stellungen mittlerweile alle fotografisch erfasst und konzentrierte sich anschließend auf die Lokalisierung feindlicher Batterien und auf das Einschießen der eigenen Artillerie:[1033]

> „Hier ist alles beim alten [sic!]. Vorwärts kommen wir sehr sehr langsam, und ich glaube, daß die Sache hier in einen netten Winterschlaf ausartet. Überall wird fieberhaft gearbeitet. Wir haben sehr wenig zu tun, da wir unsere Stellungen ganz genau kennen. Und die Infanterie hat sich ganz häus-

1025 NL. J. Suwelack, Brief an seine Eltern, Großenhain, 9.8.1914.
1026 Möser, Fahren und Fliegen, S.521.
1027 NL J. Suwelack, Feldpostkarte J. Suwelacks, Doische, undatiert [vermutlich zweite Septemberwoche 1914].
1028 Nl J. Suwelack, Feldpostbrief J. Suwelacks an seine Familie, St. Etienne, 17.9.1914.
1029 Vgl. NL J. Suwelack, Eintrag Josef Suwelacks in sein Flugbuch, Aufzeichnungen vom 20.9.1914 bis zum 4.5.1915. Das Flugbuch diente zugleich als Notizbuch. Josef vermerkte darin die Feldadressen befreundeter Flieger und anderer Bekannter. Siehe die Aufstellung im Anhang.
1030 Vgl. ebd.
1031 Vgl. ebd. und Feldpostbrief an seine Familie, St. Etienne, 21.9.1914.
1032 Vgl. NL J. Suwelack, Einträge ins Flugbuch für den Monat Oktober 1914.
1033 Vgl. NL J. Suwelack, Feldpostbrief an die Eltern, Lille, 3.12.1914.

lich eingerichtet. Ganze unterirdische Wohnungen mit Öfen und allem. Und fast alle haben gute warme Wintersachen. Hier in Lille geht alles seinen gewohnten Gang."[1034]

Bedingt durch die Kälte, das winterliche Erstarren des Krieges und den Rückzug der Soldaten in unterirdische Schlupfwinkel blieben die Frontverläufe unverändert. Seinem Bruder Richard teilte Suwelack Anfang Dezember ebenfalls von der Erstarrung und Ruhe an der Front mit:

„Wir liegen hier immer noch in Lille und unser Korps steht vor Armentières. Momentan scheint alles den Winterschlaf zu schlafen. Ein Vorkommen ist hier gänzlich unmöglich. Das Grundwasser ist auf 1 mtr und unsere Pioniere können keine Sappen treiben. Der Boden ist sehr lehmig und oftmals rutschen die Unterstände zusammen. Im Allgemeinen ist hier Ruhe und in den Schützengräben flackern lustig kleine Öfchen. Aber unsere Stellungen sind brillant ausgebaut."[1035]

Die Feldfliegerabteilung, besser gesagt das Wachpersonal samt Werkzeugen, Benzin, Öl und den Apparaten etc. war auf dem französischen Militärflugplatz (Abb. 19) unweit der Stadt Lille stationiert.[1036] Die Beobachter, Flieger, Monteure und Burschen hingegen bewohnten Zimmer in der Rue Nationale in Lille.[1037]

Zum Jahreswechsel begann man auf dem Flugplatz mit der Errichtung eines großen Holzschuppens, der Platz für sechs Maschinen bieten sollte und wohl auch als „Casino" genutzt wurde (Abb. 20).[1038]

Die Flugzeuge wurden mit der Bahn in demontiertem Zustand transportiert und dann über die Etappenflugzeugparks an die zu versorgenden Feldfliegerabteilungen verteilt. Untergebracht wurden sie in großen Zelten, die Tageslicht für Reparaturarbeiten hereinlassen und ein Gefälle zum Ablauf verschiedener Flüssigkeiten (Kühlwasser, Öl etc.) haben mussten sowie über einen Boden verfügen sollten, der Feuchtigkeit abhielt.[1039] Trotzdem wurden die Flugzeuge durch den Transport und den Einsatz stark beansprucht und wiesen rasch Verschleißerscheinungen auf. So beklagte sich Josef Suwelack in einem Brief an Alfred von Back-Begávar von Anfang April 1915 darüber, dass die Maschinen „durch den Transport und das lange Stehen im Regen auf der Bahn zu viel [leiden]"[1040] würden.

Verlief der Truppen- und Materialtransport relativ reibungslos, so funktionierte die Zustellung der Feldpost zu Beginn nicht zuverlässig. Richard erhielt die erste Post seines Bruders am 8. November 1914, der Brief war Richard zufolge jedoch bereits am 3. November in Köln abgestempelt worden: „Dieses läßt mich vermuten, daß Du denselben einem Auto, welches nach Deutschland zurückfuhr, mitgegeben hast. Ich möchte Dich deshalb bitten, in Zukunft möglichst denselben Weg

1034 NL J. Suwelack, Feldpostbrief an die Eltern, Lille, 27.11.1914.
1035 NL J. Suwelack, Feldpostbrief an Richard Suwelack, Lille, 3.12.1914.
1036 Vgl. NL J. Suwelack, Brief an seine Familie, Lille, 19.10.1914.
1037 Vgl. NL J. Suwelack, Feldpostbrief an Richard Suwelack, Lille, 1.11.1914.
1038 Vgl. NL J. Suwelack, Brief Josefs an Richard („Natz") Suwelack, Lille, 3.1.1915.
1039 BA-MA PH 9-XV/2, Ausarbeitung einer Fliegervorschrift, darin: Unterbringungsarten für Flugzeuge, Aufbau von Flugzelten, 1913.
1040 NL J. Suwelack, Brief Josefs an Alfred von Back-Begávar von den Kondor-Werken, Ostern 1915.

Abb. 19: „Unser Flugplatz in der linken Ecke des Platzes oben unsere Zelte", Lille, Januar 1915. Die Markierung „AT" in der unteren Bildhälfte diente den Fliegern als Orientierungshilfe (Foto: Familienarchiv Walter Suwelack, NL Josef Suwelack)

Verlief der Truppen- und Materialtransport relativ reibungslos, so funktionierte die Zustellung der Feldpost zu Beginn nicht zuverlässig. Richard erhielt die erste Post seines Bruders am 8. November 1914, der Brief war Richard zufolge jedoch bereits am 3. November in Köln abgestempelt worden: „Dieses läßt mich vermuten, daß Du denselben einem Auto, welches nach Deutschland zurückfuhr, mitgegeben hast. Ich möchte Dich deshalb bitten, in Zukunft möglichst denselben Weg zu benutzen, da die Beförderung der Feldpost von einem Truppenteil zum andern anscheinend mangelhaft ist."[1041]

Die private Beförderung der Post mit dem Auto, vorbei an den offiziellen Poststellen, birgt eine mögliche Erklärung dafür, weshalb Josef Suwelack zahlreiche Luftbilder, die der Geheimhaltung unterlagen, an seine Familie senden konnte, ebenso wie Fotos, die den Schrecken des Krieges in seiner ganzen Härte zeigten. Da die Flieger über Kameras verfügten, hatten sie die Möglichkeit – nicht aber die Erlaubnis –, Bilder jenseits der eigentlichen „Arbeits-Fotografie" auch für den privaten Gebrauch zu machen und zu vervielfältigen. Josef schärfte seinen Eltern

1041 NL J. Suwelack, Brief Richards an Josef, o. O. 8.11.1914.

Abb. 20: Fertiger Schuppen der FFA 24 auf dem Flugplatz in Lille, Januar 1915 (Foto: Familienarchiv Walter Suwelack, NL Josef Suwelack)

mehrmals ein, „die Luftaufnahmen sorgfältig zu verschließen und sie keinem, den Ihr nicht gut kennt, zu zeigen. Sie sind nur für mich da und Ihr dürft sie auf keinen Fall aus der Hand geben."[1042] Die privaten Aufnahmen erfüllten zwei Funktionen: Erstens sollten sie seine Familie „ins Bild" setzen und zweitens waren sie „eine wertvolle Erinnerung" für Josef Suwelack selbst, der etliche Aufnahmen aus seiner Kriegszeit in Alben beschriftet verwahrte.[1043]

Von seiner Familie in Billerbeck hatte Josef bis Mitte September nur zwei Briefe erhalten, doch beruhigte er seine Eltern, sie sollten sich keine Sorgen machen, denn die Post dringe selten bis zu seiner Abteilung durch.[1044] Am 10. Oktober hatte er nur einen weiteren Brief seiner Eltern empfangen, von seinem

1042 NL J. Suwelack, Feldpostbrief an seine Familie. Lille, undatiert [wahrscheinlich 4.4.1915, Ostern, anlässlich der Kommunion seiner Schwester Maria]. Vgl. auch den Feldpostbrief vom 20.10.1914: „Nur eins. Keines von diesen Bildern aus der Hand geben oder gar veröffentlichen. Ich habe sonst die größten Unannehmlichkeiten". Vendegies-Neuville, 20.10.1914.

1043 NL J. Suwelack, Feldpostbrief an seine Familie. Lille, undatiert [wahrscheinlich 4.4.1915]. Vgl. auch Feldpostbrief an seine Familie, Petit Rechain, 19.4.1915: „Schreibt mir bitte mal sofort, ob auch die letzten Luftaufnahmen angekommen sind. Ihr erwähnt nichts davon. Sie sind eine wertvolle Erinnerung."

1044 Vgl. NL J. Suwelack, Feldpostbrief an seine Familie, St. Etienne, 17.9.1914.

Bruder hingegen noch nichts gehört: „Hat er Euch geschrieben? Ich hab alle 4 Tage geschrieben."[1045] In einem Brief von Mitte Oktober bat Josef seine Eltern daher, die erhaltenen Briefe auf der Schreibmaschine abzutippen und sie ihm beziehungsweise Richard zu senden, damit sie mehr voneinander erführen:

> „Die Zeitung habe ich immer nach etwa 8–10 Tagen. Nur von 3 Wochen fehlt jede Post. Übermorgen schicke ich einige Photografien. Sendet doch Richard Abschriften auf der Schreibmaschine von meinen Briefen, denn es dauert etwa 1–2 Monate, bis Briefe von einem Truppenteil zum anderen kommen. Es geht schneller über Billerbeck, und auf der Schreibmaschine kann man die Briefe schnell abschreiben. Sendet mir auch die Abschriften von Richards Briefen."[1046]

Insgesamt kann dennoch festgehalten werden, dass, abgesehen von zwischenzeitlichen Informationslücken und ausbleibenden Nachrichten von der Front oder aus der Heimat, die Beförderung der Feldpost relativ gut funktionierte. In Anbetracht dessen, dass zwischen 1914 und 1918 28,7 Milliarden Briefe befördert wurden, davon 24 Prozent von Soldaten und 76 Prozent von Zivilisten, kann man von einer funktionierenden Post-Logistik sprechen.[1047]

Die Kriegserfahrungen Josef Suwelacks waren im Unterschied zu den der Grabenkämpfer am Boden geprägt von einer abstrakten distanzierten Sicht auf das Kriegsgeschehen, dem Gefühl der Kontrolle und Orientierung, aber auch der Bewegungsfreiheit in einem ansonsten der individuellen Kontrolle der Soldaten entzogenen und erstarrten Krieg. Die Feldfliegerabteilung 24 war gleich in mehrfacher Hinsicht mobil: Sie verfügte nicht nur über Flugzeuge, Automobile, Pferde und Fahrräder, ein Dach über dem Kopf und gute Kleidung, sondern hatte als eine von wenigen Einheiten die Erlaubnis, andere Truppenteile zu besuchen.[1048]

Da die Fliegerei jedoch wetterabhängig war, konnte nicht jeden Tag geflogen werden, so dass die Ruhephasen zur physischen und mentalen Regeneration genutzt wurden. Bereits die Ausstattung der Abteilung lässt auf einen hohen und luxuriösen Lebensstandard im Krieg schließen, ein Eindruck, der durch die Betrachtung der Versorgung der Fliegertruppe mit Nährmitteln und Luxusgütern noch verstärkt wird.

1045 NL J. Suwelack, Feldpostbrief an seine Familie, Vendegies-au-Bois, 10.10.1914.

1046 NL J. Suwelack, Feldpostbrief an seine Familie, Lille, undatiert [wahrscheinlich 24.10.1914]: „Nun liegen wir schon 6 Tage hier in Lille."

1047 Vgl. Ortwin Buchbender, Zur Geschichte der Rezeption und Sammlung von Feldpost in Deutschland, in: Didczuneit/Ebert/Jander (Hg.), Schreiben im Krieg – Schreiben vom Krieg, S.17–21, hier S.19.

1048 Vgl. NL J. Suwelack, Feldpostbrief an seine Familie, St. Eitenne, 3.10.1914. Siehe zu den der FfA 24 zur Verfügung stehenden Mobilitätsmaschinen den fotografischen Nachlass Josef Suwelacks, die Aufnahmen sind in der Regel nicht datiert.

Imagepflege I: Luxusleben, Freizeitaktivitäten und der Umgang mit Langeweile

> „Soeben komme ich von der Jagd nach Hause und habe 2 Hasen geschossen. Heute ist für mich Ruhetag, denn ich bin 4 Tage hintereinander geflogen. ... Unser Quartier hier ist glänzend. Wir haben auch fast immer auf Schlössern gelegen und jetzt liegen wir in einem vornehmen Patrizierhause in Lille. Wir haben Bier, Sect, Rheinwein, Rotwein, alles was man wünscht und unsere Verpflegung ist glänzend."[1049]

Sätze, wie Josef Suwelack sie seinem Bruder und auch seinen Eltern in zahlreichen Briefen immer wieder schrieb. Die Verpflegung sei glänzend und das Dienstpersonal gebe sich alle erdenkliche Mühe, ihm ein angenehmes Leben zu bereiten. „Das Leben im Krieg hat auch seine Reize", schrieb Josef Ende Oktober auf einer Feldpostkarte seinen Eltern.[1050]

Ohne das fotografische Quellenmaterial zu betrachten, mag der Eindruck entstehen, Suwelack wolle seine Eltern lediglich beruhigen und ihnen, wie so viele Soldaten es ihm in ihren Briefen nach Hause gleichtaten, sein Wohlbefinden versichern. Doch das fotografische Quellenmaterial stütz Suwelacks Aussagen. Regelmäßig schloss er seine Schreiben an seine Eltern mit der Formel, er sei „gesund und munter".[1051]

Auch Richard beklagte sich nicht über seine Verpflegung, denn er sitze „an der Quelle" und habe in den ersten Kriegswochen alkoholische Getränke in „Hülle und Fülle" zur Verfügung gehabt.[1052] Josef beschrieb die Versorgungssituation in Lille für die Armee als ausgezeichnet: „Es gibt hier viele deutsche Läden und bei den Marketendern kann man alles haben. Die Kölnische Zeitung Morgenausgabe haben wir abends 6 Uhr schon in Händen."[1053] Weitere Fotos erwecken den Eindruck eines luxuriösen Lebens im Hinterland des Krieges, das charakteristisch für die Selbstinszenierung der Fliegertruppen zu sein scheint. In seinem 1929 erschienenen Buch „Fliegerschule 4" gab Richard Euringer einige Fliegertopoi wieder, die während des Krieges entstanden waren und sich mit Suwelacks Aussagen decken:

> „Wir kennen den Stumpfsinn des Grabens nicht, noch die zermalmende Dämonie des Trommelfeuers. ... Wir Flieger schlafen in seidigen Betten, wir tafeln in Schlössern, wir trinken Sekt, wenn auch gewiss nicht alle Tage. ... Wir schießen Hasen, wir kochen Kakao, wir bändigen uns ein Reitpferd an, wir tapezieren unsere Buden, legen uns elektrisches Licht, lassen uns Grammophonplatten schicken ... [,] halten Hunde, lesen Bücher, waschen und frottieren uns, ... sausen im Auto nach

1049 NL J. Suwelack, Feldpostbrief an Richard Suwelack, Lille, 1.11.1915.
1050 Vgl. NL J. Suwelack, Feldpostkarte an seine Familie, Marche, undatiert [vermutlich Ende Oktober 1914].
1051 NL J. Suwelack, Feldpostkarte an den Vater, Vendegies-au-Bois, undatiert [Poststempel 11.10.1914]; Feldpostkarte an Helene Suwelack, Lille, undatiert [wohl Anfang November 1914]; Feldpostkarte an Richard Suwelack, undatiert [abgestempelt am 16.12.1914] u.a. Noch im September 1915 schrieb Josef, es gehe ihm sehr gut: Feldpostbrief an seine Familie, Petit Rechain, 12.9.1915.
1052 NL J. Suwelack, Brief Richards an Josef, o.O. [vermutlich Vorges], 8.11.1914.
1053 NL J. Suwelack, Brief an Richard, Lille, 26.12.1914.

der Stadt, ... sind beweglich, kommen herum, ... sind Liebkind beim Generalstab, Engel des Himmels der Infanterie, Stammgäste der Batterien, ... flott und frisch und ausgeschlafen."[1054]

Abb. 21: „Mein Beobachter Lt. Wilisch unser Verpflegungsoffizier. An der Wand abgezogene Hasen", Lille, vermutlich 1. November 1914 (Foto: Familienarchiv Walter Suwelack, NL Josef Suwelack)

Aufgrund ihrer hohen Mobilität war es den Fliegern möglich, wie auf der Aufnahme zu sehen, ihre Stellung zu verlassen und sich als mit allem versorgte, entspannte, körperlich und mental gesunde Einheit zu inszenieren. Verwegenheit, Männlichkeit und Kameradschaft waren zentrale Themen in Suwelacks Feldpostbriefen, denn sie knüpften an Fliegertopoi der Vorkriegszeit an und wurden zu einer mentalen Schutzbarriere, einer Flucht in die Vergangenheit.

Suwelacks Familie, Freunde und Bekannte aus Billerbeck und Essen trugen ihrerseits dazu bei, für sein leibliches Wohl zu sorgen. Das Einzige, um das Josef des Öfteren bat, waren Genussmittel wie Schokolade und Zigaretten, ein Wunsch, dem seine Familie und Freunde großzügig nachkamen. Seine Eltern, die durch die Molkerei und die Schweinmast materiell recht gut gestellt waren, bemühten sich sehr, ihren ältesten Sohn mit Kleidung und Lebensmitteln zu unterstützen, freilich nicht ohne ihm auch einige elterliche Mahnungen mit auf den Weg zu geben:

1054 Richard Euringer, Fliegerschule 4. Buch der Mannschaft, Hamburg 1929, S. 17f.

Abb. 22: „Ein Glas Bier bei *Vater Schmidt*", Moorslede, 5. Mai 1915
(Foto: Familienarchiv Walter Suwelack, NL Josef Suwelack)

> „Hast Du auch mal Appetit auf eine schöne Mettwurst oder sonst? Schreibe es doch, was Du haben willst. Du gibst nie Antwort, auf das, was wir schreiben. ... Und schreib mal wieder, ob Du die Sachen bekommst oder nicht. Richard macht das immer, wenn er die Sachen hat, dann schreibt er sofort. Mache Du doch dasselbe, hörst Du wohl."[1055]

Kurz vor Weihnachten 1914 erhielt Josef Liebesgaben von zu Hause, darunter zwei Pakete vom Vaterländischen Frauenverein und Zigaretten vom Schützenverein aus Billerbeck, für die er sich in einem Brief vom zweiten Weihnachtsfeiertag an seine Eltern bedankte. „Ich finde das riesig nett und aufmerksam, und jeder von den braven Billerbeckern hier vorne wird sich sehr gefreut haben darüber."[1056]

1055 NL J. Suwelack, Der einzig erhalten gebliebene Brief Josefs Mutter an ihren Sohn, Billerbeck, 18.4.1914.
1056 NL J. Suwelack, Brief an seine Familie, Lille, 26.12.1914.

Freundinnen Josef Suwelacks, eine Frau Putty Pohl aus Essen, die mit ihm in einem Mietshaus in der Zweigertstraße wohnte, sowie eine Louise Pöppinghaus,[1057] sandten ihm häufiger Zigaretten der Marke Queen und „die feinste Schokolade. Vorgestern hat sie [Louise] wieder 60 Queen geschickt, Sie ist schon bald tot vor Sehnsucht. ... Es gibt auch noch gute Seelen." Auch Familie Banholt aus Billerbeck, die mit Josef Suwelack befreundet gewesen sein muss, sowie sein Vermieter Herr Thomae aus Essen versorgten den Flieger mit Zigaretten.[1058]

Seine monatlichen Ausgaben gab Suwelack in einem Brief mit „etwa 250 Mk" an, denn die Feldfliegerabteilung habe in Lille ein Casino und müsse sich selbst versorgen. Er habe sich „im Ganzen für über 1.000 Mk. Sachen kaufen müssen", wohingegen die Offiziere alles gestellt bekämen.[1059] In einem Schreiben an seine Eltern von Anfang August 1915 berichtete er auch von einem Koch, der für das leibliche Wohl der Abteilung sorgte:

> „Er ist Chef in einem ersten Hamburger Hotel, und seine Tätigkeit wirkt überaus wohltuend auf uns. Ich glaube nicht, daß in der ganzen Gegend hier irgendwo auch nur annähernd so gut gegessen wird wie bei uns. Wir haben jeden Tag etwas anderes. Doch ich will aufhören, den Koch zu loben, sonst glaubt Ihr noch, wir seien nur des Essens halber hier."[1060]

Nicht nur physisch war Suwelack in guter Verfassung – lediglich vom 10. bis zum 17./18. März 1915 hatte er eine Grippe, die er im Bett und mit heißen Bädern auskurierte –,[1061] seine Möglichkeiten der Freizeitgestaltung boten ihm auch Gelegenheit zur mentalen Erholung. Er malte Aquarelle, schrieb zahlreiche Briefe, las, unternahm Spaziergänge und -ritte in die Umgebung und hielt seine Impressionen mit seiner Kamera, die er etwa zum Jahreswechsel bekam, fest.[1062]

Im November 1914 und Januar 1915 flog er an nur drei Tagen, im Dezember an insgesamt vier.[1063] Seine Kriegszeit war in diesen Monaten somit hauptsächlich von Langeweile bestimmt, was einige Auszüge aus Briefen verdeutlichen: „Wir sind wenig geflogen in letzter Zeit. Wetter sehr schlecht."[1064] „Hier ist alles beim alten [sic!]. Vorgestern bin ich noch geflogen, aber für die nächsten 8 Tage

1057 Über beide Frauen liegen keinerlei biografische Daten vor.

1058 NL J. Suwelack, Feldpostbrief an Richard Suwelack, Lille, 3.12.1914. Siehe zur Familie Banholt aus Billerbeck die zur Verfügung stehenden Informationen in der genealogischen Datenbank: http://familienanzeigen.genealogy.net/suchetod.php?nachname=Nordick&vorname=&ortsname=&PID=882&APID=879 (11.2.2015).

1059 NL J. Suwelack, Feldpostbrief an Richard Suwelack, Lille, 1.2.1915.

1060 NL J. Suwelack, Feldpostbrief an seine Eltern, o.O. [Petit Rechain], 8.8.1915.

1061 Vgl. NL J. Suwelack, Feldpostbrief an seinen Vater, Lille, 19.3.1915.

1062 Vgl. NL J. Suwelack, Feldpostbriefe an Richard Suwelack, Lille, 3.1.1915 und Petit Rechain, 19.4.1915 sowie an die Eltern, Lille, 27.11.1914. Zum Briefeschreiben siehe z.B. den Inhalt des Briefes aus Lille vom 13.2.1915 an Richard: „Habe heute und gestern 21 Briefe geschrieben auch an Pujol."

1063 Vgl. NL J. Suwelack, Eintragungen ins Flugbuch.

1064 NL J. Suwelack, Feldpostbrief an Richard, Lille, 3.2.1915.

habe ich wieder Ruhe, da wir momentan keine Aufträge haben."[1065] „Hier hat sich Wesentliches nicht ereignet. Immer dasselbe."[1066]

Die erlebte Eintönigkeit, das Leben von Tag zu Tag verstärkte den Wunsch nach Abwechslung und Aufträgen. In Zeiten, da Suwelack nichts Neues von sich und seiner Arbeit zu berichten hatte, ging er dazu über, die jüngst zurückliegenden spektakulärsten und andere ihm im Gedächtnis gebliebenen Ereignisse zu schildern. Gerade hier erfährt man besonders Interessantes, denn es handelt sich sowohl um die Beschreibung von banalen Alltäglichkeiten, die der „Realität" vermutlich näher kamen als die für die Öffentlichkeit bestimmten Erlebnisberichte, als auch um Außeralltägliches.

Anfang Dezember 1914 hatte Suwelack seinem Bruder geschrieben, dass sämtliche Stellungen fotografisch erfasst und die Schützengräben und Batterien deutlich zu erkennen seien: „Wir haben uns aus einem Kino einen Apparat requiriert und haben unserem Generalkommando die ganze Stellung bis ins Kleinste vorgeführt. Ab und zu suchen wir Batterien und schießen unsere Artillerie ein."[1067]. Freimütig schrieb Josef seinem Bruder auch von Unfällen:

> „Die einzige Sache, die uns aus unserer Seelenruhe etwas aufweckte war, daß gestern Abend eine von unseren Maschinen verbrannte und zwar ein DFW von Herrn von Plock [Ploetz] durch die Unvorsichtigkeit eines Landsturmmannes von der Wache, der auf eine brennende Petroleumlampe Pertroleum einfüllte. Meine Maschine ist noch mit Mühe und Not gerettet worden. Einige Brandlöcher, die zugekleistert werden können und das Höhensteuer wurde beschädigt. Sonst hat sich nennenswertes nicht ereignet."[1068]

Zeigte er sich am Schicksal des Landsturmmannes nicht interessiert, so berührte ihn aber der Unfall seines Monteurs, eines Herrn Winter, dem bei einer Benzinexplosion beide Arme verbrannt waren (Abb. 23): „Da der arme Kerl keine Eltern und Geschwister mehr hat, habe ich gut für ihn gesorgt. Er ist jetzt in Deutschland. Ein Bild von ihm liegt auch bei."[1069]

Zum Thema der Briefe machte Suwelack nun auch die zunehmende Gefahr für sein eigenes Leben, obschon er gegenüber seinen Eltern das Risiko der Kriegsfliegerei verharmloste. Richard und seinem Vater schrieb er von einem Gefecht zwischen ihm und einem deutschen (!) Zeppelin LZ 35, der ihm den Holm und die linke Tragdecke durchschossen habe.[1070] Während die Schilderung des Angriffs in beiden Briefen ungefähr gleich ausfiel, fügte Josef im Brief an seinen Vater noch hinzu, dass das Feuer sofort eingestellt worden sei, als die Besatzung

1065 NL J. Suwelack, Feldpostbrief an seine Eltern, Lille, 27.11.1914.
1066 NL J. Suwelack, Brief an seine Familie, Ostern 1915.
1067 NL J. Suwelack, Feldpostbrief an Richard, Lille, 3.12.1914.
1068 NL J. Suwelack, Feldpostbrief an Richard Suwelack, Lille, 8.1.1915.
1069 NL J. Suwelack, Feldpostbrief an seine Eltern, Petit Rechain, 19.4.1915.
1070 Vgl. NL J. Suwelack, Feldpostbrief an Richard, Lille, 1.4.1915 und Feldpostbrief an seinen Vater, Lille, 19.3.1915.

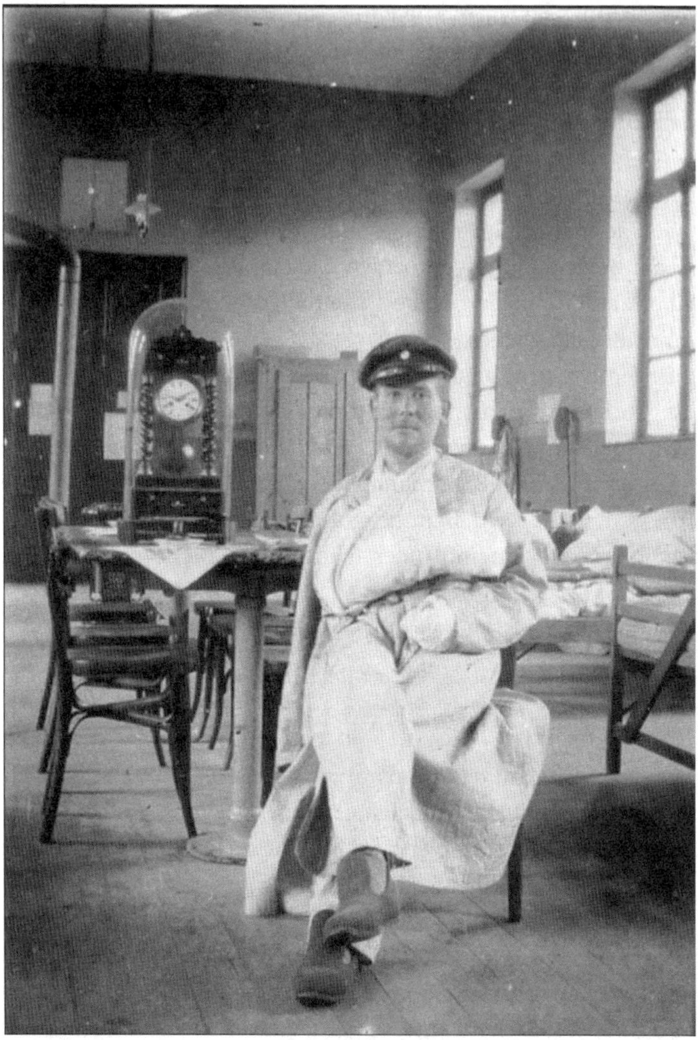

Abb. 23: Josef Suwelacks Monteur Winter, dem bei einer Benzinexplosion beide Arme verbrannt waren, vermutlich Petit Rechain, Anfang April 1915 (Foto: Familienarchiv Walter Suwelack, NL Josef Suwelack)

des Zeppelins seine Kennzeichnung am Flugzeug registriert habe: „Kurz darauf landete ich, weil ich mit dem zerbrochenen Holm nicht weiterfliegen wollte. Ich habe eine Meldung an das Hauptquartier geschickt.“[1071]

Dass sich die Gefahrenbeschreibungen in den Briefen an den Bruder und die Eltern glichen, war jedoch eher die Ausnahme, denn in der Regel verschwieg Josef die Schrecken des Krieges gegenüber seinen Eltern und brachte stattdessen seine

1071 Ebd.

Sprachlosigkeit zum Ausdruck. Gegenüber Richard, der wie er als Soldat von Verwundeten und Toten hörte und die Schrecken des Krieges sah, musste er sich nicht verstellen und die Gemüter beruhigen.

Wie die visuelle Wahrnehmung durch die Linse gerahmt war, so waren auch die Zeitspannen, in denen Suwelack dachte, äußerst begrenzt, sowohl nach hinten in die Vergangenheit als auch nach vorne in eine ungewisse Zukunft. Der Wunsch, sich profilieren und nützlich machen zu können und seinem Leben einen Sinn zu geben, bestimmte Suwelacks Handeln. In einer Zeit, in der alles auf Anfang gesetzt war, bemühte sich der Flieger, seine Zukunft zu gestalten, die ihm jedoch immer wieder entglitt.[1072] Rückhalt und Orientierung boten ihm vor allem seine Familie und die zu bewältigenden Aufgaben fernab des Krieges, wie die Begleichung von Rechnungen, Treffen mit Geschäftspartnern und der Besuch von Freunden und seiner Familie – Routinen und Handlungen, die sein Vorkriegsleben bestimmt hatten.

Familie, Freunde, Religion und zivile Arbeit als Rückhalt und Orientierungshilfe in einer Zeit ohne Zukunft

„Was die Zukunft bringt, ist schwarz und ungewiss",[1073] stellte Suwelack Weihnachten 1914 in einem Brief an seine Familie fest. Die Zeit schien auch für ihn aus den Fugen geraten, denn nichts war mehr, wie er es in Erinnerung hatte, und nur die Imagination blieb:

> „Weihnachten ist vorüber, und ich habe am Hl. Abend Euch alle, wie jedes Jahr, das schöne Fest feiern sehen. Im Geiste war ich unter Euch. Alle einzelnen habe ich mir vorgestellt, wie Ihr diesesmal ohne Richard und mir [sic!] unter dem Baum saßet. Willy und Otto haben Geige und Klavier gespielt. Aber so ganz rein war die Weihnachtsfreude nicht. Dich, mein liebes Mütterchen, habe ich gesehen, wie Du an Deine beiden Jungen gedacht hast und Vater schwirrten viel zu schwarze Gedanken durch den Kopf. Aber es ist nicht die Zeit, den Kopf hängen zu lassen."[1074]

Mit diesem Brief schlug Suwelack eine Brücke in die Heimat, knüpfte ein schriftliches Band zwischen sich und seiner Familie und wollte auf diese Weise seine und die Stimmung seiner Eltern und Geschwister heben. Gleiches tat auch sein Bruder Richard, der Josef in einem Brief frohe Weihnachten und ein erfülltes neues Jahr wünschte: „Möge es Dir das bringen, was Du davon erhoffst, einen glücklichen Abschluß des Krieges und ein fröhliches Wiedersehen zu Hause."[1075]

Feste boten stets Schreibanlässe für Josef und seine Familie, besonders der Namenstag von Josefs Vater und das Osterfest, das zugleich auch die Erstkommunion von Josefs Schwester Maria war:

1072 Vgl. dazu den Abschnitt zu den Kondor-Werken.
1073 NL J. Suwelack, Feldpostbrief an seine Familie, Lille, 26.12.1914.
1074 Ebd.
1075 NL J. Suwelack, Feldpostbrief Richard Suwelacks an Josef, o. O., 15.12.1914.

> „Zum heutigen frohen Feste, das unserer Familie beschieden ist ... meine herzlichsten Glückwünsche. Richard und ich können leider an der schönen Feier nicht teilnehmen, aber mit unseren Herzen und Gefühlen sind wir deswegen doch bei euch. Gerade hier draußen im Felde treten die Erinnerungen an die liebe Heimat und frohen und glücklichen Kindertage besonders hervor."[1076]

Die Erinnerung an bessere Zeiten, die als Topos in zahlreichen Feldpostbriefen auftaucht, bot Trost und ermöglichte es dem Schreibenden zugleich, sich für einen Moment in Erinnerungen zu flüchten, für einen Augenblick konnte er sich so der Illusion der „heilen Welt" hingeben.

Josef und seine Eltern begriffen den Krieg als göttliche Prüfung, der sie nur durch Glaubenstreue und einen starken Durchhaltewillen begegnen könnten:

> „Bete zu Gott, daß er in seinem Strafgericht, welches er offenbar über die Welt verhängt hat, unsere Familie verschone. Wir beten tagtäglich für Dich. Lieber Joseph! Was haben wir doch für einen schweren Krieg. ... Es scheint aber, daß der liebe Gott in diesem schweren Ringen mit uns ist. Die Türkei ist jetzt dazugekommen und hat ebenso mit Frankreich, England und mit Rußland Krieg. England hat es flugs verstanden, uns alles auf den Leib zu hetzen, aber die Rache wird nicht ausbleiben. Die Buren [?] in Afrika, sind schon so gut wie verloren für England. Indien und Ägypten folgen. Gottes Mühlen mahlen langsam aber fein. Für Rußland wird auch die Stunde kommen."[1077]

Wie viele andere waren weder Josef noch seine Eltern auf die Dauer des Krieges vorbereitet. Zuversichtlich, dass der Feldzug rasch beendet sein werde, notierte Suwelack auf einer Feldpostkarte von Anfang September 1914: „Wir trinken schon Champagner aus Reims. Die Rothosen haben sehr eilige Beine. Wir können kaum nachkommen. In zwei Wochen ist Schluß."[1078] Nachdem ein weiterer Monat verstrichen war, äußerte Richard Suwelack jedoch erste Zweifel an der verbreiteten und propagierten Vorstellung, Ende des Jahres wieder bei der Familie zu sein, dennoch gab er sich siegesgewiss: „In der Hoffnung Weihnachten wieder zu Hause zu sein, werden sich wohl viele getäuscht haben, denn jetzt fängt die Sache für uns an etwas schwieriger zu werden. Aber kriegen werden wir die Luders ja doch."[1079]

War anfänglich noch die Rede von wenigen Kriegswochen, verschob sich das erwartete Kriegsende immer weiter ins Ungewisse. Wie wenig die Brüder das tatsächliche Ausmaß des Krieges abschätzen konnten, zeigt sich auch daran, dass sie noch im Oktober 1914 davon ausgingen, dass der Krieg sich lediglich um ein halbes Jahr verlängern werde. Selbst als dieses halbe Jahr ergebnislos verstrichen war, kamen sie nicht über eine weitere erwartete Zeitspanne von einem halben Jahr hinaus.

1076 NL J. Suwelack, Feldpostbrief an seine Familie, Lille, undatiert [Ostern 1915, Poststempel 7.4.1915].
1077 NL J. Suwelack, Brief an Josef von seinem Vater, Billerbeck, 8.11.1914.
1078 NL J. Suwelack, Feldpostkarte an seine Familie, Reims, undatiert [Poststempel 5.9.1914].
1079 NL J. Suwelack, Feldpostbrief Richard Suwelacks an Josef, Vorges, 12.10.1914.

Josef gab sich im Ende Oktober 1914 noch optimistisch: „Na im Juli werden wir uns wiedersehen, wenn der Krieg zu Ende ist“,[1080] und Richard schloss sich dieser Meinung an, denn Deutschland werde „nicht eher ruhen, bis seine Weltmacht gesichert“ sei.[1081] Im Gegensatz zu diesen ersten Zukunftsprognosen, die noch einen konkreten Monat benannt hatten, äußerte sich Josef später immer weniger zu einem möglichen Kriegsende. Einen Brief von Ostern 1915 schloss er neben der Grußformel mit der vagen Formulierung, dass es mit seiner „Reise ... wohl noch geraume Zeit“ dauern werde.[1082]

Was blieb, war die vage Hoffnung, dass die „alte Schweinerei“[1083] bald vorbei sei und er seiner zivilen Tätigkeit als Flugzeugkonstrukteur und Flieger wieder nachgehen könne. Am Tag vor seinem Tod brachte er noch die Hoffnung zum Ausdruck, der Krieg möge enden und ihn gesund zu seinen Eltern zurückführen, doch eine tatsächliche Zukunftsperspektive hatte Josef nicht.[1084] Einzig sein Vater dachte über die Zukunft und (finanzielle) Vorsorge seines Sohnes nach:

> „Sorge dann auch [dafür], daß du an der hiesigen Sparkasse etwas abträgst. Wenn der Krieg mal zu Ende und Du für Deine Existenz bemüht bist, ist es doch sicher notwendig, daß Du von mir noch etwas zu erwarten hast, als wenn ich Deine alten Schulden alle in Anrechnung bringen müsste. Ich bin für Deine Zukunft noch sehr besorgt, solange diese nicht sichergestellt ist.“[1085]

Abgesehen von der Hoffnung auf Frieden und dem stets vagen Ausblick in die Zukunft beschränkten sich die Planungen Josef Suwelacks auf Urlaubsgesuche, um Gelegenheit zu haben, seine Familie zu sehen oder sich wegen neuer Entwürfe und Ideen mit den Kondor-Werken in Verbindung zu setzen, aber auch, um persönliche Angelegenheiten zu ordnen.

Seine Essener Wohnung hatte der Flieger noch bis Mitte des Jahres 1915 behalten und die Miete weiter gezahlt. Dann aber gab er sie zum ersten Oktober 1915 auf, hatte er wohl erkannt, dass sich sein Frontaufenthalt noch auf ungewisse Zeit verlängern würde. Seine Möbel wollte er von einem Spediteur abholen „und irgendwo gut unterstellen“ lassen. Ein Transport nach Billerbeck sei zu teuer, deshalb plante Josef, in Essen ein „billiges Zimmer, evtl. bei Thomae“ zu mieten und seine Möbel dort zu lagern.[1086]

Die Miete für die Wohnung betrug 1.620 Mark jährlich, das Mietverhältnis war 1914 auf ein Jahr festgesetzt worden und konnte nur bei eingehaltener dreimonati-

1080 NL J. Suwelack, Feldpostbrief an Richard Suwelack, Lille, 27.10.1914.
1081 NL J. Suwelack, Feldpostbrief Richard Suwelacks an Josef, o. O. [vermutlich Vorges], 8.11.1914.
1082 NL J. Suwelack. Feldpostbrief an seine Familie, Lille, undatiert [Ostern 1915].
1083 NL J. Suwelack, Feldpostbrief an Richard Suwelack, o. O., 6.7.1915.
1084 Vgl. NL J. Suwelack, Feldpostbrief an seine Eltern, Petit Rechain, 12.9.1915.
1085 NL J. Suwelack, Brief seines Vaters, Billerbeck, undatiert [vermutlich 15.3.1915].
1086 NL J. Suwelack, Feldpostbrief an Richard, Petit Rechain, 6.7.1915. W. Thomae war der Vermieter Josef Suwelacks. Vgl. NL J. Suwelack, Brief an Otto Suwelack, Lille, undatiert [Poststempel 10.7.1915].

ger Kündigungsfrist aufgelöst werden. Sein Vermieter schrieb Josef am 2. Oktober 1914 und zeigte sich flexibel:

> „Sollten Sie ... von Januar ab nur noch 1 oder 2 Monate hier in Essen bleiben, so bin ich ja auch kein Unmensch und werden wir uns wohl darüber einigen. Ich werde also vor dem 1. April 1915 keinen weiteren Schritt tun, um die Wohnung anderweitig zu vermieten. ... Ihre Sachen sowie Ihre Wohnung wird von meinem Mädchen gut in Ordnung gehalten."[1087]

Seiner Haushälterin hatte Josef bereits im August 1914 gekündigt,[1088] doch weigerte sie sich, das Haus zu verlassen: „Deine Haushälterin ist noch immer in der Wohnung. Sie will nicht heraus. Ich habe ihr schon geschrieben, daß Du keinen Pfennig bezahlst nach dem 1. August. Sie hätte keine Berechtigung, weil der Vertrag mit dem Kriege aufhörte, da Du sie zu dem Zwecke, Dir den Haushalt zu führen, nicht mehr verwenden könntest. Sie beklagte sich, daß Du noch nicht geschrieben hättest. Juli habe ich mit 35 Mark bezahlt."[1089] Erst im November habe die Haushälterin, wie ihm sein Vater schrieb, „unter großem Geschimpfe" Josefs Wohnung verlassen und aufgrund eines ausgebliebenen Gehaltes „Wertsachen und Bilder mitgenommen .... Ein *unverschämtes* Luder. Ich kann wohl nichts unternehmen?"[1090]

Obwohl Suwelack regelmäßig Geld an seinen Vermieter überwies, standen Ende Juli 1915 noch 800 Mark aus und Thomae gestattete es ihm nicht, die Möbel abtransportieren zu lassen, bevor der Geldeingang erfolgt sei. Daher schrieb Josef an seinen Vater: „Ich halte es für das beste [sic!], die Möbel nun bis Mitte September dort stehen zu lassen und sie dann nach dort zu senden."[1091]

Wesentlich dringlicher und für Suwelacks Zukunft entscheidend war jedoch eine Problematik, die ihm als Dienstuntauglichem im April/Mai 1915 entstanden war. Nachdem die Inspektion der Fliegertruppen allen Zivilfliegern den Vertrag kündigen wollte, erhielt Josef ein Schreiben des Bezirkskommandos in Essen, in dem er aufgefordert wurde, sich als Landsturmpflichtiger zu melden. Er erhob sofort Einspruch und legte seinen Ausmusterungsschein vor, woraufhin seine Abteilung telegrafisch Informationen einholte.[1092]

Zunächst aber hatte der Beschluss weitreichende Folgen für Josefs finanzielle Situation sowie für seine Karrierechancen, denn als Flieger erhielt er ein monatliches Gehalt von rund 750 Mark und Zulagen für besonders erfolgreiche Flüge, als

1087 NL J. Suwelack, Brief seines Vermieters Herrn W. Thomae, Essen, 2.10.1914.
1088 Vgl. NL J. Suwelack, Feldpostkarte an seine Familie, St. Etienne, 20.9.1914.
1089 NL J. Suwelack, Brief von Josefs Vater an seinen Sohn, Billerbeck, 29.9.1914.
1090 NL J. Suwelack, Brief an seinen Sohn, Billerbeck, 8.11.1914.
1091 NL J. Suwelack, Brief an seinen Vater, Köln (auf der Durchreise), 30.7.1915.
1092 Vgl. NL J. Suwelack, Feldpostbrief an Richard Suwelack, Petit Rechain, 9.5.1915. Die Nachforschungen in Berlin hatten ergeben, dass für Josef die Aussicht bestand, wieder als Zivilflieger an der Front tätig zu sein, da er vor dem Krieg bereits als dienstuntauglich erklärt worden war. Neben ihm betreffe diese Regelung noch Hellmuth Hirth und Hans Vollmöller. Vgl. den Brief an Richard Suwelack, o. O., 6.7.1915.

Landsturmmann hingegen hätte er nur rund 16 Mark im Monat erhalten.[1093] Außerdem konnte er als Landsturmmann keine Offizierslaufbahn einschlagen und somit keine Karriere beim Militär machen.[1094] Dies hätte eine Herabsetzung seines gewohnten Lebensstandards bedeutet, denn als Bürgerlicher, der von zu Hause ein komfortables Leben gewohnt war, sich bei den Kondor-Werken Mitbestimmungsrechte erarbeitet hatte und mit der Militärführung mehrfach in Berührung gekommen war, hatte Suwelack mittlerweile einen entsprechenden „Geschmack" entwickelt, ein Anspruchsdenken, das sich aus dem bisher Erreichten und den gesammelten Erfahrungen speiste. Allerdings gab er sein bürgerliches Leben zu keiner Zeit zugunsten eines aristokratischen Lebensstils auf, wie es die „Feudalisierungsthese"[1095] für das Bürgertum behauptet, vielmehr wollte er nach dem Krieg als Unternehmer tätig werden. Zwar machte es die zunehmende Durchlässigkeit des Offizierskorps für Bürgerliche möglich, sich nach „unten" abzugrenzen und sich zusätzlich mit symbolischem Kapital auszuzeichnen. Mit seiner technisch-taktischen Kompetenz als Aufklärer wäre es Suwelack möglich gewesen, in die Reihen des Offizierskorps vorzustoßen und sich innerhalb des Systems Freiräume zu schaffen.[1096] Doch blieb er Zeit seines Lebens Zivilflieger, auch wenn er sich im Sommer 1915 zu einer Offiziersausbildung entschlossen hatte. Als Sohn eines erfolgreichen Unternehmers und überzeugt, mit seinen Kriegserfahrungen z. B. in Spanien Fuß fassen zu können, nahm sich Suwelack als selbstbewusst und selbstbestimmt wahr.[1097]

1093 Vgl. ebd. und NL J. Suwelack, Aufstellung für den erhaltenen Sold für den Zeitraum vom 20. September 1914 bis zum 31. März 1915.

1094 Vgl. ebd. und Lothar Mertens, Das Einjährig-Freiwilligen Privileg. Der Militärdienst im Zeitgeist des deutschen Kaiserreiches, in: Zeitschrift für Religions- und Geistesgeschichte 42 (1990), S.316–329.

1095 Siehe zur Feudalisierungsthese den Überblick bei Dieter Hertz-Eichenrode, Die Feudalisierungsthese – ein Rückblick, in: Vierteljahrschrift für Sozial- und Wirtschaftsgeschichte 89 (2002), H. 3, S.265–287; Cornelius Torp/Sven-Oliver Müller, Das Bild des Deutschen Kaiserreichs im Wandel, in: dies. (Hg.), Das Deutsche Kaiserreich in der Kontroverse, Göttingen 2009, S.9–27, bes. S.12ff., sowie den Überblick zur Kontroverse um die Elitentheorie, die postulierte, das Deutsche Kaiserreich sei einzig von der preußischen Führungsschicht gelenkt worden: Roger Chickering, Zwischen Dynamik und Stillstand. Gedanken zur deutschen Innenpolitik, in: Bernd Heidenreich/Sönke Neitzel (Hg.), Das Deutsche Kaiserreich 1890–1914, Paderborn/München/Wien/Zürich 2011, S.61–73.

1096 Vgl. Wilhelm Deist, Zur Geschichte des preußischen Offizierkorps 1888–1918, in: Hanns Hubert Hofmann (Hg.), Das deutsche Offizierkorps 1860–1960, Boppard a.Rh. 1980, S.39–57, hier S.50–52. Siehe zum „Integrationsmilitarismus" im Deutschen Kaiserreich: Bernhard R. Kroener, Integrationsmiliarismus. Zur Rolle des Militärs als Instrument bürgerlicher Partizipationsbemühungen im Deutschen Reich und im 19. Jahrhundert bis zum Ausbruch des Ersten Weltkrieges, in: ders., Kriegerische Gewalt und militärische Präsenz in der Neuzeit. Ausgewählte Schriften, hg. von Ralf Pröve und Bruno Thoß, Paderborn/München/Wien/Zürich 2008, S.83–107, hier S.98ff.

1097 Siehe zur Vorstellung eines quasi-monarchischen Anspruchsdenkens des Wirtschaftsbürgertums: Peter-Christian Witt, Monarchen und Bürger. Über Untertanen und Untertänigkeit im wilhelminischen Deutschland (1890–1914), in: Horst Lademacher/Walter Mühlhausen (Hg.), Frei-

Noch bevor er zum Landsturm berufen werden sollte, war sein Sold bereits um mehr als die Hälfte gekürzt worden und er musste den vierten Teil seiner Bezüge zurückzahlen:[1098] „Den meisten Zivilfliegern ist der Vertrag gekündigt worden. ... Außerdem bekomme ich jetzt nur noch 75 Mk. [pro Woche] statt 185 wie früher."[1099] Josef bat seinen Vater daher in einem Brief vom Anfang Juni, sich um die Angelegenheit zu kümmern und seinen Ausmusterungsbeleg bei den entsprechenden Behörden vorzulegen.[1100] Da eine Vertragsverlängerung dennoch ausblieb, wollte Suwelack versuchen, als Kriegsfreiwilliger geführt zu werden, und ließ seine Kontakte spielen: „Es ist ein Herr in Berlin von uns, der alle Hebel in Bewegung setzt, andernfalls kann ich nur ... Offizier-Stellvertreter werden."[1101] Schließlich teilte er Richard am 1. September mit, dass „ein Freund von [ihm] ..., Leutnant Bonde, der bei der Inspektion" sei, „das Ding gedreht" habe.[1102] Hier zeigt sich erneut die Wichtigkeit von Beziehungen zur Militärverwaltung, von der Beförderungen, Versetzungen, Auszeichnungen und Empfehlungen abhingen, aber auch die Bedeutung der Unterstützung durch die Familie, die für Josef seine privaten Angelegenheiten regelte.

Suwelack rechnete mit erheblichen Geldnachzahlungen in einer Höhe von 3.000 bis 3.500 Mark[1103] und hoffte auf eine baldige Beförderung. Nach einem zweistündigen Flug exerziere er jetzt täglich zusätzlich anderthalb Stunden und er erwartete, dass die Ausbildung zum Unteroffizier in drei bis vier Wochen beendet sein werde: „Meinen Burschen habe ich behalten, und sonst lebe ich genau wie früher. Ich muß sagen, daß es mir so sehr gut gefällt."[1104]

Mehrmals war Josef Suwelack während seiner Kriegszeit für wenige Tage bei seinen Eltern zu Gast. Mitte November 1914 hielt er sich z. B. für einen Tag in Billerbeck auf, wo sich zu diesem Zeitpunkt anscheinend auch sein Bruder Richard befand.[1105] Von dort reiste er weiter nach Berlin, wo er „eine geheime Sache" zu erledigen hatte, und traf sich mit Herrn Spengler, dem Direktor der Aviatik AG, um für die Kondor-Werke und das Spaniengeschäft eine Geschäftsverbindung aufzubauen.[1106] Am Morgen des 20. Novembers wollte sich Josef mit Alfred von Back-Begávar im Kaiserhof zum Essen treffen, um sich über die Entwicklung bei

heitsstreben – Demokratie – Emanzipation. Aufsätze zur politischen Kultur in Deutschland und in den Niederlanden, Münster/Hamburg 1993, S.139–187.

1098 Vgl. NL J. Suwelack, Feldpostbrief an Richard, Lille, 9.4.1915.

1099 NL J. Suwelack, Feldpostbrief an Richard, Petit Rechain, 1.5.1915.

1100 Vgl. NL J. Suwelack, Feldpostbrief an seinen Vater, o. O., undatiert [Poststempel 8.6.1915].

1101 NL J. Suwelack, Feldpostbrief an Richard, Petit Rechain, 18.8.1915.

1102 NL J. Suwelack, Feldpostbrief an Richard, Petit Rechain, 1.9.1915.

1103 Vgl. NL J. Suwelack, Feldpostbrief an seinen Vater, Petit Rechain, 5.9.1915. Siehe auch die Aufstellung für die vergüteten Flüge Suwelacks.

1104 Vgl. NL J. Suwelack, Feldpostbrief an seine Familie, Petit Rechain, 12.9.1915.

1105 Vgl. NL J. Suwelack, Brief von seinem Vater, Billerbeck, undatiert [vermutlich Anfang November 1914].

1106 Vgl. NL J. Suwelack, Brief an Rirchard, Essen, 20.11.1914.

Kondor ins Bild setzen zu lassen, anschließend fuhr er mittags weiter nach Aachen, von wo er mit dem Auto von Oberleutnant Müller abgeholt wurde und Lille am 22. November erreichte:[1107] „Am folgenden Morgen noch todmüde ging's über Lüttich, Brüssel nach Lille. Eine ordentliche Freßkiste haben wir auch noch mitgenommen ... und abends wurde ich mit großem Jubel wieder aufgenommen."[1108]

Im April 1915 traf sich Josef Suwelack erneut mit von Back-Begávar, denn er wollte sich bei den Kondor-Werken ein Kampfflugzeug nach dem Vorbild eines Bristol-Doppeldeckers bauen lassen und hoffte auf einen mehrwöchigen Urlaub.[1109] Dieser wurde ungefähr für den Zeitraum vom 10. bis zum 21. Mai genehmigt und Josef hielt sich erneut für einige Tage in Billerbeck, Berlin und Essen auf. In Berlin lud ihn Hans Goldschmidt in sein Hotel Esplanade ein, in Essen traf er sich mit Karl Goldschmidt. Beide seien von seinem Vorschlag, ein Kampfflugzeug zu bauen, angetan gewesen.[1110] Suwelack konnte für die Konstrukion des Apparates nicht nur Details aus erster Hand weitergeben, sondern hatte im April 1915 zudem fotografisches Material aus einem erbeuteten englischen Bristol-Doppeldecker an von Back-Begávar gesandt. Der Flieger selbst sei nachts mit einem Auto der Nachbardivision entkommen:[1111]

> „Der Flieger mußte morgens in undurchsichtigem Nebel landen, entkam aber. Der Apparat ist ein engl. Bristol, der anscheinend nur zum Bombenwerfen gebraucht wurde. Immerhin weist er manche interessante Sachen auf. ... Das Fahrgestell ist noch vorsintflutlich, aber die Flächenkurve und Flügelkonstruktion waren ganz ausgezeichnet sowie die Beschläge verblüffend einfach und fest. Der Apparat ist nach Berlin-Adlershof geschickt. Zur Nachahmung könnte ich die Flügelkonstruktion sehr empfehlen."[1112]

Einige Angehörige der FfA 24 nahmen die Kriegsbeute sogleich in Augenschein und inszenierten sich als die überlegene Einheit. Der Erfolg war im Unterschied zu dem der Grabenkämpfer, die ihren Gegner in der Regel nicht sahen, direkt greifbar und konnte stolz präsentiert werden, ein Umstand, der für die mediale Luftkriegsberichterstattung eine wichtige Rolle spielte.

Die Technikfaszination Josef Suwelacks war auch noch während des Krieges ungebrochen. Sowohl in seinen Erlebnisberichten – davon wird noch die Rede sein – an seine Eltern, an Verwandte und an die Öffentlichkeit als auch in Briefen an seinen Bruder, in denen er auf seine Arbeit zu sprechen kam, berichtete er ausführlich von den technischen Eigenschaften, Vorzügen und Nachteilen der von ihm geflogenen Apparate:

1107 Vgl. ebd. und Feldpostbrief an seine Eltern, Lille, 23.11. und 27.11.1914.
1108 NL J. Suwelack, Feldpostbrief an seine Familie, Lille, 27.11.1914.
1109 Vgl. NL J. Suwelack, Feldpostbrief an seinen Vater vom 19.3.1915 und an Richard vom 1.5.1915.
1110 Vgl. NL J. Suwelack, Feldpostbrief an seine Familie, Petit Rechain, 22.5.1915.
1111 Vgl. NL J. Suwelack, Feldpostbrief an seine Familie, Lille, Ostern 1915.
1112 NL J. Suwelack, Feldpostbrief an Alfred von Back-Begávar, Lille, Ostern 1914.

> „Ich fliege immer noch meinen ... [D.F.W.-Doppeldecker] mit dem ich neuerdings mit voller Belastung in 9 Minuten auf 1000 mtr steige. Jetzt bekomme ich einen Fokker Eindecker. Ich werde den allerdings nicht zu Flügen über den Feind benutzen[,] sondern ich behalte meinen alten D.F.W. bei. Im Frühjahr bekomme ich eine Kampfflugmaschine.“[1113]

Im Februar schrieb er Richard: „Ich habe jetzt einen neuen Albatros, der enorm schnell steigt. Je schneller, desto besser jetzt. Nur jetzt bekomme ich ein Kampfflugzeug mit 160 PS Mercedes. Ich habe dann Sicherheitsdienst.“[1114] Schnelligkeit sei angesichts der leichten französischen Eindecker mit Maschinengewehren entscheidend, und mit „unbewaffneten Maschinen sich zu stellen“, wie es die Aufklärungsflugzeuge waren, sei „Unsinn“.[1115]

Wie in der Vorkriegszeit fotografierte Josef auch weiterhin Flugzeuge, Flugzeugabstürze und die Ausstattung der Maschinen. Das fotografische Anschauungsmaterial sollte die Grundlage für seine Tätigkeit als Konstrukteur nach dem Krieg bilden, und es ist anzunehmen, dass er auch technische Daten über Motoren, die Flügelverspannung oder das Fahrgestell notierte, gleichwohl solche Aufzeichnungen im Nachlass für die Kriegszeit fehlen. Mit der Aufrechterhaltung von habitualisierten Verhaltensmustern, der Faszination für alles Technische und der fotografischen Aneignung der technischen Seite des Krieges war es Josef zum Teil möglich, die Gewalterfahrung und das Zerstörungspotenzial der Kriegsmaschinen auszublenden beziehungsweise es aus der Sicht des interessierten Technikers und Fliegers zu betrachten.

Die Verbindung zu seiner Familie gab ihm zusätzlich Sicherheit und Orientierung. Nicht zuletzt waren seine Eltern als moralische Instanz um seine körperliche und seelische Unversehrtheit besorgt.[1116] Vor allem Josefs Mutter war in Sorge, dass Josef seinen Pflichten als Katholik nicht regelmäßig nachkomme, und auch sein Vater sah die christliche Moral durch den Krieg gefährdet.[1117] Den Forderungen seiner Eltern versuchte Josef durch entsprechend formulierte Briefe nachzukommen, gab aber auch zu bedenken, dass der Krieg ihn und seine Umwelt täglich mit außergewöhnlichen Umständen konfrontierte:

> „Liebe Mutter! Deinen Brief habe ich erhalten, und ich werde Deine guten Ratschläge getreulich so weit es geht befolgen. Du mußt Dir hier die Sache nicht so einfach vorstellen. Morgen, am 10. haben wir alle Feldgottesdienst. Das ist die einzige Gelegenheit. Ich habe vor 8 Tagen einen kath. Feldgeistlichen kennengelernt, und den will ich nachher besuchen, um zu fragen, wo und wann ich beichten und kommunizieren kann. Du mußt bedenken, daß die Sachsen alle protestantisch sind, und es gibt hier wenige kath Geistliche.“[1118]

1113 NL J. Suwelack, Feldpostbrief an Richard, Lille, 26.12.1914.
1114 NL J. Suwelack, Lille, undatiert [Februar 1915].
1115 NL J. Suwelack, Feldpostbrief an Alfred von Back-Begávar, Ostern 1915.
1116 NL J. Suwelack, Brief seines Vaters, Billerbeck, 8.11.1914.
1117 NL J. Suwelack, Brief seines Vaters, Billerbeck, undatiert [wahrscheinlich November 1914].
1118 NL J. Suwelack, Brief an seine Mutter, Lille, 9.1.1915.

Er bat um Verständnis, dass er wegen seiner dienstlichen Verpflichtungen nicht wie gewohnt regelmäßig den Gottesdienst besuchen könne: „Jetzt haben wir hier in Lille eine Garnisonskirche. Aber da wir morgens jetzt um 1/2 8 Uhr auf dem Flugplatz sein müssen, komme ich wochentags nicht dazu, in die Kirche zu gehen."[1119]

Suwelacks Briefe dienten der Beruhigung der streng katholischen Eltern und sollten ihnen die Furcht nehmen, ihr Sohn könne als Sünder sterben. Indem der Krieg als Prüfung und als über die Welt verhängtes „Strafgericht"[1120] gedeutet wurde, dem man nur durch die Einhaltung der religiösen Pflichten „reinen Gewissens" begegnen könne, war es naheliegend, dass die Religion ein bestimmendes Thema in den Briefen der Eltern war.

Zu Suwelacks religiöser Praxis während des Krieges lässt sich nur wenig mehr sagen. Lediglich ein Brief vom 5. September 1915 gewährt noch einen Einblick in seine Religiosität, hier bat er seinen Vater um ein Gebetbuch mit Texten zu Messe, Beichte, Kommunion und Kreuzweg.[1121] Ob Suwelack das Buch nutze oder betete, ist aus den Quellen nicht lesbar. Vor allem erscheint Josef als pflichtbewusster Sohn, der den Anweisungen der Eltern Folge leisten und ihnen die Zeit ohne ihn erträglicher machen wollte.

Letztmalig besuchte Suwelack seine Eltern während eines zweiwöchigen Fronturlaubs vom 15. Juli bis 1. August 1915, etwa in der Zeit vom 21. bis 27. oder 28. Juli 1915. Dort erfuhr er von jüngst gefallenen Billerbeckern und der Situation in der Heimat: „In den Städten und industriellen Betrieben macht sich der Krieg eher bemerkbar ... . In der Molkerei arbeiten jetzt schon drei Franzosen, die in Maubeuge gefangen genommen wurden."[1122] Gleichzeitig schrieb er Richard, dass er alle gesund angetroffen und man ihm zu Ehren „zwei fette Hähne gebraten" habe.[1123] Somit entsteht der Eindruck, dass Josefs Familie in Billerbeck trotz des Krieges relativ gut versorgt war. Etliche Billerbecker waren eingezogen worden, darunter auch Nachbarn und Freunde von Josef, von denen einige im Krieg verwundet wurden oder fielen.[1124]

Die Gegenwart des Todes wurde in den Briefen selten explizit thematisiert. Stattdessen beschränkten sich die Informationen auf Berichte von Begräbnissen und die Namen von persönlich bekannten, gefallenen Soldaten. Visuelle und emotionale Eindrücke von Verstümmelung, Tod und zerstörter Landschaft kamen nur

1119 NL J. Suwelack, Brief an seine Eltern, Lille, 2.3.1915.
1120 NL J. Suwelack, Brief von Josefs Vater an seinen Sohn, Billerbeck, 8.11.1914.
1121 NL J. Suwelack, Brief Josef Suwelacks an seinen Vater, Petit Rechain, 5.9.1915.
1122 NL J. Suwelack, Brief an Richard, Billerbeck, 26.7.1915.
1123 Ebd.
1124 Von Josefs Freunden aus Billerbeck sind keine Briefe oder Karten im Nachlass überliefert. Allerdings nennt Josef einige Namen in seinen Briefen und auch seine Mutter erkundigte sich, ob er an seine Freunde geschrieben habe.

zu Beginn des Krieges zur Sprache und verschwanden rasch hinter einer Fassade aus „selbst erlebten" und „selbst erzählten" Fliegerberichten, die das Bedürfnis der Heimat nach „authentischen" Heldengeschichten befriedigten und ein Publikum bedienten, das die Soldaten als tapfere und siegesgewisse Männer in den Krieg hatte ziehen sehen. Dieses verklärte Bild galt es zu Propagandazwecken, aber auch als eigenen Schutzmechanismus gegen die Kriegsrealität so lange wie möglich aufrechtzuerhalten. Möglicherweise wollte Suwelack aber auch seine Angehörigen beruhigen, indem er ihnen authentische, das heißt die Kriegsrealität schildernde Berichte ersparte.

Imagepflege II: Messbare Erfolge und Auszeichnungen in einem anonymen Krieg: Technik als Partner in einem ‚sportlichen' Wettkampf?

Erlebnisberichte von der Front, die von Siegesgewissheit und Erfolgen sprachen und die Illusion eines fairen und gerechten Kampfes Mann gegen Mann aufrechterhielten, verloren mit zunehmender Kriegsdauer nicht an Popularität, sondern blieben ein beliebter Stoff in Zeitungsartikeln, autobiografischen Texten[1125] und Zeitschriften.[1126] Zahlreiche Titel und Schriftenreihen,[1127] die zwischen 1914 und 1918 erschienen, halfen, den Kontroll- und Orientierungsverlust zu kompensieren und das Bild vom heldenhaften Kampf auf dem „Feld der Ehre" lebendig zu halten. Bis zum Oktober 1918 war in den Zeitungen nichts von militärischen Rückschlägen oder gar von Niederlagen zu lesen.[1128]

Dem Bedürfnis der Heimat nach tapferen Helden kam auch Suwelack in seinen Briefen und in einem Zeitungsartikel nach, denn seine Eltern und Verwandten verlangten ausdrücklich nach Erlebnisberichten. Bereits im September 1914

1125 Gunther Plüschow, Die Abenteuer des Fliegers von Tsingtau, Berlin 1916 (Aufl. 700 Tsd.); Erich Killinger, Die Abenteuer des Ostseefliegers im Ersten Weltkrieg, Berlin 1917 (Aufl. 420 Tsd.); Manfred von Richthofen, Der rote Kampfflieger, Berlin 1917 (Aufl. 1, 226 Mio.). Vgl. Thomas F. Schneider/Julia Heineman/Frank Hischer/Johanna Kuhlmann/Peter Puls (Hg.), Die Autoren und Bücher der deutschsprachigen Literatur zum Ersten Weltkrieg 1914–1939. Ein bio-bibliographisches Handbuch, Göttingen 2008, S.10–11.

1126 Siehe z. B. Rubrik „Aus Flieger-Feldpostbriefen": „Ein Flug auf Leben und Tod", in: Flugsport (1915), Nr. 4, S.106–107; „Im Flugzeug über Krakau", in: ebd. (1915), Nr. 2, S.51–53 oder „Fliegererlebnisse", in: ebd. (1915), Nr. 6, S.167–172.

1127 Hier nur eine Auswahl an Titeln: „Im Flugzeug über Feindesland", Dresden 1914, 32 S., online unter: http://resolver.staatsbibliothek-berlin.de/SBB00008B6C 00000000. Erschienen in der Reihe „Helden der Luft. Episoden aus Fliegerfahrten und Kämpfen in Feindesland", Berlin, Nr 10: „Der Kampf um die Telefunkenstation" (1915), 24 S., online unter: http://resolver. staatsbibliothek-berlin.de/SBB00006C4E00000000; Nr. 21 „Meldet, wir haben gesiegt!" (1915), 21 S., online unter: http:// resolver.staatsbibliothek-berlin.de/SBB00006C5E00000000 und weitere Titel in dieser Reihe (alle 15.2.2015).

1128 Vgl. Neitzel, Blut und Eisen, S.211.

war im „Essener Generalanzeiger" ein Brief Josef Suwelacks gedruckt worden, in dem er von seine Eindrücken als Feldflieger schrieb:

„Ein herrliches Schauspiel ist so ein Artilleriekampf; mit welch furchtbarer Schnelligkeit unsere Leute anrücken, ist gar nicht zu schildern. Die Artillerie ist vorn, die Infanteristen klettern auf die Geschütze, um den laufenden Rothosen noch etwas nachzuschicken. Das Rattern der Maschinengewehre und das Schießen der Infanterie vergeht in dem Kanonendonner. Herrlich ist es, wenn die weißen Wölkchen plötzlich auftauchen. Es ist ja ein etwas komisches Gefühl, wenn beim Fliegen rechts und links weiße Wölkchen erscheinen; man fliegt dann lieber 2.000 Meter hoch. In 1.200 Meter Höhe sind noch Flugzeuge getroffen, und zwar sehr gut. Ein Albatros hatte 17 Schuß durch die Flügel und einen 10 Zentimeter hinter dem Sitz. Ich habe für die nächste Zeit eine wunderbare Aufgabe, über die ich aber nichts schreiben darf. Aber wir Flieger vergehen vor Ungeduld; unsere Hauptleute haben viel zu tun, uns in Schach zu halten. Hauptsächlich werden wir zu den Artilleriekämpfen herangezogen. Neulich haben wir einen französischen Flughafen entdeckt und dann bombardiert, und zwar mit Erfolg. Heute nachmittag (der Brief ist am 1. September abgeschickt worden. Die Red.) fliege ich über Givet und, so Gott will, bald über Paris. Das wird ein lustiges Fliegen, uns alle beseelt nur der eine Gedanke: Durch! Hindernisse gibt es für unsere Truppen nicht mehr. Französische Flugzeuge haben wir noch keine gesehen. Nur Leutnant St. hatte einen Kampf mit einem französischen D.-Decker, der mit Maschinengewehr ausgerüstet war. Der Beobachter, Leutnant v. Sch., hat mit seinem Karabiner den D.-Decker zum Landen in unseren Truppen gezwungen."[1129]

Die Schilderung Suwelacks steht beispielhaft für andere Aufzeichnungen von Piloten und Beobachtern, die in ähnlicher Form vom Geschehen am Boden berichteten. In der Luft gab es noch das Gefühl von Freiheit und Überlegenheit,[1130] die Kämpfe wurden als „schönes Schauspiel" wahrgenommen und das Geschehen von oben als „Film"[1131] aus der Perspektive des Regisseurs beschrieben:[1132] „Das XIX. Korps hat seine Sache gut gemacht, und wir konnten von oben in aller Ruhe uns ansehen, wie wunderbar und furchtbar unsere Artillerie die schöne Stadt [Lille] zum dritten Teil in Schutt und Trümmer legte. Es war ein Riesenbrand."[1133]

Die maschinisierte Schlacht ließ durch das massenhafte anonyme Töten traditionelle Heldenbilder obsolet werden; an ihre Stelle traten „neue Helden"[1134] oder vielmehr reaktivierte Helden zur See oder in der Luft, die sich bereits in der Vorkriegszeit im sportlichen Wettkampf ausgezeichnet und als Beherrscher der Tech-

1129 Josef Suwelack, Ein Artilleriekampf aus der Vogelschau, in: Essener Generalanzeiger vom 13.9.1914. Der Artikel wurde außerdem abgedruckt im „Pilsener Tageblatt" vom 15.9.1914, online seit 2011 unter: http://anno.onb.ac.at/cgi-content/anno?aid=pit&datum=19140915&seite=3 (15.2.2015).

1130 Siehe auch die Aufzeichnungen im Kriegstagebuch Major Wilhelm Siegerts vom Oktober 1915: „Die am Boden verweilen / Haben am Alltag genug, / Die dem Alltag enteilen, / Heben sich hoch im Flug ... Die am Boden nur kleben, / Kriechen von Wiege zu Gruft, / Herrlich und hehr ist das Leben / Nur den Helden der Luft". Archiv DTMB, Kriegstagebuch Siegert, Eintrag von Oktober 1915, S.I 170.

1131 Dazu Paul Virilio, Krieg und Kino. Logistik der Wahrnehmung, Frankfurt a. M. 1989, bes. S.19ff.

1132 Vgl. Kehrt, Moderne Krieger, S.98; Köppen, Das Entsetzen, S.227ff.

1133 NL J. Suwelack, Brief an seine Familie, Lille, 19.10.1914.

1134 Siehe zu den „neuen Helden" die Ausführungen bei Schilling, Kriegshelden, S. 252ff. und S.286ff.

nik hervorgetan hatten. Die „neuen" Helden knüpften dabei in ihrer Selbstdarstellung an traditionelle Ehr- und Männlichkeitsvorstellungen an und versöhnten so die moderne Technik mit den traditionellen Kriegsvorstellungen. Für den Grabenkämpfer hingegen ließ sich dieses Heldenbild nicht aufrechterhalten, hier wurde es in der Weimarer Republik und dem Nationalsozialismus durch das Bild des „Neuen Menschen" mit Stahlhelm und harten Zügen ersetzt.[1135]

Die Nennung konkreter Details und „exakter" Zahlen verlieh dem Kriegsgeschehen etwas Rationales und machte es scheinbar berechenbar. Suwelack konnte seine jüngsten Erfolge benennen und in messbaren Kategorien denken.[1136] Das Flugzeug als „Distanzwaffe" war zugleich eine „Distanzierungswaffe", die einen Abstand zwischen „Täter und Opfer, zwischen Handeln und Verantworten" schuf.[1137] Das Flugzeug als „Prothese"[1138] verschmolz Flieger und Maschine zu einer Einheit, einem Cyborg, so dass ein Teil der Verantwortung auf den Apparat als Akteur und Partner übertragen und die Maschine damit „vermenschlicht" wurde.[1139] So wurden die Flugzeuge z. B. mit den Namen überflogener Orte markiert (Abb. 24) und damit eine Art Biografie der Mensch-Maschine-Interaktion geschrieben.

Die zweidimensionale Sicht aus der Luft ließ das Schlachtfeld als abstraktes Mosaik aus Linien und Rechtecken ohne Menschen erscheinen, über dem das erhabene Fliegerduo schwebte, das sogar noch die Zeit fand, sich während des Fluges selbst zu fotografieren, und damit seine Kühnheit unterstrich.

Major Wilhelm Siegert verglich das Schlachtfeld im Ersten Weltkrieg rückblickend mit „glühende[n] Ketten" und die Mündungsfeuer der Artillerie mit „Hunderttausende[n] von Leuchtkugeln aller Farben wie in Sekt tanzende Perlen."[1140] Die Gewalterfahrung wurde in der Berichterstattung verharmlost und für die Adressaten in ein Szenario transformiert, in dem das Schwarz-Weiß-Bild von Gut und Böse noch Gültigkeit besaß.[1141] Der gewaltsame Akt des Tötens aus der Dis-

1135 Bernd Hüppauf, Schlachtenmythen und die Konstruktion des „Neuen Menschen", in: Gerhard Hirschfeld/Gerd Krumeich/Irina Renz (Hg.), „Keiner fühlt sich mehr als Mensch …". Erlebnis und Wirkung des Ersten Weltkriegs, Essen 1993, S.43–83, bes. S.62ff.

1136 Vgl. Wohl, A Passion for Wings, S.203ff.

1137 Kaufmann, Kriegführung im Zeitalter technischer Systeme, S. 338.

1138 Diesen Begriff verwendet Kurt Möser, um die Interaktion zwischen Mensch und Maschine zu beschreiben. Vgl. Kurt Möser, Grauzonen der Technikgeschichte, Karlsruhe 2011, S.109.

1139 Siehe zur Akteur-Netzwerk-Theorie: Nina Degele, Einführung in die Techniksoziologie, München 2002, S.126–141; Andréa Belliger/David Krieger, Einführung in die Akteur-Netzwerk-Theorie, in: dies. (Hg.), ANThology. Ein einführendes Handbuch in die Akteur-Netzwerk-Theorie, Bielefeld 2006, S.13–50, hier S.15ff. und S.39ff.

1140 Zit nach Neumann, In der Luft unbesiegt, S.26.

1141 Vgl. Burkhard Fuhs, Fliegende Helden. Die Kultur der Gewalt am Beispiel von Kampfpiloten und ihren Maschinen, in: Rolf Brednich/Walter Hartinger (Hg.), Gewalt in der Kultur. Vorträge des 29. Deutschen Volkskongresses, Bd. 2, Passau 1992, S.705–720, hier S.714 und S.720.

tanz wurde durch die Definition der Fliegerei als „Arbeit" relativiert[1142] und mit einer Tagesleistung verbunden:

> „Gestern und vorgestern bin ich geflogen und hab an jedem Tag 10 Bomben geworfen auf die Engländer. Ihr glaubt ja gar nicht, welch furchtbar moralischen Eindruck so eine Bombe hat. ... Ein Krach und irgend ein Haus fliegt in die Luft. Oder die Bombe fällt in eine Kolonne. Alles Wirrwarr. Die Pferde werden scheu. Dann ein 2tes Mal, ein 3tes usw. Jetzt sieht man, wie der Flieger von der Artillerie beschossen wird. 10, 20, 30 Sprengpunkte, alle ganz nahe von unten gesehen. Jetzt Karabiner heraus, und es geht ein mörderisches Gewehrfeuer los. Indessen der Mann dort oben dreht den 2ten Kreis. Wiederum 5 Bomben. Dieses Experiment machen wir tagtäglich. Der engl[ische] Stab hat jeden Tag etwa 20 Bomben erhalten. Oftmals hängen wir für unsere lieben Vettern recht freundliche Grüße daran, wenn wir dann unter uns die Rauchwolken aufsteigen sehen, kehren wir befriedigt nach Hause zurück. Wir haben dann unser Tagewerk getan. Jedenfalls, wenn man so an eigenem Leibe verspürt hat, was Fliegerbomben sind, so sind wir dann abends sehr befriedigt."[1143]

Abb. 24: Maschine von Rittmeister Paul Dietze und Major Neumann mit den Namen der überflogenen Orte, undatiert, vermutlich 1915 in Lille (Foto: Familienarchiv Walter Suwelack, NL Josef Suwelack)

Das Zählen der Bombenabwürfe schien für die Flieger eine Gelegenheit gewesen zu sein, an die sportlichen Erfolge der Vorkriegszeit anzuknüpfen und sich auf das Handwerk der Fliegerei zu konzentrieren. Die Auswirkungen der Bombeneinschläge direkt sehen und hören zu können, musste zwar einerseits schockierend wirken, doch andererseits konnten die Flieger die verursachte Zerstörung auf ihr Können und ihr individuelles Handeln zurückführen. Für die Adressaten ihrer Berichte hatten sie damit einen direkten Einfluss auf den Krieg und leisteten ihren

1142 Vgl. Kehrt, Moderne Krieger, S. 96.
1143 NL J. Suwelack, Feldpostbrief Josefs an seine Familie, Lille, undatiert [vermutlich 23.10.1914].

Beitrag dazu. Krieg in diesem Sinne glich eher einem Spiel, bei dem derjenige mit den überlegenen Waffen gewann. Die Folgen der Bombenabwürfe – Tod und Zerstörung – wurden durch Phrasen nur indirekt kommuniziert, denn es war lediglich die Rede von einem „mörderischen Gewehrfeuer" oder aufsteigenden „Rauchwolken", nicht jedoch von zerfetzten Körpern oder Erdkratern, Auswirkungen, die aus der Luft ohnehin nicht zu erkennen waren.[1144] Die Beschreibung der menschlichen Katastrophe blieb der Bodenperspektive vorbehalten.

Abb. 25: „Stellungsaufnahme ca. 1.200 m", undatiert (Foto: Fotoalbum Suwelack)

Mit zahlreichen Luftaufnahmen (z. B. Abb. 25) ließ Suwelack seine Eltern an seinen Flügen teilhaben und gab ihnen Einblick in seine Aufklärungsarbeit. Die Luftaufnahmen hatten eine besondere Bedeutung für Josef und verweisen auf seine Faszination für die neue Luftbild-Technik und die technischen Möglichkeiten überhaupt. In einem Brief an seine Eltern, dem er einige Luftaufnahmen beifügte, erläuterte er ihnen die Bilder:

> „Auf Bild 29 und 30 sind verschiedene kleine weiße Wölkchen. Es sind die Schrapnells, die auf mich abgeschossen wurden. Die Sprengpunkte liegen etwa 3–500 mtr unter uns und auf Bild Nr. 30 seht Ihr 4 Stück ziemlich nahe liegen. Immerhin in 2.000 mtr ist nichts zu machen und Ihr seht auf den Bildern, was wir aus diesen Höhen sehen können. Wir haben auch noch sehr gute Gläser mit. Das schönste Bild ist Nr. 34, 33, 32."[1145]

1144 Siehe zur Ausblendung des Tötens: John Protevi, Affect, agency, and responsibility: The act of killing in the age of cyborgs, in: Kevin McSorley (Hg.), War and the Body. Militarisation, practice and experience, Abingdon/Oxon/New York 2013, S. 128–137.

1145 NL J. Suwelack, Feldpostbrief an seine Familie, St. Etienne, 3.10.1914.

Nicht nur seine Eltern forderten Josef regelmäßig auf, ausführlich von seinen Erlebnissen zu berichten, sondern auch bei Freunden und Verwandten hinterließen die Erzählungen nachhaltigen Eindruck und verstärkten den Wunsch, von neuen Heldentaten zu hören, denn Josef versicherte regelmäßig, in der Luft könne ihm nichts passieren.[1146] Das Interesse an den Fliegerberichten erstreckte sich dabei sogar auf die kirchliche Sphäre. Eine Verwandte Suwelacks, Schwester Emerentia, geborene Maria Catharina Elisabeth Brockmann (1866–1924), die 1892 in den Schwesternorden der Göttlichen Vorsehung in Münster eingetreten war, schrieb Josef im Juni 1915, dass sie sich eine schriftliche Schilderung seiner Erlebnisse als Flieger wünsche: „Hier in Cleve ist nämlich eine alte kranke Schwester, die sich immer sehr für Deine Nachrichten interessiert; Du hattest mal eine Luftfahrt in der Zeitung beschrieben, die war so schön."[1147] Gleichzeitig dankte sie Josef für seinen „aufopfernden Dienst", den er dem Deutschen Reich „und dadurch uns allen" leiste: „[W]ir sind Euch tapferen Helden zu großem Dank verpflichtet, daß Ihr den Feind von uns fernhaltet. Wäre doch der entsetzliche Krieg zu Ende."[1148]

Das Leben der Piloten, wie diese es nach außen hin schriftlich und bildlich transportierten, war geprägt und geleitet von der Aussicht auf Erfolg. Stolz berichtete Josef von erfolgreichen Flügen und Meldungen, von Treffen mit führenden Militärs und von erhaltenen Auszeichnungen. Zum „richtigen" Soldat-Sein zählte vor allem die symbolische Anerkennung der erbrachten Leistungen in Form von Auszeichnungen, die man gut sichtbar öffentlich präsentieren konnte. Schon wenige Wochen nach Kriegsbeginn wurde mit der Verleihung des Eisernen Kreuzes und weiterer Auszeichnungen an Flieger begonnen, was die Moral hochhalten und Anreize schaffen sollte, tapfer in die Schlacht zu ziehen.

Ehrgeizig und darauf bedacht, seine Position bei der Fliegerabteilung zu unterstreichen, schrieb Josef auf einer Karte an seine Essener Freundin Putty Pohl im September aus Reims:

> „Sehr geehrtes Fräulein. Vor 8 Tagen bin ich als erster deutscher Flieger in Reims gelandet und habe dort für 1 Million französische Flugzeuge und Motoren entdeckt. Heute bin ich hier in Belgien und fliege in 3–4 Tagen nach Chalons. Umstehende Villa die vom Besitzer verlassen ist, ist mein Quartier. Hoffentlich darf ich bald über Paris. Es grüßt Sie und alle Bekannte herzlichst Ihr J. Suwelack"[1149]

1146 Vgl. NL J. Suwelack, Feldpostbrief an seine Familie, Lille, Ostern 1915.

1147 NL J. Suwelack, Brief Schwester Emerentias an Josef, Cleve im Juni 1915. Wahrscheinlich bezog sie sich auf den Bericht im „Essener Generalanzeiger" vom September 1914.

1148 Ebd.

1149 NL J. Suwelack, Brief an Putty Pohl, wohnhaft Zweigerstraße Essen von Anfang September 1914. Möglicherweise bezog sich der „Flugsport" in seiner Ausgabe vom 16.9.1914 auf die Aussagen Josefs, denn die Zeitschrift berichtete in einem Artikel zum „Handstreich von Reims" vom Führer des Etappenflugzeugparks der 3. Armee, der in einer Fabrik zahlreiche französische Ein- und Doppeldecker entdeckt und beschlagnahmt habe, siehe Flugsport (1914), Nr. 19, S. 766f.

Putty Pohl gratulierte Josef daraufhin in einem Schreiben zu seinen „famosen Erfolge[n], von denen mir auch Herr Füßmann erzählt." Seine Arbeit sei „interessant, gerade weil es gefahrvoll ist."[1150] Die Beispiele verdeutlichen, dass die Heimat nach abenteuerlichen Geschichten verlangte, in denen die Flieger möglichst waghalsig und mutig erschienen und erst im letzten Moment dem Feind entkommen waren. Die Berichte über Josefs Flüge, wenn auch nicht „erfunden", dürfen trotzdem nicht mit der Realität des Luftkrieges gleichgesetzt werden. Sie geben vielmehr Auskunft über Gewaltvorstellungen und den kriegerischen Fliegerdiskurs, über die Flieger als Werbe- und Kriegsikonen, mit denen die Militärführung öffentlich vorzeigbare Helden hatte.[1151] Am 21. September teilte Suwelack seiner Familie voller Stolz mit, dass er nach drei Flügen über den feindlichen Stellungen mit seinem Beobachter „gute Meldungen gebracht" habe: „Unser Hauptmann hat mir sofort das Off[iziers] Fliegerabzeichen verliehen. Meine Maschine steigt am besten von allen der Abteilung. Und wir werden nur losgeschickt bei gutem Wetter."[1152]

Für Suwelack als Zivilflieger war es wichtig, sich gegenüber den ranghöheren Offizieren zu behaupten. Mit dem Zusatz, dass er nur bei gutem Wetter fliege, beruhigte er seine Eltern und unterstrich seine privilegierte Stellung. Das Sich-Abheben von der Masse drückte sich nicht nur in der Fliegerei selbst aus, sondern wurde öffentlich durch Auszeichnungen und Treffen mit der Militärführung sichtbar. Josef Suwelack wurde während seines Kriegseinsatzes bereits im Dezember 1914 mit dem Eisernen Kreuz und dem Albrechtsorden mit Schwertern II. Klasse ausgezeichnet, was er seiner Familie umgehend mitteilte.[1153] Bereits Anfang Oktober hatte er an seine Eltern geschrieben, dass er das Eiserne Kreuz erhalten werde: „Leider ist das Exemplar noch nicht da, aber in einigen Tagen werde ich den Orden auch tragen. ... Sagt es noch nicht der Öffentlichkeit, bevor ich offiziell dekoriert bin. Aus Essen haben mir meine Freunde geschrieben, daß ich momentan der populärste Mann in der ganzen Gegend bin."[1154]

Am 9. Dezember 1914 erfolgte die Verleihung des Albrechtsordens[1155] durch den sächsischen König.[1156] Josef sandte kurze Zeit später eine Porträtaufnahme,

1150 NL J. Suwelack, Brief von Putty Pohl, Essen, 18.9.1914. Über Putty Pohl waren keine weiteren biografischen Informationen zu finden. Auch ist dies der einzige vorhandene Brief von ihr im Nachlass. Suwelack erwähnte ihren Namen jedoch in einigen Briefen.

1151 Vgl. Möser, Fahren und Fliegen, S.554f.

1152 NL J. Suwelack, Feldpostbrief an seine Familie, St. Etienne, 21.9.1914. Diese Form der Selbstdarstellung kann als typisch für die Fliegerszene gelten. Vgl. Kehrt, Moderne Krieger, S.175.

1153 NL J. Suwelack, Feldpostbrief an Richard, Lille, 1.11.1914.

1154 NL J. Suwelack, Feldpostbrief an seine Familie, St. Etienne, 3.10.1914.

1155 Am 29. Oktober 1886 war die Stiftung des Albrechtsordens mit Schwertern für militärische Verdienste erfolgt, dem das Albrechtskreuz angeschlossen war. Es wurde für Verdienste im Feld und als militärische Auszeichnung mit Schwertern verliehen. Im Jahr 1914 gab es 470 erfasste (!) Verleihungen, zwischen 1914 und 1918 wurde das Ritterkreuz des Albrechtsordens mit Schwertern II. Klasse insgesamt 11.267 Mal verliehen. Erhard Roth, Die Verleihung des Ritter-

die der Fotograf der Abteilung gemacht hatte,[1157] nach Hause und fügte scheinbar beiläufig hinzu, dass die Abteilung „[n]eulich abends ... mit dem Prinzen Ernst A [lbrecht] von Sachsen zu Abend gegessen" habe.[1158]

Die zitierten Passagen und die Fotos belegen, dass Suwelack in jeder Hinsicht überzeugt war, als Zivilflieger zum Offizierskorps zu gehören, und sich selbst als Gleichrangigen sah. Wie in der Vorkriegszeit setzte er das Fotografieren von Generälen und Offizieren fort und brachte damit auch visuell seinen Stolz zum Ausdruck, zu diesem erlesenen Kreis zu zählen. Mit den sichtbaren Auszeichnungen als Symbolen der militärischen Wertschätzung und Anerkennung, der Beschreibung von Festessen mit Generälen und dem Königshaus sowie der bildlichen Dokumentation dieser besonderen Momente trug Suwelack zur Konstruktion eines luxuriösen, ehrenhaften und elitären Fliegerbildes bei, das bis heute die öffentliche Wahrnehmung von Fliegern prägt.[1159] Für die Mitglieder der Fliegertruppen schien der Weg stets nur nach oben zu führen und so ließ Suwelack keines der Treffen mit Angehörigen des Generalstabs unerwähnt.

Eine Woche nach der Übergabe von Lille an das deutsche Armee-Oberkommando schrieb er z. B. an seine Eltern, dass sein Beobachter Oberleutnant Müller und er über die Stadt geflogen, nachmittags über Dünkirchen gewesen seien, um die Eisenbahnlinien zu kontrollieren, und gute Meldungen zurückgebracht hätten. Daraufhin zog

> „Seine Exzellenz General von Laffert ... mich in ein längeres Gespräch und erkundigte sich über meine privaten Verhältnisse. Er sprach mir seine Freude darüber aus, daß er mir schon nach so kurzer Zeit das eiserne Kreuz hätte geben können. Sodann ließ mich der Prinz Ernst von Sachsen, der bei unserem Armeecorps ist, zu sich holen. Er lag auf einem Strohbündel und schrieb einen Brief an seinen Vater. Er erkundigte sich genau über alle Einzelheiten des Fluges, und ich sprach etwa 1/2 Stunde mit ihm."[1160]

kreuzes 2. Klasse mit Schwertern des königlich sächsischen Albrechtsordens im Ersten Weltkrieg, Konstanz am Bodensee 1998, S.10.

1156 NL J. Suwelack, Verleihungsurkunde des Albrechtsordens mit Schwerten II. Klasse durch den sächsischen König Friedrich August, Dresden, 9.12.1914. Der Orden wurde außerdem 1914 bzw. 1915 verliehen an Oberleutnant Paul Dietze, Oberleutnant Emil Clemens, Leutnant Hans Schneider, Rudolf Hasenlohr und Oskar Teichmann von der Feldfliegerabteilung 24. Siehe Roth, Die Verleihung des Ritterkreuzes, S.42ff.

1157 Vgl. NL J. Suwelack, Feldpostbrief an Richard Suwelack, Lille, 3.1.1915.

1158 NL J. Suwelack, Feldpostbrief an seine Familie, Lille, 3.1.1915.

1159 Vgl. Fuhs, Fliegende Helden.

1160 NL J. Suwelack, Feldpostbrief an seine Familie, Lille, 19.10.1914. Maximilian von Laffert (1855–1917) war seit November 1912 Kommandierender General des XIX. Armeekorps in Leipzig und wurde im Mai 1913 zum General der Kavallerie befördert. Im Ersten Weltkrieg führte er sein Korps an die Westfront von der Maas bis zur Marne, konnte sich in der Schlacht profilieren und war an der Eroberung von Lille beteiligt. Danach kämpfte er weiterhin an der Lys sowie im Frontgebiet von Lille. Vgl. Peter Mertens, Laffert, Maximilian von, in: Sächsische Biografie, hg. vom Institut für Sächsische Geschichte und Volkskunde e.V., bearb. von Martina Schattkowsky, online unter: http://www.isgv.de/saebi/ (10.3.2015).

Josef hob besonders die Ruhe der deutschen Truppen hervor, die die Wasserleitungen abgestellt und einen großen Brand in Lille gelegt hätten. Der Topos von der Nervenstärke der Truppen, der auch schon in der Vorkriegszeit einen wesentlichen Bestandteil des Fliegerdiskurses gebildet hatte, setzte sich im Krieg fort und wurde als nationaler Charakterzug gewertet, der über Sieg oder Niederlage entschied:

„Gerade fragte mich S[eine] Exzcellenz, ob der Brand wohl noch zu löschen wäre, da braust ein 100-PS-Opelwagen eine Ordonnanz vom A.O.K. heran und der Offiz[ier] meldet, daß gerade ein Parlamentär die Stadt bedingungslos übergeben habe. L[iebe] E[ltern], das war ein unvergeßlicher Moment f[ür] mich. Die Freude auf allen Gesichtern. Und dann, als ob eigentlich alles ganz selbstverständlich wäre, gings in die Quartiere."

Als Zivilflieger zählte sich Suwelack zu den „oberen Zehntausend"[1161] und profitierte durch seine Position als Flugzeugführer, in der er Hand in Hand mit dem ranghöheren Beobachter agierte, und durch seine fliegerische Qualifikation vom Krieg: Die bestehenden Rangunterschiede wurden zum Teil nivelliert beziehungsweise unterlaufen. Theoretisch waren die Rollen zwar klar definiert:

„Die gleiche militärische Ausbildung, die übereinstimmende Standesauffassung und der naturgemäß festere, kameradschaftliche Anschluß innerhalb des Offizierkorps erzeugte eine größere Einheitlichkeit der Ansichten und ein natürliches Zusammenwirken als es zwischen dem BO [= Beobachtungsoffizier] und einem Flugzeugführer der Fall sein wird, der nicht Offizier ist."[1162]

Die Fotos suggerieren jedoch einen kameradschaftlichen, respektvollen und gleichrangigen Umgang miteinander und zeigen die Flieger und Beobachter als Gemeinschaft, gleichwohl sich die Rangordnung in der Kleidung oder der Aufstellung auf dem Foto zum Teil widerspiegelt.

Eine Position bei der Fliegertruppe zu bekleiden, brachte nicht nur Privilegien sowie gesellschaftliche und militärische Wertschätzung und Bewunderung mit sich, sondern war gleichzeitig auch Ausdruck einer größeren Durchlässigkeit des Militärs für Bürgerliche und Männer mit einer technischen Ausbildung. Die hohe Motivation junger Männer, sich freiwillig zur Fliegertruppe zu melden, entsprang vermutlich dem Wunsch, in die obere Liga aufzusteigen und aufgrund der erbrachten Leistung und nicht aufgrund der Abstammung befördert zu werden. Aufzeichnungen anderer Flieger und Beobachter stützen die These, dass innerhalb der Fliegertruppe ein Wettstreit um die Anzahl der Erfolge und Auszeichnungen entstand. Ernst Eberstein, der als Beobachter den Krieg an der West- und Ostfront miterlebte, notierte in seinem Kriegstagebuch am 15. August 1914: „Die Herren sind sehr dankbar für unsere Meldungen, sie sind die besten[,] die sie haben. Sie erkennen unsere Leistungen voll an. ... [D]er Kommandierende spricht mit [Ernst]

1161 Fritzsche, A Nation of Fliers, S.84.
1162 BA-MA PH 24/259, Die Flugzeuginstrumente und ihr Einbau, um 1912 bis 1917, zusammengestellt von der physikalischen Abteilung Gruppe 5, darin: Anleitung zum Beobachtungsoffizier (Siegert).

Canter und mir sehr anerkennend, wir werden zum Eisernen Kreuz eingegeben!"[1163]

In die Erzählungen mischten sich oft auch romantische Naturbeschreibungen, wie sie in der Vorkriegszeit in Fliegerberichten zu finden waren, in denen die Piloten von einer idyllischen Landschaft und dem Vertrauen in die eigenen Fähigkeiten berichteten:

> „Das ganze Land sah aus wie eine große Stadt, die hell erleuchtet ist. Da, in der Ferne eine Leuchtrakete – unser Signal Abt. 24. Die Richtung stimmte genau. Von der Erde war mittlerweile nichts mehr zu sehen. Die Richtung hatte ich eingeschlagen nach dem Kompaß. Meine elektr(ische) Lampe versagte. Jetzt ein neues Signal. Wie mochten die Unsrigen um uns in Sorge sein? Aber ich hatte meine Maschine so ruhig und sicher in der Hand, daß mir auch nicht die geringste Sorge aufkam. Während des Gleitfluges, der 16 Minuten dauerte, genossen wir das Herrlichste, was ich je erlebt. Über uns der schöne Sternenhimmel. Hinter uns das erste Viertel des Mondes und unter uns auf Frankreichs Boden unsere tapferen Soldaten, die ermüdet von den Kämpfen des Nachmittags in den Wäldern lagen. Jetzt zeigte uns eine 3te Rakete nochmals unseren Platz, und eine halbe Minute später stand mein Apparat ruhig und sicher vor meinem Zelt. ... Rasch wurde die Maschine in das Zelt geschoben. Ein kleiner Imbiß, und dann sausten wir mit unserem 100-PS-Benz durch die Nacht zum General, dem wir unsere reichen Entdeckungen mitteilten. Ein Händedruck war die vorläufige Belohnung für meinen Flug."[1164]

In der Luft wurde eine Parallel- oder Idealwelt entworfen, in der im Gegensatz zum düsteren und „dreckigen" Boden noch Schönheit, Ruhe und vermeintliche Sicherheit herrschte. Die Konstruktion eines „anderen" Krieges, in dem noch Regeln, Individualität und persönliche Wertschätzung der erbrachten Leistung vorhanden waren, verlieh dem realen Krieg Ordnung und einen Sinn. Die Poetisierung des Kriegserlebnisses aus der Luft, bei der der Akt des Tötens und die Ausmaße der Zerstörung verharmlost oder in „beeindruckend schöne" Licht- und Farbenspiele verkehrt wurden, ließ die Leser indirekt am Krieg teilhaben und ermöglichte, das Gewaltdispositiv mithilfe der Technik zu verarbeiten, dessen Komplexität zu reduzieren und sich zu distanzieren.

Anders verhielt es sich auf dem „Boden der Tatsachen", wo die Zerstörung sichtbar wurde und sich nicht mehr beschönigen ließ. Abseits der Schlachtfelder inszenierten sich die Flieger jedoch ebenfalls als robuste und unerschrockene Einheit, die ihren Kriegseinsatz scheinbar genoss und sich auslebte. Unmittelbar konfrontiert mit Leichen, Granatsplittern, verwüsteter Landschaft und zerstörten Gebäuden begegnete Josef Suwelack, der sonst zahlreiche Worte für seine Arbeit bei den Feldfliegern fand, dieser Form des Krieges mit Schweigen. Das Narrativ vom romantischen Krieg stieß dort an seine Grenzen, wo die Waffentechnik ihre Wirkung gezeigt hatte, und fand seinen Ausdruck nur noch im Paradoxon der artikulierten Sprachlosigkeit.

1163 Archiv-DTMB NL 151, Ernst Eberstein, Kriegstagebuch, Eintrag vom 15.8.1914, S.10.
1164 NL J. Suwelack, Feldpostbrief an seine Familie, St. Etienne, 24.9.1914.

Von Kameradschaft und Männlichkeit sowie vom Umgang mit Gewalt und Tod jenseits der Fliegerei

Eine „heile Welt im Stahlgewitter“[1165] existierte bei der Fliegertruppe als Einheit mit einem Stützpunkt – in der Regel einem Militärflugplatz – nicht nur in der Phantasie, sondern wurde in zahlreichen Fotos auch bildlich konstruiert. Ein „Feldpostbrief eines Fliegers“, 1916 erschienen in „Deutsches Heldenbuch 1914/15“, beschrieb das Fronterlebnis bei der Fliegertruppe als abenteuerliche und vergnügliche Reise:

> „Tagsüber lagen wir im Freien, exerzierten, suchten Ortschaften nach versteckten Franzosen ab. Von Zeit zu Zeit wurde geflogen, manchmal kamen die Flieger mit neuen Löchern in der Maschine zurück. Wir aßen im Freien Mannschaftskost, fuhren im Auto nach der Front und ließen uns die Schrapnells um die Ohren sausen.... Hatten wir anstrengenden Dienst gehabt, dann schliefen wir im Freien im Schatten eines Brunnens. Das reine Lagerleben!“[1166]

Das „Lagerleben“ der Piloten und Beobachter thematisieren auch zahlreiche Fotos im Nachlass Josef Suwelacks. Sie stehen mit der Darstellung von Luxus, Ruhe, Entspannung und Vergnügen in starkem Kontrast zu den Bildern, die die Erinnerung an den Ersten Weltkrieg nachhaltig geprägt haben und die das massenhafte Abschlachten an der Front, die seelische und körperliche Verwundung und den Abnutzungskrieg, in dem der Einzelne bedeutungslos wurde, zeigen. Selbst in den Feuerpausen konnten sich die Soldaten, die unmittelbar an der Front wochen- und monatelang in den Schützengräben, unterirdischen Tunneln und Unterständen ausharrten, nie sicher sein, nicht doch noch von einer Kugel, Bombe oder Granate getroffen zu werden. Erschöpfung, Müdigkeit und körperliche Angeschlagenheit beziehungsweise Versehrtheit widersprachen dem propagierten stählernen und kraftvollen Soldatenbild und mündeten in einer „Krise der Männlichkeit“.[1167] Gerade an diesem Ort, an dem von Menschlichkeit und Zivilisation nicht mehr die Rede sein konnte, wirkten die „friedlichen“ und den „normalen Alltag“ zeigenden Bilder der Flieger wie vom Krieg isoliert. „Echte Männer“, so könnte

1165 So der Aufsatztitel bei: Reimann, Die heile Welt im Stahlgewitter.

1166 Feldpostbrief eines Fliegers, in: Deutsches Heldenbuch 1914/15. Feldpostbriefe, wahre Erlebnisse und Erzählungen vom Kriegsschauplatz in Ost und West, mit Abbildungen, 2. verm. und erw. Aufl., Stuttgart 1916, 228 S., S.27–32, hier S.31, online unter: http://resolver.staatsbibliothek-berlin.de/SBB0000B53300000000 (15.2.2015).

1167 Vgl. zur „Krise der Männlichkeit“ als Infragestellung des patriotischen Vaterlandsverteidigers: Ute Frevert, Gesellschaft und Militär im 19. und 20. Jahrhundert. Sozial-, kultur- und geschlechtergeschichtliche Annäherungen, in: dies. (Hg.), Militär und Gesellschaft im 19. und 20. Jahrhundert, Stuttgart 1997, S.7–16; Kühne, Männergeschichte als Geschlechtergeschichte, S.16–20; Jason Crouthamel, Deutsche Soldaten und „Männlichkeit“ im Ersten Weltkrieg, in: Aus Politik und Zeitgeschichte (2014), H. 16/17, S.39–45; Markus Funck, Bereit zum Krieg? Entwurf und Praxis militärischer Männlichkeit im preußisch-deutschen Offizierskorps vor dem Ersten Weltkrieg, in: Karen Hagemann/Stefanie Schüler-Springorum (Hg.), Heimat-Front. Militär- und Geschlechterverhältnisse im Zeitalter der Weltkriege, Frankfurt a.M. 2002, S.69–90.

man zugespitzt formulierten, die salon- und kriegsfähig zugleich waren,[1168] gab es nur noch jenseits des Gefechts, z. B. in der Luft oder im Hinterland der Front in der Kunstfigur des Ritters.

Die Inszenierung von Männlichkeit bei der Feldfliegerabteilung 24 knüpfte an den militärischen Pilotenhabitus der Vorkriegszeit an und ließ die Piloten und Beobachter auf der „Bühne" im Feindesland erneut als selbstbewusste Dandys auftreten, die sich demonstrativ auch ungeschützt im Freien von nichts aus der Ruhe und um ihre Nerven bringen ließen (Abb. 26).

Militärisch-funktionale Kleidung kombiniert mit extravaganten Mänteln und Pelzen bestätigte auch optisch die Narrenfreiheit[1169] der Fliegertruppe, die sich Freiräume schuf und relativ autonom agierte. Besonders Josef Suwelack sticht auf Fotos regelmäßig deutlich hervor: breitbeinig, mit in die Hüften gestemmten Händen, offenem Mantel und sichtlich vergnügt, inszenierte sich der Westfale vor der Linse als enthusiastischer Krieger voller Tatendrang und Energie.

Am 24. September 1914 sprach Josef in einem Brief an seine Eltern von Pflichterfüllung und der Freude darüber, dem Vaterland dienen zu können.[1170] Wenige Tage zuvor hatte er zwar noch eingestanden, dass es angesichts der Zerstörung und des menschlichen Leides „doch nicht so leicht" sei, im Krieg zu kämpfen, sich aber dennoch zuversichtlich gezeigt, dass die deutschen Truppen siegen würden: „Reims brennt im Norden. Wenn wir jetzt Truppen genügend haben, gehts auf Paris los. Bei den großen Kämpfen haben die Franzosen immer mehr Tote wie wir. Unsere Soldaten sind meist verwundet durch Schrapnell, aber die französischen Schützengräben liegen fast immer bis oben voll."[1171]

Den Zusammenhalt bei der Fliegertruppe, sei es das gemeinschaftliche Frühstück, ein Bierabend oder das „Schultern" des Kameraden, dokumentieren einige Fotos im Nachlass Suwelacks. Flugzeugführer und Beobachter treten als aufeinander eingespieltes und miteinander eng verbundenes Team auf, eine Beziehung, die über ein reines Arbeitsverhältnis hinausging. Die Zeitgenossen sprachen von „Fliegerehen" zwischen der Flugzeugbesatzung.

Ein Foto, das rückseitig mit „Freiluftspiele" beschriftet ist, zeigt Josef Suwelack (sehr wahrscheinlich) auf dem Rücken Leutnant Teichmanns, mit dem er zu Aufklärungsflügen startete, als untrennbare Einheit (Abb. 27). Winkend und in die Kamera lachend ließen die Angehörigen der Fliegertruppen den Betrachter und womöglich auch sich selbst vergessen, dass sie sich im Krieg befanden. Im Bild

1168 Vgl. Markus Funck, Vom Höfling zum soldatischen Mann. Varianten und Umwandlungen adeliger Männlichkeit zwischen Kaiserreich und Nationalsozialismus, in: Eckart Conze/Monika Wienfort (Hg.), Adel und Moderne. Deutschland im europäischen Vergleich im 19. und 20. Jahrhundert, Köln 2004, S. 205–235, bes. S. 215ff. und S. 225ff.

1169 Vgl. Fuhs, Fliegende Helden, S. 714.

1170 Vgl. NL J. Suwelack, Feldpostbrief an seine Familie, St. Etienne, 24.9.1914.

1171 NL J. Suwelack, Feldpostbrief an seine Familie, St. Etienne, 17.9.1914.

festgehaltene Momente des Vergnügens schufen ein seelisches Refugium und waren eine überlebensnotwendige Inszenierung, um die Nerven zu behalten, eine Scheinnormalität herzustellen und sich gegenseitig Halt zu geben. Da Suwelack in der Regel nur bestimmte Personen mit Namen versah, ist anzunehmen, dass er mit diesen besonders viel Zeit verbrachte und er sie zu seinen Freunden zählte.

Außerhalb dieser fliegerischen Idylle, bei Besuchen der Schlachtfelder oder der Schützengräben, wurde der Krieg mit touristisch-voyeuristischem Blick fotografiert. Bilder, die Tote zeigen, beschränken sich im Fall von Josef Suwelack stets auf gegnerische Stellungen, gefallene deutsche Soldaten sind in der visuellen Erinnerung hingegen nur in der Darstellung von Kriegsgräbern oder Bestattungen greifbar. Auf diese Weise konnten die deutschen Soldaten als Helden, die feindlichen Truppen als Verlierer erinnert und so das Schwarz-Weiß-Bild von Gut und Böse aufrechterhalten werden. Trotzdem ist auch hier festzustellen, dass der Tod auf Distanz gehalten wurde und sich unter die Faszination Furcht und vielleicht auch Ekel mischte. In den Briefen an seine Eltern berichtete Suwelack anfänglich noch von der Zerstörung: „Gewaltig war der Eindruck, den die erste Schlacht auf uns machte, besonders grauenhaft war der Besuch eines Schlachtfeldes. An einem Waldrand lagen ganze Haufen von Zuaven und auf der freien Fläche auf einer Wiese 20 mtr. vor den Laufgräben lagen die Unsern.“[1172] Insgesamt wurde der Tod jedoch weitgehend ausgeklammert. „Gestern habe ich die Schlachtfelder von Ypern besucht ... . Ich kann Euch das, was ich da gesehen habe, gar nicht schildern, es war zu gräßlich.“[1173] Stattdessen berichtete Suwelack von der Wohnsituation, der Verpflegung und der Arbeit bei der Fliegertruppe. In einem Brief von Anfang März 1915 kam er jedoch auch auf die soziale Verelendung und Mangelernährung der Bevölkerung zu sprechen: „Wenn ihr die Not hier im Lande sehen würdet! in 2 Wochen ist das Mehl hier in Lille zu Ende und nur das Militär bekommt Brot. Fleisch hat der Franzose pro Woche 300 Gramm zu beanspruchen. Was sind das für Zeiten für die Leute?“[1174]

Im Verlauf des Krieges hatte Josef Suwelack selbst ebenfalls Verluste zu beklagen. Mehrere Freunde und Verwandte waren gefallen, ebenso ein Angehöriger der Fliegertruppe, an dessen Beerdigung Josef teilnahm.[1175] Walther Schwering[1176] (1885–1915), mit dem Josef und Richard befreundet waren, war am 1. Fe-

1172 NL J. Suwelack, Feldpostbrief an seine Familie, Rocroi, 1.9.1914.
1173 NL J. Suwelack, Feldpostbrief an seine Familie, Petit Rechain, 10.5.1915.
1174 NL J. Suwelack, Feldpostbrief an seine Familie, Lille, 2.3.1915.
1175 Heute befindet sich das Grab Brandts auf dem Deutschen Soldatenfriedhof in Vladslo, Belgien.
1176 Walther Schwering war der erstgeborene Sohn des Arztes und Sanitätsrates Dr. med. Walter Schwering (1850–1936) und dessen Frau Johanna, geborene Heyl (1858–1938), aus Billerbeck. Der promovierte Altphilologe war im Ersten Weltkrieg als Kriegsfreiwilliger als Gefreiter im bayerischen Reserve-Infanterie-Regiment Nr. 18. Vgl. König-Heuer/Paulus, „In der Ferne das Donnern der Kanonen“, S. 18.

Abb. 26: „Mittagsschläfchen, Flugplatz, Obl. [Hans] Schneider", Lille, undatiert, vermutlich März 1915 – links eine Fliegerkamera (Foto: Familienarchiv Walter Suwelack, NL Josef Suwelack)

Abb. 27: Hintere Reihe (v.l.): Lt. Lothar Wilisch, Hauptmann Meinhard Rosenmüller, Hauptmann Klubert, Josef Suwelack. Vorne (v.l.): Dr. Neumeister, Lt. Rössler, Walter Hoefig, Lille, 1914/15 (Foto: Familienarchiv Walter Suwelack, NL Josef Suwelack)

bruar 1915 in einem Kriegslazarett in Tournai an Typhus gestorben.[1177] Einen Monat zuvor war Anton Suwelack (1890–1915), ein Vetter Josefs, bei Ypern als Ersatz-Reservist gefallen.[1178] Ferner erlag Edmund Ahlers (1890–1915), der Sohn des Gaststättenbesitzers und Organisators der Billerbecker Flugtage 1912 Hubert Ahlers, im Mai 1915 als Kriegsfreiwilliger Gefreiter des 1. Garde-Fußregiments bei Jaroslaw seinen Verwundungen.[1179] Den Tod Ahlers' kommentierte Suwelack mit der Feststellung, es sei „schade um den prächtigen Kerl", anschließend informierte er Richard über seine Probleme mit der Inspektion der Fliegertruppen im Zusammenhang mit seiner bevorstehenden Kündigung als Zivilflieger.[1180] Die Verlusterfahrung bildete ein Randthema in den Briefen, da Tod und Zerstörung allgegenwärtig waren und sich Verluste, so drückte es Richard in einem Brief an seinen Bruder aus, „ja tatsächlich nicht vermeiden"[1181] lassen würden. Eine vehemente Kritik am Krieg übte Josef, vermutlich aus Gründen der (Selbst-)Zensur, nicht, sondern gab stattdessen der Uneinsichtigkeit und Zähigkeit der sich widersetzenden Soldaten die Schuld am Fortgang des Krieges.

Ob Suwelack grundsätzlich gewaltbereit war, lässt sich aufgrund der wenigen Äußerungen nicht belegen. Vielmehr ist eine Angriffslust zu erkennen, die möglicherweise aus dem Gefühl der Gruppenstärke und der Solidarität untereinander erwuchs.[1182] So schrieb er:

> „Und doch ist das Volk noch frech genug. Es müßten eben jeden Tag ein paar Exemplare auf dem Grand Place an die Mauer gestellt werden. Na, wir sind ja wenig liebenswürdig mit den Herrschaften. Wenn uns jemand auf der Straße nicht ausweicht, ziehen wir ihm coram populo eines mit dem Reitstock übern Rücken, und man kennt uns Flieger schon in der ganzen Stadt."[1183]

1177 Vgl. NL J. Suwelack, Brief Martha Schwerings an Richard Suwelack, Billerbeck, 2.2.1915. Siehe auch Familienanzeigen aus dem Raum Billerbeck, bearbeitet von Rotraud Illisch und Sabine Lleshi, online unter: http://familienanzeigen.genealogy.net/anzeigen.php?bild=anzeigen/BillerbAnz/T/T–1915–2–1-schwering-walther.jpg&PID=882 (16.2.2015).

1178 Anton Suwelack stammte vom Hof in Greven-Pentrup und war der Bruder des Hofbesitzers Bernhard Suwelack und des Arztes Dr. Josef Suwelack aus Duisburg-Meiderich. Ich danke Herrn Walter Suwelack für diese Auskunft.

1179 Über das exakte Todesdatum besteht Unklarheit. In der Todesanzeige der Familie wird als Todestag der 15. Mai 1915 angegeben, in den Verlustlisten zum Ersten Weltkrieg wurde er am 9. Juni 1915 als „schwer verwundet" genannt. Da die Aktualisierung der Verlustlisten verständlicherweise einige Zeit in Anspruch nahm, ist es wahrscheinlich, dass Josef Suwelacks Angaben in einem Brief an seinen Bruder vom 9. Mai zeitlich eher zutreffen: „Du wirst inzwischen wohl erfahren haben, daß Edmund Ahlers gefallen ist". NL J. Suwelack, Feldpostbrief an Richard Suwelack, Petit Rechain, 9.5.1915.

1180 Ebd.

1181 NL J. Suwelack, Brief Richards an Josef, Vorges, 12.10.1914.

1182 Siehe zum Aspekt der Angriffslust: Norbert Elias, Über den Prozeß der Zivilisation. Soziogenetische und psychogenetische Untersuchungen, Bd. 1: Wandlungen des Verhaltens in den westlichen Oberschichten des Abendlandes, ND Frankfurt a. M. 2005, S.S.263ff. bes. S.269.

1183 NL J. Suwelack, Feldpostbrief an seine Familie, Lille, 2.3.1915.

Als Besatzer und Repräsentant des Deutschen Reiches wollte Suwelack die Überlegenheit der deutschen Einheiten unterstreichen und sich selbst als Mann inszenieren, dem man Respekt und (Ehr-)Furcht entgegenbrachte. Umgeben von Brutalität und Waffentechnik sank die Hemmschwelle, ebenfalls Gebrauch von den mitgeführten Waffen zu machen, ohne hier jedoch einen „Tötungstrieb"[1184] unterstellen zu wollen: „Da ich stets mit Karabiner und Revolver fliege, nahmen wir diese und gingen in das Dorf hinein. Wir gingen auf eine Schar 20jähriger Burschen los, die, als sie unsere Karabiner sahen, die Hände hoben."[1185] Ob diese Aussage lediglich der Ausdruck eines selbstbewussten und unerschrockenen Fliegerhabitus ist, kann rückblickend nur schwer beurteilt werden. Ähnlich äußerte sich der Flieger in einem Brief vom 2. Februar an Eduardo Pujol, in dem er dem spanischen Unternehmer von seinen Leistungen berichtete: „Ich habe ... auch schon manchen schönen Erfolg gehabt. ... Bei den Kämpfen, die ich mit französischen Fliegern hatte, sind diese meistens ausgewichen und einen hab ich in 2.400 mtr. Höhe abgeschossen."[1186] Josef Suwelack bediente in seinen Briefen die Erwartungen seiner Eltern, Freunde und Bekannten mit Berichten über Siege, Erfolge, Auszeichnungen und spannende Erlebnisse, die das Bild vom abenteuerlichen Krieg suggerierten. Mit Interesse nahm Suwelack die ihn umgebende Landschaft, die Zerstörung, die Zivilbevölkerung und fremde Soldaten als eine Art „Kriegstourist" wahr, der „seinen" Krieg in vielschichtigen Nuancen und aus unterschiedlichen Blickwinkeln fotografierte.

Neben Mannschaftsaufnahmen und Fotografien von Flugzeugen umfassen seine privaten Fliegeralben auch Sehenswürdigkeiten der von ihm besuchten Orte. Faszinierend wirkte offenbar auch Kriegsgerät wie Fliegerabwehrkanonen, Bomben oder Minen, neben denen sich die Feldfliegerabteilung in Szene setzte: „Es entstand eine Bildersprache, die auf das technologische Vernichtungspotential verwies, ohne die Destruktion selbst abzubilden"[1187].

Die Kriegserfahrung Josef Suwelacks war von Brüchen gekennzeichnet, die sich auch auf der visuellen Ebene widerspiegeln: Auf der einen Seite gab es das „Lager"- und Luxusleben hinter der Front, individuelle Freizeitgestaltung und ver-

1184 Siehe zur kritischen Auseinandersetzung mit Niall Fergusons „Der falsche Krieg": Benjamin Ziemann, Gewalt im Ersten Weltkrieg. Töten – Überleben – Verweigern, Essen 2013, S.13ff. Ziemann kritisiert die pauschalen und plakativen Thesen von der Lust am Töten oder eines Triebs, die sich weder be- noch widerlegen ließen und unhistorisch seien.

1185 NL J. Suwelack, Brief an seine Familie, Rocroi, 2.9.1914. Siehe zum Krieg gegen die Zivilbevölkerung: Anton Holzer, Das Lächeln der Henker. Der unbekannte Krieg gegen die Zivilbevölkerung 1914–1918, mit zahlreichen bisher unveröffentlichten Fotografien, Darmstadt 2008.

1186 NL J. Suwelack, Brief an Eduardo Pujol, Lille, 2.2.1915. Der Brief ist unvollständig, einige Passagen sind durchgestrichen und überarbeitet worden, denkbar ist, dass möglicherweise einige Seiten von der Zensurstelle einbehalten wurden oder dass Suwelack den Brief gar nicht erst verschickte.

1187 Köppen, Das Entsetzen, S.242.

meintliche Sicherheit, auf der anderen Seite geriet diese „heile Welt“ durch die Kriegsgräuel, den Lärm von Bombenschüssen und -explosionen sowie durch Nachrichten von Verwundeten und Toten aus den Fugen. Die Flucht in die Erzählung, in der eher banale Themen aufgegriffen wurden, blendete den Krieg weitgehend aus. In der Realität veränderten die leichten und wendigen Eindecker, die seit 1915 zum Einsatz kamen, in Kombination mit dem Maschinengewehr den Luftkrieg und mit ihm die Wahrscheinlichkeit, in einen „Luftkampf“ zu geraten. Trotz der hohen Mobilität der Fliegertruppe waren insgesamt jedoch kaum Fortschritte zu verzeichnen und die einzige Hoffnung auf ein Ende des Krieges bestand darin, dass wieder etwas in Bewegung geriet.

Am gleichen Tag, an dem Josef den Abschiedsbrief an seine Familie verfasste, schrieb er in einem weiteren Brief, dass die Schlacht weitergehe, die französische Offensive bei Arras erneut beginne und die deutsche Seite gerüstet für den Kampf sei.[1188] Die Gewaltsteigerung durch eine verbesserte (Waffen-)Technik ließ Josef Suwelack zunehmend auch über seine eigene Sicherheit nachdenken und dies bewog ihn möglicherweise, seinen Abschiedsbrief zu verfassen. In der Absicht, seiner Familie Trost zu spenden und sie in der Gewissheit zurückzulassen, dass ihr ältester Sohn als frommer Katholik und pflichtbewusster Soldat gestorben war, verlieh Josef seinem Leben Sinn und beurteilte seinen Tod im Sinne des zeitgenössischen Diskurses als von Gott gewolltes Opfer für die Nation:

> „Meine Liebe zu Euch und den Glauben an unseren lieben Gott, den Ihr mir tief ins Herz eingepflanzt habt, hat mich stets hochgehalten. Und dieser Glaube und die Hoffnung auf einen gnädigen Richter machen mir das Sterben leicht. Zudem gibt es für mich keinen schöneren Tod als den auf dem Felde der Ehre. Wie mancher läßt jetzt sein Herzblut für sein geliebtes Vaterland und ich hielt es für meine Pflicht, jetzt mit allem Können dafür einzutreten. – Also weinet nicht, denn ich habe nicht umsonst gelebt.“[1189]

Am folgenden Tag, dem 13. September 1915, fiel Josef Suwelack in Steenwerck bei Armentières und wurde nach der Mitteilung englischer Flieger mit militärischen Ehren in Erquinghem-sur-Lys begraben.

In einer Zeit, in der es keine Verbindlichkeiten mehr gab und die Sinnsuche auch zur Existenzfrage wurde, blickte der junge Josef Suwelack, dessen Karriere als Flieger nur wenige Jahre währte, auf ein dichtes und in vielerlei Hinsicht bewegtes Leben zurück. Da die genauen Todesumstände, wie in der Rezeptionsgeschichte angedeutet, nicht endgültig geklärt werden konnten, wird auf Spekulationen verzichtet und im Folgenden stattdessen die zeitgenössische Erinnerung an den westfälischen Flieger zum Thema gemacht.

1188 NL J. Suwelack, Feldpostbrief an seine Familie, Petit Rechain, 12.9.1915.
1189 NL J. Suwelack, Abschiedsbrief an seine Familie, Lille, 12.9.1915.

D. In memoriam II: Erinnerungen an Josef Suwelack

Suwelacks Tod löste Bestürzung und Anteilnahme bei seinen Freunden, Geschäftspartnern und in der Öffentlichkeit aus, seine Familie war tief getroffen. Hatte der eingangs erwähnte erste Brief Hauptmann Barths an Josefs Vater noch Hoffnungen geweckt, Josef könne unbeschadet gelandet und eventuell in Gefangenschaft geraten sein,[1] erreichte den Molkereidirektor wenige Tage später ein weiterer Brief Barths, in dem ihm dieser mitteilte, dass „die beiden braven Menschen und Soldaten nicht mehr" seien: „Neben einem kleinen Bild der beiden jungen Helden aus den letzten Tagen übergebe ich Ihnen noch den letzten Brief ihres braven Sohnes – seinem Wunsch gemäß."[2]

Der Abschiedsbrief des Piloten hatte seine Funktion, die Familie zu trösten, nicht verfehlt. Josefs Vater brachte gegenüber Richard sein Erstaunen über die Todesahnung seines Sohnes zum Ausdruck und fügte hinzu, dass ihm der Abschiedsbrief Trost spende: „Ach wie tröstet mich dieser Brief. In zahlreichen Schreiben hatte er schon immer so herzlich um Verzeihung gebeten wegen seines großen Leichtsinns. Den heutigen Brief kann man nicht lesen ohne unter Tränen."[3]

Am 3. November des Jahres folgte noch ein letzter Brief der Feldfliegerabteilung 24, der die Aussage eines englischen Gefangenen der Abteilung wiedergab, dass Suwelack und Leutnant Teichmann von einem Maschinengewehr-Schuss in der Luft getroffen worden seien. Josef habe noch „die Energie besessen ..., das Flugzeug fast glatt zu landen, dann sind beide tot im Flugzeug gefunden worden. Ein englischer General soll am Begräbnis mit teilgenommen haben."[4]

Wenngleich die Briefe, Zeitungsberichte und offiziellen Berichte zum 13. September 1915 keine „Wahrheit" verkünden konnten, war Josef Suwelack übereinstimmend den Heldentod gestorben. Er hatte in der Wahrnehmung der Zeitgenossen noch in einer scheinbar ausweglosen Situation sein fliegerisches Geschick bewiesen und die Nerven behalten. Das Mysterium seines Todes und eine Bestattung mit militärischen Ehren bildeten den Schluss- und zugleich den Ausgangspunkt für den Flieger-Mythos um Josef Suwelack.

1 Vgl. NL J. Suwelack, Brief Joseph Suwelack senior an Richard Suwelack, Billerbeck, 17.9.1915.
2 NL J. Suwelack, Brief Hauptmann Barths an Josefs Vater, Lille, 17.9.1915.
3 NL J. Suwelack, Brief Joseph Suwelack senior an Richard Suwelack, Billerbeck, 19.9.1915.
4 NL J. Suwelack, Brief Oberleutnant Kurt Röslers an Joseph Suwelack senior, o. O., 3.11.1915.

Lokalzeitungen aus dem Münsterland und dem Raum Essen beziehungsweise Gelsenkirchen erinnerten in mehrspaltigen Nachrufen an Stationen der Fliegerkarriere Josef Suwelacks und würdigten ihn als einen engagierten und passionierten Flieger, Konstrukteur und Kriegsteilnehmer, der sich ganz in den Dienst der Fliegerei und des Vaterlandes gestellt habe. Auf nationaler Ebene führte der „Flugsport" die fliegerischen Erfolge Suwelacks auf der Rumpler-Taube, dem Aviatik- und dem Kondor-Eindecker an, darunter seinen Weltrekord, seine Platzierungen auf Flugwettbewerben und seinen Flug von Essen nach England.[5] Ferner erschien in der „Kölnischen Volkszeitung" ein kurzer Artikel zum Tod des Piloten, der die Pionierarbeit und das Pflichtbewusstsein des Fliegers herausstellte:

> „Der weit über seine Heimatprovinz hinaus bekannte Flieger Joseph [sic!] Suwelack aus dem benachbarten Billerbeck, Sohn des dortigen Molkereidirektors Suwelack, hat am 13. September im schweren Luftkampfe mit englischen Fliegern den Heldentod gefunden. Suwelack, der ein Alter von nur 27 Jahren erreicht hat, war einer der ersten Pioniere der deutschen Flugkunst, versuchte sich früh in Konstruktionen und gehörte zu den ersten, die sich das Pilotenzeugnis auf der Rumpler-Taube erwarben. ... Als der Krieg ausbrach stellte sich Suwelack sofort der Militärbehörde zur Verfügung. Als Führer eines Doppeldeckers hat er manchen Flug über Feindesland gemacht, wo er nunmehr ruhm- und ehrenvoll für das Vaterland gestorben ist."[6]

Von spanischer Seite wurde an Suwelacks dortige Konstrukteurtätigkeit auf dem Flugplatz Cuatro Vientos bei der Firma Pujol Comabella y Compañia de Barcelona und an einen aufgeschlossenen jungen Flieger erinnert. Er sei ein aufrichtiger Mann gewesen, der sich stets zuerst um das Wohl des Flugzeuges und dann um das eigene gesorgt habe. Der „verdammte Krieg" („la maldita guerra") habe jedoch eine deutsch-spanische Zusammenarbeit und die Produktion zum Stillstand gebracht. Innerhalb kürzester Zeit habe Josef mit seiner offenen Art viele Freunde gefunden und seine Liebe zu Spanien entdeckt: „Möge der sympathische und grundanständige Suwelack in Frieden ruhen."[7]

Insbesondere die regional erschienenen Nachrufe zeichneten den Fliegerweg Josef Suwelacks detailliert nach und bildeten die Basis für den Mythos Suwelack, wie er bis in die Gegenwart weiterlebt. Die „Gelsenkirchener Zeitung" veröffentlichte am 23. September 1915 einen doppelseitigen aus dem Münsterland eingesandten Artikel über Josef Suwelack, der die wesentlichen Stationen des Fliegers und seine Anfänge in der Heimat nachzeichnete. Betont wurden seine westfälische Beharrlichkeit, seine Verwurzelung in der Heimat und seine Leistungen für die Fliegerei:

> „Fliegertod – Fliegerlos: Nun ist auch Suwelack auf dem Felde der Ehre gefallen, der erste Flieger des Münsterlandes, dessen Name im Münsterlande populär war, als vor einigen Jahren die Flug-

5 Vgl. „Der Tod des Fliegers Suwelack", in: Flugsport (1915), Nr. 20, S. 639f.
6 Kölnische Volkszeitung vom 23.9.1915, Rubrik „Luftfahrt".
7 Nachruf Josef Suwelack, in: Heraldo Deportivo Nr. 20 vom 5.12.1915, S. 238, online seit Mai 2009 unter: http://hemerotecadigital.bne.es/issue.vm?id=0002757794&search=&lang=en (6.10.2014). An dieser Stelle sei Rut Neuschäfer für die Übersetzung herzlich gedankt.

> kunst auch in Deutschland sich zaghaft Bahn brach … . Und als in Deutschland schon fleißig geflogen wurde, da war das Münsterland noch lange nicht der Ehre teilhaftig, einen Flieger über sich gesehen zu haben. Endlich im Frühjahr 1909 schien Münster das Glück haben zu sollen, als erste Stadt in Westfalen ein regelrechtes Schaufliegen [auf der Loddenheide] abhalten zu können. … Einer der fleißigsten Besucher des Flugschuppens, der nicht müde wurde, immer wieder zur Loddenheide herauszupilgern, war ein junger Mann aus Billerbeck, der mit heißen Wangen stundenlang wartete, um die ersten Flieger in der Luft zu sehen; er wollte nämlich auch einer werden … . Die ‚Loddenheide' und das ‚Motörken' bildeten von nun an sein Sinnen und Trachten, und als zäher Westfale ließ er von seiner Idee nicht ab, bis er dieses Ziel erreicht hatte."[8]

Hätten seine Flüge auf dem Aviatik-Eindecker „unter keinem günstigen Stern" gestanden, so habe sich Suwelack dann im Stillen an die Konstruktion seines Kondor-Eindeckers begeben, mit dem er „in der Folgezeit recht schöne Erfolge" habe vorzeigen können:

> „Längere Zeit hat damals Prinz Heinrich im Gespräch mit dem Flieger verweilt, der dem Prinzen über verschiedene der Flieger, die er hatte landen sehen, Nachricht geben konnte. Von seinen Kondorflugzeugen, die sich durch ihren eleganten hölzernen Bootskörper auszeichnen, hat Suwelack mehrere an die spanische Militärverwaltung verkauft. … Als der Krieg ausbrach, folgte Suwelack sofort der Mobilitätsbestimmung für Flieger und stellte seine Kraft in den Dienst des Vaterlandes. … Ob er über Paris gewesen ist, wissen wir nicht, aber daß [sic!] wissen wir: wenn dereinst die Geschichte des Krieges geschrieben wird und dabei die Einzelheiten des Luftkampfes enthüllt werden, die jetzt noch verschwiegen bleiben müssen, dann wird auch der Name Suwelack in Ehren genannt werden."[9]

Dem Grundsatz, über Tote nur Gutes zu schreiben, folgten auch alle anderen Nachrufe. Suwelack blieb als erfolgreicher Flieger und eifriger Konstrukteur mit westfälischen Wurzeln in Erinnerung.

Seine Familie ließ einen Totenzettel mit einem Kriegsporträt Josefs bei der Billerbecker Buchdruckerei Knüppel anfertigen, die auch den „Billerbecker Anzeiger" herausgab. Als katholischer Brauch fanden Totenzettel im Münsterland erst um 1900 größere Verbreitung, im Rheinland hatte die Entwicklung bereits Mitte des 19. Jahrhunderts eingesetzt.[10] Während der Zeit des Ersten Weltkrieges erfuhren Totenzettel dann eine qualitative Veränderung, denn während in Friedenszeiten aufgrund hoher Kosten Fotografien nur bei der Geistlichkeit oder hochrangigen Persönlichkeiten Verwendung fanden, wurden jetzt auch vermehrt großformatige Bilder der Gefallenen mit gedruckt.[11]

Als „letzte[r] Besuch der Toten"[12] waren Totenzettel eine Möglichkeit, Abschied zu nehmen und für das Seelenheil des Verstorbenen zu beten. Der Soldat, weit

8 „Zum Heldentod Suwelacks", in: Gelsenkirchener Zeitung vom 23.9.1915, 2 Seiten, hier S.1.

9 Ebd., S.2.

10 Vgl. Monika Schulte, Die Totenzettel aus dem Ersten und Zweiten Weltkrieg. Soziale Bedeutung für das Image der Männlichkeit, o. O. 2011, S.7f.

11 Vgl. ebd., S.4ff.; Beate Sturm, „Heldentod fürs Vaterland". Totenzettel als Quelle für Einzelschicksale aus der Zeit des Ersten Weltkriegs, in: Jörg Engelbrecht (†)/Simone Frank/Christian Krumm/Holger Schmenk (Hg.), Rhein-Maas. Geschichte, Sprache und Kultur, Bd. 5: Der Erste Weltkrieg, Oberhausen 2014, S.62–75, hier S.66.

12 Schulte, Die Totenzettel, S.14

entfernt gefallen in einem fremden Land, erhielt ein Gesicht und seine körperliche und mentale Unversehrtheit zurück: „Der Schrecken des Krieges ist auf den Fotos ... nicht präsent – vielmehr sehen wir den Soldaten ... wie er in den Krieg ging und wie seine Angehörigen ihn in Erinnerung behalten wollten." Die Totenzettel wurden vorrangig in katholisch geprägten Gegenden bei der Totenmesse in der Kirche oder auf dem Friedhof verteilt. Sie gaben Auskunft über bedeutende Lebensstationen des Verstorbenen, über dessen Familie, Angehörige, Todesursache und Mitgliedschaften in caritativen, politischen, christlichen oder Sportvereinen. Ferner erfuhren die Trauernden etwas über die Ausbildung und die Karriere des Toten. Ganz im Sinne des christlichen Wertekanons wurden die Verstorbenen als gute Christen und brave Söhne beziehungsweise treu sorgende Ehemänner beschrieben.[13]

Der ausgewählte Bibelspruch auf der Vorderseite des Totenzettels von Josef Suwelack aus dem zweiten Buch der Makkabäer „Sie waren bereit, für Gesetz und Vaterland zu sterben" (2 Makk 8, 21) stellte eine Verbindung zu Kaiser und Reich her. Der Spruch war in die Abbildung eine Engels eingefügt, der einen sterbenden Soldaten stützt, ihm mit dem linken Zeigefinger den Weg in den Himmel weist und im Sterben Beistand leistet. Diese Darstellung spendete Trost und diente in Verbindung mit dem abgebildeten Lorbeerblatt der Erinnerung des tapferen Soldaten als Opferheld für die Nation. Eine knappe Skizzierung der fliegerischen Etappen Suwelacks im Innenteil des Totenzettels stellte seinen Weltrekord heraus, präsentierte ihn als pflichtbewussten Deutschen und frommen Katholiken, der noch am Tag vor seinem Tod „die hl. Sakramente empfangen und mit Ergebenheit in Gottes hl. Willen seine Pflicht" erfüllt habe.[14]

Der familiäre Nachruf lobte den „außergewöhnlichen Tatendrang" und sein Engagement als Konstrukteur und Kriegsfreiwilliger. Die gewählte Fotografie zeigte Josef Suwelack mit Eisernem Kreuz, Albrechtsorden II. Klasse und Flugzeugführerabzeichen auf der Höhe seiner Kraft, gesund, ruhig, stolz und erfolgreich, so, wie er künftigen Generationen im Gedächtnis bleiben sollte (Abb. 28).

Josef Suwelack hat westfälische Geschichte geschrieben. Er ließ als Billerbecker Flugpionier und späterer Unternehmer seine Landsleute an seinem beschleunigten und modernen Leben teilhaben und lebte im regional-kollektiven Gedächtnis fort:

> „Zu diesen heldenhaften Fliegern gesellt sich nun auch der bekannte Billerbecker Joseph [sic!] Suwelack, der nicht nur im Münsterland und in der Heimatprovinz Westfalen wie im Rheinland sich außerordentlicher Beliebtheit erfreute, sondern auch in Luftfahrtkreisen allgemein Ansehen genoß. ... Mit Joseph Suwelack verliert der deutsche Flugsport einen seiner besten Vertreter. Wenn er auch nur ein Alter von 27 Jahren erreichte, so hat er doch in seinen jungen Jahren auf dem

13 Sturm, Totenzettel als Quelle, S. 63, 68, Zitat ebd.
14 NL J. Suwelack, Totenzettel Josef Suwelacks, September 1915.

> neuen, für uns Deutsche besonders wichtigen Gebiete hervorragendes geleistet, was dauernd in der Geschichte der Flugzeugkunst verzeichnet bleiben wird.“[15]

Mit seiner Teilnahme an Flugwettbewerben und seinen selbst konstruierten Flugzeugen war es ihm gelungen, Westfalen über seine Landesgrenzen hinaus bekannt zu machen und seine Landsleute mit der Flugtechnik und der Moderne zu versöhnen.

Der Aufbruch in die Moderne wurde für Josef Suwelack zu einem Leitgedanken seines Lebens, und ein hohes Maß an Anpassungsfähigkeit, Risiko und vorausschauendem Denken war der bestimmende Faktor in seinem Streben danach, sich eine dauerhafte Stellung in der Fliegerei zu sichern. In einer Zeit und in einer Branche, in der fast nichts von Dauer war, musste er sich immer wieder neu erfinden und mit der Zeit gehen. Der Kontingenzerfahrung der Fliegerei und des Ersten Weltkrieges begegnete er mit einem hedonistischen Leben im Augenblick für die und mit der Fliegerei. In Erinnerung blieb ein Mensch, der sich ausprobierte, Neues wagte, sich der Technik und seiner Zeit aufgeschlossen zeigte – ein passionierter westfälischer Flieger und Lebenskünstler.

Zur frommen Erinnerung
an den Kriegsfreiwilligen Flugzeugführer
Josef Suwelack
Feldflieger Abt. 24. Inhaber des eisernen Kreuzes und Ritter des Albrechtsordens II. Klasse mit Schwertern
geboren zu Billerbeck am 30. April 1888.

Von einem außergewöhnlichen Tatendrang beseelt, widmete er sich seit dem Jahre 1909 dem Flugwesen und baute selbst Flugapparate. Nachdem er in Berlin sein Examen als Flugzeugführer gemacht, stellte er mit einer damals noch unvollkommenen Maschine einen Weltrekord auf und erhielt bei zahlreichen Konkurrenzflügen viele Prämien und Geschenke. Er gründete dann in Essen die Gesellschaft „Kondorwerke“, deren Direktor er wurde. Gleich am ersten Mobilmachungstage stellte er sich freiwillig dem Heere infolge eines mit dem Reiche abgeschlossenen Vertrages. Nachdem er 13 Monate im Felde und fast täglich Flüge über die feindlichen Stellungen gemacht, wurde ihm sein Flug über die feindliche Linie am 13. September verhängnisvoll. Englische Flieger berichteten bei seiner Abteilung, daß er mit seinem Beobachter nach einem Kampfe mit ihren Maschinen gestürzt und in Steenwerk in Frankreich mit allen militärischen Ehren begraben sei.

Die große Gefahr bei den in letzter Zeit vermehrten harten Luftkämpfen erkennend, schrieb er am Vorabend seines Todes einen rührenden Abschiedsbrief und übergab ihn seinem Hauptmann zur Beförderung an seine Eltern; er teilte darin mit, daß er am selben Tage die hl. Sakramente empfangen und mit Ergebenheit in Gottes hl. Willen seine Pflicht erfüllen und für das Vaterland sterben würde.

Abb. 28: Innenseite des Totenzettels für Josef Suwelack, Billerbeck, September 1915 (Foto: Familienarchiv Walter Suwelack, NL Josef Suwelack)

15 „Luftfahrt. Josef Suwelack †“, in: Westfälischer Merkur vom 22.9.1915.

E. Schlussbetrachtung

Stünde man am Ende dieser Arbeit vor der Aufgabe, ein Plädoyer über Josef Suwelack und seine Zeitgenossen zu verfassen, wie sähe es aus? Die Fakten liegen längst nicht alle vor und die Beweisführung kann definitiv nicht als abgeschlossen gelten. Ferner gilt es, die „Komplizen" in die Bewertung mit einzubeziehen, da Suwelack nicht alleine für sein Handeln verantwortlich war, sondern durch äußere und personelle Strukturen beeinflusst und geformt wurde. Ohne seine Familie, seine Freunde, Geschäftspartner und das Militär, die Entwicklung der motorisierten Fliegerei, der Wettbewerbsausprägungen und der (militärischen) Richtlinien hätte sein Leben sehr wahrscheinlich einen anderen Verlauf genommen.

Eingebunden in ein privates beziehungsweise geschäftliches Feld auf der einen und ein medial-öffentliches Feld auf der anderen Seite verlief sein Leben weitgehend zwischen diesen beiden Polen der Selbst- und der Fremdwahrnehmung. Für die Öffentlichkeit und auch den Historiker greifbar wurde Josef Suwelack erst als junger Mann Anfang Zwanzig, als er mit seinen ersten heimatlichen Flugversuchen auf sich und seine Apparate aufmerksam machte. Seine Kindheit und Jugend ließen sich nur indirekt über den Kontakt zur Familie Suwelack und die Stadt Billerbeck rekonstruieren.

Als Sohn eines wohlhabenden und innovativen Molkereidirektors mit einem florierenden Unternehmen, der seinen Sohn in jeglicher Hinsicht förderte und die eigenen Talente und Neigungen herausfinden ließ, stand der junge Westfale am Ende seiner Schulzeit vor der für die Zeit eher untypischen Möglichkeit, eine freie Berufswahl treffen zu können. Dass Josef Suwelack nicht den naheliegenden Berufsweg in der Molkerei seines Vaters einschlug, sondern sich der Flugtechnik zuwandte, lässt zum einen den Rückschluss auf einen liberalen Erziehungsstil zu. Zum anderen zeigt die Hinwendung Suwelacks zur Fliegerei, dass die moderne Technik Einzug in alle Lebensbereiche hielt und auch in agrarisch geprägten Städten Aufmerksamkeit erfuhr. Der Westfale zählte zu einer neuen Generation von technikbegeisterten jungen Männern, die sich aus dem Wirtschaftsbürgertum rekrutierte, nach dem Trial-and-Error-Prinzip flug(un)fähige Apparate baute und hoffte, Geld, Ruhm und soziale Anerkennung zu erlangen.

Die literarische Phantastik, aber auch das Zeitungs- und Zeitschriftenwesen beflügelte die Vorstellungen der jungen Bastler, die in einer Gesellschaft der Auf- und Umbrüche aufwuchsen. Technik war explizit männlich konnotiert und so wurde auch die Fliegerei zu einem Feld der Männlichkeit. Obschon die ersten

Fliegerinnen als Sensation gefeiert wurden, dominierte die Vorstellung, dass zum Fliegen vermeintlich typisch männliche Eigenschaften und Charakterzüge wie Mut, Nervenstärke, Tapferkeit und physische Stärke gehörten. Fliegerinnen wie Melli Beese versuchten, sich diesem männlichen Fliegerhabitus anzupassen, mussten jedoch vor den gesellschaftlich-militärischen Zwängen kapitulieren und wurden systematisch aus der Wettbewerbsfliegerei hinausgedrängt.

Beschleunigung als Signum der Zeit Josef Suwelacks führte zu einem Nebeneinander von Optimismus und kulturellem Ideenaustausch auf der einen und Erschütterungen des Weltbildes und Unsicherheit auf der anderen Seite. Das Flugzeug wurde zu einer Chiffre für diese neue Zeit- und Raumerfahrung, denn es symbolisierte Fortschrittlichkeit, nationale Größe, Grenzenlosigkeit und individuelle Freiheit. Zugleich war es Ausdruck eines Technisierungsschubes und eines Kräftemessens der Nationen.

Als gut situierter Direktoren-Sohn in einer komfortablen, offenen Ausgangssituation war es möglicherweise Suwelacks Wunsch, gleich zu Beginn der noch unausgereiften Flugtechnik, sich selbst als Flieger zu versuchen und Teil dieser neuen Entwicklung zu werden. Konstruktionspläne waren für wenig Geld zu erstehen, von neuen Bestmarken berichteten die Zeitungen und die Preisgelder auf Flugwettbewerben waren verlockend. Sicherlich sollte auch der Aspekt der Pionierleistung nicht unterschätzt werden. Die ersten Flieger bewiesen Mut, sich in Eigenkonstruktionen vor Publikum zu präsentieren, sie nahmen Rückschläge und Spott in Kauf, in der Hoffnung, sich zu beweisen und von der Masse abzuheben. Die Aussicht auf Geldgewinne und weiteres symbolisches Kapital suggerierte die Möglichkeit eines raschen gesellschaftlichen Aufstiegs und schuf ein anderes, neuartiges Lebensgefühl.

Der Flugplatz in Berlin-Johannisthal wurde zum Versammlungsort der deutschen Fliegerszene und des gegenseitigen Erfahrungsaustausches. Bedeutende dort ansässige Unternehmen wie die Rumpler-Werke, die Albatros-Werke, die Luftverkehrsgesellschaft oder die Flugzeugbauanstalt Fokker entwickelten sich in den folgenden Jahren zu nationalen Marktführern. Der Wechsel Josef Suwelacks von der Provinz Westfalen nach Berlin wurde zum eigentlichen Beginn seiner Fliegerkarriere. Der Erwerb des Pilotenführerzeugnises im Jahr 1910 verschaffte ihm Zugang zu Flugwettbewerben, also zu den technischen Experimentierfeldern, auf denen unter reglementierten Voraussetzungen die neuesten Flugzeuge unter Praxisbedingungen getestet wurden und sich die Piloten öffentlich zur Schau stellten. Der Pilotenschein wies ihn zugleich als Experten aus und verlieh ihm eine öffentliche Fliegeridentität. Von jetzt an wurden seine Erfolge und Fehlschläge nicht nur in seiner Heimat verfolgt, sondern spätestens mit seinem Weltrekord im Dezember 1911 auch (inter-)national registriert.

Eine regelmäßige Berichterstattung über die jüngsten Flugvorstellungen Suwelacks, hier vor allem die regionale Berichterstattung in Billerbeck und Münster, stilisierte ihn zu einem Lokalhelden, dessen Leistungen und Bekanntheit positiv auf die Region zurückwirkten. Als um den Jahreswechsel 1912 ein eigener Flugtag in Billerbeck angeregt wurde, war fast die ganze Stadt an der Organisation der Veranstaltung beteiligt und bereitete „ihrem" Flieger einen herzlichen Empfang. Die kleinstädtische Flugbegeisterung ergriff Jung und Alt und das eigens zum Flugtag gestaltete Plakat, das Josef Suwelack als Hauptattraktion abbildete, symbolisierte die Offenheit der Bevölkerung für die Herausforderungen der Zukunft. Als Personifikation der (lokalen) Moderne fungierte der Flieger als Vermittler zwischen Altem und Neuem, als Bindeglied zwischen Tradition und Moderne. Der Vortrag Suwelacks „Aus meinem Fliegerleben", das Interview im Anschluss an die Umrundung des Billerbecker Ludgerus-Doms sowie sein 1913 erschienener Erlebnisbericht „Mein Flug zur Nordsee" stehen beispielhaft für die mediale Verbreitung und Aneignung der Flugtechnik und ihrer Akteure. Die Flieger gestalteten ihr öffentliches Profil aktiv mit, sie befriedigten das Leser- und Augenzeugen-Interesse nach authentischen, selbst erzählten romantisch-verklärten Erlebnis- und Abenteuerberichten. Die Konstruktion des Bildes vom unerschrockenen, aktiv zupackenden und nervenstarken Fliegerhelden kann als Versuch gewertet werden, die Flugtechnik als ein der individuellen Kontrolle der Piloten unterworfenes Artefakt zu präsentieren. Unfälle und Todesstürze wurden über den Rückgriff auf den Ikarus-Mythos zum nationalen Opfer für die „große Sache" mit Sinn aufgeladen und in den Fortschritts-Diskurs eingepasst. Eine hochkomplexe Technikerfahrung, die das Verständnis der meisten Menschen überstieg, sie verstörte und zugleich faszinierte, konnte auf diese Weise in ihre Lebenswirklichkeit integriert werden.

Josef Suwelack bewahrte sich Zeit seines – in der Tat sehr kurzen – Lebens eine positive Einstellung zur Technik. Möglicherweise inspiriert durch den unternehmerischen Erfolg des Vaters ließ er neue Impulse und Innovationen auf sich wirken und sie in sein Handeln einfließen. Der steigende Bekanntheitsgrad und die Erfolge auf Flugwettbewerben wirkten sich auf seine Selbstwahrnehmung aus, steigerten sein Selbstbewusstsein und seine Karrieremöglichkeiten. Dies führte aber auch zu der Notwendigkeit, sich immer wieder aufs Neue beweisen und profilieren zu müssen. Mit der Zeit zu gehen, wurde in der Fliegerbranche überlebenswichtig. Nur wer zum passenden Zeitpunkt in Anwesenheit der „richtigen" Leute seine Konstruktion vorstellte, Verhandlungsgeschick bewies und bereit war, in weitere Verbesserungen zu investieren, konnte sich mittel- bis langfristig auf dem Markt behaupten. Wer Flugzeuge erfolgreich verkaufen und sich einen Namen machen wollte, musste nicht nur über erhebliches ökonomisches Kapital verfügen, sondern vor allem auch sich selbst effektiv und glaubhaft vermarkten. Dazu zählte, in der Öffentlichkeit anlässlich von Wettbewerben in Erscheinung zu treten, sich durch besonders eindrucksvolle oder ausdauernde Flüge hervorzutun, mit dem Ziel, Aufträge zu erhalten, eine wohlwollende Pressekritik zu erreichen und im Gedächtnis zu bleiben. Eine Karriere als Flieger kam einem Streben nach Glück gleich und wurde gleichermaßen von persönlichen Kontakten, gesellschaftlichem Einfluss und Ansehen wie von Zufällen beeinflusst.

Parallel zu den ersten Ansätzen einer Flugzeugindustrie bildeten sich neue Organisations- und Interessenverbände heraus, eine Luftfahrtlobby, die Einfluss auf die Wettbewerbsausrichtung und die technischen Vorgaben nahm. Flugsportvereine und Fliegerverbände vertraten die Interessen einzelner Städte und Flieger, richteten Flugtage aus und luden zu Vorträgen und Versammlungen ein, so dass ein reger Austausch untereinander erfolgen konnte. Eine Vereinsmitgliedschaft war zum Teil exklusiv und kostenpflichtig. In den Flugvereinen saßen unter anderem Großindustrielle, hochrangige Militärs und Stadtoberhäupter, die wirtschaftliche, kommunale und flugsportliche Interessen vertraten und Einfluss auf die Luftfahrtentwicklung nahmen. Sich als Flieger einem solchen Verein anzuschließen bedeutete, sich aktiv am Vereinsleben zu beteiligen und die Verbindung zur Vereinsführung zu suchen. Hier wurden neue Kontakte geknüpft, es wurde über Leitlinien der Fliegerei verhandelt und über anzustrebende Entwicklungen diskutiert.

Die Flugzeugunternehmen und ihre Akteure lebten von Empfehlungen und ihrer durch die Medien verbreiteten Reputation. Die Entscheidung, eine Maschine für einsatzfähig und solide zu befinden, basierte letztlich auch auf dem Urteil der Prüfungskommission, die mit den Flugzeugfirmen in schriftlichem oder mündlichem Austausch stand. Das Erteilen einer Teilnahmegenehmigung oder der Ankauf einer Maschine hingen stark von persönlichen Beziehungen ab. Je besser der Kontakt zur Militärverwaltung oder dem Vorsitz der Flugvereine war, desto eher bestand die Aussicht auf Erfolg. Zwischen den Flugzeugunternehmen selbst herrschte ein harter Wettbewerb, ein darwinistischer Kampf ums Dasein, um Aufträge und Fürsprecher bei der Militärverwaltung. Ferner war das Flugzeuggeschäft eine Frage des gegenseitigen Vertrauens: in die Konstrukteure, die Gesellschafter, die (Test-)Piloten und in die Militärführung. Der Name Josef Suwelack, seit 1912/13 dann auch der Name „Kondor“, stand stellvertretend als Marke und Gütesiegel für die Firma und den Flieger, aber auch für die Städte Essen und Gelsenkirchen, mit denen die Kondor-Werke in Verhandlungen traten und in denen sie sich vor Ort niederließen.

Als Flieger und Verantwortlicher für die Umsetzung seiner Konstruktionspläne vertrat Josef Suwelack das Essener Unternehmen auf Wettbewerben und bei Abnahmeflügen der Militärverwaltung nach außen. Intern beschränkte sich sein Mitspracherecht hingegen auf Konstruktionsfragen, das eigentliche Geldgeschäft tangierte ihn eher weniger. Sicherlich war er als technischer Angestellter daran interessiert, ein gutes Gehalt zu bekommen und hohe Umsätze zu erwirtschaften. Allerdings wurde es immer wieder deutlich, dass Suwelack in erster Linie aus Liebe zur Fliegerei flog und die Aufmerksamkeit sowie die Perspektive aus der Luft genoss. Da er in einer wohlhabenden Familie primär sozialisiert worden war, ist es aus heutiger Sicht dennoch nachvollziehbar, dass er diesen Lebensstandard erhalten und sich, wenn möglich, sozial noch weiter verbessern wollte. Bourdieus These der Annahme des eigenen Schicksals (*amor fati*) und der Herkunft bestätigte sich bei Suwelack. Er legte Wert auf gutes Essen, leistete sich hochwertige Kleidung und Luxusgüter. Darüber hinaus präsentierte er sich als erfolgreicher Unternehmer und suchte den Kontakt zur gesellschaftlichen Elite, zu Industriellen, dem Militär und dem Kaiserhaus.

Treffen mit dem *Who-is-Who* der Fliegerei waren eine wesentliche Voraussetzung für Erfolg, aber längst keine Garantie. Verlangt wurden persönlicher Einsatz, Ausdauer und Geduld, öffentliche Präsenz, Flexibilität, Verfügbarkeit und Aktualität. Insbesondere der letzte Punkt war angesichts des raschen technischen Wandels ein entscheidendes Erfolgskriterium. Im Konkurrenzkampf mit dem In- und Ausland musste es einer Flugzeugfirma gelingen, das Militär zeitnah von ihren Apparaten und dem technischen Know-how zu überzeugen. Gelang dies nicht, galten die Maschinen bereits wieder als überholt, was zu Absatzschwierigkeiten, Personalabbau und Resignation führte. Die Erschließung neuer Absatzmärkte, wie der Kontakt der Kondor-Werke nach Spanien, war eine mögliche Option, finanzielle Engpässe zu überbrücken oder sogar zu überwinden. Angesichts der zwischenzeitlichen Absatzschwierigkeiten der Kondor-Werke ist es bemerkenswert, dass das Unternehmen sich auf dem Gebiet des Flugzeugbaus noch bis Anfang der 1920er Jahre betätigte, obwohl dort längst andere *big player* herrschten.

Josef Suwelack erschloss sich immer wieder neue Betätigungsfelder und war flexibel, beizeiten aber auch sprunghaft und orientierungslos. Der 1912 mit dem Deutschen Reich geschlossene privatrechtliche Dienstvertrag erscheint als eine seiner wenigen vorausschauenden Entscheidungen. Größtenteils lebte er jedoch im Augenblick, als Flaneur, der gleichermaßen mit der Zeit ging, aber auch, das zeigt seine Zeit bei den Kondor-Werken, zum falschen oder zu spätem Zeitpunkt in den Flugzeugbau investierte.

Der Erste Weltkrieg konfrontierte ihn mit Erfahrungen, die er in dieser Form nicht erwartet hatte und die einen Bruch in seiner Biografie bedeuteten. Dem Krieg sowie dem Schwinden von Vergangenheit und Zukunft begegnete er mit einer Flucht in Vertrautes. Die Briefe an seine Familie und die sich wiederholenden Fotomotive wurden zu einem Ritual, das kurzfristig Halt und Orientierung bot, dem Krieg einen Sinn verlieh. In seinen Darstellungen knüpfte Suwelack an die Vorkriegszeit an und verharmloste seinen Kriegseinsatz über den Rückgriff auf Sportvokabular und die Definition der Fliegerei als Arbeit beziehungsweise als Spiel.

Die Fliegerabteilungen waren eine der wenigen Möglichkeiten, sich individuell auszuzeichnen, sich aus der Distanz – in der Luft oder dem Hinterland der Front – in der Rolle des Regisseurs zu wähnen und Ruhe zu finden. Das Ideal des tapferen Einzelkämpfers in einem fairen, ehrenhaften Zweikampf und das nach außen gezeigte Luxus- und Lagerleben standen im Kontrast zum anonymen Massensterben und erlaubten eine Identifikation mit den Helden der Vorkriegszeit. Die Kluft zwischen Kriegserwartung und -erfahrung versuchten die Flieger zu überbrücken, indem sie den Krieg rationalisier- und messbar machten: Erfolge ließen sich in der Luft konkret benennen, Bombenabwürfe und die geleistete „Arbeit" in Zahlen ausdrücken, Auszeichnungen vorweisen und Überlegenheit demonstrieren. Sogar der Fliegertod konnte noch als ruhmreich und ehrenhaft gedeutet werden, denn in der zeitgenössischen Vorstellung kämpften hier ebenbürtige Männer gegeneinander, die bis zuletzt ihre Pflicht erfüllten und deren Lebensende kein anderes sein konnte, als im Dienst der Fliegerei zu sterben. Und während für „normale" Soldaten mit dem Tod häufig ein Vergessensprozess einsetzte, blieb die Erinnerung an die Flieger oftmals im kollektiven Gedächtnis bestehen. Der Mythos um die Person Josef Suwelacks, der bereits zu seinen Lebzeiten in Ansätzen ausgebildet war, erlebte nach seinem Tod eine Renaissance. Suwelack war – und ist es in Westfalen noch immer – ein Vorbild, das für politische und ideologische Zwecke vereinnahmt wurde und wird.

Abschließend soll der Versuch unternommen werden, die Frage nach der Genese und Tradierung dieses Mythos zu beantworten, also die Frage danach, weshalb Josef Suwelack zu einem Mythos wurde. Die Ergebnisse erheben keinen Anspruch auf Vollständigkeit und Allgemeingültigkeit, vielmehr greifen sie Merkmale für die Ausbildung von Mythen auf und verstehen sich somit als Grundlage einer Typologie, die durch weitere Einzelfall- oder Gruppenstudien zu anderen Fliegern zu überprüfen wäre.

Legitimation des „Lebensweges": Bedeutsam für die Entstehung eines Mythos erscheint das Ausmachen eines retrospektiv als einschneidend beurteilten (Schlüssel-)Erlebnisses oder einer besonderen stereotypen Neigung der Person. Ein Mythos braucht einen Anfang, ein Fundament, auf dem alle weiteren Erlebnisse und Erfahrungen aufbauen und auf das sie zurückgeführt werden können. Im Fall von Josef Suwelack war es das von ihm selbst verbreitete Narrativ vom „Erlebnis auf dem Tennisplatz", das Gespräch über die Gebrüder Wright, das ihn bewogen habe, Flieger zu werden. Unentschlossenheit, Zweifel und Orientierungslosigkeit als Formen der Kontingenz bleiben hier ausgeblendet. Sie werden ersetzt durch die Darstellung von Zielstrebigkeit und stereotypen Charakterzügen wie dem Topos der westfälischen Beharrlichkeit sowie eine in die Zukunft gerichtete Erzählung. Der Eigenlogik des Mythos ordnen sich alle weiteren Lebensstationen, die Mytheme, unter und alles, was die aktuelle Vorbildfunktion stören könnte, wird weggelassen. Es sind insbesondere solche Elemente, die sich über die Zeit erhalten, die einfache und plausible Erklärungsmuster für komplexe Erfahrungen bereithalten und Lösungen in Krisenzeiten versprechen. Fliegergeschichten erfüllten diese Bedingungen in besonderem Maße, denn die Piloten hoben sich durch ihre Tätigkeit von der Masse ab und führten ein freies, abenteuerliches Leben, das mit hohem Sozialprestige und Erfolg verbunden war.

Das Narrativ der Erfolgsgeschichte: Eine Voraussetzung für die Entstehung des Personen-Mythos war der Bekanntheitsgrad Josef Suwelacks in seiner Heimatstadt Billerbeck. Die Lokalpresse begleitete den Piloten von Beginn seiner Fliegertätigkeit an und informierte ausführlich und regelmäßig über den Werdegang des bekannten Direktoren-Sohnes. Unter öffentlicher Beobachtung inszenierte sich Suwelack frühzeitig als beharrlicher Konstrukteur, der seine Bau- und Flugversuche bildlich dokumentierte. Stolz präsentierte er sich in Billerbeck mit eigenem Briefbogen und suchte den Anschluss an die deutsche Fliegerszene in Berlin-Johannisthal. An der Gestaltung seines öffentlichen Profils wirkte Suwelack durch Schauflüge, Interviews und Berichte für Flugzeitschriften und Zeitungen mit.

Das leistungsorientierte Narrativ der Erfolgsgeschichte dominierte den zeitgenössischen Fliegerdiskurs, denn sich auszuzeichnen und aufzufallen war notwendig, um im Gespräch zu bleiben. Angesichts der Kurzlebigkeit von Rekorden und technischen Verbesserungen war der Übergang von Ruhm und Bedeutungslosigkeit fließend. Erfolg wurde deshalb zu einem dehnbaren Begriff, der von verschiedenen Personengruppen unterschiedlich ausgelegt werden konnte, zu einer Frage der Wahrnehmung und Perspektive. Suwelacks Erfolg ließ sich letztlich einerseits über seine sichtbaren Leistungen – Weltrekord im Passagierflug, Bau des ersten Kondor-Eindeckers, erhaltene Geldprämien und (militärische) Auszeichnungen – definieren und andererseits über seine unternehmerische Ausdauer und sein Medienecho.

Die *mythmaker* als Motor des Mythos: Ohne Interessengruppen, die von der Verbreitung eines Mythos profitieren und an dessen Gestaltung mitwirken, verliert der Mythos an Bedeutung. Nicht ohne Grund ist der Name Josef Suwelack mit Westfalen und mit Billerbeck im Speziellen verbunden. Die Kleinstadt erinnert anlässlich von Jubiläen – z. B. Suwelacks Geburtstag oder die Jahrestage des Weltrekords, der Billerbecker Flugtage etc. – an „ihren“ Flieger und macht ihn zu einem Teil ihrer Identität. Wichtigster Träger des Suwelack-Mythos ist die Familie Suwelack, die aktive Erinnerungspflege betreibt und es sich zur Aufgabe gemacht hat, ihrem Vorfahren ein bleibendes Denkmal zu setzen. Basierend auf der Familiengeschichte, spätestens mit der Veröffentlichung einer Biografie über Josef Suwelack im Jahr 1988, bildete sich ein fester Kanon heraus, auf den sich zahlreiche weitere Darstellungen berufen. Der Nachlass des Piloten und die in seine Zeit zurückreichende Berichterstattung lieferten den Stoff für seinen Mythos, der vielfach an ältere Erzähltraditionen wie den Ikarus-Mythos oder romantische Narrationen anschloss.

„Das Bedürfnis nach Helden“: Jede Zeit braucht ihre(n) Helden, um sich zu orientieren, über Wert- und Moralvorstellungen zu verhandeln und das politische, gesellschaftliche oder individuelle Handeln zu legitimieren. „Authentische“, gegenwartsbezogene Lebensgeschichten, die Parallelen zur eigenen Lebenswelt aufweisen, haben das Potenzial zum Mythos zu werden, vor allem dann, wenn ausreichend Stationen und (Helden-)Taten vorhanden sind, die einen Kanon ergeben. Vergleicht man heutige Fliegerdarstellungen miteinander, so zeigt sich, dass sie stets auf die gleichen Topoi und Deutungsmuster rekurrieren. Erzählt wird von leidenschaftlichen Pionieren, die ihr Leben der Fliegerei gewidmet hätten. Ein Treffen, ein Traum, eine Vision etc. machte sie über Nacht zu entschlossenen Fliegern. Als Draufgänger ließen sie sich nicht von ihrem Weg abbringen, stellten Rekorde auf und wurden nur kurzzeitig durch externe Strukturen gebremst, um anschließend (unbeirrt) ihren Weg fortzusetzen. Sie vollbrachten Außergewöhnliches, zeichneten sich durch ihr Handeln aus und riskierten ihr Leben für die Fliegerei. Als Technikbeherrscher hatten sie ihr und das Leben der Passagiere in der Hand und wurden an ihren Leistungen gemessen. Flieger galten und gelten als körperlich und mental belastbare Menschen. Nach wie vor ist die Passagierfliegerei eine Männerdomäne – die Frauenquote lag in Deutschland 2013 bei 6 Prozent. Dass Männer „die Geschichte machen“, lässt sich auch für die frühe Fliegerei konstatieren. Von den wenigen Fliegerinnen abgesehen, waren es Namen wie Otto Lilienthal, die Gebrüder Wright, Max Immelmann oder Manfred von Richthofen, die die öffentliche Vorstellung über die frühe Fliegerei bis heute geprägt haben.[1]

Josef Suwelack konnte zum Mythos werden, weil er privat und öffentlich schriftliche und bildliche Spuren seines Wirkens hinterlassen hat, die bereits zu seinen Lebzeiten Eingang in Zeitungs- und Zeitschriftenartikel fanden. Das gesellschaftliche Interesse an seiner Person und die Faszination für die „Helden der Lüfte“ machten Suwelack zu einer Ikone des Fortschritts. Es war ein zentrales Anliegen dieser Arbeit, den Leserinnen und Lesern einen alternativen Vorschlag für den Blick auf die ersten Piloten und ihre Geschichten zu machen und für die Mechanismen der Flieger(helden-)konstruktion zu sensibilisieren. So ist diese Biographie Suwelacks eine Art Meta-Mythos – ein neuer Teil seiner Biographie aus der Perspektive des 21. Jahrhunderts; zugleich bietet sie eine Rückschau auf die Genese der fortdauernden Narrative, die seine Person begleiten.

Hinter dem medial konstruierten selbstinszenierten Flieger verbarg sich ein Mensch, der mit der Zeit ging, der Pläne, aber keinen Lebensplan hatte, dem es

1 Siehe zu dem Bedürfnis nach Helden z. B. Tepe, Entwurf einer Theorie des politischen Mythos; Albrecht von Lucke, Neue Helden braucht das Land, in: Frankfurter Hefte 4 (2008), S. 64f.; Ralf von den Hoff u. a., Helden – Heroisierung – Heroismen. Transformationen und Konjunkturen von der Antike bis zur Moderne. Konzeptionelle Ausgangspunkte des Sonderforschungsbereichs 948, in: Ulrich Bröckling/Barbara Korte/Birgit Studt (Hg.), helden. heroes. héros. E-Journal zu Kulturen des Heroischen, Bd. 1.1 (2013), S. 9–14, bes. S. 9f., online unter: https://freidok.uni-freiburg.de/fedora/objects/freidok:10290/datastreams/FILE1/content (26.2.2018). Vgl. zu den Fliegerinnen: http://www.spiegel.de/karriere/berufsleben/flug-jubilaeum-die-ersten-pilotinnen-der-lufthansa-a-922185.html, online seit dem 16.9.2013 (17.2.2015).

jedoch immer wieder gelang, sich neu zu erfinden, zu improvisieren und sein Gesicht zu wahren. Suwelacks Sprunghaftigkeit erscheint deshalb beinahe nachvollziehbar: Er lebte in einer bewegten Zeit voller Auf- und Umbrüche, seine Stimmungen schwankten zwischen Hoffnung und Resignation. In der Luft konnte er sich frei und selbstbestimmt fühlen, am Boden sah er sich mit vielen Problemen konfrontiert, für deren Lösung er seine Freunde und Familie einspannte. Seine Förderer, allen voran sein Vater, seine Freunde bei den Kondor-Werken und Fürsprecher bei der Militärverwaltung ermöglichten ihm sein kostspieliges Fliegerleben. Es wäre jedoch zu kurz gegriffen, seine Erfolge oder Misserfolge ausschließlich anderen zuzuschreiben.

Josef Suwelack steht beispielhaft für einen Menschen des beginnenden 20. Jahrhunderts, der die Initiative ergriff, nach Möglichkeiten der Selbstverwirklichung suchte und in der Fliegerei eine Aufgabe fand, die ihn zum Gegenstand des öffentlichen Interesses machte. In einer kontingenten Zeit, die in der Zäsurerfahrung des Ersten Weltkriegs kulminierte, wurde die Fliegerei zu seinem Anker und einer Orientierung seines Lebens.

Fliegerleben scheinen prädestiniert zu sein, als heldenhafte Erfolgs- und Aufstiegsgeschichten erzählt und zu Mythen zu werden. Hundert Jahre nach seinem Tod lebt die Erinnerung an Suwelack fort, und sie wird es wohl auch weiterhin tun. Flieger haben ihre Faszination nicht verloren. Das Leben von Fliegern war, wird vermutlich auch künftig von öffentlichem und wissenschaftlichem Interesse sein. Die Geschichte Josef Suwelacks hat ihr Ende noch nicht erreicht – dies ist eine von ihnen.

F. Anhang

1. Eltern und Geschwister von Josef Suwelack

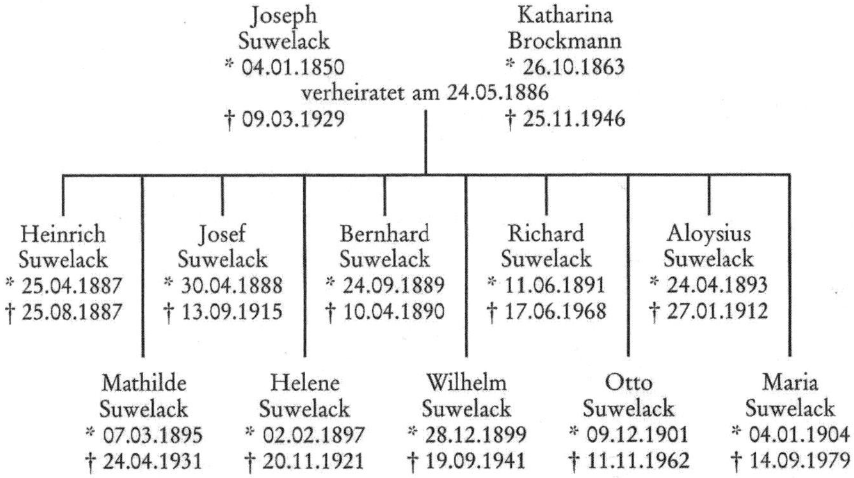

2. Stationen der Fliegerkarriere von Josef Suwelack 1909–1915

1909:	Erste Flugversuche in Billerbeck und Münster mit selbstgebauten Gleitflugzeugen
1910:	Erster Flug in Berlin-Johannisthal
2.8.1911:	Pilotenprüfung auf „Rumpler-Taube“, Pilot und Fluglehrer bei den Rumpler-Werken
8.12.1911:	Dauerweltrekord im Passagierflug mit einem Passagier
14.2.1911:	Kündigung Josef Suwelacks bei *Rumpler* zum 1.3.1912
März 1912:	Wechsel zur Aviatik A.G.
15.6.1912:	Ratifizierung des Gesellschaftsvertrages der Kondor-Werke, Ernennung Josef Suwelacks zum „technischen Direktor“
1911–1912:	Teilnahme an diversen Flugwettbewerben in Berlin, Billerbeck, Detmold, Dresden, Hannover und Münster
11.5.1913:	Teilnahme am Prinz-Heinrich-Flug
3.8.1914:	Meldung als Kriegfreiwilliger
Aug.1914 – Sept. 1915:	Einsatz als Aufklärungsflieger in Belgien und Frankreich bei der Feldfliegerabteilung 24, XIX. Armeekorps
9.12.1914:	Verleihung des Ritterkreuzes des Albrechtsordens II. Klasse mit Schwertern und des Eisernen Kreuz II. Klasse durch den Sächsischen König
13.9.1915:	Absturz und Tod in Frankreich (begraben auf dem englischen Soldatenfriedhof in Erquinghem-sur-Lys, Frankreich)

G. Abkürzungsverzeichnis

Abb.	Abbildung
AEG	Allgemeine Elektrizitäts-Gesellschaft
AG	Aktiengesellschaft
AOK	Armeeoberkommando
BA-MA	Bundesarchiv-Militärarchiv Freiburg
Bd./Bde.	Bande/Bände
BDF	Bund Deutscher Flugzeugführer
BO	Beobachtungsoffizier
DFG	Deutsche Flugmaschinenbau Gesellschaft
DFW	Deutsche Flugzeug-Werke
DLV	Deutscher Luftfahrer Verband
DLZ	Deutsche Luftschiffer/Luftfahrer Zeitschrift
DTMB	Deutsches Technikmuseum Berlin
FfA	Feldfliegerabteilung
G.m.b.H.	Gesellschaft mit beschränkter Haftung
GWU	Geschichte in Wissenschaft und Unterricht
H.	Heft
HA	Historisches Archiv
Hg./hg.	Herausgeber/herausgeben
HZ	Historische Zeitschrift
IdFlieg	Inspektion der Fliegertruppen
ILA	Internationale Luftfahrt-Ausstellung
km	Kilometer
KoGenLuft	Kommandierender General der Luftstreitkräfte
KW	Kondor-Werke
LVG	Luft-Verkehrs-Gesellschaft A.G.
M	Mark
m	Meter
MGFA	Militärgeschichtliches Forschungsamt
MGZ	Militärgeschichtliche Zeitschrift
ND	Nachdruck/Neudruck
NL	Nachlass
NVfL	Niederrheinischer Verein für Luftschiffahrt
o.D./o.O.	ohne Datum/ohne Ort
OHL	Oberste Heeresleitung
StdA	Stadtarchiv
Tab.	Tabelle
TG	Theodor Goldschmidt
TH	Technische Hochschule
VDMI	Verein Deutscher Motorluftschiff-Industrieller
v.l.	von links
VPK	Verkehrstechnische Prüfungskommission
zit. nach	zitiert nach

H. Quellen- und Literaturverzeichnis

Ungedruckte Quellen

Bundesarchiv-Militärarchiv (BA-MA), Freiburg

BA-MA PH 9-V/67 Generalinspektion des Militär-Verkehrswesens der Preußischen Armee 4 Bewaffnung und Abwehr von Luftfahrzeugen: Bericht über die Verwendung des Fliegerkommandos Blau im Kaisermanöver 1911.

BA-MA PH 9-XII/5 Inspektion der Nachrichtentruppen der Preußischen Armee: Schreiben der Inspektion des Luft- und Kraftfahrwesens vom 27. Mai 1912 betr. der Denkschrift des Deutschen Luftfahrer Verbandes von April 1912.

BA-MA PH 9-XV/2 Inspektion der Fliegertruppen der Preußischen Armee: Ausarbeitung einer Fliegervorschrift: „Besondere Bedingungen" 1913 (Flugzeugabnahme).

BA-MA PH 9-XV/2 Inspektion der Fliegertruppen der Preußischen Armee: Ausarbeitung einer Fliegervorschrift (Entwürfe).

BA-MA PH 9-XIII/126: Inspektion des Luftbildwesens: Photographieren aus Flugzeugen.

BA-MA PH 9-XX/68 Inspektion des Militär-Luft und Kraftfahrwesens der Preußischen Armee: Beschaffungsbeschluss für Tauben-Typen A 7/11 bis A 9/11 vom 28. Juli 1911.

BA-MA PH 17-I/25 Kommandierender General der Luftstreitkräfte der Preußischen Armee: Die Bildmeldung der Flieger (Druck) – Teil I: Beobachter, Teil II: Bildoffizier, 1917–1918.

BA-MA PH 19/131 Kommandostellen und Einheiten der Fliegertruppen der Preußischen Armee: Lehr- und Versuchsanstalt Döberitz, 2. Juli 1912, Betreff: Euler Maschinen Gewehrflugzeug.

BA-MA PH 24/72 Dienststellen und Einheiten der technischen Truppen der Preußischen Armee: Betr. Subvention von Flugplätzen. Schreiben an die General-Inspektion des Militär- und Verkehrsflugwesens vom 7. Juni 1912.

BA-MA PH 24/91 Dienststellen und Einheiten der technischen Truppen der Preußischen Armee: Lt. Falkenstein, Denkschrift über den weiteren Ausbau von Flugzeugen als Kampfmittel. Jüteborg, den 10. März 1914.

BA-MA PH 24/99 Dienststellen und Einheiten der technischen Truppen der Preußischen Armee: Unfälle von Flugzeugen. 1914.

BA-MA PH 24/165 Dienststellen und Einheiten der technischen Truppen der Preußischen Armee: Prüfung von Serienflugzeugen und Zubehör: Kommissionsverhandlung, Unfall Eckenbrecher, Döberitz 21.9. bis 9.10.1913.

BA-MA PH 24/259 Dienststellen und Einheiten der technischen Truppen der Preußischen Armee: Die Flugzeuginstrumente und ihr Einbau, um 1912 bis 1917, Zusammengestellt von der physikalischen Abteilung Gruppe 5.

Deutsches Technikmuseum (Archiv-DTMB), Berlin

Archiv P. M. Grosz: Schriftwechsel der K. B. Insp. d. Mil. Luft- u. Kraftf-Wesens und der K. B. Flieger-Ersatz-Abteilung vom 17., 21. und 23. Juli 1915.

DTMB III. Ssg. 2 Firmenschriften Rumpler: E. Rumpler Luftfahrzeugbau.

NL Ernst Eberstein

NL Wilhelm Siegert

NL Willy Stiasny

Familienarchiv Walter Suwelack, Warendorf

NL Josef Suwelack

Historisches Archiv Krupp (HA-Krupp), Essen

Akten zur Westdeutschen Flugplatzgesellschaft und zum Flugplatz Gelsenkirchen-Essen-Rotthausen:
HA Krupp WA 41/72–116
HA-Krupp WA 41/73–130

Privatarchiv Rotraut Illisch, Billerbeck

Jahrgänge des Billerbecker Anzeigers

Stadtarchiv Billerbeck (StdA Billerbeck)

StdA Bilerbeck: Heinrich Fasse: Geschichtliche Nachrichten aus dem Amte Billerbeck 1900–1924 (mit 60 Abblldungen) Bd. 1.
StdA Billerbeck: 250 Jahre 1734–1984 Handelshof, Gaststätte und Brennerei Franz-Josef Ahlers Billerbeck Langestraße 18.

Stadtarchiv Essen (StdA Essen)

StdAE Bestand 102 I, Akte 1059–1061

Stadtarchiv Münster (StdA Münster)

StdA Münster, Akte 1936–08–17a.
StdA Münster, Akte 1936–08–17b
Vogt, Betina: „Unsere Zukunft liegt in der Luft!“ Die Entwicklung der zivilen Luftfahrt auf dem Flugplatz Loddenheide 1909–1933. Unveröffentlichte Arbeit, hervorgegangen aus dem Geschichtswettbewerb des Bundespräsidenten 1990/91. Münster 1990/91.

Gedruckte Quellen

Zeitschriften

(Die Jahreszahlen geben nur die zitierten Jahrgänge an.)
Allgemeine Automobil-Zeitung. Offizielles Organ des Kartells Deutscher Automobil-Clubs und Vereins Deutscher Motorfahrzeug-Industrieller, Berlin 1900ff.
Deutscher Luftschiffer-Verband (Hg.), Deutsche Zeitschrift für Luftschiffahrt, Berlin 1910f.
Der Motorwagen. Automobil- und flugtechnische Zeitschrift, Berlin.
Luftschiffahrt, Flugtechnik und -Sport. Illustrierte Zeitschrift für das gesamte Ballon- und Flugwesen redigiert von Gustav Riefenstahl, Bielefeld 1911.
Polytechnische Rundschau, Stuttgart 1912f.

Sport im Bild, redigiert von Kurt Doerry (Berlin) und B. Wirth (Österreich-Ungarn). August Scherl Verlag (ab 1909), 1911ff., online für den Zeitraum von 1897–1934 (mit Lücken) unter: http://anno.onb.ac.at/cgi-content/anno-plus?aid=sib (Stand: Januar 2015).

Ursinius, Oskar (Hg.), Flugsport. Illustrirte technische Zeitschrift und Anzeiger für das gesamte „Flugwesen", Frankfurt a. M. 1909ff.

Zeitschrift für Flugtechnik und Motorluftschiffahrt offizielles Organ d. Vereins Deutscher Flugtechniker, München/Berlin 1911.

Zeitungen / Periodika

Berliner Tage- und Handelsblatt

Berliner Volks-Zeitung

Billerbecker Anzeiger

B. Z. am Mittag

Deutscher Reichsanzeiger und Königlich Preußischer Staatsanzeiger

Essener Generalanzeiger

Gelsenkirchener Zeitung

Kölnische Volkszeitung

Lippische Landeszeitung

Münsterische Zeitung

Westfälischer Merkur

Zeitgenössische Literatur / (Denk-)Schriften

Ausschreibungen für die Nationale Flugwoche Berlin-Johannisthal vom 24. September bis 1. Oktober 1911.

Endig, W., So baue ich mir eine Rumplertaube, Leipzig 1914 (Selbst ist der Mann, H. 104).

Erblich, Heinz, Moderne Flugzeuge in Wort und Bild, Berlin 1914 (Autotechnische Bibliothek Bd. 51).

Erblich, Heinz, Fliegerschule. Was muß ich wissen, wenn ich Flieger werden will? Ein Lehr- und Handbuch für den Flugschüler, 2. Aufl., Berlin 1915 (Autotechnische Bibliothek Bd. 50).

Hahn, Willy, Aus meinem Fliegerleben, in: Die Flieger. Jahrbuch des Bund Deutscher Flugzeugführer e.V. 1913, Berlin 1914.

Hildebrandt, Alfred, Flugzeug-Industrie, in: Jahrbuch der Motorluftschiff-Studiengesellschaft Bd. 5 (1912), S. 39–52.

Hirth, Hellmuth, Meine Flugerlebnisse. 20.000 Kilometer im Luftmeer, 2. erw. Aufl., Berlin 1915.

Jahrbuch des Deutschen Luftschiffer-Verbands, Berlin 1910.

Korf, Georg, So werden wir fliegen. Die natürliche Lösung des Flugproblems. Der Mensch als Segler der Lüfte in naher Zukunft. Mit einem 2. Teil: „Wenn wir fliegen", Oranienburg 1909.

Kunz, A., Der Verpflegungsbeamte im Kriege. Eine Sammlung von Bestimmungen über das Feld-Verpflegungswesen und die persönlichen Verhältnisse des Verpflegungsbeamten, nach amtlichen Quellen, 2. Aufl., Berlin 1913.

Martin, Rudolf, Berlin-Bagdad. Das Deutsche Weltreich im Zeitalter der Luftschiffahrt 1910–1931, Stuttgart/Leipzig 1907.

Martin, Rudolf, Stehen wir vor einem neuen Weltkriege?, Leipzig 1908.

Mayer, Eduard von, Technik und Kultur. Gedanken über die Verstaatlichung des Menschen, Berlin 1906 (Kulturprobleme der Gegenwart, hg. von Leo Berg, Bd. 3).

Neyen, Emil, Die Flugkunst ohne Schleier. Statistik von 1911 über die Flieger-Unfälle, Berlin 1912.

Nuntz, Nathan, Zur Physiologie und Hygiene der Luftfahrt, Berlin 1912 (Luftfahrt und Wissenschaft, Bd. 3).

Oldershausen, Frhr. von (Hg.), Denkschrift über den Prinz Heinrich Flug 1913, Straßburg 1914.

Rasch, F. (Hg.), Jahrbuch des Deutschen Luftfahrer-Verbandes 1913, Berlin 1913.

Rasch, F. (Hg.), Jahrbuch des Deutschen Luftfahrer-Verbandes 1914, Berlin 1914.

Romen, A., Die Militärpensionsgesetze vom 31. Mai 1906 nebst den Ausführungsbestimmungen, 1. Teil: Das Offizierpensionsgesetz, Berlin 1907.
Steffen, Hans, Bertram, Otto: Orientierung auf Überlandflügen, Berlin 1912.
Suwelack, Josef, Flugtechnische Fragen, in: Jahrbuch des NVfL 1912, S.126–141.
Suwelack, Richard, Festschrift anläßlich des 25jährigen Bestehens der Dampfmolkerei Billerbeck Gesellschaft mit beschränkter Haftung, o. O. 1909.
Vorreiter, Ansbert (Hg.), Jahrbuch der Luftfahrt, München 1912.
Wells, Herbert George, The War in the Air, London 1909.

Rezeption seit den 1920er Jahren

Abendgesellschaft des Zoologischen Gartens Münster (Westf.), Familie Uhlenhorst off: Buren un Kabeleeren. Großes Volksstück mit Gesang und Tanz in 4 Akten von Franz Falger, Münster 1929.
„Aus ‚Billerbecker Großflugtagen' wurden 1912 begrabene Hoffnungen", in: Billerbecker Anzeiger, Ostern 1982.
Blogger ‚Moritz': Tollkühne Kerle aus Westfalen erobern den Himmel, online seit dem 3.6.2012 unter: http://vivatwestfalia.blogspot.de/2011/09/tollkuhne-kerle-aus-westfalen-erobern.html.
Dietrich, Richard, Im Flug über ein halbes Jahrhundert, Gütersloh 1942.
Elgaß, Peter, Josef Suwelack – ein kurzes Fliegerleben, in: Fliegermagazin Nr. 4 (1989), S.86–88.
Fernagut, Alain, Josef Suwelak, un aviateur allemand inhumé au cimetière militaire britannique d'Erquinghem, in: La Voix du Nord vom 13.3.2007, 3 Teile.
Fischer, Detlef, Münster von A bis Z. Wissenswertes in 1500 Stichworten über Geschichte, Kunst und Kultur, 2. Aufl., Münster 2004.
Freitag, Werner (Hg.), Geschichte der Stadt Billerbeck, Bielefeld 2012.
Glanemann, Martina, Technikbegeisterung und kleinstädtische Strukturen. Flugtage im Münsterland 1908–1914. Unveröffentlichte Masterarbeit, Münster 2012.
Hernig, Hartmut/Kraus, Michael, Und das ist unsere Geschichte: Gelsenkirchener Lesebuch, Oberhausen 1984.
Johanterwage, Adolf, Fliegen war sein Lebensinhalt, in: Die Glocke, 27.9.2001.
„Josef Suwelack und der Traum vom Fliegen. Ein neues 135 seitiges Buch über den berühmten Billerbecker Flugpionier", in: Billerbecker Anzeiger, 28.10.1988.
Krewerth, Rainer A. (Hg.), Ein westfälisches Fliegerleben. Josef Suwelack 1888–1915, Münster 2001.
Kranzhoff, Jörg Armin, Arado-Flugzeuge: Vom Doppeldecker zum Strahlflugzeug, Bonn 2000.
Kranzhoff, Jörg Armin, Edmund Rumpler – Wegbereiter der industriellen Flugzeugfertigung, Bonn 2004.
Krystofiak, Carsten, Möönsch, he flögg!! Tüftler, Dandys, Seifenkisten: Todesmutige Luftfahrtpioniere machten zu Kaisers Zeiten Münsters Luftraum unsicher, in: Ultimo 22 (2010), S.8–9.
Lange, Bruno, Das Buch der deutschen Luftfahrttechnik, Textteil, Mainz 1970.
Langenfeld, Hans/Prange Klaus, Münster. Die Stadt und ihr Sport, Münster 2002.
Loy, Johannes, Flugpionier aus dem Münsterland, in: Westfälische Nachrichten, 27.9.2001.
Müller, H., „Ja, wenn ick un Motörken hädde ...". Josef Suwelack vor 100 Jahren geboren, in: Aerokurier 5 (1988), S.625–626.
NL J. Suwelack, Brief Hauptmann Barths an Joseph Suwelack senior, 14.9.1915, 2 Seiten.
NL J. Suwelack, „Vor 25 Jahren. Ein Streifzug durch den November des Jahres 1911", [Unbekannte Zeitung] 1936.
NL J. Suwelack, „Erstmalig um die Türme von Billerbeck. Vier Stunden bedeuteten damals noch Weltrekord", [Unbekannte Zeitung] 1961.
Radzicki, Frank, Der Traum vom Fliegen im Ruhrgebiet: Bilder der Luftfahrt, Erfurt 2006.
Rieth, Hugo, Die Kondor-Flugzeugwerke (1912–1918). Ein Kapitel aus der Pionierzeit deutscher Luftfahrtgeschichte, geschrieben in Essen und Rotthausen, in: Die Heimatstadt Essen 1975, S.121–130.
Rißmann-Ottow, Guido, Glück ab! Frühe Luftfahrt im Revier, Essen 2002.
Schröder, Gerd, 50 Jahre Flugsport in Westfalen, in: Westfalenspiegel 1960, S.21–24.
Schröder, Markus, Lokalzeitgeschichte – Flugpionier Josef Suwelack, in: Lokalzeit Münsterland des WDR vom 24.10.2001.

Schröder, Markus, Lokalzeitgeschichte – Münsterländische Flugpioniere, in: Lokalzeit Münsterland des WDR vom 23.9.2009.
„Segelflugzeug ‚Josef Suwelack'. Buersche Segelflieger ehren einen gefallenen Kriegsflieger", in: Buersche Zeitung, 22.6.1936.
„Sputnik-Sensation im Jahre 1911", in: Münsterische Zeitung, 1.5.1958.
StdA Münster Bestand Dok-Ztg. Serie Stadt-Dok. Nr. 806, 1960–1961.
StdA Münster: Amt 23, 23; Dokumentation Straßennamen, S.536f., online unter: http://www.lwl.org/westfaelische-geschichte/nstopo/strnam/Begriff_43.html.
Supf, Peter, Das Buch der deutschen Fluggeschichte, Bd. 1: Vorzeit – Wendezeit – Werdezeit, Berlin 1935.
Suwelack, Winfried/Suwelack, Walter, Josef Suwelack und der Traum vom Fliegen. Leben und Tod des Billerbecker Flugpioniers und Weltrekordlers Josef Suwelack, 30. April 1888 bis 13. September 1915, anlässlich der 100. Wiederkehr seines Geburtstages, Warendorf 1988 (Beiträge zur Landes- und Volkskunde des Kreises Coesfeld, Bd. 23).
„Vom Weltrekord im Passagierflug", in: Münsterische Zeitung, 30.3.1988.
Wittmann, Livia Käthe/Zibler, Barbara, Melli Beese und die „Flügel am Horizont". Die Geschichte der ersten deutschen Pilotin, Berlin 2009.

Literatur

Alkemeyer, Thomas/Boschert, Bernhard/Schmidt, Robert u. a. (Hg.), Aufs Spiel gesetzte Körper. Aufführungen des Sozialen in Sport und populärer Kultur, Konstanz 2003.
Anderson, Benedict, Imagined Communities: Reflections on the Origin and Spread of Nationalism, London/New York 1991.
Antonsich, Marco, Searching for Belonging – An analytical Framework, in: Geography Compass 4/6 (2010), S.644–659.
Appelgate, Celia, A Nation of Provincials. The German Idea of Heimat, Berkeley/Los Angeles/Oxford 1990.
Arand, Tobias, Der Nachlass des Leutnants Bernhard Mangels aus Münster. Exemplarische Überlegungen zur inhaltlichen und formalen Erschließung von Kriegsfotoalben deutscher Veteranen des Ersten Weltkrieges, in: Westfälische Forschungen 58 (2008), S.421–439.
Arbeitskreis für kirchliche Zeitgeschichte, Konfession und Cleavages. Ein Erklärungsmodell zur regionalen Entstehung des katholischen Milieus in Deutschland, in: Historisches Jahrbuch 120 (2000), S.358–395.
Asendorf, Christoph, Super Constellation. Das Flugzeug als kulturelle Erfahrung, in: Wolfgang Ruppert (Hg.), Fahrrad, Auto, Fernsehschrank. Zur Kulturgeschichte der Alltagsdinge, Frankfurt a. M. 1993, S.188–216.
Assmann, Aleida, Zum Problem der Identität aus kulturwissenschaftlicher Sicht, in: Rolf Lindner (Hg.), Die Wiederkehr des Regionalen. Über neue Formen kultureller Identität, Frankfurt a. M./New York 1994, S.13–35.
Assmann, Aleida, Der lange Schatten der Vergangenheit: Erinnerungskultur und Geschichtspolitik, München 2006.
Assmann, Aleida, Einführung in die Kulturwissenschaft: Grundbegriffe, Themen, Fragestellungen, 3. neu bearbeitete Aufl., Berlin 2011.
Assmann, Aleida, Ist die Zeit aus den Fugen? Aufstieg und Fall des Zeitregimes der Moderne, München 2013.
Assmann, Jan, Frühe Formen politischer Mythomotorik. Fundierende, kontrapräsentische und revolutionäre Mythen, in: Dietrich Harth/ders. (Hg.), Revolution und Mythos, Frankfurt a. M. 1992, S.39–61.
Assmann, Jan, Das kulturelle Gedächtnis. Schrift, Erinnerung und Identität in frühen Hochkulturen, 6. Aufl., München 2007.
Augsteijn, Joost/Storm, Eric (Hg.), Region and State in Nineteenth-Century Europe. Nation-Building, Regional Identities and Separatism, Basingstoke/Hampshire 2012.
Ausschreibung zum Geschichtswettbewerb des Bundespräsidenten 2008/09: Helden: verehrt – verkannt – vergessen, in: Spuren suchen 22 (2008), S.3–42.

Baumgart, Ralf/Eichener, Volker, Norbert Elias zur Einführung, Hamburg 1991.
Baumgart, Winfried, Deutsch-russische Beziehungen 1890–1914/18, in: ders. (Hg.), Preußen-Deutschland und Rußland vom 18. bis zum 20. Jahrhundert, Berlin 1991, S.105–118.
Bausinger, Hermann, Name und Stereotyp, in: Stereotypvorstellungen im Alltagsleben. Beiträge zum Themenkreis Fremdbilder – Selbstbilder – Identität. Festschrift für Georg R. Schroubek zum 65. Geburtstag, München 1988, S.13–19.
Bayer, Walter/Habersack, Mathias (Hg.), Aktienrecht im Wandel, Bd. II Grundsatzfragen des Aktienrechts, Tübingen 2007.
Becker, Frank, Strammstehen vor der Obrigkeit? Bürgerliche Wahrnehmung der Einigungskriege und Militarismus im Deutschen Kaiserreich, in: HZ 277 (2003), S.87–113.
Becker, Frank, Synthetischer Militarismus. Die Einigungskriege und der Stellenwert des Militärischen in der deutschen Gesellschaft, in: Michael Epkenhans (Hg.), Das Militär und der Aufbruch in die Moderne 1860 bis 1890. Armeen und Marinen und der Wandel von Politik, Gesellschaft und Wirtschaft in Europa, den USA sowie Japan, München 2003, S.125–141.
Becker, Frank, Perspektiven einer Carl-Diem-Biographie, in: BIOS 18 (2005), H. 2, S.157–168.
Becker, Frank, Begriff und Bedeutung des politischen Mythos, in: Barbara Stollberg-Rillinger (Hg.), Was heißt Kulturgeschichte des Politischen, Berlin 2005 (Zeitschrift für historische Forschung, Beiheft 35), S.129–148.
Behringer, Wolfgang/Ott-Koptschalijski, Constance, Der Traum vom Fliegen. Zwischen Mythos und Technik, Frankfurt a.M. 1991.
Belliger, Andréa/Krieger, David, Einführung in die Akteur-Netzwerk-Theorie, in: dies. (Hg.), ANThology. Ein einführendes Handbuch in die Akteur-Netzwerk-Theorie, Bielefeld 2006, S.13–50.
Bendick, Rainer, Kriegserwartung und Kriegserfahrung. Der Erste Weltkrieg in deutschen und französischen Schulgeschichtsbüchern (1900–1939/45), 2. Aufl., Herbolzheim 2003
Berghoff, Hartmut, „Dem Ziele der Menschheit entgegen“. Die Verheißungen der Technik an der Wende zum 20. Jahrhundert, in: Ute Frevert (Hg.), Das Neue Jahrhundert. Europäische Zeitdiagnosen und Zukunftsentwürfe um 1900, Göttingen 2000, S.47–78 (Geschichte und Gesellschaft, Sonderheft 18).
Bernig, Jörg, Anachronistisches Kriegsbild, Selbstinszenierung und posthume Heroisierung. Manfred von Richthofen: *Der rote Kampfflieger* (1917), in: Thomas F. Schneider/Hans Wagener (Hg.), Von Richthofen bis Remarque: deutschsprachige Prosa zum 1. Weltkrieg, Amsterdam/New York 2003, S.97–111 (Amsterdamer Beiträge zur neueren Germanistik, 53).
Bette, Karl-Heinrich, Sportsoziologie, Bielefeld 2010.
Biermann, Kai/Cielewicz, Erhardt, Flugplatz Döberitz. Geburtsort der militärischen Luftfahrt in Deutschland, Berlin 2005.
Bizeul, Yves, Politische Mythen im Zeitalter der „Globalisierung“, in: Klaudia Knabel (Hg.), Nationale Mythen – kollektive Symbole: Funktionen, Konstruktionen und Medien der Erinnerung, Göttingen 2005 (Formen der Erinnerung, Bd. 23), S.17–36.
Boelcke, Siegfried, Kriegsvermessungen und ihre Lehren, Berlin 1920.
Borchard, Beatrix, Lücken schreiben oder: Montage als biographisches Verfahren, in: Hans-Erich Bödeker (Hg.), Biographie schreiben, Göttingen 2003, S.211–241.
Borscheid, Peter, Das Tempo-Virus. Eine Kulturgeschichte der Beschleunigung, Frankfurt a.M./New York 2004.
Bösch, Frank/Danyel, Jürgen (Hg.), Zeitgeschichte – Konzepte und Methoden, Göttingen 2012.
Bourdieu, Pierre, Sozialer Sinn. Kritik der theoretischen Vernunft, Frankfurt a.M. 1987.
Bourdieu, Pierre, Die biographische Illusion, in: BIOS 1 (1990), S.75–81.
Bourdieu, Pierre, Die männliche Herrschaft, Frankfurt a.M. 2005.
Bourdieu, Pierre, Die gesellschaftliche Definition der Photographie, in: ders. (Hg.), Eine illegitime Kunst. Die sozialen Gebrauchsweisen der Photographie, ND Hamburg 2006, S.85–110.
Bourdieu, Pierre, Kult der Einheit und kultivierte Unterschiede, in: ders. (Hg.), Eine illegitime Kunst. Die sozialen Gebrauchsweisen der Photographie, Hamburg 2006, S.25–84.
Bourdieu, Pierre, Die Regeln der Kunst. Genese und Struktur des literarischen Feldes, ND Frankfurt a.M. 2008.
Breuer, Stefan, Bürokratie und Charisma. Zur politischen Soziologie Max Webers, Darmstadt 1994.
Buchbender, Ortwin, Zur Geschichte der Rezeption und Sammlung von Feldpost in Deutschland, in: Veit Didczuneit/Jens Ebert/Thomas Jander (Hg.), Schreiben im Krieg – Schreiben vom Krieg: Feldpost im Zeitalter der Weltkriege, Essen 2011, S.17–21.

Burleigh, Michael, Irdische Mächte, göttliches Heil. Die Geschichte des Kampfes zwischen Politik und Religion von der Französischen Revolution bis in die Gegenwart, München 2008.

Campion, Garry, The Good Fight. Battle of Britain in Wartime Propaganda and the Few, Basington 2008.

Castan, Joachim, Der Rote Baron. Die ganze Geschichte des Manfred von Richthofen, Stuttgart 2007.

Caysa, Volker, Rauschinszenierungen. Zum Verhältnis von Rausch und Ekstase im Erlebnissport, in: Thomas Alkemeyer (Hg.), Aufs Spiel gesetzte Körper. Aufführungen des Sozialen in Sport und populärer Kultur, Konstanz 2003, S.55–76.

Chickering, Roger, Zwischen Dynamik und Stillstand. Gedanken zur deutschen Innenpolitik, in: Bernd Heidenreich/Sönke Neitzel (Hg.), Das Deutsche Kaiserreich 1890–1914, Paderborn/München/Wien/Zürich 2011, S.61–73.

Cinnirella, Marco, Ethnic and national Stereotypes: A social Identity Perspective, in: C.C. Barfoot (Hg.), Beyond Pug's Tour. National and Ethnic Stereotyping in Theory and Literary Practice, Amsterdam/Atlanta 1997, S.37–51.

Clark, Christopher, Wilhelm II.: Die Herrschaft des letzten deutschen Kaisers, München 2008.

Clark, Christopher, Die Schlafwandler. Wie Europa in den Ersten Weltkrieg zog, München 2015.

Corn, Joseph J., The Winged Gospel. America's Romance with Aviation, Baltimore/London 2002.

Crouthamel, Jason, Deutsche Soldaten und „Männlichkeit" im Ersten Weltkrieg, in: Aus Politik und Zeitgeschichte 16/17 (2014), S.39–45.

Csikszentmihalyi, Mihaly, Flow: das Geheimnis des Glücks, 14. Aufl., Stuttgart 2008.

Däbritz, Walther/Paulick, Wendt, Th. Goldschmidt AG Essen. Neun Jahrzehnte Geschichte einer deutschen Fabrik, Essen 1937.

Dampfmolkerei Billerbeck Lutum-Gausing & Co., Gesamtübersichts- und Schlußbericht erstattet auf der Generalversammlung am 23.3.1964, o.O. 1964.

Degele, Nina, Einführung in die Techniksoziologie, München 2002.

Deist, Wilhelm, Zur Geschichte des preußischen Offizierkorps 1888–1918, in: Hanns Hubert Hofmann (Hg.), Das deutsche Offizierkorps 1860–1960, Boppard am Rhein 1980, S.39–57.

Deist, Wilhelm, Zensur und Propaganda in Deutschland während des Ersten Weltkrieges, in: ders. (Hg.), Militär, Staat und Gesellschaft. Studien zur preußisch-deutschen Militärgeschichte, München 1991, S.153–163.

Delvos, Hubert, Die Entwicklung des Düsseldorfer Luftverkehrs, Düsseldorf 1941.

Dewitz, Bodo von, So wird bei uns der Krieg geführt! Amateurfotografie im Ersten Weltkrieg, München 1989.

Dickhoff, Erwin, Essener Köpfe: wer war was?, Essen 1985.

Didczuneit, Veit/Ebert, Jens/Jander, Thomas (Hg.), Schreiben im Krieg. Schreiben vom Krieg. Feldpost im Zeitalter der Weltkriege, Essen 2011.

Dietrich, Richard, Im Flug über ein halbes Jahrhundert, Gütersloh 1942.

Dietschy, Paul, Kriegshelden – Sporthelden? Das Dilemma der französischen Sportpresse im Ersten Weltkrieg, in: Arnd Krüger/Swantje Scharenberg (Hg.), Zeiten für Helden – Zeiten für Berühmtheiten im Sport. Reflektionen der 9. Hoyaer Tagung Sportstars, Helden und Heldinnen. Veränderungen in der Darstellung berühmter Sportler und Sportlerinnen in den Massenmedien, Berlin 2014, S.49–64.

Dipper, Christof, Italien und Deutschland seit 1800. Zwei Gesellschaften auf dem Weg in die Moderne, in: ders./Lutz Kinkhammer/Alexander Nützenadel (Hg.), Europäische Sozialgeschichte. Festschrift für Wolfgang Schieder, Berlin 2000, S.485–503.

Ditt, Karl, Was ist ‚westfälisch'? Zur Geschichte eines Stereotyps, in: Westfälische Forschungen 52 (2002), S.45–94.

Ditt, Karl, Karl Wagenfeld (1869–1939): Dichter, Heimatfunktionär, Nationalsozialist?, in: Matthias Frese (Hg.), Fragwürdige Ehrungen!? Straßennamen als Instrument von Geschichtspolitik und Erinnerungskultur, Münster 2012, S.179–232.

Dörner, Andreas, Politischer Mythos und symbolische Politik. Sinnstiftung durch symbolische Formen am Beispiel des Hermannsmythos, Opladen 1995.

Dülffer, Jost, Die zivile Reichsleitung und der Krieg: Erwartungen und Bilder 1890–1914, in: ders., Im Zeichen der Gewalt. Frieden und Krieg im 19. und 20. Jahrhundert, hg. von Martin Kröger/Ulrich S.Soénius/Stefan Wunsch, Köln/Weimar/Wien 2003, S.124–140.

Dülffer, Jost, Kriegserwartung und Kriegsbild in Deutschland vor 1914, in: ders., Im Zeichen der Gewalt. Frieden und Krieg im 19. und 20. Jahrhundert, Köln/Weimar/Wien 2003, S.107–123.

Dunning, Eric, Sport Matters. Sociological studies of sport, violence and civilization, London/New York 1999.

Echternkamp, Jörg, Militärgeschichte, in: Frank Bösch/Jürgen Danyel (Hg.), Zeitgeschichte – Konzepte und Methoden, Göttingen 2012, S.293–312.

Edgerton, David, England and the Aeroplane. An Essay on a Militant and Technological Nation, Houndsmills/London 1992.

Eisenberg, Christiane, Arbeiter, Bürger und der ‚bürgerliche Verein' 1820–1870. Deutschland und England im Vergleich, in: Jürgen Kocka (Hg.), Bürgertum im 19. Jahrhundert. Deutschland im europäischen Vergleich, Bd. 2, Göttingen 1995, S.187–219.

Eisermann, Thilo, Pressephotographie und Informationskontrolle im Ersten Weltkrieg, Deutschland und Frankreich im Vergleich, Hamburg 2000.

Elias, Norbert/Dunning, Eric, Sport und Spannung im Prozess der Zivilisation, Frankfurt a. M. 2003 (Gesammelte Schriften, Bd. 7).

Epkenhans, Michael/Förster, Stig/Hagemann, Karen, Einführung: Biographien und Selbstzeugnisse in der Militärgeschichte – Möglichkeiten und Grenzen, in: dies. (Hg.), Militärische Erinnerungskultur. Soldaten im Spiegel von Biographien, Memoiren und Selbstzeugnissen, Paderborn 2006, S.IV-XVI.

Epkenhans, Michael/Seggern, Andreas von, Leben im Kaiserreich. Deutschland um 1900, Stuttgart 2007.

Esposito, Fernando, Mythische Moderne. Aviatik, Faschismus und die Sehnsucht nach Ordnung in Deutschland und Italien, München 2011 (Ordnungssysteme. Studien zur Ideengeschichte der Neuzeit, Bd. 32).

Etzemüller, Thomas, Biographien: lesen – erforschen – erzählen, Frankfurt a. M. 2012 (Historische Einführungen 12).

Fach, Wolfgang/Köhnke, Karl-Christian/Middell, Matthias u. a., Regionenbezogene Identifikationsprozesse. Das Beispiel „Sachsen" – Konturen eines Forschungsprogramms, in: Heinz-Werner Wollersheim/Sabine Tzschachel/Matthias Middell (Hg.), Region und Identifikation, Leipzig 1998, S.1–32.

Farrell, Joseph, Das Fliegen. D'Annunzio und die Futuristen, in: Bodo-Michael Baumunk (Hg.), Die Kunst des Fliegens (anlässlich der Ausstellung „Die Kunst des Fliegens" vom 2. Juli bis 27. Oktober 1996 im Zeppelin-Museum Friedrichshafen), Friedrichshafen 1996, S.72–83.

Felken, Detlef, Die Größe des Anderen. Anmerkungen zur Lage der Biographie, in: Non Fiktion 8 (2013), H. 1, S.13–26.

Fetz, Bernhard, Der Stoff, aus dem das (Nach-)Leben ist, in: ders. (Hg.), Die Biographie – Zur Grundlegung ihrer Theorie, Berlin/New York 2009, S.103–154.

Fetz, Bernhard/Hemecker, Wilhelm (Hg.), Theorie der Biographie. Grundlagentexte und Kommentar, Berlin/New York 2011.

Fetz, Bernhard, Struktur statt Psychologie. Die „Anti-Biographie" als biographisches Modell, in: ders./ Wilhelm Hemecker (Hg.), Theorie der Biographie. Grundlagentexte und Kommentar, Berlin/New York 2011, S.361–366.

Fink, Carl, Die Geschichte des Luftbildwesens unter besonderer Berücksichtigung der organisatorischen Arbeiten, die für den Ausbau des Luftbildwesens bis Dezember 1917 geleistet worden sind [1918], ND Hildesheim 1958.

Fischer, Christina/Weber, Florian, „Ein Bier wie seine Heimat – ursprünglich, ehrlich und charaktervoll". Eine Untersuchung der Vermarktung ‚fränkischer' Biere aus diskurstheoretischer Perspektive, in: Olaf Kühne/Florian Weber (Hg.), Bausteine der Regionalentwicklung, Wiesbaden 2015, S.73–94.

Fischer, Detlef, Chronik des Münsterlandes, Münster 2003.

Fischer, Detlef, Münster von A bis Z. Wissenswertes in 1500 Stichworten über Geschichte, Kunst und Kultur, 2. Aufl., Münster 2004.

Flender, Armin/Pfau, Dieter/Schmidt, Sebastian, Regionale Identität zwischen Konstruktion und Wirklichkeit: eine historisch-empirische Untersuchung am Beispiel des Siegerlandes, Baden-Baden 2001.

Florschütz, Gottlieb, Sport in Film und Fernsehen: Zwischen Infotainment und Spektakel, Wiesbaden 2005, S.125–130.

Focault, Michel, Was ist ein Autor?, in: ders. (Hg.), Schriften zur Literatur, Frankfurt a. M. 1988 [1969], S.7–31.

Förster, Stig, Der deutsche Generalstab und die Illusion des kurzen Krieges, 1871–1914. Metakritik eines Mythos, in: Militärgeschichtliche Mitteilungen 54 (1995), S.61–95.
Förster, Stig, Im Reich des Absurden: Die Ursachen des Ersten Weltkrieges, in: Bernd Wegner (Hg.), Wie Kriege entstehen. Zum historischen Hintergrund von Staatenkonflikten, Paderborn 2000, S.211–252.
Frank, Saskia, Zeppelin-Ereignisse: Technikkatastrophen im medialen Prozess, Marburg 2008.
Frei, Oliver, Bilder für den Krieg. Die Tagebücher des Lichtbildoffiziers Erich Ewald 1914–1918/19. Ein Beitrag zur Geschichte der Luftaufklärung und des militärischen Luftbildwesens, Berlin 2011.
Freitag, Werner (Hg.), Geschichte der Stadt Billerbeck, Bielefeld 2012.
Frevert, Ute, Gesellschaft und Militär im 19. und 20. Jahrhundert: Sozial-, kultur- und geschlechtergeschichtliche Annäherungen, in: dies. (Hg.), Militär und Gesellschaft im 19. und 20. Jahrhundert, Stuttgart 1997.
Frie, Ewald, Schauplätze des Lebens. Über das Zusammenspiel von Biographie und Region, in: Martin Dröge (Hg.), die biographische Methode in der Regionalgeschichte, Münster 2011, S.15–24.
Fritzsche, Peter, A Nation of Fliers. German Aviation and the Popular Imagination, Cambridge/Massachusetts/London 1992.
Fröhlich, Gerhard/Rehbein, Boike, Bourdieu Handbuch. Leben – Werk – Wirkung, Stuttgart/Weimar 2009.
Fuchs, Stephan, „Vom Segen des Krieges". Katholische Gebildete im Ersten Weltkrieg. Eine Studie zur Kriegsdeutung im akademischen Katholizismus, Wiesbaden 2004.
Fuhs, Burkhard, Fliegende Helden. Die Kultur der Gewalt am Beispiel von Kampfpiloten und ihren Maschinen, in: Rolf Brednich/Walter Hartinger (Hg.), Gewalt in der Kultur. Vorträge des 29. Deutschen Volkskongresses, Bd. 2, Passau 1992, S.705–720.
Funck, Markus, Bereit zum Krieg? Entwurf und Praxis militärischer Männlichkeit im preußisch-deutschen Offizierskorps vor dem Ersten Weltkrieg, in: Karen Hagemann/Stefanie Schüler-Springorum (Hg.), Heimat-Front. Militär- und Geschlechterverhältnisse im Zeitalter der Weltkriege, Frankfurt a. M. 2002, S.69–90.
Funck, Markus, Vom Höfling zum soldatischen Mann. Varianten und Umwandlungen adeliger Männlichkeit zwischen Kaiserreich und Nationalsozialismus, in: Eckart Conze/Monika Wienfort (Hg.), Adel und Moderne. Deutschland im europäischen Vergleich im 19. und 20. Jahrhundert, Köln 2004, S.205–235.
Fussell, Paul, The Great War and Modern Memory, ND Oxford 2013.
Gebauer, Gunter, Die Mythen-Maschine, in: Theodor W. Adorno/Volker Caysa (Hg.), Sportphilosophie, Leipzig 1997, S.290–317.
Gebauer, Gunter/Poser, Stefan/Schmidt, Robert Schmidt/Stern, Martin (Hg.), Kalkuliertes Risiko: Technik, Spiel und Sport an der Grenze, Frankfurt a. M./New York 2006.
Geinitz, Christian/Hinz, Uta, Das Augusterlebnis in Südbaden. Ambivalente Reaktionen der deutschen Öffentlichkeit auf den Kriegsbeginn 1914, in: Gerhard Hirschfeld/Gerd Krumeich/Dieter Langewiesche/Hans-Peter Ullmann (Hg.), Kriegserfahrungen. Studien zur Sozial- und Mentalitätsgeschichte des Ersten Weltkriegs, Essen 1997, S.20–35.
Geinitz, Christian, Kriegsfurcht und Kampfbereitschaft. Das Augusterlebnis in Freiburg. Eine Studie zum Kriegsbeginn 1914, Essen 1998.
George, Arthur/Whitehouse, Joseph, Fliegerasse 1914–1918, Stuttgart 1970.
Gerhard, Ute/Link, Jürgen, Zum Anteil der Kollektivsymbolik an den Nationalstereotypen, in: Jürgen Link/Wulf Wülfing (Hg.), Nationale Mythen und Symbole in der zweiten Hälfte des 19. Jahrhunderts. Strukturen und Funktionen von Konzepten nationaler Identität, Stuttgart 1991, S.16–54.
Gibby, Bryce D., Valiant young men: heroes of flight, Wilmington 2006.
Giesen, Bernhard, Die Aura des Helden, in: Anne Horner/Ronald Kurt/Jo Reicherts (Hg.), Diesseitsreligion: zur Deutung der Bedeutung moderner Kultur, Konstanz 1999, S.437–444.
Glanemann, Martina, Technikbegeisterung und kleinstädtische Strukturen – Flugtage im Münsterland 1908–1914. Unveröffentlichte Masterarbeit, Münster 2012.
Gmeline, Patrick de, Baron rouge et Cigogne blanche: Manfred von Richthofen et René Fonck, Paris 2011.
Grundmann, Melanie, Der Dandy: wie er wurde, was er war. Eine Anthologie, Köln/Weimar/Wien 2007.
Grütter, Heinrich Theodor (Hg.), 200 Jahre Krupp. Ein Mythos wird besichtigt, Essen 2012.

Grütter, Heinrich Theodor/Hauser, Walter (Hg.), 1914 Mitten in Europa. Die Rhein-Ruhr-Region und der Erste Weltkrieg. Katalogbuch zur Ausstellung des LVR-Industriemuseums und des Ruhr Museums auf der Kokerei Zollverein 30. April bis 26. Oktober 2014, Essen 2014.

Gundler, Bettina, „Nie wird die Fliegerei als Beruf überfüllt sein von Frauen". Geschlechterverhältnisse und berufliche Perspektiven von Pilotinnen in den Anfängen der Luftfahrt, in: Metis 4 (1995), H. 1, S.29–40.

Haak, Sebastian, „And Feel the Tingle of the Greatest Game of Them All". Über eine (verdrängte) Facette der Wahrnehmung von Krieg und Gewalt in den Briefen US-amerikanischer Soldaten, in: Veit Didczuneit/Jens Ebert/Thomas Jander (Hg.), Schreiben im Krieg — Schreiben vom Krieg: Feldpost im Zeitalter der Weltkriege, Essen 2011, S.175–184.

Hackenberger, Willi, Die Alten Adler: Pioniere der deutschen Luftfahrt, München 1960.

Hafeneder, Rudolf (Hg.), Überblick über das militärische Geowesen Deutschlands im 19. und 20. Jahrhundert, Euskirchen 2004 (Geoinformationsdienst der Bundeswehr H. 2).

Hafer, Xaver, Die Anfänge der Stabilitätsbetrachtung in der Flugmechanik, in: Zeitschrift für Flugwissenschaft und Weltraumforschung 12 (1988), S.37–44.

Haude, Rüdiger, Grenzflüge. Politische Symbolik der Luftfahrt vor dem Ersten Weltkrieg. Das Beispiel Aachen, Köln/Weimar/Wien 2007.

Heintel, Peter, Das Glück der Geschwindigkeit, in: Barbara Ränsch-Trill (Hg.), Zeit und Geschwindigkeit. Sportliches Erleben in beschleunigten Prozessen, Sankt Augustin 2002, S.100–114.

Heise, Matthias, Die Renaissance der Regionen. Neue Ansätze in den Theorien der Internationalen Beziehungen: Regionaler Sicherheitskomplex und Regionale Ordnungen, Frankfurt a.M. 2008.

Herbert, Ulrich, Best. Biographische Studien über Radikalismus, Weltanschauung und Vernunft 1903–1989, 3. Aufl., Bonn 1996.

Hersch, Jeanne, Mythos und Politik, in: Kurt Hoffmann (Hg.), Die Wirklichkeit des Mythos, München/Zürich 1965.

Hertz-Eichenrode, Dieter, Die Feudalisierungsthese — ein Rückblick, in: Vierteljahrschrift für Sozial- und Wirtschaftsgeschichte 89 (2002), H. 3, S.265–287.

Hettling, Manfred/Echernkamp, Jörg, Heroisierung und Opferstilisierung. Grundelemente des Gefallenengedenkens von 1813 bis heute, in: dies. (Hg.), Gefallenengedenken im globalen Vergleich: Nationale Tradition, politische Legitimation und Individualisierung der Erinnerung, München 2013, S.123–158.

Hildebrandt, Tina/Niejahr, Elisabeth, „Was ist ein Held?" Verteidigungsminister Thomas de Mazière und der Schriftsteller Uwe Tellkamp im Gespräch über Drohnen, gezieltes Töten, Moral und Tapferkeit, in: Die Zeit, 14.7.2013.

Hobsbawm, Eric, Introduction: Inventing Traditions, in: ders./Terence Ranger (Hg.), The Invention of Tradition, 7. Aufl., Cambridge 2009, S.1–14.

Hobsbawm, Eric, Mass-Producing Traditions: Europe, 1870–1914, in: ders./Terence Ranger (Hg.), The Invention of Tradition, 7. Aufl., Cambridge 2009, S.263–307.

Hobson, Rolf, Martimer Imperialismus: Seemachtideologie, seestrategisches Denken und der Tirpitzplan 1875 bis 1914, München 2004 (Beiträge zur Militärgeschichte, Bd. 61).

Hoegen, Jesko von, Der Held von Tannenberg: Genese und Funktion des Hindenburg-Mythos, Köln/Weimar/Wien 2007, S.182.

Höhler, Sabine, Luftfahrtforschung und Luftfahrtmythos. Wissenschaftliche Ballonfahrt in Deutschland 1880–1910, Frankfurt a.M./New York 2001.

Holtschoppen, Felix, Rebellische Technik. Maschinenphantasien in der literarischen Phantastik um 1900, Berlin 2005.

Holzer, Anton, Den Krieg sehen. Zur Bildgeschichtsschreibung des Ersten Weltkriegs, in: ders. (Hg.), Mit der Kamera bewaffnet. Krieg und Fotografie, Marburg 2003, S.57–70.

Holzer, Anton, Das fotografische Gesicht des Krieges, in: ders. (Hg.), Mit der Kamera bewaffnet. Krieg und Fotografie, Marburg 2003, S.7–20.

Hoeppner, Ernst von, Deutschlands Krieg in der Luft, Leipzig 1921 [ND Barsinghausen 2013].

Höfler, Günther A., Die Invasionen der Kriegsmaschinen in die Literatur. Der Erste Weltkrieg als literarische Technikerfahrung, in: Theo Elm/Hans H. Hiebel (Hg.), Medien und Maschinen, Freiburg 1991, S.225–244.

Horn, Eva, Erlebnis und Trauma. Die narrative Konstruktion des Ereignisses in Psychiatrie und Kriegsroman, in: Inka Mülder-Bach (Hg.), Modernität und Trauma. Beiträge zum Zeitenbruch des Ersten Weltkrieges, Wien 2000, S.131–162.

Hugger, Paul, Die Photographie als Dokument des privaten Erinnerns, in: Brigitte Bönisch-Brednich/Rolf W. Brednich/Helge Gerndt (Hg.), Erinnern und Vergessen. Vorträge des 27. Deutschen Volkskundekongresses Göttingen 1989, Göttingen 1991, S.235–242.

Humburg, Martin, Das Gesicht des Krieges. Feldpostbriefe von Wehrmachtsoldaten aus der Sowjetunion 1941–1944, Wiesbaden 1998.

Humburg, Martin, „Jedes Wort ist falsch und wahr – das ist das Wesen des Wortes". Vom Schreiben und Schweigen in der Feldpost, in: Veit Didczuneit/Jens Ebert/Thomas Jander (Hg.), Schreiben im Krieg – Schreiben vom Krieg: Feldpost im Zeitalter der Weltkriege, Essen 2011, S.75–85.

Hüppauf, Bernd, Kriegsfotografie und die Erfahrung des Ersten Weltkriegs, in: Barbara Naumann (Hg.), Vom Doppelleben der Bilder. Bildmedien und ihre Texte, München 1993, S.29–50.

Hüppauf, Bernd, Schlachtenmythen und die Konstruktion des „Neuen Menschen", in: Gerhard Hirschfeld/Gerd Krumeich/Irina Renz (Hg.), „Keiner fühlt sich mehr als Mensch …". Erlebnis und Wirkung des Ersten Weltkriegs, Essen 1993, S.43–83.

Hüppauf, Bernd, Fotografie, in: Gerhard Hirschfeld/Gerd Krumeich/Irina Renz (Hg.), Enzyklopädie Erster Weltkrieg, Paderborn/München/Wien/Zürich 2003, S.499–500.

Hüppauf, Bernd, Was ist Krieg? Zur Grundlegung einer Kulturgeschichte des Kriegs, Bielefeld 2013.

Hurrell, Andrew, Regionalism in Theoretical Perspective, in: Louise Fawcett/ders. (Hg.), Regionalism in World Politics: Regional Organisations and International Order, ND Oxford 2004, S.37–73.

Ilisch, Peter, Bischöfliche Siedlung und Wigbold der Bürger Billerbeck 809–1803, in: Werner Freitag (Hg.), Geschichte der Stadt Billerbeck, Bielefeld 2012, S.21–108.

Ingold, Felix Philipp, Literatur und Aviatik. Europäische Flugdichtung 1909–1927, Frankfurt a.M. 1980.

Jäger, Georg, Das Zeitschriftenwesen, in: ders. (Hg.), Geschichte des deutschen Buchhandels. Frankfurt a.M. 2003, S.368–389 (Das deutsche Kaiserreich 1871–1918, Teil 2).

Jäger, Helmut, Erkundung mit der Kamera. Die Entwicklung der Photographie zur Waffe und ihr Einsatz im 1. Weltkrieg, 2. Aufl., München 2007.

Jäger, Jens, Fotografiegeschichte(n). Stand und Tendenzen der historischen Forschung, in: Archiv für Sozialgeschichte 48 (2008), S.511–537.

Jäger, Jens, Fotografie und Geschichte, Frankfurt a.M./New York 2009 (Campus Historische Einführungen, Bd.7).

Karstens, Simon, Die Summe aller Wahrheiten und Lügen. Ein Erfahrungsbericht zur geschichtswissenschaftlichen Biographie, in: BIOS 24 (2011), H. 1, S.78–97.

Kaschuba, Wolfgan, Die Überwindung der Distanz. Zeit und Raum in der europäischen Moderne, Frankfurt a.M. 2004.

Kaufmann, Stefan, Kriegführung im Zeitalter technischer Systeme – Zur Maschinisierung militärischer Operationen im Ersten Weltkrieg, in: MGZ 61 (2002), S.337–367.

Kauther, Alexander/Wirtz, Paul, Robert Thelen – ein Alter Adler in Berlin-Johannisthal, Berlin/Jülich 2011 (Dokumentationsreihe über den Flugplatz Johannisthal 1909–1914, H. 27).

Kehrt, Christian, „Schneid, Takt und gute Nerven". Der Habitus deutscher Militärpiloten und Beobachter im Kontext technisch strukturierter Handlungszusammenhänge, 1914–1918, in: Technikgeschichte 72 (2005), S.177–201.

Kehrt, Christian, „Das Fliegen ist immer noch ein gefährliches Spiel" – Risiko und Kontrolle der Flugzeugtechnik von 1908 bis 1914, in: Gunter Gebauer/Stefan Poser/Robert Schmidt/Martin Stern (Hg.), Kalkuliertes Risiko: Technik, Spiel und Sport an der Grenze, Frankfurt a.M./New York 2006, S.199–224.

Kehrt, Christian, „A horribly beautiful picture". Aesthetic dimensions of aerial warfare, in: Reinhard Heil/Andreas Kaminiski/Markus Stippak u.a. (Hg.), Tensions and Convergences. Technological and Aestetic Transformations of Society, Bielefeld 2007, S.337–349.

Kehrt, Christian, Moderne Krieger. Die Technikerfahrungen deutscher Militärpiloten 1910–1945, Paderborn/München/Wien/Zürich 2010 (Krieg in der Geschichte, Bd.58).

Kehrt, Christian, Vom Luftakrobaten zum Cyborg? Technikerfahrungen deutscher Militärpiloten im 20.Jahrhundert, in: Eberhard Birk/Heiner Möllers/Wolfgang Schmidt (Hg.), Die Luftwaffe zwischen Politik und Technik, Berlin 2012, S.15–37 (Schriften zur Geschichte der Deutschen Luftwaffe, Bd.2).

Kennan, George F., The Decline of Bismarck's European Order. Franco-Russian Relations, 1875–1890, Princeton 1979.

Kern, Stephen, The Culture of Time and Space, 1880–1918, Cambridge/Massachusetts 1983.

Keupp, Heiner, Diskursarena Identität. Lernprozesse in der Identitätsforschung, in: ders./Renate Höfer (Hg.), Identitätsarbeit heute. Klassische und aktuelle Perspektiven der Identitätsforschung, ND Frankfurt a. M. 2001, S.11–39.

Klein, Christian, Einleitung: Biographik zwischen Theorie und Praxis. Versuch einer Bestandsaufnahme, in: ders. (Hg.), Grundlagen der Biographik, Stuttgart 2002, S.1–22.

Klein, Christian, „Biographismus ist eine andere Art der Kontingenzbekämpfung“. Der Literaturkritiker Hubert Winkels über die „gute Biographie“, ihr Verhältnis zu Roman und Realität sowie den Erfolg des Genres, in: Non Fiktion. Arsenal der anderen Gattungen 8 (2013), H. 1, S.105–111.

Klüting, Hermann, Soldaten in Westfalen und am Niederrhein. Das Königlich Preußische VII. Armeekorps, Beckum 1982.

Knightley, Philip, The Eye of War. Words and Photographs from the Front Line, London 2003.

König, Wolfgang/Weber, Wolfhard, Netzwerke, Stahl und Strom: 1840 bis 1914, Berlin 1990 (Propyläen-Technikgeschichte 4).

König-Heuer, Ursula/Paulus, Julia (Hg.), „In der Ferne das Donnern der Kanonen“. Briefwechsel eines Billerbecker Soldaten mit seiner Mutter im Ersten Weltkrieg, Münster 2014.

Köppen, Manuel, Das Entsetzen des Beobachters. Krieg und Medien im 19. und 20. Jahrhundert, Heidelberg 2005.

Köppen, Manuel, Luftbilder. Die Medialisierung des Blicks, in: Gerhard Paul (Hg.), Das Jahrhundert der Bilder 1900 bis 1949, Göttingen 2009, S.180–187.

Koselleck, Reinhart, Darstellung, Ereignis und Struktur, in: Gerhard Schulz (Hg.), Geschichte heute. Positionen, Tendenzen und Probleme, Göttingen 1973, S.307–317.

Koselleck, Reinhart, „Erfahrungsraum“ und „Erwartungshorizont“ – zwei historische Kategorien, in: ders., Vergangene Zukunft. Zur Semantik geschichtlicher Zeiten, Frankfurt a. M. 1995, S.349–375.

Krais, Beate/Gebauer, Gunter, Habitus, 3. Aufl., Bielefeld 2010.

Krais, Beate, Die männliche Herrschaft: ein somatisiertes Herrschaftsverhältnis, in: Österreichische Zeitschrift für Soziologie 36 (2011), H. 4, S.33–50.

Kranzhoff, Jörg Armin, Die Arado-Flugzeuge: Vom Doppeldecker zum Strahlflugzeug, Bonn 2000.

Kranzhoff, Jörg Armin, Edmund Rumpler – Wegbereiter der industriellen Flugzeugfertigung, Bonn 2004.

Krause, Detlef, Zur Geschichte der Mitteldeutschen Creditbank, in: Bankhistorisches Archiv. Zeitschrift für Bankengeschichte (1991), H. 17, S.17–41.

Krauss, Martina, Kleine Welten. Alltagsfotografie – die Anschaulichkeit einer „privaten Praxis“, in: Gerhard Paul (Hg.), Visual History. Ein Studienbuch, Göttingen 2006, S.57–75.

Krewerth, Rainer A. (Hg.), Ein westfälisches Fliegerleben. Josef Suwelack 1888–1915, Münster 2001.

Kroener, Bernhard R., Integrationsmilitarismus. Zur Rolle des Militärs als Instrument bürgerlicher Partizipationsbemühungen im Deutschen Reich und in Preußen im 19. Jahrhundert bis zum Ausbruch des Ersten Weltkrieges, in: ders./Ralf Pröve/Bruno Thoß (Hg.), Kriegerische Gewalt und militärische Präsenz in der Neuzeit. Ausgewählte Schriften, Paderborn/München/Wien/Zürich 2008, S.83–107.

Krumeich, Gerd, Die Dolchstoß-Legende, in: Etienne François/Hagen Schulze (Hg.), Deutsche Erinnerungsorte 1, München 2001, S.585–599.

Kruse, Wolfgang, Die Kriegsbegeisterung in Deutschland zu Beginn des Ersten Weltkriegs. Entstehungszusammenha?nge, Grenzen und ideologische Strukturen, in: Marcel van der Linden/Gottfried Mergner (Hg.), Kriegsbegeisterung und mentale Kriegsvorbereitung, Berlin 1991, 73–87.

Kühl, Ernst, Das schöne Münster: 50 Jahre Sportfliegen in Münster, Münster 1960.

Kühne, Thomas, Männergeschichte als Geschlechtergeschichte, in: ders. (Hg.), Männergeschichte – Geschlechtergeschichte. Männlichkeit im Wandel der Moderne, Frankfurt a. M./New York 1996, S.7–30.

Küster, Thomas, „Regionale Identität“ als Forschungsproblem. Konzepte und Methoden im Kontext der modernen Regionalgeschichte, in: Westfälische Forschungen 52 (2002), S.1–44.

Kutz, Martin, Das Flugzeug als Waffe – der Erste Weltkrieg als Experimentierfeld des Luftkriegs. Museum für Verkehr und Technik (Hg.), Hundert Jahre deutsche Luftfahrt. Lilienthal und seine Erben, Gütersloh/München 1991, S.39–59.

Landwehr, Achim, Historische Diskursanalyse, 3. überarb. Aufl., Frankfurt a. M. 2008 (Historische Einführungen, Bd. 4).

Lange, Bruno, Das Buch der deutschen Luftfahrttechnik, 2 Bde., Mainz 1970.

Lange, Klaus, „Regionen“, in: Handwörterbuch der Raumforschung und Raumordnung, Bd. 3, 2. Aufl., Hannover 1970, Sp. 2705–2719.
Langenfeld, Hans/Prange, Klaus, Münster. Die Stadt und ihr Sport, Münster 2002.
Lässig, Simone, Die historische Biographie auf neuen Wegen?, in: GWU 60 (2009), S. 540–553.
Latour, Bruno, Wir sind nie modern gewesen. Versuch einer symmetrischen Anthropologie, Berlin 1995.
Le Goff, Jacques, Ludwig der Heilige, Stuttgart 2000.
Leonhard, Jörn, Die Büchse der Pandora. Geschichte des Ersten Weltkriegs, München 2014.
Lewis, David W., Eddie Rickenbacker, an American hero in the twentieth century, Baltimore 2008.
Lindner-Wirsching, Almut, Patrioten im Pool. Deutsche und französische Kriegsberichterstatter im Ersten Weltkrieg, in: Ute Daniel (Hg.), Augenzeugen. Kriegsberichterstattung vom 18. zum 21. Jahrhundert, Göttingen 2006, S. 113–140.
Loewenstein, Bedrich, Der Fortschrittsglaube. Geschichte einer europäischen Idee, Göttingen 2009.
Lüdtke, Alf, Ikonen des Fortschritts. Eine Skizze zu Bild-Symbolen und politischen Orientierungen in den 1920er und 1930er Jahren in Deutschland, in: ders./Inge Marßolek/Adelheid von Saldern (Hg.), Amerikanisierung: Traum und Alptraum in Deutschland des 20. Jahrhunderts, Stuttgart 1996, S. 199–210.
LWL-Industriemuseum (Hg.), Helden. Von der Sehnsucht nach dem Besonderen. Katalog zur Ausstellung des LWL-Industriemuseum Heinrichshütte Hattingen 12.3.–31.10.2010, Essen 2010.
Makropoulos, Michael, Modernität und Kontingenz, München 1997.
Marschik, Matthias, „Flieger grüss mir die Sonne ...“. Eine kleine Kulturgeschichte der Luftfahrt, Wien 2000.
Masse, Kaspar, Nachwelten zwischen „Heimat“ und „Kulisse“. Anmerkungen zur volkskundlich-kulturwissenschaftlichen Regionalitätsforschung, in: Zeitschrift für Volkskunde 94 (1998), S. 53–70.
Meier, Christian, Von der Schwierigkeit, ein Leben zu erzählen. Zum Projekt einer Caesar-Biographie, in: Jürgen Kocka/Thomas Nipperdey (Hg.), Theorie und Erzählung in der Geschichte. Beiträge zur Historik, Bd. 3, München 1979, S. 229–258.
Meier, Christian, Die Faszination des Biographischen, in: Frank Niess (Hg.), Interesse an der Geschichte, Frankfurt a. M. 1989, S. 100–111.
Meighörner, Jeannine, „Die Essenz dieses Mannes ist Mut“. Graf Zeppelin und die Presse: ein Heldenepos, in: Wolfgang Meighörner (Hg.), Der Graf. 1838–1917 (Ausstellungskatalog), Friedrichshafen 2000, S. 204–227.
Mertens, Lothar, Das Einjährig-Freiwilligen Privileg. Der Militärdienst im Zeitgeist des deutschen Kaiserreiches, in: Zeitschrift für Religions- und Geistesgeschichte 42 (1990), H. 4, S. 316–329.
Militärgeschichtliches Forschungsamt (Hg.), Die Militärluftfahrt bis zum Beginn des Ersten Weltkrieges, 2 Bde., Frankfurt a. M. 1965/66.
Militärgeschichtliches Forschungsamt (Hg.), Die Entwicklung der Heeres- und Marineflugzeuge (technischer Band), 2. Aufl., Frankfurt a. M. 1966.
Milrach, Ernst, Die Geschichte des Niederrheinischen Vereins für Luftschiffahrt, in: Festschrift zum XXI. Deutschen Luftfahrer-Tag anläßlich des 25jährigen Jubiläums des Deutschen Luftfahrt-Verbandes und des Niederrheinischen Vereins für Luftschiffahrt: Sektionen Wuppertal und Bonn, Juni 1927, Barmen 1927, S. 26–44.
Mommsen, Wolfgang J., Großmachtstellung und Weltpolitik. Die Außenpolitik des Deutschen Reiches 1870–1914, Frankfurt a.M./Berlin 1993.
Morrow, John H., Building German Air Power, 1909–1914, Knoxville 1976.
Morrow, John H., German Air Power in World War I, Lincoln/Nebraska/London 1982.
Morrow, John H., Knights of the Sky. The Rise of Military Aviation, in: Frans Coetzee/Marilyn Shevin-Coetzee (Hg.), Authority, Identity and the Social History of the Great War, Oxford/Providence 1995, S. 305–324.
Möser, Kurt, Fliegerblick 1914, in: Journal of New Frontiers in Spatial Concepts 1 (2009), S. 99–106.
Möser, Kurt, Fahren und Fliegen in Frieden und Krieg. Kulturen individueller Mobilitätsmaschinen 1880–1930, Heidelberg/ Ubstadt-Weiher/Neustadt a. d.W./Basel 2009 (Technik + Arbeit 13).
Möser, Kurt, Schlachtflieger 1918 – ein technisches Waffensystem im Kontext, in: Technikgeschichte 77 (2010), S. 185–230.
Möser, Kurt, Grauzonen der Technikgeschichte, Karlsruhe 2011 (Technikdiskurse. Karlsruher Studien zur Technikgeschichte).
Möser, Kurt, Technische Zukunftskriege vor dem Großen Krieg, in: Simone Finkele/Burkhardt Krause

(Hg.), Technikfiktionen und Technikdiskurse. Ringvorlesung des Instituts für Literaturwissenschaft im Sommersemester 2009, Karlsruhe 2012, S.209–244 (Karlsruher Studien Technik und Kultur, Bd.4).
Mosse, George L., Gefallen für das Vaterland. Nationales Heldentum und namenloses Sterben, Stuttgart 1993.
Mülder-Bach, Inka (Hg.), Modernität und Trauma. Beiträge zum Zeitenbruch des Ersten Weltkrieges, Wien 2000.
Münkler, Herfried, Politische Mythen und nationale Identität. Vorüberlegungen zu einer Theorie politischer Mythen, in: Wolfgang Frindte/Harald Pätzholt (Hg.), Mythen der Deutschen. Deutsche Befindlichkeiten zwischen Geschichten und Geschichte, Opladen 1994, S.21–27.
Münkler, Herfried, Die Deutschen und ihre Mythen, Berlin 2009.
Münkler, Herfried, Der große Krieg. Die Welt 1914–1918, 4. Aufl., Berlin 2014.
Nalepka, Cornelia, Postmoderne Biographik. Dieter Kühns *N* und Hans Magnus Enzensbergers *Der kurze Sommer der Anarchie*, in: Wilhelm Hemecker (Hg.), Die Biographie – Beiträge zu ihrer Geschichte, Berlin/New York 2009, S.393–421.
Naumann, Michael, Strukturwandel des Heroismus: vom sakralen zum revolutionären Heldentum, Königstein/Ts. 1984.
Neff, Bernhard, „Dekorationsmilitarismus". Die sozialdemokratische Kritik eines vermeintlich nicht kriegsmäßigen Militärwesens (1890–1911), in: Wolfram Wette (Hg.), Schule der Gewalt. Militarismus in Deutschland 1871–1945, Berlin 2005, S.91–110.
Neitzel: Sönke, Zum strategischen Mißerfolg verdammt? Die deutschen Luftstreitkräfte in beiden Weltkriegen, in: Bruno Thoß/Hans-Erich Volkmann (Hg.), Erster Weltkrieg – Zweiter Weltkrieg. Ein Vergleich. Krieg, Kriegserlebnis, Kriegserfahrung in Deutschland, Paderborn 2002, S.167–192.
Neitzel, Sönke, Blut und Eisen: Deutschland und der Erste Weltkrieg, Zürich 2003.
Neumann, Georg Paul, Die gesamten deutschen Luftstreitkräfte im 1. Weltkrieg, ND Bremen 2011 [1920].
Ní Dhúill, Caitríona, Widerstand gegen die Biographie: Sigrid Weigels Ingeborg-Bachmann-Studie, in: Wilhelm Hemecker (Hg.), Die Biographie – Beiträge zu ihrer Geschichte, Berlin/New York 2009, S.43–68.
Núnes, Xosé-Manoel, Historiographical Approaches to Sub-national Identities in Europe: A Reappraisal and Some Suggestions, in: Eric Storm (Hg.), Region and State in Nineteenth-Century Europe. Nation-Building, Regional Identities and Separatism, Basingstoke/Hampshire 2012, S.13–35.
Nye, David E., The Invented Self. An Anti-Biography, from documents of Thomas A. Edison, Odense 1983.
Nye, David E., Nach Thomas Edison. Rückblick auf eine Anti Biographie [2003], in: Bernhard Fetz/Wilhelm Hemecker (Hg.), Theorie der Biographie. Grundlagentexte und Kommentar, Berlin/New York 2011, S.347–360.
Osiezki, Maria, „Der Flug ist das Leben wert". Entgrenzung weiblicher Lebensentwürfe in der Luftfahrt des 20. Jahrhunderts, in: Wolfgang Meighörner (Hg.), Frau und Flug: die Schwestern des Ikarus, Marburg 2004 S.8–25.
Osterhammel, Jürgen, Die Verwandlung der Welt, 3. Aufl., München 2009.
Otto, Walter F., Mythos und Welt, Stuttgart 1962.
Pagé, Georges, L'aviation francaise, 1914–1918, Paris 2011.
Palmen, Polo (Hg.), Flugplatz Großenhain. Historischer Abriss, o. O. 2007.
Pardoe, Blaine Lee, Terror of the Autumn Skies. The true Story of Frank Luke, America's rogue Ace of World War I, New York 2008.
Peters, Ralf/Zehnter, Annette, Grenzen überwinden. 150 Jahre Th. Goldschmidt, Bottrop 1997.
Pfleger, Arthur, Franz im Feuer. Vier Jahre Flugzeugbeobachter, Regensburg 1930.
Pieters, Walter M., Above Flanders' Fields: a complete record of the the Belgian fighter pilots and their units during the Great War, 1914–1918, London 1998.
Pohl, Hans, Zur Entwicklung der Formen der Betriebs- und Unternehmensorganisation, insbesondere der Großorganisation im Verhältnis zum persönlich geführten Geschäft, in: ders., Wirtschaft, Unternehmen, Kreditwesen, soziale Probleme. Ausgewählte Aufsätze Teil 1, Stuttgart 2005, S.536–568.
Pohl, Jürgen, Regionalbewußtsein als Thema der Sozialgeographie. Theoretische Überlegungen und empirische Untersuchungen am Beispiel Friaul, Kallmünz 1993 (Münchener Geographische Hefte Nr.70).

Pohl, Jürgen, Raumbezogene Identität als geographisches Problem, in: Heiner Meulemann/Agnes Elting-Camus (Hg.), 26. Deutscher Soziologentag. Lebensverhältnisse und soziale Konflikte im neuen Europa (Tagungsband II), Opladen 1993, S.380–383.

Pöppinghege, Rainer, Geschichtspolitik per Stadtplan. Kontroversen zu historisch-politischen Straßennamen, in: Matthias Frese (Hg.), Fragwürdige Ehrungen!? Straßennamen als Instrument von Geschichtspolitik und Erinnerungskultur, Münster 2012, S.21–40.

Poser, Stefan, „Kannst du bremsen, Geliebter?" Wechselbeziehungen zwischen Technik, Spiel und Risiko, in: Gunter Gebauer/Stefan Poser/Robert Schmidt/Martin Stern (Hg.), Kalkuliertes Risiko: Technik, Spiel und Sport an der Grenze, Frankfurt a. M./New York 2006, S.15–37.

Potempa, Harald, Die Königlich-Bayerische Fliegertruppe: 1914–1918, Frankfurt a. M. 1997.

Potempa, Harald, Die Königlich-Bayerische Fliegertruppe als Teil der Deutschen Luftstreitkräfte bis 1918, in: Blätter zur Geschichte der Deutschen Luft- und Raumfahrt XIX (2013), S.11–22.

Potempa, Harald, Krieg in der Luft, in: Markus Po?hlmann/Harald Potempa/Thomas Vogel (Hg.), Der Erste Weltkrieg 1914–1918. Der deutsche Aufmarsch in ein kriegerisches Jahrhundert, München 2013, S.89–109.

Prinz, Joseph, Greven an der Ems. Die Geschichte der Stadt und des Amtes Greven, ND Greven 2005.

Protevi, John, Affect, Agency, and Responsibility: the Act of Killing in the Age of Cyborgs, in: Kevin McSorley (Hg.), War and the Body. Militarisation, Practice and Experience, Abingdon/Oxon/New York 2013, S.128–137.

Pyta, Wolfram, Biographisches Arbeiten als Methode. Geschichtswissenschaft, in: Christian Klein (Hg.), Handbuch Biographie. Methoden, Traditionen, Theorien, Stuttgart 2009, S.331–338.

Radkau, Joachim, Das Zeitalter der Nervosität. Deutschland zwischen Bismarck und Hitler, Wien 1998.

Radkau, Joachim, Technik in Deutschland. Vom 18. Jahrhundert bis heute, Frankfurt a. M./New York 2008.

Radzicki, Frank, Der Traum vom Fliegen im Ruhrgebiet: Bilder der Luftfahrt, Erfurt 2006.

Reitmayer, Morten, Bankiers im Kaiserreich. Sozialprofil und Habitus der deutschen Hochfinanz, Göttingen 2011.

Reimann, Aribert, Die heile Welt in Stahlgewittern: Deutsche und englische Feldpost aus dem Ersten Weltkrieg, in: Gerhard Hirschfeld/Gerd Krumeich/Dieter Langewiesche, Hans-Peter Ullmann (Hg.), Kriegserfahrungen. Studien zur Sozial- und Mentalitätsgeschichte des Ersten Weltkriegs, Essen 1997, S.129–145.

Reimann, Aribert, Der große Krieg der Sprachen. Untersuchungen zur historischen Semantik in Deutschland und England zur Zeit des Ersten Weltkrieges, Essen 2000.

Reimann, Aribert, Der Erste Weltkrieg – Urkatastrophe oder Katalysator?, in: Aus Politik und Zeitgeschichte 29/30 (2004), S.30–38.

Reinicke, Helmut, Deutsche Auffahrten Mentalitätsformen der Technikbegeisterung in der Entwicklung der Aviation, in: Wechselwirkung 54 (1992), H. 4, S.22–28.

Rhiemeier, Günter, Nichts dauert ewig. Vom Aufstieg und Niedergang des Kondor-Werkes, in: Lemgoer Hefte 3 (1995), S.22–28.

Rieger, Bernhard, Technology and the Culture of Modernity in Britain and Germany, 1890–1945, Cambridge 2005.

Rieth, Hugo, Die Kondor-Flugzeugwerke (1912–1918). Ein Kapitel aus der Pionierzeit deutscher Luftfahrtgeschichte, geschrieben in Essen und Rotthausen, in: Die Heimatstadt Essen 1975, S.121–130.

Rieth, Hugo, Der Wind bestimmt die Richtung, in: Jahrbuch Essen 1988, S.91–104.

Rißmann-Ottow, Guido, Glück ab! Frühe Luftfahrt im Revier, Essen 2002.

Rohkrämer, Thomas, Eine andere Moderne? Zivilisationskritik, Natur und Technik in Deutschland 1880–1933, Paderborn 1999.

Rohkrämer, Thomas, Der Militarismus der „kleinen Leute". Die Kriegervereine im deutschen Kaiserreich 1871–1914, München 1990.

Rolf, Malte, Einführung: Imperiale Biographien. Lebenswege imperialer Akteure in Groß- und Kolonialreichen (1850–1918), in: Geschichte und Gesellschaft 40 (2014), S.5–21.

Rosa, Hartmut, Beschleunigung. Die Veränderung der Zeitstruktur in der Moderne, ND Frankfurt a. M. 2010.

Rose, Andreas, Zwischen Empire und Kontinent. Britische Außenpolitik vor dem Ersten Weltkrieg, München 2011.

Rosenboom, Sebastian, Im Einsatz über der „vergessenen Front". Der Luftkrieg an der Ostfront im Ersten Weltkrieg, Potsdam 2013 (Potsdamer Schriften zur Militärgeschichte, Bd. 23).
Rösler, Katja, Die Sammel-Molkerei in Billerbeck im Kontext kleinstädtischer Entwicklung und Agrarmodernisierung um 1900. Unveröffentlichte Masterarbeit, Münster 2012.
Rother, Rainer (Hg.), Die letzten Tage der Menschheit. Bilder des Ersten Weltkrieges, Berlin 1994.
Runge, Anita, Wissenschaftliche Biographik, in: Christian Klein (Hg.), Handbuch Biographie: Methoden, Traditionen, Theorien, Stuttgart 2009, S.113–121.
Salewski, Michael/Stölken-Fitschen, Ilona (Hg.), Moderne Zeiten. Technik und Zeitgeist im 19. und 20. Jahrhundert, Stuttgart 1994.
Saupe, Lothar, Palästina von oben. Luftbilder der bayerischen Fliegerstaffel 304 im Ersten Weltkrieg aus Beständen des Bayerischen Hauptstaatsarchivs. Begleitausstellung des Bayerischen Hauptstaatsarchivs zur Ausstellung „Mission Palästina", 18. Juni bis 15. September 2006, München 2006.
Scharte, Sebastian, Westfalenbilder und Westfalenbewußtsein. Die Herausforderungen ‚Regionale Kulturpolitik' und ‚Verwaltungsstrukturreform', Münster 2003.
Schellack, Fritz, Nationalfeiertage in Deutschland von 1871 bis 1945, Frankfurt a. M./Bern/New York/Paris 1990.
Scherke, Katharina, Die These von der ‚Ästhetisierung der Lebenswelt' als eine Form der Analyse des Modernisierungsprozesses, in: Sabine A. Haring/dies. (Hg.), Analyse und Kritik der Modernisierung um 1900 und um 2000, Wien 2000, S.109–131 (Studien zur Moderne 12).
Scherstjanoi, Elke, Als Quelle nicht überfordern! Zu Besonderheiten und Grenzen der wissenschaftlichen Nutzung von Feldpostbriefen in der (Zeit-)Geschichte, in: Veit Didczuneit/Jens Ebert/Thomas Jander (Hg.), Schreiben im Krieg — Schreiben vom Krieg: Feldpost im Zeitalter der Weltkriege, Essen 2011, S.117–125.
Schikorsky, Isa, Kommunikation über das Unbeschreibbare. Beobachtungen zum Sprachstil von Kriegsbriefen, in: Wirkendes Wort 42 (1992), H. 2, S.295–315.
Schilling, René, Der Körper des „Helden". Deutschland 1813–1945, in: Bielefelder Graduiertenkolleg Sozialgeschichte (Hg.), Körper Macht Geschichte, Geschichte Macht Körper. Körpergeschichte als Sozialgeschichte, Bielefeld 1999, S.119–140.
Schilling, René, „Kriegshelden". Deutungsmuster heroischer Männlichkeit in Deutschland 1813–1945, Paderborn/München/Wien/Zürich 2002.
Schmetzke, Rudolf, Luftfahrt und Versicherung. Eine historische Betrachtung über die Entstehung der deutschen Luftfahrtversicherung, Berlin 1954.
Schmidt, Svenja, Kassandra – ein Mythos im Wandel der Zeit. Antiker Mythos und moderne Literatur am Beispiel der „Kassandra" von Christa Wolf, Marburg 2004 (diplomica, Bd. 13).
Schmidt-Gernig, Alexander, Zukunftsmodell Amerika? Das europäische Bürgertum und die amerikanische Herausforderung um 1900, in: Ute Frevert (Hg.), Das neue Jahrhundert. Europäische Zeitdiagnosen und Zukunftsentwürfe um 1900, Göttingen 2000, S.79–112 (Geschichte und Gesellschaft, Sonderheft 18).
Schmiechen-Ackermann, Detlef (Hg.), „Volksgemeinschaft": Mythos, wirkmächtige soziale Verheißung oder soziale Realität im „Dritten Reich"? Zwischenbilanz einer kontroversen Debatte, Paderborn/München/Wien/Zürich 2012 (Nationalsozialistische „Volksgemeinschaft", Bd. 1).
Schmitt, Günter, Fliegende Kisten von Kitty Hawk bis Kiew, 2. Aufl., Berlin (Ost) 1990.
Schmitt, Günter/Schwipps, Werner, 20 Kapitel frühe Luftfahrt, Planegg 1990.
Schmitt, Günter, Als die Oldtimer flogen. Die Geschichte des Flugplatzes Berlin-Johannisthal, 3. Aufl., Oberhaching 1995.
Schneider, Thomas F./Heinemann, Julia u. a. (Hg.), Die Autoren und Bücher der deutschsprachigen Literatur zum Ersten Weltkrieg 1914–1939. Ein bio-bibliographisches Handbuch, Göttingen 2008.
Schneider, Ute, Spurensuche: Reinhart Koselleck und die „Moderne", in: dies./Lutz Raphael (Hg.), Dimensionen der Moderne: Festschrift für Christof Dipper, Frankfurt a. M./Berlin/Bern u. a. 2008, S.61–72.
Schnürer, Florian, „Nun hat der große Flieger dich geholt ..." – Die Totenfeiern für die ‚Ritter der Lüfte' als transnationales Medienereignis, in: Kulturwissenschaften exemplarisch: Gießener Forschungsbeiträge zu acht Kernkompetenzen, Trier 2011, S.167–192.
Schrijver, Frans, Regionalism after Regionalisation: Spain, France and the United Kingdom, Amsterdam 2006.
Schröder, Ernst August, Die Sportluftfahrt im Ruhrland, in: Ruhrland Almanach der Lüfte, Essen 1929, S.19–23.

Schröder, Ernst August jun., Ernst August Schröder 1873–1949. Ein deutsches Bürgerleben vom Ausgang des 19. in das folgende schicksalhaft-wechselvolle Halbjahrhundert hinein aus eigener Kraft gestaltet, Essen/Nümbrecht-Benroth 1975–1977.

Schröder, Ernst August jun., Von den Wehen in der Geburtsstunde des Luftverkehrs, in: Beiträge zur Geschichte von Stadt und Stift Essen 104 (1991/92), S.101–122.

Schubert, Werner, Die Abschaffung des Konzessionssystems durch die Aktienrechtsnovelle von 1870, in: Zeitschrift für Unternehmens- und Gesellschaftsrecht 10 (1981), H. 2, S.285–317.

Schüler-Springorum, Stefanie, Vom Fliegen und Töten. Militärische Männlichkeit in der deutschen Fliegerliteratur, 1914–1939, in: Karen Hagemann (Hg.), Heimat – Front. Militär und Geschlechterverhältnisse im Zeitalter der Weltkriege, Frankfurt a.M. 2002, S.208–233.

Schulte, Monika, Die Totenzettel aus dem Ersten und Zweiten Weltkrieg: Soziale Bedeutung für das Image der Männlichkeit, o.O. 2011.

Schumann, Dirk, Der brüchige Frieden. Kriegserinnerungen, Kriegsszenarien und Kriegsbereitschaft, in: Ute Frevert (Hg.), Das neue Jahrhundert. Europäische Zeitdiagnosen und Zukunftsentwürfe um 1900, Göttingen 2000, S.113–145 (Geschichte und Gesellschaft, Sonderheft 18).

Schwarz, Angela, Der Schlüssel zur modernen Welt. Wissenschaftspopularisierung in Großbritannien und Deutschland im Übergang zur Moderne (ca. 1870–1914), Stuttgart 1999, S.164–172.

Schwarz, Angela, Wie uns die Stunde schlägt. Zeitbewußtsein und Zeiterfahrungen im Industriezeitalter als Gegenstand der Mentalitätsgeschichte, in: Archiv für Kulturgeschichte 83 (2001), S.451–479.

Schweiger, Günter, Image und Imagetransfer, in: Bruno Tietz/Richar Köhler (Hg.), Handwörterbuch des Marketings, 2. Aufl., Stuttgart 1995, Sp.915–928.

Schwingel, Markus, Pierre Bourdieu zur Einführung, 6. Aufl., Hamburg 2009.

Schwipps, Werner, Schwerer als Luft: Die Frühzeit der Flugtechnik in Deutschland, Koblenz 1984 (Die deutsche Luftfahrt, Bd.8).

Schwipps, Werner, Riesenzigarren und fliegende Kisten. Bilder aus der Frühzeit der Luftfahrt in Berlin, Berlin 1984 (Berliner Beiträge, Bd.3).

Seifert, Karl-Dieter, Hans Grade. Ingenieur – Flugpionier – Automobilbauer. Ein Leben in stürmischen Zeiten, Magdeburg 2008.

Seifert, Karl-Dieter, Mit Sammelgroschen zur deutschen Luftmacht – Die Nationalflugspende 1912–1914, Berlin 2014.

Sieger, Constanze, Die Ludgerusstadt im 19.Jahrhundert, in: Werner Freitag (Hg.), Geschichte der Stadt Billerbeck, Bielefeld 2012, S.111–260.

Siegrist, Hannes, Regionalisierung im Medium des Konsums, in: Comparativ 11 (2001), H. 1, S.7–26.

Siemer, Stefan, Bildgelehrte Geotechniker. Luftbild und Karthographie um 1900, in: Alexander Gall (Hg.), Konstruieren, kommunizieren, präsentieren. Bilder von Wissenschaft und Technik, Göttingen 2007 (Deutsches Museum, Abhandlungen und Berichte. Neue Folge, Bd.23).

Simmel, Georg, Die Großstädte und das Geistesleben, in: Klaus Lichtblau (Hg.), Georg Simmel, Soziologische Ästhetik, Darmstadt 1998, S.103–114.

Skinner, Stephen, The Stand. The final Flight of Lt. Frank Luke jr., o.O. 2008.

Smouts, Marie-Claude, The Region as the New Imagined Community?, in: Patrick Le Galès/Christian Lequesne (Hg.), Regions in Europe: The Paradox of Power, New York 1998, S.30–38.

Speth, Rudolf, Nation und Revolution. Politische Mythen im 19.Jahrhundert, Opladen 2000.

Stachel, Peter, Stadtpläne als politische Zeichensysteme. Symbolische Einschreibungen in den öffentlichen Raum, in: Rudolf Jaworski (Hg.), Die Besetzung des öffentlichen Raumes. Politische Pläne, Denkmäler und Straßennamen im europäischen Vergleich, Berlin 2007, S.13–60.

Stein, Oliver, Das Kriegsministerium und der Ausbau des deutschen Heeres 1871–1914, in: Christian Th. Mülller/Matthias Rogg (Hg.), Das ist Militärgeschichte!, Paderborn/München/Wien/Zürich 2013, S.48–62.

Steiner-Welz, Sonja, Schütte-Lanz-Luftfahrzeuge aus Mannheim, Bd.1: Die Luftschiffe, Mannheim 2006.

Stern, Martin, Heldenfiguren im Wagnissport. Zur medialen Inszenierung wagnissportlicher Erlebnisräume, in: Thomas Alkemeyer/Bernhard Boschert/Robert Schmidt u.a. (Hg.), Aufs Spiel gesetzte Körper. Aufführungen des Sozialen in Sport und populärer Kultur, Konstanz 2003, S.37–54.

Stern, Martin, Der moderne Ikarus: Technik, Spiel und Risiko in der Luft, in: Gunter Gebauer/Stefan Poser/Robert Schmidt/Martin Stern (Hg.), Kalkuliertes Risiko: Technik, Spiel und Sport an der Grenze, Frankfurt a.M./New York 2006, S.138–158.

Stierle, Karlheinz, Text als Handlung. Perspektiven einer systematischen Literaturwissenschaft, München 1975.

Storz, Dieter, Kriegsbild und Rüstung vor 1914. Europäische Landstreitkräfte vor dem Ersten Weltkrieg, Herford/Berlin/Bonn 1992.

Strauss, Claude Lévi, The Structural Study of Myth, in: The Journal of American Folklore 68 (1955), S.428–444.

Sturm, Beate, „Heldentod fürs Vaterland". Totenzettel als Quelle für Einzelschicksale aus der Zeit des Ersten Weltkriegs, in: Rhein-Maas Geschichte, Sprache und Kultur: Der Erste Weltkrieg, Bd. 5, hg. von Jörg Engelbrecht †/Simone Frank, Christian Krumm und Holger Schmenk, Oberhausen 2014, S.62–75.

Supf, Peter, Das Buch der deutschen Fluggeschichte: Vorzeit, Wendezeit, Werdezeit, Bd. 1, 2. Aufl., Stuttgart 1956.

Supf, Peter, Das Buch der deutschen Fluggeschichte: Vorkriegszeit, Kriegszeit, Nachkriegszeit bis 1932, Bd. 2, Stuttgart 1958.

Süß, Dietmar, Tod aus der Luft. Kriegsgesellschaft und Luftkrieg in Deutschland und England, München 2011.

Suwelack, Winfried/Suwelack, Walter, Josefs Suwelack und der Traum vom Fliegen. Leben und Tod des Billerbecker Flugpioniers und Weltrekordlers Josef Suwelack, 30. April 1888 bis 13. September 1915, anlässlich der 100. Wiederkehr seines Geburtstages, Warendorf 1988 (Beiträge zur Landes- und Volkskunde des Kreises Coesfeld, Bd. 23).

Syon, Guillaume de, Zeppelin! Germany and the Airship, 1900–1939, Baltimore 2002.

Taylor, John W. R./Mondey, Davin, Spione am Himmel über uns. Die Geschichte der Luftaufklärung, Stuttgart 1974.

Tepe, Peter, Entwurf einer Theorie des politischen Mythos. Mit einem Analysemodell für politische Helden, in: ders. (Hg.), Mythos No. 2: Politische Mythen, Würzburg 2006.

Thamer, Hans-Ulrich, Eine Ausstellung und ihre Folgen. Impulse der „Wehrmachtsausstellung" für die historische Forschung, in: Ulrich Bielefeld/Heinz Bude/Bernd Greiner (Hg.), Gesellschaft – Gewalt – Vertrauen. Jan Philipp Reetsma zum 60. Geburtstag, Hamburg 2012, S.489–503.

Thiele, Jörg/Hunger, Ina, Zeiterleben im Erlebnissport – Versuch einer Annäherung, in: Barbara Ränsch-Trill (Hg.), Zeit und Geschwindigkeit. Sportliches Erleben in beschleunigten Prozessen, Sankt Augustin 2002, S.124–142.

Thies, Kristina, Das Augusterlebnis in Münster im Spiegel der Kriegschronik Eduard Schultes, in: Tobias Arand: Die „Urkatastrophe" als Erinnerung – Geschichtskultur des Ersten Weltkriegs, Münster 2006, S.99–131.

Torp, Cornelius/Müller, Sven-Oliver, Das Bild des Deutschen Kaiserreichs im Wandel, in: dies. (Hg.), Das Deutsche Kaiserreich in der Kontroverse, Göttingen 2009, S.9–27.

Trischler, Helmuth, Luft- und Raumfahrtforschung in Deutschland 1900–1970, Frankfurt a. M./New York 1992.

Trischler, Helmuth, Die neue Räumlichkeit des Krieges: Wissenschaft und Technik im Ersten Weltkrieg, in: Berichte zur Wissenschaftsgeschichte 19 (1996), S.95–103.

Ulrich, Bernd, Die Desillusionierung der Kriegsfreiwilligen von 1914, in: Wolfram Wette (Hg.), Der Krieg des kleinen Mannes. Eine Militärgeschichte von unten, München/Zürich 1992, S.110–127.

Ulrich, Bernd, „Militärgeschichte von unten". Anmerkungen zu ihren Ursprüngen, Quellen und Perspektiven im 20. Jahrhundert, in: Geschichte und Gesellschaft 22 (1996), S.473–503.

Ulrich, Bernd/Vogel, Jakob/Ziemann, Benjamin (Hg.), Untertan in Uniform. Militär und Militarismus im Deutschen Kaiserreich. Quellen und Dokumente, Frankfurt a. M. 2001.

Veltzke, Veit, Die Stiftung des Eisernen Kreuzes 1813. Ein „neuer preußisch-deutscher Orden" für den Volkskrieg, in: Winfried Heinemann (Hg.), Das Eiserne Kreuz. Die Geschichte eines Symbols im Wandel der Zeit, Potsdam 2014, S.9–20.

Verhey, Jeffrey, Der „Geist von 1914" und die Erfindung der Volksgemeinschaft, Hamburg 2000.

Vierhaus, Rudolf (Hg.), Deutsche biographische Enzyklopädie, 2. Ausg., München 2008.

Villard, Henry Serrano, Contact! The story of the early birds, Washington D.C. 2002.

Virilio, Paul, Krieg und Kino. Logistik der Wahrnehmung, Frankfurt a. M. 1989.

Vogel, Jakob, Nationen im Gleichschritt. Der Kult der ‚Nation in Waffen' in Deutschland und Frankreich, 1871–1914, Göttingen 1997.

Vogt, Betina Vogt, „Unsere Zukunft liegt in der Luft!". Die Entwicklung der zivilen Luftfahrt auf dem

Flugplatz Loddenheide 1909–1933. Unveröffentlichte Arbeit, Geschichtswettbewerb des Bundespräsidenten 1990/91, Münster 1990/91.

Vogt, Hans, Seidene Kugel und Fliegende Kiste, Krefeld 1993.

Vries, John A. de, Taube, Dove of War. Historical Aviation Album Publication, o. O. 1974.

Wagner, Wolfgang, Die Flugzeuge der Kondor Werke, in: Aerokurier 1 (1974), S.67–69.

Wagner, Wolfgang, Der deutsche Luftverkehr – Die Pionierjahre 1919–1925, Koblenz 1987.

Walkenhorst, Peter, Der „Daseinskampf des deutschen Volkes": Nationalismus, Sozialdarwinismus und Imperialismus im wilhelminischen Deutschland, in: MGFA/Jörg Echternkamp/Sven Oliver Müller (Hg.), Die Politik der Nation. Deutscher Nationalismus in Krieg und Krisen 1760–1960, München 2002, S.131–148.

Walkenhorst, Peter, Nation – Volk – Rasse: Radikaler Nationalismus im Deutschen Kaiserreich 1890–1914, Göttingen 2007.

Walther, Christine, Siegertypen. Zur fotografischen Vermittlung eines gesellschaftlichen Selbstbildes um 1900, Würzburg 2007.

Watson, Nigel, The Scareship Mystery. A Survey of Worldwide Phantom Airship Scare (1909–1918), Corby 2000.

Weichlein, Siegfried, Regionalism, Federalism and Nationalism in the German Empire, in: Joost Augusteijn/Eric Storm (Hg.), Region and State in Nineteenth-Century Europe. Nation-Building, Regional Identities and Separatism, Basingstoke/Hampshire 2012, S.93–110.

Weidner, Marcus, „Wir beantragen ... unverzüglich umzubenennen". Die Straßenbenennungspraxis in Westfalen und Lippe im Nationalsozialismus, in: Matthias Frese (Hg.), Fragwürdige Ehrungen!? Straßennamen als Instrument von Geschichtspolitik und Erinnerungskultur, Münster 2012, S.41–98.

Welskopp, Thomas, Die Sozialgeschichte der Väter. Grenzen und Perspektiven der Historischen Sozialwissenschaft, in: Geschichte und Gesellschaft 24 (1998), S.169–194.

Wernitz, Frank, Die Erneuerungen des Eisernen Kreuzes 1870 und 1914. Betrachtungen der Verleihungspraxis eines „deutschen" Ehrenzeichens, in: Winfried Heinemann (Hg.), Das Eiserne Kreuz. Die Geschichte eines Symbols im Wandel der Zeit, Potsdam 2014, S.21–29.

Westfälischer Heimatbund (Hg.), Jahrbuch Westfalen, Jg. 54, Münster 2000, S.26–39.

Wette, Wolfram (Hg.), Der Krieg des kleinen Mannes. Eine Militärgeschichte von unten, München/Zürich 1992.

White, Hayden, Der historische Text als literarisches Kunstwerk, in: ders., Auch Klio dichtet oder die Fiktion des Faktischen. Studien zur Tropologie des historischen Diskurses, Stuttgart 1986, S.101–122.

Witt, Peter-Christian, Monarchen und Bürger. Über Untertanen und Untertänigkeit im wilhelminischen Deutschland (1890–1914), in: Horst Lademacher/Walter Mühlhausen (Hg.), Freiheitsstreben – Demokratie, Emanzipation. Aufsätze zur politischen Kultur in Deutschland und in den Niederlanden, Münster/Hamburg 1993, S.139–187.

Wittmann, Livia Käthe/Zibler, Barbara, Melli Beese und die „Flügel am Horizont". Die Geschichte der ersten deutschen Pilotin, Berlin 2009.

Wohl, Robert, A Passion for Wings: Aviation and the Western Imagination, 1908–1918, Yale 1994.

Wolter, Gundula, Sensationelle Sportkostüme. Die Kleidung der Flugpionierinnen, in: Wolfgang Meighörner (Hg,), Frau und Flug: die Schwestern des Ikarus, Marburg 2004, S.211–237.

Wülfing, Wulf, Die heilige Luise von Preußen, in: Jürgen Link/ders. (Hg.), Bewegung und Stillstand in Metaphern und Mythen, Stuttgart 1984, S.233–275.

Wülfing, Wulf/Bruns, Karin/Parr, Rolf, Historische Mythologie der Deutschen 1798–1918, München 1991.

Zegenhagen, Evelyn, Zwischen Herd und Himmel, zwischen Kochbuch und Flugbuch. Frauen in der deutschen Luftfahrt, in: Helmuth Trischler u. a. (Hg.), Ein Jahrhundert im Flug: Luft- und Raumfahrtforschung in Deutschland 1907–2007, Frankfurt a. M./New York 2007, S.70–87.

Ziemann, Benjamin, Zum ländlichen Augusterlebnis 1914 in Deutschland, in: Bedrich Loewenstein (Hg.), Geschichte und Psychologie. Annäherungsversuche, Pfaffenweiler 1992, S.193–203.

Ziemann, Benjamin, Sozialmilitarismus und militärische Sozialisation im deutschen Kaiserreich 1870–1914. Desiderate und Perspektiven in der Revision eines Geschichtsbildes, in: GWU 3 (2002), S.148–164.

Personenregister

(zeitgenössische Personen, ohne Josef Suwelack)

Band 70: Anselm Heinrich
Theater in der Region
Westfalen und Yorkshire 1918–1945
2012, 350 Seiten, gebunden, ISBN 978-3-506-77631-0

Band 71: Wilfried Reininghaus/Bernd Walter (Hg.)
Westfalen als Gegenstand landes- und regionalgeschichtlicher Forschung
2013, 302 Seiten, gebunden, ISBN 978-3-506-77771-3

Band 72: Karl Ditt/Cordula Obergassel (Hg.)
Vom Bildungsideal zum Standortfaktor
Städtische Kultur und Kulturpolitik in der Bundesrepublik
2012, 454 Seiten, gebunden, ISBN 978-3-506-77632-7

Band 73: Thomas Küster (Hg.)
Medien des begrenzten Raumes
Landes- und regionalgeschichtliche Zeitschriften im 19. und 20. Jahrhundert
2013, 374 Seiten, gebunden, ISBN 978-3-506-77730-0

Band 74: Jan Nikolas Dicke
Reform und Protest
Konflikte um die Neugliederung des Kreises Borken in den 1960er und 1970er Jahren
2013, 488 Seiten, gebunden, ISBN 978-3-506-77859-8

Band 75: Bärbel Sunderbrink
Revolutionäre Neuordnung auf Zeit
Gelebte Verfassungskultur im Königreich Westphalen:
Das Beispiel Minden-Ravensberg 1807–1813
2015, 412 Seiten, gebunden, ISBN 978-3-506-78150-5

Band 76: Julia Paulus/Marion Röwekamp (Hg.)
Eine Soldatenheimschwester an der Ostfront
Briefwechsel von Annette Schücking mit ihrer Familie (1941–1943)
2015, 654 Seiten, gebunden, ISBN 978-3-506-78151-2

FERDINAND SCHÖNINGH

Band 77: Franz-Werner Kersting/Clemens Zimmermann (Hg.)
Stadt-Land-Beziehungen im 20. Jahrhundert
Geschichts- und kulturwissenschaftliche Perspektiven
2015, 330 Seiten, gebunden, ISBN 978-3-506-78152-9

Band 78: Martin Dröge
Männlichkeit und ‚Volksgemeinschaft'
Der westfälische Landeshauptmann Karl Friedrich Kolbow (1899–1945):
Biographie eines NS-Täters
2015, 440 Seiten, gebunden, ISBN 978-3-506-78289-2

Band 79: Michael Prinz (Hg.)
Die vielen Gesichter des Konsums
Westfalen, Deutschland und die USA 1850–2000
2016, 296 Seiten, gebunden, ISBN 978-3-506-78415-5

Band 80: Lena Krull (Hg.)
Westfälische Erinnerungsorte
Beiträge zum kollektiven Gedächtnis einer Region
2017, 590 Seiten, gebunden, ISBN 978-3-506-78607-4

Band 81: Cordula Obergassel
„Das ist doch keine Kunst!"
Kulturpolitik und kultureller Wandel in Dortmund und Münster (1960–1985)
2017, 370 Seiten, gebunden, ISBN 978-3-506-78689-0

Band 82: Matthias Frese/Marcus Weidner (Hg.)
Verhandelte Erinnerungen
Der Umgang mit Ehrungen, Denkmälern und Gedenkorten nach 1945
2018, 386 Seiten, gebunden, ISBN 978-3-506-78798-9

Band 83: Julia Paulus (Hg.)
‚Bewegte Dörfer'
Neue soziale Bewegungen in der Provinz 1970–1990
2018, 242 Seiten, gebunden, ISBN 978-3-506-78804-7